Vector Targeting for Therapeutic Gene Delivery

Vector Targeting for Therapeutic Gene Delivery

Edited by

DAVID T. CURIEL

JOANNE T. DOUGLAS

Division of Human Gene Therapy
Departments of Medicine, Pathology and Surgery,
and the Gene Therapy Center
The University of Alabama at Birmingham

This book is printed on acid-free paper. ♾

Published by Wiley-Liss, Inc., Hoboken, New Jersey.

Published simultaneously in Canada.

For general information on our other products and services please contact our Customer Care Department within the U.S. at 877-762-2974, outside the U.S. at 317-572-3993 or fax 317-572-4002.

Wiley also publishes its books in a variety of electronic formats. Some content that appears in print, however, may not be available in electronic format.

Library of Congress Cataloging-in-Publication Data:

Vector targeting for therapeutic gene delivery/edited by David T. Curiel, Joanne T. Douglas.
p. cm.
Includes bibliographical references and index.
ISBN 0-471-43479-5 (cloth:alk. paper)
1. Gene therapy. 2. Gene targeting. 3. Genetic vectors. I. Curiel, David.
II. Douglas, Joanne T.

RB155.8 V434 2002
616'.042—dc21 2002071310

Printed in the United States of America.

10 9 8 7 6 5 4 3 2 1

CONTENTS

PREFACE

The basic mandates for gene therapy were formulated before the existence of any practical basis for the application of the concept. In this regard, early proponents of genetic therapy defined the basic criteria to be met prior to actual implementation of any human gene therapy approach. Specifically, it was required that the therapeutic gene be efficiently delivered to the relevant target cells, that the gene be expressed at an appropriate level, and that both of these ends be achieved with an acceptable margin of safety. It was thus implicit even at the earliest stage of field conceptualization that targeted gene delivery would need to be accomplished.

Despite this recognition, the earliest human gene therapy approaches proceeded without explicit attention to targeting per se. From the vector standpoint, the earliest issues addressed were related to achieving basic efficacy parameters needed to make practical gene therapy feasible. In the context of nonviral vectors, exploitation of nonspecific cellular transport processes provided the basis of genetic transduction. These methods were inefficient and, by their very design, nontargeted in intent. As an alternative, viral vectors were developed that incorporate a heterologous transgene within the viral genome and exploit the relatively more efficient processes of viral infection of a target cell. Such virus-mediated delivery, however, was restricted by the native tropism of the parent virus.

Early generations of vector systems thus prioritized the development of efficient gene delivery. By virtue of the design logic for both nonviral and viral approaches, specific targeting capacities were not embodied. In fact, the situation predicated vectors having tropism capacities frequently at odds with cell-specific targeting goals. Recognition of these limits led to initial gene therapy approaches whereby target cells could be modified ex vivo. In this schema, cell-specific transduction was achieved by a priori selection and isolation of target cells, followed by vector-mediated transduction. This schema implicitly recognized the limitations of available vector systems for achieving targeting goals. Unfortunately, only a few diseases are amenable to ex vivo gene therapy interventions, reflecting the limited repertoire of parenchymal cells that can be manipulated via extracorporeal methods. These considerations thus highlight the degree to which vector limitations, specifically limitations in vector targeting, restricted the practical implementation of the range of candidate gene therapies.

The goal of extending the range of gene therapy disease targets was fostered by the development of vector systems capable of in vivo gene delivery. In this regard, the capacity of various viral vector systems to achieve in situ gene transfer allowed the conceptualization of gene therapy approaches unconstrained by extracorporeal modifications. On this basis, gene therapy approaches for a variety of inherited and acquired disorders advanced to animal model systems and human clinical trials. Consideration of the results of these various in vivo gene therapy approaches defined the limitations of current vector systems and thereby established the clear rationale for targeting strategies.

In many respects, cancer gene therapy has illustrated key issues with respect to targeting, reflecting the fact that antitumor strategies frequently involve the delivery of toxin genes. In this scenario, the consequences of ectopic, nontargeted delivery would be manifested most prominently as therapy-related toxicity. Thus, the early clinical trials for cancer provided key insights as to the requirements for targeting and potential gains therein. In this regard, the obvious lack of targeting capacity of available vectors mitigated against approaches for disseminated neoplastic disease. Such diseases would have required tumor-selective gene delivery following systemic vector administration. The lack of any vector with such attributes meant it was necessary to address disease contexts in which targeting stringency was not such a paramount consideration. To this end, tumors localized to natural body compartments (central nervous system, thoracic cavity, peritoneal cavity) appeared to offer the ideal scenario—a space-confined tumor allowing vector concentration and containment. Thus, initial in vivo anticancer gene therapy approaches were endeavored for glioma, pleural mesothelioma, and peritoneal carcinomatosis.

The results of these trials were highly disappointing. First, extremely low levels of tumor cell transduction were achieved. Thus, despite apparent vector efficacy in model system studies, target cell transduction rates following in vivo gene delivery were limited. Secondly, ectopic gene delivery occurred in various in vivo delivery schemas, irrespective of theoretical vector containment based on anatomic aspects. Further, ectopic gene deliveries were associated with vector-related toxicities. Thus, whereas the overall profile of anticancer gene therapy has suggested an acceptable safety/toxicity profile, the occurrence of suboptimal target transduction rates might logically predicate advanced dosing schemas, a strategy countermanded by the observed phenomenon of ectopic gene delivery. In vivo gene delivery, even in the optimized scenario of compartment-based models, therefore exhibits all of the limitations involved in countervailing systemic delivery schemas—limited tumor cell transduction and ectopic gene delivery. These considerations thus clearly establish the universal applicabilities that might derive from targeting, irrespective of delivery route.

The very recent aspect of these findings explains the only recent development, and application, of targeting for gene therapy applications. In this regard, the basic ideas of retargeting vectors for selective gene delivery have been previously studied as a means to improve vector efficiency per se, as in the context of receptor-mediated conjugate vectors. In other words, the idea of targeting to improve gene therapy outcomes has been most generally recognized as a consequence of these disappointing results in human clinical trials. However, basic field paradigms have only been recently established. Further, the actual translation of targeting paradigms into the clinical context has awaited these key proof-of-principle studies whereby direct gene therapy gain via targeting has been established.

For selected viral and nonviral vector systems it has now been demonstrated that targeting strategies can allow targeted, cell-specific gene delivery. However, this has largely been demonstrated only in in vitro proof-of-principle studies. A much smaller subset of studies has demonstrated the capacity to alter vectors in the context of in vivo gene delivery schemas. Such studies have also allowed demonstration of valid functional gene therapy endpoints including increased target cell transduction for enhancement of phenotype correction and mitigation of vector-mediated toxicities. Such powerful results have predicated consideration of translating such approaches into the clinical context to validate the human therapeutic uses embodied in these targeting approaches.

The unfortunate demise of a young man in a human clinical gene therapy trial represented a field landmark. In addition to the obvious field setback provoked, basic limits inherent in current vector systems became apparent. Although a temporary retrenchment in clinical activities could have been predicted, in fact, the longer term effects have been to positively radicalize the field and regulatory agencies with respect to considering vector redesign as of paramount importance to the field. Further, the positive findings in three recent human trials (for X-linked severe combined immunodeficiency, hemophilia B, and ischemic heart disease) have been generally recognized to have been gained via vector improvements. Thus, the formula that vector gain equals gene therapy gain has clearly been established. On this basis, recognition of the need for vector targeting strategies has allowed recent approval by the National Institutes of Health Recombinant DNA Advisory Committee of a variety of human clinical trials that embody targeting principles. The use of tropism-modified viral vectors represents a fundamental paradigm shift of the basic concept of exploitation of viruses for gene therapy applications. The realization of direct gene therapy gains to these trials—that is, an improved therapeutic outcome and/or a reduction of treatment-associated toxicities—will constitute a critical validation of the targeting principle with wide implications for the field.

It is against this historical background that this book has been conceptualized. The first and second sections focus on transductional targeting strategies designed to achieve the selective delivery of the therapeutic gene by both nonviral and viral vectors, respectively. The third section discusses the alternative, but complementary, approach of transcriptional targeting, in which the therapeutic gene is placed under the control of transcriptional regulatory sequences activated in the disease cells but not in normal cells and therefore target expression selectively to the tumor cell. Any transductional targeting approach mandates ligands that can be exploited to achieve cell-specific gene delivery. Therefore, the fourth section is dedicated to the consideration of a variety of strategies that can be employed to define appropriate cellular targeting moieties. Finally, it is becoming increasingly recognized that therapeutic gene delivery in the clinical setting could greatly benefit from strategies to monitor the extent of gene expression. Accordingly, the last section of the book is dedicated to this topic. Whereas the gains of targeting have begun to become apparent in model systems, it is clear that additional, and profound, gains may yet be realized by further endeavors of this type. However, the true gains have yet to be defined in the ultimate context—human clinical gene therapies.

DAVID T. CURIEL
JOANNE T. DOUGLAS

CONTRIBUTORS

W. FRENCH ANDERSON, M.D., Gene Therapy Laboratories and the departments of Surgery and Biochemistry, Keck School of Medicine of the University of Southern California, Los Angeles, California

JOZEF ANNÉ, Ph.D., Laboratory of Bacteriology, Rega Institute, K.U. Leuven, Belgium

DAVID C. ANSARDI, Ph.D., Replicon Technologies Incorporated, OADI Technology Center, Birmingham, Alabama

QING BAI, Ph.D., Department of Molecular Genetics and Biochemistry, University of Pittsburgh School of Medicine, Pittsburgh, Pennsylvania

ANDREW BAIRD, Ph.D., Selective Genetics, Inc., San Diego, California

JORGE R. BARRIO, Ph.D., UCLA School of Medicine, Los Angeles, California

MICHAEL A. BARRY, Ph.D., Center for Cell and Gene Therapy, Department of Molecular and Human Genetics, Baylor College of Medicine, and Department of Bioengineerng, Rice University, Houston, Texas

ROBERT W. BEART, JR., M.D., The Department of Surgery, Keck School of Medicine of the University of Southern California, Los Angeles, California

KATIE BINLEY, Ph.D., Oxford BioMedica (UK) Ltd, Medawar Centre, Oxford, United Kingdom

ANDREA BLEDSOE, Ph.D., Department of Microbiology, University of Alabama at Birmingham, Birmingham, Alabama

ABRAHAM BOUT, Ph.D., Crucell Holland BV, Leiden, The Netherlands

HILDEGARD BUNING, Ph.D., Genzentrum Ludwig-Maximilians-Universitat Munchen, Munchen, Germany

Edward A. Burton, M.D., Ph.D., Department of Clinical Neurology, University of Oxford, Oxford, England

Roberto Cattaneo, Ph.D., Molecular Medicine Program, Mayo Foundation, Rochester, Minnesota

Esther H. Chang, Ph.D., Department of Oncology, Georgetown University Medical Center, Lombardi Cancer Center, Washington, D.C.

Simon R. Cherry, Ph.D., UCLA School of Medicine, Los Angeles, California

Charles N. Cobbs, M.D., Department of Surgery, University of Alabama at Birmingham, Birmingham, Alabama

David T. Curiel, M.D., Division of Human Gene Therapy, Departments of Medicine, Pathology and Surgery and the Gene Therapy Center, The University of Alabama at Birmingham, Birmingham, Alabama

Patrick S. Daugherty, Ph.D., Department of Chemical Engineering, University of California at Santa Barbara, Santa Barbara, California

Pamela B. Davis, M.D., Ph.D., Division of Pediatric Pulmonology, Rainbow Babies and Childrens Hospital, Case Western Reserve University School of Medicine, Cleveland Ohio

John Dileo, M.S., Center for Pharmacogenetics, Department of Pharmaceutical Sciences, University of Pittsburgh School of Pharmacy, Pittsburgh, PA

David Dingli, M.D., Molecular Medicine Program, Mayo Clinic, Rochester, Minnesota

Joanne T. Douglas, Ph.D., Division of Human Gene Therapy, Departments of Medicine, Pathology, and Surgery, and the Gene Therapy Center, The University of Alabama at Birmingham, Birmingham, Alabama

Stefan Dübel, Ph.D., Institut für Molekulare Genetik, Universität Heidelberg, Germany

Mario Fernandez, Ph.D., ICRF Mol Oncology Unit, London, United Kingdom

Theodore Friedmann, M.D., Center for Molecular Genetics and Department of Pediatrics, UCSD School of Medicine, La Jolla, California

Sanjiv S. Gambhir, M.D., Ph.D., UCLA School of Medicine, Los Angeles, California

JOSEPH C. GLORIOSO, Ph.D., Department of Molecular Genetics and Biochemistry, University of Pittsburgh School of Medicine, Pittsburgh, Pennsylvania

WILLIAM F. GOINS, Ph.D., Department of Molecular Genetics and Biochemistry, University of Pittsburgh School of Medicine, Pittsburgh, Pennsylvania

ERLINDA M. GORDON, M.D., Gene Therapy Laboratories and the Department of Pediatrics, Keck School of Medicine of the University of Southern California, Los Angeles, California

HIDDE J. HAISMA, Ph.D., Department of Therapeutic Gene Modulation, University Center for Pharmacy, University of Groningen, The Netherlands

FREDERICK L. HALL, Ph.D., Gene Therapy Laboratories and the Department of Surgery, Keck School of Medicine of the University of Southern California, Los Angeles, California

MICHAEL HALLEK, M.D., Genzentrum Ludwig-Maximilians-Universitat Munchen, Munchen, Germany

ANTHEA L. HAMMOND, Ph.D., Molecular Medicine Program, Mayo Foundation, Rochester, Minnesota

MENZO J. E. HAVENGA, Ph.D., Crucell Holland BV, Leiden, The Netherlands

HARVEY R. HERSCHMAN, Ph.D., UCLA School of Medicine, Los Angeles, California

LEAF HUANG, Ph.D., Center for Pharmacogenetics, Department of Pharmaceutical Sciences, University of Pittsburgh School of Pharmacy, Pittsburgh, PA

CHERYL A. JACKSON, Ph.D., Department of Physiological Optics, University of Alabama at Birmingham, Birminghan, Alabama

LISA K. JOHANSEN, Ph.D., Department of Microbiology, University of Alabama at Birmingham, Birmingham, Alabama

SAMUEL C. KAYMAN, Ph.D., Laboratory of Retroviral Biology, Public Health Research Institute, New York, New York

JÜRGEN KLEINSCHMIDT, Ph.D., Deutsches Krebsforschungszentrum, Forschungs-schwerpunkt Angewandte Tumorvirologie, Heidelberg, Germany

ROLAND E. KONTERMANN, Ph.D., Vectron Therapeutics AG, Marburg, Germany

PHILIPPE LAMBIN, M.D., Ph.D., Department of Radiation Oncology, RTIL/U.H. Maastricht, The Netherlands

WILLY LANDUYT, Ph.D., Department of Radiation Oncology, Laboratory of Experimental Radiobiology, K.U. Leuven, Belgium

DAVID LAROCCA, Ph.D., Selective Genetics, Inc., San Diego, California

NICK LEMOINE, Ph.D., ICRF Mol Oncology Unit, London, United Kingdom

BRANDI LEVIN, B.A., Department of Pathology and Kaplan Comprehensive Cancer Center, New York University Medical Center, New York, New York

SONG LI, M.D., Ph.D., Center for Pharmacogenetics, Department of Pharmaceutical Sciences, University of Pittsburgh School of Pharmacy, Pittsburgh, PA

QIANWA LIANG, Ph.D., UCLA School of Medicine, Los Angeles, California

FENG LIU, Ph.D., Center for Pharmacogenetics, Department of Pharmaceutical Sciences, University of Pittsburgh School of Pharmacy, Pittsburgh, PA

ZHENG MA, M.D., Ph.D., Center for Pharmacogenetics, Department of Pharmaceutical Sciences, University of Pittsburgh School of Pharmacy, Pittsburgh, PA

DUNCAN C. MACLAREN, Ph.D., UCLA School of Medicine, Los Angeles, California

GIANDHAM MAHENDRA, Ph.D., Department of Pathology, The University of Alabama at Birmingham, Birmingham, AL

MAJID MAHTALI, Ph.D., GIANDHAM MAHENDRA, Ph.D., Deltagen Europe S. A., Leiden, The Netherlands

DANIEL MERUELO, Ph.D., Department of Pathology and Kaplan Comprehensive Cancer Center, New York University Medical Center, New York, New York

ATSUSHI MIYANOHARA, Ph.D., Center for Molecular Genetics and Department of Pediatrics, UCSD School of Medicine, La Jolla, California

CASEY D. MORROW, Ph.D., Department of Cell Biology, University of Alabama at Birmingham, Birmingham, Alabama

ROLF MÜLLER, Ph.D., Institute of Molecular Biology and Tumor Research, Philipps-University, Marburg, Germany

DIRK M. NETTELBECK, Ph.D., Division of Human Gene Therapy and Gene Therapy Center, University of Alabama at Birmingham, Birmingham, Alabama

SANDRA NUYTS, M.D., Laboratory of Bacteriology, Rega Institute and Department of Radiation Oncology, Laboratory of Experimental Radiobiology, K. U. Leuven, Leuven, Belgium

MATTHEW T. PALMER, B.S., Department of Cell Biology, University of Alabama at Birmingham, Birmingham, Alabama

CHRISTINE PAMPENO, Ph.D., Department of Pathology and Kaplan Comprehensive Cancer Center, New York University Medical Center, New York, New York

M. BRANDON PARROTT, B.S., Department of Immunology, Baylor College of Medicine, Houston, Texas

JEAN D. PEDUZZI, Ph.D., Department of Physiological Optics, University of Alabama at Birmingham, Birmingham, Alabama

ALEKSANDR PEREBOEV, M.D., Ph.D., Division of Human Gene Therapy, Departments of Medicine, Pathology, and Surgery and The Gene Therapy Center, The University of Alabama at Birmingham, Birmingham, AL

MICHAEL E. PHELPS, Ph.D., UCLA School of Medicine, Los Angeles, California

KATHLEEN F. PIROLLO, Ph.D., Department of Oncology, Georgetown University Medical Center, Lombardi Cancer Center, Washington, D.C.

RICHARD K. PLEMPER, Ph.D., Molecular Medicine Program, Mayo Foundation, Rochester, Minnesota

SELVARANGAN PONNAZHAGAN, Ph.D., Department of Pathology and The Gene Therapy Center, The University of Alabama at Birmingham, Birmingham, Alabama

DONNA C. PORTER, Ph.D., Replicon Technologies Incorporated, OADI Technology Center, Birmingham, Alabama

MARTIN ULRICH RIED, Ph.D., Genzentrum Ludwig-Maximilians-Universitat Munchen, Munchen, Germany

PETRA ROHRBACH, Ph.D., Institut für Molekulare Genetik, Universität Heidelberg, Germany

MARIANNE G. ROTS, Ph.D., Department of Therapeutic Gene Modulation, University Center for Pharmacy, University of Groningen, The Netherlands

STEPHEN J. RUSSELL, M.D., Ph.D., Molecular Medicine Program, Mayo Clinic, Rochester, Minnesota

ISABELLA SAGGIO, Ph.D., Department of Genetics and Molecular Biology,

University of Rome “La Sapienza” and Parco Scientifico Biomedico di Roma, S. Raffaele, Italy

NAGICHETTIAR SATYAMURTHY, Ph.D., UCLA School of Medicine, Los Angeles, California

NANCY SMYTH TEMPLETON, Ph.D., Center for Cell and Gene Therapy, Department of Molecular and Cellular Biology, Baylor College of Medicine, Houston, Texas

SATOSHI TAKAHASHI, M.D., Center for Gene Therapy, Baylor College of Medicine, Houston, Texas

YADI TAN, Ph.D., Center for Pharmacogenetics, Department of Pharmaceutical Sciences, University of Pittsburgh School of Pharmacy, Pittsburgh, PA

JAN THEYS, Ph.D., Department of Radiation Oncology, RTIL/U.H. Maastricht, The Netherlands

TATSUSHI TOYOKUNI, Ph.D., UCLA School of Medicine, Los Angeles, California

LIEVE VAN MELLAERT, Ph.D., Laboratory of Bacteriology, Rega Institute, K.U. Leuven, Belgium

RONALD VOGELS, Ph.D., Crucell Holland BV, Leiden, The Netherlands

THOMAS J. WICKHAM, Ph.D., GenVec, Inc., Gaithersburg, Maryland

LIANG XU, M.D., Ph.D., Department of Oncology, Geogetown University Medical Center, Lombardi Cancer Center, Washington, D.C.

SHAHRIAR YAGHOUBI, B.S., UCLA School of Medicine, Los Angeles, California

JOHN A. T. YOUNG, Ph.D., McArdle Laboratory for Cancer Research, University of Wisconsin-Madison, Madison, Wisconsin

ASSEM G. ZIADY, Ph.D., Department of Pediatrics at Rainbow Babies and Childrens Hospital, Case Western Reserve University School of Medicine, Cleveland, Ohio

PART I

TRANSDUCTIONALLY TARGETED VECTORS—NONVIRAL

1

ALTERNATIVE STRATEGIES FOR TARGETED DELIVERY OF NUCLEIC ACID–LIPOSOME COMPLEXES

NANCY SMYTH TEMPLETON, PH.D.

INTRODUCTION

Delivery of nucleic acids using liposomes holds great promise as a safe and nonimmunogenic approach to gene therapy. Furthermore, gene therapies that use these artificial reagents can be standardized and regulated as drugs rather than as biologics. Much effort has been devoted to the development of non-viral delivery due to the disadvantages of viruses used for gene delivery. The disadvantages of viral delivery include:

- Generation of immune responses to expressed viral proteins that subsequently kill the target cells required to produce the therapeutic gene product
- Random integration of some viral vectors into the host chromosome
- Clearance of viral vectors delivered systemically
- Difficulties in engineering viral envelopes or capsids to achieve specific delivery to cells other than those with natural tropism for the virus
- Potential recombination of the viral vector with DNA sequences in the host chromosome that generates a replication-competent, infectious virus
- Inability to administer certain viral vectors more than once

Vector Targeting for Therapeutic Gene Delivery, Edited by David T. Curiel and Joanne T. Douglas
ISBN 0-471-43479-5

- High costs for producing large amounts of high-titer viral stocks for use in the clinic
- Limited size of the nucleic acid that can be packaged and used for viral gene therapy

The advantages in using liposomes for gene therapy are several and include:

- Lack of immunogenicity
- Lack of clearance by complement using improved formulations
- Unlimited size of nucleic acids that can be delivered (from single nucleotides to large mammalian artificial chromosomes)
- Ability to perform repeated administrations in vivo without adverse consequences
- Low cost and relative ease in creating nucleic acid–liposome complexes in large scale for use in the clinic
- Relative ease in creating targeted complexes for delivery and gene expression in specific cell types, organs, or tissues
- Greater safety for patients due to few or no viral sequences present in nucleic acids used for delivery, thereby precluding generation of an infectious virus

The disadvantage of nonviral delivery systems had been the low levels of delivery and gene expression produced by first generation complexes. However, recent advances have been made that dramatically improve transfection efficiencies of nonviral vectors.

Delivery of nucleic acid-based therapeutics to specific target tissues, organs, or cells is desirable or required for certain applications. For example, lower amounts of complexes could be administered intravenously if the bulk of the injected material is delivered to target cells that are solely responsible for producing the therapeutic gene product. Injection of lower amounts of complexes would be most cost-effective, would provide another level of safety to the patient, and may produce greater efficacy for the treatment of certain diseases. Furthermore, strategies such as suicide gene approaches that are designed to kill target cells, such as tumor cells, require targeted delivery or gene expression to avoid killing normal cells. Cell or tissue-specific gene expression can also be achieved by creating plasmids containing specific promoters to produce expression exclusively in the target cells.

Nonviral delivery vehicles have no target specificity, and therefore, retargeting is not required. Basically, ligands are used to coat nucleic acid–liposome complexes to achieve specific delivery to cell surface receptors. To efficiently coat the surface of complexes, nucleic acids must be encapsulated within the delivery vehicle so that the ligand does not interfere with nucleic acid condensation. In addition, the encapsulated nucleic acid does not prevent the attachment of the ligand on the surface of the complexes.

Methods used to date for targeted delivery of nonviral delivery vehicles have produced inefficient gene expression in the target cells. Although gene expression is apparently produced primarily in the target cell, the levels of expression are lower in these cells than those produced using the nontargeted vehicle for delivery. Therefore, we have designed alternative methods to avoid this major disadvantage. These novel approaches involve targeted delivery of liposomes that are optimized for charge on the surface of complexes, cell entry by fusion with the membrane, and penetration across tight barriers in vivo to reach and diffuse through the target tissue/organ efficiently. These alternative strategies are the focus of this chapter.

ENCAPSULATION OF NUCLEIC ACIDS

Liposomes have different morphologies based on their composition and the formulation method. Formulations frequently used for the delivery of nucleic acids are small unilamellar vesicles (SUVs), multilamellar vesicles (MLVs), or bilamellar invaginated structures recently developed in our laboratory (Fig. 1.1). SUVs condense nucleic acids on the surface and form "spaghetti and meatballs" structures (Sternberg, 1996). DNA–liposome complexes made

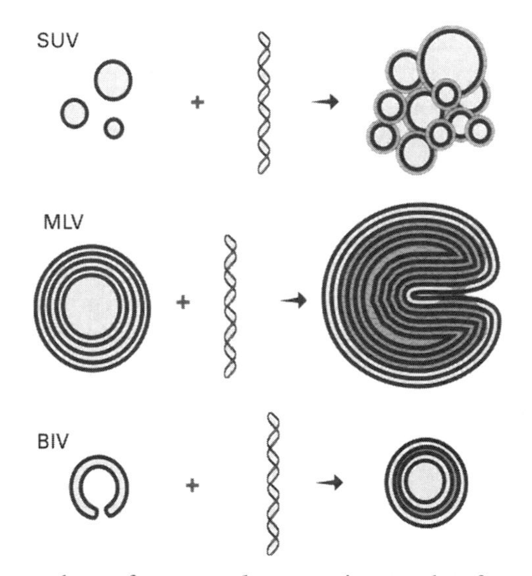

Figure 1.1 Diagrams drawn from cryoelectron micrographs of cross sections through vitrified films of various types of liposomes and DNA–liposome complexes. SUVs are small unilamellar vesicles that condense nucleic acids on the surface and produce "spaghetti and meatballs" structures. MLVs are multilamellar vesicles that appear as "Swiss rolls" after mixing with DNA. BIVs are bilamellar invaginated vesicles produced using a formulation developed in our laboratory (Templeton et al., 1997). Nucleic acids are efficiently encapsulated between two bilamellar invaginated structures.

using SUVs produce little or no gene expression on systemic delivery although these complexes transfect numerous cell types efficiently in vitro (Felgner et al., 1987; Felgner et al., 1994). Furthermore, SUV liposome–DNA complexes cannot be targeted efficiently. At best, the ligand could be added to the SUV prior to mixing with DNA. However, data from our laboratory and others show that liposome+ligand conjugates condense nucleic acids inefficiently compared to liposomes alone. In addition, the nucleic acids are exposed and not protected within the liposome.

SUV liposome–DNA complexes also have a short half-life within the circulation, generally about 5 to 10 minutes. Polyethylene glycol (PEG) has been added to liposome formulations to extend the half-life (Papahadjopoulos et al., 1991; Senior et al., 1991; Gabizon et al., 1994), however, PEGylation created other problems that have not been solved. PEG seems to hinder delivery of cationic liposomes into cells due to its sterically hindering ionic interactions that are discussed further in the section Charge Interactions. Furthermore, an extremely long half-life in the circulation (i.e., several days) has caused problems for patients because the bulk of the formulation accumulates in the skin, hands, and feet. For example, patients contract mucositis and hand-and-foot syndrome (Gordon et al., 1995; Uziely et al., 1995) that cause extreme discomfort to the patient. Addition of PEG into formulations developed in our laboratory also caused steric hindrance in the bilamellar invaginated structures that did not encapsulate DNA efficiently, and gene expression was diminished.

Some investigators have loaded nucleic acids within SUVs using a variety of methods; however, the bulk of the DNA does not load or stay within the liposomes. Furthermore, most of the processes used for loading nucleic acids within liposomes are extremely time consuming and not cost-effective. Therefore, SUVs are not the ideal liposomes for creating nonviral vehicles for targeted delivery.

Complexes made using MLVs appear as "Swiss rolls" (Fig. 1.1) when viewing cross sections by cryoelectron microscopy (Gustafsson et al., 1995). These complexes can become too large for systemic administration or deliver nucleic acids inefficiently into cells due to inability to unravel at the cell surface. Addition of ligands onto MLV liposome–DNA complexes further aggravates these problems. Therefore, MLVs are not useful for the development of targeted delivery of nucleic acids.

Using a formulation developed in our laboratory, nucleic acids are efficiently encapsulated between two bilamellar invaginated structures (BIVs) (Fig. 1.1) (Templeton et al., 1997). Figure 1.2 shows the assembly of the nucleic acid–liposome complexes. We created these structures using 1,2-dioleoyl-3-trimethylammonium-propane (DOTAP) and cholesterol, and a novel formulation procedure. This procedure is different because it includes a brief, low frequency sonication followed by manual extrusion through filters of decreasing pore size. The 0.1 and 0.2 μm filters used are made of aluminum oxide and not the polycarbonate typically used by other protocols. Aluminum oxide membranes contain more pores per surface area, evenly spaced and sized pores,

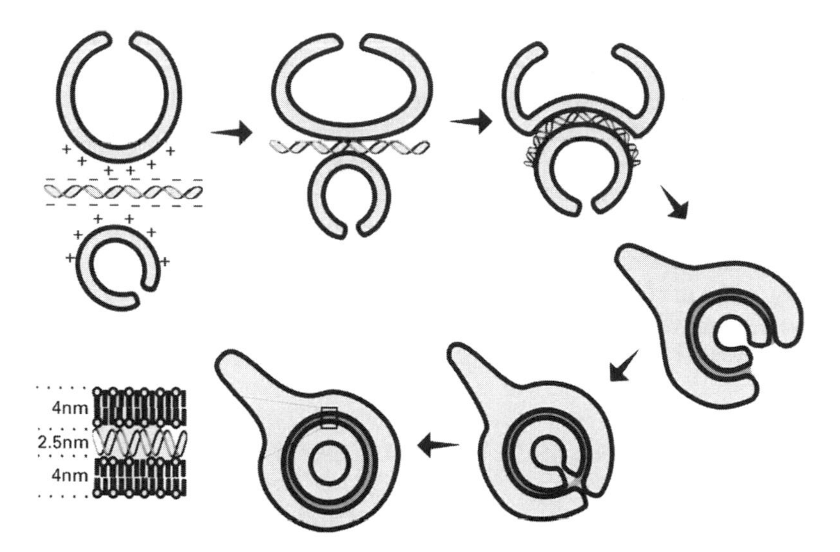

Figure 1.2 Proposed model showing cross sections of extruded DOTAP : Chol liposomes (BIVs) interacting with nucleic acids. Nucleic acids adsorb onto a BIV via electrostatic interactions. Attraction of a second BIV to this complex results in further charge neutralization. Expanding electrostatic interactions with nucleic acids cause inversion of the larger BIV and total encapsulation of the nucleic acids. Inversion can occur in these liposomes because of their excess surface area, which allows them to accommodate the stress created by the nucleic acid–lipid interactions. Nucleic acid binding reduces the surface area of the outer leaflet of the bilayer and induces the negative curvature due to lipid ordering and reduction of charge repulsion between cationic lipid headgroups. Condensation of the internalized nucleic acid–lipid sandwich expands the space between the bilayers and may induce membrane fusion to generate the apparently closed structures. The enlarged area shows the arrangement of nucleic acids condensed between two 4 nm bilayers of extruded DOTAP : Chol.

and pores with straight channels. During the manual extrusion process the liposomes are passed through each of four different size filters only once. Use of high frequency sonication and/or mechanical extrusion produces only SUVs.

The BIVs produced condense unusually large amounts of nucleic acids of any size. Furthermore, addition of other DNA condensing agents including polymers is not necessary. For example, condensation of plasmid DNA onto polymers before encapsulation in the BIVs did not increase condensation or subsequent gene expression after transfection in vitro or in vivo. Encapsulation of nucleic acids by these BIVs alone is spontaneous and immediate, and, therefore, cost effective, requiring only one step of simple mixing.

The extruded DOTAP : Chol-nucleic acid complexes are also large enough so that they are not cleared rapidly by Kupffer cells in the liver and yet extravasate across tight barriers and diffuse through the target organ efficiently. Further addition of ligands to the surface of extruded DOTAP : Chol-nucleic acid complexes does not significantly increase the mean particle size. Extravasation and penetration through the target organ and gene expression produced after transfection are not diminished. These modified formulations are positively charged and deliver nucleic acids efficiently into cells in vitro and in vivo. Extruded DOTAP : Chol-nucleic acid complexes with or without ligands also have a 5-hour half-life in the circulation, and do not accumulate in the skin, hands, or feet. Extended half-life in the circulation is provided primarily by the formulation, preparation method, injection of optimal colloidal suspensions, and optimal nucleic acid : lipid ratio used for mixing complexes, serum stability, and size (200–450 nm). Therefore, these bilamellar invaginated liposomes are ideal for use in the development of effective, targeted nonviral delivery systems that clearly require encapsulation of nucleic acids.

LIGANDS USED FOR TARGETED DELIVERY

Using liposomes that encapsulate nucleic acids, ligands can be coated onto the surface of the complexes formed (Fig. 1.3). We have added monoclonal antibodies, Fab fragments, proteins, partial proteins, peptides, peptide mimetics, small molecules, and drugs to the surface of the complexes after mixing (Templeton et al., 1997; Templeton and Lasic, 1999). Ligands are chosen by their ability to bind efficiently to a target cell surface receptor while maintaining entry into the cell by direct fusion. Entry into the cell is presented further in the section Mechanisms of Cell Entry. The ligands most useful for gene therapeu-

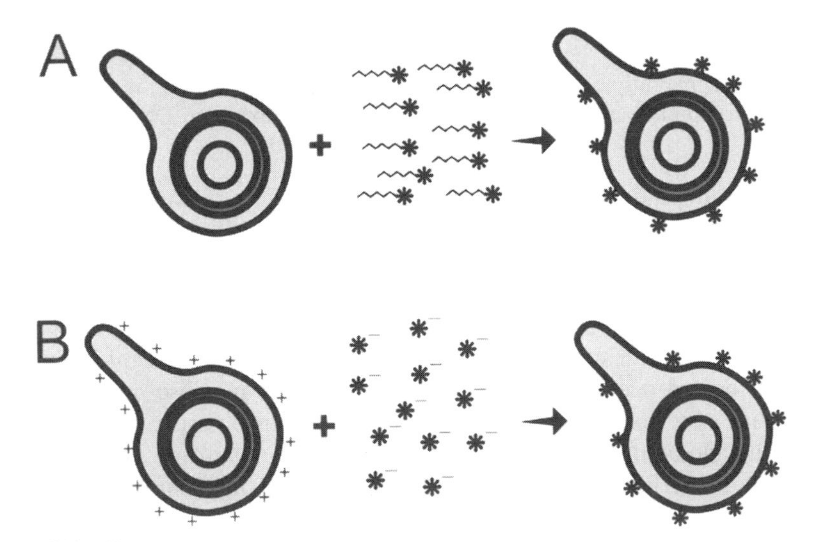

Figure 1.3 Cross sections of extruded DOTAP : Chol-nucleic acid liposome complexes (left) that are coated with ligands (right). Ligands are attached to the surface of preformed nucleic acid–liposome complexes by covalent attachments through linker lipids (*A*) or by ionic interactions (*B*).

tics in humans are those that are smallest and possess high affinity for the target receptor. Nonviral systems are desirable because they can be repeatedly administered. Therefore, immune responses may be generated in animals or people on repeated administration of complexes containing too much ligand or too large a ligand on the surface. These immune responses could cause the targeted therapeutics to be unsafe and/or ineffective for treatments in the clinic.

However, often the best ligand is not always available immediately. An investigator could also wait for years for the most appropriate ligand to be generated or produced in the amounts required for large experiments. Our experience shows that much useful information can be generated concerning targeting of a particular cell surface receptor of interest using a less-than-ideal ligand in vitro and in vivo in pilot experiments, while creating the best ligand concurrently.

ATTACHMENT OF LIGANDS

Generally investigators attach ligands to PEG for incorporation into liposomes and other conjugates or for coating onto the surface of complexes after mixing. For more information on this topic, see other chapters, particularly Chapters 2 and 3, of this volume. After extensive work with PEG in our lab, we have chosen or created alternative methods for the attachment of ligands and have avoided the use of PEG because of its numerous disadvantages presented in this chapter.

Alternative strategies to the use of PEG include attachment of ligands through ionic interactions or by covalent attachment to linker lipids. The ligands listed in the previous section are generally negatively charged. Therefore, the ligands can simply be adsorbed onto the surface of complexes with encapsulated nucleic acids after mixing (Fig. 1.3B). Additional moieties can be added to the ligand to increase the amount of negative charge and yet do not interfere with the ability of the modified ligand to bind efficiently to the appropriate cell surface receptor. For example, we used succinylated asialofetuin to target delivery of DNA–liposome complexes to the asialoglycoprotein receptor on hepatocytes in the liver. The succinic acid amides provided greater negative charge to asialofetuin and, therefore, bound to the surface of complexes more efficiently than asialofetuin alone. The amount used for adsorption onto the surface of complexes is ligand dependent. Titration studies must be performed to determine the optimal amount of ligand to coat onto the surface of complexes. Ultimately, in vivo transfection experiments must be performed to verify the optimal amount of ligand to use to provide delivery to the target cells and the highest levels of gene expression in these cells with no generation of an immune response.

Ligands or modified ligands containing reactive groups can be covalently attached to linker lipids (Fig. 1.3A). These ligand–lipid conjugates must be checked for optimal activity of the ligand to bind to its receptor. Furthermore, the covalent linkage must not be immunogenic in animals or people after repeated administration. Ligand–lipid conjugates can be spontaneously inserted into the outside membrane of complexes in which the nucleic acids are encapsulated within liposomes (Fig. 1.3A). The amount of ligand–lipid used for insertion into the surface of complexes is also ligand dependent. Titration studies must be performed to determine the optimal amount of ligand–lipid to insert into the surface of complexes. Again, in vivo transfection experiments must be performed to verify the optimal amount of ligand–lipid to use to provide delivery to the target cells and the highest levels of gene expression in these cells with no generation of an immune response.

Using the alternative approaches described above, we have produced complexes that provide delivery of nucleic acids to target cells. Furthermore, the gene expression in the target cells using the targeted complexes is higher than that using the generic extruded DOTAP : Chol complexes.

CHARGE INTERACTIONS

Cells are negatively charged on the surface with specific cell types varying in their density of negative charge. These differences in charge density can influence the ability of cells to be transfected. Cationic complexes have nonspecific ionic charge interactions with cell surfaces. Adequate charge interactions contribute, in part, to efficient transfection of cells by cationic complexes. In addition, recent publications report that certain viruses have a partial positive charge around key subunits of viral proteins on the viral surface responsible for binding to cell surface receptors. Furthermore, this partial positive charge is required for virus entry into the cell. Therefore, maintenance of an adequate positive charge on the surface of targeted liposome complexes is essential for optimal delivery into the cell.

PEGylation shields the positive charge on cationic complexes and is unable to deshield upon contact with the target cell surface. Therefore, the PEGylated complexes cannot utilize critical charge interactions for optimal transfection of cells. This topic is discussed further in Mechanisms of Cell Entry. Loss of positive charge on the surface of complexes also results in extremely low levels of gene expression in the target cells that have been transfected. Furthermore, fewer cells are transfected. As discussed above, PEGylation was first used to increase the half-life of complexes in the circulation and to avoid uptake in the lung. However, this technology also destroys the ability to efficiently transfect cells.

In using the extruded DOTAP : Chol-nucleic acid : liposome complexes, we produced an optimal half-life in the circulation without the use of PEG (Templeton et al., 1997). As mentioned above, the extended half life was produced primarily by the formulation, preparation method, injection of optimal colloidal suspensions, serum stability, optimal nucleic acid : lipid ratio used for mixing complexes, and size (200–450 nm). Furthermore, we avoid uptake in the lungs by using the negative charge of the ligands and shielding/deshielding compounds that can be added to the complexes used for targeting just prior to injection or administration in vivo. By adding ligands using the novel approaches that we have developed, adequate overall positive charge on the surface of complexes is preserved. In summary, we achieve optimal circulation time of the complexes, reach and deliver to the target organ, avoid uptake in nontarget tissues, and efficiently interact with the cell surface to produce optimal transfection.

MECHANISMS OF CELL ENTRY

Different formulations of liposomes interact with cell surfaces via a variety of mechanisms. Two major pathways for interaction are by endocytosis (Fig. 1.4A) or by direct fusion with the cell membrane (Fig. 1.4B) (Behr et al., 1989; Felgner and Ringold, 1989; Pinnaduwage and Huang, 1989; Leventis

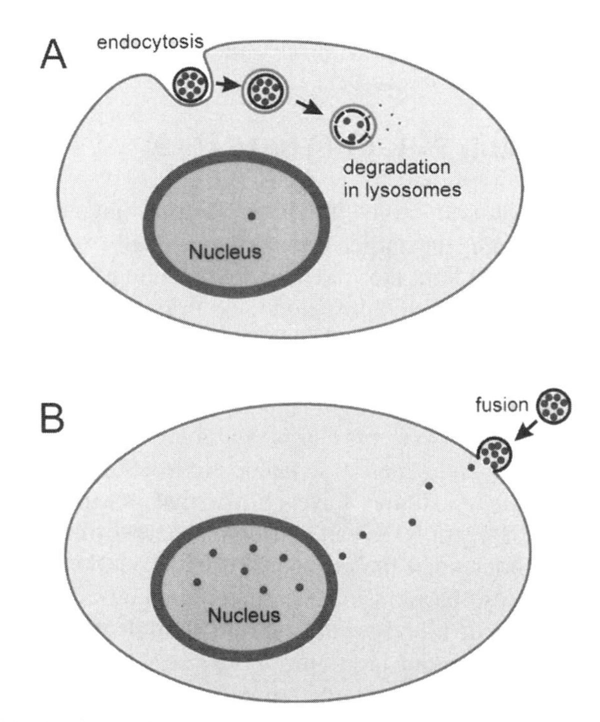

Figure 1.4 Mechanisms for cell entry of nucleic acid–liposome complexes. Two major pathways for interaction are by endocytosis (*A*) or by direct fusion with the cell membrane (*B*). Complexes that enter the cell by direct fusion allow delivery of more nucleic acids to the nucleus because the bulk of the nucleic acids do not enter endosomes.

and Silvius, 1990; Rose, Buonocore, and Whitt, 1991; Loeffler and Behr, 1993; Gustafsson et al., 1995). Many formulations, including PEGylated complexes with attached ligands used for targeted delivery, enter cells through the endocytic pathway. Unfortunately, the endosomes rapidly fuse with lysosomes and the bulk of the nucleic acid is degraded, so only a small fraction of the nucleic acid enters the nucleus. Therefore, the levels of gene expression in the target cell after transfection using ligand-associated PEGylated complexes are low.

Preliminary data suggest that nucleic acids delivered in vitro and in vivo using complexes developed in our laboratory enter the cell by direct fusion. Apparently, the bulk of the nucleic acids do not enter endosomes (Fig. 1.4B), and, therefore, far more nucleic acid enters the nucleus. Cell transfection by direct fusion produced orders of magnitude higher levels of gene expression and numbers of transfected cells versus cells transfected through the endocytic pathway. Use of the novel methods for addition of ligands to the complexes for targeted delivery results in further increased gene expression in the target cells after transfection. Therefore, we design targeted liposomal delivery systems that retain predominant entry into cells by direct fusion versus the endocytic pathway. We believe that maintenance of an adequate positive charge on the surface of complexes is essential to drive cell entry by direct fusion.

EXTRAVASATION AND TARGET PENETRATION

One of the primary goals in targeted delivery is to achieve extravasation into, and penetration through, the target organ/tissue ideally by noninvasive systemic administration. Without these events, therapeutic efficacy is highly compromised for any treatment including gene and drug therapies. Achieving this goal is difficult due to the many tight barriers that exist in animals and people. Furthermore, many of these barriers become tighter in the transition from neonates to adults.

We believe that nonviral systems can play a pivotal role in achieving the target organ extravasation and penetration needed to treat or cure certain diseases. Many of our preliminary studies have shown that extruded DOTAP : Chol-nucleic acid : liposome complexes can extravasate across tight barriers and penetrate evenly throughout entire target organs, whereas viral vectors cannot cross identical barriers. These barriers include the endothelial cell barrier in a normal mouse, the posterior blood retinal barrier in adult mouse eyes, complete and even diffusion throughout large tumors (Ramesh et al., 2001), and penetration through several tight layers of smooth muscle cells in the arteries of pigs (Templeton et al., 1999).

Currently, we are investigating the mechanisms used by extruded DOTAP : Chol-nucleic acid : liposome complexes to cross barriers and penetrate throughout target organs. By knowing more about these mechanisms, we can further develop more robust nonviral gene therapeutics.

OPTIMIZATION OF IN VIVO DELIVERY

Issues unrelated to liposome formulations must be addressed in order to produce optimal delivery, gene expression, and efficacy, including that in target organs/tissues. Based on our experience from several collaborative studies covering a wide variety of target cells and delivery routes, we have identified the major points to consider for optimization of nonviral gene therapy and delivery. The points to consider are plasmid design, plasmid DNA preparation, delivery vehicle formulation, detection of gene expression, route of administration, dosing, and administration schedule. To achieve the greatest clinical efficacy, each of these issues must be addressed and thoroughly investigated, preferably in the initial research phase and before much time and money are spent on animal studies.

Delivery of DNA and subsequent gene expression may be poorly correlated. Sometimes the delivery of DNA into the nucleus of a particular cell type may be efficient, although little or no gene expression is achieved. The causes of poor gene expression can be numerous, and other issues should be considered independent of the delivery formulation. The plasmid expression cassette may not have been optimized for animal studies. For example, many plasmids lack a full-length cytomegalovirus (CMV) promoter-enhancer, and approximately 135 variations of the CMV promoter-enhancer exist. Plasmids containing suboptimal CMV promoters-enhancers, including commercially available plasmids, produce greatly reduced or no gene expression in certain cell types or in animals. In addition, plasmids that have been optimized for overall efficiency in animals may not be best for transfection of certain cell types in vitro or in vivo. Ideally, investigators should design custom promoter-enhancer chimeras that produce the highest levels of gene expression in their target cells of interest.

Plasmids can be engineered to provide for specific or long-term gene expression, replication, or integration. Persistence elements, such as the inverted terminal repeats from adeno-associated virus, have been added to plasmids to prolong gene expression in vitro and in vivo. Apparently, these elements bind to the nuclear matrix, thereby retaining the plasmid in cell nuclei. For regulated gene expression, many different inducible promoters are used that promote expression only in the presence of a positive regulator or in the absence of a negative regulator. Tissue-specific promoters have been used for the production of gene expression exclusively in the target cells. Replication-competent plasmids or plasmids containing sequences for autonomous replication can be included to provide prolonged gene expression and sometimes produce increased levels of expression over time. Other plasmid-based strategies produce site-specific integration or homologous recombination within the host cell genome. Integration of a cDNA into a specific silent site in the genome could provide long-term gene expression without disruption of normal cellular functions. Homologous recombination could correct genetic mutations on integration of wild-type sequences that replace mutations in the genome. Plasmids that contain fewer bacterial sequences and that produce a high yield on growth in *Escherichia coli* are also desirable.

The transfection quality of plasmid DNA is dependent on the preparation protocol and training of the person preparing the DNA. In addition, optimized methods to detect and remove contaminants from plasmid DNA preparations have not been available to date. We have identified large amounts of contaminants that exist in laboratory and clinical grade preparations of plasmid DNA. These contaminants copurify with DNA by anion exchange chromatography and by cesium chloride density gradient centrifugation. Endotoxin removal does not remove these contaminants. We have developed three proprietary methods for the detection of these contaminants in plasmid DNA preparations. We can now make clinical grade (GMP) DNA that does not contain these contaminants. To provide the greatest efficacy and levels of safety, these contaminants must be assessed and removed from plasmid DNA preparations. These contaminants belong to a class of molecules known to inhibit both DNA and RNA polymerase activities. Therefore, gene expression posttransfection can be increased by orders of magnitude if these contaminants are removed from DNA preparations. The presence of these contaminants in DNA also precludes high-dose delivery of DNA–liposome complexes intravenously.

Thought should also be given to choosing the most sensitive detection method for every application of nonviral delivery rather than using the method that seems simplest. For example, detection of β-galactosidase expression is far more sensitive than detection of green fluorescent protein (GFP). Specifically, 500 molecules of β-galactosidase per cell are required for detection using X-gal staining, whereas about one million molecules of GFP per cell are required for direct detection. Furthermore, detection of GFP may be impossible if the fluorescence background of the target cell or tissue is too high. Further work is needed to develop novel in vivo detection systems that have high sensitivity and low background.

To establish the maximal efficacy for the treatment of certain diseases or for the creation of robust vaccines, injections or administrations of the nonviral gene therapeutic via different routes may be required. For particular treatments, one should not assume that one delivery route is superior to others without performing the appropriate animal experiments. In addition, people with the appropriate expertise should perform the injections and administrations.

The optimal dose should be determined for each therapeutic gene or other nucleic acid administered. The investigator should not assume that the highest tolerable dose is optimal for producing maximal efficacy.

The optimal administration schedule should also be determined for each therapeutic gene or other nucleic acid. For example, to progress faster, some investigators have simply used the same administration schedule of DNA : liposome complexes that they used for administering chemotherapeutics. In vivo experiments should be performed to determine when gene expression and/or efficacy drops significantly. Most likely, readministration of the nonviral gene therapeutic is not necessary until this drop occurs.

CONCLUSION

Despite the complexities discussed in this chapter and further challenges to be overcome for broad application of targeted nonviral vectors, we have made great progress in creating useful vectors for some target organs. In addition, we have identified important issues that require further investigation in order to produce robust targeted nonviral gene therapeutics for organs and tissues that have no useful targeted delivery vehicles at the present time. Nonviral delivery systems have clear advantages for use in targeted delivery, particularly for noninvasive systemic administration.

ACKNOWLEDGMENT

I thank Dr. David D. Roberts at the National Cancer Institute, National Institutes of Health, Bethesda, MD for preparation of the figures.

REFERENCES

Behr J-P, Demeneix B, Loeffler JP, Perez-Mutul J (1989): Efficient gene transfer into mammalian primary endocrine cells with lipopolyamine-coated DNA. *Proc Natl Acad Sci USA* 86:6982–6986.

Felgner PL, Ringold GM (1989): Cationic liposome-mediated transfection. *Nature* 337:387–388.

Felgner PL, Gadek TR, Holm M, Roman R, Chan HW, Wenz M, et al. (1987): Lipofection: a highly efficient lipid-mediated DNA transfection procedure. *Proc Natl Acad Sci USA* 84:7413–7417.

Felgner JH, Kumar R, Sridhar CN, Wheeler CJ, Tsai YJ, Border R, et al. (1994): Enhanced gene delivery and mechanism studies with a novel series of cationic lipid formulations. *J Biol Chem* 269:2550–2561.

Gabizon A, Catane R, Uziely B, Kaufman B, Safra T, Cohen R, Martin F, Huang A, Barenholz Y. (1994): Prolonged circulation time and enhanced accumulation in malignant exudates of doxorubicin encapsulated in polyethylene-glycol coated liposomes. *Cancer Res* 54:987–992.

Gordon KB, Tajuddin A, Guitart J, Kuzel TM, Eramo LR, VonRoenn J (1995): Hand-foot syndrome associated with liposome-encapsulated doxorubicin therapy. *Cancer* 75:2169–2173.

Gustafsson J, Arvidson G, Karlsson G, Almgren M (1995): Complexes between cationic liposomes and DNA visualized by cryo-TEM. *Biochim Biophys Acta* 1235:305–312.

Leventis R, Silvius JR (1990): Interactions of mammalian cells with lipid dispersions containing novel metabolizable cationic amphiphiles. *Biochim Biophys Acta* 1023:124–132.

Loeffler JP, Behr J-P (1993): Gene transfer into primary and established mammalian cell lines with lipopolyamine-coated DNA. *Methods Enzymol* 217:599–618.

Papahadjopoulos D, Allen TM, Gabizon A, Mayhew E, Matthay K, Huang SK, Lee K, Woodle MC, Lasic DD, Redemann C, Martin FJ (1991): Sterically stabilized liposomes: improvements in pharmacokinetics and antitumor therapeutic efficacy. *Proc Natl Acad Sci USA* 88:11460–11464.

Pinnaduwage P, Huang L (1989): The role of protein-linked oligosaccharide in the bilayer stabilization activity of glycophorin A for dioleoylphosphatidylethanolamine liposomes. *Biochim Biophys Acta* 986:106–114.

Ramesh R, Saeki T, Templeton NS, Ji L, Stephens LC, Ito I, Wilson DR, Wu Z, Branch CD, Minna JD, Roth JA (2001): Successful treatment of primary and disseminated human lung cancers by systemic delivery of tumor suppressor genes using an improved liposome vector. *Molecular Therapy* 3:337–350.

Rose JK, Buonocore L, Whitt MA (1991): A new cationic liposome reagent mediating nearly quantitative transfection of animal cells. *Biotechniques* 10:520–525.

Senior J, Delgado C, Fisher D, Tilcock C, Gregoriadis G (1991): Influence of surface hydrophilicity of liposomes on their interaction with plasma protein and clearance from the circulation: studies with poly(ethylene glycol)-coated vesicles. *Biochim Biophys Acta* 1062:77–82.

Sternberg B (1996): Morphology of cationic liposome/DNA complexes in relation to their chemical composition. *J Liposome Res* 6: 515–533.

Templeton NS, Lasic DD (1999): New directions in liposome gene delivery. *Molecular Biotechnology* 11:175–180.

Templeton NS, Lasic DD, Frederik PM, Strey HH, Roberts DD, Pavlakis GN (1997): Improved DNA: liposome complexes for increased systemic delivery and gene expression. *Nature Biotechnology* 15:647–652.

Templeton NS, Alspaugh E, Antelman D, Barber J, Csaky KG, Fang B, Frederik P, Honda H, Johnson D, Litvak F, Machemer T, Ramesh R, Robbins J, Roth JA, Sebastian M, Tritz R, Wen SF, Wu Z (1999): Non-viral vectors for the treatment of disease. Keystone Symposia on Molecular and Cellular Biology of Gene Therapy, Salt Lake City, Utah.

Uziely B, Jeffers S, Isacson R, Kutsch K, Wei-Tsao D, Yehoshua Z, Libson E, Muggia FM, Gabizon A (1995): Liposomal doxorubicin: antitumor activity and unique toxicities during two complementary phase I studies. *J Clin Oncol* 13:1777–1785.

2

TARGETED GENE DELIVERY VIA LIPIDIC VECTORS

SONG LI, M.D., PH.D., ZHENG MA, M.D., PH.D., YADI TAN, PH.D., FENG LIU, PH.D., JOHN DILEO, M.S., AND LEAF HUANG, PH.D.

INTRODUCTION

The success of gene therapy largely depends on the development of suitable vectors or vehicles that can deliver a gene(s) to specific target tissue with minimal toxicity. A number of methods have been developed for transfecting eukaryotic cells. These can be classified as either viral or nonviral. Viral vectors are highly efficient at transducing cells. However, safety concerns regarding their use in humans make nonviral delivery systems an attractive alternative. Nonviral vectors are particularly suitable due to their simplicity, ease of large-scale production, and lack of specific immune response in the host. A number of nonviral systems have been developed to meet gene therapy needs in different clinical settings (Li and Ma, 2001). The simplest approach to nonviral delivery systems is direct gene transfer with naked plasmid DNA.

Following the landmark discovery by Wolff and colleagues (Wolff et al., 1990) that intramuscularly injected naked plasmid DNA can be efficiently expressed in myofibers, there has been increasing interest in gene transfer via direct intratissue injection of naked DNA (Danko and Wolff, 1994). Recent studies have shown that other tissues besides muscle, such as skin, liver, kidney, and some tumors, are also susceptible to naked DNA-mediated gene transfer (Vile and Hart, 1993; Yang and Huang, 1996). Gene transfer via naked DNA can be further enhanced by electroporation (Heller et al., 1996; Zhang et al., 1996; Rols

Vector Targeting for Therapeutic Gene Delivery, Edited by David T. Curiel and Joanne T. Douglas
ISBN 0-471-43479-5

et al., 1998; Rizzuto et al., 1999; Bettan et al., 2000). Despite these encouraging results, systemic administration of free plasmid DNA under physiological conditions generally results in little, if any, gene expression in major organs. Targeted gene delivery via the systemic route requires the use of a delivery vehicle. Two major synthetic vectors have been developed to achieve this goal, namely, polymer- and lipid-based vectors (Li and Ma, 2001). In this chapter, we focus on lipidic vectors for targeted gene delivery. Emphasis is placed on the understanding of in vivo barriers in targeted gene delivery via the vascular route.

DRUG DELIVERY VERSUS GENE DELIVERY VIA LIPIDIC VECTORS

Liposomes are lipid vesicles and were initially used as a model system in membrane biophysics. Early studies on liposomes as a drug delivery system were based on their innate properties such as self-assembly to a closed, relatively permeable membrane system and recognition by the reticuloendothelial system (RES) in the blood, resulting in antigen presentation, macrophage activation, macrophage killing, and elimination of intracellular parasitic infections. Later on, it was established that a variety of other properties can be introduced by design. One such liposome, namely the pH-sensitive liposome, is designed to inhibit the degradation of encapsulated agents inside the endosome/lysosome and enhance their cytoplasmic release (Huang, Connor, and Wang, 1987). This type of liposome is useful in delivering agents that are sensitive to enzymatic degradation and whose targets reside in the cytosol or nucleus (Collins, Maxfield, and Huang, 1989). The efficiency of drug delivery via this liposome can be further enhanced via the incorporation of a targeting ligand (Wang and Huang, 1989). This new type of liposome is called a pH-sensitive immunoliposome.

A number of other types of liposomes have also been developed to suit drug delivery for different clinical settings, such as temperature-sensitive liposomes and light-sensitive liposomes. Yet, the most dramatic advancement in liposomal drug development came after the discovery that grafting of flexible water soluble polymers such as polyethylene glycol (PEG) on the surface of liposomes dramatically decreases the uptake of liposomes by the RES and prolongs their circulation times (Klibanov and Huang, 1992; Woodle and Lasic, 1992; Allen et al. 1992). This type of liposome is named a stealth liposome. It has led to a significant improvement in liposomal drug delivery to solid tumors via the systemic route. Given a sufficient circulation time, liposomal drugs are preferentially taken up by tumors due to the leaky vasculature in the tumor (Gabizon and Papahadjopoulos, 1988; Papahadjopoulos et al., 1991; Li et al., 1995). Currently, several liposomal drug formulations are being used clinically for the treatment of cancer and infectious diseases.

Recent studies have shown that the targeting efficiency of long-circulating liposomes can be further enhanced via the incorporation of a tissue-specific ligand (Lopes de Menezes et al., 2000). The experience accumulated in liposomal drug targeting has proved to be helpful in guiding the design of lipidic

TABLE 2.1 Drug Delivery Versus Gene Delivery via Lipidic Vectors

	Liposomal Drugs	Liposomal Genes
Size	Small	Large
Charge	Neutral, negative, or positive	Highly negative
Cationic lipid	Not required for achieving high encapsulation efficiency	Required for achieving high encapsulation efficiency
Cellular target	Membrane, cytosol, or nucleus	Nucleus
Cytoplasmic release	May or may not be required	Essential
Nuclear transport	Efficient	Inefficient
Aggregation in blood	No	Yes
RES uptake	Yes, requiring PEG to prolong circulation times	Yes, requiring PEG to prolong circulation times

vectors for gene delivery. However, due to the large differences between genes and chemical drugs in size and structure as well as the mechanism of action, the formulations designed for each differ in several ways, including optimal lipid composition and preparation protocol. Table 2.1 summarizes the similarities and differences between gene vectors and drug vehicles with respect to the cellular and in vivo barriers as well as the strategies to overcome these problems. The development of gene vectors is largely based on our accumulated experience in liposomal drug technology and also our understanding of how viral vectors overcome the cellular and molecular barriers in achieving efficient gene transfer.

TARGETED GENE DELIVERY VIA NEUTRAL OR ANIONIC LIPOSOMES

In the late 1970s and early 1980s, Fraley and colleagues used a reverse phase evaporation method to incorporate SV40 DNA into liposomes and demonstrated gene expression in transfected cells (Fraley et al., 1980). To improve gene delivery efficiency, Wang and Huang (1987) used pH-sensitive immunoliposomes to introduce foreign genes into target cells. Although the level of gene expression was relatively low, specific gene transfer was demonstrated in vitro and also in an ascites tumor model.

These early studies with neutral or anionic liposomes demonstrate the feasibility of using liposomal vectors for gene delivery. However, there has been a lack of progress in this field due to the technical difficulties in encapsulating a sufficient amount of DNA into the vesicles. For the past decade, attention has been primarily focused on cationic liposomes (Li and Huang, 2000). It should be noted that neutral and anionic liposomes are less toxic and more compatible with biological fluids and, therefore, may be more suitable for systemic gene delivery. Also, targeted gene delivery using these vectors might be achieved via incorporation of a tissue-specific ligand. Neutral and anionic lipidic vectors are currently being revisited and improved in several labs to achieve targeted gene delivery. These vectors are discussed in Self-Assembled Lipidic Vectors for Targeted Gene Delivery.

SYSTEMIC GENE DELIVERY USING CATIONIC LIPIDIC VECTORS: A PASSIVE TARGETING

Intravenous gene delivery via cationic liposomes was first reported by Zhu and colleagues in 1993 (Zhu et al., 1993). However, only recently were researchers able to obtain high and reproducible levels of gene expression by intravenous injection (Hong et al., 1997; Liu et al., 1997a,b; Li and Huang, 1997; Templeton et al., 1997; Wang et al., 1998). The major application of this method lies largely in pulmonary gene transfer. As discussed in the following text, pulmonary gene transfer via cationic lipidic vectors is largely due to a passive mechanism of vector aggregation. Most of the discussion is focused on studies using cationic liposome-entrapped, polycation-condensed DNA (LPD-I), a vector developed in this laboratory.

Development of LPD-I for Efficient Gene Transfer

Cationic lipids, particularly monovalent cationic lipids, are inefficient in condensing DNA. Mixing of DNA with cationic liposomes generally leads to formation of spaghetti-like structures accompanied by large aggregates that are heterogeneous in sizes. In contrast, polycations such as poly-L-lysine are highly efficient in condensing DNA. Polycation/DNA complexes are toroid in shape and are about 50 nm in size. Most cationic polymer/DNA complexes, however, have no transfection activity due to their lack of an endosome-disruption mechanism. Based on these studies, Gao and Huang (Gao and Huang, 1996) hypothesized that the introduction of cationic polymers at appropriate ratios to a mixture of cationic liposomes and DNA might lead to the formation of a new type of complexes that are not only small in size and, therefore more efficient in cellular uptake, but also efficient in cytoplasmic release following endocytosis. The initial LPD-I was composed of polylysine (PLL), 3β[N-(N′,N′-dimethylaminoethane)carbamoyl] cholesterol (DC-chol): dioleoylphosphatidylethanolamine (DOPE) liposomes, and DNA. This formulation is much more efficient than the corresponding liposome/DNA complexes in transfecting cells in vitro. This formulation is also highly efficient in transfecting neurons following intracranial injection and is currently being used in a clinical trial for the treatment of Canavan disease, a leukodystrophy caused by mutation in the aspartoacylase (ASPA) gene (Leone et al., 2000). Recently, LPD-I has been reoptimized and improved for achieving systemic gene transfer (Li and Huang, 1997; Li et al., 1998). The new formulation is composed of protamine sulfate, 1,2-dioleoyl-3-trimethylammonium-propane (DOTAP): cholesterol liposomes, and DNA. Examination of the new formulation by cryo-electron microscopy reveals viruslike particles that contain a core of protamine-condensed DNA coated with lipid bilayers. These particles are highly stable and can be lyophilized and kept at room temperature for several months without any significant changes in either biophysical properties or transfection efficiency (Li, B. et al., 2000). Systemic administration of these particles results in gene expression in all major organs including the lung, heart, liver, spleen, and kidney. The lung is the organ with the highest level of gene expression and endothelial cells appear to be the major cell type that is transfected (Li and Huang, 1997; Li et al., 1998). Similar to cationic lipid/DNA complexes, the efficiency of gene transfer by LPD-I via the vascular route is affected greatly by lipid composition. There is about a 100- to 1000-fold difference between DOPE-containing LPD and cholesterol-containing LPD with respect to the level of gene expression. Delivery of a therapeutic gene such as the Rb gene by cholesterol containing-LPD-I holds promise for the treatment of pulmonary diseases such as pulmonary metastases (Nikitin et al., 1999).

Interactions of LPD-I with Mouse Serum

As an approach to understanding why LPD or cationic lipid/DNA complexes of different lipid compositions have a dramatic difference in the efficiency of gene delivery, we have studied the interaction of LPD-I with mouse serum with an emphasis on how serum affects its biophysical and biological properties (Li et al., 1998; Li et al., 1999a). Exposure of LPD-I to mouse serum results in an immediate size increase, suggesting serum-induced aggregation of LPD particles. Interactions of LPD with mouse serum also lead to changes in surface charge: all LPDs become negatively charged following exposure to serum. Interestingly, DOPE-containing LPD recruits more serum proteins than other LPD formulations. LPD-associated serum proteins are mainly albumin and some other proteins of high molecular weight (Li et al., 1998; Li et al., 1999a). Prolonged interactions of LPD with serum lead to disintegration of the vector. Figure 2.1 shows the change of turbidity of LPDs of different lipid compositions following exposure to serum. Exposure of LPD to serum results in an immediate increase in turbidity. The DOPE-containing LPD is the most turbid one. Prolonged incubation of LPD with serum is then associated with a decrease in turbidity. The rate of decrease in turbidity is much faster for the DOPE formulation than for the other two formulations. This decrease in turbidity suggests a process of vector disintegration, which is further confirmed by other biophysical studies including sucrose density gradient ultracentrifugation analysis and fluorescence resonance energy transfer (FRET) assay (Li et al., 1999a).

Table 2.2 summarizes how serum affects the biophysical properties of LPD-I and its transfection efficiency. The immediate effect of serum is aggregation. Aggregation plays a key role in determining the initial accumulation of LPD-I in the pulmonary microvasculature. Subsequent interactions of LPD with serum

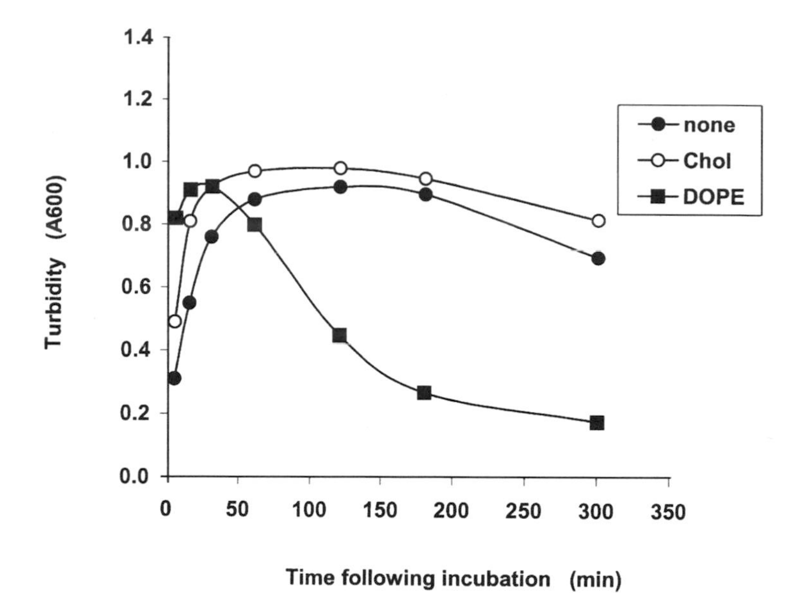

Figure 2.1 Dynamic changes in the turbidity of LPD-I after exposure to mouse serum. LPD-I containing different helper lipids were mixed with mouse serum at a ratio of 1/2 (v/v) and the mixtures were incubated at 37°C with gentle shaking. The absorbance of the mixtures at 600 nm was recorded at different times with serum alone as a blank control. From Li et al., 1999a, by permission.

lead to vector disintegration, the consequences of which are the clearance of LPD-I from the lung, DNA release, and degradation. The balance between the rates of the initial aggregation and the subsequent disintegration plays an important role in determining the amount of intact DNA that interacts with target cells (endothelial cells). The DOPE formulation, in spite of its rapid rate of initial aggregation, is inefficient in pulmonary gene transfer due to its rapid rate of disintegration. The formulations without any helper lipid, although highly resistant to serum-mediated disintegration, have a relatively slow rate

TABLE 2.2 Effect of Serum on Cationic Lipidic Vectors of Different Lipid Compositions

Helper Lipid	Aggregation Rate	Disassembly/DNA Degradation Rate	Lung		
			Accumulation	Retention	Transfection
None	+	+	++	+++	+++
Cholesterol	+++	++	+++	+++	++++
DOPE	++++	++++	++++	+	+

Source: From Li et al., 1999a, by permission.

of initial aggregation. These formulations have an intermediate level of gene expression. Cholesterol-containing formulations show the best balance between the rapid initial aggregation and the subsequent slow disintegration, and are the most efficient vector in pulmonary gene transfer. This study therefore explains why cationic lipidic vectors of different lipid compositions have a dramatic difference in their in vivo transfection efficiency. These studies also clearly demonstrate the critical role of vector aggregation in cationic lipid-mediated pulmonary gene transfer via the vascular route.

Efficient Pulmonary Gene Transfer via Sequential Injection of Cationic Liposomes and Plasmid DNA

Although the new lipidic vectors demonstrated improved efficiency in pulmonary gene transfer, their clinical applications are limited by their toxicities including the induction of proinflammatory cytokines. We and others have recently shown that systemic administration of cationic liposome/DNA complexes can trigger the production of large amounts of proinflammatory cytokines including interferon γ(IFN-γ), tumor necrosis factor α(TNF-α), interleukin 12(IL-12) and others (Freimark et al., 1998; Li et al., 1999b; Yew et al., 1999; Lanuti et al., 2000). These cytokines not only are toxic at high doses but also inhibit transgene expression (Li et al., 1999b). Cytokine induction is largely due to the unmethylated CpG motifs in bacteria plasmid DNA, although cationic lipids may play a synergistic role. Several strategies have been proposed to overcome this problem, including modification of plasmid DNA to decrease the CpG content (Yew et al., 1999; Yew et al., 2000), improvement of the specificity of the vectors for endothelial cells (Li, S. et al., 2000), and the use of a general immunosuppressant (Tan et al., 1999). Recently, we have found that the cytokine toxicity can be significantly reduced via a simple, sequential injection protocol, that is, injection of cationic liposomes followed by plasmid DNA (Tan et al., 2001). This protocol was initially shown by Song and co-workers (Song et al., 1998) to efficiently transfect the lung. They speculated that the free DNA is responsible for lung gene transfer. Cationic liposomes probably enhance gene transfer by prolonging the residency time of DNA in the pulmonary microcirculation and facilitating its interaction with endothelial cells. This hypothesis was supported by their observation that intrapulmonary arterial injection of free DNA can efficiently transfect isolated mouse lungs (Song et al., 1998). Based on these studies, we proposed that the sequential injection may be associated with a minimal cytokine response, because the majority of DNA will stay in the free form and will have a much reduced interaction with immune cells as compared to aggregated cationic lipid/DNA complexes. Indeed, sequential injection led to significantly reduced levels of a number of proinflammatory cytokines (IFN-γ, TNF-α, and IL-12) as compared to complex injection (Tan et al., 2001). Furthermore, this new protocol brought about improved pulmonary gene transfer, probably due to a decreased inhibitory effect of cytokines on gene transfer (Tan et al., 2001).

More recently, we have found that a sequential injection protocol can also improve pulmonary gene transfer by adenovirus via the vascular route (Ma et al., 2001). It is well known that adenoviral vectors, although highly efficient in liver gene transfer, have proved to be limited for pulmonary gene transfer with respect to efficiency, in part because of difficulty in assuring significant residence time in the lung and/or a paucity of cellular receptors for adenovirus on the endothelium. One approach to improve the efficiency of gene transfer to pulmonary endothelium is genetic modification of adenoviral vectors to render them endothelium-specific as discussed in other chapters of this book. Our study provides a different approach to improve pulmonary gene transfer using adenovirus. The improvement in pulmonary gene transfer was associated with a decrease in gene expression in the liver (Fig. 2.2). Similar to the sequential protocol for plasmid DNA-mediated gene transfer, the improvement in pulmonary gene transfer via the adenovirus might be due to an improved interaction of the adenoviral vectors with the pulmonary endothelium. This was supported by the observation that more luciferase cDNA was detected in the lungs in the sequential group than in the lungs of the group receiving the adenoviral vector alone. Interestingly, premixing of cationic liposomes with the adenovirus resulted in a decrease in gene expression in all major organs in a dose-dependent manner. The CpG-cytokine response that was associated with plasmid DNA was not observed in mice receiving either adenovirus alone or liposomes followed by adenovirus. This protocol, together with the genetic modification of adenoviral

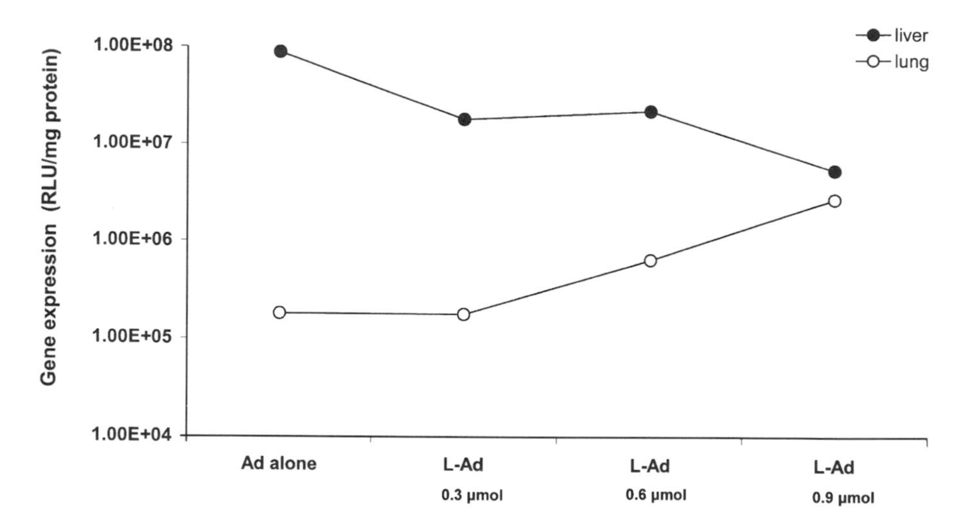

Figure 2.2 Gene expression in mice following tail vein injection of adenovirus containing luciferase cDNA (AdCMVLuc) or DOTAP: cholesterol liposomes followed by AdCMVLuc (sequential protocol). DOTAP: cholesterol liposomes (1 : 1; m/m) were prepared by the extrusion method with a final size of 100 nm. The amount of AdCMVLuc injected was 4×10^{10} particles/mouse. Gene expression in lung and liver was assayed 72 h following the injection ($n = 3$).

vectors, may prove to be useful for pulmonary gene transfer for the treatment of pulmonary diseases. This method may also be extended to pulmonary gene transfer using other types of viral vectors via the vascular route.

SELF-ASSEMBLED LIPIDIC VECTORS FOR TARGETED GENE DELIVERY

While the preceding studies indicate the potential of cationic lipidic vectors in gene transfer for the treatment of pulmonary diseases, they also support the long-held notion that positively charged lipidic vectors are not suitable for gene delivery in a tissue-specific manner. Active targeting requires the use of a vector with a neutral or anionic surface, or a cationic lipidic vector with a shielded surface. Several such vectors have been developed. These vectors differ from traditional neutral or anionic liposomes as they include a cationic component to improve the DNA entrapment efficiency. They also differ from the traditional cationic liposome/DNA complexes because their preparation follows a new self-assembling process that does not involve preformed cationic liposomes. Development of these vectors is largely based on our understanding of how viruses bypass the cellular and molecular barriers in gene transfer and our accumulated experience in targeted drug delivery using stealth liposomes.

Development of LPD-II for Targeted Gene Delivery

Preparation of LPD-II involves a process similar to that for preparation of LPD-I except that anionic liposomes instead of cationic liposomes are used (Fig. 2.3). Similar to LPD-I, these structures contain a core of PLL-condensed DNA surrounded by a lipid shell. Therefore, this novel vector has been named LPD-II (Lee and Huang, 1996).

The net charge of LPD-II particles is related to the lipid/DNA ratio. At low lipid/DNA ratios, the particles carry a net positive charge. Transfection and uptake of cationic LPD-II are independent of a targeting ligand and do not require a pH-sensitive lipid composition. When a sufficient amount of lipid is used, the particles become slightly negatively charged. Transfection and uptake of anionic LPD-II require the use of a targeting ligand, and are also dependent on the pH-sensitive lipid composition (Lee and Huang, 1996).

The advantages of LPD-II are that, besides being highly compact, it does not require purification and is a single-vial formulation. Compared with traditional anionic and neutral liposomal vectors, DNA is highly condensed in LPD-II and is quantitatively encapsulated without requiring excess amounts of lipids. Because anionic liposomes are compatible with biological fluids, LPD-II potentially can be used for tissue-specific gene delivery. The major problem with the initial LPD-II formulations is their high degree of serum-sensitivity, as incorporation of serum proteins into LPD-II stabilizes the bilayer structure,

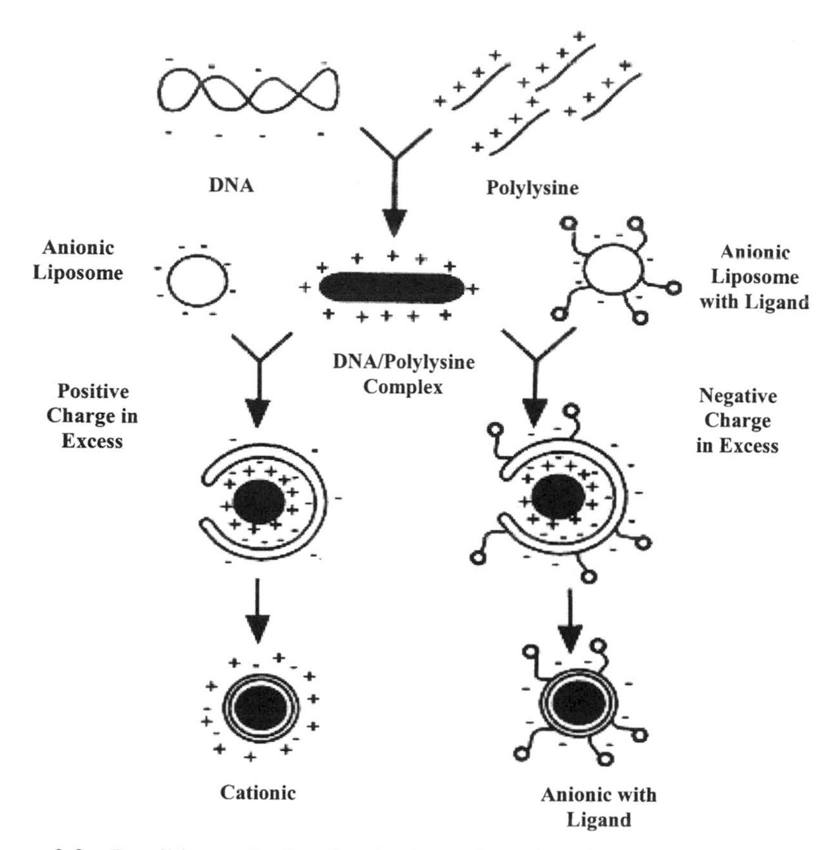

Figure 2.3 Possible mechanism for the formation of LPD-II particles. The targeting ligand is folate. From Lee and Huang, 1996, by permission.

resulting in the loss of pH-dependent fusogeneic activity. This problem can be resolved, at least partially, by the use of pH-insensitive liposomes together with a cationic polymer such as polyethylenimine (PEI), which can effectively buffer endosomal acidification. It remains to be determined whether LPD-II is effective in targeted gene delivery in vivo.

Other Lipidic Self-Assembling Systems for Targeted Gene Delivery

Several studies have shown that a hydrophobic lipid/DNA complex can be prepared in the absence of preformed liposomes (Reimer et al., 1995; Wong, Reimer, and Bally, 1996). Cationic lipid and plasmid DNA are solubilized in a Bligh and Dyer monophase consisting of chloroform/methanol/water (1 : 2.1 : 1). Subsequently, the sample is partitioned into an aqueous phase and an organic phase by the further addition of chloroform and water. The binding of DNA to cationic lipids results in charge neutralization and extraction of the complex into the organic phase. This hydrophobic lipid/DNA complex can be isolated and used as a unique intermediate for the preparation of well-defined particles. For example, the complex can be used in the preparation of oil/water emulsions. The complex can also be dissolved in alternative solvents for the preparation of membrane structures via a reverse-phase evaporation technique. One of the major advantages of this system is the flexibility in modifying the surface properties of the resulting particles. Neutral or anionic particles can be prepared by incorporating appropriate lipids into the complex. A targeting ligand can also be used for targeted in vivo gene delivery. These properties are not shared by cationic liposome/DNA complexes. The reconstituted chylomicrons recently developed by Hara and colleagues (Hara, Tan, and Huang, 1997) demonstrate the use of a hydrophobic lipid/DNA complex for preparation of well-defined particles. 3β-[N′,N′,N′-trimethylaminoethane)]cholesterol iodide (TC-chol), a quaternary ammonium derivative of DC-chol was employed to form a hydrophobic complex with DNA. The hydrophobic TC-chol/DNA complex extracted from the organic phase was then incorporated into reconstituted chylomicrons by emulsification with appropriate amounts of triglyceride, l-α-phosphatidylcholine (PC), lysophosphatidylcholine (lyso PC), cholesterol, and cholesteryl oleate. After extrusion, the size of the reconstituted chylomicrons was about 100 nm, with a DNA incorporation efficiency of greater than 60% (Hara, Tan, and Huang, 1997). Injection of reconstituted chylomicrons into mice through the portal vein resulted in gene expression in all major organs including the lung, heart, liver, spleen, and kidney, with the highest level of gene expression found in the liver. The level of gene expression by reconstituted chylomicrons was about 100-fold higher than that when naked plasmid DNA was used (Hara, Tan, and Huang, 1997). Recently, this formulation has been improved to incorporate phosphatidylethanolamine conjugated polyethylene glycol (PE-PEG). This modification significantly prolongs the circulation times of chylomicrons in the blood (Chesnoy et al., 2001). It remains to be determined whether targeted gene delivery can be achieved using this vector following incorporation of a tissue-specific ligand.

Cullis' group has reported a detergent dialysis procedure which allows encapsulation of plasmid DNA within a lipid envelope, whereby the resulting particle is stabilized in aqueous media by the presence of a PEG coating (Zhang et al., 1999; Tam et al., 2000). These stabilized plasmid-lipid particles (SPLPs) exhibit an average size of 70 nm in diameter, contain one plasmid per particle and fully protect the encapsulated plasmid from digestion by serum nucleases and *Escherichia coli* DNase I. Encapsulation is a sensitive function of the cationic lipid content, with maximum entrapment observed at dioleoyldimethylammonium chloride (DODAC) contents of 5 to 10 mol%. The formulation process results in plasmid-trapping efficiencies of up to 70% and permits inclusion of fusigenic lipids such as DOPE. The blood clearance and the biodistribution of the SPLPs can be modulated by varying the acyl chain length of the ceramide (Cer) group used as a lipid anchor for the PEG polymer. Circulation lifetimes (t1/2) observed for SPLPs with PEG-CerC14 and PEG-CerC20 were approximately 1 h and approximately 10 h, respectively (Zhang et al., 1999). The SPLPs are stable while circulating in the blood and the encapsulated DNA is fully protected from degradation by serum nucleases. Significant accumulation (approximately 10% of injected dose) of the long-circulating SPLPs with PEG-CerC20 in a distal tumor (Lewis lung tumor in the mouse flank) was observed following intravenous application (Zhang et al., 1999). The efficient accumulation, however, was associated with only a low level of gene expression in the tumor. This level might be due to the poor interaction of these particles with tumor cells in the absence of a targeting ligand. Another possibility is the difficulty in DNA release from the particles following cellular uptake due to the compact structures. These problems might be resolved by the incorporation of a targeting ligand and the use of a lipid composition that results in the formation of less compact complexes with plasmid DNA.

A different protocol for the preparation of long-circulating and targeted cationic lipid/DNA complexes involves the preparation of cationic liposome/DNA complexes followed by incubation with PE-PEG micelles. Previous studies have shown that PE-PEG in micellar form can spontaneously transfer to preformed liposomes composed of neutral lipids and impart long in vivo circulation half-life to an otherwise rapidly cleared lipid composition. The study from Huang and colleagues (Huang et al., 2001) suggested that stealth cationic lipid/DNA complexes could be prepared in a similar manner. Furthermore, ligand-conjugated PE-PEG can be incorporated to render the vectors target specific. One liver-specific vector thus prepared demonstrated a 10-fold higher level of gene expression in the liver as compared to the lung. Better characterization of the vectors may lead to a further improvement in tissue specificity and in transfection efficiency.

An interesting observation by Cheng (1996) showed that noncovalent attachment of a targeting ligand to cationic liposome/DNA complexes improves the transfection of cells that express the corresponding receptors. Later on, extended studies by Chang and colleagues (Xu et al., 1999) showed that such a system can efficiently mediate targeting of a gene to distant subcutaneous tumors via the vascular route. A detailed description of this system can be found in Chapter 3. The advantages of this system lie in its simplicity and apparent high efficiency. It remains to be addressed how this system bypasses the in vivo barriers such as nonspecific uptake by RES or extravasation at the tumor site.

CONCLUSION

The past decade has seen substantial progress in gene therapy using lipidic vectors. Several lipidic formulations are currently under clinical evaluation for local gene delivery for the treatment of cancers and cystic fibrosis. Recent studies have also led to the development of several novel lipidic formulations that show promise for targeted gene delivery via the vascular route. However, many barriers still need to be overcome before they can be used clinically. Their efficiency still needs to be improved for patients to benefit from the gene therapy. Also, the short- and long-term toxicity of lipids, especially cationic lipids, needs to be more rigorously addressed, particularly when repeated dosing is required to achieve or maintain a therapeutic effect. Further advancements in this field will continue to rely on a better understanding about the cellular and in vivo barriers in gene transfer. Experience accumulated in gene delivery using other vectors will also help to advance the development of lipidic vectors. Furthermore, hybrid systems that combine lipidic vectors with other nonviral systems or viral systems can be developed to improve the efficiency of gene transfer while overcoming the limitations of individual vectors. Finally, targeting efficiency can be further refined by improvements in gene expression systems as discussed in other chapters of this book. These advancements will help to maximize the therapeutic effect while minimizing the undesirable side effects.

ACKNOWLEDGMENT

The authors are supported by NIH grants HL 63080 (to Li), AI 48851, DK 54225, CA 74918, DK 44935, and AR 45925 (to Huang), and DOD PC 001525 (to Li).

REFERENCES

Allen TM, Mehra T, Hansen C, Chin YC (1992): Stealth liposomes: An improved sustained release system for 1-beta-D-arabinofuranosylcytosine. *Cancer Res* 52:2431–2439.

Bettan M, Emmanuel F, Darteil R, Caillaud JM, Soubrier F, Delaere P, Branelec D, Mahfoudi A, Duverger N, Scherman D (2000): High-level protein secretion into blood circulation after electric pulse-mediated gene transfer into skeletal muscle. *Mol Ther* 2:204-210.

Cheng PW (1996): Receptor ligand-facilitated gene transfer: Enhancement of liposome-mediated gene transfer and expression by transferrin. *Hum Gene Ther* 7:275–282.

Chesnoy S, Durand D, Doucet J, Stolz DB, Huang L (2001): Improved DNA/emulsion complex stabilized by polyethylene-glycol conjugated phospholipid. *Pharm Res* 18:1480–1484.

Collins D, Maxfield F, Huang L (1989): Immunoliposomes with different acid sensitivities as probes for the cellular endocytic pathway. *Biochim Biophys Acta* 987:47–55.

Danko I, Wolff JA (1994): Direct gene transfer into muscle. *Vaccine* 12:1499–1502.

Fraley R, Subramani S, Berg P, Papahadjopoulos D (1980): Introduction of liposome-encapsulated SV40 DNA into cells. *J Biol Chem* 255:10431–10435.

Freimark BD, Blezinger HP, Florack VJ, Nordstrom JL, Long SD, Deshpande DS, Nochumson S, Petrak KL (1998): Cationic lipids enhance cytokine and cell influx levels in the lung following administration of plasmid: cationic lipid complexes. *J Immunol* 160:4580–4586.

Gabizon A, Papahadjopoulos D (1988): Liposome formulations with prolonged circulation time in blood and enhanced uptake by tumors. *Proc Natl Acad Sci USA* 85:6949–6953.

Gao X, Huang L (1996): Potentiation of cationic liposome-mediated gene delivery by polycations. *Biochemistry* 35:1027–1036.

Hara T, Tan Y, Huang L (1997): In vivo gene delivery to the liver using reconstituted chylomicron remnants as a novel nonviral vector. *Proc Natl Acad Sci USA* 94:14547–14552.

Heller R, Jaroszeski M, Atkin A, Moradpour D, Gilbert R, Wands J, Nicolau C (1996): In vivo gene electroinjection and expression in rat liver. *FEBS Lett* 389:225–228.

Hong K, Zheng W, Baker A, Papahadjopoulos D (1997): Stabilization of cationic liposome-plasmid DNA complexes by polyamines and poly(ethylene glycol)-phospholipid conjugates for efficient in vivo gene delivery. *FEBS Lett* 400:233–237.

Huang L, Connor J, Wang CY (1987): pH-sensitive immunoliposomes. *Methods Enzymol* 149:88–99.

Huang SK, Jin B, Zhang W, Mullah N, Zalipsky S (2001): Efficient systemic delivery of factor VIII gene by poly(ethylene glycol)-grafted, galactose-conjugated neutral-cationic lipid/DNA complex. *Mol Ther* 3:S10–S11.

Klibanov AL, Huang L (1992): Long-circulating liposomes: development and perspectives. *J Liposome Res* 2:321–334.

Lanuti M, Rudginsky S, Force SD, Lambright ES, Siders WM, Chang MY, Amin KM, Kaiser LR, Scheule RK, Albelda SM (2000): Cationic lipid: bacterial DNA complexes elicit adaptive cellular immunity in murine intraperitoneal tumor models. *Cancer Res* 60:2955–2963.

Lee RJ, Huang L (1996): Folate-targeted, anionic liposome-entrapped polylysine-condensed DNA for tumor cell-specific gene transfer. *J Biol Chem* 271:8481–8487.

Leone P, Janson CG, Bilaniuk L, Wang Z, Sorgi F, Huang L, Matalon R, Kaul R, Zeng Z, Freese A, McPhee SW, Mee E, During MJ (2000): Aspartoacylase gene transfer to the mammalian central nervous system with therapeutic implications for Canavan disease. *Ann Neurol* 48:27–38.

Li B, Li S, Tan Y, Stolz DB, Watkins SC, Block LH, Huang L (2000): Lyophilization of cationic lipid-protamine-DNA (LPD) complexes. *J Pharm Sci* 89:355–364.

Li S, Huang L (1997): In vivo gene transfer via intravenous administration of cationic lipid-protamine-DNA (LPD) complexes. *Gene Ther* 4:891–900.

Li S, Huang L (2000): Nonviral gene therapy: promises and challenges. *Gene Ther* 7:31–34.

Li S, Ma Z (2001): Nonviral gene therapy. *Current Gene Ther* 1:201–226.

Li S, Khokhar AR, Perez-Soler R, Huang L (1995): Improved antitumor activity of cis-Bis-neodecanoato-trans-R,R-1,2-diaminocyclohexaneplatinum (II) entrapped in long circulating liposomes. *Oncol Res* 7:611–617.

Li S, Rizzo MA, Bhattacharya S, Huang L (1998): Characterization of cationic lipid-protamine-DNA (LPD) complexes for intravenous gene delivery. *Gene Ther* 5:930–937.

Li S, Tseng WC, Stolz DB, Wu SP, Watkins SC, Huang L (1999a): Dynamic changes in the characteristics of cationic lipidic vectors after exposure to mouse serum: implications for intravenous lipofection. *Gene Ther* 6:585–594.

Li S, Wu SP, Whitmore M, Loeffert EJ, Wang L, Watkins SC, Pitt BR, Huang L (1999b): Effect of immune response on gene transfer to the lung via systemic administration of cationic lipidic vectors. *Am J Physiol* 276:L796–804.

Li S, Tan Y, Viroonchatapan E, Pitt BR, Huang L (2000): Targeted gene delivery to pulmonary endothelium by anti-PECAM antibody. *Am J Physiol Lung Cell Mol Physiol* 278:L504–11.

Liu F, Qi H, Huang L, Liu D (1997a): Factors controlling the efficiency of cationic lipid-mediated transfection in vivo via intravenous administration. *Gene Ther* 4:517–523.

Liu Y, Mounkes LC, Liggitt HD, Brown CS, Solodin I, Heath TD, Debs RJ (1997b): Factors influencing the efficiency of cationic liposome-mediated intravenous gene delivery. *Nat Biotechnol* 15:167–173.

Lopes de Menezes DE, Pilarski LM, Belch AR, Allen TM (2000): Selective targeting of immunoliposomal doxorubicin against human multiple myeloma in vitro and ex vivo. *Biochim Biophys Acta* 1466:205–220.

Ma Z, Mi Z, Robbinson PD, Pitt B, Li S (2001): Redirecting adenovirus to pulmonary endothelium by cationic liposome. *Mol Ther* 3:S198–S199.

Nikitin AY, Juarez-Perez MI, Li S, Huang L, Lee WH (1999): RB-mediated suppression of spontaneous multiple neuroendocrine neoplasia and lung metastases in Rb+/- mice. *Proc Natl Acad Sci USA* 96:3916–3921.

Papahadjopoulos D, Allen TM, Gabizon A, Mayhew E, Matthay K, Huang SK, Lee KD, Woodle MC, Lasic DD, Redemann C, et al. (1991): Sterically stabilized liposomes: improvements in pharmacokinetics and antitumor therapeutic efficacy. *Proc Natl Acad Sci USA* 88:11460–11464.

Reimer DL, Zhang Y, Kong S, Wheeler JJ, Graham RW, Bally MB (1995): Formation of novel hydrophobic complexes between cationic lipids and plasmid DNA. *Biochemistry* 34:12877–12883.

Rizzuto G, Cappelletti M, Maione D, Savino R, Lazzaro D, Costa P, Mathiesen I, Cortese R, Ciliberto G, Laufer R, La Monica N, Fattori E (1999): Efficient and regulated erythropoietin production by naked DNA injection and muscle electroporation. *Proc Natl Acad Sci USA* 96:6417–6422.

Rols MP, Delteil C, Golzio M, Dumond P, Cros S, Teissie J (1998): In vivo electrically mediated protein and gene transfer in murine melanoma. *Nat Biotechnol* 16:168–171.

Song YK, Liu F, Liu D (1998): Enhanced gene expression in mouse lung by prolonging the retention time of intravenously injected plasmid DNA. *Gene Ther* 5:1531–1537.

Tam P, Monck M, Lee D, Ludkovski O, Leng EC, Clow K, Stark H, Scherrer P, Graham RW, Cullis PR (2000): Stabilized plasmid-lipid particles for systemic gene therapy. *Gene Ther* 7:1867–1874.

Tan Y, Li S, Pitt B, Huang L (1999): The inhibitory role of CpG immunostimulatory motifs in cationic lipid vector-mediated transgene expression in vivo. *Hum Gene Ther* 10:2153–2161.

Tan Y, Liu F, Li Z, Li S, Huang L (2001): Sequential injection of cationic liposome and plasmid DNA effectively transfects the lung with minimal inflammatory toxicity. *Mol Ther* 3:673–682.

Templeton NS, Lasic DD, Frederik PM, Strey HH, Roberts DD, Pavlakis GN (1997): Improved DNA: liposome complexes for increased systemic delivery and gene expression. *Nat Biotechnol* 15:647–652.

Vile RG, Hart IR (1993): Use of tissue-specific expression of the herpes simplex virus thymidine kinase gene to inhibit growth of established murine melanomas following direct intratumoral injection of DNA. *Cancer Res* 53:3860–3864.

Wang CY, Huang L (1987): pH-sensitive immunoliposomes mediate target-cell-specific delivery and controlled expression of a foreign gene in mouse. *Proc Natl Acad Sci USA* 84:7851–7855.

Wang CY, Huang L (1989): Highly efficient DNA delivery mediated by pH-sensitive immunoliposomes. *Biochemistry* 28:9508–9514.

Wang J, Guo X, Xu Y, Barron L, Szoka FC Jr (1998): Synthesis and characterization of long chain alkyl acyl carnitine esters. Potentially biodegradable cationic lipids for use in gene delivery. *J Med Chem* 41:2207–2215.

Wolff JA, Malone RW, Williams P, Chong W, Acsadi G, Jani A, Felgner PL (1990): Direct gene transfer into mouse muscle in vivo. *Science* 247:1465–1468.

Wong FM, Reimer DL, Bally MB (1996): Cationic lipid binding to DNA: characterization of complex formation. *Biochemistry* 35:5756–5763.

Woodle MC, Lasic DD (1992): Sterically stabilized liposomes. *Biochim Biophys Acta* 1113:171–199.

Xu L, Pirollo KF, Tang WH, Rait A, Chang EH (1999): Transferrin-liposome-mediated systemic p53 gene therapy in combination with radiation results in regression of human head and neck cancer xenografts. *Hum Gene Ther* 10:2941–2952.

Yang JP, Huang L (1996): Direct gene transfer to mouse melanoma by intratumor injection of free DNA. *Gene Ther* 3:542–548.

Yew NS, Wang KX, Przybylska M, Bagley RG, Stedman M, Marshall J, Scheule RK, Cheng SH (1999): Contribution of plasmid DNA to inflammation in the lung after administration of cationic lipid : pDNA complexes. *Hum Gene Ther* 10:223–234.

Yew NS, Zhao H, Wu IH, Song A, Tousignant JD, Przybylska M, Cheng SH (2000): Reduced inflammatory response to plasmid DNA vectors by elimination and inhibition of immunostimulatory CpG motifs. *Mol Ther* 1:255–262.

Zhang L, Li L, Hoffmann GA, Hoffman RM (1996): Depth-targeted efficient gene delivery and expression in the skin by pulsed electric fields: an approach to gene therapy of skin aging and other diseases. *Biochem Biophys Res Commun* 220:633–636.

Zhang YP, Sekirov L, Saravolac EG, Wheeler JJ, Tardi P, Clow K, Leng E, Sun R, Cullis PR, Scherrer P (1999): Stabilized plasmid-lipid particles for regional gene therapy: formulation and transfection properties. *Gene Ther* 6:1438–1447.

Zhu N, Liggitt D, Liu Y, Debs R (1993): Systemic gene expression after intravenous DNA delivery into adult mice. *Science* 261:209–211.

3

IMMUNOLIPOSOMES: A TARGETED DELIVERY TOOL FOR CANCER TREATMENT

KATHLEEN F. PIROLLO, PH.D., LIANG XU, M.D., PH.D. AND ESTHER H. CHANG, PH.D.

INTRODUCTION

Despite significant advances in delivery of therapeutic agents, genes, proteins, and drugs, the ability to target the agent specifically to the site of interest remains one of medicine's greatest challenges. One approach, the use of antibody molecules to target defined cell types, was actually proposed over one hundred years ago (reviewed in Erlich, 1957). It took the successful development of hybridoma technology by Köhler and Milstein (1975), with the resulting ability to produce monoclonal antibodies, to make this proposal a reality. Monoclonal antibodies (Mabs) have become accepted tools for the detection and, in limited situations, the treatment of cancer. Mab-based radioimmunodetection systems for both colon (Oncoscint®) and prostate (Prostascint®) cancers are on the market. Antibodies have also been approved by the Food and Drug Administration (FDA) for the treatment of lymphoma (Rituxan®) and breast cancer (Herceptin®) (Adams and Schier, 1999). However, there are drawbacks and limitations to the use of Mab-based agents. Full-sized Mabs exhibit a prolonged circulation time which can, particularly when radioisotopes are involved, result in bone marrow exposure producing unacceptable myelotoxicities. Adverse toxic reactions can also occur due to interactions between Fc receptors on normal tissues and the

Vector Targeting for Therapeutic Gene Delivery, Edited by David T. Curiel and Joanne T. Douglas
ISBN 0-471-43479-5

Mab carrying radioisotopes or toxins. Moreover, with regard to use as therapeutic agents in the treatment of cancer, the large size of the intact Mabs (approximately 155 kDa) limits their ability to diffuse from the capillaries into the tumor (Jain and Baxter, 1988) and thus their potential efficacy.

In the late 1980s, the Mab field took a leap forward when methodologies were developed by groups led by Skerra (Skerra and Pluckthun, 1988) and Better (Better et al., 1988) to express active antibody fragments in *Escherichia coli*. Advances in recombinant DNA technology, along with a better understanding of the genetics and structure of immunoglobulins, led to the production of recombinant antibodies that were reduced in size, dissected into minimal fragments that retained binding activity, and engineered into multivalent high avidity reagents (reviewed in Hudson and Kortt, 1999; Charnes and Baty, 2000). These smaller engineered Mab-based complexes can overcome some of the problems associated with large IgG molecules.

Various different monomeric Ig fragments, and multiples thereof, have been engineered and are currently in use (see Fig. 3.1) (reviewed in Adams and Schier, 1999; Hudson and Kortt, 1999). The smallest of these fragments that still has clinical application is the 25 kDa Fv fragment. This molecule is derived by association of the variable domain of the heavy (V_H) and light (V_L) chains. The hydrophobic interaction between these two domains is not very strong. Thus, to achieve a stable molecule, it is necessary to include a covalent link between V_H and V_L, usually accomplished through engineering a flexible 15–20 residue peptide linker to join the V_H and V_L domains. The Fv can also be stabilized by the inclusion of cysteines to form a disulfide bridge (dsFv). The first scFv molecules, developed by both Bird et al. (1988) and Huston et al. (1988), were capable of binding to target antigens with affinities similar to those of their parental Mab. Although they readily penetrate solid tumors, because their size of 25–27 kDa is below the renal threshold, they were also rapidly eliminated through the kidney, restricting their accumulation in tumors. Moreover, a decreased affinity and poor stability of scFvs, compared to their parental antibody, has often been observed (Huston et al., 1988). This decreased ability to interact with the target antigen is believed to result from the switch from the bivalent parental Mab to the smaller monovalent scFv, or to structural alterations between the IgG and the scFv, such as the presence of the peptide spacer and possible exposure of hydrophobic residues (Adams and Schier, 1999; Adams et al., 2001). Multivalent scFv-based structures (Fig. 3.1) have been engineered in an attempt to improve the functional affinity of the scFv molecule. However, a number of studies have indicated that the ability of antibody-based proteins to penetrate solid tumors is actually hindered by extremely high affinities (Weinstein et al., 1987; Fujimori et al., 1989; Juweid et al., 1992; Adams, 1998; Adams et al., 1998; Jackson et al., 1998; Adams and Schier, 1999, 2001). This has been termed the *binding site barrier* effect by Weinstein, where antibodies with an extremely high affinity will bind in an irreversible manner to the first rank of tumor cells they encounter, resulting in antibody localization only in highly vascularized regions of the tumor. For

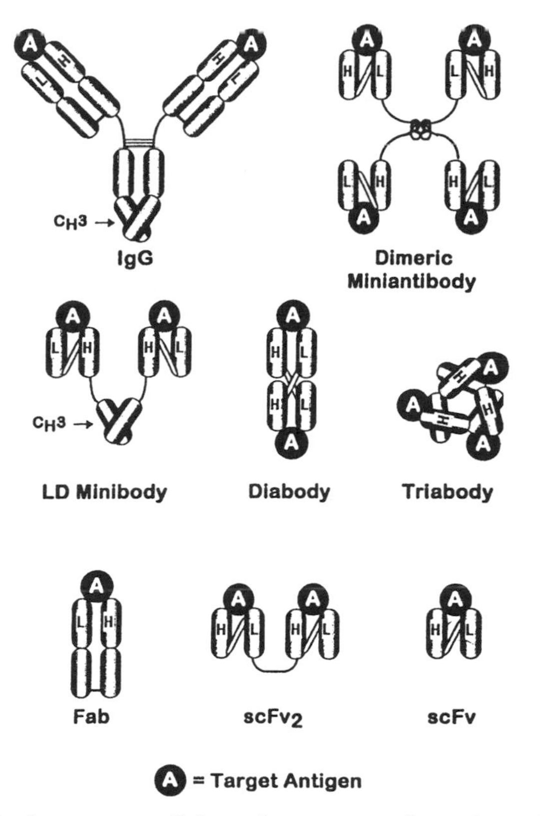

Figure 3.1 Antibody structures. Schematic structures of a variety of antibody-based molecules ranging in size from ~24 to ~155 kDa and valence from one to four binding sites. (Reprinted from Adams and Schier, 1999, by permission from Elsevier Science.)

example, the observation by Allen et al. (1995) that a sterically stabilized liposome complex bearing doxorubicin was actually more effective than its Mab (B43.13)-targeted counterpart in reducing tumor growth in a human ovarian cancer xenograft mouse model was ascribed to this binding site barrier phenomenon. Thus, an ultrahigh affinity molecule may not be the most effective delivery agent for cancer treatment, and the lower binding of the monovalent scFv fragment may be more beneficial.

Antibody fragments, including scFvs, are being used in the clinic through fusion to radioisotopes for cancer imaging, enzymes for prodrug therapy, toxins for targeted cell killing, viruses for gene therapy, biosensors for real-time detection of target molecules, and even liposomes for improved systemic delivery and gene therapy. Liposomes are essentially phospholipid bilayer vesicles with an internal space that shields the encapsulated agent from degradation in the plasma. Thus they are potential vehicles for gene/drug delivery (Gregoriadis, 1993). The bases for their attractiveness as delivery vehicles are: the low intrinsic level of toxicity of the lipid components; their versatility (Gregoriadis, 1976); simplicity of preparation; the ability to complex large amounts of DNA; versatility of use with any type and size of DNA or RNA; the ability to transfect many different types of cells (including nondividing cells); and a relative lack of immunogenicity or biohazardous activity (Farhood et al., 1994; Felgner et al., 1995). Liposomes can also be formulated in a large range of sizes and chemical components (Gregoriadis, 1976). However, conventional neutral liposomes have generally proved to be inefficient for delivery of anticancer drug to cells. Therefore, specialized liposomes are currently being developed to improve efficiency and cell-specific targeting. Improved variants include cationic liposomes as well as sterically stabilized, fusogenic, and pH- or thermo-sensitive liposomes (Felgner, 1999; Lian and Ho, 2001). Combinations and permutations of these specialized types of liposomes are being tested to achieve higher efficiency of delivery and better tumor targeting (Felgner, 1999; Xu, Pirollo, and Chang, 2001). From the perspective of human cancer therapy, liposomes have also proved safe and effective for the in vivo delivery of various types of therapeutic molecules including genes, oligonucleotides, drug, enzymes, peptides, and hormones. More than 75 clinical trials using liposomes for delivery have already been approved. At least six liposome products are already on the market (Lian and Ho, 2001). These include AmBisome® (amphotericin B lipid complex) for treatment of fungal infection (Hann and Prentice, 2001), and Doxil® (liposome-encapsulated doxorubicin) and DorinoXome® (liposome-encapsulated daunorubicin) for treatment of ovarian, breast, and other cancers (Muggia, 2001; Ranson et al., 2001) and Kaposi's sarcoma (Janknegt, 1996; Muggia, 2001), respectively. The major drawback to the use of liposomes as a carrier system is that they lack target-binding specificity. The targeting of cancer cells by liposomes can be achieved by modifying the liposomes, through addition of molecules such as antibodies and antibody fragments, so that they selectively deliver their payload to tumor cells (Bendas, 2001).

The remainder of this chapter focuses on the use of antibody-directed liposome molecules (immunoliposomes) in cancer treatment, with particular emphasis on scFv-liposomes for gene therapy.

The Evolution of Immunoliposomes

Much effort has gone into developing ways to enhance the specificity of liposomal delivery systems, most commonly by conjugating ligands to the liposome surface that will produce a specific interaction with the target cell. Ligands include vitamins (Lee and Low, 1997), glycoproteins (Ishii et al., 1989; Kikuchi et al., 1996), peptides (Oku et al., 1996; Gyongyossy-Issa et al., 1998) and oligonucleotides aptamers (Willis et al., 1998), in addition to the most commonly used ligands, antibodies or antibody fragments.

The first report of Mabs conjugated to liposomes was by Torchilin et al. (1979), who demonstrated that antimyosin-immunoliposome retained its ability to specifically bind to the receptor on the target cells. Soon after, using varied coupling techniques, other groups also reported covalent coupling of Mabs to liposomes for specific cell targeting *in vitro* (Huang et al., 1980; Barbet et al., 1981; Heath et al., 1981; Martin et al., 1981; Martin and Papahadjopoulos, 1982; Huang et al., 1983). Various attempts have since been made to improve the efficiency of delivery of the liposome payload including development of pH-sensitive immunoliposomes (Wang and Huang, 1987; Akhtar et al., 1991; Ma and Wei, 1996). However, despite promising *in vitro* results, the *in vivo* use of these Mab-immunoliposomes has been hampered by lack of stability in circulation and their rapid clearance by the mononuclear phagocyte system (MPS) (liver and spleen) (reviewed in Mastrobattista and Stoter, 1999). An example of this is the report by Matzku et al. (1990), who found virtually no uptake of Mab-immunoliposomes in either a human melanoma xenograft model or a syngeneic murine lymphoma model that spontaneously metastasizes to the liver. These investigators speculated that lack of uptake was the result of limited availability of the complex due to only moderate stability in circulation as well as inability of the immunoliposome to extravasate.

A similar problem with the liposomal delivery itself has been circumvented by the insertion of molecules such as ganglioside G_{M1} (Allen and Chonn, 1987; Gabizon and Papahadjapoulos, 1992), phosphatidylinositol (Gabizon and Papahadjapoulos, 1988), or polyethylene glycol (PEG) (Klibanov et al., 1990) into the lipid bilayer. The steric barrier of the long PEG chains and surface hydrophobicity may reduce the total amount of blood proteins bound to the liposomes. Based on the numerous reports of the advantages of PEG-coated liposomes in reducing uptake by organs of the MPS, prolonging circulation time and improving tumor uptake (Papahadjopoulos et al., 1991; Klebanov et al., 1991; Woodle and Lasic, 1992; Gabizon et al., 1994), attempts have also been made to use PEG to form sterically stabilized immunoliposomes (reviewed in Zalipsky et al., 1996). Two main strategies have been employed: (1) coupling of the antibody directly to the liposome, and (2) attachment of the antibody to the terminal end of the PEG molecule, resulting in proteins at the surface of the PEG coating (pendant type) (Fig. 3.2). The latter coupling has been accomplished by covalent methodologies that link the Mab to the PEG via a thioether or hydrazine bond (Hansen et al., 1995). Coupling to the distal end of the PEG molecule seems to be preferred, because it should minimize interference of the PEG with the antibody–antigen interaction (steric hindrance) and with coupling of the antibody to the liposome.

Most recently, recombinant antibody fragments have also been incorporated into stabilized immunoliposomes. Maruyama et al. (1999) tested whether PEG-immunoliposomes conjugated with a Fab′ fragment at the distal end of the PEG could still specifically bind the target MKN-45 (human gastric cancer) cells *in vitro* and in an *in vivo* xenograft mouse model. They reported that the Fab′-PEG-immunoliposome could not only bind to the cells *in vitro* as well as the parental Mab, but also showed equivalent tumor accumulation *in vivo*. Moreover, the Fab′-PEG-immunoliposome was also superior to the parent complex with respect to prolonged circulation time and decreased liver uptake.

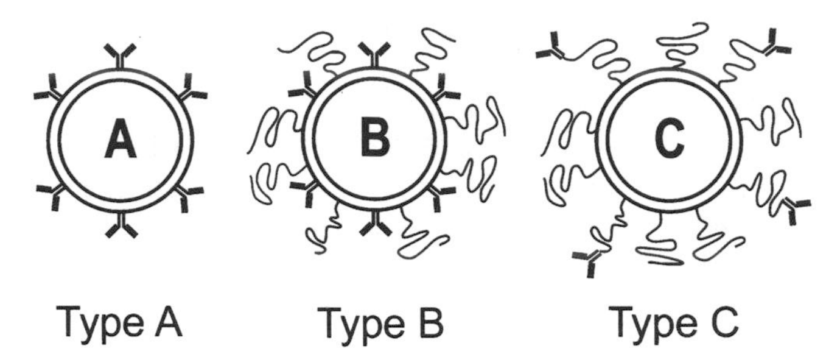

Figure 3.2 Schematic illustration of immobilization of antibody on liposome. *Type A:* PEG-free immunoliposome with the antibody covalently linked directly to the liposome; *Type B:* PEG immunoliposome with the antibody covalently linked directly to the liposome; *Type C:* Pendant type immunoliposome with the antibody attached to the distal end of the PEG molecule. (Reprinted from Maruyama et al., 1995, by permission from Elsevier Science.)

Various scFv recombinant fragments have also been used in immunoliposomes. Although not as yet complexed into sterically stabilized immunoliposomes, a lipid-tagged version of an scFv has been successfully incorporated into liposomes (Laukkanen et al., 1993, 1994; de Kruif and Storm, 1996) and used to target B-lymphocytes *in vitro* (de Kruif and Storm, 1996). Xu et al. (2001) have also used scFv-immunoliposomes to successfully target and treat solid tumors in *in vivo* mouse models. These small antibody fragments have a number of advantages over the intact Mab. In addition to the fact that they are less immunogenic, the absence of the Fc constant region will prevent the rapid Fc receptor–mediated clearance of scFv-liposomes. Moreover, lipid-tagged scFvs self-insert into the lipid bilayer molecules and thus do not need to be chemically modified for conjugation to the liposome. Therefore, reduction of binding activity and/or the induction of immunogenicity due to chemical modification are avoided. Also, the fact that the scFvs are recombinant molecules produced from phage display or *E. coli* libraries has practical advantages with respect to production for human clinical use.

THERAPEUTIC USES OF IMMUNOLIPOSOMES

The Payload: Drugs Versus Genes

Immunoliposomes are currently being developed for a variety of therapeutic uses. For example, viral replication was inhibited in acutely human immunodeficiency virus type 1 [HIV-1]-infected T-lymphoblast (CEM) cells treated with anti-HIV antibody immunoliposomes carrying antisense oligonucleotides against the *rev* and *tat* genes (Zelphati et al., 1993). Selvam et al. (1996) also used an anti-CD4 Mab-immunoliposome to deliver an antisense *rev* oligonucleotide to HIV-infected H9 cells and peripheral blood T lymphocytes *in vitro*.

They not only demonstrated specific binding of the complex to the target cells, but also an antiviral effect of the antisense oligomer with minimal toxicity. In a similar vein, an antisense RNA directed against the viral *env* region of HIV-1 displayed an anti-HIV-1 effect in the HTLV-IIIB/H9 system *in vitro* only when incorporated in immunoliposomes targeted by an antibody to CD3. The immunoliposome-delivered antisense molecule completely suppressed gp160 production and inhibited *tat* gene expression by approximately 90%, resulting in inhibition of HIV-1 production. No anti-HIV-1 activity was detected when the antisense *env* region RNA was administered free in solution or encapsulated in liposomes lacking the targeted antibody (Renneisen et al., 1990). Dufresne et al. (1999) demonstrated that an anti-HLA-DR Fab′ fragment could target immunoliposomes to lymph modes in C3H mice, with reduced uptake in the spleen, as compared to conventional liposomes. As HIV is known to accumulate and actively replicate in lymphoid tissue in the early stages of infection and throughout latency, they proposed that immunoliposomes might be useful to target drug delivery to these viral reservoirs.

Bendas et al. (1998) reported the specific binding of immunoliposomes directed against endothelial (E)-selectins *in vitro*, proposing this as a potential new type of anti-inflammatory drug therapy. Immunoliposomes carrying Mabs against the oral bacterium *Streptococcus oralis* have even been used specifically to deliver antimicrobial agents as a means to prevent tooth decay (Robinson et al., 1998). Immunoliposomes conjugated with an antibody targeting laminin B_2 were also used to deliver reporter genes (β-galactosidase and luciferase) to mature nonproliferating myotubes *in vitro* (Watanabe, 2000). However, in this instance it was found that the transfection efficiency of the immunoliposomes was not better than that of the cationic liposomes without the laminin B antibody. They suggested that this particular Mab may not be useful for myotube targeting.

In addition to these varied applications, numerous studies have been and are being performed to determine the feasibility of using immunoliposomes for drug delivery in cancer treatment. The toxic side effects of many chemotherapeutic agents on normal tissue during systemic delivery seriously limits their dosage and effectiveness. Encapsulation of the drug in a complex, such as an immunoliposome, which has specificity for the target tissue can help overcome this limitation by selective concentration of the cytotoxic agent in the tumor, thus protecting the normal cells. Various approaches using immunoliposomes for drug therapy have been employed. These include using immunoliposomes to deliver prodrug molecules and enzymes capable of activating prodrugs (Vingerhoeds et al., 1996a), as well as to deliver the chemotherapeutic agent itself. Representative examples of these studies are listed in tables 3.1 and 3.2. It is clear from these reports that targeting ability alone is not always sufficient to ensure increased efficiency of drug delivery and efficacy *in vivo*. For example, using various tumor models neither Allen et al. (1995), Vingerhoeds et al. (1996b), nor Moase et al. (2001) saw any increase in antitumor effect on established solid tumors by Mab-targeted doxorubicin liposomes as

TABLE 3.1 Representative Examples of Immunoliposomes for *In Vitro* Drug Delivery

Antibody Type	Drug	Target Cell	Results	Reference
G-22 (Mab) (Glioma associated antigen)	MTX	Glioma cells	↑ Uptake ↑ Cytotoxicity	Kito *et al.*, 1989
OV-TL3 (Fab′) (OA3 antigen) 323/A3 (Fab′) (43 kDa carcinoma glycoprotein)	DT	Ovarian carcinoma	↑ Uptake ↑ Antitumor effect	Vingerhoeds *et al.*, 1996c
H18/7 (Mab) (Murine E-selectin)	DXR	IL-1 B activated Endothelial (HUVEC)	↑ Cytotoxicity	Spragg *et al.*, 1997
OKT9 (Mab) (Transferrin receptor)	DXR	Leukemia (Parental and DXR-resistant)	↑ Uptake ↑ Accumulation ↑ Drug sensitivity	Suzuki *et al.*, 1997
My-10 (Mab) (CD-34)	Marker	Haematopietic stem cells	↑ Binding specificity	Mercadal *et al.*, 1998
14.G2a (Mab) (GD2 disialoganglioside on neuroectodermal tumors)	HPR	Melanoma	↑ Binding ↑ Growth inhibition	Pagnan *et al.*, 1999
CC52 (Mab) (colon adenocarcinoma)	FUdR Prodrug	Colon carcinoma	↑ Drug delivery No uptake of immunoliposomes	Koning *et al.*, 1999
Anti-CD19 (Mab) (CD19)	DXR	Multiple myeloma	↑ Selective cytotoxicity	Lopes de Menezes *et al.*, 2000
LL2 (Mab) (B-cell)	DXR	Lymphoma	↑ Cell association	Lundberg *et al.*, 2000

MTX = Methotrexate; DT = Diphtheria toxin; DXR = Doxorubicin; HPR = Fenretinide; FUdR = 5-fluoro-2′-deoxyuridine.

compared with their nontargeted counterparts and/or free drug. This result was despite the fact that a high level of specific binding and increased cytotoxicity was observed *in vitro* with the immunoliposomes in all three reports.

However, this is not always the case as is evidenced in Table 3.2. One factor that appears to play a role in the effectiveness of immunoliposomal drug delivery is the form of the targeting antibody, whole Mabs versus Fab′ or scFv fragments. The large size of the Mabs can interfere with penetration into larger solid tumors. In contrast, the smaller size of the antibody fragments permits them to pass more readily through the tumor vasculature. This is exemplified by comparing the effect of immunoliposomes targeted by an anti-*erbB*-2/HER-2 Mab (Goren et al., 1996) on gastric cancer with the effect of those targeted by a humanized recombinant HER-2 Fab′ fragment on breast tumors (Park et

TABLE 3.2 Representative Examples of Immunoliposomes for In Vivo Drug Delivery in Mouse Models

Antibody (Type)	Drug	Target Tumor	Treatment Route	Results	Reference
Anti-mIg[5] (Mab) (Immunoglobulin receptor)	Ara-C	B-cell lymphoma	IP	↓ Metastatic growth No effect on 1° tumor	Bankert *et al.*, 1989
174H.64 (Mab) Squamous carcinoma)	DXR	Lung carcinoma	IV	↑ Tumor regression	Ahmad *et al.*, 1993
174H.64 (Mab) (Squamous carcinoma)	DXR	Lung carcinoma (metastasis model)	IV	↑ Survival (early treatment)	Allen *et al.*, 1995
B43.13 (Mab) (Ovarian)	DXR	Ovarian	IV	↓ Antitumor efficacy	Allen *et al.*, 1995
HBJ127 (Mab) (gp125)	DXR	Colon	IV	↑ Circulation time (no efficacy reported)	Suzuki *et al.*, 1995
N-12A5 (Mab) (ErbB-2)	DXR	Gastric carcinoma	IV	No increase in antitumor effect	Goren *et al.*, 1996
OV-TL3 (Mab) (OA3)	DXR	Ovarian	IP	No increase in antitumor effect	Vingerhoeds *et al.*, 1996b
AR-3 (Mab) (CAR-3)	5-FUR Prodrug	Colon	IP	↑ Cytotoxicity ↑ Antitumor activity	Crosasso *et al.*, 1997
OX26 (Mab) (Rat transferrin)	DAU	Rat brain	IV	↑ Brain accumulation	Huwyler *et al.*, 1996, 1997

TABLE 3.2 (*Continued*)

Antibody (Type)	Drug	Target Tumor	Treatment Route	Results	Reference
rhuMab HER-2 (Fab′) (HER-2)	DXR	Breast	IV	↑ Antitumor cytotoxicity ↓ Systemic toxicity	Park *et al.*, 1997
Anti-CD19 (Mab) (CD19)	DXR	LB-cell lymphoma	IV/IP	↑ Survival ↑ Antitumor effect	Lopes de Menezes *et al.*, 1998
DH2 (Fab) (Ganglioside Gm3)	DXR	Murine melanoma	IV	↑ Antitumor effect	Nam *et al.*, 1999
SH1 (Fab) (Lex)	DXR	Murine melanoma	IV	↑ Antitumor effect	Nam *et al.*, 1999
S5A8 (Mab) (38C13)	DXR	Murine lymphoma	IV	↑ Circulation time ↑ Survival	Tseng *et al.*, 1999
IFII (Fab′) (Beta- Integrin)	DXR	Lung	IV	↓ Tumor growth ↓ Metastatic spread	Sugano *et al.*, 2000
B27.29 (Mab) (MUC)	DXR	Murine breast	IV	↑ Survival No improvement in metastasis development	Moase *et al.*, 2001

Note: Ara-C = Cytosine arabinonucleoside; DXR = Doxorubicin; 5-FUR = 5-flurouridine; DAU = Daunomycin; IV = intravenous; IP = intraperitoneal.

al., 1997). In the former case, the addition of the targeting antibody did not result in improved antitumor efficacy, while in the latter, a marked increase in tumor growth inhibition and even tumor regression was observed. However, it should be noted that different tumor models were employed in the two studies. Furthermore, the use of a Fab′ fragment in place of the intact IgG molecule reduces MPS uptake by elimination of the Fc receptor-mediated mechanisms. The binding site barrier effect of the intact Mab, as discussed previously, also needs to be taken into account. The ability of a particular immunoliposome composition to extravasate has also been shown to have a significant effect on the accumulation of the complex in the tumor (Matzku et al., 1990; Goren et al., 1996), as does the number of antibody molecules per liposome as presented in the following discussion.

It has also been postulated, but not experimentally demonstrated, that similar antitumor activity between targeted and nontargeted liposomes can result from premature leakage of the drug from the complexes (both targeted and nontargeted) prior to binding to the target cell, resulting in an increased local concentration of the drug in the vicinity of the tumor, irrespective of target cell binding (Matzku et al., 1990; Vingerhoeds et al., 1996b; Moase et al., 2001).

Perhaps the most important factor in determining the efficacy of immunoliposome drug treatment is the target itself. Because circulating tumor cells are accessible to intravenously administered immunoliposomes, hematological malignancies such as lymphomas and leukemias are most likely to respond to this form of therapy, which indeed seems to be the case. Most, if not all, of the studies reported in the literature using immunoliposomes targeting circulating tumor cells show good efficacy. The study by Lopes de Menezes et al. (1998) with human B-cell lymphoma, as well as that of Tseng et al. (1999) in a murine lymphoma model, demonstrates increased survival of animals treated with the Mab-targeted immunoliposome-doxorubicin complex when compared to those receiving either free drug or untargeted complex. Moreover, in both studies efficient internalization of the immunoliposomes by receptor-mediated endocytosis was suggested as one reason for the increased efficacy. This point is important, because once the immunoliposome complex has reached the tumor (or target) cell it must be able to efficiently release its payload into the cell. It has been shown that the liposome ratio, formulation, and procedure used to form the complex are critical for internalization and affect both the level and cellular specificity of, for example, reporter gene expression (Thierry et al., 1997). The importance of liposome formulation is discussed in detail in the following section. Tseng and colleagues (1999) also proposed that binding of the immunoliposomes to the tumor cells leads to the rapid uptake and clearance of the tumor cells by hepatosplenic macrophages, resulting in increased survival. One potential drawback to the use of immunoliposomes for treatment of circulating tumor cells is the possibility of agglutination of the tumor cells due to the interaction of more than one tumor cell with the immunoliposomes. The immunoliposomes could also agglutinate by binding to multivalent target antigens. This problem was observed by Bankert et al. (1989) with a B-cell lymphoma mouse model, resulting in animal deaths attributed to occlusion of the small capillaries by agglutination of the immunoliposomes. However, this serious side effect was not observed in either of the other two previously referenced studies.

In contrast to the hemotologic malignancies, solid tumors present a more difficult treatment problem, namely access to the target tumor cells. This issue is evident when comparing the effectiveness of immunoliposome treatment in metastasis models versus treatment of large established tumors (Bankert et al., 1989; Allen et al., 1995; Moase et al., 2001). Treatment with immunoliposomes targeted by whole Mabs in various mouse tumor models ranging from breast to lung, ovarian and even lymphoma was found to be most effective with metastases, or early treatment of small tumors, such as those produced by intraperitoneal (IP) inoculation of tumor cells. In contrast, in most instances, the immunoliposome-drug treatment had very little effect on primary or large established tumors. However, using an anti-beta 1 integrin Fab′-targeted immunoliposome for delivery of doxorubicin, Sugano et al. (2000) saw significant suppression of primary tumor growth, as well as inhibition of metastatic spread, illustrating the effect size has on the ability of the complex to reach its target effectively. Although the newly established tumor vasculature has increased permeability, as previously discussed, the size of the targeting antibody molecule can greatly affect the ability of the complex to perfuse into the interior of the tumor. Thus, the size of the Fab′ and scFv recombinant antibody fragments [approximately 55–60 kDa and 25–27 kDa, respectively (Wu and Yazaki, 2000)] would make them more useful than the intact Mab [approximately 155 kDa (Adams and Schier, 1999)] for targeting solid tumors. This was shown to be the case by Maruyama et al., (1999), who compared the target cell binding, extravasation, tissue distribution, and tumor accumulation of an immunoliposome targeted by either an intact anti-CEA antibody or its Fab′ fragment in a human gastric cancer mouse xenograft model. They found that the Fab′ pendant type immunoliposome had not only increased circulation time and decreased liver uptake, but also demonstrated a significantly higher level of tumor accumulation compared to the complex containing the intact IgG molecule. Yuan et al. (1995) showed that the vascular permeability of 25 kDa molecules (e.g., scFvs) is twice as high as for 155 kDa molecules (e.g., intact Mabs), giving the smaller scFv fragment an advantage in reaching the target cells. The interstitial hypertension within the tumor also favors scFv molecules. Intravenously administered radiolabeled scFv was found to be more evenly distributed throughout a mouse xenograft tumor, while the large IgG was concentrated near the blood vessels, and the Fab′ was intermediately distributed (Yokota et al., 1992). Early research demonstrated that liposome size affects vesicle distribution and clearance after systemic administration (Lian and Ho, 2001). Maruyama (2000) also found that long circulating PEG-liposomes with an average diameter of 100–200 nm accumulated efficiently in tumor tissue. These reports support the importance of size in targeted delivery.

Although the majority of reports in the literature deal with immunoliposome use in drug therapy, immunoliposomes have also been used successfully to deliver nucleic acid molecules, mostly in the form of oligonucleotides. We have earlier described the use of immunoliposome complexes as vehicles for delivery of antisense molecules directed against HIV (Renneisen et al., 1990; Zelphati, Zon, and Leserman, 1993; Selvam et al., 1996; Dufresne et al., 1999). In addition, Leonetti et al. (1990) used an antimouse major histocompatibility complex antibody-targeted immunoliposome carrying anti-VSV oligomers to inhibit vesicular stomatitis virus replication by greater than 95% in mouse L929 cells. Ma and Wei (1996) showed increased delivery and a twofold increase in uptake by human leukemia cells of anti-*myb* oligomers encapsulated in anti-CD32 or anti-CD2 pH-sensitive immunoliposomes when compared to oligomers encapsulated in untargeted liposomes.

Exogenous genes have also recently been efficiently delivered by immunoliposomes. However, at this time there are far fewer of these studies, the majority of which involve only reporter genes. In an *in vivo* mouse model, Wang and Huang (1987) used pH-sensitive immunoliposomes targeted by the mouse H-2K^k Mab to efficiently deliver the *CAT* gene i.p. Such pH sensitive liposomes become destabilized at pH values below the physiological blood levels such as those found in certain endosome compartments. They reported a significant increase in the level of tumor (lymphoma)-specific CAT expression with decreased nonspecific uptake by the spleen. This same group later also reported an eightfold increase in transfection efficiency of non-Ltk$^-$ cells by the herpes simplex virus thymidine kinase gene delivered by pH-sensitive immunoliposomes *in vitro* over that observed with their pH-insensitive counterparts (Wang and Huang, 1989).

While still in its early stages, the potential usefulness of this delivery system for gene therapy is exemplified by reports using immunoliposomes to carry genes to the brain. A major obstacle in the use of gene therapy for the treatment of brain tumors is the difficulty in crossing the blood–brain barrier. Thus, only highly invasive routes of administration are currently being successfully used to administer exogenous genes to the brain. However, to be efficacious, gene therapy of the brain will most likely require multiple treatments, necessitating development of delivery vehicles that can be administered intravenously and cross the blood–brain barrier. A number of investigators have explored the use of immunoliposomes as a means to penetrate the blood–brain barrier and deliver the therapeutic agent to brain cells. Mizuno and Yoshida (1996) transfected malignant glioma cells *in vitro* with the LacZ gene using immunoliposomes targeted by either the intact G-22 Mab or Fab′2 fragment (a dimer of Fab′). In both instances they observed a twofold increase in β-galactosidase activity as compared to the untargeted cationic liposome. This difference in reporter gene expression was increased further twofold upon repeated treatment with the immunoliposome. Thus, in the *in vivo* situation, multiple treatments could likely demonstrate an increased therapeutic effect. Attempts have also been made to use immunoliposomes targeted to an endogenous transport system within the blood–brain barrier, such as transferrin or insulin (Pardridge, 1997). Using immunoliposomes conjugated with OX26, an Mab to the rat transferrin receptor, Cerletti et al. (2000) investigated the uptake and transcytosis of this complex across the blood–brain barrier *in vitro* and *in vivo*. Accumulation of the immunoliposome within the endosome, and endosomal release, were observed *in vitro*. Transcytosis was confirmed *in vivo* and the immunoliposome complex was found in the postvascular compartment of brain parenchyma and not associated with the brain microvasculature. In an effort to avoid the uptake by the lung seen with the more commonly used cationic liposomal formulations, Shi and Pardridge (2000) employed PEGylated neutral liposomes targeted by the OX26 antibody to intravenously deliver either the LacZ or luciferase gene to the rat brain *in vivo*. Histological expression of β-galactosidase was observed in the brain of the rat receiving the OX26-immunoliposome. There was also a significant increase in luciferase activity in the brain after immunoliposome administration when compared to that observed after unliganded liposome, indicating that this complex can cross the blood–brain barrier after noninvasive intravenous (IV) administration. However, while very limited luciferase activity was observed in the heart and kidneys with the immunoliposome, the level of expression in the liver hepatocytes was almost sixfold higher than in the brain. Moreover, based on luciferase activity, accumulation of the immunoliposome complex was comparable in both lung and brain. Therefore, while the level in lungs may be reduced compared to cationic immunoliposomes, these neutral immunoliposomes with increased binding to the liver may not be significantly better overall than cationic immunoliposomes.

The use of scFv molecules to target liposomes for gene delivery is still in its infancy. However, an scFv fragment of the antihuman polymeric immunoglobulin receptor (pIgR) has recently been used to successfully deliver the luciferase gene to human epithelial cells, including primary human tracheal epithelial cells, *in vitro* (Gupta et al., 2001). Here also, the untargeted liposome complex displayed a far lower transfection efficiency.

Therefore, immunoliposomes targeted by either an intact Mab or a recombinant fragment (Fab′ or scFv) have the potential to become useful cancer therapeutics delivering genes and chemotherapeutic agents.

OPTIMIZATION OF LIPOSOME FORMULATION FOR TUMOR TARGETING

Over the past decade, cationic liposomes have been developed for the transfer into cells of drugs and DNA (Felgner, 1999). The cationic liposome-DNA complex (lipoplex) is formed by a combination of electrostatic attraction and hydrophobic interaction (Felgner et al., 1987; Huang and Viroonchatapan, 1999). The excess positive charge of the complex allows DNA transfection of the cells to take place (Wong et al., 1996). Cationic liposomes are composed of positively charged lipid bilayers and can be complexed to negatively charged, naked DNA by simple mixing of lipids and DNA such that the resulting complex has a net positive charge (reviewed in Clark and Hersh, 1999). The complex is easily bound and taken up by cells with high transfection efficiency (Felgner, 1999). More importantly, from the perspective of human cancer therapy, cationic liposomes have been proved to be safe and efficient for *in vivo* gene delivery (Bendas, 2001). The composition of lipids in liposomes varies widely and is critical to vector targeting and efficiency (Liu, Qi, and Huang, 1997). For antibody-targeted immunoliposomes, the antibody-to-lipid ratio is one of the most important factors for their *in vivo* targeted binding (Maruyama et al., 1999; Maruyama, Kennel, and Huang, 1990). The amount of antibody bound per vesicle depends on the Mab/liposome concentration in the incubation mixture (Nassander et al., 1995; Mercadal et al., 1998). Maruyama and colleagues (Maruyama et al., 1999; Maruyama, Kennel, and Huang, 1990) tested a series of immunoliposomes with various antibody-to-lipid ratios, from 0 to 74 antibody molecules per liposome particle. The size of the immunoliposomes remained the same, 90–130 nm in diameter, over this range. The target-binding efficiency increased with the antibody density up to about 30 antibody molecules per liposome, then reached a plateau. Further increase in antibody density only resulted in the increased liver uptake (Maruyama, Kennel, and Huang, 1990). Similarly, Huwyler et al. (1996) titrated the amount of OX26 Mab conjugated per liposome for the rate of plasmid clearance and brain accumulation, identifying a value of 29 antibody molecules per liposome particle as optimal for binding. Bendas et al. (1998) also observed selective accumulation when a specific liposome composition and a certain balance in the Mab/liposome ratio was maintained. But, as discussed in the preceding section, specific target cell binding alone is not sufficient to result in a target cell response (efficacy). Also critical are the size of the complex, the ability to extravasate, and the liposome formulation.

One factor contributing to the difference between the target cell response observed *in vitro* and *in vivo* in the studies described in the previous section is the finding that *in vitro* transfection efficiencies of lipoplex formulations cannot be easily extrapolated to the *in vivo* setting. Using a severe combined immune deficiency (SCID) mouse xenograft model, Egilmez et al. (1996) evaluated five commonly used cationic liposome formulations for *in vitro* and *in vivo* gene transfection efficiencies and found that the optimal lipid to DNA ratios *in vivo* were markedly different than those observed *in vitro* for each liposome formulation. The optimal *in vitro* transfection efficiencies varied between 12 and 55%, with 1,2-dimyristyloxy-propyl-3-dimethylhydroxy ethylammonium bromide (DMRIE)-dioleolyphosphatidylethanolamine (DOPE) being the most efficient formulation. The *in vivo* optimal ratios and overall transfection efficiencies were drastically different. The lipid to DNA optima were lower and the efficiency of transfection for individual formulations varied between 0.01 and 0.3%. The highest efficiency was obtained with 3β-N(N′,N′-dimethylaminoethane)-carbamyol-cholesterol (DC-Cholesterol) which was 7- to 8-fold more efficient than Lipofectin™ and 20- to 25-fold more efficient than Cellfectin™, Lipofectamine™, and DMRIE-DOPE. This discrepancy is attributed to differences in the microenvironment, such as the presence of serum proteins and extracellular matrix, that liposome : DNA complexes encounter *in vivo* versus *in vitro* (Egilmez et al., 1996). The changing conditions could also affect the activity of individual lipids to various degrees depending on their structure and charge. Interestingly, for intratumor injection, tumor size at the time of injection also affected transfection efficiency significantly (Egilmez et al., 1996). The authors observed a nearly exponential tumor size-dependent effect. An increase in tumor size from 20 mg to 100 mg resulted in a 4- to 5-fold decrease in transfection efficiency. This may be because there is a decreased amount of the lipoplex available per cell with increasing tumor size, or more significantly, because of changing physiological conditions within a tumor as it grows, such as pH, oxygen$_2$, vascularization, and interstitial hypertension as discussed previously. Optimization of the formulation is also critical for intravenous administration in order to obtain a defined structure, which is necessary to obtain a preparation with high reproducibility and stability, greater homogeneity of particle size, and high efficiency in systemic gene transfer (Gao and Huang, 1995; Thierry et al., 1997).

The same is true for antibody-targeted immunoliposomes (Bendas, 2001). Shi and Pardridge (2000) described an antitransferrin receptor antibody-targeted neutral immunoliposome with encapsulated plasmid DNA. A single IV administration of a low dose of DNA (10 μg per rat) in the immunoliposome resulted in luciferase and β-galactosidase reporter gene expression in the brain, which peaked 48 hours after injection. The luciferase gene expression in brain was comparable to that of lung or spleen but minimal in heart and kidney. They found that the liposome formulation and the ratios of lipid to DNA and lipid to protein were critical (Shi and Pardridge, 2000). Similar to the reports mentioned previously, their optimal antibody density was found to be 39 antibody molecules per liposome.

In our laboratory we have recently established an immunolipoplex system for tumor-targeted systemic gene delivery (Xu et al., 2001). We employed a recombinant scFv antibody fragment derived from the antitransferrin receptor (TfR) monoclonal antibody 5E9, TfRscFv, to target cationic liposomes. Our previous studies have shown that the inclusion of the transferrin ligand (Tf) in an optimized liposome-DNA preparation results in a complex that can not only be systemically delivered, but is also tumor-specific for both primary and metastatic disease (Xu et al., 1997, 1999). Moreover, the intravenously administered complex demonstrated a high level of transfection efficiency *in vivo*. Delivery of the human wild-type p53 gene via the Tf-targeting liposome complex resulted in sensitization of established tumors to conventional radiation or chemotherapy leading to long-term tumor regression (Xu et al., 1999). The TfRscFv has advantages over the Tf molecule itself. First, the size of the TfRscFv (~30 kDa with lipid tag) is much smaller than that of the Tf molecule (~80 kDa). Thus, the TfRscFv-liposome-DNA complex may exhibit better penetration into tumor vasculature. Second, as a recombinant protein, the scFv has practical advantages related to large scale production required for use in the clinic. Third, the scFv is not a blood product like human Tf, and therefore presents fewer regulatory issues related to potential contamination with blood-borne pathogens. Additional advantages of using the scFv relate to the fact that Tf interacts with the TfR with high affinity only after the ligand is saturated with iron. Large-scale production of liposomes containing iron-loaded Tf may present practical challenges. Thus the use of the scFv also enables the TfR on tumor cells to be targeted by a liposomal therapeutic complex that does not contain iron. The lipid-tagged TfRscFv was incorporated into cationic liposomes with ease via either a lipid-film solubilization method or a direct anchoring method (Bendas, 2001; Xu et al., 2001).

Optimization of the scFv Immunoliposomes for In Vitro Transfection

The *in vitro* transfection efficiency of the anti-TfR scFv-liposome complex was determined in a squamous cell carcinoma of the head and neck cell line (JSQ-3) using β-galactosidase as the reporter gene. The level of β-gal expression in the transfected cells (correlating with the transfection efficiency) was assessed by an enzymatic assay. The attachment of the anti-TfR scFv to the liposome resulted in a doubling of the enzyme activity in the immunoliposome transfected cells, as compared to the untargeted liposome complex. This level of expression was found to be comparable to that observed when transferrin itself was used as the targeting ligand. Moreover, this increase in gene expression was shown to be reporter gene DNA dose-dependent (Xu et al., 2001).

To examine the effects of the DNA-to-lipid ratio and lipid composition on the transfection efficiencies of the immunolipoplexes, JSQ-3 cells were transfected with TfRscFv-targeted cationic lipoplexes prepared with different lipid compositions and at various DNA-to-lipid ratios (Fig. 3.3). TfRscFv-targeted LipA [dioleoyltrimethylammoniumpropane : dioleoylphosphatidylethanola-

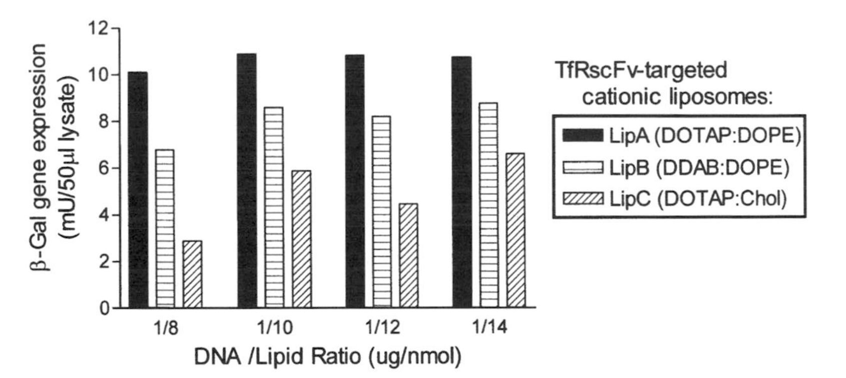

Figure 3.3 The effects of lipid composition and DNA/lipid ratio on the transfection efficiency of the TfRscFv-immunoliposome. JSQ-3 cells were transfected with TfRscFv-targeted cationic liposomes prepared with different lipid compositions and at various DNA-to-lipid ratios, using β-galactosidase as the reporter gene. TfRscFv-LipA showed superior transfection activity over TfRscFv-targeted LipB or LipC. DNA/lipid ratios of 1/10 to 1/14 μg/nmol gave the best result and had similar efficiencies *in vitro*.

mine (DOTAP : DOPE), molar ratio 1 : 1] showed superior transfection activity over other liposome compositions, that is, LipB [dimethyldioctadecylammonium bromide (DDAB) : DOPE (molar ratio 1 : 1)] or LipC (DOTAP : Cholesterol) (molar ratio 1 : 1). The liposome composition LipA was therefore chosen for further optimization. With this composition, DNA-to-lipid ratios of 1 : 10 to 1 : 14 μg/nmol gave the best result and had similar efficiencies *in vitro* (Xu et al., 2001). Moreover, there was no significant difference between either preparation method.

To compare the transfection efficiency of the immunoliposome complex prepared with different antibody-to-lipid ratios, different amounts of TfRscFv were incorporated into cationic liposomes by direct anchoring, with final protein-to-lipid ratios of 16–22 μg/μmol. Both the human breast cancer cell line MDA-MB-435 and the head and neck cell line JSQ-3 were transfected with the TfRscFv-immunoliposome carrying the *LacZ* gene. The results showed that the immunoliposomes with 22 μg scFv protein/μmol lipids resulted in higher transfection efficiency. However, these complexes tended to precipitate and were not stable when complexed with DNA (Xu et al., 2001). Therefore, as found for whole Mabs by other investigators, there is an optimal ratio of antibody to liposome. In this study, 18–20 μg/μmol of TfRscFv : lipids were determined as the optimal ratios.

To confirm this optimized formulation for TfRscFv-immunolipoplexes, a second reporter gene, the firefly luciferase gene in the plasmid pLuc, was used in the *in vitro* transfection assay. Addition of TfRscFv significantly increased the transfection efficiency of cationic liposomes, in both breast cancer (MDA-MB-435) and prostate cancer (DU145) cell lines. For both cell lines, the TfRscFv-immunolipoplexes were 10-fold more efficient than nontargeted liposomes and 2- to 3-fold more efficient than the Tf-targeted liposomes. Use of a control scFv-immunoliposome demonstrated that this increased efficiency is via the specific TfRscFv targeting ligand (Xu et al., 2001).

Optimization for In Vivo Tumor Targeting

Examination of the literature, as discussed in the preceding section, demonstrates that the *in vitro* optimized liposome formulation may not always be the best for *in vivo* gene delivery (Huang and Viroonchatapan, 1999), because the microenvironment that lipoplexes encounter *in vivo* is very different from that experienced *in vitro* (Egilmez et al., 1996). To determine if there were differences between optimal liposome formulations for *in vitro* and *in vivo* delivery, the ability of IV-administered TfRscFv-immunoliposomes, prepared using two different liposome formulations, to target and deliver the wild-type *p53* (*wtp53*) gene to prostate xenograft tumors was determined in our laboratory. Although both compositions were able to deliver the *wtp53* to the tumor, liposome composition A was somewhat more efficient, resulting in a higher level of expression (Fig. 3.4). Thus, in this instance, the lipid formulation optimized for *in vitro* use (Liposome A) was also superior *in vivo*. Ratios of scFv to

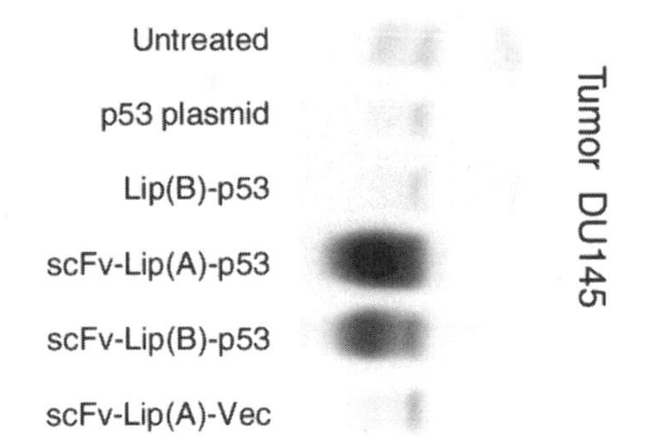

Figure 3.4 Exogenous wtp53 expression in DU145 xenograft tumors after IV injection of TfRscFv-Lip-p53. Athymic nude mice carrying human prostate tumor cell line DU145 subcutaneous xenograft tumors were injected IV with TfRscFv-targeted Lip(A)-*p53* or Lip(B)-*p53*. The animals were euthanized 48 hours later, the tumor was excised, and protein was isolated for Western analysis. The protein isolated from the tumor of an untreated animal was also included. As controls, animals were injected IV with *p53* plasmid alone, untargeted Lip(B)-p53, or TfRscFv-Lip(A) carrying empty vector. The p53 protein bands were detected using the monoclonal anti-p53 antibody Ab-2 (Oncogene Research) and the ECL Western blot kit.

total lipid were also compared for *in vivo* targeting ability. Varying the ratio of single-chain protein to lipid *in vivo*, we found that the optimal *in vivo* ratio was somewhat higher than that found to be optimal in the *in vitro* setting. The optimal DNA-to-lipid ratio of the TfRscFv-immunolipoplex is 1 μg/0.012 μmol *in vitro* (Xu et al., 2001). However, *in vivo* optimization data showed that a slightly higher DNA-to-lipid ratio gave better reporter gene expression in the tumor after systemic administration in a nude mouse tumor model. This observation supports the notion that *in vivo* optimization is critical and the *in vivo* optimal formulation cannot be predicted merely from *in vitro* results.

Our findings also emphasize the fact that for immunoliposome therapy to be effective, other factors besides antibody selection and targeting ability, such as liposome composition and antibody-to-lipid ratios, are critical and must be taken into account when designing the therapeutic complex.

ANTITUMOR EFFICACY OF IMMUNOLIPOSOMES

Immunoliposomes can, under certain conditions, be useful anticancer tools for drug and gene delivery. However, as was evident in Table 3.2, these complexes do not always demonstrate antitumor efficacy. As we discussed earlier, this inconsistency may be related to a number of factors, including differences between the antibody molecules used to target the complexes. In general, it seems that recombinant fragments have consistently better penetration with resulting increased antitumor effect. Accordingly, small scFv fragments should result in even greater efficacy (Adams and Schier, 1999). These recombinant scFv fragments have been successfully used to target radioisotopes and toxins (Colcher et al., 1999; Kang et al., 2000; Wang et al., 2001). However, despite their promise, there have been very few reports on their use for gene delivery *in vitro*. Moreover, only one report has shown tumor-specific targeting and efficacy of an scFv-targeted immunoliposome for *in vivo* gene therapy (Xu et al., 2001).

We examined the ability of TfRscFv-targeted liposomes to systemically deliver the *p53* gene specifically to tumor tissue *in vivo* (Xu et al., 2001). TfRscFv-LipA-p53 or untargeted LipA-p53 was injected via the tail vein into nude mice bearing JSQ-3 subcutaneous xenograft tumors. Two days after injection, as shown in Fig. 3.4, the tumor from the mouse injected with the TfRscFv-LipA-*p53* immunoliposomes displayed a strong exogenous wtp53 signal in Western blot analysis, while only very limited expression of exogenous p53 is evident in skin and only endogenous mouse p53 was evident in the liver of the same animal. In contrast, a much lower level of exogenous p53 expression was observed in the tumor from a mouse injected IV with unliganded lipoplex LipA-p53 (Fig. 3.5). The TfRscFv-immunolipoplex-mediated systemic gene delivery was tumor specific in that the liver showed no obvious exogenous *p53* gene expression, nor did other organs (data not shown). This experiment was reproducible with representative results shown in Fig. 3.5, thus demonstrating that the exogenous p53 gene can be systemically delivered to, and efficiently expressed in, tumors *in vivo* by the TfRscFv-targeted immunoliposome.

Of most significance, however, would be the proven efficacy of gene therapy using an scFv-immunoliposome-DNA complex. We employed a human breast cancer mouse metastasis model to test the effectiveness of the TfRscFv-immunoliposome carrying *wtp53* in sensitizing these lung metastases to the

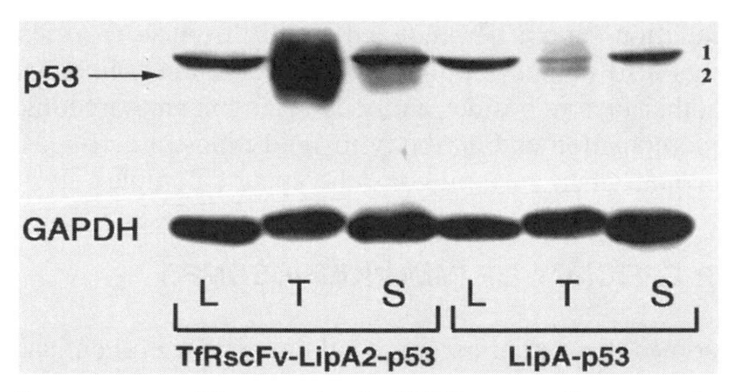

Figure 3.5 Exogenous wtp53 expression in JSQ-3 xenograft tumors after IV injection of TfRscFv-LipA2-p53. Athymic nude mice carrying JSQ-3 xenograft tumors were injected IV with either TfRscFv-LipA2-p53 or the unliganded LipA-p53. After 48 hrs the animals were euthanized, the tumor, liver, and skin excised, and protein isolated for Western analysis. T = tumor; L = liver; S = skin; 1 = endogenous p53; 2 = exogenous p53.

chemotherapeutic agent docetaxel (Xu et al., 2001). Efficacy here was measured by prolonged survival. In two separate experiments, systemic administration of the TfRscFv-liposome-p53 complex in combination with docetaxel significantly extended the life span of the animals compared to the untargeted complex plus drug (Fig. 3.6).

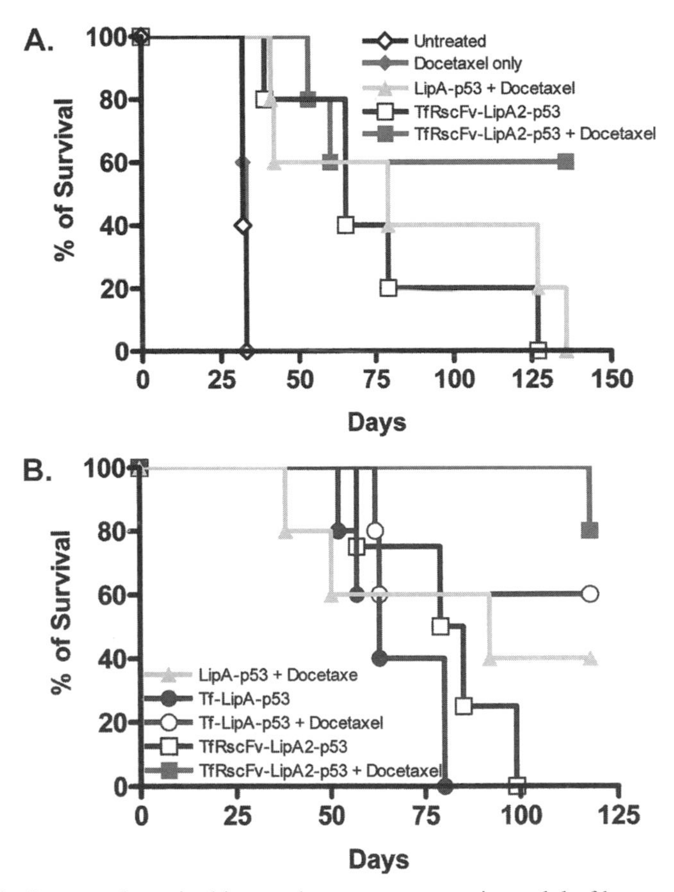

Figure 3.6 Increased survival in a nude mouse metastasis model of human breast cancer after treatment with TfRscFv-LipA2-p53 in combination with docetaxel. Beginning 10 days after tumor cell injection, twice weekly doses of TfRscFv-LipA2-*p53* (20 μg *p53* DNA/0.3 ml) with or without docetaxel (7.5 mg/kg) were intravenously administered, for a total of 10 injections. *A:* Survival curves of the mice after various treatments; *B:* A repeat experiment focusing on comparison of the two targeting ligands, TfRscFv and Tf. The systemic administration of TfRscFv-LipA2-p53 in combination with docetaxel significantly extended the life span of the animals. Long-term survivors were observed in the animals receiving the combination of TfRscFv-LipA2-*p53* and docetaxel, with 60% in the first study (*A*) and 80% in the second study (*B*), while no long-term survivors were observed in animals receiving either treatment alone.

The size of the TfRscFv-immunoliposome-*p53* complex was also assessed by dynamic laser light scattering. The size of the immunoliposome without DNA was approximately 25 nm in diameter, equivalent to that of the liposome itself. More significantly, when complexed with the *p53* plasmid DNA the size was found to be approximately 90 mm in diameter, below the permeable size of the tumor capillaries [about 100 mm in diameter (Huang et al., 1997; Feng et al., 1997)]. Thus, this small size, permitting passage through thc capillary wall, in conjunction with the tumor specificity of the scFv, may account for the antitumor efficacy of this gene therapy treatment.

CONCLUSIONS

Antibody-targeted liposomal complexes have been shown to have the potential to be effective in delivering drugs to tumor and other target cells. While drugs are currently the major beneficiary of this technology, the successful use of immunoliposomes for gene delivery shown in recent studies is likely to encourage further advances in this direction. In particular, it would seem that with the advantages recombinant scFv fragments have over intact IgG or even Fab′ fragments, as demonstrated by Xu et al. (2001), these molecules have the greatest potential for development into new improved molecular therapeutics for the treatment of cancer.

REFERENCES

Adams GP (1998): Improving the tumor specificity and retention of antibody-based molecules. *In Vivo* 12:11–21.

Adams GP, Schier R (1999): Generating improved single-chain Fv molecules for tumor targeting. *J Immunol Methods* 231:249–260.

Adams GP, Schier R, Marshall K, Wolf EJ, McCall AM, Marks JD, Weiner LM (1998): Increased affinity leads to improved selective tumor delivery of single-chain Fv antibodies. *Cancer Res* 58:485–490.

Adams GP, Schier R, McCall AM, Simmons HH, Horak EM, Alpaugh RK, Marks JD, Weiner LM (2001): High affinity restricts the localization and tumor penetration of single-chain fv antibody molecules. *Cancer Res* 61:4750–4755.

Ahmad I, Longenecker M, Samuel J, Allen TM (1993): Antibody-targeted delivery of doxorubicin entrapped in sterically stabilized liposomes can eradicate lung cancer in mice. *Cancer Res* 53:1484–1488.

Akhtar S, Basu S, Wickstrom E, Juliano RL (1991): Interactions of antisense DNA oligonucleotide analogs with phospholipid membranes (liposomes). *Nucleic Acids Res* 19:5551–5559.

Allen TM, Chonn A (1987): Large unilamellar liposomes with low uptake into the reticuloendothelial system. *FEBS Letters* 223:42–46.

Allen TM, Ahmad I, Lopes de Menezes DE, Moase EH (1995): Immunoliposome-mediated targeting of anti-cancer drugs in vivo. *Biochemical Soc Trans* 23:1073–1079.

Bankert RB, Yokota S, Ghosh SK, Mayhew E, Jou YH (1989): Immunospecific targeting of cytosine arabinonucleoside-containing liposomes to the idiotype on the surface of a murine B-cell tumor in vitro and in vivo. *Cancer Res* 49:301–308.

Barbet J, Machy P, Leserman LD (1981): Monoclonal antibody covalently coupled to liposomes: specific targeting to cells. *J Supramol Struct Cell Biochem* 16:243–258.

Bendas G, Krause A, Schmidt R, Vogel J, Rothe U (1998): Selectins as new targets for immunoliposome-mediated drug delivery. A potential way of anti-inflammatory therapy. *Pharm Acta Helv* 73:19–26.

Bendas G. (2001): Immunoliposomes: A promising approach to targeting cancer. *BioDrugs* 15:215–224.

Better M, Chang CP, Robinson RR, Horwitz AH (1988): *Escherichia coli* secretion of an active chimeric antibody fragment. *Science* 240:1041–1043.

Bird RE, Hardman KD, Jacobson JW, Johnson S, Kaufman BM, Lee SM, Lee T, Pope SH, Riordan GS, Whitlow M (1988): Single-chain antigen-binding proteins. *Science* 242:423–426.

Cerletti A, Drewe J, Fricker G, Eberle AN, Huwyler J (2000): Endocytosis and transcytosis of an immunoliposome-based brain drug delivery system. *J Drug Target* 8:435–446.

Chames P, Baty D. (2000): Antibody engineering and its applications in tumor targeting and intracellular immunization. *FEMS Microbiol Lett* 189:1–8.

Clark PR, Hersh EM (1999): Cationic lipid-mediated gene transfer: current concepts. *Curr Opin Mol Ther* 1:158–176.

Colcher D, Pavlinkova G, Beresford G, Booth BJ, Batra SK (1999): Single-chain antibodies in pancreatic cancer. *Ann NY Acad Sci* 880:263–280.

Crosasso P, Brusa P, Dosio F, Arpicco S, Pacchioni D, Schuber F, Cattel L (1997): Antitumoral activity of liposomes and immunoliposomes containing 5-fluorouridine prodrugs. *J Pharm Sci* 86:832–839.

de Kruif J, Storm G (1996): Biosynthetically lipid-modified human scFv fragments from phage display libraries as targeting molecules for immunoliposomes. *FEBS Lett* 399:232–236.

Dufresne I, Desormeaux A, Bestman-Smith J, Gourde P, Tremblay MJ, Bergeron MG (1999): Targeting lymph nodes with liposomes bearing anti-HLA-DR Fab′ fragments. *Biochim Biophys Acta* 1421:284–294.

Egilmez NK, Iwanuma Y, Bankert RB (1996): Evaluation and optimization of different cationic liposome formulations for in vivo gene transfer. *Biochem Biophys Res Comm* 221:169–173.

Ehrlich, P (1957): *Immunology and Cancer Research.* Oxford: Pergamon Press.

Farhood H, Gao X, Son K, Yang YY, Lazo JS, Huang L, Barsoum J, Bottega R, Ep RM, (1994): Cationic liposomes for direct gene transfer in therapy of cancer and other diseases. *Ann NY Acad Sci* 716:23–34.

Felgner PL (1999): Progress in Gene Delivery Research and Development. In Huang L, Hung MC, Wagner E, eds. *Non-viral Vectors for Gene Therapy.* San Diego, CA: Academic Press, pp 26–38.

Felgner PL, Gadek TR, Holm M, Roman R, Chan HW, Wenz M, Northrop JP, Ringold GM, Danielsen M (1987): Lipofection: a highly efficient, lipid-mediated DNA-transfection procedure. *Proc Nat Acad Sci USA* 84:7413–7417.

Felgner PL, Tsai YJ, Sukhu L, Wheeler CJ, Manthorpe M, Marshall J, Cheng SH (1995): Improved cationic lipid formulations for in vivo gene therapy. *Ann NY Acad Sci* 772:126–139.

Feng D, Nagy JA, Hipp J, Pyne K, Dvorak HF, Dvorak AM (1997): Reinterpretation of endothelial cell gaps induced by vasoactive mediators in guinea-pig, mouse and rat: many are transcellular pores. *J Physiol* 504:747–761.

Fujimori K, Covell DG, Fletcher JE, Weinstein JN (1989): Modeling analysis of the global and microscopic distribution of immunoglobulin G, F(ab′)2, and Fab in tumors. *Cancer Res* 49:5656–5663.

Gabizon A, Papahadjopoulos D (1988): Liposome formulations with prolonged circulation time in blood and enhanced uptake by tumors. *Proc Nat Acad Sci USA* 85:6949–6953.

Gabizon A, Papahadjopoulos D (1992): The role of surface charge and hydrophilic groups on liposome clearance in vivo. *Biochim Biophys Acta* 1103:94–100.

Gabizon A, Catane R, Uziely B, Kaufman B, Safra T, Cohen R, Martin F, Huang A, Barenholz Y (1994): Prolonged circulation time and enhanced accumulation in malignant exudates of doxorubicin encapsulated in polyethylene-glycol coated liposomes. *Cancer Res* 54:987–992.

Gao X, Huang L (1995): Cationic liposome-mediated gene transfer. *Gene Ther* 2:710–722.

Goren D, Horowitz AT, Zalipsky S, Woodle MC, Yarden Y, Gabizon A (1996): Targeting of stealth liposomes to erbB-2 (Her/2) receptor: in vitro and in vivo studies. *Br J Cancer* 74:1749–1756.

Gregoriadis G (1976): The carrier potential of liposomes in biology and medicine. *New Engl J Med* 295:765–770.

Gregoriadis G (1993): Liposomes, a tale of drug targeting. *J Drug Target* 1:3–6.

Gupta S, Eastman J, Silski C, Ferkol T, Davis PB (2001): Single chain Fv: a ligand in receptor-mediated gene delivery. *Gene Ther* 8:586–592.

Gyongyossy-Issa MI, Muller W, Devine DV (1998): The covalent coupling of Arg-Gly-Asp-containing peptides to liposomes: purification and biochemical function of the lipopeptide. *Arch Biochem Biophys* 353:101–108.

Hann IM, Prentice HG (2001): Lipid-based amphotericin B: A review of the last 10 years of use. *Int J Antimicrob Agents* 17:161–169.

Hansen CB, Kao GY, Moase EH, Zalipsky S, Allen TM (1995): Attachment of antibodies to sterically stabilized liposomes: evaluation, comparison and optimization of coupling procedures. *Biochim Biophys Acta* 1239:133–144.

Heath TD, Macher BA, Papahadjopoulos D (1981): Covalent attachment of immunoglobulins to liposomes via glycosphingolipids. *Biochim Biophys Acta* 640:66–81.

Huang L, Viroonchatapan E (1999): Introduction. In Huang L, Hung M, Wagner E., eds. *Non-viral Vectors for Gene Therapy*. San Diego, CA: Academic Press, pp 3–22.

Huang A, Huang L, Kennel SJ (1980): Monoclonal antibody covalently coupled with fatty acid. A reagent for in vitro liposome targeting. *J Biol Chem* 255:8015–8018.

Huang A, Kennel SJ, Huang L (1983): Interactions of immunoliposomes with target cells. *J Biol Chem* 258:14034–14040.

Huang SK, Martin FJ, Friend DS, Papahadjopoulos D (1997): Mechanism of Stealth Liposome Accumulation in Some Pathological Tissues. In Lasic, Martin F eds. *Stealth Liposomes*. Boca Raton, FL: CRC Press Inc. pp 119–125.

Hudson PJ, Kortt AA (1999): High avidity scFv multimers; diabodies and triabodies. *J Immunol Methods* 231:177–189.

Huston JS, Levinson D, Mudgett-Hunter M, Tai MS, Novotny J, Margolies MN, Ridge RJ, Bruccoleri RE, Haber E, Crea R (1988): Protein engineering of antibody binding sites: recovery of specific activity in an anti-digoxin single-chain Fv analogue produced in *Escherichia coli*. *Proc Nat Acad Sci USA* 85:5879–5883.

Huwyler J, Wu D, Pardridge WM (1996): Brain drug delivery of small molecules using immunoliposomes. *Proc Nat Acad Sci USA* 93:14164–14169.

Huwyler J, Yang J, Pardridge WM (1997): Receptor mediated delivery of daunomycin using immunoliposomes: pharmacokinetics and tissue distribution in the rat. *J Pharmacol Exp Ther* 282:1541–1546.

Ishii Y, Aramaki Y, Hara T, Tsuchiya S, Fuwa T (1989): Preparation of EGF labeled liposomes and their uptake by hepatocytes. *Biochem Biophys Res Commun* 160:732–736.

Jackson H, Bacon L, Pedley RB, Derbyshire E, Field A, Osbourn J, Allen D (1998): Antigen specificity and tumour targeting efficiency of a human carcinoembryonic antigen-specific scFv and affinity-matured derivatives. *Br J Cancer* 78:181–188.

Jain RK, Baxter LT (1988): Mechanisms of heterogenous distribution of monoclonal antibodies and other macro-molecules in tumors: significance of elevated interstitial pressure. *Cancer Res* 48:7022–7032.

Janknegt R (1996): Liposomal formulations of cytotoxic drugs. *Suppor Care Cancer* 4:298–304.

Juweid M, Neumann R, Paik C, Perez-Bacete MJ, Sato J, van Osdol W, Weinstein JN (1992): Micropharmacology of monoclonal antibodies in solid tumors: direct experimental evidence for a binding site barrier. *Cancer Res* 52:5144–5153.

Kang N, Hamilton S, Odili J, Wilson G, Kupsch J (2000): In vivo targeting of malignant melanoma by 125Iodine- and 99mTechnetium-labeled single-chain Fv fragments against high molecular weight melanoma-associated antigen. *Clinical Cancer Res* 6:4921–4931.

Kikuchi A, Sugaya S, Ueda H, Tanaka K, Aramaki Y, Hara T, Arima H, Tsuchiya S, Fuwa T (1996): Efficient gene transfer to EGF receptor overexpressing cancer cells by means of EGF-labeled cationic liposomes. *Biochem Biophys Res Commun* 227:666–671.

Kito A, Yoshida J, Kageyama N, Kojima N, Yagi K (1989): Liposomes coupled with monoclonal antibodies against glioma-associated antigen for targeting chemotherapy of glioma. *J Neurosurg* 71:382–387.

Klibanov AL, Maruyama K, Torchilin VP, Huang L (1990): Amphipathic polyethyleneglycols effectively prolong the circulation time of liposomes. *FEBS Lett* 268:235–237.

Klibanov AL, Maruyama K, Beckerleg AM, Torchilin VP, Huang L (1991): Activity of amphipathic poly(ethylene glycol) 5000 to prolong the circulation time of liposomes depends on the liposome size and is unfavorable for immunoliposome binding to target. *Biochim Biophys Acta* 1062:142–148.

Köhler G, Milstein C (1975): Continuous cultures of fused cells secreting antibody of predefined specificity. *Nature* 256:495–497.

Koning GA, Morselt HW, Velinova MJ, Donga J, Gorter A, Allen TM, Zalipsky S, Kamps JA, Scherphof GL (1999): Selective transfer of a lipophilic prodrug of 5-fluorodeoxyuridine from immunoliposomes to colon cancer cells. *Biochim Biophys Acta* 1420:153–167.

Laukkanen ML, Teeri TT, Keinanen K (1993): Lipid-tagged antibodies: bacterial expression and characterization of a lipoprotein-single-chain antibody fusion protein. *Protein Eng* 6:449–454.

Laukkanen ML, Alfthan K, Keinanen K (1994): Functional immunoliposomes harboring a biosynthetically lipid-tagged single-chain antibody. *Biochemistry* 33:11664–11670.

Lee RJ, Low PS (1997): Folate-targeted liposomes for drug delivery. *J Liposome Res* 455–466.

Leonetti JP, Machy P, Degols G, Lebleu B, Leserman L (1990): Antibody-targeted liposomes containing oligodeoxyribonucleotides complementary to viral RNA selectively inhibit viral replication. *Proc Nat Acad Sci USA* 87:2448–2451.

Lian T, Ho RJ (2001): Trends and developments in liposome drug delivery systems. *J Pharm Sci* 90:667–680.

Liu F, Qi H, Huang L (1997): Factors controlling the efficiency of cationic lipid-mediated transfection in vivo via intravenous administration. *Gene Ther* 4:517–523.

Lopes de Menezes DE, Pilarski LM, Allen TM (1998): In vitro and in vivo targeting of immunoliposomal doxorubicin to human B-cell lymphoma. *Cancer Res* 58:3320–3330.

Lopes de Menezes DE, Pilarski LM, Belch AR, Allen TM (2000): Selective targeting of immunoliposomal doxorubicin against human multiple myeloma in vitro and ex vivo. *Biochim Biophys Acta* 1466:205–220.

Lundberg BB, Griffiths G, Hansen HJ (2000): Specific binding of sterically stabilized anti-B-cell immunoliposomes and cytotoxicity of entrapped doxorubicin. *Int J Pharm* 205:101–108.

Ma DD, Wei AQ (1996): Enhanced delivery of synthetic oligonucleotides to human leukaemic cells by liposomes and immunoliposomes. *Leuk Res* 20:925–930.

Martin FJ, Papahadjopoulos D (1982): Irreversible coupling of immunoglobulin fragments to preformed vesicles. An improved method for liposome targeting. *J Biol Chem* 257:286–288.

Martin FJ, Hubbell WL, Papahadjopoulos D (1981): Immunospecific targeting of liposomes to cells: a novel and efficient method for covalent attachment of Fab′ fragments via disulfide bonds. *Biochemistry* 20:4229–4238.

Maruyama K (2000). In vivo targeting by liposomes. *Biol Pharm Bull* 23:791–799.

Maruyama K, Kennel SJ, Huang L (1990): Lipid composition is important for highly efficient target binding and retention of immunoliposomes. *Proc Natl Acad Sci USA:* 5744–5748.

Maruyama K, Takizawa T, Yuda T, Kennel SJ, Huang L, Iwatsuru M (1995): Targetability of novel immunoliposomes modified with amphipathic poly(ethylene glycol)s conjugated at their distal terminals to monoclonal antibodies. *Biochim Biophys Acta* 1234:74–80.

Maruyama K, Ishida O, Takizawa T, Moribe K (1999): Possibility of active targeting to tumor tissues with liposomes. *Adv Drug Delivery Rev* 40:89–102.

Mastrobattista KGA, Stoter G (1999): Immunoliposomes for the targeted delivery of antitumor drugs. *Adv Drug Delivery Rev* 92:103–127.

Matzku S, Krempel H, Weckenmann HP, Schirrmacher V, Sinn H, Stricker H (1990): Tumour targeting with antibody-coupled liposomes: failure to achieve accumulation in xenografts and spontaneous liver metastases. *Cancer Immunol Immunother* 31:285–291.

Mercadal M, Carrion C, Domingo JC, Petriz J, Garcia J, de Madariaga MA (1998): Preparation of immunoliposomes directed against CD34 antigen as target. *Biochim Biophys Acta* 1371:17–23.

Mizuno M, Yoshida J (1996): Repeated exposure to cationic immunoliposomes activates effective gene transfer to human glioma cells. *Neurol Med Chir* (Japan) 36:141–144.

Moase EH, Qi W, Ishida T, Gabos Z, Longenecker BM, Zimmermann GL, Ding L, Krantz M, Allen TM (2001): Anti-MUC-1 immunoliposomal doxorubicin in the treatment of murine models of metastatic breast cancer. *Biochim Biophys Acta* 1510:43–55.

Muggia FM (2001): Liposomal encapsulated anthracyclines: New therapeutic horizons. *Curr Oncol Rep* 3:156–162.

Nam SM, Kim HS, Ahn WS, Park YS (1999): Sterically stabilized anti-G(M3), anti-Le(x) immunoliposomes: targeting to B16BL6, HRT-18 cancer cells. *Oncol Res* 11:9–16.

Nassander UK, Steerenberg PA, De Jong WH, Van Overveld WO, Te Boekhorst CM, Poels LG, Jap PH, Storm G (1995): Design of immunoliposomes directed against human ovarian carcinoma. *Biochim Biophys Acta* 1235:126–139.

Oku N, Tokudome Y, Koike C, Nishikawa N, Mori H, Saiki I, Okada S (1996): Liposomal Arg-Gly-Asp analogs effectively inhibit metastatic B16 melanoma colonization in murine lungs. *Life Sci* 58:2263–2270.

Pagnan G, Montaldo PG, Pastorino F, Raffaghello L, Kirchmeier M, Allen TM, Ponzoni M (1999): GD2-mediated melanoma cell targeting and cytotoxicity of liposome-entrapped fenretinide. *Int J Cancer* 81:268–274.

Papahadjopoulos D, Allen TM, Gabizon A, Mayhew E, Matthay K, Huang SK, Lee KD, Woodle MC, Lasic DD, Redemann C (1991): Sterically stabilized liposomes: improvements in pharmacokinetics and antitumor therapeutic efficacy. *Proc Nat Acad Sci USA* 88:11460–11464.

Pardridge WM (1997): Drug delivery to the brain. *J Cereb Blood Flow Metab* 17:713–731.

Park JW, Hong K, Kirpotin DB, Meyer O, Papahadjopoulos D, Benz CC (1997): Anti-HER2 immunoliposomes for targeted therapy of human tumors. *Cancer Lett* 118:153–160.

Ranson MR, Cheeseman S, White S, Margison J (2001): Caelyx (stealth liposomal doxorubicin) in the treatment of advanced breast cancer. *Crit Rev Oncol Hematol* 37:115–120.

Renneisen K, Leserman L, Matthes E, Schroder HC, Muller WE (1990): Inhibition of expression of human immunodeficiency virus-1 in vitro by antibody-targeted liposomes containing antisense RNA to the env region. *J Biol Chem* 265:16337–16342.

Robinson AM, Creeth JE, Jones MN (1998): The specificity and affinity of immunoliposome targeting to oral bacteria. *Biochim Biophys Acta* 1369:278–286.

Selvam MP, Buck SM, Blay RA, Mayner RE, Mied PA, Epstein JS (1996): Inhibition of HIV replication by immunoliposomal antisense oligonucleotide. *Antiviral Res* 33:11–20.

Shi N, Pardridge WM (2000): Noninvasive gene targeting to the brain. *Proc Nat Acad Sci USA* 97:7567–7572.

Skerra A, Pluckthun A (1988): Assembly of a functional immunoglobulin Fv fragment in *Escherichia coli*. *Science* 240:1038–1041.

Spragg DD, Alford DR, Greferath R, Larsen CE, Lee KD, Gurtner GC, Cybulsky MI, Tosi PF, Nicolau C, Gimbrone MA (1997): Immunotargeting of liposomes to activated vascular endothelial cells: a strategy for site-selective delivery in the cardiovascular system. *Proc Nat Acad Sci USA* 94:8795–8800.

Sugano M, Egilmez NK, Yokota SJ, Chen FA, Harding J, Huang SK, Bankert RB (2000): Antibody targeting of doxorubicin-loaded liposomes suppresses the growth and metastatic spread of established human lung tumor xenografts in severe combined immunodeficient mice. *Cancer Res* 60:6942–6949.

Suzuki S, Watanabe S, Masuko T, Hashimoto Y (1995): Preparation of long-circulating immunoliposomes containing adriamycin by a novel method to coat immunoliposomes with poly(ethylene glycol). *Biochim Biophys Acta* 1245:9–16.

Suzuki S, Inoue K, Hongoh A, Hashimoto Y, Yamazoe Y (1997): Modulation of doxorubicin resistance in a doxorubicin-resistant human leukaemia cell by an immunoliposome targeting transferring receptor. *B J Cancer* 76:83–89.

Thierry AR, Rabinovich P, Peng B, Mahan LC, Bryant JL, Gallo RC (1997): Characterization of liposome-mediated gene delivery: expression, stability and pharmacokinetics of plasmid DNA. *Gene Ther* 4:226–237.

Torchilin VP, Khaw BA, Smirnov VN, Haber E (1979): Preservation of antimyosin antibody activity after covalent coupling to liposomes. *Biochem Biophys Res Commun* 89:1114–1119.

Tseng YL, Hong RL, Tao MH, Chang FH (1999): Sterically stabilized anti-idiotype immunoliposomes improve the therapeutic efficacy of doxorubicin in a murine B-cell lymphoma model. *Int J Cancer* 80:723–730.

Vingerhoeds MH, Steerenberg PA, Hendriks JJ, Dekker LC, Van Hoesel QG, Crommelin DJ, Storm G (1996a): Immunoliposome-mediated targeting of doxorubicin to human ovarian carcinoma in vitro and in vivo. *Br J Cancer* 74:1023–1029.

Vingerhoeds MH, Haisma HJ, Belliot SO, Smit RH, Crommelin DJ, Storm G (1996b): Immunoliposomes as enzyme-carriers (immuno-enzymosomes) for antibody-directed enzyme prodrug therapy (ADEPT): optimization of prodrug activating capacity. *Pharm Res* 13:604–610.

Vingerhoeds MH, Steerenberg PA, Hendriks JJ, Crommelin DJ, Storm G (1996c): Targeted delivery of diphtheria toxin via immunoliposomes: efficient antitumor activity in the presence of inactivating anti-diphtheria toxin antibodies. *FEBS Lett* 395:245–250.

Wang CY, Huang L (1987): pH-sensitive immunoliposomes mediate target-cell-specific delivery and controlled expression of a foreign gene in mouse. *Proc Nat Acad Sci USA* 84:7851–7855.

Wang CY, Huang L (1989): Highly efficient DNA delivery mediated by pH-sensitive immunoliposomes. *Biochemistry* 28:9508–9514.

Wang L, Liu B, Schmidt M, Lu Y, Wels W, Fan Z (2001): Antitumor effect of an HER2-specific antibody-toxin fusion protein on human prostate cancer cells. *Prostate* 47:21–28.

Watanabe Y, Sawaishi Y, Tada H, Sato E, Suzuki T, Takada G (2000): Immunoliposome-mediated gene transfer into cultured myotubes. *Tohoku J Expos Med* 192:173–180.

Weinstein JN, Eger RR, Covell DG, Black CD, Mulshine J, Carrasquillo JA, Larson SM, Keenan AM (1987): The pharmacology of monoclonal antibodies. *Ann NY Acad Sci* 507:199–210.

Willis MC, Collins BD, Zhang T, Green LS, Sebesta DP, Bell C, Kellogg E, Gill SC, Magallanez A, Knauer S, Bendele RA, Gill PS, Janjic N, Collins B (1998): Liposome-anchored vascular endothelial growth factor aptamers. *Bioconjug Chem* 9:573–582.

Wong FM, Reimer DL, Bally MB (1996): Cationic lipid binding to DNA: characterization of complex formation. *Biochemistry* 35:5756–5763.

Woodle MC, Lasic DD (1992): Sterically stabilized liposomes. *Biochim Biophys Acta* 1113:171–199.

Wu AM, Yazaki PJ (2000): Designer genes: recombinant antibody fragments for biological imaging. *Quart J Nucl Med* 44:268–283.

Xu L, Pirollo KF, Chang EH (1997): Transferrin-liposome-mediated p53 sensitization of squamous cell carcinoma of the head and neck to radiation in vitro. *Hum Gene Ther* 8:467–475.

Xu L, Pirollo KF, Chang EH (1999): Transferrin-liposome-mediated systemic p53 gene therapy in combination with radiation results in regression of human head and neck cancer xenografts. *Hum Gene Ther* 10:2941–2952.

Xu L, Pirollo KF, Chang EH (2001a): Tumor-targeted p53 gene therapy enhances the efficacy of conventional chemo/radiotherapy. *J Control Release* 6:115–128.

Xu L, Tang W, Huang C, Alexander W, Xiang L, Pirollo K, Rait A, Chang EH (2001): Systemic p53 gene therapy of cancer with immunolipoplexes targeted by anti-transferrin receptor scFv. *Mol Med* 7:723–734.

Yokota T, Milenic DE, Whitlow M, Schlom J (1992): Rapid tumor penetration of a single-chain Fv and comparison with other immunoglobulin forms. *Cancer Res* 52:3402–3408.

Yuan F, Dellian M, Fukumura D, Leunig M, Berk DA, Torchilin VP, Jain RK (1995): Vascular permeability in a human tumor xenograft: Molecular size dependence and cutoff size. *Cancer Res* 55:3752–3756.

Zalipsky S, Hansen CB, Oaks JM, Allen TM (1996): Evaluation of blood clearance rates and biodistribution of poly(2-oxazoline)-grafted liposomes. *J Pharm Sci* 85:133–137.

Zelphati O, Zon G, Leserman L (1993): Inhibition of HIV-1 replication in cultured cells with antisense oligonucleotides encapsulated in immunoliposomes. *Antisense Res Dev* 3:323–338.

4

RECEPTOR-DIRECTED GENE DELIVERY USING MOLECULAR CONJUGATES

ASSEM G. ZIADY, PH.D. AND PAMELA B. DAVIS, M.D., PH.D.

INTRODUCTION

In an effort to sidestep the safety limitations of viral vectors and the cytotoxicity of liposomal carriers, investigators have used receptor-targeted molecular conjugates to direct gene transfer into mammalian cells in vitro (Wu and Wu, 1987; Zatoukal et al., 1992; Ferkol, Kaetzel, and Davis, 1993; Monsigny et al., 1994; Midoux et al., 1993; Chen, Stickles, and Daiscendt, 1994; Perales et al., 1997; Erbacher et al., 1996; Schwarzenberger et al., 1996; Ross et al., 1995; Hart et al., 1995; Ziady et al., 1997; Ziady et al., 1998; Batra et al., 1994; Rojanasakul et al., 1994; Gottschalk et al., 1994; Buschle et al., 1995; Cotten, Wagner, and Birnstiel, 1993a; Thurher et al., 1990; Wagner, Curiel, and Cotten, 1994; Wagner et al., 1990; Gottschalk et al., 1996; Ferkol et al., 1996a; Zenke et al., 1990; Chen et al., 1994; Chen et al., 1995; Fisher et al., 2000; Hart et al., 1998) and in vivo (Wu, Wilson, and Wu, 1989; Wu et al., 1991; Wilson et al., 1992; Perales et al., 1994a; Chowdhury et al., 1993; Ferkol et al., 1995; Christiano, Smith, and Woo, 1993; Cotten et al., 1993b; Christiano et al., 1993; Ferkol et al., 1998; Ziady et al., 1999; Ferkol et al., 1993; Stankovics et al., 1994; Gao et al., 1993; Ohtake et al., 1999; Hoganson et al., 1998; Jenkins et al., 2001). Complexes consisting of DNA, noncovalently bound to a poly-

Vector Targeting for Therapeutic Gene Delivery, Edited by David T. Curiel and Joanne T. Douglas
ISBN 0-471-43479-5

cation polymer that is chemically conjugated to a ligand, can bind a cell surface receptor and be internalized. Some receptors targeted for this purpose (e.g., the asialoglycoprotein receptor [reviewed in Stockert, 1995]) traffic their cargo to lysosomes for degradation, while others recycle to the cell surface (e.g., the transferrin receptor [reviewed in Ponko and Lok, 1999]) or transport their ligands across the cell (e.g., the polymeric immunoglobulin receptor [reviewed in Kaetzel et al., 1997]). The receptors most successful in gene delivery are constitutively expressed, abundant, and exhibit high specificity for the ligand but low selectivity for the attached cargo. Immunogenicity is usually low for polycations (Maurer, 1962) and DNA (Kantor, Ojeda, and Benacerraf, 1962), and the packaging capacity is high (Cotten et al., 1992), allowing for the inclusion of native promoters, enhancer sequences, and intronic sequences that augment gene expression (Yew et al., 1997). Furthermore, targeting a select receptor can provide exquisite specificity for a target cell population. Once perfected in animal models, this method may provide a specific, noninfectious, and minimally toxic alternative to present techniques employed in human gene therapy.

The basic design of molecular conjugates consists of two elements: a receptor ligand coupled with a polycation (e.g., polylysine) capable of condensing DNA (Fig. 4.1). Receptor ligands, ranging from simple sugars to complex antibodies have been used to target molecular conjugate DNA complexes to specific cell types. Once constructed, this molecular conjugate should interact electrostatically with the negatively charged phosphates of the DNA backbone and produce a stable conjugate-DNA complex appropriate for gene transfer (Ziady et al., 1999; Adami et al., 1998). This DNA complex alone should be capable of cell entry and nuclear localization, but various reagents have been used to aid in the internalization, endosomal escape, and nuclear translocation. These reagents are necessary for efficient gene transfer via some receptor systems in vitro (Gottschalk et al., 1994; Cotten et al., 1990), however, in vivo they have

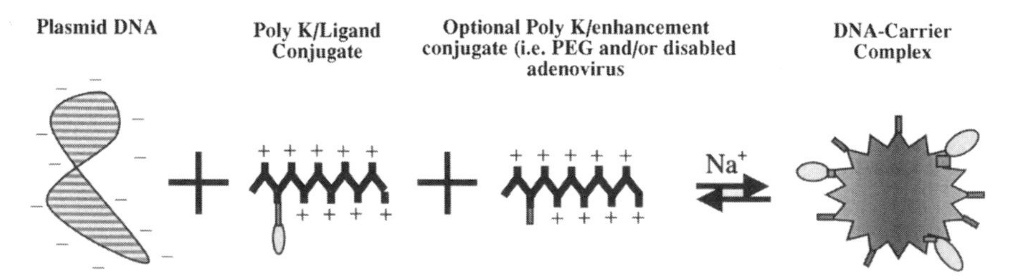

Figure 4.1 General scheme of receptor targeted DNA complex construction. DNA complexes are formed by mixing plasmid DNA with the molecular conjugate under the proper salt conditions. Molecular conjugates consist of a polycation coupled to a receptor ligand. Polycations modified with enhancers (e.g., PEG or adenovirus) may also be included in the DNA complex.

little effect. Initial studies reported that the efficacy of targeted molecular conjugate gene delivery was low and inconsistent, especially in animals. However, recent developments in the field have improved a number of factors including homogeneity and stability, allowing for the design of superior second and third generation targeted molecular conjugates capable of consistently efficient gene transfer in vivo. Although in vitro models are essential in the characterization and initial testing of efficacy, duration, and toxicity of these vectors, in this chapter we focus mostly on developments that have resulted in higher efficacy in vivo, a necessary step in the direction of human gene therapy.

The novelty of this method of gene transfer is its ability to introduce DNA plasmids into specific cell types and to access the nuclei of nondividing cells. Selective targeting of cells may be important, because indiscriminate transgene expression may be disadvantageous. Targeted molecular conjugate DNA delivery depends on several factors, including the presence and number of cell surface receptors on the target cell, receptor–conjugate affinity, access and interaction, the stability of the conjugate-DNA complex, and endocytosis of the complex (Fig. 4.2).

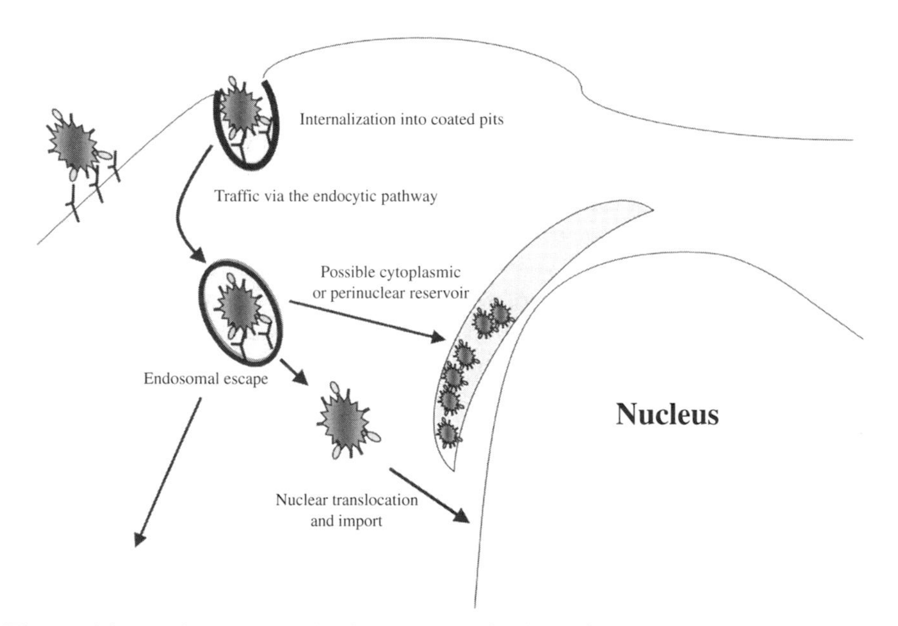

Figure 4.2 Cellular internalization and localization of receptor targeted DNA complexes. Once in contact with the cell surface receptor, the complex is internalized through the endocytic pathway and translocates to the nucleus by either an active or passive mechanism. Endosomal escape of DNA complexes is necessary prior to nuclear import. Alternatively, some receptors may route their cargo to a perinuclear reservoir where DNA complexes are retained and slowly released for prolonged duration.

TARGETED MOLECULAR CONJUGATE TECHNOLOGY

Selection of the Protein Carrier for DNA

The negatively charged phosphate groups on DNA interact with polycations, which can serve as DNA carriers. Under the proper ionic conditions, DNA can be compacted with polycations into a form that resists nuclease degradation. For gene transfer, it is important that the polycation has sufficient affinity for DNA so that the complex remains intact until it enters the cell. However, polycations that bind DNA very tightly might decrease the efficiency of gene transfer duc to the unavailability of the DNA during transcription. Poly-L-amino acids, such as poly-L-lysine (polyK), poly-L-arginine, and poly-L-ornithene have the advantage of limited immunogenicity and prompt metabolism. In contrast, peptides consisting of D-isomers of amino acids are relatively resistant to degradation (Cortes et al., 1993). Poly-L-arginine, though it will compact DNA into small complexes, is less efficient than polyK. It has been postulated that polyK could serve as a nuclear localization signal and may be more efficient in targeting the conjugate-DNA complex to the nucleus once it has been internalized, though this has been contested.

The level of substitution of the polycation with linker and ligand, as well as the length and type of polycation, markedly affect the ability of these complexes to efficiently transfer genes (Ziady et al., 1998; Wagner et al., 1991). Ziady et al., targeting the sec-R, reported that polycations of longer chain length (256 vs. 36 a.a.), modified at less than 10% of the residues, gave more protracted expression and protected DNA from degradation to a greater extent (Ziady et al., 1999). Similar results were reported for complexes directed at the epidermal growth factor receptor (EGF-R). However, high concentrations of NaCl were necessary to prevent aggregation. Heavy substitution of the polymer resulted in less compact DNA complexes, probably due to steric hindrance, though for glycosylated carriers, heavy substitution seems permissible.

A variety of synthetic cationic polymers have been shown to mediate gene delivery into cells. Cascade polymers, or starburst dendramers, have transferred reporter genes to different cell types in vitro. Producing complexes with a net positive charge, Szoka and colleagues achieved optimal transfection efficiency using dendramers 68 Å in diameter (Haensler and Szoka, 1993a). Conjugation of a membrane destabilizing protein to this dendramer further enhanced gene delivery. Polyethylenimine (PEI), an organic polymer synthesized by polymerizing aziridine, has been employed as a DNA carrier due to a superior ability to disrupt endosomes during acidification, causing rupture and more efficient release. This enabled Boussif and co-workers to successfully introduce plasmids containing the luciferase reporter gene into cells of the central nervous system, which are typically difficult to transfect (Boussif et al., 1995). However, although efficient, PEI is quite toxic and is not biodegradable (Godbey, Wu, and Mikos, 1999).

Other DNA-binding molecules have been used to link expression plasmids to ligands and deliver genes to mammalian cells. Nucleoproteins, like histones (Chen, Stickles, and Daiscendt, 1994; Wagner et al., 1991) and protamines (Wagner et al., 1990; Chen et al., 1995), have been employed as the polycation in several molecular conjugates. The intercalating agent bisacridine, chemically modified to contain terminal galactose residues, has been used to direct gene transfer via the asialoglycoprotein receptor (Haensler and Szoka, 1993b). Using a novel variation, Fominaya and Wels exploited the high affinity of the yeast transcriptional activator GAL4 for specific DNA recognition sequences to bind expression plasmids to a molecular conjugate and transfect cells in vitro (Fominaya and Wels, 1996).

Selection of the Targeting Ligand

Construction of the molecular conjugate begins with the selection of an appropriate ligand to target a suitable receptor on a specific cell type. The choice of ligand depends on the receptor to be targeted. Natural as well as synthetic ligands have been used by a number of investigators. Technologies such as peptide phage display have been useful in identifying ligands that are bound and internalized by specific cell types. Because the ligand is the portion of the complex with the greatest potential for immunogenicity and toxicity, ligand size, structure, and design should be carefully considered. Ligands should have high affinity for the receptor, low immunogenicity, initiate minimal cell signaling, and be easy to couple, with no loss of binding affinity, to the polycation. Ligands tested for gene transfer include mono- and disaccharides (Monsigny et al., 1994; Midoux et al., 1993; Chen, Stickles, and Daiscendt, 1994; Perales et al., 1997; Erbacher et al., 1996; Ferkol et al., 1996, 1998), peptides/proteins (Zatoukal et al., 1992; Schwarzenberger et al., 1996; Ross et al., 1995; Hart et al., 1995; Ziady et al., 1997, 1998, 1999), folate (Gottschalk et al., 1994), glycoproteins (Wu and Wu, 1987; Zatoukal et al., 1992), lectins (Batra et al., 1994), and antibodies (Ferkol, Kaetzel, and Davis, 1993; Rojanasakul et al., 1994; Buschle et al., 1995; Chen et al., 1994, 1995; Ferkol et al., 1995). Generally, shorter ligands have been less immunogenic (Haselden, Kay, and Larche, 2000) and easier to synthesize. Furthermore, different ligands against the same receptor may elicit different signaling and trafficking responses. For example, the Fab fragments of an antibody against the EGF-R, which bind but fail to activate the receptor, have been used to target the EGF-R without initiating the signaling cascades or receptor trafficking caused by the native ligand, EGF (Chen et al., 1994). This may be desirable if activation of the target receptor is disadvantageous. Furthermore, varying receptor trafficking may enhance or fine-tune the intensity and duration of transgene expression (Hoganson et al., 2001).

DNA

DNA purity directly influences the ability to produce a homogenous population of targeted complexes. DNA should be purified by double CsCl gradient centrifugation or an equivalent high quality method. Endo- and exotoxin contamination of plasmid preparations has been shown by Cotten and co-workers to injure primary cells in vitro and affect the expression of reporter genes (Cotten et al., 1994). Thus, the use of pure, pharmaceutical grade DNA is essential. Although plasmids as large as 48 kb have been successfully delivered to cells with receptor-targeted molecular conjugates, DNA size affects the mobility of free DNA within the cytoplasm (Lukacs et al., 2000), and even for compacted DNA, larger plasmids complexed with molecular conjugates are less efficient in reaching the nucleus following direct injection into the cytosol. Futhermore, although molecular conjugates can compact large DNA molecules, the larger the plasmid, the larger the complex size, so minimizing plasmid size may be advantageous for enhancing efficacy, especially in vivo.

For molecular conjugate DNA delivery, the plasmid DNA transgenes remain episomal, but expression from them can be extended by selection of promoter and enhancer, and by targeting long-lived cells. Self-replicating episomal vectors that contain viral sequences that act as an origin of replication could allow the transgene to persist in dividing cells (Cooper, 1996). In addition, artificial chromosomes are being developed that should permit the delivery and continued expression of not only the transgene (cDNA or intact gene) but its promoter and enhancer elements as well (Huxley, 1994).

The transgene product can provoke host cellular immune responses, with destruction of the gene-expressing cells. The CpG sequences in plasmids grown in bacteria can also incite host immune responses (Yew et al., 1999), and inflammation can be reduced and duration of expression enhanced by specifically removing as many of these sequences as possible from the expression plasmids (Yew et al., 2000).

Molecular Conjugate Condensation of DNA

In initial reports, the polycation was simply viewed as a method to adhere the ligand to the plasmid DNA. Consistent with this concept, gene transfer mediated through the asialoglycoprotein receptor showed little dependence on the size of the condensed DNA particles (Wu and Wu, 1987; Wu et al., 1991). In these experiments, DNA was annealed to excess polycation, then dialyzed against a sodium chloride gradient. This technique was capable of compacting 200–500 μg of DNA with short length polyK (M_r = 3,800) in 1 ml low salt (~150 mM NaCl) solution. The DNA was gradually condensed into multimolecular positively charged particles with an apparent size distribution ranging between 50 and 200 nm in diameter (Wu and Wu, 1987; Wu et al., 1991; Wu, Wilson, and Wu, 1989).

Subsequent studies revealed that the size of the conjugate DNA complexes may be important, particularly in vivo (Perales et al., 1997; Ziady et al., 1999; Cortes et al., 1993), possibly due to size restrictions of the endocytic pathway (Neutra et al., 1985), on the movement of free DNA in cytoplasm once endosomal escape has occurred, or on nuclear import. In addition, Ziady et al.

(1999) have shown that more tightly condensed complexes are more resistant to nuclease degradation. Wagner and co-workers have shown that the degree of condensation of the complexes correlated with the level of transgene expression (Wagner et al., 1991). Specifically, conjugate-DNA complexes condensed into toroid structures measuring 80–100 nm in diameter were most effective for gene transfer into cells grown in culture. Condensation with longer polymers further complicated this method, resulting in heterogeneous precipitates ranging in size between 50 and 1000 nm, unless the molecular conjugate was modified with polyethylene glycol (PEG) or smaller concentrations of DNA (30–60 μg/ml) were used. Filtration could isolate the smaller particles from this mixed population but reduced the effective DNA concentration.

Perales and colleagues developed an alternative method of condensing expression plasmids into compact particles suitable for gene transfer and have successfully stabilized these complexes, even at high concentrations of DNA (Perales et al., 1997, 1994a). Up to 12 mg DNA/ml could be compacted with any size polycation (Gedeon et al., 2000). This technique involved the gradual addition of small aliquots of the ligand-polycation conjugate to plasmid DNA. Adjusting the concentration of sodium chloride in the solution yielded "unimolecular" complexes measuring as small as 10–20 nm in diameter by electron microscopy, depending on plasmid size (Perales et al., 1997; Ziady et al., 1999; Perales et al., 1994a; Gedeon et al., 2000). The complexes had neutral net surface charge, critical for avoiding complement activation and providing for a more efficient vector for gene transfer in vivo (Plank et al., 1996).

The length of the polylysine affects how much sodium chloride is necessary to form and maintain these complexes, with longer polymers requiring higher salt concentrations to prevent aggregation. The secondary structure of the polycation also influenced the structure of the complexes. Some synthetic peptides arranged in different conformations (i.e., random coil, α-helix, and β-sheet structures) produced multimolecular aggregates, unsuitable for receptor-mediated uptake, when bound to DNA (Perales et al., 1994b). Furthermore, the DNA condensation was destabilized by increased substitution of the polycation during conjugate synthesis. Extensive substitution lessens the polycation affinity to DNA, resulting in less tightly packed complexes that may be too large for internalization (Ziady et al., 1998). Substitution usually eliminates the positive charge on the polycation and thus lessens its affinity for DNA. Steric hindrance by bulky ligands also results in less tightly packed complexes. Ziady et al. reported more effective gene transfer via the serpin-enzyme complex (sec-R) in vitro with the less substituted polycations (Ziady et al., 1998). Poly-L-lysine alone, however, produced the smallest complexes but failed to transfect cells. In addition, duration of transgene expression was prolonged with longer chain length polylysine, which compacted DNA to a greater extent (Ziady et al., 1999). Wagner and colleagues discovered that molecular conjugates with fewer ligand moieties are more effective at delivering reporter genes by receptor-mediated endocytosis (Wagner et al., 1991). Conjugates containing approximately 1 transferrin per 100 lysine residues resulted in maximal transgene expression in human erythroid cells. Moreover, the partial replacement of the transferrin-based conjugate with free polylysine produced smaller toroidal structures and higher levels of luciferase activity.

Enhancing Elements

Endosomal Escape Once inside the cell, these DNA must reach the nucleus, where it is transcribed. For many receptors, it appears that the route to the nucleus is via endosomal escape and passage through the cytoplasm (Fig. 4.2), though this suggestion is based on pharmacological data rather than direct observation. Few studies have examined the trafficking of receptor-targeted DNA complexes. Scanlin and colleagues reported that lactosylated DNA molecular conjugate complexes administered to airway epithelial cells in culture are internalized, traffic to the perinuclear region, and are internalized through the nuclear pore. Blockage of the nuclear pore protein complex blocked the translocation of plasmid DNA across the nuclear membrane (Klink et al., 2001).

Szoka and colleagues (Haensler and Szoka, 1993a) reported that polycations that can be protonated (such as polyethylenimine) may uncouple the endosomal proton pump, resulting in endosome lysis due to an influx of water. Other investigators have used a variety of agents to enhance endosomal escape, either incorporated into the molecular conjugate-DNA complexes or administered separately (Gottschalk et al., 1994; Adami et al., 1998; Curiel et al., 1991; Plank et al., 1994). Cotten and associates have shown that chloroquine, a lysosomotropic agent that interferes with the acidification of lysosomes and inhibits hydrolytic enzymes, greatly improves the expression of transgenes in human erythroid cells (K562 cell line) transfected with the transferrin-polylysine conjugate (Adami et al., 1998). Chloroquine has also been shown to augment the expression of transgenes delivered to other cell types, like hepatoma cells and human macrophages, by receptor-mediated endocytosis (Midoux et al., 1993; Erbacher et al., 1996). Colchicine, which disrupts microtubules, has also been shown to improve the survival of transgenes in vivo by interfering with the trafficking of endosomes (Chowdhury et al., 1996).

Although the efficiency of gene transfer in many systems has been enhanced through disruption of the endocytic pathway by pharmacological agents, this is not the case for all receptor systems. Gene transfer via some receptor systems, such as the sec-R, is not enhanced by endosomal rupture (unpublished observations, A. Ziady, T. Ferkol, and P. Davis). DNA complexes targeted to this receptor rapidly traffic to the nuclear periphery, where within 15 minutes DNA dissociates from polylysine and transverses the nuclear membrane (Funke, Ziady, and Davis, 2000). It is thought that prolonged gene expression via this system results from a nonlysosomal pool of complexes that remains at the nuclear periphery as an intracellular reservoir after a single dose administration. Since endolysotropic agents are often toxic, their use should be balanced with the maintenance of the safety advantages of molecular conjugates. Indeed, as in the case of the sec-R, these agents may not be necessary.

Several investigators have employed adenoviruses to improve the release of macromolecules taken up by receptor-mediated endocytosis. The regulation of endosomal pH is critical for the appropriate trafficking of ligands along endocytotic pathways, and the pH of the endosomal vesicle becomes more acidic as it routes to the lysosome. Acidification of the late endosomes produces conformational changes in the adenoviral capsid proteins and causes an interaction between the protein and vesicle membrane, thus disrupting the endosome. This observation led to the hypothesis that endosomal disruption caused by the adenovirus could improve the survival and expression of transgenes. Curiel and associates have exploited this property of the virus and have coupled replication-defective adenovirus to transferrin-based molecular conjugates (Curiel et al., 1975). These adenovirus-linked molecular conjugates greatly enhanced the expression of different transgenes in a variety of cell types in vitro. Even addition of free adenovirus to the complexes was effective in enhancing transgene expression. However, complexes containing adenoviruses may be retargeted toward adenovirus receptors, and the inflammatory and immunogenic properties of the adenovirus (or other viruses, which have also been tested) may be deleterious for gene transfer, especially if it needs to be repeated. Moreover, adenovirus may increase the size of the complex.

To circumvent some of these problems, investigators have included individual proteins, mostly viral in origin, in molecular conjugates to enhance endosomal release of DNA. Wagner and associates have shown that peptide sequences derived from the influenza haemagglutinin HA-2 bound to the transferrin-polylysine molecular conjugate markedly increased the level of transgene expression in cells grown in culture (Wagner et al., 1992). Other investigators have constructed synthetic peptides that imitate the endosomolytic functions of viral proteins such as the fusion protein of the respiratory syncytial virus (Gottschalk et al., 1996). The inclusion of such proteins may add to the immunogenicity of the complexes.

Nuclear Localization The ultimate goal of gene transfer is the delivery of exogenous DNA to the nucleus where transgenes can undergo transcription (Fig. 4.2). The process of nuclear trafficking of DNA complexes once they escape the endosome, however, is perhaps the least understood process in receptor-mediated gene transfer. Nuclear entry of the conjugate-DNA complexes may be a passive, random process and may simply relate to the total number of copies of the transgene present in the cytoplasm and the mobility of DNA in the cell. Access of transgenes to the nucleus is enhanced by cell division, during the disintegration of the nuclear envelope. The efficiency of gene transfer via the asialoglycoprotein receptor increased when hepatic regeneration was induced by subtotal hepatectomy (Wu, Wilson, and Wu, 1989; Wu et al., 1991; Ferkol et al., 1993). Nevertheless, reporter genes can be effectively delivered to nondividing cells (Ferkol, Kaetzel, and Davis, 1993; Ferkol et al., 1995), especially if DNA is compacted (Gedeon et al., 2000). Thus, the role of cell division may differ from system to system and will depend on the size of the DNA complexes and characteristics of the targeted receptor.

Transgene expression in the liver was also remarkably prolonged in the animals that underwent subtotal hepatectomy, though it is uncertain if the exogenous genes were integrated into the host cell's genome (Wu, Wilson, and Wu, 1989; Wu et al., 1991; Ferkol et al., 1993). It is likely that the majority of DNA transferred by these molecular conjugates in cells and tissues exists as episomes in the nucleus and will not persist in the nuclei of transfected cells. The prolonged survival of the reporter gene may be related to the transgene escaping the asialoglycoprotein receptor pathway, thus avoiding degradation in lysosomes and allowing the DNA to persist in nonhydrolytic cytoplasmic vesicles that serve as a protective pool of these complexes (Chowdhury et al., 1993).

Another compelling hypothesis is that the polylysine component of the carrier may assist in nuclear trafficking of the DNA. Several viral proteins responsible for translocation of the viral genome to the nucleus are highly basic and have sequences rich in lysine. The amino acid sequence Phe-Lys-Lys-Lys-Arg-Lys-Val has been shown to transport the simian virus-40 (SV-40) large T antigen to the nuclei of mammalian cells (Kalderon et al., 1984), and the sequence Lys-Lys-Lys-Tyr-Lys-Leu-Lys (Bukrinsky et al., 1993) serves as a specific nuclear localization signal for the matrix protein for the human immunodeficiency virus type-1.

Nucleoproteins, such as histones and protamine that are also highly basic, have been employed as the polycation or nuclear localization signal in several molecular conjugates. The addition of histone H4 to transferrin-based conjugates enhanced transgene expression in cells in vitro (Wagner et al., 1991). Histone H1, chemically modified to contain terminal galactose residues, was also used to introduce reporter genes into hepatoma cells via the asialoglycoprotein receptor (Midoux et al., 1993). The mechanisms by which these proteins enhance gene transfer still need to be determined.

Serum Stability Although molecular conjugate condensation of DNA protects it against degradation temporarily, prolonged survival in blood is not achieved. Long chain polymers, shown to protect DNA against degradation in vitro (Ziady et al., 1999), have also been shown to prolong DNA stability and circulation time (Dash et al., 1999). Nevertheless, DNA complexes not internalized in the first few passes rapidly aggregate, bind to serum albumin, and are eliminated (Dash et al., 1999). To circumvent this problem, investigators have substituted the polycation portion of the conjugate with PEG. Wagner and colleagues reported that such modification not only reduced aggregation, but also stabilized high concentrations of DNA molecular conjugate [illegible] in solution (Katayose and Kataoka, 1997). Rice and co-workers demonstrated that PEGylation of receptor-targeted molecular conjugates resulted in a prolonged circulation time of the complexes after intravenous administration (Kwok et al., 1999). Rice (Adami and Rice, 1999) and Seymour (Oupicky et al., 2000) have also cross-linked the polycation moieties in the complex together after complex formation to enhance stability. These cross-links are designed for reduction once inside the cell to allow access of the DNA for transcription. Although these enhancements may facilitate delivery of the entire application to the target tissue in vivo, oversubstitution of the molecular conjugate may affect the interaction with DNA (Ziady et al., 1998). Shielding provided by the PEG substitution, which protects against aggregation, may also interfere with the availability of the ligand for receptor binding. Furthermore, overstabilization of the DNA molecular complex is possible and leads to decreased efficiency due to the unavailability of the DNA for transcription.

Modifications for stability in blood may not be essential for some receptor systems. A receptor that is highly expressed on the target tissue and is rapidly internalized once ligated may clear sufficient quantities of the DNA complex before inactivation by serum. DNA complexes targeted to the sec-R, a receptor that rapidly internalizes its ligand, are internalized within 2 minutes and trafficked to the nucleus in as little as 10–15 minutes, in vitro (Funke, Ziady, and Davis, 2000). This may explain why serum stabilization was not necessary to achieve considerable gene transfer in vivo (Ziady et al., 1999). Thus, although possibly advantageous, PEGylation should be carried out under controlled conditions and the resulting receptor-targeted complexes should be fairly characterized.

TARGETED RECEPTORS

The novelty of receptor-mediated gene transfer is its selective nature. Indiscriminate transgene delivery and expression may be disadvantageous. A variety of cell surface receptors have been targeted for gene delivery. The most successful receptors are abundant, constitutively expressed, and exhibit a high specificity for a given ligand but a low selectivity for the attached cargo. In this section we discuss only those tested in vivo as well as in vitro, although the others are listed in Table 4.1.

The Asialoglycoprotein Receptor

In their seminal work, Wu and colleagues described a soluble, targetable DNA carrier that delivered expression plasmids to hepatocytes via the asialoglycoprotein receptor (Wu and Wu, 1987). The asialoglycoprotein receptor is an integral membrane glycoprotein that clears galactose-terminal glycoproteins from the blood by endocytosis through the coated pit/coated vesicle pathway in hepatocytes (reviewed in Stockert, 1995) and routes the cargo to lysosomes. However, the trafficking of the asialoglycoproteins is leaky, and DNA transferred via this asialoglycoprotein receptor can escape. Because expression of this receptor is limited to hepatocytes, this system has excellent cell selectivity.

The ligand was prepared by desialylation of orosomucoid, which exposed the terminal galactose. The ligand was linked to poly-L-lysine and com-

TABLE 4.1 Receptors Targeted with Molecular Conjugate Vectors

Receptor	Ligand	Trafficking
Asialoglycoprotein	Asialoorsomucoid galactose	Lysosomal
Transferrin	Transferrin	Recycled
Polymeric immunoglobulin	Anti-pIgR antibody	Transcytotic
Serpin enzyme complex	Peptide ligands	Lysosomal
Epidermal growth factor (EGF)	Epidermal growth factor and anti-EGF-R antibody	Lysosomal/perinuclear/recycled
Fibroblast growth factor (FGF)	Fibroblast growth factor 2 and basic FGF	Lysosomal/perinuclear/recycled
Vascular endothelial growth factor (VEGF)	Vascular endothelial growth factor	Lysosomal/perinuclear/recycled
Integrin	Peptide ligand	Lysosomal
Folate	Folate	Lysosomal
Mannose	Mannose	Lysosomal
Cell surface glycocalyx	Lectins	Lysosomal/recycled
Surfactant A	Surfactant protein A	Lysosomal
C-kit	Anti-CD3 antibody	Lysosomal
Carbohydrate	Anti-T_n antibody	Lysosomal
CD3	Steel factor (SLF)	Lysosomal

[1] Wu and Wu, 1987; Monsigny *et al.*, 1994; Midoux *et al.*, 1993; Chen, Stickles, and Daiscendt, 1994; Perales *et al.*, 1997; Wu, Wilson, and Wu, 1989; Wu *et al.*, 1991; Wilson *et al.*, 1992; Perales *et al.*, 1994a; Chowdhury *et al.*, 1993; Christiano, Smith, and Woo, 1993; Christiano *et al.*, 1993; Ferkol *et al.*, 1993; Stankovics *et al.*, 1994.
[2] Zatoukal *et al.*, 1992; Wagner, Curiel, and Cotten, 1994; Wagner *et al.*, 1990; Zenke *et al.*, 1990; Cotten *et al.*, 1992, 1993b; Gao *et al.*, 1993; Wagner *et al.*, 1991.
[3] Ferkol, Kaetzel, and Davis, 1993; Ferkol *et al.*, 1995, 1996.
[4] Ziady *et al.*, 1997, 1998, 1999.
[5] Chen *et al.*, 1994; Ohtake *et al.*, 1999; Wells, 1999.
[6] Fisher *et al.*, 2000; Hoganson *et al.*, 1998; Powers, McLeskey, and Wellstein, 2000.

plexed with DNA. These complexes specifically targeted human hepatoma cells (HepG2) in vitro (Wu and Wu, 1987), and in vivo (Wu, Wilson, and Wu, 1989; Wu et al., 1991; Ferkol et al., 1993) systemic infusions of complexes in rats in vivo produced reporter gene activity only in the liver. The gene expression persisted for weeks when the animals underwent subtotal hepatectomy at the time of injection. Using this approach, Nagase analbuminemic rats received intravenous administration of complexes containing the structural gene for human albumin. The transgene was detected as an episome in the liv-

Target Cells	Successful Transfection	Co-transfer Elements Used	Refs.
Hepatocytes	Moderate in vitro and low in vivo	Endosomolytic agents (unless partial hepatectomy)	[1]
Ubiquitous	Moderate in vitro and low in vivo	Endosomolytic agents/PEG adenovirus particles	[2]
Respiratory and intestinal epithelia, hepatocytes	Moderate in vitro and in vivo	None	[3]
Hepatocytes, glial neurons, macrophages, respiratory epithelia	High in vitro and moderate in vivo	None	[4]
Ubiquitous, tumor tissue	Moderate in vitro and low in vivo	Adenovirus particles	[5]
Epithelium, tumor tissue	High in vitro and low low in vivo	PEG	[6]
Vascular endothelium, smooth, cardiac, skeletal muscle, and tumor disease	Moderate in vitro	None	[7]
Ubiquitous	Moderate in vitro	None	[8]
Ubiquitous	Moderate in vitro	None	[9]
Macrophages	Moderate in vitro and low in vivo	None	[10]
Ubiquitous	Moderate in vitro	None	[11]
Respiratory epithelia, and alveoli	Moderate in vitro	None	[12]
Hematopoietic stem cells	Moderate in vitro	None	[13]
Carcinoma cells and lymphocytes	Moderate in vitro	None	[14]
Lymphocytes	Moderate in vitro	Adenovirus particles	[15]

[7] Fisher *et al.*, 2000
[8] Hart *et al.*, 1995, 1998; Jenkins *et al.*, 2001.
[9] Gottschalk *et al.*, 1994
[10] Erbacher *et al.*, 1996; Ferkol *et al.*, 1996, 1998.
[11] Batra *et al.*, 1994.
[12] Ross *et al.*, 1995.
[13] Schwarzenberger *et al.*, 1996.
[14] Thurher *et al.*, 1990.
[15] Buschle *et al.*, 1995.

ers of the animals 2 weeks after treatment, and low levels of human albumin were measured in the blood for 4 weeks (Wu et al., 1991).

Using a similar system, Wilson and co-workers delivered a chimeric gene containing the low-density lipoprotein receptor cDNA to the livers of Watanabe rabbits, an animal model for familial hypercholesterolemia (Wilson et al., 1992). Low density lipoprotein receptor mRNA was isolated one day after administration of the conjugate-DNA complexes, but no transcripts were detected 3 days posttransfection. The total cholesterol levels in the blood of transfected rabbits were also reduced by 30% 2 days after treatment, but returned to pretreatment levels by 5 days after transfection.

Stankovics and co-workers reported that an expression plasmid containing the gene encoding methylmalonyl CoA mutase, bound to an asialoorosomucoid-polylysine molecular conjugate, increased the enzyme's activity in the liver to a level that could be therapeutic in patients with the inborn error of metabolism, methylmalonic aciduria (Stankovics et al., 1994). Expression was short-lived, lasting less than 2 days after treatment, and the complexes elicited an antibody response against the ligand, asialoorosomucoid, which may limit its use in chronic diseases. No serologic response to the plasmid DNA was detected in the mice after injection, which is encouraging, because antibodies directed against DNA could preclude such an approach for gene transfer.

Other investigators have used alternative ligands to successfully target the asialoglycoprotein receptor. Molecular conjugates consisting of lactosylated polylysine (Midoux et al., 1993), galactosylated histones (Chen, Stickles, and Daiscendt, 1994), and galactosylated albumin-polylysine galactosylated albumin-polylysine (Ferkol et al., 1993) have been used to transfer reporter genes to hepatoma cells in vitro. A molecular conjugate consisting of a polylysine chemically modified with α-D-galactopyranosyl phenylisothiocyanate introduced functional genes to hepatocytes in vitro and in vivo (Perales et al., 1994a). This receptor has also been successfully targeted in vivo by investigators who delivered chimeric oligonucleotides, which specifically altered a single base pair in genomic DNA, resulting in genetic correction of deficiency of glucuronyltransferase activity in Gunn rats.

The Transferrin Receptor

The transferrin receptor, a dimeric glycoprotein 180 kDa in size (reviewed in Ponka and Lok, 1999), binds to its natural ligand, transferrin, rapidly internalizes the complex, and then recycles the ligand back to the cell surface. The transferrin receptor is ubiquitous, present on the cell membranes of most dividing eukaryotic cells, including erythroblasts, hepatocytes, and tissue macrophages; this receptor system has been exploited to deliver drugs and toxins to tumor cells in vitro, as reviewed by Wagner and colleagues (Wagner, Curiel, and Cotten, 1994).

Complexes directed at the transferrin receptor transfect avian and human erythroid cells in vitro (Zatoukal et al., 1992; Wagner et al., 1990; Zenke et al., 1990), as well as other cells in primary culture. In some cell lines, gene expression is augmented by co-treatment with lysosomotropic agents, blocked by excess free transferrin to the medium, and enhanced by agents (e.g., by desferroxamine) that increase expression of the transferrin receptor (Thurher et al., 1990). Inactivated adenovirus particles included in the conjugate DNA complexes also augmented transgene expression (Cotten et al., 1992). Plasmid size, up to 48 kb, had little effect of the efficacy of transfer. Complexes less than 100 nm in diameter were optimal for gene delivery.

Systemic injection of transferrin receptor-directed complexes, including those containing an inactivated adenovirus, has had limited success, as reviewed by Cotten (1995). It is possible that the inflammatory responses associated with adenovirus entry interfered with the survival of transfected cells. These same immune responses, however, could potentially be useful in a vaccine application. Zatoukal and colleagues have used the transferrin-based conjugate to deliver the interleukin-2 gene to murine melanoma cells, and high levels of the cytokine were produced in culture (Zatoukal et al., 1995). Based on this data, tumor cells transfected ex vivo have been reimplanted in animal models and act as a vaccine by inducing an immune response against the tumor.

Local injections of transferrin-based conjugate-DNA complexes into the liver have given high levels of reporter gene expression (Cotten, 1995). Intratracheal instillation of DNA bound to human transferrin-polylysine and transferrin-adenovirus-polylysine molecular conjugates resulted in transient low-level expression of the reporter gene, which peaked one day after transfection and returned to pretreatment levels by 7 days.

The Serpin Enzyme Complex Receptor

Ziady et al. have used the serpin enzyme complex receptor (sec-R) to examine a number of characteristics of receptor-mediated molecular conjugate gene transfer. The sec-R, originally described as a binding site on human hepatoma cells and blood monocytes, recognizes a sequence in α_1-antitrypsin, which is exposed only when it is complexed with a serine protease such as neutrophil elastase or modified by either metalloelastase or by the collaborative action of active oxygen intermediates and neutrophil elastase (reviewed in Perlmutter, 1994). The receptor is present on such cell types as mononuclear phagocytes, neutrophils, myeloid cell lines U937 and HL60, human intestinal epithelial cell line CaCo2, mouse fibroblast L cells, rat neuronal cell line PC12, and human glial cell line U373MG. Using synthetic peptides based in sequence on α_1-antitrypsin, Ziady and colleagues targeted reporter genes specifically to receptor bearing cells in vitro (Ziady et al., 1997, 1998) and in vivo (Ziady et al., 1999). Studies detailing the effects of substitution of polyK with receptor ligands on expression (Ziady et al., 1998) demonstrated that sparsely substituted polyK of various lengths could be used to extend or shorten the duration of expression as well as affect the intensity of expression in vitro and in vivo (Ziady et al., 1999). More recently, Funke and co-workers have examined this receptor's ability to rapidly clear DNA complexes and transport them to the nuclear periphery (Funke, Ziady, and Davis, 2000).

The Polymeric Immunoglobulin Receptor

The lungs play a pivotal role in a number of genetic and nongenetic diseases and have been proposed as potential organs for gene therapy. Gene transfer to the respiratory epithelial cells has been mediated by different receptors, including the polymeric immunoglobulin receptor (reviewed in Kaetzel et al., 1997), which is specifically adapted for the internalization and nondegradative transport of macromolecules (e.g., dimeric immunoglobulin A and pentameric immunoglobulin M) across epithelia and hepatocytes, and appears to be well suited for receptor-mediated gene transfer. The distribution of this receptor in humans in the airway epithelium and the serous cells of the submucosal glands has led to speculation that this receptor may be an attractive target for the treatment of lung diseases, such as cystic fibrosis.

DNA has been directed to human tracheal epithelial cells in vitro using the Fab fragment of an antibody directed against the human secretory component, the ectoplasmic domain of the receptor, as ligand. Gene expression was specific for cells that expressed the receptor and persisted in the presence of an excess of dIgA, the natural ligand for the receptor, but was inhibited by excess soluble receptor, which presumably occupies the recognition site on the Fab fragment. However, expression was transient. In vivo, Fab fragments of antibodies directed against the rat polymeric immunoglobulin receptor introduced expression plasmids into tissues that express the receptor when complexes were injected intravenously (Ferkol et al., 1995). The pattern of reporter gene expression conformed to the spatial distribution of cells that express the receptor. Expression lasted less than 12 days after injection. However, a neutralizing antibody response directed against the Fab portion of the complexes, which was associated with reduced gene delivery by the complexes, developed after the third injection (Ferkol et al., 1996b). Clearly, these immunologic obstacles must be dealt with before such vectors could be used clinically.

The Mannose Receptor

Tissue macrophages have been targeted via the mannose receptor (reviewed in Stahl and Ezekowitz, 1998), which is abundant in macrophages, and internalizes glycoproteins with mannose, glucose, fucose, and N-acetylglucosamine residues in exposed, nonreducing positions. Ligands internalized by the mannose receptor are trafficked to lysosomes. Polylysine, glycosylated with various monosaccharides, was used for receptor-mediated gene transfer to murine and human macrophages in primary culture (Erbacher et al., 1996; Ferkol et al., 1996a). The transfection efficiency varied, but transgene expression co-localized with monocyte/macrophage markers (Ferkol et al., 1998) and could be blocked by excess mannosylated bovine serum, suggesting receptor specificity.

The mannose-terminal glycoprotein conjugate directs gene transfer into tissue macrophages in rodents in vivo following systemic injection (Erbacher et al., 1996). Reporter genes were successfully delivered to reticuloendothelial organs in mice, though transfection efficiency was low and transgene expression was short-lived, peaking 4 days after transfection.

Other Receptors

The epidermal growth factor receptor (EGF-R) (reviewed in Wells, 1999), has been a target for exogenous gene delivery. Due to increased expression levels of this receptor in certain types of cancer cells, this system has been attractive. However, although quite successful in vitro (Chen et al., 1994), transgene delivery by this receptor in vivo has been inefficient. More recently, a Fab fragment of the monoclonal antibody B4G7 against the human EGF-R has been conjugated to polylysine and used to direct suicide gene delivery to melanoma cells that were then implanted in nude mice (Ohtake et al., 1999). Suppression of tumor growth indicated successful gene delivery, confirming previous studies that characterized EGF-R as a suitable receptor for molecular conjugate cancer therapy.

The fibroblast growth factor receptor (FGF-R) (reviewed in Powers, McLeskey, and Wellstein, 2000) has also been used to a limited extent both in vitro and in vivo. Sosnowski and colleagues established that basic fibroblast growth factor (FGF2) was an efficient ligand for targeting genes to the FGF receptor in vitro and in vivo (Hoganson et al., 1998). Seymour and colleagues demonstrated the efficiency of gene delivery via this receptor in a study of nuclear import of transferrin, vascular endothelial growth factor (VEGF), and basic FGF containing DNA complexes (Fisher et al., 2000). Analysis using fluorescence microscopy showed enhanced uptake of bFGF-targeted complexes. Although VEGF and transferrin-targeted complexes were restricted to cytoplasmic or perinuclear compartments, bFGF-targeted complexes showed efficient delivery into the nucleus, with accumulation of more than 10^5 plasmids per cell within distinct intranuclear compartments. Thus, the FGF receptor has been promising in vitro, but this efficacy remains to be seen in animals.

Integrin targeted molecular conjugates have also been successful in delivering transgenes in vivo. Small peptides, which minimize the immune response, are capable of targeting integrins and there is less size constraint on the attached complex (Hart et al., 1995). Coutelle and colleagues added an integrin-binding domain to their polycation-DNA complexes and achieved efficient gene transfer in vitro (Hart et al., 1995). Kinnon and co-workers further enhanced this technique in cultured corneal endothelium by including liposomes in the DNA complex to augment endosomal escape (Hart et al., 1998). This maneuver increased luciferase expression by 2 orders of magnitude while remaining relatively specific, as evidenced by the reduction in expression observed when competing antibodies were added at the time of transfection. Jenkins et al. reported that integrin-targeted molecular conjugates delivered the bacterial β-galactosidase gene to bronchial epithelium and parenchymal cells in vivo, with high efficacy (Jenkins et al., 2001). Vector administration was repeatable and resulted in no inflammation.

CONCLUSION

A variety of molecular conjugates have been used to transfer exogenous, functional genes into mammalian cells by exploiting receptor-mediated endocytosis. This method of nonviral gene transfer provides a flexible, noninfectious method for DNA delivery to specific target cells in vitro and in vivo. The incorporation of enhancing agents to stabilize the DNA complex or aid in endosomal release and nuclear targeting greatly enhances transgene expression. However, despite their potential advantages (Table 4.1), these gene delivery systems have produced variable results in vivo. Reporter genes introduced into target cells by certain receptor-targeted molecular conjugates (i.e., asialoglycoprotein, sec-R) have resulted in prolonged expression. This may be related to the size and conformation of the plasmid DNA, the method of administration, and/or the DNA complex size and stability. Other molecular conjugates have resulted in transient expression in tissues at moderate levels in vivo, despite modifications that enhance stability in the recipient cell. Extensive characterization of receptor-targeted molecular conjugates should shed light on these intricacies. Rigorous analysis will lead the way for the development of second and third generation molecular conjugates embodying the safety advantages of nonviral gene delivery in the context of a stable and efficient vector.

Presently, some investigators are addressing the limitations of receptor-targeted molecular conjugates by incorporating features of molecular complexes into other gene transfer vehicles, such as liposomes. Condensation of plasmid DNA with basic proteins (e.g., polylysine) has been shown to augment gene transfer by cationic liposomes due to improved encapsulation of the DNA and possibly improved cell trafficking. Liposomes tend to indiscriminately fuse with cells, and the addition of a ligand to the lipid-DNA complex may permit the selective delivery of genes to target cells (Miller and Vile, 1995). Chimeric vectors are also being designed to combine elements of both viral and nonviral systems, where investigators have included viral proteins into molecular conjugates to enhance transgene endosomal escape and nuclear uptake.

The converse should also be true, and recombinant viruses could be modified to include ligands that should permit their targeting to specific cells that express the receptor. An ecotropic Moloney murine leukemia virus that was chemically modified with lactose specifically targeted the asialoglycoprotein receptor on the surface of HepG2 hepatoma cells (Wu and Wu, 1987). Alternatively, Kasahara and co-workers have genetically engineered the sequence for erythropoietin into the envelope of an ecotropic retrovirus (Kasahara, Dozy, and Kan, 1994). This hybrid vector was used to target transgenes to murine and human cells that express the erythropoietin receptor. Avian retroviruses have also been modified to express an integrin on their surface and used to infect eukaryotic cells (Valsesia-Whitmann et al., 1994). These chimeras could potentially permit the tissue-specific targeting of recombinant viruses to cell surface receptors, and may represent the future of receptor-mediated gene transfer.

REFERENCES

Adami RC, Rice KG (1999): Metabolic stability of glutaraldehyde cross-linked peptide DNA condensates. *J Pharm Sci* 88(8):739–746.

Adami RC, Collard WT, Gupta SA, Kwok KY, Bonadio, J, Rice KG (1998): Stability of peptide-condensed plasmid DNA formulations. *J Pharm Sci* 87:678–683.

Batra RK, Wang-Johanning F, Wagner E, Garver RI, Curiel DT (1994): Receptor-mediated gene delivery employing lectin-binding specificity. *Gene Ther* 1:255–260.

Boussif O, Lezoualc'h F, Zanta MA, Mergny MD, Scherman D, Demeneix B, Behr JP (1995): A versatile vector for gene and oligonucleotide transfer into cells in culture and in vivo: polyethylenimine. *Proc Natl Acad Sci USA* 92:7297–7301.

Bukrinsky MI, Haggerty S, Dempsey MP, Sharova N, Adzhubel A, Spitz L, Lewis P, Goldfarb D, Emerman M, Stevenson M (1993): A nuclear localization signal within HIV-1 matrix protein that governs infection of non-dividing cells. *Nature* 365:666–669.

Buschle M, Cotten M, Kirlappos H, Mechtler K, Schaffner G, Zauner W, Birnstiel ML, Wagner E (1995): Receptor-mediated gene transfer into human T lymphocytes via binding of DNA/CD3 antibody particles to the CD3 T cell receptor complex. *Hum Gene Ther* 6:753–761.

Chen J, Stickles RJ, Daiscendt KA (1994): Galactosylated histone-mediated gene transfer and expression. *Hum Gene Ther* 5:429–435.

Chen J, Gamau S, Takayanagi A, Shimizu N (1994): A novel gene delivery system using EGF receptor-mediated endocytosis. *FEBS Lett* 338:167–169.

Chen SY, Zani C, Khouri Y, Warasco WA (1995): Design of a genetic immunotoxin to eliminate immunogenicity. *Gene Ther* 2:116–123.

Chowdhury NR, Wu CH, Wu GY, Yerneni PC, Bommineni VR, Chowdhury JR (1993): Fate of DNA targeted to the liver by asialoglycoprotein receptor-mediated endocytosis in vivo. *J Biol Chem* 268:11265–11271.

Chowdhury NR, Hays RM, Bommineni VR, Frank N, Chowdhury JR, Wu CH, Wu GY (1996): Microtubular disruption prolongs the expression of human bilirubin-uridinediphosphoglucuronate-glucuronosyltransferase-1 gene transferred into Gunn rat livers. *J Biol Chem* 271:2341–2346.

Christiano RJ, Smith LC, Woo SL (1993): Hepatic gene therapy: adenovirus enhancement of receptor-mediated gene delivery and expression in primary hepatocytes. *Proc Natl Acad Sci USA* 90:2122–2126.

Christiano RJ, Smith LC, Kay MA, Brinkley BR, Woo SL (1993): Hepatic gene therapy: efficient gene delivery and expression in primary hepatocytes utilizing a conjugated adenovirus DNA complex. *Proc Natl Acad Sci USA* 90:11548–11552.

Cooper MJ (1996): Noninfectious gene transfer and expression systems for cancer gene therapy. *Semin Oncol* 23:172–187.

Cortes F, Panneerselvam N, Mateos S, Ortiz T (1993): Poly-D-lysine enhances the genotoxicity of bleomycin in cultured CHO cells. *Carcinogenesis* 14:2543–2546.

Cotten M (1995): The entry mechanism of adenovirus and some solutions to the toxicity problems associated with adenovirus-augmented, receptor-mediated gene delivery. In The Molecular Repertoire of Adenoviruses III (eds., Doerfler W, Bšhm P). *Current Topics in Microbiology and Immunology* 199:283–295.

Cotten M, Wagner E, Birnstiel ML (1993a): Receptor-mediated transport of DNA into eukaryotic cells. *Methods Enzymol* 217:618–644.

Cotten M, Laengle-Rouault F, Kirlappos H, Wagner E, Mechtler K, Zenke M, Beug H, Birnstiel ML (1990): Transferrin-polycation mediated introduction of DNA into human leukemic cells: stimulation by agents that affect survival of transfected DNA or modulate transferrin receptor levels. *Proc Natl Acad Sci USA* 87:4033–4037.

Cotten M, Wagner E, Zatloukal K, Phillips S, Curiel DT, Birnstiel ML (1992): High-efficiency receptor-mediated delivery of small and large (48 kilobase) gene constructs using endosome-disruption activity of defective or chemically inactivated adenovirus particles. *Proc Natl Acad Sci USA* 89:6094–6098.

Cotten M, Wagner E, Zatloukal K, Birnstiel ML (1993b): Chicken adenovirus (CELO) particles augment receptor-mediated DNA delivery to mammalian cells and yield exceptional levels of stable transformants. *J Virol* 67:3777–3785.

Cotten M, Baker A, Saltik M, Wagner E, Buschle M (1994): Lipopolysaccharide is a frequent contaminant of plasmid DNA preparations and can be toxic to primary human cells in the presence of adenovirus. *Gene Ther* 1:239–246.

Curiel DT, Agarwal S, Wagner E, Cotten M (1991): Adenovirus enhancement of transferrin-polylysine-mediated gene delivery. *Proc Natl Acad Sci USA* 88:8850–8854.

Dash PR, Read ML, Barrett LB, Wolfert MA, Seymour LW (1999): Factors affecting blood clearance and in vivo distribution of polyelectrolyte complexes for gene delivery. *Gene Ther* 6(4):643–650.

Erbacher P, Bousser MT, Raimond J, Monsigny M, Midoux P, Roche AC (1996): Gene transfer by DNA/glycosylated polylysine complexes into human blood monocyte derived macrophages. *Hum Gene Ther* 7:721–729.

Ferkol T, Kaetzel CS, Davis PB (1993): Gene transfer into respiratory epithelial cells by targeting the polymeric immunoglobulin receptor. *J Clin Invest* 92:2394–2400.

Ferkol T, Lindberg GL, Perales JC, Chen J, Ratnoff OD, Hanson RW (1993): Regulation of the phosphoenolpyruvate carboxykinase/human factor IX gene introduced into the livers of adult rats by receptor-mediated gene transfer. *FASEB J* 7:1081–1091.

Ferkol T, Perales JC, Kaetzel CS, Eckman E, Hanson RW, Davis PB (1995): Gene transfer into airways in animals by targeting the polymeric immunoglobulin receptor. *J Clin Invest* 95:493–502.

Ferkol T, Pellicena-Palle A, Eckman E, Perales JC, Tosi M, Redline R, Hanson RW, Davis PB (1996): Immunologic responses of gene transfer via the polymeric immunoglobulin receptor in mice. *Gene Ther* 3(8):669–678.

Ferkol T, Perales JC, Mularo F, Hanson RW (1996a): Receptor-mediated gene transfer into macrophages. *Proc Natl Acad Sci USA* 93:101–105.

Ferkol T, Mularo F, Hilliard J, Lodish S, Perales JC, Ziady A, Konstan M (1998): Transfer of the gene encoding human $alpha_1$ antitrypsin into pulmonary macrophages in vivo. *Am J Respir Cell Mol Biol* 18:591–601.

Fisher KD, Ulbrich K, Subr V, Ward CM, Mautner V, Blakey D, Seymour LW (2000): A versatile system for receptor-mediated gene delivery permits increased entry of DNA into target cells, enhanced delivery to the nucleus and elevated rates of transgene expression. *Gene Ther* 7(15):1337–1343.

Fominaya J, Wels W (1996): Target cell-specific DNA transfer mediated by a chimeric multidomain protein: novel non-viral gene delivery system. *J Biol Chem* 271:10560–10568.

Funke M, Ziady A, Davis P (2000): Fluorescent analysis of endocytosis of sec-R receptor targeted DNA complexes. *Pediatr Pulmonol* 20(Suppl):A207.

Gao L, Wagner E, Cotten M, Agarwal S, Harris C, Romer M, Miller L, Hu PC, Curiel D (1993): Direct *in vivo* gene transfer to airway epithelium employing adenovirus-polylysine-DNA complexes. *Hum Gene Ther* 4:17–24.

Gedeon C, Ziady A, Miller TJ, Payne JM, Kowalczyk TH, Pasumarthy MK, Davis PB, Moen RC, Cooper MJ (2000): High-level expression of compacted DNA complexes following intratracheal administration. *Mol Ther* 1(Suppl):S78.

Godbey WT, Wu KK, Mikos AG (1999): Poly(ethylenimine) and its role in gene delivery. *J Control Release* 60(2-3):149–160.

Gottschalk S, Christiano RJ, Smith LC, Woo SL (1994): Folate receptor mediated DNA delivery into tumor cells: potosomal disruption results in enhanced gene expression. *Gene Ther* 1:185–191.

Gottschalk S, Sparrow JT, Hauer J, Mims MP, Leland FE, Woo SLC, Smith LC (1996): Synthetic vehicles for efficient gene transfer and expression in mammalian cells. *Gene Ther* 3:448–457.

Haensler J, Szoka FC (1993): Polyamidoamine cascade polymers mediate efficient transfection of cells in culture. *Bioconjug Chem* 4:372–379.

Haensler J, Szoka FC (1993): Synthesis and characterization of a trigalactosylated bisacridine compound to target DNA to hepatocytes. *Bioconjug Chem* 4:85–93.

Hart SL, Arancibia-Carcamo CV, Wolfert MA, Mailhos C, O'Reilly NJ, Ali RR, Coutelle C, George AJ, Harbottle RP, Knight AM, Larkin DF, Levinsky RJ, Seymour LW, Thrasher AJ, Kinnon C (1998): Lipid-mediated enhancement of transfection by a nonviral integrin-targeting vector. *Hum Gene Ther* 9:575–585.

Hart SL, Harbottle RP, Cooper R, Miller A, Williamson R, Coutelle C (1995): Gene delivery and expression mediated by an integrin-binding peptide. *Gene Ther* 2:552–554.

Haselden BM, Kay AB, Larche M (2000): Peptide-mediated immune responses in specific immunotherapy. *Int Arch Allergy Immunol* 122:229–237.

Hoganson DK, Chandler LA, Fleurbaaij GA, Ying W, Black ME, Doukas J, Pierce GF, Baird A, Sosnowski BA (1998): Targeted delivery of DNA encoding cytotoxic proteins through high-affinity fibroblast growth factor receptors. *Hum Gene Ther* 9(17):2565–2575.

Hoganson DK, Sosnowski BA, Pierce GF, Doukas J (2001): Uptake of adenoviral vectors via fibroblast growth factor receptors involves intracellular pathways that differ from the targeting ligand. *Mol Ther* 3:105–112.

Huxley C (1994): Mammalian artificial chromosomes: a new tool for gene therapy. *Gene Ther* 1:7–12.

Jenkins RG, Herrick SE, Meng QH, Kinnon C, Laurent GJ, McAnulty RJ, Hart SL (2001): An integrin-targeted non-viral vector for pulmonary gene therapy. *Gene Ther* 7:393–400.

Kaetzel CS, Blanch VJ, Hempen PM, Phillips KM, Piskurich JF, Youngman KR (1997): The polymeric immunoglobulin receptor: structure and synthesis. *Biochem Soc Trans* 25:475–480.

Kalderon D, Richardson WD, Markham AF, Smith AE (1984): Sequence requirements for nuclear location of simian virus 40 large-T antigen. *Nature* 311:33–38.

Kantor FS, Ojeda A, Benacerraf B (1962): Studies on artificial antigens: antigenicity of DNP-polylysine and DNP copolymer of lysine and glutamic acid in Guinea pigs. *Hum Exp Med* 117:55–69.

Kasahara N, Dozy AM, Kan YW (1994): Tissue specific targeting of retroviral vectors through ligand-receptor interactions. *Science* 266:1373–1376.

Katayose S, Kataoka K (1997): Water-soluble polyion complex associates of DNA and poly(ethylene glycol)-poly(L-lysine) block copolymer. *Bioconjug Chem* 8:702–707.

Klink DT, Chao S, Glick MC, Scanlin TF (2001): Nuclear translocation of lactosylated poly-l-lysine/cdna complex in cystic fibrosis airway epithelial cells. *Mol Ther* 3(6):831–841.

Kwok KY, McKenzie DL, Evers DL, Rice KG (1999): Formulation of highly soluble poly(ethylene glycol)-peptide DNA condensates. *J Pharm Sci* 88:996–1003.

Lukacs GL, Haggie P, Seksek O, Lechardeur D, Freedman N, Verkman AS (2000): Size-dependent DNA mobility in cytoplasm and nucleus. *J Biol Chem* 275:1625–1629.

Maurer PH (1962): Antigenicity of polypeptides (poly-alpha amino acids). *J Immunol* 88:330–345.

Midoux P, Mendes C, Legrand A, Raimond J, Mayer R, Monsigny M, Roche AC (1993): Specific gene transfer mediated by lactosylated poly-L-lysine into hepatoma cells. *Nucleic Acids Res* 21:871–878.

Miller N and Vile R (1995): Targeted vectors for gene therapy. *FASEB J* 9:190–199.

Monsigny M, Roche AC, Midoux P, Mayer R (1994): Glycoconjugates as carriers for specific delivery of therapeutic drugs and genes. *Adv Drug Delivery Rev* 14:1–24.

Neutra MR, Ciechanover A, Owen LS, Lodish HF (1985): Intracellular transport of transferrin- and asialoorosomucoid-colloidal gold conjugates to lysosomes after receptor-mediated endocytosis. *J Histochem Cytochem* 33:1134–1144.

Ohtake Y, Chen J, Gamou S, Takayanagi A, Mashima Y, Oguchi Y, Shimizu N (1999): Ex vivo delivery of suicide genes into melanoma cells using epidermal growth factor receptor-specific Fab immunogene. *Jpn J Cancer Res* 90:460–468.

Oupicky D, Howard KA, Konak C, Dash PR, Ulbrich K, Seymour LW (2000): Steric stabilization of poly-L-Lysine/DNA complexes by the covalent attachment of semitelechelic poly[N-(2-hydroxypropyl)methacrylamide]. *Bioconjug Chem* 11:492–501.

Perales JC, Ferkol T, Beegen, H, Ratnoff OD, Hanson RW (1994a): Gene transfer *in vivo:* Sustained expression and regulation of genes introduced into the livers by receptor targeted uptake. *Proc Natl Acad Sci USA* 91:4086–4090.

Perales JC, Ferkol T, Molas M, Hanson RW (1994b): Receptor-mediated gene transfer. *Eur J Biochem* 226:255–266.

Perales JC, Grossman GA, Molas M, Liu G, Ferkol T, Harpst J, Oda H, Hanson RW (1997): Biochemical and functional characterization of DNA complexes capable of targeting genes to hepatocytes via the asialoglycoprotein receptor. *J Biol Chem* 272:11, 7398–7407.

Perlmutter DH (1994): The SEC receptor: A possible link between neonatal hepatitis in α_1-antitrypsin deficiency and Alzheimer's disease. *Pediatric Res* 36:271–277.

Plank C, Oberhauser B, Mechtler K, Koch C, Wagner E (1994): The influence of endosome-disruptive peptides on gene transfer using synthetic virus-like gene transfer systems *J Biol Chem* 269:12918–12924.

Plank C, Mechtler K, Szoka F, Wagner E (1996): Activation of the complement system by synthetic DNA complexes: a potential barrier for intravenous gene delivery. *Hum Gene Ther* 7:1437–1446.

Ponka P, Lok CN (1999): The transferrin receptor: role in health and disease. *Int J Biochem Cell Biol* 31:1111–1137.

Powers CJ, McLeskey SW, Wellstein A (2000): Fibroblast growth factors, their receptors and signaling. *Endocr Relat Cancer* 7:165–197.

Rojanasakul Y, Wang LJ, Malanga CJ, Ma JKH, Liaw JH (1994): Targeted gene delivery to alveolar macrophages via Fc receptor-mediated endocytosis. *Pharm Res* 11:1731–1736.

Ross GF, Morris RE, Ciraolo G, Huelsman K, Bruno MD, Whitsett JA, Baatz JE, Korfhagen TR (1995): Surfactant proteinA-polylysine conjugates for delivery of DNA to airway cells in culture. *Hum Gene Ther* 6:31–40.

Schwarzenberger P, Spence SE, Gooya JM, Michiel D, Curiel DT, Ruscetti FW, Keller JR (1996): Targeted gene transfer to human hematopoietic progenitor cell lines through the c-kit receptor. *Blood* 87:472–478.

Stahl PD, Ezekowitz RA (1998): The mannose receptor is a pattern recognition receptor involved in host defense. *Curr Opin Immunol* 10:50–55.

Stankovics J, Crane AM, Andrews E, Wu CH, Wu GY, Ledley FD (1994): Overexpression of human methylmalonyl CoA mutase in mice after *in vivo* gene transfer with asialoglycoprotein/polylysine/DNA complexes. *Hum Gene Ther* 5:1095–1104.

Stockert RJ (1995): The asialoglycoprotein receptor: relationships between structure, function, and expression. *Physiol Rev* 75:591–609.

Thurher M, Wagner E, Clausen H, Mechtler K, Rusconi S, Dinter A, Birnstiel ML, Berger EG, Cotten M (1990): Carbohydrate receptor mediated gene transfer to human T leukemic cells. *Glycobiology* 4:429–435.

Valsesia-Whitmann S, Drynda A, Deleage G, Aumailley M, Heard JM, Danos O, Verdier G, Cosset FL (1994): Modifications in the binding domain of avian retrovirus envelope protein to redirect the host range of retroviral vectors. *J Virol* 68:4609–4619.

Wagner E, Curiel DT, Cotten M (1994): Delivery of drugs, proteins, and genes into cells using transferrin as a ligand for receptor-mediated endocytosis. *Adv Drug Del Rev* 14:113–135.

Wagner E, Zenke M, Cotten M, Beug H, Birnstiel ML (1990): Transferrin-polycation conjugates as carriers for DNA uptake into cells. *Proc Natl Acad Sci USA* 87:3410–3414.

Wagner E, Cotten M, Foisner R, and Birnstiel MT (1991): Transferrin-polycation-DNA complexes: the effect of polycations on the structure of the complex and DNA delivery to cells. *Proc Natl Acad Sci USA* 88:4255–4259.

Wagner E, Plank C, Zatloukal K, Cotten M, Birnstiel ML (1992): Influenza virus hemaglutinin HA-2 N-terminal fusogenic peptides augment gene transfer by transferrin-polylysine-DNA complexes: toward a synthetic virus-like gene-transfer vehicle. *Proc Natl Acad Sci USA* 89:7934–7938.

Wells A (1999): EGF receptor. *Int J Biochem Cell Biol* 31:637–643.

Wilson JM, Grossman M, Wu CH, Chowdhury NR, Wu GY, Chowdhury JR (1992): Hepatocyte-directed gene transfer in vivo leads to transient improvement of hypercholesterolemia in low-density lipoprotein receptor-deficient rabbits. *J Biol Chem* 267:963–967.

Wu GY, Wu CH (1987): Receptor-mediated *in vitro* gene transformation by a soluble DNA carrier system. *J Biol Chem* 262:4429–4432.

Wu CH, Wilson JM, Wu GY (1989): Targeting genes: delivery and persistent expression of a foreign gene driven by mammalian regulator elements *in vivo. J Biol Chem* 264:16985–16987.

Wu GY, Wilson JM, Shalaby F, Grossman M, Shafritz DA, Wu CH (1991): Receptor-mediated gene delivery *in vivo:* Partial correction of genetic analbuminemia in Nagase rats. *J Biol Chem* 266:14338–14342.

Yew NS, Wysokenski DM, Wang KX, Ziegler RJ, Marshall J, McNeilly D, Cherry M, Osburn W, Cheng SH (1997): Optimization of plasmid vectors for high-level expression in lung epithelial cells. *Hum Gene Ther* 8:575–584.

Yew NS, Wang KX, Przybylska M, Bagley RG, Stedman M, Marshall J, Scheule RK, Cheng SH (1999): Contribution of plasmid DNA to inflammation in the lung after administration of cationic lipid : pDNA complexes. *Hum Gene Ther* 10:223–234.

Yew NS, Zhao H, Wu IH, Song A, Tousignant JD, Przybylska M, Cheng SH (2000): Reduced inflammatory response to plasmid DNA vectors by elimination and inhibition of immunostimulatory CpG motifs. *Mol Ther* 1:255–262.

Zatoukal K, Wagner E, Cotten M, Phillips S, Plank C, Steinheim P, Curiel D, Birnstiel ML (1992): Transferrinfection: a highly efficient way to express gene constructs in eukaryotic cells. *Ann NY Acad Sci* 660:136–153.

Zatoukal K, Schneeberger A, Berger M, Schmidt W, Koszik F, Kutil R, Cotten M, Wagner E, Buschle M, Maass G, Payer E, Stingl G, Birnstiel ML (1995): Elicitation of a systemic and protective anti-melanoma immune response by an IL-2-based vaccine: assessment of critical cellular and molecular parameters. *J Immunol* 154:3406–3419.

Zenke M, Steinlein P, Wagner E, Cotten M, Beug H, Birnstiel ML (1990): Receptor-mediated endocytosis of transferrin-polycation conjugates: an efficient way to introduce DNA into hematopoietic cells. *Proc Natl Acad Sci USA* 87:3655–3659.

Ziady A, Perales JC, Ferkol T, Gerken T, Beegen H, Perlmutter DH, Davis PB (1997): Gene transfer into hepatocyte cell lines via the serpin enzyme complex (SEC) receptor. *Am J Physiol* 273:G545–G552.

Ziady A, Ferkol T, Gerken T, Dawson DV, Perlmutter DH, Davis PB (1998): Ligand substitution of receptor targeted DNA complexes affects gene transfer into hepatoma cells. *Gene Ther* 5:1685–1697.

Ziady A, Ferkol T, Dawson DV, Perlmutter DH, Davis PB (1999): Chain length of the polymer portion of receptor-targeted DNA complexes modulates gene transfer both in vitro and in vivo. *J Biol Chem* 274:4908–4916.

PART II

TRANSDUCTIONALLY TARGETED VECTORS—VIRAL

5

PSEUDOTYPING OF ADENOVIRAL VECTORS

MENZO J. E. HAVENGA, PH.D. RONALD VOGELS, PH.D. ABRAHAM BOUT, PH.D. AND MAJID MEHTALI, PH.D.

INTRODUCTION

The development of safer and more efficient gene delivery systems offers potential preventive or therapeutic solutions to a wide variety of medical problems. Several promising gene transfer technologies, viral or nonviral, are currently under investigation. Nonviral gene delivery is being pursued using naked DNA or cationic lipids and cationic polymers complexed with DNA (reviewed by Nishikawa and Huang, 2001). Although relatively safe, nonviral vectors have been usually characterized by poor in vivo gene transfer efficiency and transient gene expression. In general, vectors derived from viruses constitute more efficient in vivo gene delivery systems but are associated with specific limitations such as toxicity and strong antiviral host immune responses (reviewed by Jolly, 1994). Recombinant vectors have been derived from a variety of viruses including retroviruses, adenoviruses, adeno-associated viruses, herpes simplex viruses, poxviruses, polioviruses, baculoviruses, and Sindbis viruses. There are several considerations for the selection of an optimal gene delivery vehicle: (1) the accessibility of the target tissue, which dictates whether host cell transduction should be performed ex vivo or in vivo; (2) the desired persistence of expression of the therapeutic gene, which can be

Vector Targeting for Therapeutic Gene Delivery, Edited by David T. Curiel and Joanne T. Douglas
ISBN 0-471-43479-5

sustained or transient depending on the selected vector and on the disease indication; (3) the susceptibility of the target tissue to viral transduction.

Adenovirus-based vectors have been widely studied as one of the most promising gene transfer systems. Adenoviruses are nonenveloped double-strand DNA viruses with spiked icosahedral morphology. The virion particle consists of 252 capsomers per nucleocapsid (240 hexons and 12 pentons) with the fiber protein projecting from each of the twelve vertices. The fiber, the predominant determinant of the viral tissue tropism, is linked to the viral capsid via the penton base protein (Fig. 5.1). Adenovirus serotypes 2 (Ad2) and 5 (Ad5) are commonly used as gene delivery vehicles for a number of reasons:

- Their genome is fully sequenced and their genomic organization well characterized (Chroboczek, Bieber, and Jacrot, 1992).
- Nonreplicative E1-deleted viruses can be produced at high yields on E1-expressing complementation cells such as the 293 (Graham et al., 1977) or the PER.C6 (Fallaux et al., 1998) cell lines. The latter cell system also prevents the generation of replication-competent adenoviruses (RCA) during manufacturing, which is a major safety issue (Zhu et al., 1999; Hehir et al., 1996). In addition, PER.C6 cells have been adapted to grow in suspension in the absence of animal sera thus paving the road for large-scale productions, a prerequisite for the development of recombinant adenoviruses as pharmaceutical drugs.
- In vivo gene delivery is relatively efficient in a wide variety of tissues.
- In contrast to retroviral vectors, adenoviruses are not susceptible to inactivation by human complement.

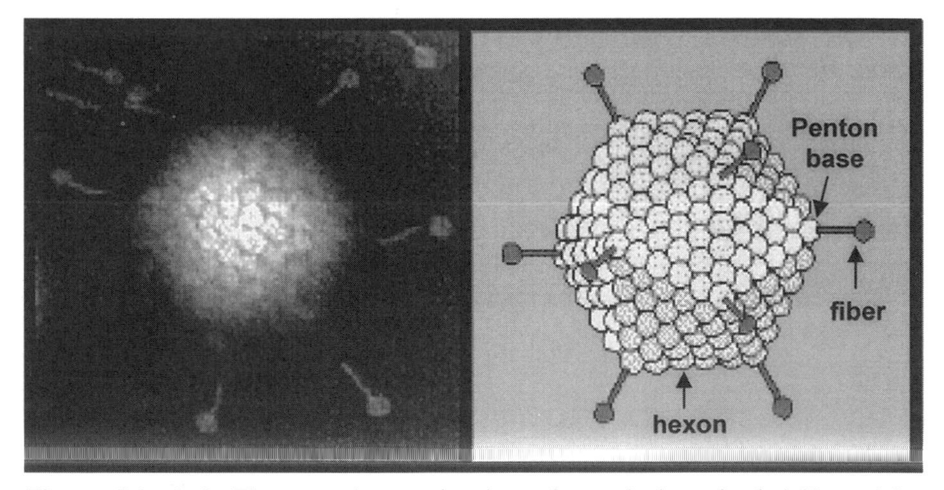

Figure 5.1 *Left:* Electron microscopic view of negatively stained Ad5 particles (133000x). *Right:* Cartoon showing the adenovirus icosahedral capsid built up from the fiber, penton base, and hexon structural proteins.

- Expression of therapeutic genes is strong and transient, which is an advantage for the treatment of diseases such as cancers or cardiac ischemia, and for applications in vaccination.
- The adenoviral particle is relatively stable and can be lyophilized and stored for prolonged periods (Bout, 1996; Croyle et al., 1998).

These attractive features were strong incentives for academic and industrial investigators to use recombinant Ad5- or Ad2-derived vectors for the tentative treatment of various inherited or acquired human diseases. Although numerous preclinical studies in nonhuman primates as well as clinical trials in humans have shown safety at moderate viral doses, therapeutic benefit from an adenovirus-mediated gene therapy approach has yet to be demonstrated. Current limitations of the adenovirus vectors and strategies to improve them are discussed in detail in this chapter.

LIMITATIONS OF CURRENT RECOMBINANT ADENOVIRUS VECTORS

Several hurdles are associated with the use of Ad2 and Ad5 viruses as gene delivery vehicles. In particular, several tissues of major therapeutic interest have been found to be refractory to transduction by Ad2- and Ad5-derived vectors. High dosages of recombinant vectors are therefore necessary to genetically engineer the target cells in vivo, which may in turn compromise the safety of the treated patients. Overcoming these hurdles requires the development of vectors that are (1) able to transduce specifically and efficiently the desired human cells while bypassing nontarget tissues; (2) not prone to neutralization by human serum; and (3) characterized by an acceptable toxicity profile.

Vector Binding and Entry

Both Ad5 and Ad2 viruses require the presence of the coxsackies virus and adenovirus receptor (CAR) for efficient binding to the cell surface (Bergelson et al., 1997). However, CAR was recently shown to be absent on several primary human cell types such as tumor cells (Cripe et al., 2001), smooth muscle cells (Wickham et al., 1996), liver sinusoidal endothelial cells (Hegenbarth et al., 2000), synoviocytes (Goossens et al., 1999), airway epithelial cells (Zabner et al., 1997), and dendritic cells (Rea et al., 1999). These cell types constitute important targets for the treatment of cancers, cardiovascular disease, rheumatoid arthritis, cystic fibrosis, and for the development of adenovirus-based vaccines, respectively. Although lacking CAR, such cells can nonetheless be genetically modified in vitro and in vivo by using high multiplicities of infection and/or prolonged virus exposure times (Zabner et al., 1997). However, administration of high vector dosages hampers the therapeutic use of recombinant adenoviruses given their strong tropism to the liver, where they may induce undesired toxicity as witnessed by elevations of liver enzymes

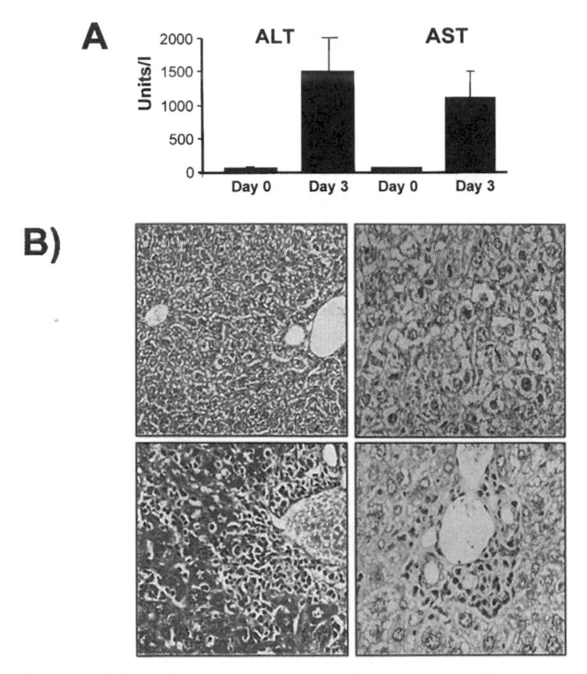

Figure 5.2 Toxicity of E1-deleted adenovirus in mice. *A*. Three days after systemic administration of 5×10^{11} virus particles of Ad5 carrying no transgene, detection of elevated liver enzymes (ALT, AST) in serum (day 0 as control) indicates severe liver damage. *B*. Microscopic section of the liver taken 3 days after administration of phosphate-buffered saline PBS (upper left panel) or after administration of 5×10^{11} virus particles (upper right panel). Clear cytoplasmic vacuolar changes, dense chromatin, and mitosis can be observed in the vector-treated animals, indicative of tissue destruction and regeneration. Anisokaryosis, cellular hypertrophy, periportal inflammation, and cell death is observed in the liver at 14 days (lower left panel) and 28 days (lower right panel) after vector administration.

(alanine aminotranserase, ALT and aspartate aminotransferase, AST), necrosis, and increased mitosis (Fig. 5.2). Therefore, the development of a vector with a more restricted tropism that selectively and efficiently transduces the target tissue would allow the use of lower therapeutic doses and hence reduce the vector toxicity.

Vector Neutralization by Preexisting Antibodies

Both Ad5 and Ad2 wild-type viruses are associated with respiratory illness and are the cause of the common cold in humans. Therefore, most individuals have developed neutralizing antibodies directed against these viral serotypes (D'Ambrosio et al., 1982; Flomenberg et al., 1995; Schulick et al., 1997). We and others have shown in animal models that the presence of neutralizing anti-

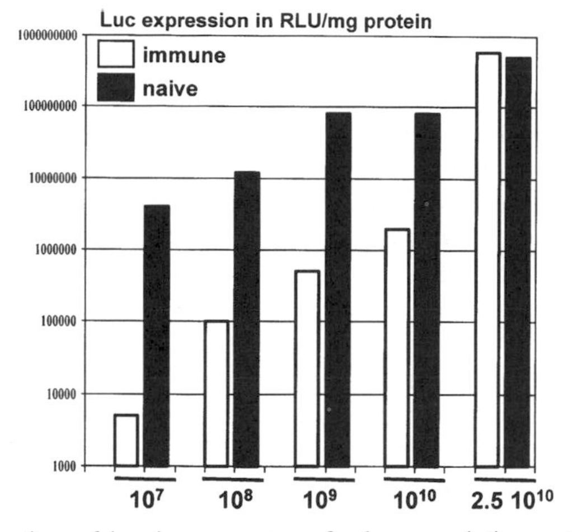

Figure 5.3 Prevention of in vivo gene transfer by preexisting anti-Ad5 neutralizing antibodies. Mice bearing syngeneic tumors were either nontreated (naíve) or immunized (immune) at day 0 by intravenous injection of 5×10^{10} virus particles of an Ad5 vector expressing no transgene. Two weeks later, increasing dosages of a vector expressing the luciferase transgene were administered to the mice by intra-tumoral injection. Transgene expression was in the tumor monitored 48 hours later.

adenoviral antibodies in the serum prevents local (Fig. 5.3) and systemic gene transfer (Schulick et al., 1997; Kuriyama et al., 1998; Dai et al., 1995; Kass-Eisler et al., 1994). These observations were confirmed in phase I clinical trials where adenovirus-mediated gene transfer was prevented in patients with high preexisting titers of neutralizing antibodies (Gahery-Segard et al., 1997). Inhibition of transduction can be overcome by the administration of higher vector doses (Fig. 5.3). However, it is also known that neutralizing antibody titers can vary significantly among individuals (D'Ambrosio et al., 1982). As a consequence, a standard vector dose administered to a group of patients can lead to huge differences in clinical outcome, ranging from no evident pathology and gene transfer to severe toxicity and efficient gene transfer (Chen et al., 2000). Such a variation in clinical outcome is not acceptable and adenoviral vectors which are less prone to neutralization are therefore required for a safe and uniform dosing in all patients.

Vector Toxicity

The severity of the virus-mediated toxicity critically depends on the viral dose. Approximately 90% of the virus particles are cleared from the organs during the first 3 days after vector injection, irrespective of the route of delivery (Lusky et al., 1998; Worgall et al., 1997; Brough et al., 1997). The finding that this process is identical for psoralen- and ultraviolet (UV)-inactivated viruses indicates that the acute toxicity is independent of viral replication or transgene expression (Cotten et al., 1994; McCoy et al., 1995; Schnell et al., 2001). This rapid clearance of recombinant adenovirus is caused by nonspecific defense mechanisms involving the uptake of the virions by dendritic cells and macrophages and the subsequent release of cytokines such as tumor necrosis factors-α (TNF-α), interleukin-6 (IL-6), and IL-12 (Benihoud et al., 1998; Elkon et al., 1997; Worgall et al., 1997; Zhang et al., 2001; Sung, Qin, and Bromberg, 2001; Verma and Somia, 1997; McCoy et al., 1995; Muruve et al., 1999; Schnell et al., 2001). These findings were further confirmed by the demonstration of a significant reduction in viral clearance in animals experimentally depleted for macrophages (Stein et al., 1998). From these studies, it can be deduced that the rapid clearance during the first few days after vector administration is caused by the adenovirus particles which, upon binding to cells, cause an acute inflammatory reaction. To circumvent the massive elimination of the virions and the toxicity observed in the first few days after vector administration, transient suppression of the immune system has been investigated in animal models using either immunosuppressive agents such as corticosteroids (Lawrence et al., 1999; Elshami et al., 1995), cyclosporin (Russi, Hirschowitz, and Crystal, 1997), FK506 (Lochmuller et al., 1995) and cyclophosphamide (Jooss, Yang, and Wilson, 1996) or chemotherapeutic and biological agents such as etoposide (Yang, Greenough, and Wilson, 1996; Smith et al., 1996), anti-CD40 antibodies (Kay et al., 1997) or CTLA4-Ig (Jooss, Turka, and Wilson, 1998; Ideguchi et al., 2000). In most cases, an increase in the level and duration of transgene expression and a reduction of the local inflammation in the treated organ was observed in the immunosuppressed animals. However, the effect of such drugs on the acute toxicity was not carefully investigated. In one of the very few clinical studies addressing the same question, no significant differences in initial gene transfer were observed in patients with malignant pleural mesothelioma treated by systemic corticosteroid administration immediately before and after vector injection (Sterman et al., 2000). However, the acute clinical toxicity was significantly decreased in the patients treated with steroids, with reduced serum levels of IL-6 and TNF. All these studies suggest that the acute vector toxicity is predominantly triggered by the nonspecific uptake of the viral particles by antigen-presenting cells (e.g., Kupffer cells, macrophages, and dendritic cells), leading to the production of high systemic levels of inflammatory cytokines. Reducing the therapeutic viral dosages and/or developing vectors less susceptible to nonspecific uptake by antigen-presenting cells could therefore minimize this major drawback.

Antiadenoviral Cellular Immunity

Ad2- and Ad5-derived vectors yield short-term (2–3 weeks) transgene expression in immune competent animals, whereas expression of the transgene can be monitored for more than a year in immunodeficient mice (Dai et al., 1995, Barr et al., 1995; Kass-Eisler et al., 1994; Smith et al., 1993; Li et al., 1993; Michou et al., 1997; Christ et al., 1997). Also, agammaglobulinemic patients infected with adenoviruses were shown to be able to successfully battle the viral infection, confirming the importance of the antiadenoviral cellular immune response in controlling the spread of the virus, although exceptions were noted (Siegal et al., 1981). These preclinical and clinical observations support the notion that the host antiviral cellular immune response can induce the progressive elimination of the transduced cells. Short-term duration of transgene expression is probably sufficient to achieve therapeutic benefit for many indications such as cancer or cardiovascular diseases. However, other applications require longer-term expression of the transgene. Therefore, emphasis has been given in recent years to research studying the interaction between the host immune system and the recombinant vectors. Adenovirus clearance is considered to be a biphasic process with, as indicated previously, a rapid nonspecific elimination of the adenoviral capsids by the antigen-presenting cells. Then, de novo synthesis of adenoviral proteins in transduced cells might either directly interfere with cell viability (Bal et al., 2000) or trigger the cytotoxic T-cell (CTL)-mediated destruction of the transduced cells (Tripathy et al., 1996; Zsengeller et al., 1995; Yei et al., 1994; Yang, Ertl, and Wilson, 1994; Yang et al., 1995; Yang et al., 1996). These hypotheses led to the development of a new generation of vectors with multiple regulatory genes deleted in an attempt to fully suppress expression of the viral proteins and hence reduce the host antiviral cellular immunity (Gao, Yang, and Wilson, 1996; Engelhardt, Litzky, and Wilson, 1994; Brough et al., 1996; Dedieu et al., 1997; Gorziglia et al., 1996; Lusky et al., 1998; Raper et al., 1998). Studies performed by us and others with vectors deleted or attenuated in E1, E2A, and/or E4 regulatory regions have clearly shown a dramatic reduction in expression of the viral proteins in transduced cells (Fig. 5.4), associated with reduced hepatotoxicity in mice (unpublished data, Christ et al., 2000).

A further vector development was the recent generation of so-called high-capacity or gutless viruses (Parks et al., 1996; Sandig et al., 2000; Morsy et al., 1998; Fisher et al., 1996; Kochanek et al., 1996). These vectors are deleted of all coding sequences and are characterized in vivo by a reduced toxicity profile. The immunogenicity of such high-capacity vectors and vectors carrying multiple deletions has been investigated in numerous animal models, revealing serious discrepancies among the studies. Whereas several studies have reported a similar host antiviral immune response and persistence of transduced cells for E1-deleted and multideleted vectors (Morral et al., 1997; Lieber et al., 1996; Lusky et al., 1998), other studies have shown prolonged transgene expression and persistence of transduced cells when using multiple-deleted or gutless vectors (Engelhardt, Litzky, and Wilson, 1994; Dedieu et al., 1997; Brough et al., 1996, 1997; Gorziglia et al., 1996). The reasons for these discrepancies remain unclear but might be due to the use of different mouse strains and vectors. In addition, it has also been reported that, in the absence of any de novo synthesis of viral proteins, antiviral CTLs can be induced (Kafri et al., 1998) and viral epitopes can be efficiently presented by major histocompatability complex

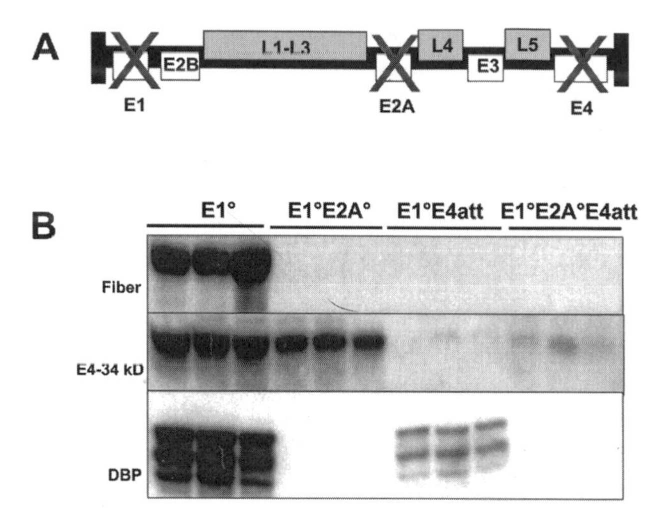

Figure 5.4 Prevention of viral gene expression by multiple deletions of the Ad5 regulatory genes. *A*. Schematic representation of the Ad5 genome structure showing the proximate locations of the viral genes encoding the early (E1, E2A, E2B, E3, and E4) and late proteins (L1-3, L4, L5). Crosses indicate the deleted (E1 or E2A) or attenuated (E4) regulatory regions. *B*. Expression of the viral early (E4-34kD, DBP) and late (Fiber) proteins in human A549 cells infected with Ad5 vectors deleted in E1 (E1°), in both E1 and E2A (E1°E2A°), in both E1 and E4 (E1°E4att) or in E1, E2A, and E4 (E1°E2A°E4att).

class I (MHC-I) antigens (Molinier-Frenkel et al., 2000). These data suggest that both newly synthesized viral antigens and the injected capsids are able to efficiently elicit an antiviral CTL response. However, the impact of this cellular immune response on the persistence of the transduced cells is still controversial.

GENERATION OF IMPROVED ADENOVIRUS VECTORS

From the previously listed limitations, it is clear that the use of Ad2- or Ad5-derived vectors for human gene transfer poses a dilemma. On one hand, the therapeutic viral dose must be elevated to overcome the presence of neutralizing antibodies and to compensate for the low efficiency of transduction of the target cells, in particular for CAR-deficient cells. On the other hand, a high viral dose causes undesired local and systemic toxicity and possibly results in the eradication of the transduced cells by the host immune response. To overcome this dilemma, adenoviral vectors must be generated that are not neutralized by the antiviral antibodies preexisting in the serum and that are able to specifically and efficiently transduce the tissue of interest and not undesired organs.

Several strategies have been explored to develop vectors that specifically target the cell types of interest, among which two have been more intensively investigated. In the first approach, bispecific antibodies have been developed that can simultaneously bind to the adenovirus fiber, preventing the binding to CAR, and to a molecule present specifically on the targeted cells (Wickham et al., 1996; Douglas et al., 1996; Haisma et al., 2000). Such adenovirus-antibody complexes were shown in vitro to specifically bind and infect the targeted cells, but a detailed in vivo analysis of the biological and pharmacological properties of the complexes is still lacking. The second strategy encompasses specific genetic modifications of the viral capsid proteins to ablate the CAR and integrin binding motifs located in the fiber and penton base, respectively, and to insert alternative binding ligands in the penton base, hexon, or fiber (Wickham, Carrion, and Kovesdi, 1995; Vigne et al., 1999; Krasnykh et al., 1998; Kasono et al., 1999; Dmitriev et al., 1998; Leissner et al., 2001). While preliminary results indicate that such engineered viruses can indeed transduce more efficienctly and specifically the targeted cells, they remain fully sensitive to neutralization by the anti-Ad5 antibodies present in most human sera.

Attempts to generate adenoviral vectors that are less prone to neutralization have been explored by several groups, mainly by investigating the possibility of shielding the viral neutralization epitopes with inert chemical components such as polyethylene glycol (PEG) (Fisher et al., 2001; O'Riordan et al., 1999; Croyle, Yu, and Wilson, 2000). These PEGylated vectors were shown in vitro to be less sensitive to the presence of neutralizing antibodies. However, they were also partially impaired in their ability to enter cells, probably as a consequence of the steric hindrance of the fiber knob domain by the PEG moieties. In addition, in vivo entrapment of the PEGylated vector was observed in certain organs (Croyle, Yu, and Wilson, 2000). Recently, Blackwell et al. (2000) investigated an alternative strategy to overcome the anti-Ad5 neutralizing activity found in cancer patients' ascites fluids. They generated an Ad5 vector carrying an insertion of an Arg-Gly-Asp (RGD) peptide sequence in the fiber knob and demonstrated that this virus can efficiently enter cell lines and primary ovarian cancer cells in the presence of ascites fluid containing high titers of neutralizing anti-Ad5 antibodies. These promising results indicate that precise genetic modifications of the virus particle may allow enhanced gene transfer in the presence of neutralizing antibodies. However, the anti-Ad5 antibodies present in the ascites fluid were predominantly directed against the fiber. It is therefore unclear whether knob modifications may also reduce neutralization by antibodies directed against other structural proteins such as the hexon and penton base.

While the above strategies may succeed in generating vectors with restricted tropism or reduced sensitivity to anti-Ad5 antibodies, they result in most cases in the development of complex multicomponent systems (i.e., virus plus bispecific antibody or virus plus PEG) that do not combine the desired biological features of an optimal vector (i.e., specific tropism without neutralization by human sera). In order to generate such a single optimal vector that naturally combines the desired properties, an alternative strategy exploits the wide diversity of the human adenovirus family. Given the large number of reported human adenoviruses that differ in symptoms, tissue tropism, and immunogenicity, theoretically it should be possible to identify a virus that is not neutralized by human sera but can efficiently infect tissues of therapeutic importance.

The Natural Diversity of the Adenoviridae

To date, 51 different human adenovirus serotypes have been identified (De Jong et al., 1999) and grouped into six subgroups (A, B, C, D, E, and F) on the basis of (1) the lack of serological cross-neutralization; (2) unrelated hemagglutination patterns; and (3) substantial differences in DNA (Francki et al., 1991). Besides human adenoviruses, approximately 50 nonhuman adenoviruses have also been isolated from different species, including mice, frogs, cats, deer, macaques, sheep, cows, and chimpanzees (Wigand, Mauss, and Adrian, 1989; Clark et al., 1973; Sorden, Woods, and Lehmkuhl, 2000; Lapointe et al., 2000; Lakatos et al., 2000). The disease association of most of the human and some of the animal adenoviruses has been documented and is summarized in Table 5.1. Altogether, more than one hundred different adenoviruses are at the disposal of scientists to derive an adenoviral vector less prone to neutralization and more infectious toward Ad5-refractory cells. In addition, this diversity also allows individual capsid subunits (i.e., fiber, hexon, penton base) to be selected from particular serotypes and used as building blocks to generate chimeric or pseudotyped viruses that combine the specific properties conferred by the selected proteins.

Molecular Determinants of the Viral Tropism

Mechanisms for cellular entry have been described best for subgroup C adenoviruses and consist of two distinct processes. The virus first binds to CAR via the knob domain of the fiber protein and then is internalized through the interaction of the penton base protein with $\alpha v\beta 3$ and $\alpha v\beta 5$ cellular integrins (Greber et al., 1993; Wickham et al., 1993; Goldman and Wilson, 1995). While at least some members of subgroup A, C, D, E, and F adenoviruses are able to bind to CAR (Roelvink et al., 1998), interaction with an alternative receptor has also been reported. For instance, productive infection with Ad8, Ad19a, and Ad37 (subgroup D) is dependent on sialic acids present on cellular membrane proteins, whereas infection with Ad9 and Ad19, also subgroup D members, is not (Arnberg et al., 1999, 2000). In addition, infection with subgroup B viruses has been shown to be independent of the presence of CAR (Stevenson et al., 1995). The receptor for subgroup B viruses is still not characterized but studies have shown that a 130 kDa protein is involved in the binding of Ad3, a subgroup B member, to HeLa cells (Di Guilmi et al., 1995). Moreover, individual members of subgroup B are most likely also able to enter cells using different attachment molecules, as suggested by results from competition experiments between Ad3 and Ad35 viruses (Shayakhmetov et al., 2000). Viruses that use the same

TABLE 5.1 Mastadenoviridae and Possible Disease Association

Serotypes	Native Host	Disease Association
HAV (1–7, 11a, 14, 21, 31, 34, 35, 39, 42–48, 51)	Human	Respiratory illness
HAV (8, 19, 37, 50)	Human	Keratoconjunctivitis
HAV (5, 7, 11, 14, 16, 21, 34, 35, 39, 42–47)	Human	Hemorrhagic cystitis and urogenital tract infections
HAV (1–3, 5, 28, 31, 40, 41)	Human	Gastroenteritis
HAV (2, 3, 5–7, 12, 31, 32, 49)	Human	Central nervous system disease
HAV (1, 2, 5, 30, 31)	Human	Hepatitis
HAV (9, 10, 13, 15, 17, 18, 20, 22–29, 33, 36, 38)	Human	None
FAV-1	Frog	None
BAV1/2/3/6/9	Cow	None
BAV4/5	Cow	Pneumoenteritis
BAV7/8	Cow	Respiratory illness
BAV-10	Cow	Enteric disease
OAV1/2/3/4	Sheep	None
OAV5	Sheep	Respiratory illness
OAV6	Sheep	Respiratory illness
PAV1/2/3/4/5	Pig	None
GAV1	Goat	Encephalitis
DAV1	Mule deer	Systemic vasculitis, pulmonary edema, and hemorrhagic enteropathy
MAV1/2	Mouse	Unknown
CAV1	Dog	Ocular disease
CAV2	Dog	Respiratory illness
ChAV (pan 11)	Chimpanzee	Chronic interstitial nephritis
ChAV (pan 5–7/9)	Chimpanzee	None
ChAV (C1)	Chimpanzee	Respiratory illness
SAV 16–18	Vervet monkey	None
SAV (V340)	African green monkey	Pneumoenteritis
SAV (1)	Cynomolgus monkey	None
SAV (11, 15, 17, 20, 23, 25, 27, 30–34, 36–39)	Rhesus monkey	Unknown
SAV (AA153)	Baboon	None

Note: Disease association has been well documented and proven for most human adenovirus strains and was experimentally established for several nonhuman adenoviruses by administration of purified virus in healthy animals. In several of these studies disease was seen only in newborns and not in adults. Some viruses were isolated from cell culture experiments only and were not tested for disease association (unknown).

fiber receptor may also display markedly different binding characteristics and differ in their internalization strategy, as shown for Ad2 and Ad9 (Roelvink, Kovesdi, and Wickham, 1996). The shorter length of the Ad9 fiber (11 nm) relative to the Ad2 fiber (37 nm) probably allows the fiber-independent penton base-dependent binding of Ad9 to the cellular integrins and therefore modulates the binding characteristics of the virus (Roelvink, Kovesdi, and Wickham, 1996). The RGD motif located in the penton base is necessary for the successful internalization of several adenoviruses (Mathias et al., 1994; Hidaka et al., 1999; Pearson et al., 1999). However, mutations in the RGD motif do not prevent transduction of mouse hepatocytes in vitro and in vivo (Hautala et al., 1998). Also, Ad40 and Ad41 (subgroup F) have no RGD motifs in the penton base, suggesting other internalization mechanisms (Mautner, Steinthorsdottir, and Bailey, 1995; Albinsson and Kidd, 1999). Finally, although CAR and integrins are important for Ad5 infection in vitro, there is no apparent correlation between expression of these molecules in tissues and the biodistribution of Ad5 vectors injected intravenously in mice (Fechner et al., 1999). Taken together, these results show that the viral fiber and penton base play a critical role in the determination of the viral tissue tropism, but suggest that anatomical barriers may also modulate the ability of adenoviral vectors to efficiently transduce certain tissues in vivo.

Pseudotyping of Ad5 Vectors for Improved Target Transduction and Reduced Toxicity

The detailed knowledge of the molecular determinants of viral tropism allows the development of vectors with new cell specificities by the precise replacement of the Ad5 fiber- or knob-coding sequences with the corresponding sequences from viral serotypes selected for their defined tissue tropism (Shayakhmetov et al., 2000; Stevenson et al., 1997; Havenga et al., 2001; Goossens et al., 2001; Rea et al., 2001). Several investigators have reported the generation of chimeric viruses that efficiently transduce human cell types known to be refractory to Ad5-based viruses. For instance, transduction of human $CD34^+$ cells (Shayakhmetov et al., 2000; Yotnda et al., 2001), human airway epithelia cells (Zabner et al., 1999), or human tumor cell lines (Stevenson et al., 1997) is very significantly enhanced when using Ad5.Fb35 (Ad5 vectors carrying the fiber from Ad35), Ad5.Fb17, and Ad5.Fb3 vectors, respectively, instead of Ad5 vectors. To take full advantage of the complex natural diversity of human adenoviruses, we have undertaken in our laboratory the generation and characterization of a library of Ad5 vectors carrying fibers from alternative human serotypes. For that purpose, fiber-coding sequences from all human adenoviral serotypes, with the exception of serotypes 1, 6, 10, 18, and 26, were cloned, sequenced, and analyzed for the presence of conserved functional domains. As shown in Figure 5.5A, the phylogenetic tree of the fiber knob domains closely matches the six subgroups previously characterized by classical serotyping methods, such as hemagglutination and cross-

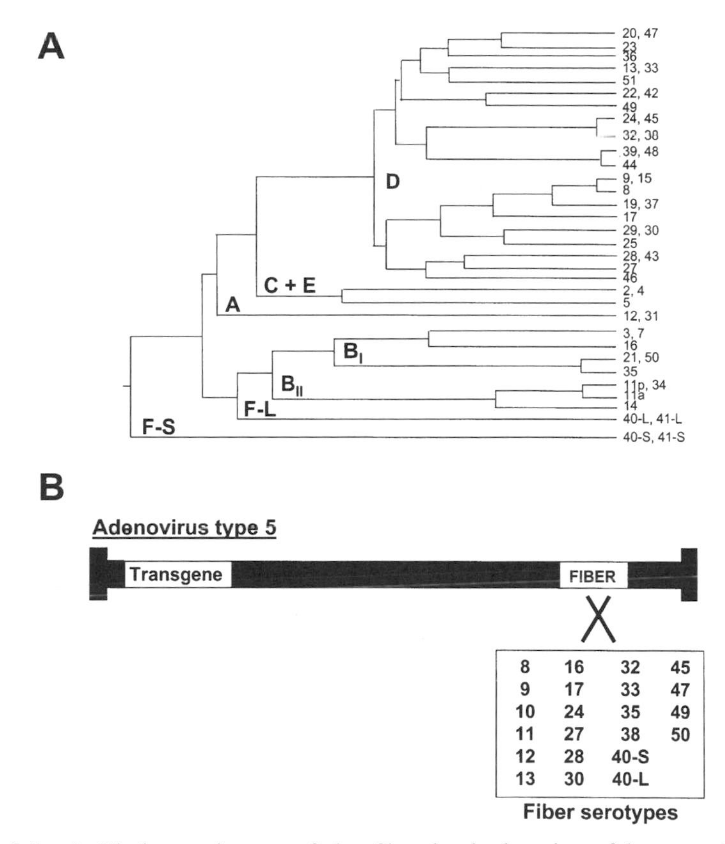

Figure 5.5 *A*. Phylogenetic tree of the fiber knob domains of human adenovirus serotypes. Most fiber domains were sequenced in our laboratory but several sequences were also obtained from Genbank. *B*. A library of fiber-chimeric vectors was generated by deletion of the Ad5 fiber sequence (with the exception of the tail), and substitution by the fiber sequence isolated from the alternative human adenovirus serotypes. At present, the library consists of 22 different fiber-chimeric vectors (numbers in the box indicate the viral serotype from which the fiber in the chimeric vector was isolated).

neutralization assays. As expected, sequence homologies of the knob domains were higher within a subgroup (A: 81.7% (n = 2); B: 53.1% ± 20.1% (n = 10); C: 63.9% (n = 2); D: 57.4 ± 10.9% (n = 30); Short fiber F: 96.9% (n = 2); Long fiber F: 91.0% (n = 2)) than between subgroups (34.4% ± 11.3%). Our chimeric vector library was generated by deleting the entire Ad5 fiber-coding sequence, except for part of the tail of the Ad5 fiber, and replacing it with the fiber sequence from the other human viral serotypes. At present, the vector library consists of 22 different chimeric Ad5 vectors, each expressing the luciferase, β-galactosidase, or green fluorescent protein (GFP) marker genes (Fig. 5.5B). The impact of the fiber replacement on the tropism of the Ad5 chimeric vectors is illustrated in Figure 5.6 and Table 5.2. A representa-

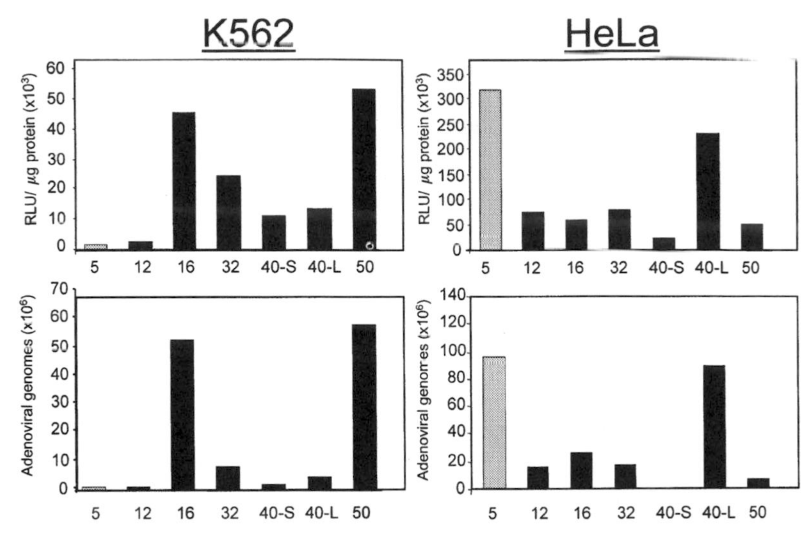

Figure 5.6 Modulation of the viral tropism by fiber swapping. Chimeric Ad5 vectors carrying fiber molecules from different viral serotypes (5, 12, 16, 32, 40-S, 40-L, 50) were used to infect 5×10^5 human K562 or HeLa cells at a defined multiplicity of infection (1000 virus particles per cell). After a 1 h exposure, the virus was removed and 48 h later half of the cell population was used to determine the level of expression of the luciferase reporter gene (upper graphs: luciferase activity is expressed in relative light units (RLU) per μg total protein). The other half of the cell population was used to quantify by real-time polymerase chain reaction (PCR) analysis the number of adenoviral genome copies transduced into the target cells.

tive panel of vectors was selected and used to transduce human cells sensitive (HeLa) and refractory (human erythroid leukemia K562) to infection by Ad5 vectors. Analysis of transgene expression and quantification of the transduced viral genomes confirm that the efficiency of transduction can be significantly modulated by fiber replacement. In addition, these results also demonstrate that increased levels of transgene expression correlate with an increased viral genome copy number. Although we cannot exclude an influence on transgene expression of the fiber involvement in viral intracellular trafficking (Miyazawa et al., 1999; Miyazawa, Crystal, and Leopold, 2001), these observations support the critical role of fiber in vector binding to the target cell. The results also demonstrate the complexity of receptor recognition because K562 cells, which do not express CAR, are poorly transduced with Ad5 and Ad5.Fb12 but well transduced with Ad5.Fb16 and Ad5.Fb50, while Ad5.Fb32, Ad5.Fib40-S and Ad5.Fib40-L show variable levels of transgene expression. In contrast, Ad5 and Ad5.Fb40-L but not Ad5.Fb16 or Ad5.Fb50 efficiently transduce HeLa cells (Fig. 5.6). Interestingly, in most cases, the vector identified as transducing the tested target cell type most efficiently carries a fiber from a subgroup

TABLE 5.2 Sensitivity of Primary Cells and Established Tumor Cell Lines to Transduction by Fiber-Chimeric Ad5 Vectors

Primary Cells	Tissue	CAR	$\alpha_V\beta 3$	$\alpha_V\beta 5$	Best Vector	Fold Increase of Transduction[(a)]
Endothelial	Umbilical vein	+	++	+	Ad5.Fb16	8 ± 3x ($n = 9$)
Smooth muscle	Umbilical vein	–	++	++	Ad5.Fb16	64 ± 20x ($n = 8$)
Synoviocytes	Synovium	–	+	+	Ad5.Fb16	154 ± 68x ($n = 4$)
Fibroblasts	Skin	–	++	–	Ad5.Fb16	64 ± 58x ($n = 4$)
Amniocytes	Amniotic fluid	–	++	+	Ad5.Fb50	7x ($n = 1$)
Hepatocytes	Liver	ND	ND	ND	Ad5	– ($n = 2$)
Mature dendritic	Hemopoietic	–	–	+	Ad5.Fb35	21 ± 5x ($n = 5$)
Immature dendritic	Hemopoietic	–	–	+	Ad5.Fb35	9 ± 5x ($n = 3$)
HBSC	Hemopoietic	–	+	+	Ad5.Fb16	138 ± 86x ($n = 3$)
Chondrocytes	Hemopoietic	–	+	+	Ad5.Fb16	13 ± 5x ($n = 3$)
Myoblasts	Skeletal muscle	ND	ND	ND	Ad5.Fb50	18 ± 8x ($n = 3$)
Melanocytes	Skin	ND	ND	ND	Ad5.Fb35	17x ($n = 1$)
FDPC	Skin	ND	ND	ND	Ad5.Fb35	36x ($n = 1$)
Established Cell Lines						
A549	Lung	++	+	++	Ad5	– ($n = 12$)
K562	Erythroid	–	+	+	Ad5.Fb16	814 ± 176x ($n = 4$)
TF-1	Erythroid	–	++	+	Ad5.Fb35	60 ± 18x ($n = 3$)
CEM	Lymphoid	–	–	–	Ad5.Fb16	11x ($n = 1$)
MCF-7	Breast	ND	ND	ND	Ad5.Fb50	42x ($n = 2$)
HeLa	Cervix	++	+	+	Ad5	– ($n = 4$)
HepG2	Liver	++	–	+	Ad5	– ($n = 3$)
CAPAN-A	Pancreas	–	–	+	Ad5.Fb16	5x ($n = 3$)
Mz-Cha-1	Cholangio	–	–	+	Ad5.Fb50	107x ($n = 1$)

Notes: Expression of CAR and integrins was monitored with flow cytometry. Cells were scored as –, +, ++ when less than 2%, between 2%–50%, or more than 50% of the cell population express the molecule, respectively.[a] Fold increase of luciferase activity obtained with the best vector when compared to a luciferase expression in cells transduced with the Ad5 vector. Luciferase activity was expressed in RLU per microgram total cellular protein. Total cellular protein as measured 48 h after a 2 h vector exposure of 1000 virus particles per cell.
ND = not determined. HBSC = human bone marrow stroma cells; FDPC = follicular dermal papilla cells. The number of independent experiments performed (each in triplicate) is listed in parentheses.

B virus. This is illustrated in Table 5.2, which summarizes the results of an in vitro study in which a panel of human primary cells and tumor cell lines was tested for expression of CAR and $\alpha v\beta 3/\alpha v\beta 5$ integrins and for the efficiency of transduction by the chimeric vectors in the library. The vectors identified in such in vitro studies can then be selected for further evaluation in vivo. In this respect, it is noteworthy that vectors identified as being superior to Ad5 for gene transfer into human and nonhuman primate cells may not necessarily be superior in other animals used classically as preclinical models such as mice or pigs. Such a case is illustrated in Figure 5.7, where human and rhesus monkey carotid artery smooth muscle cells are better transduced with Ad5.Fb16 vectors whereas in contrast, cells from rat, mouse, rabbit, and pig are much more sensitive to Ad5 than Ad5.Fb16 viruses. Although this phenomenon has not been completely elucidated, we hypothesize that the cellular attachment molecules utilized for infection by subgroup B viruses are not conserved between species. Nonhuman primates may therefore in such cases be the only appropriate preclinical models in studies using fiber-chimeric vectors. As mentioned earlier, adenoviruses have also been isolated from many different animal species. Little is known about their biology and their cellular receptors but it has been reported that some of these nonhuman viruses can infect human cells. For instance, recombinant vectors have been derived from ovine adenovirus OAV287, bovine adenovirus type 3 (BAV-3), porcine adenovirus type 3 (PAV-3), avian adenovirus (CELO), or canine adenovirus type 2 and shown to transduce human cell lines (Khatri, Xu, and Both, 1997; Kremer et al., 2000; Zakhartchouk et al., 1998; Rasmussen et al., 1999; Reddy et al., 1999; Michou et al., 1999; Tan et al., 2001; Klonjkowski et al., 1997). Fibers isolated from

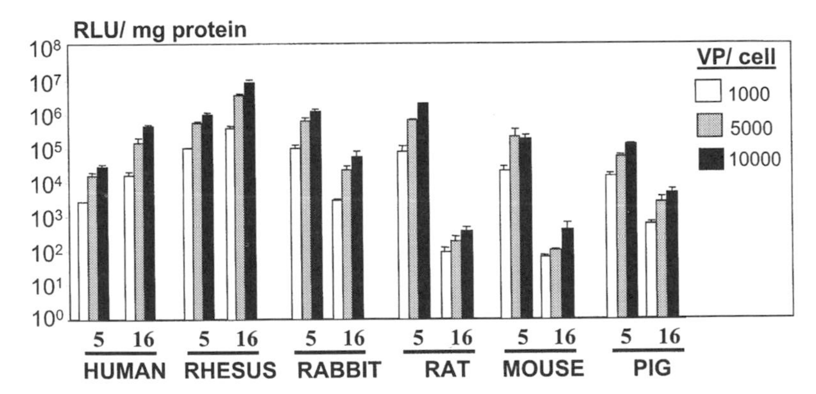

Figure 5.7 Viral tropism is modulated by the species origin of the tested tissues. Smooth muscle cells isolated from the carotid artery of different species were infected with 1000, 5000, or 10,000 virus particles per cell of Ad5 or Ad5.Fb16 expressing the luciferase reporter gene. Virus was removed 1 h later and luciferase activity was determined 48 h later. Luciferase activity is expressed as relative light units (RLU) per mg protein.

such nonhuman viruses can therefore be similarly used to enrich and expand the Ad5-fiber chimeric library with vectors that can efficiently infect tissues from both human and animal origins.

Altogether, data from the available studies support the notion that vectors with improved transduction efficiencies for important human target tissues such as the lung or the lymphoid organs can be generated through fiber swapping. Although in vivo data are still scarce, such vectors may allow the reduction of the therapeutic viral doses currently used for Ad5 vectors, with the associated advantage of a possible reduction of the toxicity risks. However, such fiber-chimeric viruses may still remain fully susceptible to neutralization by human sera since hexon, penton base, and fiber together constitute the major epitopes recognized by the neutralizing antibodies (see following section). Additional engineering must therefore be considered to overcome this drawback. Again, the wide natural diversity of human adenoviruses may be exploited to derive vectors that can better transduce the major human target tissues and in addition are not neutralized by the antiadenoviral antibodies present in most human sera. For that purpose, a better understanding of the overall prevalence in the human population of the neutralizing antibodies directed against all human adenovirus serotypes is required.

Antiadenoviral Neutralizing Antibodies in the Human Population

An extensive study was performed in 1982 in Italy by D'Ambrosio et al. to determine the presence in the sera from 453 children and 51 young adults of neutralizing antibodies to 33 different human adenovirus serotypes. This survey showed that 74.6% of the children contained neutralizing antibodies directed against at least one serotype, going up to 98% for the young adults. The viral serotypes most often recognized in the children and young adults were subgroup C serotypes 2 (41%), 5 (33%), and 1 (29%), followed by subgroup B serotype 3 (22%). Serotypes to which there was virtually no neutralizing activity included 8, 9, 20, 24–26, 28, and 32. Another finding of this study was that the number of individuals with neutralizing activity against the most frequent types rises with age while it was steady for the less frequent ones. The results obtained from Italy for serotypes 1 to 8 were similar to results from a survey performed in the United States and Sweden (Tai and Graystorm, 1962), but differed from data obtained in Taiwan and Japan (Tai and Graystorm, 1962), suggesting geographical variations in viral epidemiology. In addition, the distribution of the adenovirus serotypes, and hence of the corresponding neutralizing antibody titers, may also differ according to the health status of the individuals. For instance, some adenovirus serotypes are found more frequently in immunocompromised patients than in healthy individuals (Hierholzer et al., 1988; Hierholzer, 1992).

In an attempt to identify an adenovirus serotype less prone to neutralization by most human sera, we have recently conducted a large screening program in which sera from 560 healthy volunteers between the ages of 20 and 70 years

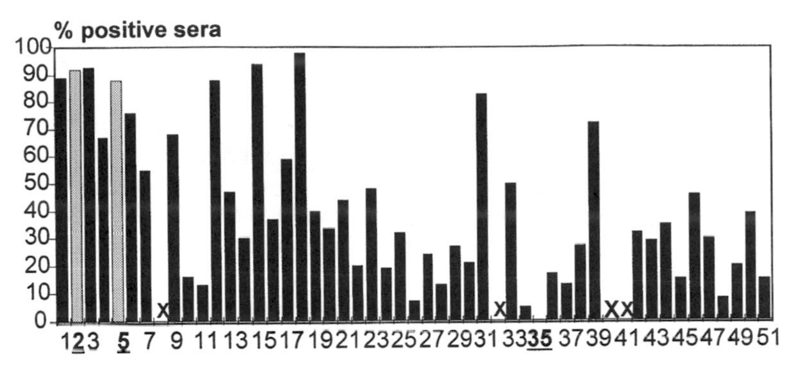

Figure 5.8 Prevalence of antiadenoviral neutralizing antibodies in human sera. One hundred sera from healthy volunteers were collected in Belgium and were tested for the presence of neutralizing activity against all 51 human adenovirus serotypes identified to date, with the exception of serotypes 8, 32, 40, and 41 (marked with X). As expected, Ad2 and Ad5 (light grey bars) were neutralized efficiently (>80%) by the sera. Ad35 was neutralized by none of the tested sera.

were collected from different locations in Europe (the Netherlands, United Kingdom, and Belgium), Japan, and the United States (East Coast and West Coast) and tested for the presence of neutralizing antibodies against almost all human adenovirus serotypes (Fig. 5.8 summarizes the data obtained with the sera from 100 Belgian volunteers). While high antibody titers against Ad5 and Ad2 viruses were found in the majority of the volunteers (>80%), a few viral serotypes were identified that were barely or not neutralized by the tested sera (Fig. 5.8). Roughly similar results were obtained with sera collected in the other geographical locations. Altogether, Ad35 was identified as the viral serotype against which most healthy individuals have developed no or only very low neutralizing antibody titers (an average of 7.0% of the sera scored positive for the presence of neutralizing anti-Ad35 antibodies). As an extension of this study, we also analyzed sera collected from patients suffering from cardiovascular diseases and found similarly low antibody titers against Ad35. This striking property of the Ad35 serotype can therefore be exploited to derive vectors that are less prone to neutralization by the preexisting antibodies present in the sera of the patients to be treated. Two major strategies can be considered for that purpose: the use of the individual structural components of Ad35 that confer to the virus its resistance to neutralization or the development of vectors fully derived from Ad35.

Pseudotyping of Adenoviral Vectors to Circumvent Neutralization

The structure of the adenovirus particle predicts that there are predominantly three structural viral proteins toward which neutralizing antibodies can be directed, namely, penton base, hexon, and fiber (Fig. 5.1). Wohlfart (1988) suggested that an adenovirus may be neutralized by two pathways. The first pathway, commonly referred to as extracellular neutralization, occurs via aggregation of the free virions or via inhibition of virus binding to the cell surface by antifiber or antipenton base antibodies. The second pathway, referred to as intracellular neutralization, occurs via inhibition of the interaction between the adenovirus and the endosomal membrane by the antipenton base or antihexon antibodies. However, the respective impacts on vector neutralization of antibodies directed against the three different structural proteins remains unclear. Whereas some studies have reported that the hexon constitutes the primary target for neutralization (Philipson, Longberg-Holm, and Pattersson, 1968; Haase and Pereira, 1972; Toogood, Crompton, and Hay, 1992; Wohlfart, 1988), other investigators have shown in patients treated with recombinant Ad5 vectors that antibodies against the fiber and penton base have a strong synergistic effect in neutralization (Gahery-Segard et al., 1998). In addition, the route of vector administration or virus infection may also influence the nature of the antibody response (Gahery-Segard et al., 1997). In these studies, intraperitoneal injection of Ad5 vectors into mice was shown to induce predominantly antihexon IgG2a neutralizing antibodies. In contrast, mice treated intravenously developed predominantly neutralizing IgM antibodies recognizing several capsid proteins. In another study, passive immunization experiments using hyperimmune mouse sera specifically depleted of antibodies directed against either fiber, penton base, or hexon have demonstrated that antifiber antibodies were the main component involved in viral neutralization (Rahman et al., 2000).

Despite these contradictory results, most data suggest that hexon proteins play the major role in the induction of neutralizing antibodies. Analysis of the primary sequence and tertiary structure of the hexon protein has revealed the existence of seven hypervariable domains (Crawford-Miksza and Schnurr, 1996) located on two distinct loops (L1 and L2) orientated outward (Athappilly et al., 1994; Roberts et al., 1986; Eiz, Adrian, and Pring-Akerblom, 1995; Pring-Akerblom et al., 1995; Pring-Akerblom and Adrian, 1993; Toogood, Crompton, and Hay, 1992; Rux and Burnet, 2000). In vitro and in vivo neutralization studies performed with chimeric Ad5 vectors in which the hexon L1 or L2 domains have been replaced by the L1 or L2 domains from Ad2 have shown that L1 is the predominant target for neutralization (Gall, Crystal, and Falck-Pedersen, 1998). These results were supported by an independent study using a chimeric Ad5 vector in which a larger part of the Ad5 hexon (spanning loops L1 to L4) was deleted and replaced by the corresponding hexon sequence from Ad12 (Subgroup A) (Roy et al., 1998). In vitro analysis of the sensitivity of this Ad5.hexon12 virus toward anti-Ad12 antibodies revealed a neutralization profile similar to Ad12 (Roy et al., 1998). Furthermore, the chimeric Ad5.hexon12 vector could be efficiently administered in mice preimmunized with Ad5 and therefore containing high serum levels of anti-Ad5 neutralizing antibodies. Interestingly, the authors also tested 23 different human serum samples for neutralization of Ad5.hexon12, Ad5 and Ad12. Nine sera were able to neutralize efficiently Ad5, Ad12, and Ad5.hexon12 whereas 7 sera neutralized only Ad12 and Ad5.hexon12 but not

Ad5. Altogether, these results strongly support the hypothesis that the viral hexon protein contains the major neutralization epitopes.

To take advantage of these observations and of our finding that the Ad35 serotype is not neutralized by most human sera, we have undertaken the generation and analysis of chimeric Ad5 vectors that bear the fiber, penton base, and/or hexon proteins from Ad35. Our results confirm the notion that fiber and penton base are not the major targets for neutralization since a chimeric Ad5 vector carrying simultaneously the Ad35 penton base and fiber proteins (Ad5.Fb35Pb35) was still efficiently neutralized by all tested human sera (Fig. 5.9). However, the fiber and penton base do contain minor neutralization epitopes, since neutralization by most sera was slightly attenuated when using Ad5.Fb35Pb35 as compared to Ad5 (Fig. 5.9). Such results suggest that a chimeric Ad5 vector carrying simultaneously the fiber, penton base, and hexon from Ad35 should be fully insensitive to neutralization. Unfortunately, while Ad5.Fb35Pb35 can be produced at high titers, replacement of the hexon gene did not result in the generation of viable viruses. Similarly, an independent attempt to develop a chimeric Ad5 vector carrying a subgroup B hexon (Ad5.hexon7) also failed to generate viable viruses (Gall et al., 1998). The reason for this failure is unclear because successful hexon swapping between members of the same subgroup or between members from subgroup A, D, and C viruses is feasible (unpublished results; Roy et al., 1998).

Recombinant Adenoviral Vectors Derived from Rare Serotypes

Given the technical difficulties in generating a chimeric vector that is not neutralized by human sera and that better transduces Ad5-refractory tissues, the development of a new vector that is fully derived from an adenoviral serotype that naturally combines both properties constitutes an attractive strategy. As mentioned earlier, Ad35 viruses are not neutralized by most of the human sera (Fig. 5.8) and the Ad35 fiber binds to an unknown receptor present in several human tissues that do not express CAR (Table 5.2). The availability of a gene transfer vector derived from Ad35 would theoretically constitute a significant improvement over current Ad5-based vectors. We have therefore cloned and sequenced the Ad35 genome and developed a plasmid-based system for the generation of recombinant Ad35 vectors, allowing us to successfully generate various recombinant E1A-, E1-, and/or E3-deleted Ad35 vectors. In vitro neutralization experiments using an E1A-deleted Ad35 vector expressing the luciferase gene confirmed the insensitivity of this novel recombinant vector to a panel of sera collected from healthy volunteers while an Ad5-luciferase vector was neutralized under similar experimental conditions (Fig. 5.10).

These results clearly indicate that the natural diversity of the human adenovirus serotypes can be usefully exploited to identify viruses with features advantageous for the development of novel gene transfer vectors with improved properties. In addition, specific structural or regulatory components of these human viruses may also be selected for the generation of pseudotyped vectors that com-

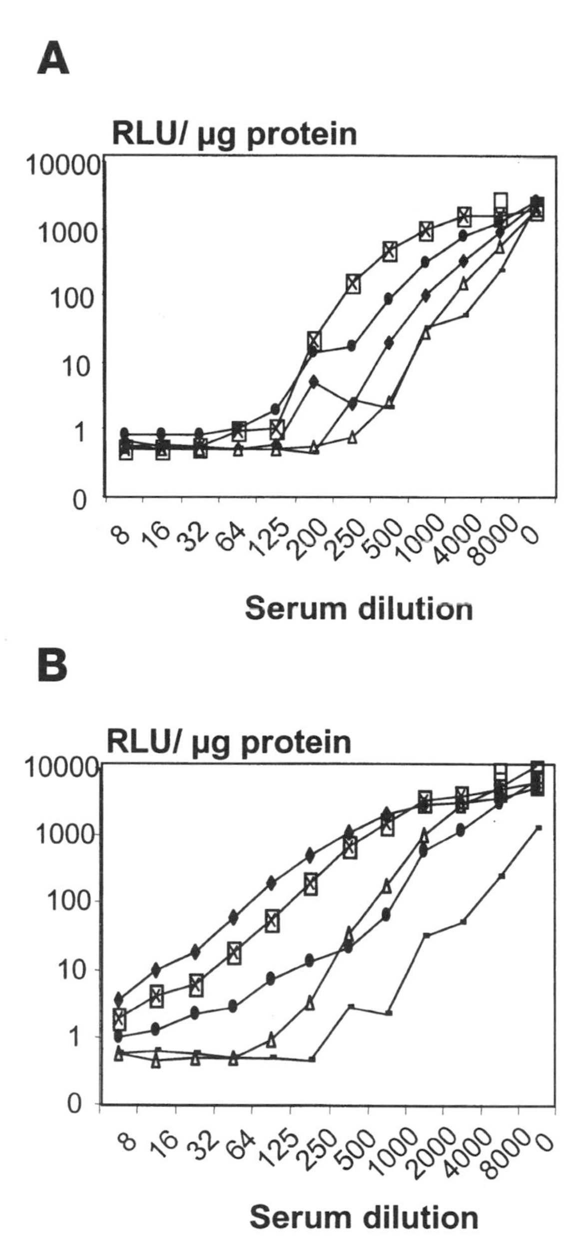

Figure 5.9 Sensitivity of Ad5 and Ad5-Fb35-Pb35 vectors to neutralization by human sera. Serial dilutions of five individual human sera (50 μl) were added to the vectors (10^7 virus particles in 50 μl) and the mixture was then added to 10^4 human A549 cells in 96-well plates. Cells were lysed 48 h later and luciferase activity (expressed as RLU/μg protein) was determined. The inhibition of gene transfer by the neutralizing antibodies is reflected in the loss of luciferase activity upon increasing serum concentration. *A*. Neutralization of Ad5. *B*. Neutralization of the chimeric Ad5-Fb35-Pb35 vector carrying both the penton base and the fiber proteins from Ad35.

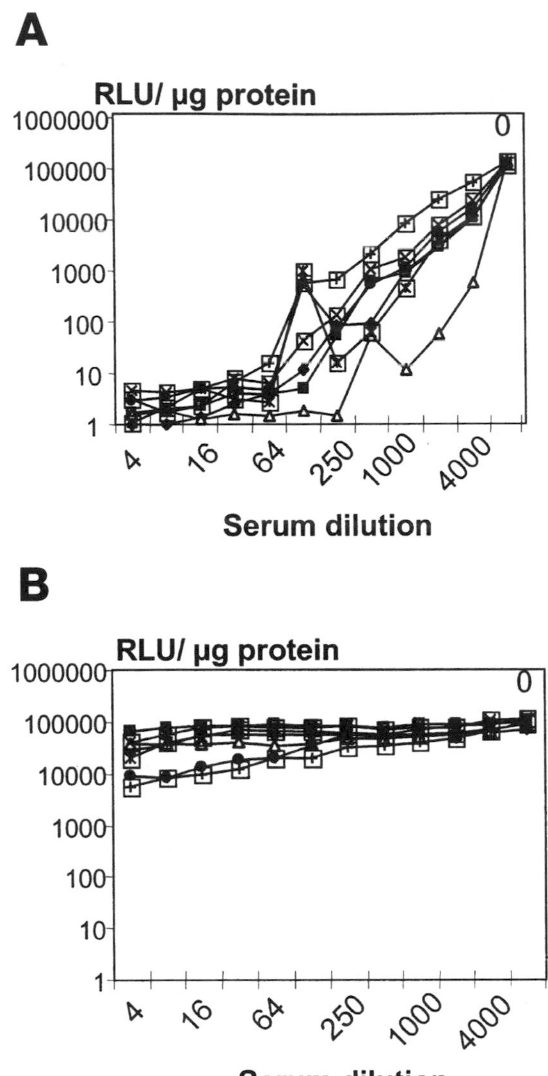

Figure 5.10 Sensitivity of Ad5 and Ad35 vectors to neutralization by human sera. A neutralization assay was performed for Ad5 (*A*) and Ad35 (*B*) vectors expressing the luciferase reporter gene, as shown in Figure 5.9.

bine the attractive properties of the individual components. Finally, the list of adenovirus serotypes available for investigation can be tremendously expanded by also considering all nonhuman adenoviruses. However, basic knowledge of the biological and immunological properties, molecular structure, and genetic organization of these human and nonhuman viruses, a prerequisite for the generation of recombinant vectors, is in most cases still lacking.

CONCLUSION

The realization that the Adenoviridae constitutes a very large and complex family that comprises members characterized by a huge natural diversity in pathological, immunological, and biological properties suggests that, besides Ad2 and Ad5, other human or animal viruses may be selected for the generation of gene transfer vectors with more favorable characteristics. Already, several studies on Ad5 vectors pseudotyped with fiber proteins from other human adenovirus serotypes have illustrated the potential strength of such a strategy. Despite the current technical limitations, vectors pseudotyped for several capsid proteins from different origins (e.g., fiber, penton base, and hexon) may constitute an important step forward in the generation of vectors that are not neutralized by human sera and that better transduce the target tissue. One may even envision that not only structural proteins, but also specific regulatory proteins selected from other human or animal serotypes for their attractive biological properties (e.g., immunomodulatory function), may in the future be used to generate improved (e.g., less immunogenic) chimeric vectors. Our limited knowledge on the biology of most human and nonhuman adenoviruses constitutes today the most serious limitation in our ability to take full advantage of the possibilities offered by the natural diversity of the Adenoviridae.

REFERENCES

Albinsson B, Kidd AH (1999): Adenovirus type 41 lacks an RGD alpha(v)-integrin binding motif on the penton base and undergoes delayed uptake in A549 cells. *Virus Res* 64:125–136.

Arnberg N, Edlund K, Kidd AH, Wadell G (1999): Adenovirus type 37 uses sialic acid as a cellular receptor. *J Virol* 74:42–48.

Arnberg N, Kidd AH, Edlund K, Olfat F, Wadell G (2000): Initial interactions of subgenus D adenoviruses with A549 cellular receptors: sialic acid versus alpha(v) integrins. *J Virol* 74:7691–7693.

Athappilly FK, Murali R, Rux JJ, Cai Z, Burnett RM (1994): The refined crystal structure of hexon, the major coat protein of adenovirus type 2, at 2.9 A resolution. *J Mol Biol* 242:430–455.

Bal HP, Chroboczek J, Schoehn G, Ruigrok RW, Dewhurst S (2000): Adenovirus type 7 penton purification of soluble pentamers from *Escherichia coli* and development of an integrin-dependent gene delivery system. *Eur J Biochem* 267:6074–6081.

Barr D, Tubb J, Ferguson D, Scaria A, Lieber A, Wilson C, Perkins J, Kay MA (1995): Strain related variations in adenovirally mediated transgene expression from mouse hepatocytes in vivo: comparisons between immunocompetent and immunodeficient inbred strains. *Gene Ther* 2:151–155.

Benihoud K, Saggio I, Opolon P, Salone B, Amiot F, Connault E, Chianale C, Dautry F, Yeh P, Perricaudet M (1998): Efficient, repeated adenovirus-mediated gene transfer in mice lacking both tumor necrosis factor alpha and lymphotoxin alpha. *J Virol* 72:9514–9525.

Bergelson JM (1999): Receptors mediating adenovirus attachment and internalization. *Biochem Pharmacol* 57:975–979.

Bergelson JM, Cunningham JA, Droguett G, Kurt-Jones EA, Krithivas A, Hong JS, Horwitz MS, Crowell RL, Finberg RW (1997): Isolation of a common receptor for Coxsackie B viruses and adenoviruses 2 and 5. *Science* 275:1320–1323.

Blackwell JL, Li H, Gomez-Navarro J, Dmitriev I, Krasnykh V, Richter CA, Shaw DR, Alvarez RD, Curiel DT, Strong TV (2000): Using a tropism-modified adenoviral vector to circumvent inhibitory factors in ascites fluid. *Hum Gen Ther* 11:1657–1669.

Bout A (1996): Prospects for human gene therapy. *Eur J Drug Metab Pharmacokinet* 21:175–179.

Brough DE, Lizonova A, Hsu C, Kulesa VA, Kovesdi I (1996): A gene transfer vector-cell line system for complete functional complementation of adenovirus early regions E1 and E4. *J Virol* 70:6497–6501.

Brough DE, Hsu C, Kulesa VA, Lee GM, Cantolupo LJ, Lizonova A, Kovesdi I (1997): Activation of transgene expression by early region 4 is responsible for a high level of persistent transgene expression from adenovirus vectors in vivo. *J Virol* 71:9206–9213.

Chen Y, Yu DC, Charlton D, Henderson DR (2000): Pre-existent adenovirus antibody inhibits systemic toxicity and antitumor activity of CN706 in the nude mouse LNCaP xenograft model: implications and proposals for human therapy. *Hum Gene Ther* 11:1553–1559.

Christ M, Lusky M, Stoeckel F, Dreyer D, Dieterle A, Michou AI, Pavirani A, Mehtali M (1997): Gene therapy with recombinant adenovirus vectors: evaluation of the host immune response. *Immunol Lett* 57:19–25.

Chroboczek J, Bieber F, Jacrot B (1992): The sequence of the genome of adenovirus type 5 and its comparison with the genome of adenovirus type 2. *Virology* 186:280–285.

Clark HF, Michalski F, Tweedell KS, Yohn D, Zeigel RF (1973): An adenovirus, FAV-1, isolated from the kidney of a frog (*Rana pipiens*). *Virology* 51:392–400.

Cotten M, Saltik M, Kursa M, Wagner E, Maass G, Birnstiel ML (1994): Psoralen treatment of adenovirus particles eliminates virus replication and transcription while maintaining the endosomolytic activity of the virus capsid. *Virology* 205:254–261.

Crawford-Miksza L, Schnurr DP (1996): Analysis of 15 adenovirus hexon proteins reveals the location and structure of seven hypervariable regions containing serotype-specific residues. *J Virol* 70:1836–1844.

Cripe TP, Dunphy EJ, Holub AD, Saini A, Vasi NH, Mahller YY, Collins MH, Snyder JD, Krasnykh V, Curiel DT, Wickham TJ, DeGregori J, Bergelson JM, Currier MA (2001): Fiber knob modifications overcome low, heterogeneous expression of the coxsackievirus-adenovirus receptor that limits adenovirus gene transfer and oncolysis for human rhabdomyosarcoma cells. *Cancer Res* 61:2953–2960.

Croyle MA, Roessler BJ, Davidson BL, Hilfinger JM, Amidon GL (1998): Factors that influence stability of recombinant adenoviral preparations for human gene therapy. *Pharm Dev Technol* 3:373–378.

Croyle MA, Yu QC, Wilson JM (2000): Development of a rapid method for the PEGylation of adenoviruses with enhanced transduction and improved stability under harsh storage conditions. *Hum Gene Ther* 11:1713–1722.

Dai Y, Schwarz EM, Gu D, Zhang WW, Sarvetnick N, Verma IM (1995): Cellular and humoral immune responses to adenoviral vectors containing factor IX gene: tolerization of factor IX and vector antigens allows for long-term expression. *Proc Natl Acad Sci USA* 92:1401–1405.

D'Ambrosio E, Del Grosso N, Chicca A, Midulla M (1982): Neutralizing antibodies against 33 human adenoviruses in normal children in Rome. *J Hyg (Lond)* 89:155–161.

De Jong JC, Wermenbol AG, Verweij-Uijterwaal MW, Slaterus KW, Wertheim-Van Dillen P, Van Doornum GJ, Khoo SH, Hierholzer JC (1999): Adenoviruses from human immunodeficiency virus-infected individuals, including two strains that represent new candidate serotypes Ad50 and Ad51 of species B1 and D, respectively. *J Clin Microbiol* 37:3940–3945.

Dedieu JF, Vigne E, Torrent C, Jullien C, Mahfouz I, Caillaud JM, Aubailly N, Orsini C, Guillaume JM, Opolon P, Delaere P, Perricaudet M, Yeh P (1997): Long-term gene delivery into the livers of immunocompetent mice with E1/E4-defective adenoviruses. *J Virol* 71:4626–4637.

Di Guilmi AM, Barge A, Kitts P, Gout E, Chroboczek J (1995): Human adenovirus serotype 3 (Ad3) and the Ad3 fiber protein bind to a 130-kDa membrane protein on HeLa cells. *Virus Res* 38:71–81.

Dmitriev I, Krasnykh V, Miller CR, Wang M, Kashentseva E, Mikheeva G, Belousova N, Curiel DT (1998): An adenovirus vector with genetically modified fibers demonstrates expanded tropism via utilization of a coxsackievirus and adenovirus receptor-independent cell entry mechanism. *J Virol* 72:9706–9713.

Douglas JT, Rogers BE, Rosenfeld ME, Michael SI, Feng M, Curiel DT (1996): Targeted gene delivery by tropism-modified adenoviral vectors. *Nat Biotechnol* 14:1574–1578.

Eiz B, Adrian T, Pring-Akerblom P (1995): Immunological adenovirus variant strains of subgenus D: comparison of the hexon and fiber sequences. *Virology* 213:313–320.

Elkon KB, Liu CC, Gall JG, Trevejo J, Marino MW, Abrahamsen KA, Song X, Zhou JL, Old LJ, Crystal RG, Falck-Pederson E (1997): Tumor necrosis factor alpha plays a central role in immune-mediated clearance of adenoviral vectors. *Proc Natl Acad Sci USA* 94:9814–9818.

Elshami AA, Kucharczuk JC, Sterman DH, Smythe WR, Hwang HC, Amin KM, Litzky LA, Albelda SM, Kaiser LR (1995): The role of immunosuppression in the efficacy of cancer gene therapy using adenovirus transfer of the herpes simplex thymidine kinase gene. *Ann Surg* 222:298–307.

Engelhardt JF, Litzky L, Wilson JM (1994): Prolonged transgene expression in cotton rat lung with recombinant adenoviruses defective in E2a. *Hum Gene Ther* 5:1217–1229.

Fallaux FJ, Bout A, van der Velde I, van den Wollenberg DJ, Hehir KM, Keegan J, Auger C, Cramer SJ, van Ormondt H, van der Eb AJ, Valerio D, Hoeben RC (1998): New helper cells and matched early region 1-deleted adenovirus vectors prevent generation of replication-competent adenoviruses. *Hum Gene Ther* 9:1909–1917.

Fechner H, Haack A, Wang H, Wang X, Eizema K, Pauschinger M, Schoemaker R, Veghel R, Houtsmuller A, Schultheiss HP, Lamers J, Poller W (1999): Expression of coxsackie adenovirus receptor and alphav-integrin does not correlate with adenovector targeting in vivo indicating anatomical vector barriers. *Gene Ther* 6:1520–1535.

Fisher KD, Stallwood Y, Green NK, Ulbrich K, Mautner V, Seymour LW (2001): Polymer-coated adenovirus permits efficient retargeting and evades neutralizing antibodies. *Gene Ther* 8:341–348.

Fisher KJ, Choi H, Burda J, Chen SJ, Wilson JM (1996): Recombinant adenovirus deleted of all viral genes for gene therapy of cystic fibrosis. *Virology* 217:11–22.

Flomenberg P, Piaskowski V, Truitt RL, Casper JT (1995): Characterization of human proliferative T cell responses to adenovirus. *J Infect Dis* 171:1090–1096.

Francki RIB, Fauquet CM, Knudson L, Brown F (1991): Archives of virology supplementum 2: Classification and nomenclature of adenoviruses. In: *Fifth Report of the International Committee on Taxonomy of Viruses*, Vienna: Springer-Verlag Wien, pp. 140–144.

Gahery-Segard H, Juillard V, Gaston J, Lengagne R, Pavirani A, Boulanger P, Guillet JG (1997): Humoral immune response to the capsid components of recombinant adenoviruses: routes of immunization modulate virus-induced Ig subclass shifts. *Eur J Immunol* 27:653–659.

Gahery-Segard H, Farace F, Godfrin D, Gaston J, Lengagne R, Tursz T, Boulanger P, Guillet JG (1998): Immune response to recombinant capsid proteins of adenovirus in humans: antifiber and anti-penton base antibodies have a synergistic effect on neutralizing activity. *J Virol* 72:2388–2397.

Gall JG, Crystal RG, Falck-Pedersen E (1998): Construction and characterization of hexon-chimeric adenoviruses: specification of adenovirus serotype. *J Virol* 72:10260–10264.

Gao GP, Yang Y, Wilson JM (1996): Biology of adenovirus vectors with E1 and E4 deletions for liver-directed gene therapy. *J Virol* 70:8934–8943.

Goldman MJ, Wilson JM (1995): Expression of alpha v beta 5 integrin is necessary for efficient adenovirus-mediated gene transfer in the human airway. *J Virol* 69:5951–5958.

Goossens PH, Schouten GJ, ’t Hart BA, Bout A, Brok HP, Kluin PM, Breedveld FC, Valerio D, Huizinga TW (1999): Feasibility of adenovirus-mediated nonsurgical synovectomy in collagen-induced arthritis-affected rhesus monkeys. *Hum Gene Ther* 10:1139–1149.

Goossens PH, Havenga MJ, Pieterman E, Lemckert AA, Breedveld FC, Bout A, Huizinga TW (2001): Infection efficiency of type 5 adenoviral vectors in synovial tissue can be enhanced with a type 16 fiber. *Arthritis Rheum* 44:570–577.

Gorziglia MI, Kadan MJ, Yei S, Lim J, Lee GM, Luthra R, Trapnell BC (1996): Elimination of both E1 and E2 from adenovirus vectors further improves prospects for in vivo human gene therapy. *J Virol* 70:4173–4178.

Graham FL, Smiley J, Russell WC, Nairn R (1977). Characterization of a human cell line transformed by DNA from human adenovirus type 5. *J Gen Virol* 36:59–65.

Greber UF, Willettes M, Webster P, Helenius A (1993). Stepwise dismantling of adenovirus 2 during entry into cells. *Cell* 75:477–483.

Haase AT, Pereira HG (1972): The purification of adenovirus neutralizing antibody: adenovirus type 5 hexon immunoadsorbent. *J Immunol* 108:633–636.

Haisma HJ, Grill J, Curiel DT, Hoogeland S, van Beusechem VW, Pinedo HM, Gerritsen WR (2000): Targeting of adenoviral vectors through a bispecific single-chain antibody. *Cancer Gene Ther* 7:901–904.

Hautala T, Grunst T, Fabrega A, Freimuth P, Welsh MJ (1998): An interaction between penton base and alpha v integrins plays a minimal role in adenovirus-mediated gene transfer to hepatocytes in vitro and in vivo. *Gene Ther* 5:1259–1264.

Havenga MJ, Lemckert AA, Grimbergen JM, Vogels R, Huisman LG, Valerio D, Bout A, Quax PH (2001): Improved adenovirus vectors for infection of cardiovascular tissues. *J Virol* 75:3335–3342.

Hegenbarth S, Gerolami R, Protzer U, Tran PL, Brechot C, Gerken G, Knolle PA (2000): Liver sinusoidal endothelial cells are not permissive for adenovirus type 5. *Hum Gene Ther* 11:481–486.

Hehir KM, Armentano D, Cardoza LM, Choquette TL, Berthelette PB, White GA, Couture LA, Everton MB, Keegan J, Martin JM, Pratt DA, Smith MP, Smith AE, Wadsworth SC (1996): Molecular characterization of replication-competent variants of adenovirus vectors and genome modifications to prevent their occurrence. *J Virol* 70:8459–8467.

Hidaka C, Milano E, Leopold PL, Bergelson JM, Hackett NR, Finberg RW, Wickham TJ, Kovesdi I, Roelvink P, Crystal RG (1999): CAR-dependent and CAR-independent pathways of adenovirus vector-mediated gene transfer and expression in human fibroblasts. *J Clin Invest* 103:579–587.

Hierholzer JC (1992): Adenoviruses in the immunocompromised host. *Clin Microbiol Rev* 5:262–274.

Hierholzer JC, Wigand R, Anderson LJ, Adrian T, Gold JW (1988): Adenoviruses from patients with AIDS: A plethora of serotypes and a description of five new serotypes of subgenus D (types 43–47). *J Infect Dis* 158:804–813.

Ideguchi M, Kajiwara K, Yoshikawa K, Uchida T, Ito H (2000): Local adenovirus-mediated CTLA4-immunoglobulin expression suppresses the immune responses to adenovirus vectors in the brain. *Neuroscience* 95:217–226.

Jolly DJ (1994): Viral vector systems for gene therapy. *Cancer Gene Ther* 1:51–64.

Jooss K, Yang Y, Wilson JM (1996): Cyclophosphamide diminishes inflammation and prolongs transgene expression following delivery of adenoviral vectors to mouse liver and lung. *Hum Gene Ther* 7:1555–1566.

Jooss K, Turka LA, Wilson JM (1998): Blunting of immune responses to adenoviral vectors in mouse liver and lung with CTLA4Ig. *Gene Ther* 5:309–319.

Kafri T, Morgan D, Krahl T, Sarvetnick N, Sherman L, Verma IM (1998): Cellular immune response to adenoviral vector infected cells does not require de novo viral gene expression: Implications for gene therapy. *Proc Natl Acad Sci USA* 95:11377–11382.

Kasono K, Blackwell JL, Douglas JT, Dmitriev I, Strong TV, Reynolds P, Kropf DA, Carroll WR, Peters GE, Bucy RP, Curiel DT, Krasnykh V (1999): Selective gene delivery to head and neck cancer cells via an integrin targeted adenoviral vector. *Clin Cancer Res* 5:2571–2579.

Kass-Eisler A, Falck-Pedersen E, Elfenbein DH, Alvira M, Buttrick PM Leinwand LA (1994): The impact of developmental stage, route of administration and the immune system on adenovirus-mediated gene transfer. *Gene Ther* 1:395–402.

Kay MA, Meuse L, Gown AM, Linsley P, Hollenbaugh D, Aruffo A, Ochs HD, Wilson CB (1997): Transient immunomodulation with CD40-ligand antibody and CTLA4Ig enhances persistance and secondary adenovirus-mediated gene transfer into mouse liver. *Proc Natl Acad Sci USA* 94:4686–4691.

Khatri A, Xu ZZ, Both GW (1997): Gene expression by atypical recombinant ovine adenovirus vectors during abortive infection of human and animal cells in vitro. *Virology* 239:226–237.

Klonjkowski B, Gilardi-Hebenstreit P, Hadchouel J, Randrianarison V, Boutin S, Yeh P, Perricaudet M, Kremer EJ (1997): A recombinant E1-deleted canine adenoviral vector capable of transduction and expression of a transgene in human-derived cells and in vivo. *Hum Gene Ther* 8:2103–2115.

Kochanek S, Clemens PR, Mitani K, Chen HH, Chan S, Caskey CT (1996): A new adenoviral vector: Replacement of all viral coding sequences with 28 kb of DNA independently expressing both full-length dystrophin and beta-galactosidase. *Proc Natl Acad Sci USA* 93:5731–5736.

Krasnykh V, Dmitriev I, Mikheeva G, Miller CR, Belousova N, Curiel DT (1998): Characterization of an adenovirus vector containing a heterologous peptide epitope in the HI loop of the fiber knob. *J Virol* 72:1844–1852.

Kremer EJ, Boutin S, Chillon M, Danos O (2000): Canine adenovirus vectors: an alternative for adenovirus-mediated gene transfer. *J Virol* 74:505–512.

Kuriyama S, Tominaga K, Kikukawa M, Nakatani T, Tsujinoue H, Yamazaki M, Nagao S, Toyokawa Y, Mitoro A, Fukui H (1998): Inhibitory effects of human sera on adenovirus-mediated gene transfer into rat liver. *Anticancer Res* 18:2345–2351.

Lakatos B, Farkas J, Adam E, Dobay O, Jeney C, Nasz I, Ongradi J (2000): Serological evidence of adenovirus infection in cats. *Arch Virol* 145:1029–1033.

Lapointe JM, Woods LW, Lehmkuhl HD, Keel MK, Rossitto PV, Swift PK, and Maclachlan NJ (2000): Serologic detection of adenoviral hemorrhagic disease in black-tailed deer in California. *J Wildl Dis* 36:374–377.

Lawrence MS, Foellmer HG, Elsworth JD, Kim JH, Leranth C, Kozlowski DA, Bothwell AL, Davidson BL, Bohn MC, Redmond RD (1999): Inflammatory responses and their impact on beta-galactosidase transgene expression following adenovirus vector delivery to the primate caudate nucleus. *Gene Ther* 6:1349–1351.

Leissner P, Legrand V, Schlesinger Y, Ali-Hadji D, van Raaij M, Cusak S, Pavirani A, Mehtali M (2001). Influence of adenoviral fiber mutations on viral encapsidation, infectivity and in vivo tropism. *Gene Ther* 8:49–55.

Li Q, Kay MA, Finegold M, Stratford-Perricaudet LD, Woo SL (1993): Assessment of recombinant adenoviral vectors for hepatic gene therapy. *Hum Gene Ther* 4:403–409.

Lieber A, He CY, Kirillova I, Kay MA (1996): Recombinant adenoviruses with large deletions generated by Cre mediated excision exhibit different biological properties compared with first-generation vectors in vitro and in vivo. *J Virol* 70:8944–8960.

Lochmuller H, Petrof BJ, Allen C, Prescott S, Massie B, Karpati G (1995): Immunosuppression by FK506 markedly prolongs expression of adenovirus-delivered transgene in skeletal muscles of adult dystrophic [mdx] mice. *Biochem Biophys Res Commun* 213:569–574.

Lusky M, Christ M, Rittner K, Dieterle A, Dreyer D, Mourot B, Schultz H, Stoeckel F, Pavirani A, Mehtali M. (1998): In vitro and in vivo biology of recombinant adenovirus vectors with E1, E1/E2A, or E1/E4 deleted. *J Virol* 72:2022–2032.

Mathias P, Wickham T, Moore M, Nemerow G (1994): Multiple adenovirus serotypes use alpha v integrins for infection. *J Virol* 68:6811–6814.

Mautner V, Steinthorsdottir V, Bailey A (1995): Enteric adenoviruses. In: *The Molecular Reportoire of Adenoviruses* III, Doerfler, W., Bohm, P. (eds). Berlin-Heidelberg: Springer-Verlag.

McCoy RD, Davidson BL, Roessler BJ, Huffnagle GB, Janich SL, Laing TJ, Simon RH (1995): Pulmonary inflammation induced by incomplete or inactivated adenoviral particles. *Hum Gene Ther* 6:1553–1560.

Michou AI, Santoro L, Christ M, Julliard V, Pavirani A, Mehtali M (1997): Adenovirus-mediated gene transfer: influence of transgene, mouse strain and type of immune response on persistence of transgene expression. *Gene Ther* 4:473–482.

Michou AI, Lehrmann H, Saltik M, Cotton M (1999): Mutational analysis of the avian adenovirus CELO, which provides a basis for gene delivery vectors. *J Virol* 73:1399–1410.

Miyazawa N, Crystal RG, Leopold PL (2001): Adenovirus serotype 7 retention in a late endosomal compartment prior to cytosol escape is modulated by fiber protein. *J Virol* 75:1387–1400.

Miyazawa N, Leopold PL, Hackett NR, Ferris B, Worgall S, Falck-Pedersen E, Crystal RG (1999): Fiber swap between adenovirus subgroups B and C alters intracellular trafficking of adenovirus gene transfer vectors. *J Virol* 73:6056–6065.

Molinier-Frenkel V, Gahery-Segard H, Mehtali M, Leboulaire C, Ribeault S, Boulanger P, Tursz T, Guillet JG, Farace F (2000). Immune response to recombinant adenovirus in humans: capsid components from viral input are targets for vector-specific cytotoxic T lymphocytes. *J Virol* 74:7678–7682.

Morral N, O'Neal W, Zhou H, Langston C, Beaudet A (1997): Immune responses to reporter proteins and high viral dose limit duration of expression with adenoviral vectors: comparison of E2a wild type and E2a deleted vectors. *Hum Gene Ther* 8:1275–1286.

Morsy MA, Gu M, Motzel S, Zhao J, Lin J, Su Q, Allen H, Franlin L, Parks RJ, Graham FL, Kochanek S, Bett AJ, Caskey CT (1998): An adenoviral vector deleted for all viral coding sequences results in enhanced safety and extended expression of a leptin transgene. *Proc Natl Acad Sci USA* 95:7866–7871.

Muruve DA, Barnes MJ, Stillman IE, Libermann TA (1999): Adenoviral gene therapy leads to rapid induction of multiple chemokines and acute neutrophil-dependent hepatic injury in vivo. *Hum Gene Ther* 10:965–976.

Nishikawa M, Huang L (2001): Nonviral vectors in the new millennium: delivery barriers in gene transfer. *Hum Gene Ther* 12:861–870.

O'Riordan CR, Lachapelle A, Delgado C, Parkes V, Wadsworth SC, Smith AE, Francis GE (1999): PEGylation of adenovirus with retention of infectivity and protection from neutralizing antibody in vitro and in vivo. *Hum Gene Ther* 10:2575–2576.

Parks RJ, Chen L, Anton M, Sankar U, Rudnicki MA, Graham FL (1996): A helper-dependent adenovirus vector system: removal of helper virus by Cre-mediated excision of the viral packaging signal. *Proc Natl Acad Sci USA* 93:13565–13570.

Pearson AS, Koch PE, Atkinson N, Xiong M, Finberg RW, Roth JA, Fang B (1999): Factors limiting adenovirus-mediated gene transfer into human lung and pancreatic cancer cell lines. *Clin Cancer Res* 5:4208–4213.

Philipson L, Lonberg-Holm K, Pettersson U (1968): Virus-receptor interaction in the adenovirus system. *J Virol* 2:1064–1070.

Pring-Akerblom P, Adrian T (1993): The hexon genes of adenoviruses of subgenus C: comparison of the variable regions. *Res Virol* 144:117–127.

Pring-Akerblom P, Trijssenaar FE, Adrian T (1995): Sequence characterization and comparison of human adenovirus subgenus B and E hexons. *Virology* 212:232–236.

Rahman A, Tsai V, Ramachandra M, Maneval D, Laface D, Shabram P (2000): Immunodepletion of adenoviral antibodies to increase transduction efficiency in a mouse model. *Keystone Symposia: Gene Therapy: The Next Millenium*, Colorado, USA abstract 425:97.

Raper SE, Haskal ZJ, Ye X, Pugh C, Furth EE, Gao GP, Wilson JM (1998): Selective gene transfer into the liver of non-human primates with E1-deleted, E2A-defective, or E1-E4 deleted recombinant adenoviruses. *Hum Gene Ther* 9:671–679.

Rasmussen UB, Benchaibi M, Meyer V, Schlesinger Y, Schughart K (1999): Novel human gene transfer vectors: evaluation of wild-type and recombinant animal adenoviruses in human-derived cells. *Hum Gene Ther* 10:3587–3599.

Rea D, Schagen FH, Hoeben RC, Mehtali M, Havenga MJ, Toes RE, Melief CJ, Offringa R (1999): Adenoviruses activate human dendritic cells without polarization toward a T-helper type 1-inducing subset. *J Virol* 73:10245–10253.

Rea D, Havenga MJ, van Den Assem M, Sutmuller RP, Lemckert A, Hoeben RC, Bout A, Melief CJ, Offringa R (2001): Highly efficient transduction of human monocyte-derived dendritic cells with subgroup B fiber-modified adenovirus vectors enhances transgene-encoded antigen presentation to cytotoxic T cells. *J Immunol* 166:5236–5244.

Reddy PS, Idamakanti N, Hyun BH, Tikoo SK, Babiuk LA (1999): Development of porcine adenovirus-3 as an expression vector. *J Gen Virol* 80:563–570.

Roberts MM, White JL, Grutter MG, Burnett RM (1986): Three-dimensional structure of the adenovirus major coat protein hexon. *Science* 232:1148–1151.

Roelvink PW, Kovesdi I, Wickham TJ (1996): Comparative analysis of adenovirus fiber-cell interaction: adenovirus type 2 (Ad2) and Ad9 utilize the same cellular fiber receptor but use different binding strategies for attachment. *J Virol* 70:7614–7621.

Roelvink PW, Lizonova A, Lee JG, Li Y, Bergelson JM, Finberg RW, Brough DE, Kovesdi I, Wickham TJ (1998): The coxsackievirus-adenovirus receptor protein can function as a cellular attachment protein for adenovirus serotypes from subgroups A, C, D, E, and F. *J Virol* 72:7909–7915.

Roy S, Shirley PS, McClelland A, Kaleko M (1998): Circumvention of immunity to the adenovirus major coat protein hexon. *J Virol* 72:6875–6879.

Russi TJ, Hirschowitz EA, Crystal RG (1997): Delayed-type hypersensitivity response to high doses of adenoviral vectors. *Hum Gene Ther* 8:323–330.

Rux JJ, Burnet RM (2000): Type-specific epitope locations revealed by X-ray crystallographic study of adenovirus type 5 hexon. *Mol Ther* 1:18–30.

Sandig V, Youil R, Bett AJ, Franlin LL, Oshima M, Maione D, Wang F, Metzker ML, Savino R, Caskey CT (2000): Optimization of the helper-dependent adenovirus system for production and potency in vivo. *Proc Natl Acad Sci USA* 97:1002–1007.

Schnell MA, Zhang Y, Tazelaar J, Gao GP, Yu QC, Qian R, Chen SJ, Varnavski AN, LeClair C, Raper SE, Wilson JM (2001): Activation of innate immunity in nonhuman primates following intraportal administration of adenoviral vectors. *Mol Ther* 3:708–722.

Schulick AH, Vassalli G, Dunn PF, Dong G, Rade JJ, Zamarron C, Dichek DA (1997): Established immunity precludes adenovirus-mediated gene transfer in rat carotid arteries. Potential for immunosuppression and vector engineering to overcome barriers of immunity. *J Clin Invest* 99:209–219.

Shayakhmetov DM, Papayannopoulou T, Stamatoyannopoulos G, Lieber A (2000): Efficient gene transfer into human CD34(+) cells by a retargeted adenovirus vector. *J Virol* 74:2567–2583.

Siegal FP, Dikman SH, Arayata RB, Bottone EJ (1981): Fatal disseminated adenovirus 11 pneumonia in an agammaglobulinemic patient. *Am J Med* 71:1062–1067.

Smith TA, Mehaffey MG, Kayda DB, Saunders JM, Yei S, Trapnell BC, McClelland A, Kaleko M (1993): Adenovirus mediated expression of therapeutic plasma levels of human factor IX in mice. *Nat Genet* 5:397–402.

Smith TA, White BD, Gardner JM, Kaleko M, Mcclelland A (1996): Transient immunosuppression permits successful repetitive intravenous administration of an adenovirus vector. *Gene Ther* 3:496–502.

Sorden SD, Woods LW, Lehmkuhl HD (2000): Fatal pulmonary edema in white-tailed deer (Odocoileus virginianus) associated with adenovirus infection. *J Vet Diagn Invest* 12:378–380.

Stein CS, Pemberton JL, van Rooijen N, Davidson BL (1998): Effects of macrophage depletion and anti-CD40 ligand on transgene expression and redosing with recombinant adenovirus. *Gene Ther* 5:431–439.

Sterman DH, Molnar-Kimber K, Iyengar T, Chang M, Lanuti M, Amin KM, Pierce BK, Kang E, Treat J, Recio A, Litzky L, Wilson JM, Kaiser LR, Albelda SM (2000): A pilot study of systemic corticosteroid administration in conjunction with intrapleural adenoviral vector administration in patients with malignant pleural mesothelioma. *Cancer Gene Ther* 7:1511–1518.

Sterner G (1962): Adenovirus infection in childhood: an epidemiological and clinical survey among Swedish children. *Acta Paediatr Scand* 142:1–30.

Stevenson SC, Rollence M, White B, Weaver L, McClelland A (1995): Human adenovirus serotypes 3 and 5 bind to two different cellular receptors via the fiber head domain. *J Virol* 69:2850–2857.

Stevenson SC, Rollence M, Marshall-Neff J, McClelland A (1997): Selective targeting of human cells by a chimeric adenovirus vector containing a modified fiber protein. *J Virol* 71:4782–4790.

Sung RS, Qin L, Bromberg JS (2001): TNFalpha and IFNgamma induced by innate anti-adenoviral immune responses inhibit adenovirus-mediated transgene expression. *Mol Ther* 3:757–767.

Tai FH, Graystorm JT (1962): Adenovirus neutralizing antibodies in persons on Taiwan. *Proc Soc Exp Biol Med* 109:881–884.

Tan PK, Michou AI, Bergelson JM, Cotton M (2001): Defining CAR as a cellular receptor for the avian adenovirus CELO using a genetic analysis of the two viral fibre proteins. *J Gen Virol* 82:1465–1472.

Toogood CI, Crompton J, Hay RT (1992): Antipeptide antisera define neutralizing epitopes on the adenovirus hexon. *J Gen Virol* 73:1429–1435.

Tripathy SK, Black HB, Goldwasser E, Leiden JM (1996): Immune responses to transgene-encoded proteins limit the stability of gene expression after injection of replication-defective adenovirus vectors. *Nat Med* 2:545–550.

Verma IM, Somia N (1997): Gene therapy-promises, problems and prospects. *Nature* 389:239–242.

Vigne E, Mahfouz I, Dedieu JF, Brie A, Perricaudet M, Yeh P (1999): RGD inclusion in the hexon monomer provides adenovirus type 5-based vectors with a fiber knob-independent pathway for infection. *J Virol* 73:5156–5161.

Wadell G (1984): Molecular epidemiology of human adenoviruses. *Curr Top Microbiol Immunol* 110:191–220.

Wickham TJ, Mathias P, Cheresh DA, Nemerow GR (1993): Integrins alpha v beta 3 and alpha v beta 5 promote adenovirus internalization but not virus attachment. *Cell* 73:309–319.

Wickham TJ, Carrion ME, Kovesdi I (1995): Targeting of adenovirus penton base to new receptors through replacement of its RGD motif with other receptor-specific peptide motifs. *Gene Ther* 2:750–756.

Wickham TJ, Segal DM, Roelvink PW, Carrion ME, Lizonova A, Lee GM, Kovesdi I (1996): Targeted adenovirus gene transfer to endothelial and smooth muscle cells by using bispecific antibodies. *J Virol* 70:6831–6838.

Wigand R, Mauss M, Adrian T (1989): Chimpanzee adenoviruses are related to four subgenera of human adenoviruses. *Intervirology* 30:1–9.

Wohlfart C (1988): Neutralization of adenoviruses: kinetics, stoichiometry, and mechanisms. *J Virol* 62:2321–2328.

Worgall S, Leopold PL, Wolff G, Ferris B, Van Roijen N, Crystal RG (1997): Role of alveolar macrophages in rapid elimination of adenovirus vectors administered to the epithelial surface of the respiratory tract. *Hum Gene Ther* 8:1675–1684.

Yang Y, Wilson JM (1995): Clearance of adenovirus-infected hepatocytes by MHC class I-restricted CD4+ CTLs in vivo. *J Immunol* 155:2564–2570.

Yang Y, Ertl HC, Wilson JM (1994): MHC class I-restricted cytotoxic T lymphocytes to viral antigens destroy hepatocytes in mice infected with E1-deleted recombinant adenoviruses. *Immunity* 1:433–442.

Yang Y, Greenough K, Wilson JM (1996): Transient immune blockade prevents formation of neutralizing antibody to recombinant adenovirus and allows repeated gene transfer to mouse liver. *Gene Ther* 3:412–420.

Yang Y, Li Q, Ertl HC, Wilson JM (1995): Cellular and humoral immune responses to viral antigens create barriers to lung-directed gene therapy with recombinant adenoviruses. *J Virol* 69:2004–2015.

Yang Y, Su Q, Grewal IS, Schilz R, Flavell RA, Wilson JM (1996): Transient subversion of CD40 ligand function diminishes immune responses to adenovirus vectors in mouse liver and lung tissues. *J Virol* 70:6370–6377.

Yei S, Mittereder N, Tang K, O'Sullivan C, Trapnell BC (1994): Adenovirus-mediated gene transfer for cystic fibrosis: quantitative evaluation of repeated in vivo vector administration to the lung. *Gene Ther* 1:192–200.

Yotnda P, Onishi H, Heslop HE, Shayakhmetov D, Lieber A, Brenner M, Davis A (2001): Efficient infection of primitive hematopoietic stem cells by modified adenovirus. *Gene Ther* 8:930–937.

Zabner J, Freimuth P, Puga A, Fabrega A, Welsh MJ (1997): Lack of high affinity fiber receptor activity explains the resistance of ciliated airway epithelia to adenovirus infection. *J Clin Invest* 100:1144–1149.

Zabner J, Chillon M, Grunst T, Moninger TO, Davidson BL, Gregory R, Armentano D (1999): A chimeric type 2 adenovirus vector with a type 17 fiber enhances gene transfer to human airway epithelia. *J Virol* 73:8689–8695.

Zakhartchouk AN, Reddy PS, Baxi M, Baca-Estrada ME, Mehtali M, Babiuk LA, Tikoo SK (1998): Construction and characterization of E3-deleted bovine adenovirus type 3 expressing full-length and truncated form of bovine herpesvirus type 1 glycoprotein gD. *Virology* 250:220–229.

Zhang Y, Chirmule N, Gao GP, Qian R, Croyle M, Joshi B, Tazelaar J, Wilson JM (2001): Acute cytokine response to systemic adenoviral vectors in mice is mediated by dendritic cells and macrophages. *Mol Ther* 3:697–707.

Zhu J, Grace M, Casale J, Chang AT, Musco ML, Bordens R, Greenberg R, Schaefer E, Indelicato SR (1999): Characterization of replication-competent adenovirus isolates from large-scale production of a recombinant adenoviral vector. *Hum Gene Ther* 10:113–121.

Zsengeller ZK, Wert SE, Hull WM, Hu X, Yei S, Trapnell BC, Whitsett JA (1995): Persistence of replication-deficient adenovirus-mediated gene transfer in lungs of immune-deficient (nu/nu) mice. *Hum Gene Ther* 6:457–467.

6

TARGETING OF ADENOVIRAL GENE THERAPY VECTORS: THE FLEXIBILITY OF CHEMICAL AND MOLECULAR CONJUGATION

HIDDE J. HAISMA, PH.D. AND MARIANNE G. ROTS, PH.D.

INTRODUCTION

Recombinant adenoviral vectors are promising reagents for therapeutic interventions in humans, including gene therapy for biologically complex diseases like cancer and cardiovascular diseases. In this regard, the major advantage of adenoviral vectors is their superior in vivo gene transfer efficiency on a wide spectrum of both dividing and non-dividing cell types. However, this broad tropism also represents an important limitation for their use in therapeutic applications where specific gene transfer is required. In addition, several potential target cells for gene therapy are poorly transduced by adenoviral vectors due to the scarcity of an appropriate cell surface receptor (Nalbantoglu et al., 1999; Wickham et al., 1996a, 1996b; Zabner et al., 1997).

Adenoviral infection is initiated by the recognition of target cells by the C-terminal domain of the fiber protein, termed the *knob*, and the primary cellular receptor, the coxsackie B virus and adenovirus receptor (CAR) (Bergelson

Vector Targeting for Therapeutic Gene Delivery, Edited by David T. Curiel and Joanne T. Douglas
ISBN 0-471-43479-5

et al., 1997). An alternative receptor, the major histocompatibility complex (MHC) class I 2 domain, has also been proposed (Hong et al., 1997), however, recent data indicate a lack of correlation between MHC class I heavy chain expression and susceptibility to Ad infection, confirming an exclusive role of CAR as the primary receptor (Davison et al., 1999; Hong et al., 1997; McDonald et al., 1999). After binding of the knob domain of the fiber, entry of the virus into the cell occurs via interaction of the Arg-Gly-Asp (RGD) sequence located in the viral penton base protein with cellular $\alpha v\beta 3$ and $\alpha v\beta 5$ integrins (Wickham et al., 1993). Many primary tumors express low levels of CAR resulting in low levels of gene delivery into these cancer cells (Cripe et al., 2001; Grill et al., 2001; Li et al., 1999; Miller et al., 1998). For ex vivo applications, such as transduction of dendritic cells for vaccination strategies, it may be sufficient to enhance transduction efficiency, without the need for specificity. In this case, targeting is obtained by isolating the specific cell type, that is, dendritic cells. Similarly, vector targeting can also be achieved by local administration of the gene therapy vector in, for instance, cardiac or tumor tissue.

Selective targeting of adenoviral vectors obviously provides a number of advantages. First, cell type-specific targeting will result in transgene expression only in the cell type of choice, thus permitting systemic administration whereby the virus would be localized to specific cell types, tissues, or organs. Second, the inflammatory and immune responses against the vector may be reduced, as these are thought to derive from uptake into antigen-presenting cells and from virus binding to native receptors, promoting the concomitant stimulation of cytokines. Since infection efficiency is increased by targeting, toxicity from the administration of the viral vector may be decreased, as the dose of virus can be reduced.

Targeting adenoviral vectors toward alternative surface receptors on specific cell types requires both the ablation of native viral tropism and the introduction of a novel binding specificity. Two general strategies are currently being considered to target adenoviral vectors in order to enhance vector infectivity and specificity (Fig. 6.1). In the first approach, adenoviral vectors are genetically modified to alter the binding specificity of the viral capsid thus creating a stable single-reagent genetic medicine (Krasnykh et al., 2000). In the second—physical—approach, a new ligand is directly chemically coupled onto the viral capsid (Romanczuk et al., 1999) or the virus is complexed with bispecific molecules that on one side bind to the viral capsid and on the other side redirect the virus to a novel receptor (Dmitriev et al., 2000; Douglas et al., 1996; Doukas et al., 1999; Haisma et al., 1999, 2000a; Hong et al., 1999; Miller et al., 1998; Watkins et al., 1997b; Wickham et al., 1996b).

The major advantage of the physical retargeting strategy is its flexibility. First, the continuous identification of high-affinity peptide ligands and antibodies with cell specificity vastly increases the number of potential targets for vector targeting. This mandates a rapid evaluation of these targets as potential targets for specific gene delivery. Such evaluations would more easily be

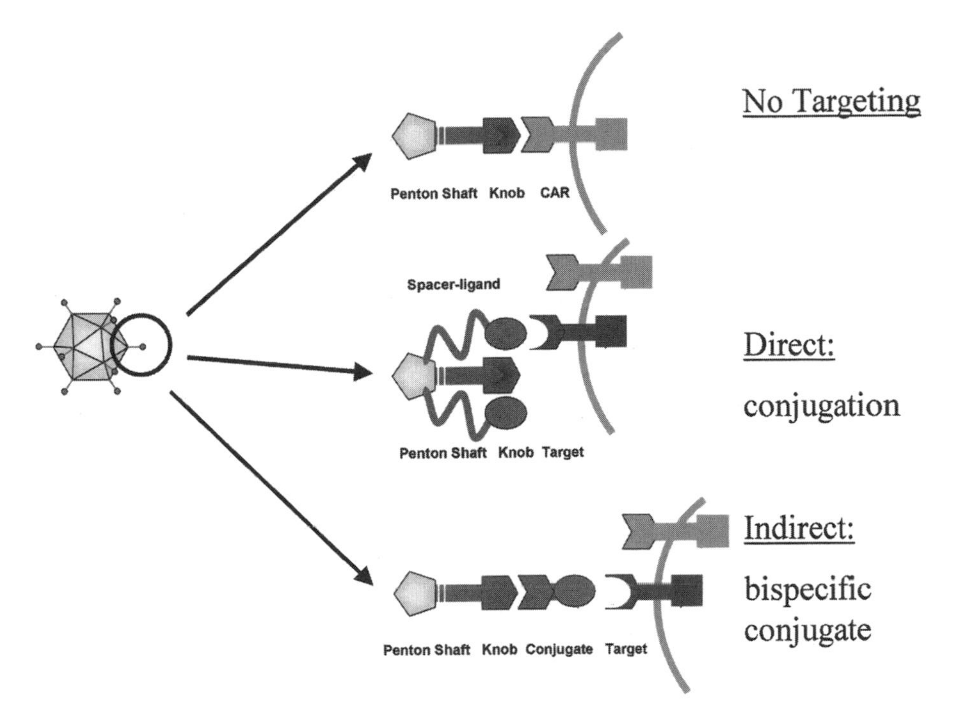

Figure 6.1 Physical approaches toward adenoviral targeting.

accomplished by preparing conjugates with the newly identified molecules than by using a genetic approach in which one of the proteins of the capsid of the adenovirus is modified. The genetic approach has been successful in only a limited number of cases in which small peptides have been inserted in either the HI loop of the fiber knob or in the RGD motif in the penton base (Krasnykh et al., 1998; Wickham, Carrion, and Kovesdi, 1995; Xia et al., 2000). So far, it has not been possible to include single-chain antibodies (scFvs) in a genetically modified adenovirus. Second, the use of conjugates allows the rapid evaluation of adenoviral vectors containing different transgenes, targeted to specific cells using the same conjugates. Third, as adenoviral vectors with genetically modified capsid proteins are being developed to enhance overall infectivity, conjugates may be used to study the effects of targeting with these new reagents.

In this chapter, physical approaches toward adenoviral targeting are reviewed.

DIRECT CHEMICAL CONJUGATION OF ADENOVIRUS

One method to achieve targeting of adenoviral vectors has been by the physical modification of adenovirus. In this regard, studies have been directed toward increased transduction without specificity for a particular cell type. This has mainly been accomplished via cationic liposomes, calcium phosphate precipitates, and polyethylene glycol. Only a few studies have incorporated a targeting ligand using either bifunctional polyethylene glycol or a biotin-avidin system. Alternatively, a number of different groups have reported the use of polymers to produce coated adenoviral particles. These complexes have been used either to target adenovirus, or adenoviral vectors have been incorporated into DNA/polymer complexes to improve cellular transfection efficiency. The use of adenoviral particles to improve the transfection of adipocytes, monocytes, and hepatoma cells has been described (Meunier-Durmort et al., 1996, 1997). In these studies, DNA/polymer complexes were produced using the polymers lipofectamine and polyethylenimine (PEI) followed by addition of adenovirus to the transfecting mixture. Adenoviral particles improved transfection some 140- to 300-fold when compared to PEI alone. The formation of DNA/polymer/Ad complexes using PEI has also been described by Baker et al. (1997). The authors concluded that involvement of the adenovirus receptor and/or integrins was probably responsible for the observed improvements, by promoting endosomal release. Clearly, the formation of complexes between cationic polymers and adenovirus does not impart any specific targeting ability to the virus. In the cases just described, cell type-specific promoters were used to achieve tissue-specific gene expression. For example, Meunier-Durmort (Meunier-Durmort et al., 1996, 1997) used the phosphoenolpyruvate carboxykinase promoter fused to the CAT gene which was shown to be cAMP-responsive in adipocytes using this system.

Cross-linking complexes between cationic polymers and adenoviral vectors have also been used to improve transfection of airway epithelial cells in cystic fibrosis (CF) gene therapy (Fasbender et al., 1997). The formation of DNA/Ad complexes was reported with a variety of different cationic polymers and lipids. Using primary cultures of human airway epithelial cells, the complexes were shown to correct the electrophysiological abnormalities that characterize CF epithelia more efficiently than adenovirus alone. Once again, no element of direct targeting is conferred by the complex formation with polymers, but targeting is achieved simply by local administration into the airways.

Besides improving transfection efficiency, the use of polymers to coat virus particles has two distinct advantages. First, it may be used to mask capsid proteins and thereby prevent interaction with CAR and/or integrins as well as the stimulation of a humoral immune response. Second, coating with an activated polymer provides a mechanism by which novel targeting ligands may be attached to the virus, thus enabling its redirection.

O'Riordan and colleagues described the use of activated polyethylene glycol (PEG), which was covalently linked to the surface of the adenovirus via reaction with surface-exposed lysine residues (O'Riordan et al., 1999). Results indicate that PEGylation is sufficient to protect adenovirus particles from the effects of neutralizing antibodies in the lungs of mice with high antibody titers to adenovirus (O'Riordan et al., 1999). Coupling of adenoviral particles to bifunctional PEG allowed the introduction of a peptide derived from urokinase plasminogen activator to target the apical membrane of airway epithelium (Drapkin et al., 2000), resulting in enhanced gene transfer. A novel method of redirecting adenoviral particles is the use of targeting peptides derived from a phage display library. The peptide with the most effective binding to human airway epithelial cells was coupled to the surface of an adenovirus using bifunctional PEG molecules (Romanczuk et al., 1999). The chemically modified adenoviral vector was able to effect gene transfer to well-differentiated human airway epithelial cells by an alternative pathway dependent on the incorporated peptide. Incorporation of the cystic fibrosis transmembrane conductance regulator (CFTR) gene in a similarly modified vector resulted in correction of defective chloride ion transport in well-differentiated epithelial cultures established from human CF donors. In addition, the presence of PEG molecules on the surface of the virus reduced antibody neutralization in vitro (Romanczuk et al., 1999).

Another approach uses the high affinity biotin-avidin system to expand the tropism of adenoviral vectors (Smith et al., 1999). Photoactivable Biotin was employed to attach the targeting ligand stem cell factor (SCF) via an avidin bridge to the capsid of the adenoviral vector. Gene transfer was directed specifically to hematopoietic cell lines, resulting in up to a 2,440-fold increase in luciferase expression with efficiencies equivalent to recombinant virus infection of permissive cells. In addition, the flexibility of this approach was demonstrated by substitution of other biotinylated antibodies directed against CD44 and IL-2.

INDIRECT CONJUGATION OF ADENOVIRUS

An alternative to the direct chemical modification of adenovirus is the use of bispecific molecules, which contain a first specificity for the virus whereas the second specificity redirects the virus to a novel receptor. In this regard, the virus-binding molecule can be an antibody or peptide directed against a capsid protein or a soluble form of the adenovirus receptor CAR. It is important to note that the virus-binding ligand should preferably be monovalent to circumvent the problem of viral clustering, which may occur with bivalent molecules. Also, in vivo trapping in the reticuloendothelial system through the Fc domain of whole immunoglobulins (Igs) should preferably be avoided if systemic administration is considered.

The receptor ligand can be an antibody, peptide or other receptor-binding molecule. The bispecific molecule can be made by chemical cross-linking or by expressing a recombinant molecule containing both binding specificities. Chemical cross-linking has the advantage that, if both ligands are available in a relatively pure form, conjugates can be prepared with relative ease according to standardized techniques, allowing the rapid exchange of ligands. A disadvantage of chemical conjugates is the heterogeneity of the end product, which usually contains conjugates of different sizes that need to be purified in order to obtain a reasonably pure product free of aggregates and unreacted reagents.

Chemical Antibody–Ligand Conjugates

Cell surface receptors with known ligands were among the first targets to be used for redirecting adenovirus. The high affinity folate receptor is highly expressed on several malignant cell lines. A conjugate prepared from folate and a Fab fragment of a neutralizing antiknob monoclonal antibody was used by Douglas et al. (1996). The conjugate was complexed with adenovirus and shown to redirect infection to target cells. The conjugate mediated specific killing of folate receptor-positive cells when an adenoviral vector containing the suicide gene thymidine kinase was employed in combination with the prodrug ganciclovir. This study showed, for the first time, that adenovirus could be specifically redirected toward an alternative cellular receptor with retention of gene transduction capacity.

The same group reported the use of another ligand, basic fibroblast growth factor, FGF2. A bifunctional conjugate consisting of the blocking antiadenoviral knob Fab linked to FGF2 enhanced the transduction of Kaposi's sarcoma cells 7.7–44 fold. In this regard, two Kaposi's sarcoma cell lines that were previously refractory to native adenoviral transduction could be successfully transduced by the addition of the conjugate (Goldman et al., 1997). Studies aimed at understanding the mechanism of enhanced gene transduction by FGF2-targeted adenovirus have shown that FGF2 retargeting results in increased adenovirus entry. Nuclear delivery is also increased, but to a level that is directly proportional to viral entry. In addition, after entry, the retargeted particle rapidly localizes to the nucleus in a time frame similar to that of adenovirus alone. Through unknown mechanisms, transgene expression is always enhanced with FGF2-mediated delivery, whether overall transduction of the cell population is increased, equivalent, or decreased relative to nontargeted adenoviral vectors (Hoganson et al., 2001).

Several other ligands have been coupled to antibodies directed against adenovirus to produce bispecific targeting molecules. The Hc fragment of tetanus toxin has specific nerve cell binding and transport properties, but lacks any toxicity. The Fab fragment of the neutralizing antiknob antibody covalently bound to Hc was attached to an adenovirus. Infection of neuronal and non-neuronal cell lines with this retargeted virus showed highly increased neuronal cell selectivity but no significant enhancement of gene delivery into these cells. Intramuscular injection of retargeted virus into mouse tongues resulted in selective gene transfer to the neurons of the hypoglossal nucleus, where no pathological changes were observed (Schneider et al., 2000).

A lung-homing peptide was identified by in vivo phage display and chemically linked to a neutralizing antifiber knob antibody (Trepel et al., 2000). Cells expressing the receptor for the peptide that were previously refractory to adenoviral infection were sensitized to the Ad5 vector in the presence of the peptide conjugate. Interestingly, direct chemical conjugation of the peptide onto the virus capsid abolished the infectivity of the adenovirus.

Chemical Bispecific Antibody Conjugates

Many cell-specific cell surface proteins have been defined by monoclonal antibodies. The abundance of readily available high-affinity antibodies to virtually any cell type is one of the major advantages of this approach. These antibodies are versatile reagents that have been employed in adenoviral retargeting strategies.

The bispecific antibody (bsAb) conjugate approach has been used with viral attachment directed against different viral proteins. In early studies, Wickham (Wickham et al., 1996b, 1997) employed a bispecific antibody with specificities for a FLAG epitope genetically introduced into the penton base of Ad5 (Ad.FLAG) and for α integrins or human CD3 to redirect the virus to endothelial and smooth muscle cells or human T cells, respectively. Complexing Ad.FLAG with the bsAb directed against α v integrins increased the transduction of human venule endothelial cells and human intestinal smooth muscle cells by 7- to 9-fold compared with transduction by Ad.FLAG alone. Similarly, the anti-FLAG x anti-CD3 bsAb increased Ad.FLAG binding 30-fold, induced the efficient uptake of Ad.FLAG into the cells, and led to a 100- to 500-fold increase in the transduction of resting T cells. Moreover, 25–90% of the T cells were transduced by the bsAb-complexed Ad.FLAG at multiplicities of infection between 20 and 100 active particles per cell compared to 10% for Ad.FLAG alone. The same approach was used for targeting E-selectin, a surface adhesion molecule that is only expressed by activated endothelial cells. An anti-E-selectin mAb, 1.2B6, was complexed with the adenovirus vector AdLacZ.FLAG by conjugating it to an anti-FLAG mAb. Gene transduction of cultured endothelial cells was increased 20-fold compared with AdZ.FLAG complexed with a control bispecific antibody providing endothelial cells were activated by cytokines. The anti-E-selectin-complexed vector transduced 29 +/– 9% of intimal endothelial cells in segments of pig aorta cultured with cytokines ex vivo, compared with less than 0.1% transduced with the control construct.

These studies showed that bispecific antibody conjugates could target adenovirus to alternate receptors, but did not address the issue of specificity, as in these studies native adenovirus binding through CAR was not blocked. This issue was addressed in a study by Miller et al. (1998), who used a neutralizing antiknob antibody to block native CAR binding of the adenovirus. A bispecific antibody conjugate was prepared with this antibody and an antibody directed against the epidermal growth factor receptor (EGFR), a (brain) tumor-associated marker negligibly expressed in normal, mitotically quiescent neural tissues. The results demonstrate that EGFR-targeted adenovirus gene transfer was EGFR-specific and independent of fiber–CAR interactions. Furthermore, EGFR, targeting significantly enhanced adenovirus gene delivery to 7 of 12 established glioma cell lines and to 6 of 8 cultured primary gliomas. Interestingly, EGFR-targeted adenovirus gene transfer enhancement did not correlate with EGFR expression across cell lines, suggesting the importance of other factors. In a subsequent study by Blackwell et al., retargeting of Ad to EGFR via this bispecific conjugate enhanced the selectivity of Ad infection for squamous cell carcinoma of the head and neck cells relative to normal tissue from the same patient (Blackwell et al., 1999).

Similarly, Haisma (Haisma et al., 1999) and Heideman (Heideman et al., 2001) used this neutralizing antifiber antibody conjugated to an antibody against the epithelial cell adhesion molecule (EpCAM) to target the adenovirus to the EpCAM antigen present on tumor cells. The EpCAM antigen was chosen as the target because this antigen is highly expressed on a variety of adenocarcinomas of different origin such as stomach, esophagus, breast, ovary, colon, and lung, whereas EpCAM expression is limited in normal tissues. In these studies, the EpCAM-targeted adenovirus was shown to infect specifically cancer cell lines of different origin expressing EpCAM. Interestingly, the EpCAM molecule is a noninternalizing cell surface protein. Gene transfer was dramatically reduced in EpCAM-negative cell lines, thus showing the specificity of the EpCAM-targeted adenovirus. Again, the infection with targeted adenovirus was shown to be independent of CAR. The bispecific antibody conjugate could also successfully mediate gene transfer to primary human colon cancer cells, whereas it almost completely abolished infection of liver cells.

Another marker explored for targeting adenovirus is TAG-72, which is overexpressed on most ovarian cancers (Kelly et al., 2000). A monoclonal antibody (Mab) that has been investigated clinically for immunotherapy and immunodetection of ovarian carcinomas, namely CC49, was used to construct a bispecific conjugate with the Fab fragment of a neutralizing antiknob Mab to target adenovirus binding via TAG-72. This conjugate facilitated TAG-72-specific, CAR-independent adenovirus reporter gene transfer to both ovarian cancer cell lines and primary ovarian cancer cells cultured from malignant ascites fluid. Fab-CC49 was selective for tumor cells, augmenting adenovirus gene transfer to primary ovarian cancer cells while decreasing gene transfer to autologous cultured mesothelial cells.

The same neutralizing antiknob antibody was used to generate a CD40-directed bispecific antibody for transduction of dendritic cells, to be used in a cancer vaccination strategy. Adenovirus targeted to CD40 demonstrated dramatic improvements in gene transfer towards dendritic cells (DCs) relative to untargeted adenovirus vectors. Importantly, this efficient gene transfer was accompanied by the maturation of the DCs and an enhanced allostimulatory capacity in a mixed lymphocyte reaction (Tillman et al., 1999), and resulted in an enhancement in the efficacy of DC-based vaccination against human papilloma virus 16-induced tumor cells in a murine model (Tillman et al., 2000).

Molecular Conjugates

The expression of recombinant bispecific fusion proteins in bacteria, yeast, or mammalian cells provides a simple and efficient way to produce a targeting moiety. It allows the production of bispecific molecules in a single step, without the need for chemical linkage of two components, which results in heterogeneous conjugates. It is amenable to upscaling to clinical grade production.

Antibody-Ligand Fusion Proteins

The ability to engineer recombinant Abs has facilitated the production of bispecific Abs or fusion proteins. This has resulted in the construction of an "adenobody" reported by Watkins (Watkins et al., 1997). A virus neutralizing single chain Fv (scFv) antibody fragment designated S11e was isolated from a phage library and a C-terminal fusion protein with epidermal growth factor (EGF) was constructed. This fusion protein, or adenobody, bound both to the fiber protein of the adenovirus and to EGFR on human cells, and was able to direct adenoviral binding to the new receptor. Using this system, the efficiency of viral infection was markedly enhanced and was targeted to the EGFR.

Taking this approach one step further, a peptide-antibody fusion protein was constructed (Nicklin et al., 2000). Phage display was used to isolate peptides that bind selectively and efficiently to quiescent human umbilical vein endothelial cells (HUVECs) with reduced or negligible binding to nonendothelial cells, including vascular smooth muscle cells and hepatocytes. The peptide SIGYPLP or the positive control peptide KKKKKKK (polylysine) were cloned upstream of the S11e scFv directed against the knob domain of the adenovirus to create fusion proteins. Adenovirus-mediated gene transfer via fiber-dependent infection was blocked with S11e, whereas inclusion of the KKKKKKK peptide retargeted gene transfer. The peptide SIGYPLP, however, retargeted gene delivery specifically to endothelial cells with a significantly enhanced efficiency over nontargeted adenovirus and without transduction of nontarget cells in vitro.

An antibody directed against the adenovirus penton base was fused to tumor necrosis factor alpha (TNF α), insulin-like growth factor (IGF)-1 or EGF by Li (Li et al., 2000b). Adenoviral vectors complexed with these bifunctional antibodies increased gene delivery 10- to 50-fold to human melanoma cells lacking alpha v integrins. The bifunctional antibodies also enhanced gene delivery by fiberless adenoviral particles, which cannot bind to CAR. Improved gene delivery correlated with increased virus internalization and attachment.

sCAR Fusion Proteins

An alternative to using antibodies for viral attachment is the use of a soluble form of the extracellular domain of the cellular receptor CAR, sCAR. The ectodomain of CAR was genetically fused to human EGF and expressed in insect cells. The sCAR-EGF protein was capable of binding to adenoviral vectors and directing them to EGFR, thereby achieving targeted delivery of reporter genes (Dmitriev et al., 2000). This construct was used by Wesseling et al. (2001) who showed that sCAR-EGF redirects adenovirus to the EGFR, leading to an enhanced in vitro gene transfer efficiency to pancreatic carcinoma cells, which were otherwise resistant to adenoviral gene transfer. Similarly, Ebbinghaus et al. (2001) produced a bispecific hybrid adapter protein consisting of the amino-terminal extracellular domain of the human CAR protein (CARex) and the Fc region of the human immunoglobulin. CARex-Fc was purified from COS7 cell supernatants and mixed with adenoviral particles, thus blocking adenovirus infection of CAR-positive but Fc receptor-negative cells. The functionality of the CARex domain was further confirmed by successful immunization of mice with CARex-Fc followed by selection of a monoclonal antihuman CAR antibody, which blocked adenovirus infection of CAR-positive cells. When mixed with adenovirus expressing green fluorescent protein, CARex-Fc mediated an up to 250-fold increase of transgene expression in CAR-negative human monocytic cell lines expressing the high-affinity Fcgamma receptor I (CD64) but not in cells expressing the low-affinity Fcgamma receptor II (CD32) or III (CD16).

An interesting new approach was used by Hemminki et al. (2001) who constructed an adenovirus expressing an sCAR-EGF fusion protein. Retargeting adenovirus to EGFR resulted in a more than 150-fold increase in gene transfer. When the virus expressing the retargeting molecule was combined with an oncolytic adenovirus, an increased oncolytic potency in vitro and therapeutic gain in vivo was observed.

Bispecific Antibody Fusion Proteins

Given the versatility of antibodies as reagents for adenoviral retargeting strategies, it comes as no surprise that antibodies directed against adenoviral capsid proteins have been combined with antibodies against a targeting moiety for cell-specific gene delivery. The construction and expression of such bispecific scFv fusion proteins provides a simple and efficient way to produce an immunological targeting moiety.

A construct was made that encodes the neutralizing anti-adenovirus fiber Ab (S11) fused to an scFv Ab (425) directed against EGFR. The fusion protein was produced in an eukaryotic cell line to allow correct folding and disulfide bond formation. The bispecific scFv markedly enhanced the infection efficiency of adenoviral vectors in EGFR-expressing cell lines, such as gliomas (Haisma et al., 2000). Combining targeting of adenoviruses to integrins and EGFRs further increases gene transfer into primary glioma cells and spheroids, as was shown by Grill et al. (2001). Targeting to the EGFR was performed with the single-chain bispecific antibody described previously, whereas targeting to the αv integrins was performed by insertion of an integrin-binding sequence, RGD-4C, in the HI-loop of the knob domain of the adenoviral fiber protein (Krasnykh et al., 1998). Increased luciferase gene transfer in primary glioma cells was observed in 8 of 13 samples with EGFR-targeting (2–11 times enhancement) and in all of the samples with RGD-targeting (2–42 times enhancement). Combining the two targeting motifs further enhanced the gene transfer in primary glioma cells in an additive manner (3–56 times). The double-targeted adenovirus also strongly augmented gene transfer into organotypic glioma spheroids. Conversely, gene transfer into normal brain explants was reduced dramatically using the double targeted adenovirus.

Van Beusechem (2000) used doubly ablated adenoviral vectors, lacking Coxsackie adenovirus receptor and αv integrin binding capacities (Roelvink et al., 1999), together with bispecific single-chain antibodies targeted toward human EGFR or EpCAM. Targeted doubly ablated adenoviral vectors showed very efficient and specific gene transfer in cell lines and primary human tumor specimens. On primary glioma cell cultures, EGFR-targeting augmented the median gene transfer efficiency of doubly ablated adenoviral vectors 123-fold. Moreover, EGFR-targeted doubly ablated vectors were selective for human brain tumors versus the surrounding normal brain tissue. They transduced organotypic glioma and meningioma spheroids with similar efficiency as did native adenoviral vectors, while exhibiting more than 10-fold reduced background on normal brain explants from the same patients. As a result, EGFR-targeted doubly ablated adenoviral vectors had a 5- to 38-fold improved tumor over normal brain targeting index compared to native vectors.

Nettelbeck et al. (2001) developed a recombinant bispecific antibody to the endothelial cell surface protein endoglin, for vascular targeting of adenoviruses. Endoglin (CD105), a component of the transforming growth factor beta receptor complex, represents a promising target for antivascular cancer therapy. Endoglin is expressed predominantly on endothelial cells and is upregulated in angiogenic areas of tumors. A single-chain Fv fragment directed against human endoglin was isolated from a human semisynthetic antibody library and used to construct a bispecific single-chain diabody directed against endoglin and the adenovirus fiber knob domain. This diabody mediated enhanced and selective adenovirus transduction of HUVECs, which was independent from binding to CAR and αv integrins.

The aforementioned in vitro studies have shown that a wide variety of cell surface molecules can function as a surrogate cell surface receptor for adenoviral infection. Targeting can be achieved through direct coupling of ligands or by indirect binding through a knob-binding moiety. It is recognized that the native adenovirus entry pathway comprises two distinct steps: the primary cellular receptor for Ad5, CAR, serves purely as a docking site for the virion on the cell surface (Leon et al., 1998; Wang and Bergelson, 1999), with internalization being mediated by the interaction of the RGD motif in the penton base with cellular αv integrins. By analogy, it is therefore possible to retarget an adenoviral vector simply by redirecting binding to an alternative cellular receptor, with subsequent internalization mediated by the interaction between the penton base and cellular integrins. Thus, a wide range of cell surface molecules can be exploited as alternative binding sites for targeted adenoviral vectors, including molecules that do not normally serve as cellular receptors.

IN VIVO STUDIES

In vivo retargeting of adenoviral vectors aims at selective gene expression after systemic administration. In this regard, not only should the vector be targeted toward a new cell surface specific ligand, but the natural tropism of the vector should be ablated also. Adenoviruses need to cross the vasculature in order to reach target cells located outside the blood compartment. Normal vascular endothelium consists of a continuous lining of endothelial cells, which are connected with each other by tight junctions. Beneath this cellular layer is the basement membrane and, in larger vessels, an additional layer of smooth muscle cells is present. This tight barrier prevents the passage of large molecules and particulates. In solid tumors, the vessels formed by the process of angiogenesis often show an increased permeability due to large fenestrae (up to 400 nm) and an irregularly formed basement membrane (Dvorak et al., 1988). At tumor sites with an increased vascular permeability, particles with a size similar to that of adenovirus (90 nm) can extravasate (Yuan et al., 1994), allowing adenoviral vectors to reach the tumor cells. However, at certain sites in the body, such as the liver and spleen, the vascular anatomy is different. At these sites, the endothelial lining contains fenestrations that are approximately 80–120 nm in diameter. The nonspecific sequestration/infection in the liver is a major concern in adenoviral gene therapy and warrants strong research efforts in retargeting these vectors.

Initial in vivo studies for retargeting adenoviral vectors were carried out in intraperitoneal models of human cancer in mice. This approach allows the evaluation of targeting ligand specificity without concerns of sequestration of adenovirus in liver, which normally accounts for >90% of the vector localization. Using adenovirus coated with the polymer poly-[N-(2-hydroxypropyl)methacrylamide] (pHPMA), Fisher et al. (2001) showed that incorporation of the targeting ligand FGF2 onto the polymer-coated virus produces ligand-mediated, CAR-independent binding and uptake into cells bearing appropriate receptors. Retargeted virus was resistant to antibody neutralization and infected receptor-positive target cells selectively in xenografts in vivo in an intraperitoneal model of pancreatic cancer. Similarly, an FGF2 antibody conjugate showed enhanced adenovirus-mediated gene transfer to a human ovarian cancer cell line, permitting the transduction of a greater number of target cells to be achieved by a given dose of virus (Rancourt et al., 1998). In a murine intraperitoneal model of human ovarian carcinoma, an FGF2-redirected adenoviral vector carrying the gene for herpes simplex virus thymidine kinase (AdCMVHSV-TK) was shown to result in a significant prolongation of survival compared with the same number of particles of unmodified AdCMVHSV-TK. In addition, equivalent survival rates were achieved with a 10-fold lower dose of the FGF2-redirected AdCMVHSV-TK compared with the unmodified vector. In this study, no sign of treatment-related toxicity was observed in any treatment group.

The main challenge for adenovirus targeting obviously is systemic administration. FGF2-Ad was administered intravenously to normal mice by Gu et al. (Gu et al., 1999, 1998) to determine whether increased selectivity could be observed in vivo. FGF2-Ad demonstrated markedly decreased hepatic toxicity and liver transgene expression (7- to 20-fold) compared with untargeted adenovirus treatment. Importantly, FGF2-Ad carrying the herpes simplex virus thymidine kinase (TK) gene transduced adenovirus-resistant FGFR-positive tumor cells both ex vivo and in vivo, which resulted in substantially enhanced survival (180–260%) when the prodrug ganciclovir was administered (Gu et al., 1999). In a subsequent study, Printz et al. (2000) analyzed the distribution of adenovirus after intravenous administration and found by semiquantitative polymerase chain reaction (PCR) analyses that the liver uptake of FGF2-Ad vector genome sequences was 10- to 20-fold reduced in the case of FGF2-Ad when compared with native adenovirus. This decrease in liver deposition translated into a significant reduction in subsequent toxicity as measured by serum transaminases. True targeting after systemic administration was first shown by Reynolds et al. (2000). They used a bispecific antibody to target adenovirus specifically to angiotensin-converting enzyme (ACE), which is preferentially expressed on pulmonary capillary endothelium and which may thus enable gene therapy for pulmonary vascular disease. Administration of retargeted adenovirus via tail vein injection into rats resulted in at least a 20-fold increase in both adenovirus DNA localization and luciferase transgene expression in the lungs, compared to the untargeted vector. Furthermore, targeting led to reduced transgene expression in nontarget organs, especially the liver, where the reduction was more than 80%. However, significant transgene expression was found in the liver where absolute transgene expression was still higher than in lung. This problem was addressed by the same authors in a subsequent study in which they combined the transductional targeting strategy with a transcriptional targeting approach in which expression of the transgene was placed under the control of the endothelial-specific promoter for vascular endothelial growth factor receptor type 1, flt-1 (Reynolds et al., 2001). The endothelial-specific promoter led to greatly reduced gene expression in the liver on systemic vector administration, while permitting efficient gene expression in the pulmonary endothelium. Consequently, the combined transductional and transcriptional targeting strategy resulted in a synergistic, 300,000-fold improvement in the selectivity of transgene expression for lung versus the usual site of vector sequestration, the liver. Immunohistochemical and immunoelectron microscopy analysis confirmed that the pulmonary transgene expression was specifically localized to the target endothelial cells. This study thus proved that targeting of adenovirus to a selective, well-accessible cell type close to the vasculature results in enhanced specific gene transfer after intravenous administration. These impressive results have encouraging implications for the use of targeted adenoviral vectors in vivo.

CONCLUSIONS

Physically targeted adenoviruses have been very valuable tools in validating receptor molecules as targets and establishing several key concepts with respect to the goal of achieving systemic adenovirus administration: (1) targeted adenoviral vectors can achieve effective and specific gene delivery via CAR-independent cellular entry pathways. Thus, the interaction of the targeted virus with its native receptor CAR is not obligatory for effective cellular entry; (2) CAR-independent cell transduction allows enhanced levels of gene transfer, especially in CAR-deficient cells. This increased gene transfer is of therapeutic relevance, as shown in suicide gene therapy approaches for cancer. It can be achieved with knob-binding neutralizing molecules containing a targeting ligand; (3) the intrinsic internalization capacity of the target receptor on cells is not a relevant factor predicating its utility for adenovirus retargeting. In this regard, cross-linking of adenovirus to internalizing, as well as noninternalizing, receptors allowed CAR-independent gene transfer with comparable enhancement of efficiency; (4) the physically targeted adenoviruses retain their specificity in vivo, as shown in animal models after both intraperitoneal and intravenous injection; (5) in vivo targeting of adenovirus results in reduced hepatic uptake and toxicity; and (6) increased efficiency of infection of target cells allows lower dosages of adenovirus to be given.

Despite the key concepts shown by physically targeted adenovirus, a major disadvantage of these vectors is that they consist of at least two components, that need to be combined. This is a major drawback for manufacturing, validation, and the use in clinical trials. For cancer gene therapy, the use of replicative adenoviruses expressing their own targeting molecules as recently described (Hemminki et al., 2001) would possibly be a solution to this problem. Nonetheless, the regulatory issues posed by the use of bispecific conjugates to retarget Ad vectors are not insurmountable. In this regard, an Ad vector retargeted by a bispecific molecule consisting of the neutralizing antiknob Fab fragment conjugated to FGF2 has recently been approved for a clinical trial for ovarian cancer at the University of Alabama at Birmingham.

REFERENCES

Baker A, Saltik M, Lehrmann H, Killisch I, Mautner V, Lamm G, Christofori G, Cotten M (1997): Polyethylenimine (PEI) is a simple, inexpensive and effective reagent for condensing and linking plasmid DNA to adenovirus for gene delivery. *Gene Ther* 4:773–782.

Bergelson JM, Cunningham JA, Droguett G, Kurt-Jones EA, Krithivas A, Hong JS, Horwitz MS, Crowell RL, Finberg RW (1997): Isolation of a common receptor for Coxsackie B viruses and adenoviruses 2 and 5. *Science* 275:1320–1323.

Blackwell JL, Miller CR, Douglas JT, Li H, Reynolds PN, Carroll WR, Peters GE, Strong TV, Curiel DT (1999): Retargeting to EGFR enhances adenovirus infection efficiency of squamous cell carcinoma. *Arch Otolaryngol—Head & Neck Surg* 125:856–863.

Cripe TP, Dunphy EJ, Holub AD, Saini A, Vasi NH, Mahller YY, Collins MH, Snyder

JD, Krasnykh V, Curiel DT, Wickham TJ, DeGregori J, Bergelson JM, Currier MA (2001): Fiber knob modifications overcome low, heterogeneous expression of the coxsackievirus-adenovirus receptor that limits adenovirus gene transfer and oncolysis for human rhabdomyosarcoma cells. *Cancer Res* 61:2953–2960.

Davison E, Kirby I, Elliott T, Santis G (1999): The human HLA-A*0201 allele, expressed in hamster cells, is not a high-affinity receptor for adenovirus type 5 fiber. *J Virol* 73:4513–4517.

Dmitriev I, Kashentseva E, Rogers BE, Krasnykh V, Curiel DT (2000): Ectodomain of coxsackievirus and adenovirus receptor genetically fused to epidermal growth factor mediates adenovirus targeting to epidermal growth factor receptor-positive cells. *J Virol* 74:6875–6884.

Douglas JT, Rogers BE, Rosenfeld ME, Michael SI, Feng M, Curiel DT (1996): Targeted gene delivery by tropism-modified adenoviral vectors. *Nat Biotechnol* 14:1574–1578.

Doukas J, Hoganson DK, Ong M, Ying W, Lacey DL, Baird A, Pierce GF, Sosnowski BA (1999): Retargeted delivery of adenoviral vectors through fibroblast growth factor receptors involves unique cellular pathways. *FASEB J* 13:1459–1466.

Drapkin PT, O'Riordan CR, Yi SM, Chiorini JA, Cardella J, Zabner J, Welsh MJ (2000): Targeting the urokinase plasminogen activator receptor enhances gene transfer to human airway epithelia. *J. Clin. Invest.* 105:589–596.

Dvorak HF, Nagy JA, Dvorak JT, Dvorak AM (1988): Identification and characterization of the blood vessels of solid tumors that are leaky to circulating macromolecules. *Am J Pathol* 133:95–109.

Ebbinghaus C, Al Jaibaji A, Operschall E, Schoffel A, Peter I, Greber UF, Hemmi S (2001): Functional and selective targeting of adenovirus to high-affinity Fcgamma receptor I-positive cells by using a bispecific hybrid adapter. *J Virol* 75:480–489.

Fasbender A, Zabner J, Chillon M, Moninger TO, Puga AP, Davidson BL, Welsh MJ (1997): Complexes of adenovirus with polycationic polymers and cationic lipids increase the efficiency of gene transfer in vitro and in vivo. *J Biol Chem* 272:6479–6489.

Fisher KD, Stallwood Y, Green NK, Ulbrich K, Mautner V, Seymour LW (2001): Polymer-coated adenovirus permits efficient retargeting and evades neutralising antibodies. *Gene Ther* 8:341–348.

Goldman CK, Rogers BE, Douglas JT, Sosnowski BA, Ying W, Siegal GP, Baird A, Campain JA, Curiel DT (1997): Targeted gene delivery to Kaposi's sarcoma cells via the fibroblast growth factor receptor. *Cancer Res* 57:1447–1451.

Grill J, van Beusechem VW, Van Der Valk P, Dirven CM, Leonhart A, Pherai DS, Haisma HJ, Pinedo HM, Curiel DT, Gerritsen WR (2001) Combined targeting of adenoviruses to integrins and epidermal growth factor receptors increases gene transfer into primary glioma cells and spheroids. *Clin Cancer Res* 7:641–650.

Gu DL, Gonzalez AM, Printz MA, Doukas J, Ying W, D'Andrea M, Hoganson DK, Curiel DT, Douglas JT, Sosnowski BA, Baird A, Aukerman SL, Pierce GF (1999): Fibroblast growth factor 2 retargeted adenovirus has redirected cellular tropism: evidence for reduced toxicity and enhanced antitumor activity in mice. *Cancer Res* 59:2608–2614.

Haisma HJ, Grill J, Curiel DT, Hoogeland S, van Beusechem VW, Pinedo HM, Gerritsen WR (2000a): Targeting of adenoviral vectors through a bispecific single-chain antibody. *Cancer Gene Ther* 7:901–904.

Haisma HJ, Pinedo HM, Rijswijk A, der Meulen-Muileman I, Sosnowski BA, Ying W, Beusechem VW, Tillman BW, Gerritsen WR, Curiel DT (1999): Tumor-specific gene transfer via an adenoviral vector targeted to the pan-carcinoma antigen EpCAM. *Gene Ther* 6:1469–1474.

Heideman DA, Snijders PJ, Craanen ME, Bloemena E, Meijer CJ, Meuwissen SG, van Beusechem VW, Pinedo HM, Curiel DT, Haisma HJ, Gerritsen WR (2001): Selective gene delivery toward gastric and esophageal adenocarcinoma cells via EpCAM-targeted adenoviral vectors. *Cancer Gene Ther* 8:342–351.

Hemminki A, Dmitriev I, Liu B, Desmond RA, Alemany R, Curiel DT (2001): Targeting oncolytic adenoviral agents to the epidermal growth factor pathway with a secretory fusion molecule. *Cancer Res* 61:6377–6381.

Hoganson DK, Sosnowski BA, Pierce GF, Doukas, J. (2001): Uptake of adenoviral vectors via fibroblast growth factor receptors involves intracellular pathways that differ from the targeting ligand. *Mol Ther* 3:105–112.

Hong SS, Galaup A, Peytavi R, Chazal N, Boulanger P (1999): Enhancement of adenovirus-mediated gene delivery by use of an oligopeptide with dual binding specificity. *Hum Gene Ther* 10:2577–2586.

Hong SS, Karayan L, Tournier J, Curiel DT, Boulanger PA (1997): Adenovirus type 5 fiber knob binds to MHC class I alpha2 domain at the surface of human epithelial and B lymphoblastoid cells. *EMBO J* 16:2294–2306.

Kelly FJ, Miller CR, Buchsbaum DJ, Gomez-Navarro J, Barnes MN, Alvarez RD, Curiel DT (2000): Selectivity of TAG-72-targeted adenovirus gene transfer to primary ovarian carcinoma cells versus autologous mesothelial cells in vitro. *Clin Cancer Res* 6:4323–4333.

Krasnykh V, Dmitriev I, Mikheeva G, Miller CR, Belousova N, Curiel DT (1998): Characterization of an adenovirus vector containing a heterologous peptide epitope in the HI loop of the fiber knob. *J Virol* 72:1844–1852.

Krasnykh V, Dmitriev I, Navarro JG, Belousova N, Kashentseva E, Xiang J, Douglas JT, Curiel DT (2000): Advanced generation adenoviral vectors possess augmented gene transfer efficiency based upon coxsackie adenovirus receptor-independent cellular entry capacity. *Cancer Res* 60:6784–6787.

Leon RP, Hedlund T, Meech SJ, Li S, Schaack J, Hunger SP, Duke RC, DeGregori J (1998): Adenoviral-mediated gene transfer in lymphocytes. *Proc Natl Acad Sci U.S.A.* 95:13159–13164.

Li D, Duan L, Freimuth P, O'Malley BW, Jr. (1999): Variability of adenovirus receptor density influences gene transfer efficiency and therapeutic response in head and neck cancer. *Clin Cancer Res* 5:4175–4181.

Li E, Brown SL, Von Seggern DJ, Brown GB, Nemerow GR (2000): Signaling antibodies complexed with adenovirus circumvent CAR and integrin interactions and improve gene delivery. *Gene Ther* 7:1593–1599.

McDonald D, Stockwin L, Matzow T, Blair Zajdel ME, Blair GE (1999): Coxsackie and adenovirus receptor (CAR)-dependent and major histocompatibility complex (MHC) class I-independent uptake of recombinant adenoviruses into human tumour cells. *Gene Ther* 6:1512–1519.

Meunier-Durmort C, Ferry N, Hainque B, Delattre J, Forest C (1996): Efficient transfer of regulated genes in adipocytes and hepatoma cells by the combination of liposomes and replication-deficient adenovirus. *Eur J Biochem* 237:660–667.

Meunier-Durmort C, Grimal H, Sachs LM, Demeneix BA, Forest C (1997): Adenovirus enhancement of polyethylenimine-mediated transfer of regulated genes in differentiated cells. *Gene Ther* 4:808–814.

Miller CR, Buchsbaum DJ, Reynolds PN, Douglas JT, Gillespie GY, Mayo MS, Raben D, Curiel DT (1998) Differential susceptibility of primary and established human glioma cells to adenovirus infection: targeting via the epidermal growth factor receptor achieves fiber receptor-independent gene transfer. *Cancer Res* 58:5738–5748.

Nalbantoglu J, Pari G, Karpati G, Holland PC (1999) Expression of the primary coxsackie and adenovirus receptor is downregulated during skeletal muscle maturation and limits the efficacy of adenovirus-mediated gene delivery to muscle cells. *Hum Gene Ther* 10:1009–1019.

Nettelbeck DM, Miller DW, Jerome V, Zuzarte M, Watkins SJ, Hawkins RE, Muller R, Kontermann RE (2001): Targeting of adenovirus to endothelial cells by a bispecific single-chain diabody directed against the adenovirus fiber knob domain and human endoglin (CD105). *Mol Ther* 3:882–891.

Nicklin SA, White SJ, Watkins SJ, Hawkins RE, Baker AH (2000): Selective targeting of gene transfer to vascular endothelial cells by use of peptides isolated by phage display. *Circulation* 102:231–237.

O'Riordan CR, Lachapelle A, Delgado C, Parkes V, Wadsworth SC, Smith AE, Francis GE (1999): PEGylation of adenovirus with retention of infectivity and protection from neutralizing antibody in vitro and in vivo. *Hum Gene Ther* 10:1349–1358.

Printz MA, Gonzalez AM, Cunningham M, Gu DL, Ong M, Pierce GF, Aukerman SL (2000): Fibroblast growth factor 2-retargeted adenoviral vectors exhibit a modified biolocalization pattern and display reduced toxicity relative to native adenoviral vectors. *Hum Gene Ther* 11:191–204.

Rancourt C, Rogers BE, Sosnowski BA, Wang M, Piche A, Pierce GF, Alvarez RD, Siegal GP, Douglas JT, Curiel DT (1998): Basic fibroblast growth factor enhancement of adenovirus-mediated delivery of the herpes simplex virus thymidine kinase gene results in augmented therapeutic benefit in a murine model of ovarian cancer. *Clin Cancer Res* 4:2455–2461

Reynolds PN, Zinn KR, Gavrilyuk VD, Balyasnikova IV, Rogers BE, Buchsbaum DJ, Wang MH, Miletich DJ, Grizzle WE, Douglas JT, Danilov SM, Curiel DT (2000): A targetable, injectable adenoviral vector for selective gene delivery to pulmonary endothelium in vivo. *Mol Ther* 2:562–578.

Reynolds PN, Nicklin SA, Kaliberova L, Boatman BG, Grizzle WE, Balyasnikova IV, Baker AH, Danilov SM, Curiel DT (2001): Combined transductional and transcriptional targeting improves the specificity of transgene expression in vivo. *Nat Biotechnol* 19:838–842.

Roelvink PW, Mi LG, Einfeld DA, Kovesdi I, Wickham TJ (1999): Identification of a conserved receptor-binding site on the fiber proteins of CAR-recognizing adenoviridae. *Science* 286:1568–1571.

Romanczuk H, Galer CE, Zabner J, Barsomian G, Wadsworth SC, O'Riordan CR

(1999): Modification of an adenoviral vector with biologically selected peptides: A novel strategy for gene delivery to cells of choice. *Hum Gene Ther* 10:2615–2626.

Schneider H, Groves M, Muhle C, Reynolds PN, Knight A, Themis M, Carvajal J, Scaravilli F, Curiel DT, Fairweather NF, Coutelle C (2000): Retargeting of adenoviral vectors to neurons using the Hc fragment of tetanus toxin. *Gene Ther* 7:1584–1592.

Smith JS, Keller JR, Lohrey NC, McCauslin CS, Ortiz M, Cowan K, Spence SE (1999): Redirected infection of directly biotinylated recombinant adenovirus vectors through cell surface receptors and antigens. *Proc Natl Acad Sci USA* 96:8855–8860.

Tillman BW, de Gruijl TD, Luykx-de Bakker SA, Scheper RJ, Pinedo HM, Curiel TJ, Gerritsen WR, Curiel DT (1999): Maturation of dendritic cells accompanies high-efficiency gene transfer by a CD40-targeted adenoviral vector. *J Immunol* 162:6378–6383.

Tillman BW, Hayes TL, DeGruijl TD, Douglas JT, Curiel DT (2000): Adenoviral vectors targeted to CD40 enhance the efficacy of dendritic cell-based vaccination against human papillomavirus 16-induced tumor cells in a murine model. *Cancer Res* 60:5456–5463.

Trepel M, Grifman M, Weitzman MD, Pasqualini R (2000): Molecular adaptors for vascular-targeted adenoviral gene delivery. *Hum Gene Ther* 11:1971–1981.

van Beusechem VW, van Rijswijk AL, van Es HH, Haisma HJ, Pinedo HM, Gerritsen WR (2000): Recombinant adenovirus vectors with knobless fibers for targeted gene transfer. *Gene Ther* 7:1940–1946.

Wang X, Bergelson JM (1999): Coxsackievirus and adenovirus receptor cytoplasmic and transmembrane domains are not essential for coxsackievirus and adenovirus infection. *J Virol* 73:2559–2562.

Watkins SJ, Mesyanzhinov VV, Kurochkina LP, Hawkins RE (1997): The 'adenobody' approach to viral targeting: specific and enhanced adenoviral gene delivery. *Gene Ther* 4:1004–1012.

Wesseling JG, Bosma PJ, Krasnykh V, Kashentseva EA, Blackwell JL, Reynolds PN, Li H, Parameshwar M, Vickers SM, Jaffee EM, Huibregtse K, Curiel DT, Dmitriev I (2001): Improved gene transfer efficiency to primary and established human pancreatic carcinoma target cells via epidermal growth factor receptor and integrin-targeted adenoviral vectors. *Gene Ther* 8:969–976.

Wickham TJ, Carrion ME, Kovesdi I (1995): Targeting of adenovirus penton base to new receptors through replacement of its RGD motif with other receptor-specific peptide motifs. *Gene Ther* 2:750–756.

Wickham TJ, Mathias P, Cheresh DA, Nemerow GR (1993): Integrins alpha v beta 3 and alpha v beta 5 promote adenovirus internalization but not virus attachment. *Cell* 73:309–319.

Wickham TJ, Roelvink PW, Brough DE, Kovesdi I (1996a): Adenovirus targeted to heparan-containing receptors increases its gene delivery efficiency to multiple cell types. *Nat Biotechnol* 14:1570–1573.

Wickham TJ, Segal DM, Roelvink PW, Carrion ME, Lizonova A, Lee GM, Kovesdi I (1996b): Targeted adenovirus gene transfer to endothelial and smooth muscle cells by using bispecific antibodies. *J Virol* 70:6831–6838.

Wickham TJ, Lee GM, Titus JA, Sconocchia G, Bakacs T, Kovesdi I, Segal DM

(1997): Targeted adenovirus-mediated gene delivery to T cells via CD3. *J Virol* 71:7663–7669.

Xia H, Anderson B, Mao Q, Davidson BL (2000): Recombinant human adenovirus: targeting to the human transferrin receptor improves gene transfer to brain microcapillary endothelium. *J Virol* 74:11359–11366.

Yuan F, Leunig M, Huang SK, Berk DA, Papahadjopoulos D, Jain RK (1994): Microvascular permeability and interstitial penetration of sterically stabilized (stealth) liposomes in a human tumor xenograft. *Cancer Res* 54:3352–3356.

Zabner J, Freimuth P, Puga A, Fabrega A, Welsh MJ (1997): Lack of high affinity fiber receptor activity explains the resistance of ciliated airway epithelia to adenovirus infection. *J Clin Invest* 100:1144–1149.

7

GENETIC TARGETING OF ADENOVIRAL VECTORS

THOMAS J. WICKHAM, PH.D.

INTRODUCTION

Gene therapy pharmaceutical products using the adenoviral vector continue to show promise, with some currently entering pivotal clinical trials. Adenoviral vectors containing a variety of transgenes have been shown to have a good safety profile. For these and future adenoviral products, a limiting feature of the vector is efficient gene delivery to the appropriate target cells. Current generations of adenoviral vectors are limited because the primary receptors used to enter cells are often low or absent in many tumors and clinically relevant tissues. In contrast, these same receptors are highly expressed in nontarget organs leading to inappropriate vector delivery to these organs following regional or systemic administration. Due to these limitations, the current generations of adenoviral vectors are largely restricted only to those situations where the vector can be administered by direct injection or where the consequences of nontarget tissue transduction are minimal. The development of adenoviral vectors that target tumors or specific tissues would improve the safety, efficiency, and flexibility of adenoviral vectors for the treatment of diverse diseases. For example, tumor-specific vectors could dramatically change the treatment paradigm for cancer by allowing clinicians to better treat disseminated cancers rather than be limited to locally advanced disease.

Vector Targeting for Therapeutic Gene Delivery, Edited by David T. Curiel and Joanne T. Douglas
ISBN 0-471-43479-5

CREATING A TURN-KEY APPROACH TO THE DESIGN OF TARGETED VECTORS

Since the first publications on adenovirus targeting, numerous investigators have shown that targeting is technically feasible and functionally important. There is improved understanding of the virus/receptor interactions that facilitate the entry of native adenoviral vectors into cells, how the adenoviral coat proteins can be engineered to contain peptide ligands, and the diversity of receptors that can be successfully targeted using adenovirus. This knowledge provides us with the molecular understanding to retarget adenovirus. The challenge of future targeting efforts is to assemble the basic components for a turn-key approach to targeted vector design. The ultimate goal of this turn-key approach is the development of safer and more effective gene therapy-based pharmaceutical products. Programs to develop these products should combine the targeted vector backbones with the evaluation of the therapeutic gene at an early research stage. With such a turn-key system, the vector and therapeutic gene are designed and optimized with the specific application in mind.

Elements of a Turn-Key Approach

To implement a turn-key approach to targeting and drug discovery three basic components are needed. First, a base vector is necessary in which all of the native adenovirus receptor interactions have been stripped away. This base vector then serves as a platform for the addition of tissue-specific ligands. Second, special production cell lines are required for the propagation of targeted vectors. Since the native receptor interactions have been ablated, the targeted vector is unable to attach and enter into standard packaging cell lines. The final component is a rapid ligand identification system that enables tissue-specific vectors to be rapidly identified and then modified for testing with a therapeutic gene. Each of these three elements of a turn-key drug discovery system is reviewed in detail.

In the generation of a turn-key drug discovery system, other characteristics need to be considered that enable the targeted vectors to be rapidly taken into clinical and commercial applications. The vector should be a single component to reduce the complexity of production. This is most efficiently accomplished through genetic, rather than nongenetic, modifications. There should be a single packaging cell line so that diverse vectors can be efficiently produced. There should be a single, scalable purification process so that all targeted vectors can take advantage of the processes already developed to purify adenoviral vectors. Finally, the modifications in the vector that enable it to be targeted should not compromise its inherent particle activity.

BACKGROUND: EFFICIENCY VERSUS SPECIFICITY

Adenoviral tropism depends on the native receptors that the virus uses to enter cells. Adenovirus utilizes two of its coat proteins to interact with two distinct

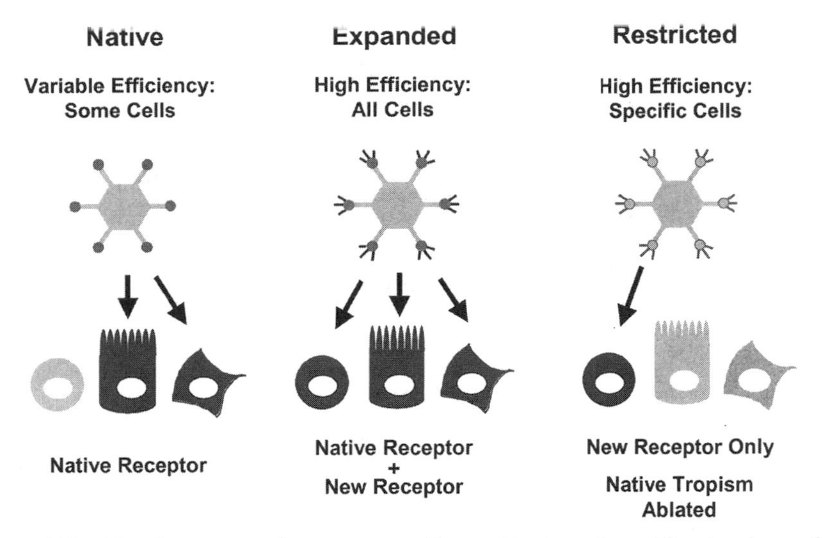

Figure 7.1 Tropism expansion versus cell-specific targeting. The tropism of an unmodified adenovirus is determined by the biodistribution of its native receptors. Since the native receptors are not expressed in all tissues, this results in variable transduction efficiency of adenovirus. Adenovirus tropism can be expanded through the incorporation of ligands that bind to broadly expressed receptors. Generally, the native receptor binding capabilities of the vector are left intact. Adenovirus tropism can be restricted through the ablation of native tropism and redirection to a specific cell type through the introduction of a tissue or tumor-specific ligand into the coat protein of the vector.

receptors during entry into cells. The adenoviral fiber protein binds with high affinity to the Coxsackie adenovirus receptor (CAR) (Bergelson et al., 1997; Bergelson 1999; Tomko et al., 2000). Following attachment via CAR, the penton base protein then binds to αv integrins, which mediate internalization and may also have a role in escape from the endosome (Wickham et al., 1993, 1994; Bai, Harfe, and Freirmuth, 1993).

Adenoviral tropism can be altered in two basic ways (Fig. 7.1). First, the vector can be modified to transduce cell types beyond its native tropism. Such a modification does not prevent entry into nontarget cells but can increase the efficiency of entry into cells lacking high levels of adenoviral receptors. Tropism expansion is useful since many cells, including many types of tumor cells, have been shown to be low or devoid of CAR expression (Table 7.1). In addition, although integrins are more broadly expressed than CAR, there are also cell types, notably certain hematopoietic cells, that express neither CAR nor αv integrins.

A second approach to alter tropism is by modifying the vector to transduce only a particular cell type with high efficiency (Fig. 7.1). Such a modification requires at least two basic elements. First, the native tropism of the vector must be blocked or ablated to prevent entry into nontarget cells via native

TABLE 7.1 Cells and Tissues with Low Expression of Adenovirus Receptors

	Comments	References
	Cancer Cells	
Bladder cancer	Little or no CAR on 2/7 Low CAR on additional 2/7	Li et al., 1990a
Glioma	Little or no CAR on 8/12	Miller et al., 1998
Glioma	Low CAR on majority of 15 cell lines	Asaoka et al., 2000
Head and neck cancer	Low CAR on 3/3	Kasono et al., 1999; Blackwell et al., 1999
Head and neck cancer	Variable CAR. Little or no CAR on 1/3	Li et al., 1999
Lung and pancreatic cancer	Low CAR in 2/8	Pearson et al., 1999
Melanoma	Low CAR on 10/14 Poor transduction in same 10/14	Hemmi et al., 1998
Mesothelioma	Low CAR on 2/4	Lanuti et al., 1999
NSCLC adenocarcinoma	5/5 refractory 3/3 NSCLC susceptible	Batra et al., 1998
Rhabdomyosarcoma	Little CAR in 13 primary tumors	Cripe et al., 2001
Head and neck	CAR expressed in undifferentiated. Not available in differentiated	Hutchin, Pickles, and Yarbrough, 2000
	Noncancer Cells	
Airway epithelia	Basolateral localization	Walters, et al., 1999; Pickles et al., 1998
Cardiomyocytes	Very low in normal adult High in newborns or in myocarditis	Ito et al., 2000; Noutsias et al., 2001
Dendritic cells	No detectable CAR	Rea et al., 1999
Endothelial	Cell density dependent	Carson et al., 1999
Macrophage	Little or no CAR	Kaner et al., 1999
Skeletal muscle	Barely detectable in adult	Nalbantoglu et al., 1999, 2001
Smooth muscle	Little or no CAR	Lanuti et al., 1999
Synoviocytes	Little or no CAR	Goossens et al., 2001
Bone marrow mesenchymal progenitor cells	CAR only expressed on subpopulation	Conget and Minguell, 2000
Lymphocytes	Expression of CAR on lymphocytes renders them transducible	Leon et al., 1998; Schmidt et al., 2000

CAR or integrins. Second, a cell-specific ligand must be added to the vector to permit its selective entry into only those cells that express the receptor for the ligand. This type of tropism modification represents true targeting of the vector. The rationales for targeting gene transfer by this approach are numerous and discussed throughout this book.

INCREASING EFFICIENCY: TROPISM EXPANSION

Tropism expansion has been achieved through the genetic addition of ligands to the fiber or other coat proteins. The types of ligands that have been used to date are limited with the most common being polylysine Arg-Gly-Asp (RGD) motifs, or knob switching (Tables 7.2 and 7.3). However, the broad success of approaches for targeting diverse receptors using nongenetic approaches strongly indicates that numerous receptors can be targeted as long as small, high-affinity peptide ligands can be identified for genetic insertion into the adenovirus coat proteins (Table 7.4). Polylysine or other polybasic amino acid stretches are general motifs for binding to heparan sulfate. Heparan sulfate is a sugar moiety linked to a broad variety of heparan sulfate proteoglycan receptors, which are broadly expressed on most cell types. The RGD-4C motif, CDCRGDCFC, is a high affinity ligand for αv integrins (Koivunen, Wang, and Ruoslahti, 1995; Pasqualini et al., 1997). Despite the presence of an RGD motif in penton base, the additional presence of the RGD-4C motif in the fiber or hexon has been shown to increase the transduction of cells that express low levels of CAR (Wickham et al., 1997a; Vigne et al., 1999). Most of the data with tropism-expanded vectors has demonstrated improved delivery efficiency in cell culture. However, a number of reports have now demonstrated improved gene transfer by tropism-expanded vectors when used in vivo (Bouri et al., 1999; Wickham et al., 1997a; Shinoura et al., 1999; Douglas et al., 2001; Suzuki et al., 2001).

TURN-KEY TISSUE-SPECIFIC VECTORS

Elimination of Native Receptor Binding

Genetic Approaches to CAR Ablation Three basic approaches have been taken to genetically ablate CAR binding (Fig. 7.2). The first, and seemingly most straightforward, approach has been to remove the fiber entirely from the virus and then target it via the penton base or other coat protein. Adenoviral particles that lack fiber have been produced through this approach (Legrand et al., 1999; Von Seggern et al., 1999). However, the applicability of these particles remains to be determined due to the unusually high particle to plaque-forming unit (pfu) ratios of these vectors. One reason for these high particle to pfu ratios appears to stem from the incomplete processing of certain structural

TABLE 7.2 Tropism Expansion Using Genetic Modification

Ligand	Application or Target Cell	References
Polylysine	Acute myeloid leukemia	Gonzalez et al., 1999a
Polylysine	B lymphocytes and lymphoma	Li, Wickham, and Keegan, 2001
Polylysine	Bone marrow, myeloma	Gonzalez et al., 1999b
Polylysine	Broad	Wickham et al., 1996
Polylysine	Broad	Wickham et al., 1997a
Polylysine	Glioma	Staba et al., 2000
Polylysine	Glioma	Shinoura, et al., 1999, 2000
Polylysine	Glioma	Yoshida et al., 1998
Polylysine	Muscle	Bouri et al., 1999
Polylysine	Rhabdomyosarcoma	Cripe et al., 2001
Polylysine	Vasculature	Kibbe et al., 2000
RGD-4C	Biodistribution	Reynolds et al., 1999
RGD-4C	Broad	Wickham et al., 1997a
RGD-4C	Broad	Vigne et al., 1999
RGD-4C	Broad	Dmitriev et al., 1998
RGD-4C	Broad	Krasnykh et al., 1998
RGD-4C	Cancer	Suzuki et al., 2001
RGD-4C	Dendritic cells	Asada-Mikami et al., 2001
RGD-4C	Dendritic cells	Okada et al., 2001
RGD-4C	Glioma	Grill et al., 2001
RGD-4C	Glioma	Staba et al., 2000
RGD-4C	Glioma	Miller et al., 1998
RGD-4C	Head and neck cancer	Kasono et al., 1999
RGD-4C	Kidney vasculature	McDonald et al., 1999b
RGD-4C	Myelomonocytic leukemia	Garcia-Castro et al., 2001
RGD-4C	Ovarian cancer	Blackwell et al., 2000; Vanderkwaak et al., 1999
RGD-4C	Rhabdomyosarcoma	Cripe et al., 2001
RGD-4C	Vasculature	Wickham et al., 1997a; Hay et al., 2001
NGR	Tumor vasculature	Mizuguchi et al., 2001
Transferrin receptor	Brain endothelium	Xia et al., 2000

proteins (Legrand et al., 1999). In addition, significant changes in the nuclear architecture during the replication of fiberless vectors compared to fiber-containing vectors suggest that the fiber may play additional roles in the maturation of the vector particle (Fuvlon-Dutheul et al., 1999).

A second approach to ablate CAR binding has been to remove the knob portion of the fiber, which is necessary for fiber trimerization, and replace it

TABLE 7.3 Changing Adenovirus Tropism via Pseudotyping

Fiber Subgroup	Fiber Serotype	Application or Target Cell	References
B	3	EBV-transformed B cells	Von Seggern et al., 2000
B	3	Head and neck, fibroblast, monocyte	Stevenson et al., 1997, 1995
B	7	Diverse	Gall et al., 1996
B	16	Synovial tissue	Goossens et al., 2001
B	16	Endothelial cells Smooth muscle	Havenga et al., 2001
B	35	Stem cells	Shayakhmetov et al., 2000; Yotnda et al., 2001
B	35	Dendritic	Rea et al., 2001
D	17	Airway epithelia	Zabner et al., 1999
D	17	Central nervous system cells	Chillon et al., 1999
D and B	9 and 35	Broad	Shayakhmetov and Lieber, 2000

TABLE 7.4 Receptors Targeted Using Nongenetic Approaches

Target Receptor	Application or Target Cell	References
N-acetylglucosamine receptor	Liver	Thoma et al., 2000
ACE	Lung endothelium	Reynolds et al., 2000
CD105	Tumor endothelium	Nettelbeck et al., 2001
CD3	T cells	Wickham et al., 1997b
CD40	Dendritic cells	Tillman et al. 1999, 2000
CD70	B cells	Israel et al., 2001
EGF receptor	Cancer, EGF receptor-expressing cells	Dmitriev et al., 2000; Miller et al., 1998; Haisma et al., 2000
Endothelial receptor	Endothelial cells	Nicklin et al., 2001
EpCAM	Carcinoma	Haisma et al., 1999
E-selectin	Endothelial cells, inflammation	Wickham et al., 1997c; Harari et al., 1999
Fcg receptors	B cells, B lymphoma	Li et al., 2001
FcgRI (CD64)	Monocytic cells	Ebbinghaus et al., 2001
FGF receptor	Cancer, broad	Rancourt et al., 1998; Doukas et al., 1999; Sosnowski et al., 1999; Gu et al., 1999; Printz et al., 2000; Chandler, et al., 2000

TABLE 7.4 *(Continued)*

Target Receptor	Application or Target Cell	References
Folate receptors	Cancer	Douglas et al., 1996
P2Y2 receptor	Differentiated epithelial cells	Kreda et al., 2000
HCCR	Hepatocellular carcinoma	Yoon et al., 2000
GFE-1 ligand receptor	Receptor-expressing cells	Trepel et al., 2000
TAG-72	Ovarian cancer	Kelly et al., 2000
Tetanus toxin receptor	Neurons	Schneider et al., 2000
TNF, IGF-1, and EGF receptors	Melanoma	Li et al., 2000
VEGF receptor	Endothelium	Watkins et al., 1997
αv integrins	Broad	Wickham et al., 1996b

Notes: EGF = epidermal growth factor; FGF = fibroblast growth factor; IGF = insulin-like growth factor; VEGF = vascular endothelial growth factor.

with an alternative trimerization domain. Using this approach it is possible to produce chimeric proteins that fold and trimerize correctly (van Beusechem et al., 2000; Krasnykh et al., 2001; Magnusson et al., 2001). Viral particles containing these fibers are able to bind and enter cells via the chimeric fiber. An advantage of this approach is that it could permit ligands that are larger than short peptides to be incorporated into the chimera. However, it remains to be determined whether these particles maintain the inherent particle of an

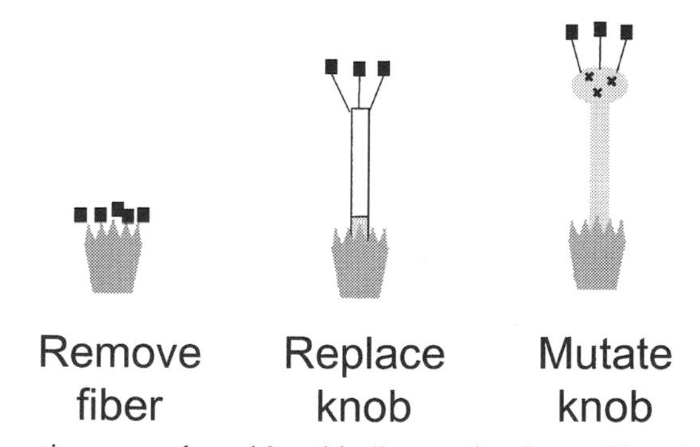

Figure 7.2 Genetic approaches ablate binding to the Coxsackie adenovirus receptor (CAR). CAR binding can be ablated [illegible]. The fiber gene can be genetically removed so that the resultant particles are fiberless. The knob, which is necessary for trimerization of the fiber, can be removed and replaced with a nonnative trimerization domain such as a coiled coil. Finally, the knob can be mutated so that the molecular interactions that permit CAR binding are destroyed.

unmodified adenovirus. Vectors containing a polyhistidine tag that recognizes an artificial single-chain antibody receptor were not able to transduce cells expressing the artificial receptor as efficiently as a vector with a native fiber (Krasnykh et al., 2001). A similar decrease in the transduction of 293 cells was observed with a similar vector containing the RGD-4C sequence (Magnusson et al., 2001). These results indicate that either the receptor affinities are not as optimal as the native fiber/CAR interaction or that the particle activity of these vectors is somehow negatively affected by the deletion of the knob.

A final genetic approach to ablate CAR binding has been to identify specific amino acid mutations in the knob that ablate binding to CAR (Figure 7.3). The crystal structure of the Ad5 knob has been solved, revealing those exposed amino acids that would be available for receptor binding (Xia et al., 1994). Four β-sheets are present at the top of each knob monomer and face toward the cell. These sheets were originally termed the R-sheet indicating a potential role in receptor binding. Interestingly, extensive mutational analysis of the fiber knob together with the crystal structure of the Ad12 fiber knob complexed with CAR have shown that CAR binds on the side of the knob across the AB loop rather than on top of the knob via the R-sheet (Bewley et al., 1999; Roelvink et al., 1999). A number of mutations in and around the AB loop have been shown to ablate binding to CAR (Roelvink et al., 1999; Kirby et al., 1999, 2000; Leissner et al., 2001; Jakubczak et al., 2001). Several groups have now made CAR-ablated vectors and have shown reduced ability of these vectors to transduce CAR-expressing cells (Leissner et al., 2001; Jakubczak et al., 2001; Roelvink et al., 1999). The inherent particle activity of these vectors appears to

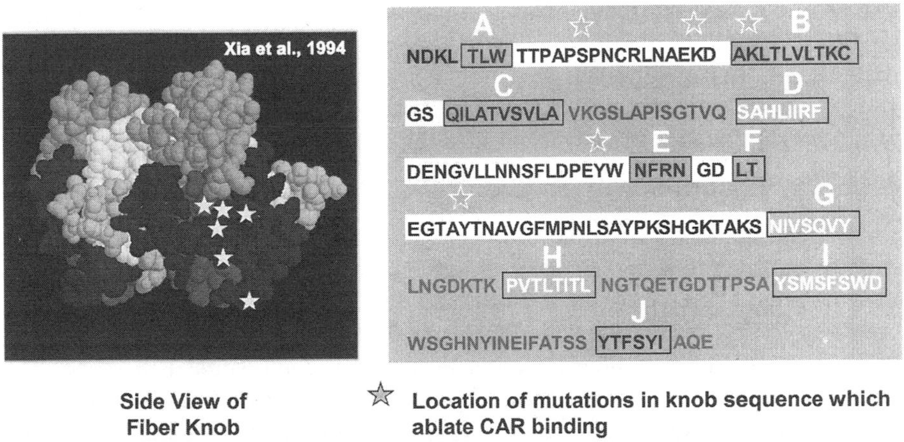

Xia et al., 1994

NDKL TLW TTPAPSPNCRLNAEKD AKLTLVLTKC
GS QILATVSVLA VKGSLAPISGTVQ SAHLIIRF
DENGVLLNNSFLDPEYW NFRN GD LT
EGTAYTNAVGFMPNLSAYPKSHGKTAKS NIVSQVY
LNGDKTK PVTLTITL NGTQETGDTTPSA YSMSFSWD
WSGHNYINEIFATSS YTFSYI AQE

Side View of Fiber Knob

☆ Location of mutations in knob sequence which ablate CAR binding

Figure 7.3 Locations of mutations that ablate binding of the Ad5 knob to CAR. CAR-ablating mutations (indicated by a star) are mostly localized in and around the AB loop. The majority of these mutations cluster on the side of the knob and have been shown via crystal structure analysis to directly interact with CAR.

be maintained. CAR-ablated vectors expressing hemagglutinin A (HA) peptide tag in the HI loop of the fiber have been shown to transduce cells expressing an artificial single-chain antibody receptor specific for HA as efficiently as a vector with a native fiber (Roelvink et al., 1999). In addition, the particle to pfu ratio of these vectors is similar to a vector with a native fiber (unpublished observations).

A natural hypothesis on the effect of CAR ablation in the fiber is that transduction of the liver would be dramatically reduced, because since CAR is highly expressed in the mouse liver. However, recent data has shown the somewhat unexpected result that ablation of CAR results in little, if any, reduction of mouse liver transduction (Leissner, 2001; unpublished observations). These data clearly indicate that other interactions are able to mediate liver transduction in the absence of CAR binding.

Ablation of CAR and Integrins One plausible hypothesis for the remaining liver transduction by the CAR-ablated vector is that the penton base mediates liver transduction in the absence of a CAR interaction. To test this hypothesis, our group constructed a panel of vectors in which CAR, integrin, or CAR plus integrin interactions were genetically ablated from the virus. Using this panel of vectors, the individual roles of CAR and integrins in adenovirus-mediated transduction of a variety of tissues in vivo could be dissected. Transduction of CAR and integrin-expressing cells with the panel of vectors demonstrated that CAR was the predominant receptor interaction governing transduction. However, a significant residual transduction of cells by CAR-ablated vectors often remains. This residual transduction has been found to be integrin-mediated, because ablation of both CAR and integrin interactions knocks out the majority of the residual transduction observed with CAR-ablated vectors. The finding that penton base can directly mediate transduction in the absence of CAR interactions is not surprising. It has been shown previously that the penton base can mediate adenovirus binding to macrophages via an $\alpha m\beta 2$ integrin-dependent interaction (Huang et al., 1996). Furthermore, ablation of the integrin interaction alone has been shown to result in significant reductions of transduction of CAR-negative cells (Hidaka et al., 1999).

The most dramatic reductions in gene transfer in vivo have been observed following ablation of both the CAR and integrin interactions. These findings support the idea that the ablation of both interactions is necessary to create a truly targeted adenoviral vector. While the ablation of CAR has only a modest effect on liver transduction following intrajugular administration, the ablation of both the CAR and integrin interactions generally results in a more than 100-fold reduction in liver transduction (Figure 7.4). In addition, the doubly ablated vector demonstrates significant reductions in all tissues examined. This result is in contrast to the CAR-ablated vector, which surprisingly demonstrates a trend toward increased transduction of certain tissues. A similar pattern of transduction has generally been found for the transduction of most tissues that we have investigated following various routes of administration of the panel of

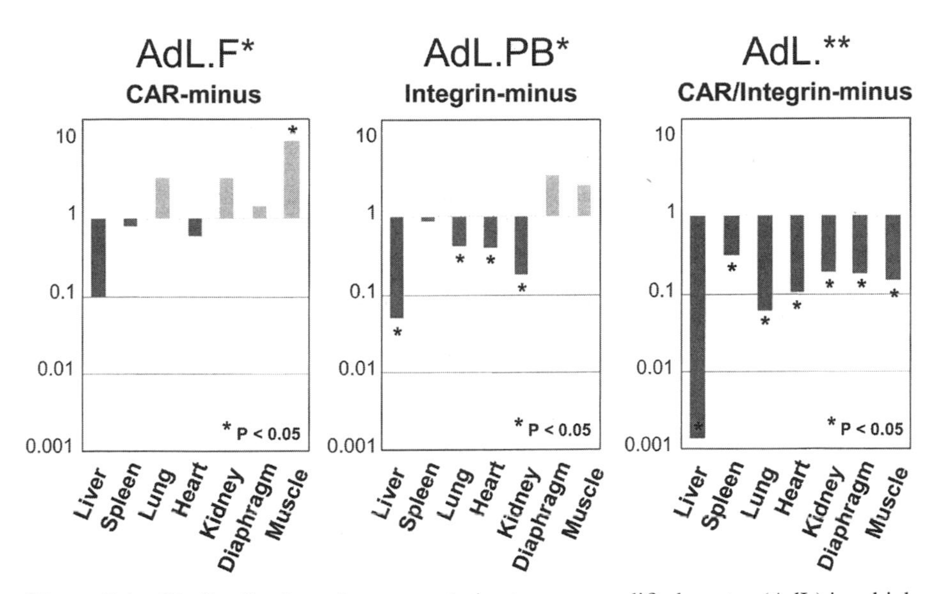

Figure 7.4 Biodistribution of vectors relative to an unmodified vector (AdL) in which CAR (AdL.F*), integrin (AdL.PB*), or CAR plus integrin (AdL.**) binding have been ablated. AdL, AdL.F*, AdL.PB*, and AdL.** were each administered at 1×10^{11} vector particles to mice by intrajugular injection at day 0. The mice were then sacrificed at day 1 and the luciferase activities, normalized for protein content from each organ, were measured and plotted relative to the unmodified vector, AdL.

vectors. These results strongly support the use of a base targeting vector that is ablated for both CAR and integrin interactions.

Remaining Interactions Our results with CAR- and integrin-ablated vectors indicate that most of the native tropism of adenovirus has been removed. However, other receptor interactions for adenovirus have been reported, including interactions of fiber with major histocompatibility complex class I (MHC I) and heparan sulfate (Dechecchi et al., 2000). We have not been able to demonstrate any involvement of heparan sulfate-containing receptors in the entry of native adenovirus and have found that high levels of heparin do not reduce the very low levels of residual transduction observed for the CAR- and integrin-ablated vectors in vitro. Similarly, MHC I has been shown not to be directly involved in adenovirus attachment to cells (Davison et al., 1999; McDonald et al., 1999a). However, it is still not possible to rule out the possibility that receptors other than CAR or integrins play some role in the residual transduction of CAR- and integrin-ablated adenovirus vectors. In fact, the rapid uptake of adenovirus by macrophages in the lung following intranasal administration, and by Kupffer cells in the liver following intravenous injection, suggests that adenovirus may enter certain cells via innate mechanisms that are used to clear pathogens (Harrod et al., 1999; Zsengeller et al., 2000).

Cell Lines for Production of Targeted Vectors

Fiber-Complementing Cell Lines Two general approaches have been devised to produce genetically-modified targeted vectors. The first general approach has been to create a packaging cell line that complements for fiber expression (Legrand et al., 1999; Von Seggern et al., 1998, 1999, 2000). For example, the fiber gene could be deleted from the vector backbone, so that in the absence of a proper packaging cell line, the resultant vector particles would be fiberless. A fiber-complementing cell line would thereby provide the modified fiber protein required for targeting and produce vector particles carrying the modified fiber protein. Such an approach as been successfully used to produce the fiberless adenoviral vectors discussed previously as well as vectors that are pseudotyped with another fiber protein (Legrand et al., 1999; Von Seggern et al., 1999, 2000). A similar approach has been used as a novel system to rapidly evaluate the function of adenoviral vectors with modified fibers (Jakubczak et al., 2001). This approach uses a transient transfection/infection system that combines transfection of cells with plasmids that express high levels of the modified fiber protein and infection with a fiber-deleted adenoviral vector. The resultant vectors contain the modified fiber proteins, thereby allowing the mutations to be rapidly evaluated in the context of the virus particle. A drawback to the fiberless backbone/fiber complementing cell line approach is that a new cell line is required to produce each different targeted vector. However, similar approaches can be easily envisioned whereby only a single cell line would be required.

Pseudoreceptor-Expressing Cell Lines A second approach to produce targeted adenovirus vectors has been to create cell lines that express an artificial, nonnative receptor that binds to a common epitope present on all targeted vectors. This artificial receptor, or pseudoreceptor, allows the vector to bind and enter the packaging cell line (Douglas et al., 1999; Einfeld et al., 1999) (Fig. 7.5). One advantage of this approach is that virtually any targeted vector can be produced in a single cell line. Pseudoreceptor cell lines have been used to propagate vectors containing chimeric fibers with nonnative trimerization domains (Krasnykh et al., 2001) as well as vectors that have been ablated in both CAR- and integrin-binding (unpublished observations). Single-chain antibody-based pseudoreceptors have been designed to recognize either the fiber knob or short peptide tags (Douglas et al., 1999; Einfeld et al., 1999). An advantage of peptide tag-recognizing pseudoreceptors is that they can be used to produce any type of modified vector as long as the vector contains the peptide tag. The disadvantage of this approach is the requirement of the vector to express both the targeting ligand and the peptide tag. Pseudoreceptors recognizing a coat protein epitope common to all targeted vectors have the advantage of not requiring the incorporation of a peptide tag. However, in the case of a fiber knob-recognizing pseudoreceptor, the pseudoreceptor cannot be utilized to produce vectors in which the fiber knob has been replaced with a

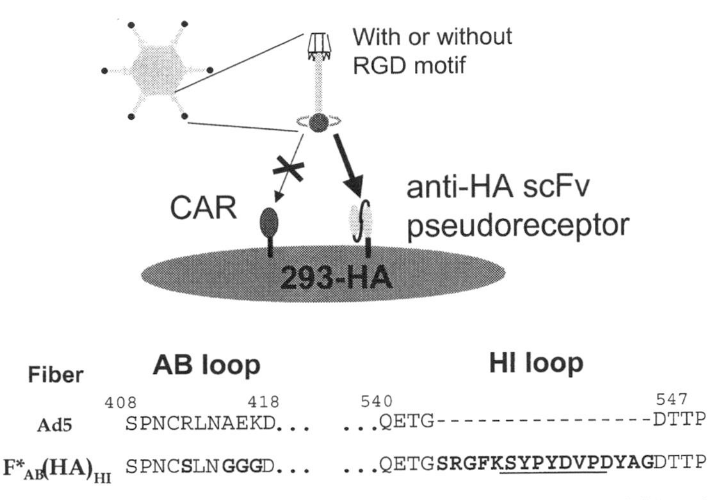

Figure 7.5 Schematic depicting the growth of CAR or CAR and integrin-ablated vectors in the 293-HA cell line. The cell line expresses an artificial receptor comprised of a transmembrane protein domain fused to a single chain antibody recognizing the HA peptide epitope. The HA peptide epitope (underlined) is genetically incorporated into the HI loop of a fiber knob containing a CAR-ablating mutation in the AB loop (emboldened amino acids).

nonnative trimerization domain. Using such pseudoreceptor cell lines, vectors ablated in both CAR- and integrin-binding can be grown to high titers that are nearly identical to unmodified vectors (unpublished observations).

Redirection: Tissue-Specific Ligands

As discussed previously, vectors with all known native tropism completely ablated and production cell lines for growth of targeted vectors have been developed. The ligand remains as the last critical path item for development of highly selective adenovirus vectors. While ligand identification technology has been around as long as, or longer than, the field of gene therapy, the identification of ligands that function properly within the adenovirus vector has remained a key hurdle. Fortunately, this situation is rapidly changing. More focused efforts toward the identification of ligands using traditional methods such as phage display are accelerating. In addition, advances in adenovirus-compatible ligand identification technology promise to dramatically increase the speed and efficiency for identifying tissue-selective vectors.

Ligand Location Fiber has been the major coat protein used for redirecting the adenovirus to new receptors; however, evidence now indicates that other coat proteins can also be used for inserting ligands (Fig. 7.6). In the fiber,

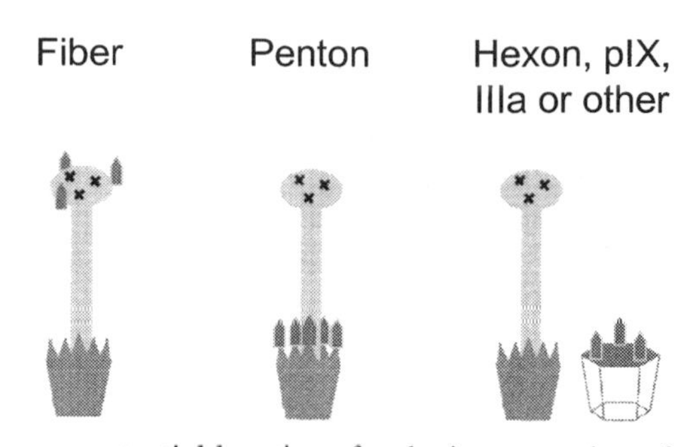

Figure 7.6 Known or potential locations for the incorporation of peptide ligands into adenovirus.

peptides have been put either at the C-terminus or in the HI-loop of the fiber knob (Michael et al., 1995; Wickham et al., 1997a; Krasnykh et al., 1998). Peptides inserted at the C-terminus have typically contained a flexible amino acid linker of approximately 10 amino acids. The linker presumably functions to bring the ligand away from the fiber knob in order to be better available to bind the receptor. The C-terminus of the fiber emerges from the knob in close proximity to where the fiber shaft enters the knob so the idea of a linker appears to have merit (Xia et al., 1994). Sequence comparisons of knobs from different adenovirus serotypes reveals that the HI loops of adenovirus knobs are hypervariable. This finding indicates that the knob would be amenable to changes or insertions in this region. This idea is supported by data showing that diverse types of peptides can be inserted into the HI loop and produce correctly folded and functional protein (Krasnykh et al., 1998; Roelvink et al., 1999; Mizuguchi et al., 2001; Xia et al., 2000). It is not yet clear whether either the C-terminal or HI loop location is better for ligand placement. It is most likely that the optimal location is peptide dependent. It is possible that other locations in the fiber protein could be used for inserting ligands; however, no data have been published on this idea.

In addition to fiber, other coat proteins have also been shown to permit the functional incorporation of ligands. The HA peptide ligand has been functionally incorporated into the penton base to create a vector that can enter cells via an anti-HA artificial receptor (Einfeld et al., 1999). Additionally, incorporation of the RGD-4C ligand into the hexon has been shown to result in vector particles with properties very similar to vectors containing the RGD-4C ligand incorporated into the fiber (Vigne et al., 1999). The hexon may have additional advantages for ligand location because hexon is present at a higher copy number (720/virion) than the fiber (36), penton base (60), or pIX (240) (Stewart et al., 1993). Relative to the fiber, the higher copy number of the other coat proteins may permit lower affinity ligands to function better in the context of adenovirus due to increased avidity. It remains to be determined whether the nonfiber coat proteins have advantages for ligand placement compared to fiber.

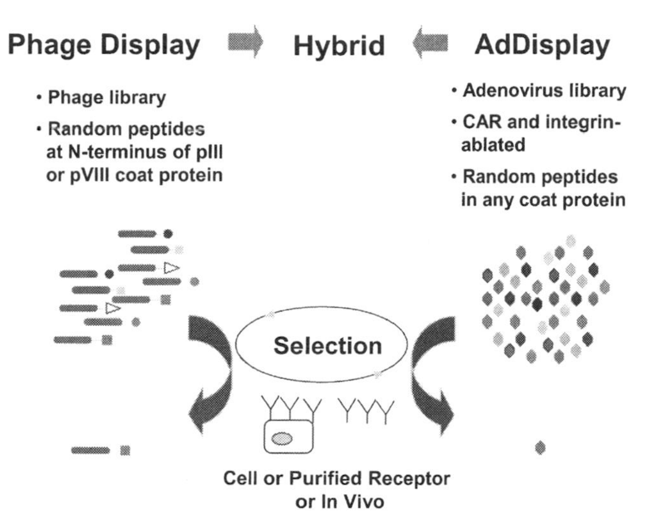

Figure 7.7 Potential approaches for the identification of ligands for use in targeted adenovirus vectors. Phage display has been the standard approach for identifying small, high-affinity peptide ligands for incorporation into adenovirus vectors. More advanced ligand identification techniques have begun to address some of the limitations in compatibility between adenovirus and phage. These techniques include hybrid approaches, where adenovirus coat proteins are incorporated into phage for selection. In addition, AdDisplay is being developed in which adenovirus itself is used as the vehicle.

Ligand Identification—Phage Display Phage display has been used to identify a large number of peptide and single-chain antibody ligands (Fig. 7.7). The advantage of this system for identifying ligands is that it is rapid and well tested. Selections against targets can be performed in a matter of weeks and numerous ligands have been identified and characterized. While it is apparent that peptide ligands can be rapidly identified, it is not always clear how well these ligands will function in the context of the adenovirus. The affinity of the ligand, maintenance of its receptor-binding activity once incorporated into an adenoviral coat protein, and its impact on adenoviral coat protein folding complicate the simple movement of phage-derived peptides into adenovirus. Since phage display is extensively discussed elsewhere in this book (chapters 25–27), it is discussed here only briefly in terms of its utility for identifying peptide ligands for genetically targeted adenoviral vectors.

Phage display technology has been used to isolate a number of potentially useful receptor-specific, peptide ligands. Some of these ligands include peptides that bind to, αv integrins (Koivunen et al., 1995), $\alpha5\beta1$ integrin (Koivunen et al., 1994), E-selectin (Fukuda et al., 2000), CD13 (Pasqualini et al., 2000), the gelatinases MMP2 and MMP9 (Koivunen et al., 1999), and the melanoma-associated antigen, NG2 (Burg et al., 1999). The high-affinity peptide motif, ACDCRGDCFCG, which binds to αv integrins has an affinity of approximately 100-fold higher than the RGD motif in fibronectin used by the $\alpha v\beta 3$ integrin (Koivunen et al., 1995). An adenoviral vector containing the high affinity RGD motif incorporated into a modified fiber coat protein is redirected toward binding to αv integrin receptors on proliferating endothelial cells (Wickham et al., 1997a; Xia et al., 2000). More recently, phage-derived peptide ligands specific for CD13 and the transferrin receptor have been reported to function in the context of adenovirus (Mizuguchi et al., 2001; Xia et al., 2000). These results demonstrate the great potential of peptide phage display technology in obtaining high affinity ligands that specifically target viral particles to specific tissues.

The use of phage display to identify ligands is not limited to the foregoing situation where the target receptor must be known and purified. Phage display screening technologies have also been developed to target specific cells and tissues without prior knowledge of the target receptor. Rather than using purified receptor, the phage library can be screened against whole cells, which permits the isolation of cell-specific phage. This method has been used to isolate peptide sequences specific for macrophage and dendritic cells (Barry, Dower, and Johnston, 1996).

Another novel phage selection technique is simply to inject the entire phage library into the vascular system of an animal and subsequently select and isolate phage clones that home to endothelial cells within specific tissues (Pasqualini and Ruoslahti 1996a, 1996b; Pasqualini, Koivunen, and Ruoslahti, 1997; Pasqualini 1999). The basis for this technique is that endothelial cells express certain receptors that are specific to a tissue, an organ, or a disease state. In vivo selection of phage has been used to isolate peptide sequences that localize phage specifically to tumors, the brain, or the kidney (Pasqualini and Ruoslahti, 1996a, 1996b). A great advantage of this technique is that it generally enriches only those ligand sequences that are specific and eliminates most nonspecific sequences. Furthermore, the use of a viral particle during the selection most closely simulates the use of a gene delivery vehicle injected in vivo.

The major disadvantage of the phage display system for the development of genetically targeted adenoviral vectors is that the peptides identified via phage are not necessarily compatible with, or functional in, the context of an adenoviral vector. Despite the identification of multiple receptor-specific peptide ligands using phage display in the past 5 years, only a handful of the ligands have been successfully incorporated into adenovirus. While this situation appears to be changing, the lag in successful reports can be at least partially attributed to difficulties related to ligand compatibility. The first hurdle encountered when attempting to transfer a phage-derived peptide into adenovirus is that the resulting chimeric coat protein may not fold properly. It is not easily predicted whether the incorporation of the phage-derived peptide sequence into the adenoviral coat protein will result in a correctly folded coat protein. If the chimeric protein does fold properly, a further complication relates to differences in how the peptide is displayed on the phage versus the adenovirus. The M13 phage pIII or pVIII systems utilize the N-terminus of the proteins to identify peptides. When these types of peptides are incorporated into either the C-terminus or the HI loop of the adenovirus fiber, the free N-terminus is lost. This incompatibility of transfer may result in a peptide that is no longer functional in the context of adenovirus. Furthermore, it is not necessarily easy to predict, *a priori*, whether a peptide will be functional. Finally, if the peptide does retain function in the context of adenovirus, depending on how the phage selection was performed, the peptide affinity may be too low to permit adequate adenovirus binding to the target receptor on the cell.

Advanced Phage Display Systems Rapid advances are now being made in ligand identification that will likely overcome the hurdles to obtaining peptide ligands that function in the context of adenovirus. One straightforward approach is to utilize display technologies that use a loop or C-terminal extension rather than an N-terminal peptide extension. For example, the T7 phage display system utilizes a C-terminal random peptide. Such peptides derived from the T7 system are more likely to function in the context of a C-terminal addition to fiber as compared to peptides derived from the N-terminus of pIII on M13. Another approach would be to identify and utilize exposed N-termini on the coat proteins of adenovirus. It may be, for example, that the N-terminus of either pIX or IIIa are exposed on the adenovirus, which may permit these coat proteins to be more compatible with M13-derived peptides. A phage-based approach being attempted is to display the entire trimeric fiber protein on the phage particle. A basic proof of principle for this approach has recently been demonstrated by the successful display of a functional fiber knob on the M13 phage (Pereboev et al., 2001).

AdDisplay The foregoing problems related to the compatibility of phage display-derived peptides and adenovirus could be avoided if adenovirus itself could be made into a display system. In fact, such a system would not only avoid the disadvantages of phage display but would also have additional advantages over the previous phage display systems. Our group has been building such a system called AdDisplay (Fig. 7.7).

AdDisplay combines the CAR and integrin-ablated vector backbone and the pseudoreceptor production cell line as described previously, together with technology to create libraries of adenoviruses. Libraries of adenoviral vectors are made using cosmid cloning as a first step. The entire adenovirus genome ablated for CAR- and integrin-binding is put into a cosmid. Unique restriction sites are inserted into any coat protein that then permit the insertion of random oligonucleotides encoding potential peptide ligands. The ablation of CAR and integrin interactions allows the library to be efficiently selected *in vivo* or on any cell type regardless of CAR expression. Following construction of the cosmid library, it is then transfected into packaging cell lines and converted into vector. The conversion step is the key rate-limiting step to obtaining high-complexity libraries; however, certain approaches can be used to increase the conversion rate in order to obtain reasonable complexities of vector particles for use in selections.

The AdDisplay system is expected to permit the rapid identification of ligands without the issues of compatibility. The ligands identified are already optimized within the context of adenovirus. The vectors identified can then be rapidly evaluated and converted to express therapeutic proteins. In addition, the selections are not limited to any particular protein. Presumably ligands could be selected in the context of fiber, penton base, hexon, pIX, or any other exposed coat protein. Another attractive feature of the system is that selections could be done *in vivo*. Perhaps the most attractive feature of a selectable system is that the desired phenotype is selected without the need for any a priori knowledge of the target or targeting mechanism. For example, selection for an adenoviral vector that homes to a tumor may identify ligands that not only function in target cell binding and entry. It may also be possible to select for other ligands that improve vector pharmacology through mechanisms that are distinct from the binding via a tumor-specific receptor.

CONCLUSION

With the two major hurdles of native tropism ablation and vector growth largely solved, it is clear that the last remaining hurdle is the rapid identification of ligands that function in adenovirus. From the recent advances made in more advanced display methodologies it is highly likely that this last hurdle will be overcome. This achievement will open up a whole new era in adenovirus-mediated gene therapy where the vector is selectively designed for the particular application or indication. In this way, the clinical use of targeted vectors that utilize receptor-based targeting together with promoter-based targeting will become the norm rather than the exception.

REFERENCES

Asada-Mikami R, Heike Y, Kanai S, Azuma M, Shirakawa K, Takaue Y, Krasnykh V, Curiel DT, Terada M, Abe T, Wakasugi H (2001): Efficient gene transduction by RGD-fiber modified recombinant adenovirus into dendritic cells. *Jpn J Cancer Res* 92:321–327.

Asaoka K, Tada M, Sawamura Y, Ikeda J, Abe H (2000): Dependence of efficient adenoviral gene delivery in malignant glioma cells on the expression levels of the Coxsackievirus and adenovirus receptor. *J Neurosurg* 92:1002–1008.

Bai M, Harfe B, Freimuth P (1993): Mutations that alter an Arg-Gly-Asp (RGD) sequence in the adenovirus type 2 penton base protein abolish its cell-rounding activity and delay virus reproduction in flat cells. *J Virol* 67:5198–5205.

Barry MA, Dower WJ, Johnston SA (1996): Toward cell-targeting gene therapy vectors: Selection of cell-binding peptides from random peptide-presenting phage libraries. *Nat Med* 2:299–305.

Batra RK, Olsen JC, Pickles RJ, Hoganson DK, Boucher RC (1998): Transduction of non-small cell lung cancer cells by adenoviral and retroviral vectors. *Am J Respir Cell Mol Biol* 18:402–410.

Bergelson, JM (1999): Receptors mediating adenovirus attachment and internalization. *Biochem Pharmacol* 57:975–979.

Bergelson JM, Cunningham JA, Droguett G, Kurt-Jones EA, Krithivas A, Hong JS, Horwitz MS, Crowell RL, Finberg RW (1997): Isolation of a common receptor for Coxsackie B viruses and adenoviruses 2 and 5. *Science* 275:1320–1323.

Bewley MC, Springer K, Zhang YB, Freimuth P, Flanagan JM (1999): Structural analysis of the mechanism of adenovirus binding to its human cellular receptor, CAR. *Science* 286:1579–1583.

Blackwell JL, Miller CR, Douglas JT, Li H, Reynolds PN, Carroll WR, Peters GE, Strong TV, Curiel DT (1999): Retargeting to EGFR enhances adenovirus infection efficiency of squamous cell carcinoma. *Arch Otolaryngol Head Neck Surg* 125:856–863.

Blackwell JL, Li H, Gomez-Navarro J, Dmitriev I, Krasnykh V, Richter CA, Shaw DR, Alvarez RD, Curiel DT, Strong TV (2000): Using a tropism-modified adenoviral vector to circumvent inhibitory factors in ascites fluid. *Hum Gene Ther* 11:1657–1669.

Bouri K, Feero WG, Myerburg MM, Wickham TJ, Kovesdi I, Hoffman EP, Clemens PR (1999): Polylysine modification of adenoviral fiber protein enhances muscle cell transduction. *Hum Gene Ther* 10:1633–1640.

Burg MA, Pasqualini R, Arap W, Ruoslahti E, Stallcup WB (1999): NG2 proteoglycan-binding peptides target tumor neovasculature. *Cancer Res* 59:2869–2874.

Carson SD, Hobbs JT, Tracy SM, Chapman NM (1999): Expression of the coxsackievirus and adenovirus receptor in cultured human umbilical vein endothelial cells: Regulation in response to cell density. *J Virol* 73:7077–7079.

Chandler LA, Doukas J, Gonzalez AM, Hoganson DK, Gu DL, Ma C, Nesbit M, Crombleholme TM, Herlyn M, Sosnowski BA, Pierce GF (2000): FGF2-Targeted adenovirus encoding platelet-derived growth factor-B enhances de novo tissue formation. *Mol Ther* 2:153–160.

Chillon M, Bosch A, Zabner J, Law L, Armentano D, Welsh MJ, Davidson BL (1999): Group D adenoviruses infect primary central nervous system cells more efficiently than those from group C. *J Virol* 73:2537–2540.

Conget PA, Minguell JJ (2000): Adenoviral-mediated gene transfer into ex vivo expanded human bone marrow mesenchymal progenitor cells. *Exp Hematol* 28:382–390.

Cripe TP, Dunphy EJ, Holub AD, Saini A, Vasi NH, Mahller YY, Collins MH, Snyder JD, Krasnykh V, Curiel DT, Wickham TJ, DeGregori J, Bergelson JM, Currier MA (2001): Fiber knob modifications overcome low, heterogeneous expression of the coxsackievirus-adenovirus receptor that limits adenovirus gene transfer and oncolysis for human rhabdomyosarcoma cells. *Cancer Res* 61:2953–2960.

Davison E, Kirby I, Elliott T, Santis G (1999): The human HLA-A*0201 allele, expressed in hamster cells, is not a high-affinity receptor for adenovirus type 5 fiber. *J Virol* 73:4513–4517.

Dechecchi MC, Tamanini A, Bonizzato A, Cabrini G (2000): Heparan sulfate glycosaminoglycans are involved in adenovirus type 5 and 2-host cell interactions. *Virology* 268:382–390.

Dmitriev I, Krasnykh V, Miller CR, Wang M, Kashentseva E, Mikheeva G, Belousova N, Curiel DT (1998): An adenovirus vector with genetically modified fibers demonstrates expanded tropism via utilization of a coxsackievirus and adenovirus receptor-independent cell entry mechanism. *J Virol* 72:9706–9713.

Dmitriev I, Kashentseva E, Rogers BE, Krasnykh V, Curiel DT (2000): Ectodomain of coxsackievirus and adenovirus receptor genetically fused to epidermal growth factor mediates adenovirus targeting to epidermal growth factor receptor-positive cells. *J Virol* 74:6875–6884.

Douglas JT, Rogers BE, Rosenfeld ME, Michael SI, Feng M, Curiel DT (1996): Targeted gene delivery by tropism-modified adenoviral vectors. *Nat Biotechnol* 14:1574–1578.

Douglas JT, Miller CR, Kim M, Dmitriev I, Mikheeva G, Krasnykh V, Curiel DT (1999): A system for the propagation of adenoviral vectors with genetically modified receptor specificities. *Nat Biotechnol* 17:470–475.

Douglas JT, Kim M, Sumerel LA, Carey DE, Curiel DT (2001): Efficient oncolysis by a replicating adenovirus (Ad) in vivo is critically dependent on tumor expression of primary Ad receptors. *Cancer Res* 61:813–817.

Doukas J, Hoganson DK, Ong M, Ying W, Lacey DL, Baird A, Pierce GF, Sosnowski BA (1999): Retargeted delivery of adenoviral vectors through fibroblast growth factor receptors involves unique cellular pathways. *FASEB J* 13:1459-1466.

Ebbinghaus C, Al-Jaibaji A, Operschall E, Schoffel A, Peter I, Greber UF, Hemmi S (2001): Functional and selective targeting of adenovirus to high-affinity Fcgamma receptor I-positive cells by using a bispecific hybrid adapter. *J Virol* 75:480–489.

Einfeld DA, Brough DE, Roelvink PW, Kovesdi I, Wickham TJ (1999): Construction of a pseudoreceptor that mediates transduction by adenoviruses expressing a ligand in fiber or penton base. *J Virol* 73:9130–9136.

Fukuda MN, Ohyama C, Lowitz K, Matsuo O, Pasqualini R, Ruoslahti E, Fukuda M (2000): A peptide mimic of E-selectin ligand inhibits sialyl Lewis X-dependent lung colonization of tumor cells. *Cancer Res* 60:450–456.

Gall J, Kass-Eisler A, Leinwand L, Falck-Pedersen E (1996): Adenovirus type 5 and 7 capsid chimera: Fiber replacement alters receptor tropism without affecting primary immune neutralization epitopes. *J Virol* 70:2116–2123.

Garcia-Castro J, Segovia JC, Garcia-Sanchez F, Lillo R, Gomez-Navarro J, Curiel DT, Bueren JA (2001): Selective transduction of murine myelomonocytic leukemia cells (WEHI-3B) with regular and RGD-adenoviral vectors. *Mol Ther* 3:70–77.

Gonzalez R, Vereecque R, Wickham TJ, Vanrumbeke M, Kovesdi I, Bauters F, Fenaux P, Quesnel B (1999a): Increased gene transfer in acute myeloid leukemic cells by an adenovirus vector containing a modified fiber protein. *Gene Ther* 6:314–320.

Gonzalez R, Vereecque R, Wickham TJ, Facon T, Hetuin D, Kovesdi I, Bauters F, Fenaux P, Quesnel B (1999b): Transduction of bone marrow cells by the AdZ.F(pK7) modified adenovirus demonstrates preferential gene transfer in myeloma cells. *Hum Gene Ther* 10:2709–2717.

Goossens PH, Havenga MJ, Pieterman E, Lemckert AA, Breedveld FC, Bout A,

Huizinga TW (2001): Infection efficiency of type 5 adenoviral vectors in synovial tissue can be enhanced with a type 16 fiber. *Arthritis Rheum* 44:570–577.

Grill J, Van Beusechem VW, Van Der Valk P, Dirven CM, Leonhart A, Pherai DS, Haisma HJ, Pinedo HM, Curiel DT, Gerritsen WR (2001): Combined targeting of adenoviruses to integrins and epidermal growth factor receptors increases gene transfer into primary glioma cells and spheroids. *Clin Cancer Res* 7:641–650.

Gu DL, Gonzalez AM, Printz MA, Doukas J, Ying W, D'Andrea M, Hoganson DK, Curiel DT, Douglas JT, Sosnowski BA, Baird A, Aukerman SL, Pierce GF (1999): Fibroblast growth factor 2 retargeted adenovirus has redirected cellular tropism: evidence for reduced toxicity and enhanced antitumor activity in mice. *Cancer Res* 59:2608-2614.

Haisma HJ, Pinedo HM, Rijswijk A, der Meulen-Muileman I, Sosnowski BA, Ying W, Beusechem VW, Tillman BW, Gerritsen WR, Curiel DT (1999): Tumor-specific gene transfer via an adenoviral vector targeted to the pan-carcinoma antigen EpCAM. *Gene Ther* 6:1469–1474.

Haisma HJ, Grill J, Curiel DT, Hoogeland S, van Beusechem VW, Pinedo HM, Gerritsen WR (2000): Targeting of adenoviral vectors through a bispecific single-chain antibody. *Cancer Gene Ther* 7:901–904.

Harari OA, Wickham TJ, Stocker CJ, Kovesdi I, Segal DM, Huehns TY, Sarraf C, Haskard DO (1999): Targeting an adenoviral gene vector to cytokine-activated vascular endothelium via E-selectin. *Gene Ther* 6:801–807.

Harrod KS, Trapnell BC, Otake K, Korfhagen TR, Whitsett JA (1999): SP-A enhances viral clearance and inhibits inflammation after pulmonary adenoviral infection. *Am J Physiol* 277:L580–L588.

Havenga MJ, Lemckert AA, Grimbergen JM, Vogels R, Huisman LG, Valerio D, Bout A, Quax PH (2001): Improved adenovirus vectors for infection of cardiovascular tissues. *J Virol* 75:3335–3342.

Hay CM, De Leon H, Jafari JD, Jakubczak JL, Mech CA, Hallenbeck PL, Powell SK, Liau G, Stevenson SC (2001): Enhanced gene transfer to rabbit jugular veins by an adenovirus containing a cyclic RGD motif in the HI loop of the fiber knob. *J Vasc Res* 38:315–323.

Hemmi S, Geertsen R, Mezzacasa A, Peter I, Dummer R (1998): The presence of human coxsackievirus and adenovirus receptor is associated with efficient adenovirus-mediated transgene expression in human melanoma cell cultures. *Hum Gene Ther* 9:2363–2373.

Hidaka C, Milano E, Leopold PL, Bergelson JM, Hackett NR, Finberg RW, Wickham TJ, Kovesdi I, Roelvink P, Crystal RG (1999): CAR-dependent and CAR-independent pathways of adenovirus vector-mediated gene transfer and expression in human fibroblasts. *J Clin Invest* 103:579–587.

Huang S, Kamata T, Takada Y, Ruggeri ZM, Nemerow GR (1996): Adenovirus interaction with distinct integrins mediates separate events in cell entry and gene delivery to hematopoietic cells. *J Virol* 70:4502–4508.

Hutchin ME, Pickles RJ, Yarbrough WG (2000): Efficiency of adenovirus-mediated gene transfer to oropharyngeal epithelial cells correlates with cellular differentiation and human coxsackie and adenovirus receptor expression. *Hum Gene Ther* 11:2365–2375.

Israel BF, Pickles RJ, Segal DM, Gerard RD, Kenney SC (2001). Enhancement of adenovirus vector entry into CD70-positive B-cell lines by using a bispecific CD70-adenovirus fiber antibody. *J Virol* 75:5215–5221.

Ito M, Kodama M, Masuko M, Yamaura M, Fuse K, Uesugi Y, Hirono S, Okura Y, Kato K, Hotta Y, Honda T, Kuwano R, Aizawa Y (2000): Expression of coxsackievirus and adenovirus receptor in hearts of rats with experimental autoimmune myocarditis. *Circ Res* 86:275–280.

Jakubczak JL, Rollence ML, Stewart DA, Jafari JD, Von Seggern DJ, Nemerow GR, Stevenson SC, Hallenbeck PL (2001): Adenovirus type 5 viral particles pseudotyped with mutagenized fiber proteins show diminished infectivity of coxsackie B-adenovirus receptor-bearing cells. *J Virol* 75:2972–2981.

Kaner RJ, Worgall S, Leopold PL, Stolze E, Milano E, Hidaka C, Ramalingam R, Hackett NR, Singh R, Bergelson J, Finberg R, Falck-Pedersen E, Crystal RG (1999): Modification of the genetic program of human alveolar macrophages by adenovirus vectors in vitro is feasible but inefficient, limited in part by the low level of expression of the coxsackie/adenovirus receptor. *Am J Respir Cell Mol Biol* 20:361–370.

Kasono K, Blackwell JL, Douglas JT, Dmitriev I, Strong TV, Reynolds P, Kropf DA, Carroll WR, Peters GE, Bucy RP, Curiel DT, Krasnykh V (1999): Selective gene delivery to head and neck cancer cells via an integrin targeted adenoviral vector. *Clin Cancer Res* 5:2571–2579.

Kelly FJ, Miller CR, Buchsbaum DJ, Gomez-Navarro J, Barnes MN, Alvarez RD, Curiel DT (2000): Selectivity of TAG-72-targeted adenovirus gene transfer to primary ovarian carcinoma cells versus autologous mesothelial cells in vitro. *Clin Cancer Res* 6:4323–4333.

Kibbe MR, Murdock A, Wickham T, Lizonova A, Kovesdi I, Nie S, Shears L, Billiar TR, Tzeng E (2000): Optimizing cardiovascular gene therapy: increased vascular gene transfer with modified adenoviral vectors. *Arch Surg* 135:191–197.

Kirby I, Davison E, Beavil AJ, Soh CP, Wickham TJ, Roelvink PW, Kovesdi I, Sutton BJ, Santis G (1999): Mutations in the DG loop of adenovirus type 5 fiber knob protein abolish high-affinity binding to its cellular receptor CAR. *J Virol* 73:9508-9514.

Kirby I, Davison E, Beavil AJ, Soh CP, Wickham TJ, Roelvink PW, Kovesdi I, Sutton BJ, Santis G (2000): Identification of contact residues and definition of the CAR-binding site of adenovirus type 5 fiber protein. *J Virol* 74:2804–2813.

Koivunen E, Wang B, Ruoslahti E (1994): Isolation of a highly specific ligand for the alpha 5 beta 1 integrin from a phage display library. *J Cell Biol* 124:373-380.

Koivunen E, Wang B, Ruoslahti E (1995): Phage libraries displaying cyclic peptides with different ring sizes: ligand specificities of the RGD-directed integrins. *Biotechnology* (*NY*) 13:265–270.

Koivunen E, Arap W, Valtanen H, Rainisalo A, Medina OP, Heikkila P, Kantor C, Gahmberg CG, Salo T, Konttinen YT, Sorsa T, Ruoslahti E, Pasqualini R (1999): Tumor targeting with a selective gelatinase inhibitor. *Nat Biotechnol* 17:768–774.

Krasnykh V, Dmitriev I, Mikheeva G, Miller CR, Belousova N, Curiel DT (1998): Characterization of an adenovirus vector containing a heterologous peptide epitope in the HI loop of the fiber knob. *J Virol* 72:1844–1852.

Krasnykh V, Belousova N, Korokhov N, Mikheeva G, Curiel DT (2001): Genetic targeting of an adenovirus vector via replacement of the fiber protein with the phage T4 fibritin. *J Virol* 75:4176–4183.

Kreda SM, Pickles RJ, Lazarowski ER, Boucher RC (2000): G-protein-coupled receptors as targets for gene transfer vectors using natural small-molecule ligands. *Nat Biotechnol* 18:635–640.

Lanuti M, Kouri CE, Force S, Chang M, Amin K, Xu K, Blair I, Kaiser L, Albelda S (1999): Use of protamine to augment adenovirus-mediated cancer gene therapy. *Gene Ther* 6:1600-1610.

Legrand V, Spehner D, Schlesinger Y, Settelen N, Pavirani A, Mehtali M (1999): Fiberless recombinant adenoviruses: virus maturation and infectivity in the absence of fiber. *J Virol* 73:907–919.

Leissner P, Legrand V, Schlesinger Y, Hadji DA, van Raaij M, Cusack S, Pavirani A, Mehtali M (2001): Influence of adenoviral fiber mutations on viral encapsidation, infectivity and in vivo tropism. *Gene Ther* 8:49-57.

Leon RP, Hedlund T, Meech SJ, Li S, Schaack J, Hunger SP, Duke RC, DeGregori J (1998): Adenoviral-mediated gene transfer in lymphocytes. *Proc Natl Acad Sci USA* 95:13159–13164.

Li L, Wickham TJ, Keegan AD (2001): Efficient transduction of murine B lymphocytes and B lymphoma lines by modified adenoviral vectors: enhancement via targeting to FcR and heparan-containing proteins. *Gene Ther* 8:938–945.

Li Y, Pong RC, Bergelson JM, Hall MC, Sagalowsky AI, Tseng CP, Wang Z, Hsieh JT (1999a): Loss of adenoviral receptor expression in human bladder cancer cells: a potential impact on the efficacy of gene therapy. *Cancer Res* 59:325–330.

Li D, Duan L, Freimuth P, O'Malley BW, Jr. (1999b): Variability of adenovirus receptor density influences gene transfer efficiency and therapeutic response in head and neck cancer. *Clin Cancer Res* 5:4175–4181.

Li E, Brown SL, Von Seggern DJ, Brown GB, Nemerow GR (2000): Signaling antibodies complexed with adenovirus circumvent CAR and integrin interactions and improve gene delivery. *Gene Ther* 7:1593-1599.

Magnusson MK, Hong SS, Boulanger P, Lindholm L (2001): Genetic retargeting of adenovirus: novel strategy employing "deknobbing" of the fiber. *J Virol* 75:7280–7289.

McDonald D, Stockwin L, Matzow T, Blair Zajdel ME, Blair GE (1999a): Coxsackie and adenovirus receptor (CAR)-dependent and major histocompatibility complex (MHC) class I-independent uptake of recombinant adenoviruses into human tumour cells. *Gene Ther* 6:1512–1519.

McDonald GA, Zhu G, Li Y, Kovesdi I, Wickham TJ, Sukhatme VP (1999b): Efficient adenoviral gene transfer to kidney cortical vasculature utilizing a fiber modified vector. *J Gene Med* 1:103–110.

Michael SI, Hong JS, Curiel DT, Engler JA (1995): Addition of a short peptide ligand to the adenovirus fiber protein. *Gene Ther* 2:660–668.

Miller CR, Buchsbaum DJ, Reynolds PN, Douglas JT, Gillespie GY, Mayo MS, Raben D, Curiel DT (1998): Differential susceptibility of primary and established human glioma cells to adenovirus infection: targeting via the epidermal growth factor receptor achieves fiber receptor-independent gene transfer. *Cancer Res* 58:5738–5748.

Mizuguchi H, Koizumi N, Hosono T, Utoguchi N, Watanabe Y, Kay MA, Hayakawa

T (2001): A simplified system for constructing recombinant adenoviral vectors containing heterologous peptides in the HI loop of their fiber knob. *Gene Ther* 8:730–735.

Nalbantoglu J, Pari G, Karpati G, Holland PC (1999): Expression of the primary coxsackie and adenovirus receptor is downregulated during skeletal muscle maturation and limits the efficacy of adenovirus-mediated gene delivery to muscle cells. *Hum Gene Ther* 10:1009–1019.

Nalbantoglu J, Larochelle N, Wolf E, Karpati G, Lochmuller H, Holland PC (2001): Muscle-specific overexpression of the adenovirus primary receptor CAR overcomes low efficiency of gene transfer to mature skeletal muscle. *J Virol* 75:4276–4282.

Nettelbeck DM, Miller DW, Jerome V, Zuzarte M, Watkins SJ, Hawkins RE, Muller R, Kontermann RE (2001): Targeting of adenovirus to endothelial cells by a bispecific single-chain diabody directed against the adenovirus fiber knob domain and human endoglin (CD105). *Mol Ther* 3:882–891.

Nicklin SA, Reynolds PN, Brosnan MJ, White SJ, Curiel DT, Dominiczak AF, Baker AH (2001): Analysis of cell-specific promoters for viral gene therapy targeted at the vascular endothelium. *Hypertension* 38:65–70.

Noutsias M, Fechner H, de Jonge H, Wang X, Dekkers D, Houtsmuller AB, Pauschinger M, Bergelson J, Warraich R, Yacoub M, Hetzer R, Lamers J, Schultheiss HP, Poller W (2001): Human coxsackie-adenovirus receptor is colocalized with integrins alpha(v)beta(3) and alpha(v)beta(5) on the cardiomyocyte sarcolemma and upregulated in dilated cardiomyopathy: implications for cardiotropic viral infections. *Circulation* 104:275–280.

Okada N, Tsukada Y, Nakagawa S, Mizuguchi H, Mori K, Saito T, Fujita T, Yamamoto A, Hayakawa T, Mayumi T (2001): Efficient gene delivery into dendritic cells by fiber-mutant adenovirus vectors. *Biochem Biophys Res Commun* 282:173–179.

Pasqualini R (1999): Vascular targeting with phage peptide libraries. *Q J Nucl Med* 43:159–162.

Pasqualini R, Ruoslahti E (1996a): Organ targeting in vivo using phage display peptide libraries. *Nature* 380:364–366.

Pasqualini R, Ruoslahti E (1996b): Tissue targeting with phage peptide libraries. *Mol Psychiatry* 1:423.

Pasqualini R, Koivunen E, Ruoslahti E (1997): Alpha v integrins as receptors for tumor targeting by circulating ligands. *Nat Biotechnol* 15:542–546.

Pasqualini R, Koivunen E, Kain R, Lahdenranta J, Sakamoto M, Stryhn A, Ashmun RA, Shapiro LH, Arap W, Ruoslahti E (2000): Aminopeptidase N is a receptor for tumor-homing peptides and a target for inhibiting angiogenesis. *Cancer Res* 60:722–727.

Pearson AS, Koch PE, Atkinson N, Xiong M, Finberg RW, Roth JA, Fang B (1999): Factors limiting adenovirus-mediated gene transfer into human lung and pancreatic cancer cell lines. *Clin Cancer Res* 5:4208–4213.

Pereboev A, Pereboeva L, Curiel DT (2001): Phage display of adenovirus type 5 fiber knob as a tool for specific ligand selection and validation. *J Virol* 75:7107–7113.

Pickles RJ, McCarty D, Matsui H, Hart PJ, Randell SH, Boucher RC (1998): Limited entry of adenovirus vectors into well-differentiated airway epithelium is responsible for inefficient gene transfer. *J Virol* 72:6014–6023.

Printz MA, Gonzalez AM, Cunningham M, Gu DL, Ong M, Pierce GF, Aukerman SL (2000): Fibroblast growth factor 2-retargeted adenoviral vectors exhibit a modified biolocalization pattern and display reduced toxicity relative to native adenoviral vectors. *Hum Gene Ther* 11:191–204.

Puvion-Dutilleul F, Legrand V, Mehtali M, Chelbi-Alix MK, de The H, Puvion E (1999): Deletion of the fiber gene induces the storage of hexon and penton base proteins in PML/Sp100-containing inclusions during adenovirus infection. *Biol Cell* 91:617–628.

Rancourt C, Rogers BE, Sosnowski BA, Wang M, Piche A, Pierce GF, Alvarez RD, Siegal GP, Douglas JT, Curiel DT (1998): Basic fibroblast growth factor enhancement of adenovirus-mediated delivery of the herpes simplex virus thymidine kinase gene results in augmented therapeutic benefit in a murine model of ovarian cancer. *Clin Cancer Res* 4:2455–2461.

Rea D, Schagen FH, Hoeben RC, Mehtali M, Havenga MJ, Toes RE, Melief CJ, Offringa R (1999): Adenoviruses activate human dendritic cells without polarization toward a T-helper type 1-inducing subset. *J Virol* 73:10245–10253.

Rea D, Havenga MJ, van Den Assem M, Sutmuller RP, Lemckert A, Hoeben RC, Bout A, Melief CJ, Offringa R (2001): Highly efficient transduction of human monocyte-derived dendritic cells with subgroup B fiber-modified adenovirus vectors enhances transgene-encoded antigen presentation to cytotoxic T cells. *J Immunol* 166:5236–5244.

Reynolds P, Dmitriev I, Curiel D (1999): Insertion of an RGD motif into the HI loop of adenovirus fiber protein alters the distribution of transgene expression of the systemically administered vector. *Gene Ther* 6:1336–1339.

Reynolds PN, Zinn KR, Gavrilyuk VD, Balyasnikova IV, Rogers BE, Buchsbaum DJ, Wang MH, Miletich DJ, Grizzle WE, Douglas JT, Danilov SM, Curiel DT (2000): A targetable, injectable adenoviral vector for selective gene delivery to pulmonary endothelium in vivo. *Mol Ther* 2:562–578.

Roelvink PW, Mi Lee G, Einfeld DA, Kovesdi I, Wickham TJ (1999): Identification of a conserved receptor-binding site on the fiber proteins of CAR-recognizing adenoviridae. *Science* 286:1568–1571.

Schmidt MR, Piekos B, Cabatingan MS, Woodland RT (2000): Expression of a human coxsackie/adenovirus receptor transgene permits adenovirus infection of primary lymphocytes. *J Immunol* 165:4112–4119.

Schneider H, Groves M, Muhle C, Reynolds PN, Knight A, Themis M, Carvajal J, Scaravilli F, Curiel DT, Fairweather NF, Coutelle C (2000): Retargeting of adenoviral vectors to neurons using the Hc fragment of tetanus toxin. *Gene Ther* 7:1584–1592.

Shayakhmetov DM, Lieber A (2000): Dependence of adenovirus infectivity on length of the fiber shaft domain. *J Virol* 74:10274–10286.

Shayakhmetov DM, Papayannopoulou T, Stamatoyannopoulos G, Lieber A (2000): Efficient gene transfer into human CD34(+) cells by a retargeted adenovirus vector. *J Virol* 74:2567–2583.

Shinoura N, Yoshida Y, Tsunoda R, Ohashi M, Zhang W, Asai A, Kirino T, Hamada H (1999): Highly augmented cytopathic effect of a fiber-mutant E1B-defective adenovirus for gene therapy of gliomas. *Cancer Res* 59:3411–3416.

Shinoura N, Sakurai S, Asai A, Kirino T, Hamada H (2000): Transduction of a fiber-mutant adenovirus for the HSVtk gene highly augments the cytopathic effect towards gliomas. *Jpn J Cancer Res* 91:1028–1034.

Sosnowski BA, Gu DL, D'Andrea M, Doukas J, Pierce GF (1999): FGF2-targeted adenoviral vectors for systemic and local disease. *Curr Opin Mol Ther* 1:573–579.

Staba MJ, Wickham TJ, Kovesdi I, Hallahan DE (2000): Modifications of the fiber in adenovirus vectors increase tropism for malignant glioma models. *Cancer Gene Ther* 7:13–19.

Stevenson SC, Rollence M, White B, Weaver L, McClelland A (1995): Human adenovirus serotypes 3 and 5 bind to two different cellular receptors via the fiber head domain. *J Virol* 69:2850–2857.

Stevenson SC, Rollence M, Marshall-Neff J, McClelland A (1997): Selective targeting of human cells by a chimeric adenovirus vector containing a modified fiber protein. *J Virol* 71:4782–4790.

Stewart PL, Fuller SD, Burnett RM (1993): Difference imaging of adenovirus: bridging the resolution gap between X-ray crystallography and electron microscopy. *Embo J* 12:2589–2599.

Suzuki K, Fueyo J, Krasnykh V, Reynolds PN, Curiel DT, Alemany R (2001): A conditionally replicative adenovirus with enhanced infectivity shows improved oncolytic potency. *Clin Cancer Res* 7:120–126.

Thoma C, Wieland S, Moradpour D, von Weizsacker F, Offensperger S, Madon J, Blum HE, Offensperger WB (2000): Ligand-mediated retargeting of recombinant adenovirus for gene transfer in vivo. *Gene Ther* 7:1039–1045.

Tillman BW, de Gruijl TD, Luykx-de Bakker SA, Scheper RJ, Pinedo HM, Curiel TJ, Gerritsen WR, Curiel DT (1999): Maturation of dendritic cells accompanies high-efficiency gene transfer by a CD40-targeted adenoviral vector. *J Immunol* 162:6378–6383.

Tillman BW, Hayes TL, DeGruijl TD, Douglas JT, Curiel DT (2000): Adenoviral vectors targeted to CD40 enhance the efficacy of dendritic cell-based vaccination against human papillomavirus 16-induced tumor cells in a murine model. *Cancer Res* 60:5456–5463.

Tomko RP, Johansson CB, Totrov M, Abagyan R, Frisen J, Philipson L (2000): Expression of the adenovirus receptor and its interaction with the fiber knob. *Exp Cell Res* 255:47–55.

Trepel M, Grifman M, Weitzman MD, Pasqualini R (2000): Molecular adaptors for vascular-targeted adenoviral gene delivery. *Hum Gene Ther* 11:1971–1981.

van Beusechem VW, van Rijswijk AL, van Es HH, Haisma HJ, Pinedo HM, Gerritsen WR (2000): Recombinant adenovirus vectors with knobless fibers for targeted gene transfer. *Gene Ther* 7:1940–1946.

Vanderkwaak TJ, Wang M, Gomez-Navarro J, Rancourt C, Dmitriev I, Krasnykh V, Barnes M, Siegal GP, Alvarez R, Curiel DT (1999): An advanced generation of adenoviral vectors selectively enhances gene transfer for ovarian cancer gene therapy approaches. *Gynecol Oncol* 74:227–234.

Vigne E, Mahfouz I, Dedieu JF, Brie A, Perricaudet M, Yeh P (1999): RGD inclusion in the hexon monomer provides adenovirus type 5-based vectors with a fiber knob-independent pathway for infection. *J Virol* 73:5156–5161.

Von Seggern DJ, Kehler J, Endo RI, Nemerow GR (1998): Complementation of a fibre mutant adenovirus by packaging cell lines stably expressing the adenovirus type 5 fibre protein. *J Gen Virol* 79:1461–1468.

Von Seggern DJ, Chiu CY, Fleck SK, Stewart PL, Nemerow GR (1999): A helper-independent adenovirus vector with E1, E3, and fiber deleted: structure and infectivity of fiberless particles. *J Virol* 73:1601–1608.

Von Seggern DJ, Huang S, Fleck SK, Stevenson SC, Nemerow GR (2000): Adenovirus vector pseudotyping in fiber-expressing cell lines: improved transduction of Epstein-Barr virus-transformed B cells. *J Virol* 74:354–362.

Walters RW, Grunst T, Bergelson JM, Finberg RW, Welsh MJ, Zabner J (1999): Basolateral localization of fiber receptors limits adenovirus infection from the apical surface of airway epithelia. *J Biol Chem* 274:10219–10226.

Watkins SJ, Mesyanzhinov VV, Kurochkina LP, Hawkins RE (1997): The 'adenobody' approach to viral targeting: Specific and enhanced adenoviral gene delivery. *Gene Ther* 4:1004–1012.

Wickham TJ, Mathias P, Cheresh DA, Nemerow GR (1993): Integrins alpha v beta 3 and alpha v beta 5 promote adenovirus internalization but not virus attachment. *Cell* 73:309–319.

Wickham TJ, Filardo EJ, Cheresh DA, Nemerow GR (1994): Integrin alpha v beta 5 selectively promotes adenovirus mediated cell membrane permeabilization. *J Cell Biol* 127:257–264.

Wickham TJ, Roelvink PW, Brough DE, Kovesdi I (1996a): Adenovirus targeted to heparan-containing receptors increases its gene delivery efficiency to multiple cell types. *Nat Biotechnol* 14:1570–1573.

Wickham TJ, Segal DM, Roelvink PW, Carrion ME, Lizonova A, Lee GM, Kovesdi I (1996b): Targeted adenovirus gene transfer to endothelial and smooth muscle cells by using bispecific antibodies. *J Virol* 70:6831–6838.

Wickham TJ, Tzeng E, Shears LL, 2nd, Roelvink PW, Li Y, Lee GM, Brough DE, Lizonova A, Kovesdi I (1997a): Increased in vitro and in vivo gene transfer by adenovirus vectors containing chimeric fiber proteins. *J Virol* 71:8221–8229.

Wickham TJ, Lee GM, Titus JA, Sconocchia G, Bakacs T, Kovesdi I, Segal DM (1997b): Targeted adenovirus-mediated gene delivery to T cells via CD3. *J Virol* 71:7663–7669.

Wickham TJ, Haskard D, Segal D, Kovesdi I (1997c): Targeting endothelium for gene therapy via receptors up-regulated during angiogenesis and inflammation. *Cancer Immunol Immunother* 45:149–151.

Xia D, Henry LJ, Gerard RD, Deisenhofer J (1994): Crystal structure of the receptor-binding domain of adenovirus type 5 fiber protein at 1.7 A resolution. *Structure* 2:1259–1270.

Xia H, Anderson B, Mao Q, Davidson BL (2000): Recombinant human adenovirus: targeting to the human transferrin receptor improves gene transfer to brain microcapillary endothelium. *J Virol* 74:11359–11366.

Yoon SK, Mohr L, O'Riordan CR, Lachapelle A, Armentano D, Wands JR (2000): Targeting a recombinant adenovirus vector to HCC cells using a bifunctional Fab-antibody conjugate. *Biochem Biophys Res Commun* 272:497–504.

Yoshida Y, Sadata A, Zhang W, Saito K, Shinoura N, Hamada H (1998): Generation of fiber-mutant recombinant adenoviruses for gene therapy of malignant glioma. *Hum Gene Ther* 9:2503–2515.

Yotnda P, Onishi H, Heslop HE, Shayakhmetov D, Lieber A, Brenner M, Davis A (2001): Efficient infection of primitive hematopoietic stem cells by modified adenovirus. *Gene Ther* 8:930–937.

Zabner J, Chillon M, Grunst T, Moninger TO, Davidson BL, Gregory R, Armentano D (1999): A chimeric type 2 adenovirus vector with a type 17 fiber enhances gene transfer to human airway epithelia. *J Virol* 73:8689–8695.

Zsengeller Z, Otake K, Hossain SA, Berclaz PY, Trapnell BC (2000): Internalization of adenovirus by alveolar macrophages initiates early proinflammatory signaling during acute respiratory tract infection. *J Virol* 74:9655–9667.

8

STRATEGIES TO ALTER THE TROPISM OF ADENOVIRAL VECTORS VIA GENETIC CAPSID MODIFICATION

DAVID T. CURIEL, M.D.

INTRODUCTION

Adenoviral vectors have emerged as highly useful agents for a variety of gene therapy applications for a number of reasons. Foremost among the recommending factors is the unparalleled in vivo gene delivery capacity of adenoviral vectors. Indeed, of all currently available vector systems, the potential efficacy rates obtainable for in situ gene delivery via adenoviral vectors far exceed those of alternative viral and nonviral vector systems. This unique feature of adenoviral vectors has thus recommended their use for a variety of gene therapy applications that exploit this delivery capacity. As well, the utility potentials embodied in the adenoviral vector have rationalized the high level of development endeavors which have sought to derive further benefit via direct engineering of the vector to address other gene therapy goals.

These considerations notwithstanding, clinical experience with adenoviral vectors in human trials has revealed limits not apparent in earlier studies with model systems. In vivo gene delivery rates, albeit maximal by viral vector standards, have proven to be suboptimal in applied settings. Specifically, doses

Vector Targeting for Therapeutic Gene Delivery, Edited by David T. Curiel and Joanne T. Douglas
ISBN 0-471-43479-5

of adenoviral vectors required to achieve physiologically meaningful target cell genetic modification have frequently been associated with limiting toxicities. In addition, the promiscuous tropism of the virus mitigates against the central field mandate of cell-specific gene delivery. Both of these considerations are consistent with the concept that the tropism of the parent adenovirus confers key limits on the utility of adenoviral vectors—limits from the standpoints of efficiency and specificity.

This chapter relates to strategies which have been undertaken to address these limits by approaches to alter adenoviral tropism via genetic capsid modification. In this regard, methods to alter adenoviral tropism have been proposed via physical agents designed to cross-link adenoviral vectors to target cells (Douglas et al., 1996). As well, genetic capsid modifications of adenoviral vectors have been endeavored which exploit chimerism of key capsid proteins with respect to alternate adenoviral serotypes (Krasnykh et al., 1996; Stevenson et al., 1997; Von Seggern et al., 2000). Herein we address genetic capsid modifications that have been attempted to directly engineer adenoviral vectors to achieve the goal of cell-specific targeting. Nonetheless, the historically earlier strategies for tropism alteration, just noted, have established key biological principles that have been embodied in the genetic capsid modification approaches. We thus initially consider these earlier approaches from the perspective of their biologic relevance to genetic capsid modification approaches.

BIOLOGY OF TROPISM-MODIFIED ADENOVIRAL VECTORS

The strategic intent of tropism modification of adenoviral vectors is to exploit a cellular receptor allowing efficient and specific gene delivery to target cells. Implicit in this scenario is the notion of exploiting target cell entry aspects distinct from the native primary adenovirus receptor, the coxsackie virus and adenovirus receptor (CAR). For the current purposes it is sufficient to appreciate several key factors with respect to the native adenovirus entry pathway. First, the major adenoviral serotypes employed for vectors, 2 and 5, employ a two-step entry pathway whereby binding to an attachment receptor and an internalization receptor are distinct events. In this regard, adenovirus attaches to its primary receptor CAR by virtue of interactions with specific domains of the knob portion of the fiber capsid protein (Bergelson et al., 1997; Tomko, Xu, and Philipson, 1997; Roelvink et al., 1998). Subsequent to this anchoring step, cellular αv integrins interact with an Arg-Gly-Asp (RGD) motif in the adenovirus capsid penton base to trigger internalization via a clathrin-coated pit, receptor-mediated endocytosis mechanism. From the strategic standpoint, this two-step entry pathway suggests that binding and entry may be functionally uncoupled toward the goal of tropism alteration. The concept embodied herein is that primary cellular attachment would be necessary and sufficient to trigger cellular internalization, irrespective of the specific nature of this initial attachment.

Initial studies exploring this concept were carried out by Douglas and co-workers employing a retargeting complex which served to cross-link an adenoviral vector to the receptor for folate. In this instance, the retargeting complex consisted of the Fab fragment of an antifiber knob antibody, which was chemically coupled to folate (Douglas et al., 1996). This study established several key principles with respect to the biology of tropism-modified adenoviral vectors: (1) that gene delivery could be achieved following adenoviral attachment to a nonnative receptor, and (2) that the levels of gene transfer obtainable were nearly comparable to adenoviral infection via the native pathway. This latter point was critical in consideration of historically earlier attempts to achieve tropism alteration of retroviral vectors. In these studies, altered cell attachment could be achieved via genetic engineering of the retroviral envelope proteins. Vector transduction rates of the targeted retrovirus however, were dramatically decreased (Russell and Cosset, 1999). This loss of vector efficacy was hypothesized to reflect the fact that retroviral binding and entry were functionally linked. Thus, engineered alterations designed to modify retroviral attachment had the unintended consequence of confounding key entry functions. Here the advantages accrued by exploiting the two-step entry pathway of the adenovirus emerge; tropism alterations can be achieved without loss in overall efficiency of gene delivery. This study thus established the key principle that tropism-modified adenoviral vectors could accomplish efficient gene delivery in the context of CAR-independent target cell attachment.

Whereas this study established the feasibility of retargeted adenovirus-mediated gene transfer, the underlying biologic principles relating to the entry dynamics of tropism-modified adenoviral vectors were not fully apparent. From a practical standpoint, the functional capacity of the attachment receptors to internalize is consequential for utility in several described vector contexts (Russell and Cosset, 1999). Specifically, we wondered if anchoring adenoviral vectors to cellular receptors that lacked an intrinsic internalization capacity would allow retargeted adenovirus-mediated gene transfer at high efficiencies. The practical relevance of this consideration relates to the fact that cellular receptors approached via targeted adenoviral vectors could represent molecules that lacked internalization properties. To address this issue, a strategy to target the receptor for epidermal growth factor (EGF-R) was endeavored employing native, internalizing EGF-R versus EGF-R truncation mutants lacking internalization capacity. In these studies, retargeted adenovirus-mediated gene transfer was comparable, irrespective of the internalization biology of the target EGF-R (C. R. Miller, unpublished observation). These studies thus establish a key biological principle with respect to the entry of retargeted adenoviral vectors—anchoring to a nonnative receptor, irrespective of its intrinsic internalization biology allows infection via retargeted adenovirus. This functional uncoupling of binding and entry in the two-step adenoviral entry pathway thus accrues practical benefits for strategies to alter adenoviral tropism in that target cell receptors may be exploited irrespective of their internalization capacity. This recognition expands the nature of targets, and targeting strategies, which may be considered to achieve tropism modifications of adenoviral vectors.

The recognition that CAR-independent target cell attachment could allow cell-specific gene delivery via tropism-modified adenoviral vectors provoked a large number of studies seeking to evaluate this concept in the context of various cellular substrates (Wickham et al., 1996b; Goldman et al., 1997; Rogers et al., 1997; Watkins et al., 1997; Wickham et al., 1997a; Miller et al., 1998; Rancourt et al., 1998; Reynolds et al., 1998; Rogers et al., 1998; Blackwell et al., 1999; Gu et al., 1999; Haisma et al., 1999; Harari et al., 1999; Hong et al., 1999; Tillman et al., 1999; Chandler et al., 2000; Haisma et al., 2000; Kelly et al., 2000; Printz et al., 2000; Schneider et al., 2000; Tillman et al., 2000; Trepel et al., 2000; Yoon et al., 2000; Ebbinghaus et al., 2001). These studies have validated the universality of the biologic principles of retargeted adenoviral infection and have formally established a range of cellular targets exploitable for targeted gene delivery. In the course of these studies, it was noted that retargeted adenoviral infection often achieved levels of gene transfer efficiency of even greater magnitude than the untargeted adenoviral vector. This was most noteworthy in the context of primary tissue targets, a substrate observed to be relatively refractory to adenoviral infection (Miller et al., 1998; Kasono et al., 1999; Vanderkwaak et al., 1999). In this regard, the means to evaluate target cell levels of the native adenovirus attachment receptor CAR provided the basis of correlating target cell adenoviral susceptibility with levels of CAR (Bergelson et al., 1997). Such analysis of a variety of primary tumor cells demonstrated profound deficiencies of CAR which were frequently not noted in the context of the immortalized cell line counterparts of the study tissue. As noted, such adenovirus-resistant targets could frequently be infected with augmented efficiency via retargeted adenoviral infection. These augmentations achieved via retargeted adenoviral infection could be of the magnitude of 1000- to 10,000-fold (Kasono et al., 1999; Vanderkwaak et al., 1999). Several key biologic principles thus derived from these studies, with direct consequence for retargeted adenoviral infection. First, immortalized cell lines provide a spurious substrate for analysis of vector susceptibility; primary material more correctly reflects the biologic dictates of vector tropism relevant to retargeted adenoviral infection. Second, many cellular targets exhibit profound resistance to adenoviral vectors on the basis of CAR deficiency. Of note, retargeted adenoviral infection is able to circumvent this resistance via exploiting CAR independent attachment biology. This CAR independence is relevant not only with respect to the goal of cell-specific targeting, it likewise provides the basis of enhancement of gene delivery to target cells.

From the historical perspective, most strategies to retarget adenoviral infection have employed retargeting complex-based approaches. These approaches have demonstrated gene therapy utility in both in vitro and in vivo model systems. On this basis, modification of current adenovirus-based human clinical trials has been proposed to incorporate these vector modifications. For reasons that will be advanced, these same human clinical considerations have spurred the development of genetic capsid modification approaches to achieve adenoviral vector retargeting as an alternative to the retargeting complex-based methods. Nonetheless, these earlier approaches have established critical biologic principles with respect to retargeted adenoviral infection with direct relevance to retargeting strategies based on genetic capsid modification.

GENETIC CAPSID MODIFICATION—UNDERLYING PRINCIPLES

The practical translation of tropism-modified adenoviral vectors based on retargeting complexes is limited by several considerations. In this regard, a variety of retargeting complex species have been employed to achieve retargeted adenoviral infection. This diversity has reflected the fact that a species with optimal properties has not emerged. Specifically, issues of upscaling may limit current retargeting complex utilities for human translation. In addition, issues of purity and homogeneity have represented a confounding factor. These considerations have led to the concept that the optimal scenario would be the development of an adenoviral vector embodying the aforementioned principles of retargeting, but retaining the single-unit configuration of the native adenoviral particle. Indeed, the experience of numerous investigative groups, working in direct contact with the United States Food and Drug Administration (FDA), has clearly established the desirability of this type of vector configuration.

On the basis of these considerations, strategies have been attempted to derive retargeted adenoviral vectors via genetic capsid modifications. Consideration of the native entry pathway of the adenovirus provided a logical framework for retargeting; the two-step entry pathway allows retargeting strategies based on the functionally uncoupling of binding and entry for targeting purposes. The other major lesson of the adenoviral entry pathway provides the basis for advancing the goal of retargeted adenoviral vectors via genetic capsid modification. Specifically, the key steps of adenovirus target cell binding and entry are mediated by viral capsid proteins (Nemerow 2000). It is therefore logical that these steps could be altered by modification of the corresponding capsid proteins.

The primacy of the adenovirus fiber protein in the initial attachment process logically suggests that modification of fiber could provide the basis of tropism alteration, therefore most initial work focused on this hypothesis. Nonetheless, other capsid proteins are involved in the process of adenovirus entry and these have likewise been exploited for the achievement of tropism alteration. Additionally, adenoviral structural proteins, not known to be involved in attachment or entry have been studied. The diversity of approaches has reflected the nontrivial nature of endeavoring genetic capsid engineering, an approach that must embody consideration of adenoviral biosynthesis and particle assembly.

STRATEGIES TO GENETICALLY ALTER FIBER FOR TROPISM MODIFICATION

Michael and co-workers were the first to attempt to genetically modify fiber for targeting ends. They constructed genetic fusions incorporating the gastrin-releasing protein (GRP) at the carboxy (C) terminus of the fiber and studied them from the standpoint of fiber biosynthesis. This strategy was dictated by the fact that the C terminus corresponds to the knob protein of fiber. Fiber is normally synthesized in the cellular cytosol, where it trimerizes by virtue of elements in the knob domain. It was thus apparent that any strategy to genetically modify the fiber could not perturb this interaction in order to allow for this key biosynthetic aspect. These studies established that short peptides could be added to fiber and that they did not perturb fiber biosynthesis. Analysis with anti-GRP antibodies demonstrated that the incorporated heterologous peptide was localized in a surface-exposed configuration. These studies thus made feasible the approach of exploiting the C-terminus of the fiber as a locale for direct incorporation of targeting ligands.

Independently, Wickham et al. endeavored to incorporate candidate ligands at the fiber C-terminus for retargeting purposes (Wickham et al., 1996a, 1997b). In two separate reports, the candidate ligands polylysine and the peptide RGD were evaluated. The former peptide motif is capable of recognizing cell surface heparan sulfate receptors, while the latter is capable of binding to integrins $\alpha v\beta 3$ and $\alpha v\beta 5$. The authors were able to derive viruses that incorporated these fiber additions. Importantly, the incorporated ligands were capable of conferring CAR-independent gene delivery on the derived vectors. This capacity allowed augmented infection efficiency to a series of target cells known to be relatively refractory to adenovirus. These studies thus formally established that retargeted adenoviral vectors could be obtained via genetic capsid modifications. Further, direct accrual of vectorologic gain could be demonstrated—CAR-independent gene delivery and augmented vector efficacy for refractory targets.

As noted, any genetic fiber modifications must allow retention of physiologic fiber biosynthesis. In this regard, mutations known to ablate fiber trimerization abrogate particle assembly. These considerations were highly consequential for the aforementioned studies, as attempts to incorporate relatively larger ligands proved unsuccessful (Wickham et al., 1996a, 1997b). These studies seemed to establish a size limit of amino acids with respect to compatibility with fiber biosynthesis. These results were paralleled by Hong and Engler (1996) who demonstrated a similar upper limit for incorporated heterologous peptides employing an in vitro biosynthesis system. Although neither of these studies evaluated the upper limits of the incorporated peptide with respect to qualitative aspects, they did at least suggest a general range of ligand size potentially incorporable at this locale. Of note, the apparent upper limit defined was well below the size of readily exploited targeting moieties, such as single-chain antibodies (scFvs).

Direct attempts have nevertheless been made to exploit the available capacities of the C-terminal site of fiber. Approaches were explored that sought to exploit targeting ligands of a size compatible with the apparent limits of the fiber C-terminus. In this regard, Mehtali incorporated GRP (M. Mehtali, unpublished observations). Whereas a virus could be rescued that incorporated this ligand, functional retargeting could not be demonstrated. In consideration of this result, it is noteworthy that many small peptide ligands require terminal amidation for efficient receptor binding. The absence of such processing, in this instance, could have confounded the utility of the candidate ligand. In any event, these results suggested that the issue of ligand fidelity in the vector context might be more complex than initially conceptualized.

Despite the limited utility of this site with respect to the type of ligands that could be exploited for targeting purposes, attempts to employ the derived vectors with incorporated polylysine motifs have been endeavored. In vitro analysis of these vectors on various cellular substrates has confirmed the utility of CAR-independent attachment as a means to circumvent CAR-deficiency (Gonzalez et al., 1999a, 1999b). As well, model systems of in vivo gene therapy have demonstrated practical gain with respect to gene transfer to target cells. This has been accomplished in a model of in situ gene delivery to vascular endothelium (McDonald et al., 1999) and for genetic modification of muscle cells (Bouri et al., 1999). The former study was accomplished in an isolated vascular graft, purged and devoid of serum. Direct extrapolation of these findings to in vivo gene delivery via the vascular route have not been reported. Both of these studies thus highlight the basic reality of the polylysine modification—in vitro augmentations of gene transfer are not paralleled by significant augmentations in vivo. Nonetheless, the fiber C-terminus ligand approach has demonstrated the key principle that genetic capsid modification can accomplish tropism modification of adenoviral vectors. However, issues related to the apparent size restriction imposed by this site have not allowed direct exploitation of this locale. These considerations have led to the exploration of alternative capsid sites for incorporation of heterologous targeting ligands for derivation of practically useful retargeted adenoviral vectors.

Studies exploring the utility of the fiber C-terminus established a number of key concepts with respect to the criteria for an exploitable capsid locale for ligand incorporation. In this regard, ligands should be of a size to be compatible with the structural restraints of their intended capsid site. Second, capsid incorporation should not alter configuration aspects of the targeting motifs that would confound its fidelity. As well, the targeting ligand should be of a composition and nature to be compatible with the biosynthesis of capsid proteins. The initial candidate ligands considered for capsid incorporation were thus small physiologic peptide ligands of native cellular receptor pathways. Whereas these ligands fit the aforementioned criteria, they represented an extremely limited repertoire for practical utilization. This recognition has led to the consideration of methods to directly derive nonnative ligands capable of recognizing target cell surface markers. In this regard, the recently developed technique of phage display library biopanning offers a means to define candidate ligands for this application. Specifically, this method can define target cell-specific signatures recognized by selected peptides or scFvs (O'Neil et al., 1992; Doorbar and Winter, 1994; Goodson et al., 1994; Barry et al., 1996; Pasqualini, Koivunen, and Ruoslahti, 1997; Arap et al., 1998; Pasqualini et al., 2000). The advent of this methodology thus has established a source of unlimited potential for the generation of targeting ligands compatible with the dictates of adenovirus capsid incorporation.

It is noteworthy that the fiber C-terminus studies were all endeavored without a priori knowledge of the structure of the fiber knob. The proposal of a three-dimensional model of the adenovirus fiber knob by Xia and colleagues provided the basis to explore alternate locales for ligand incorporation (Xia et al., 1994). In this regard, this model proposes specific loops within the context of the knob structure. A number of considerations suggested that these loops could function as locales for ligand incorporation. First, the model localizes these loops to the surface of the knob, which means they are theoretically accessible for direct interaction with target cell surface receptors. Second, these loops are not involved in intermolecular interactions within the fiber knob, and therefore alterations within their structure would not appear deleterious to the overall configuration of fiber. Third, significant primary amino acid sequence variability of the loops is noted between human adenovirus serotypes, suggesting that these structures do not subserve highly critical functions. These considerations predicated the evaluation of the HI loop of the fiber knob as a candidate site for ligand incorporation toward the goal of adenoviral tropism modification.

An initial proof-of-principle study was endeavored by Krasnykh and co-workers employing a peptide tag in the HI loop (Krasnykh et al., 1998) (Fig. 8.1).

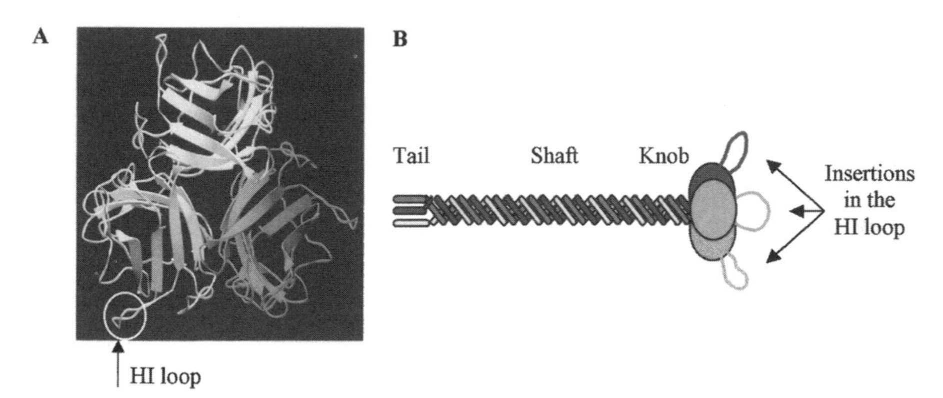

Figure 8.1 Structure of adenovirus fiber protein and its knob domain. *A*. According to the three-dimensional model of the fiber knob domain, it resembles a three-bladed propeller formed by two sheets of ß-strands connected with loops and turns. The flexible HI loop (red circle), which connects strands H and I, is exposed outside the knob and, therefore, provides a convenient locale for incorporation of targeting ligands. *B*. The fiber protein incorporated into each of the 12 vertices of the icosahedral adenoviral capsid is a homotrimeric molecule, which consists of three distinct structural domains: the *tail*, the *shaft*, and the *knob*. The tail [illegible] of the fiber in the adenoviral capsid via noncovalent association with the penton base protein, whereas the rodlike shaft serves to extend the globular knob domain away from the virion, thereby facilitating interaction between the fiber and CAR. The knob domain fulfils double duties by maintaining trimerization of the fiber and binding to CAR.

Specifically, the octapeptide FLAG tag was genetically incorporated into the HI loop of the fiber knob. Analysis of the tagged fiber, in recombinant form, demonstrated clearly that the incorporated tag did not affect the native trimeric configuration of the molecule. As well, the recombinant fiber retained the ability to interact with its native receptor, CAR. These feasibilizing studies rationalized the construction of an adenoviral vector containing the FLAG peptide at the HI loop site. The rescued virus containing the FLAG peptide within the HI loop retained the ability to accomplish CAR-dependent gene transfer. Importantly, antibody probe analysis of the virion confirmed the surface locale of the peptide tag. These studies thus established that the HI loop possesses key attributes rationalizing its development as a locale for targeting ligand incorporation.

The same investigative group attempted a follow-up analysis whereby integrin-targeting peptide was configured at the HI loop locale (Dmitriev et al., 1998). Specifically, the peptide RGD-4C had been defined by in vivo phage display biopanning by Ruoshahti and shown to recognize integrins of various classes (Pasqualini, Koivunen, and Ruoslahti, 1997; Arap et al., 1998). Importantly, this peptide had a number of features recommending its use for targeting purposes. First, its relatively small size was consistent with ligands tested within the HI loop context. Second, its targeting capacity was highly relevant to cancer applications, as the target integrins were known to be upregulated in tumor endothelium and in selected tumor cell targets. On this basis, an RGD-4C-containing virus was rescued and analyzed for its function capacities. Most noteworthy in this regard, the virus Ad5lucRGD was capable of accomplishing gene transfer via CAR-independent mechanisms. This finding is consistent with the concept that incorporation of the targeting ligand had effectively expanded the tropism of the adenoviral vector. These studies thus established the key concept that ligands incorporated at the fiber knob HI loop locale could function to alter adenoviral tropism. This finding warranted exploration of both the direct utility of the derived virus, Ad5lucRGD, as well as the potential of this site for additional adenovirus targeting applications.

As noted earlier, one rationale for modifying the tropism of adenoviral vectors is the relative paucity of the primary receptor CAR on many target cells. This recognition invokes the requirement to accomplish CAR-independent gene transfer as an approach to achieve effective adenovirus-mediated gene transfer to such targets. On this basis, direct analysis of Ad5lucRGD was endeavored in a series of CAR-deficient target cells. These studies demonstrated that augmented gene transfer could be accomplished to glioma cells (Grill et al., 2001), myelomonocytic leukemia cells (Garcia-Castro et al., 2001), head and neck cancer cells cells (Kasono et al., 1999), pancreatic cancer cells (Wesseling et al., 2001), rhabdomyosarcoma cells (Cripe et al., 2001) and various other types of neoplastic targets. Gene transfer efficacy augmentations of up to three orders of magnitude could be noted in selected contexts (Kasono et al., 1999). This was especially noteworthy for primary tumor material. Indeed the universality of augmented gene transfer accomplished by Ad5lucRGD to primary tumor cells strongly supports the concept of CAR deficiency as an axiomatic aspect of tumor biology, as has been suggested by others (Okegawa et al., 2000).

The wherewithal to increase gene transfer to target cells has naturally led to studies to establish the practical gene therapy benefits accrued via this type of tropism modification. In this regard, the RGD-4C modification allows for an infectivity enhancement of tumor cells potentially allowing an improved therapeutic index for cancer gene therapy applications. This has been confirmed in studies whereby RGD-4C-modified versions of vectors encoding various anticancer toxin genes (cytosine deaminase, herpes simplex thymidine kinase) have shown improved therapeutic capacities compared to unmodified adenoviral vector counterparts in model systems (Blackwell et al., 2000a; Hemminki et al., 2001). In addition, infectivity-enhanced versions of conditionally replicative adenoviral agents (CRAds) exhibit enhanced antitumor potency (Cripe et al., 2001; Suzuki et al., 2001). Clearly, the incorporation of RGD-4C in the HI loop does not create a vector that exhibits true tumor targeting. Nonetheless, the tropism expansion achieved provides an augmented potency, and in selected cases, a targeting advantage, which recommends this modification. On this basis, a number of proposed human clinical trials have been advanced that employ adenoviral agents with RGD-4C HI loop modifications. The validity of therapeutic goals obtainable via this new class of infectivity-enhanced adenoviruses remains to be confirmed in these human trials.

A noteworthy aspect of the HI loop-modified vectors has been their performance in vivo. In this regard, a significant drawback of the fiber knob C-terminus as a locale for ligand insertion was the apparent disconnect between the gene transfer augmentations noted in vitro and in vivo. Indeed, it has been suggested that humoral factors may directly inactivate fiber C-terminus modified viruses, thus confounding their relevance to in vivo gene therapy applications (T. J. Wickham, unpublished observations). In marked contrast, HI loop-modified vectors containing RGD-4C have demonstrated significant augmentations in gene transfer efficiency when applied in various in vivo contexts (Reynolds et al., 1999; Blackwell et al., 2000b; Bauerschmitz et al., 2001). In this regard, direct intratumoral injection with Ad5lucRGD induces significant enhancements of in situ gene transfer compared to an unmodified adenoviral vector (Blackwell et al., 2000a). In addition, systemic vascular injection of Ad5lucRGD resulted in a biodistribution pattern of gene transfer distinct from that noted with an unmodified adenoviral vector (Reynolds et al., 1999). The significantly augmented levels of gene transfer noted in selected integrin-expressing contexts is consistent with the notion that the tropism expansion capacities of Ad5lucRGD are operative in this most stringent in vivo delivery context.

The vector gains noted in the context of the HI loop strategy might logically predict the development of a host of vectors for specific targeting applications. The absence of further examples of this vector class has suggested that simple configuration of phage-derived peptides into the HI loop might not be as straightforward as noted for the peptide RGD-4C. Whereas there are several reports of other phage-derived peptides predicating corresponding targeted adenoviral vectors, the paucity of such successful applications belies the notion that a simple functional link exists between phage display biopanning target definition and vector targeting capacities based on exploitation of the HI loop locale. Indeed, a large number of physiologic peptides, as well as phage-defined peptide ligands, have not allowed for the construction of the corresponding targeted adenoviral vector (T. J. Wickham, unpublished observation). In these instances, the large majority of these peptides do not retain targeting capacity at the new locale within the adenoviral capsid. Alternatively, direct HI loop incorporation of larger targeting ligands, such as scFvs, has not been feasible. These findings highlight the key issue of ligand fidelity in the vector context. As well, these findings suggest that even a locale with the apparent utilities of the HI loop may have practical limits; the size and structural constraints imposed by the fiber knob may ultimately limit exploitation of important classes of targeting ligands.

As has been apparent up to this point, there exist a relative paucity of natural targeting ligands exploitable for vector-targeting purposes. This situation has predicated the nearly universal recognition of the enormous potentials embodied in the technique of ligand definition via phage display biopanning. Indeed, this method would appear to offer the highly desirable ability to identify unique, cell-specific markers for any given candidate target cell. In combination with the identification of an optimal capsid locale for ligand incorporation, the capacity would thus theoretically exist for a direct target definition/vector targeting link. The overriding desirability of this goal highlights the significance of the finding that numerous phage-defined peptides appear to be nonoperational in the HI loop, despite the fact that structural incompatibilities between phage peptides and the HI loop site do not impact negatively on viral structure. Thus, the issue of fidelity of a targeting ligand at phage and viral locales is of overriding significance. This issue has recently been addressed by Pereboev and co-workers who have developed a novel phage display system designed to address the fidelity issue (Pereboev, Pereboeva, and Curiel, 2001). Specifically, they have developed a phage display method whereby the fiber knob of adenovirus is displayed on the phage surface. At this locale the knob trimerizes and retains the ability to interact with CAR. These key findings suggest that the phage–knob display system is capable of presenting the knob in a functional context with direct relevance to intact adenoviral particles. Albeit early in development, this system would appear to address the concept of ligand fidelity, which has proven to be so crucial in the context of attempts to establish a truly functional target definition/vector targeting link.

The issue of exploitable ligands has also led to the exploration of the size constraints of the HI loop vis à vis incorporation of heterologous peptides. Specifically, an understanding of the upper limits of the HI loop's accommodation capacity would logically predicate the consideration of candidate targeting ligands of potentially greater size/complexity. Additionally, the creation of a locale with structural flexibilities could potentially allow ligands to function with improved fidelity. On this basis, Belousova and co-workers have endeavored an analysis of the ligand incorporation capacities at the HI loop. They have generated a series of adenoviral vectors whereby the RGD-containing loop of the penton base has been engrafted into the HI loop of the fiber knob with variable lengths of linker peptide. These studies have demonstrated that heterologous sequences of up to 83 amino acids may be configured in the HI loop of fiber knob without significant loss of vector function or infectivity (N. Belousova, unpublished observation). This expanded understanding of the ligand incorporation capacity of the fiber knob HI loop may thus allow consideration of candidate ligands of increased size. This advancement notwithstanding, the basic issue of size restriction will likely remain consequential in consideration of ligands of the size of scFvs. Indeed, this recognition highlights the basic fact that working within the constraints imposed by the structure of the fiber knob may ultimately impose severe practical limits on the targeting potentials achievable.

STRATEGIES TO GENETICALLY REPLACE FIBER FOR TROPISM MODIFICATION

From a conceptual standpoint, all of the fiber modifications strategies noted previously have proceeded from a central assumption, that is, that the knob is essential for fiber trimerization and thus incorporated targeting ligands should function within this framework. As we have seen, this approach imposes specific limits which, despite elegant viral engineering, functionally exclude a variety of targeting ligand types of high potential utility. On this basis, a strategy to circumvent this limit might involve replacement of the knob entirely. Such an approach would necessarily be required to compensate for the key structural contribution of knob to the quaternary configuration of the fiber. Specifically, the deletion of fiber knob would require an alternate means to retain trimerization of the fiber protein. Realization of this goal would potentially allow for the incorporation of a wider range of targeting ligands on the basis of removal of the structural constraints imposed by the fiber knob. These considerations have thus led to the development of methods to replace the fiber, or fiber knob, as an approach to achieve adenoviral tropism modifications.

Two critical considerations relate to the realization of such an approach. First, the N-terminus tail of the fiber is required for association with the penton base in the context of capsid assembly. Thus, any derived chimeric fiber would be required to incorporate this requisite component of fiber related to assembly. Second, a trimerization motif would be required to substitute for the functions normally provided by the knob. To this end, a variety of naturally occurring trimeric molecules have been reported that potentially could fulfill this requirement. Krasnykh et al. have pursued such a strategy exploring the utility of the bacteriophage T4 protein fibritin as a candidate heterologous trimerizer (Krasnykh et al., 2001) (Fig. 8.2). In this regard, fibritin pos-

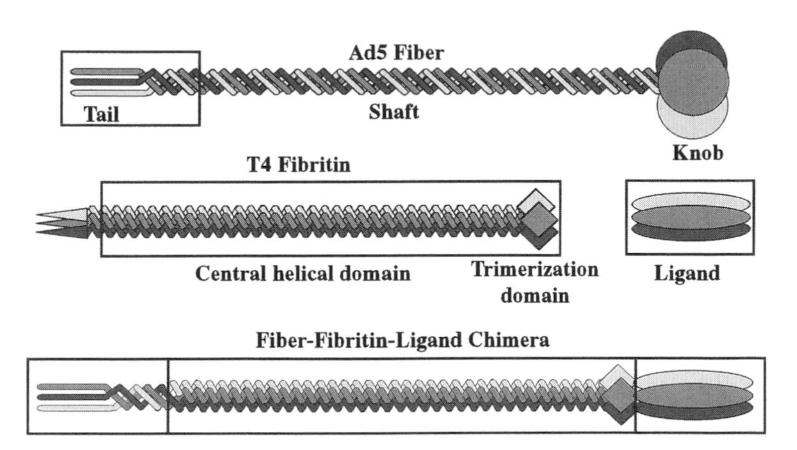

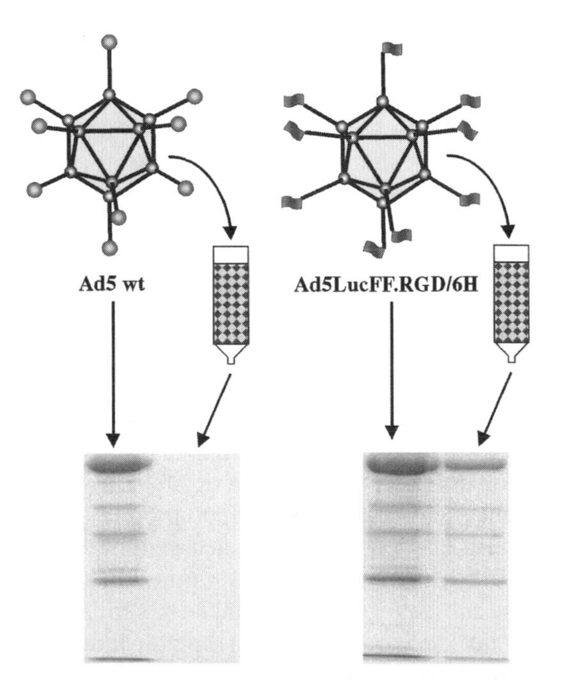

Figure 8.2 Fiber replacement strategy for genetic capsid modification tropism-modification. *A*. Schema of strategy for construction of chimeric fiber molecule whereby Ad5 fiber shaft and knob are replaced by trimerization domain derived from T4 fibritin. *B*. Demonstration of capsid incorporation of chimeric fiber-fibritin molecule. Adenovirus 5 wild-type (Ad5 wt) does not exhibit binding to a NiNTA column. An adenovirus containing chimeric fiber-fibritin with 6His Tag, is specifically bound, and elutable from the NiNTA column, indicating its incorporation into capsid.

sesses a number of attributes recommending it for this application. First, it is capable of forming stable homotrimers on the basis of its C-terminus foldon domain. Second, this trimeric structure can be retained in the context of N-terminus deletions and C-terminus additions. Additionally, fibritin fusions can be readily produced in recombinant systems, allowing for their facile characterization.

On this basis, a chimeric fiber protein was designed consisting of the native adenovirus fiber N-terminus, fibritin as a shaft/knob substitute, and a 6His peptide tag. The latter motif could function as a model ligand for target cells expressing an "artificial receptor" containing a transmembrane anti-His scFv displayed on the cell surface (Douglas et al., 1999). Rescue of this virus employed a technique described by von Seggern et al., and allowed for the derivation of a homogenous population of viral particles containing the chimeric fiber substitute protein. Gene transfer experiments demonstrated that the vector was capable of specific gene transfer via the targeted artificial receptor pathway, and that such gene transfer was accomplished in a CAR-independent manner. Thus, this strategy allowed for highly selective gene transfer without native tropism contributions. Whereas gene transfer efficiencies were less than for adenoviral vectors with wild-type fiber, this difference was within one to two orders of magnitude. These findings must be understood in the context of the fact that the receptor-ligand affinity of this model virus may have been substantially less than that of fiber knob for CAR. As well, the important parameter of shaft length had not been optimized. It will thus be critical to determine if optimization of these aspects of virus–cell interaction will allow the vector to function at efficiency levels fully comparable to adenoviral vectors containing wild-type fibers. Nonetheless, the potentials embodied in this system appear significant: the ability to accomplish truly targeted gene transfer, in combination with an expanded capacity for ligand incorporation, will address many of the limits noted in the fiber modification schemas.

Additional strategies to substitute for the trimerization function of knob have also been endeavored by other groups. Van Beusechem et al. have sought to replace the fiber knob domain with the trimeric alpha-helical coil domain of the Moloney murine leukemia virus p15 envelope protein (van Beusechem et al., 2000). These chimeric fibers appeared to lack stability and the resultant adenoviral vector particles exhibited only very limited incorporation of this fiber species. Magnusson and colleagues have taken a similar approach, endeavoring to employ the neck region peptide of human lung surfactant protein D for trimerization function (Magnusson et al., 2001). Again, the very limited vector data presented did not allow an understanding of the targeting capacities and efficiencies of this system. Thus, in these instances, the rationalizing principle of this line of investigation has not been established—that is, the incorporation of larger/more complex targeting ligands made possible by knob removal.

STRATEGIES TO ALTER ALTERNATIVE CAPSID PROTEINS FOR TROPISM MODIFICATIONS

Attempts to alter the Ad fiber capsid for tropism alteration logically derived from considerations of the major role fiber plays in target cell recognition. In addition to the aforementioned strategies to modify fiber for tropism alteration purposes, there have also been attempts to delete this protein entirely. This strategy sought to eliminate the contribution of native tropism in undermining true cell-specific targeting. Implicit in this approach was the concept that fiber ablation could be combined with the incorporation of targeting ligands at alternate capsid locales, such as the penton base and hexon. The removal of fiber would thus eliminate its contribution to ectopic delivery, as well as compensate for any steric effects it might have vis à vis cellular recognition of targeting ligands at these alternate locales. Initial attempts at derivation of fiberless adenoviral constructions demonstrated the ability to rescue such species (Nemerow, 2000). These agents appeared to possess defects in proteolytic processing of capsid proteins, in addition to the intended fiber deletion. Most importantly, gene transfer frequencies associated with fiberless adenoviral vectors have been observed to be profoundly decreased compared to adenoviral vectors containing fibers.

Whereas these studies defined unanticipated processing defects associated with fiber deletion, they did provoke consideration of alternate capsid locales as candidates for ligand incorporation (Fig. 8.3). Such a strategy was also sug-

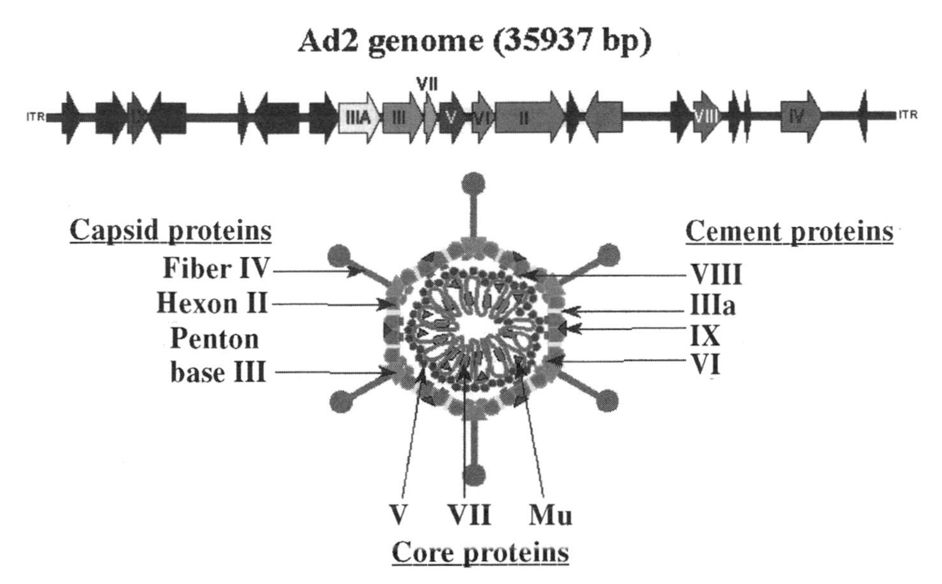

Figure 8.3 Major adenovirus structural proteins potentially exploitable for incorporation of heterologous ligands.

gested by other lines of data. In this regard, adenovirus serotype 9 appears to exploit penton base binding to cellular integrins as a major determinant of its tropism. In addition, heterologous ligands have been configured into penton base loop structures (Wickham, Carrion, and Kovesdi, 1995). Such incorporated motifs at penton base have been exploited as anchors for antibody bridge-based retargeting complexes to achieve tropism alterations. The foregoing studies have thus established that the penton base determinants may play a major role in adenoviral tropism and also that the penton base may be genetically modified to incorporate heterologous ligands. On this basis it is logical to consider further approaches designed to exploit the potential utility of penton base for tropism modification endeavors.

The other major capsid protein, hexon, is principally thought to function as scaffolding for the viral capsid structure and does not have a known role in adenoviral target cell attachment or entry. Therefore, scant attention has been paid to this locale for tropism modification purposes. However, its relative abundance within the capsid context, as well as its significant contribution to the capsid exterior, may make this an attractive locale for such endeavors. Vigne and co-workers have replaced hypervariable region 5 (HVR5) in the adenoviral hexon protein with an RGD-containing peptide with flexible peptide linkers (Vigne et al., 1999). This vector was reported to be capable of accomplishing CAR-independent gene transfer to CAR-deficient smooth muscle cells. A subsequent study by Rux and Burnett provided an additional rationale for exploiting hexon for genetic capsid modification based on their crystal structure analysis, which demonstrated a surface locale for several of the hexon HVR regains (Crawford-Miksza and Schnurr, 1996; Rux and Burnett, 2000). It is thus apparent that the hexon may offer significant potential as a locale for incorporation of heterologous targeting ligands for tropism modification purposes.

In addition to the aforementioned major capsid proteins, the adenovirus also contains a number of so-called minor capsid proteins. This group includes the capsid proteins pIIIa and pIX, termed *cement proteins*, which are thought to foster intermolecular interactions of the hexon proteins. Dimitriev and co-workers have explored this locale and have demonstrated that both the pIIIa and pIX capsid proteins can accommodate the addition of terminal extensions of heterologous peptides (Fig. 8.4). These extensions are presented on the capsid exterior and are capable, in selected instances, of allowing CAR-independent gene transfer (Fig. 8.5). These studies are at present in their early stages; however, they do establish the candidacy of additional capsid proteins for altering adenoviral tropism via genetic capsid modifications (Dmitriev et al., 2001).

Each of the aforementioned strategies clearly embodies potentials that have not been fully developed to date. These approaches may offer a direct means to achieve targeting, or may provide adjunctive methods that operate in combination with fiber modification/replacement methods.

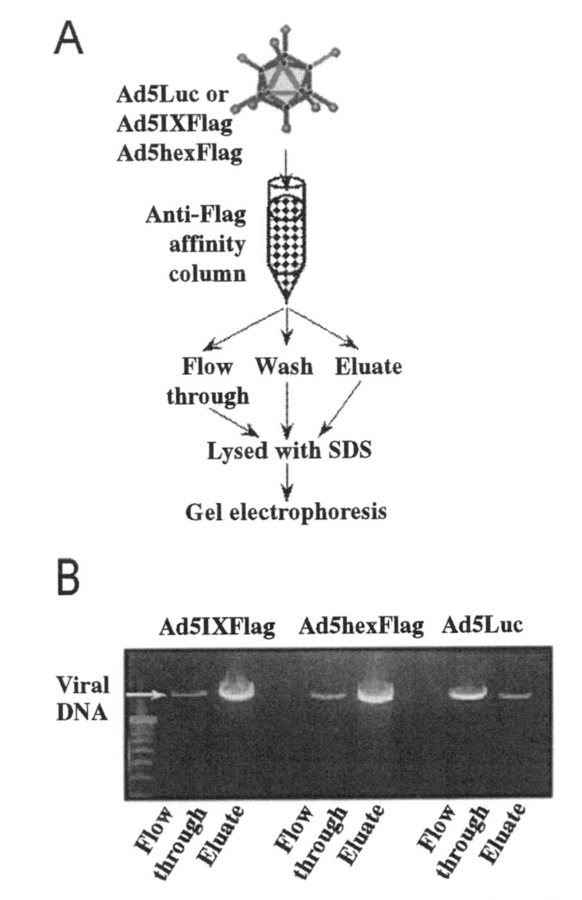

Figure 8.4 Incorporation of heterologous peptide at C-terminus of adenovirus protein IX. *A*. Schema for analysis of adenovirus species by affinity chromatography. Unmodified adenovirus (Ad5Luc), adenovirus modified to contain the Flag peptide at protein IX (Ad5IXFlag) and adenovirus modified to contain the Flag peptide at the hexon (Ad5hexFlag) were subject to column chromatography, as indicated. *B*. Analysis of adenovirus variants by column chromatography. Results demonstrate that Ad5IXFlag exhibited specific binding to column, indicating accessibility of the Flag peptide at the C-terminus of protein IX.

CONSIDERATIONS IN THE CLINICAL APPLICATION OF TROPISM-MODIFIED ADENOVIRAL VECTORS

An increasing proportion of cancer gene therapy approaches currently employ adenoviral vectors. This recognition stems from the fact that in vivo approaches to cancer gene therapy may be most practically endeavored by direct tumor cell transduction. These considerations relate especially to the anticancer gene ther-

A

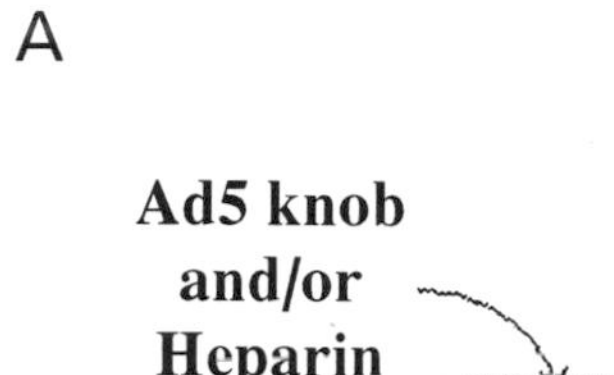
Ad5 knob
and/or
Heparin

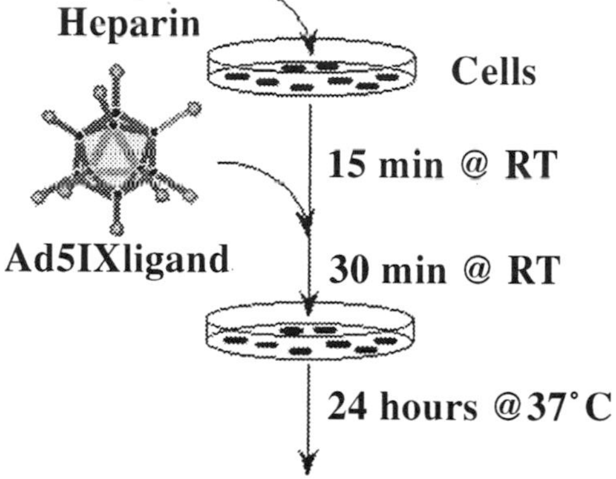
Cells
15 min @ RT
Ad5IXligand
30 min @ RT
24 hours @37°C
Luciferase assay

B

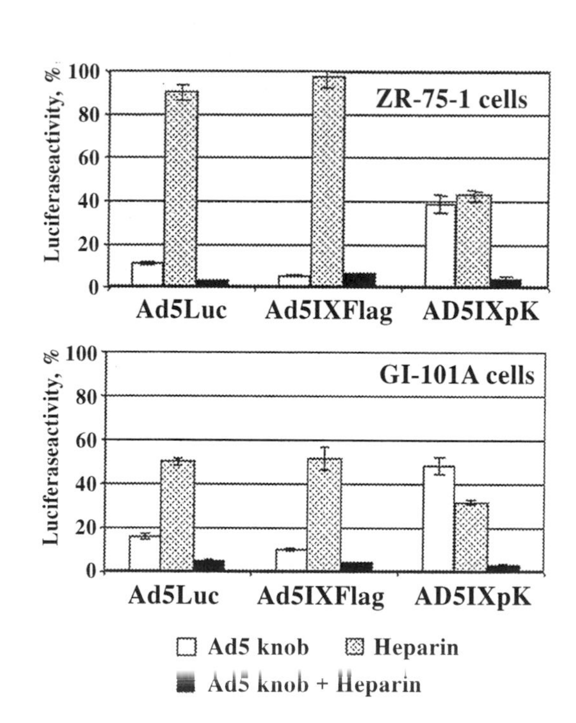
Luciferaseactivity, %
100
80
60
40
20
0
ZR-75-1 cells
Ad5Luc
Ad5IXFlag
AD5IXpK
Luciferaseactivity, %
100
80
60
40
20
0
GI-101A cells
Ad5Luc
Ad5IXFlag
AD5IXpK
Ad5 knob
Heparin
Ad5 knob + Heparin

apy approaches of molecular chemotherapy and mutation compensation, which involve direct in situ gene delivery to tumor cells in patients (Gomez-Navarro, Curiel, and Douglas, 1999). On this basis, the unparalleled in vivo gene transfer efficiency of the adenoviral vector has predicated its emergence as the vector most capable of achieving the requisite levels of tumor cell transduction. Nonetheless, the key limits of adenoviral vectors, noted previously, have enormous bearing on the results obtained in these trials. In the first instance, the promiscuous tropism of adenoviral vectors would lead to widespread ectopic gene transfer if vector was administered in a noncontained fashion. On this basis, in vivo adenoviral protocols have been largely confined to disease contexts whereby the vector could be delivered within a body compartment, such as the central nervous system or peritoneal cavity. Thus, the lack of a cell-specific targeting capacity severely restricts the application of cancer gene therapies to only a highly defined set of disease types and stages. Studies carried out to date have established that such containment approaches, as developed, are not adequate to fully mitigate the possibility of ectopic gene delivery. Both murine model systems and human clinical trials have highlighted that such ectopic gene delivery may be the basis of toxicities (Lambright et al., 2001). Thus, targeting is required not only to allow application of gene therapy to noncompartment contexts of neoplastic disease, but also to provide a safety margin even for application to clinical settings where intrinsic targeting appears to be operative.

In addition, human trials endeavored to date have demonstrated a disappointingly low magnitude of in situ gene transfer accomplished via adenoviral vectors (Krasnykh et al., 1996), even in optimized settings whereby vector is administered directly to a body compartment containing the tumor targets. Of note, such human interventions were frequently based on model system studies in which the adenoviral vector appeared to be highly efficacious in accomplishing gene transfer to tumor targets. As noted earlier, there frequently exists a major disparity between levels of the primary adenovirus receptor CAR asso-

Figure 8.5 Utilization of adenovirus capsid protein IX for incorporation of a heterologous ligand. Genetic modification of adenovirus allowed incorporation of polylysine peptide at the C-terminus of protein IX. *A*. Schema for gene transfer via protein IX-modified adenoviral vector encoding luciferase reporter gene. *B*. adenoviral vectors with unmodified capsid (Ad5Luc), Flag peptide at protein IX (Ad5IX Flag), or polylysine peptide at protein IX (Ad5IXpK) were delivered to ZR-75-1 or GI-101A cells in the presence of various blocking agents (Ad5 knob and/or heparin) with luciferase activity measured as a function of unblocked adenovirus of the corresponding species. Ad5IXpK exhibits nonblockable gene transfer in the presence of knob and/or heparin, indicating polylysine at the protein IX site functions to allow CAR-independent gene transfer based on pK interaction with corresponding cellular receptors.

ciated with immortalized human cells and their primary tumor counterparts. This adenovirus-refractory phenotype of primary tumors has led to the employment of elevated adenoviral doses as a means to achieve an effective level of gene transfer relevant to the endeavored genetic intervention. In this regard, a very steep toxicity curve has been noted with respect to human responses to administered adenoviral vectors. Further, all major adenovirus-associated toxicities are clearly linked to adenoviral dose. In the aggregate, these considerations suggest that adenoviral vectors that circumvent CAR deficiency will be required for future cancer gene therapy approaches, both to achieve a potentially effective gene delivery to tumor cells to produce a therapeutic effect and to reduce the administered dose to avoid untoward toxicities (Krasnykh et al., 2000).

It should be readily apparent from some of the earlier considerations that available technologies could potentially address some of these human trial-related vector limits. However, it must be recognized that any engineered alterations of basic vector function, such as attachment or entry, renders the vector effectively a new biologic agent. In this regard, such new agents require a level of safety and analysis of a greater order than the simple transition of a new transgene to an existing vector. To this end, the Recombinant DNA Advisory Committee (RAC) of the National Institutes of Health has as one of its primary missions to consider such issues in the context of a new vector type for human application.

Work by Rancourt and colleagues demonstrated that an HSV-tk-encoding adenovirus, retargeted via a fibroblast growth factor (FGF) ligand, could achieve improved survival in a murine model of cancer of the ovary (Rancourt et al., 1998). On this basis, a human trial was proposed by Alvarez and co-workers that embodies this retargeting principle in its protocol design. Close interactions between the NIH RAC, the FDA, and the investigative group allowed the definition of the key safety issues potentially stemming from this first use of a tropism-modified viral vector. This trial is currently in the process of regulatory analysis, and once begun, will represent the first use of a tropism-modified viral agent in a human trial. It is hypothesized that this strategy will allow improved specific targeting to human cells. This parameter may be directly tested in the human system and will thus provide important proof-of-principle data with broad relevance to the field.

The concept of vector potency enhancement is likewise being addressed via tropism-modified adenoviral agents in a human trial. Specifically, Peters and colleagues will employ an adenoviral vector carrying the toxin gene cytosine deaminase (CD) and containing an HI loop RGD-4C. It is anticipated that this infectivity-enhanced adenoviral vector will accomplish higher levels of CD gene transfer at comparable vector doses. This finding would validate the employment of infectivity-enhanced adenoviral vectors as a generalized method to accomplish a desired level of transgene expression at lower vector doses. A highly novel aspect of this trial is that the expression unit of the infectivity-enhanced adenoviral vector also encodes a gene employed for imaging analysis. Specifically, the expression unit of the somatostatin receptor 2 (SSTR-2) induces infected cells to express cell surface SSTR-2 (Zinn et al.,

2001) This expression sensitizes target cells to binding by labeled octreotide, a frequently employed radiotracer compound. By virtue of this sensitization, direct monitoring of vector transductional efficacy, and localization, may be achieved. This is the first ever use of an imaging methodology in a human clinical gene therapy trial. Clearly, the type of data it will generate will determine the direct value of this type of tropism modification for adenoviral utility in the clinical context.

Many of the key biologic determinants relevant to reticuloendothelial clearance of adenoviral vectors in vivo have not been defined. These characterizations will provide the relevant database for Ad engineering efforts that embody such biologic considerations. Both of these endeavors—defining the behavior of adenoviral vectors in vivo and circumventing biologic limits to targeting—will require analytic methods appropriate to the in vivo context of such studies. On this basis, the development and application of imaging modalities will be critical for this analysis to progress along the most rational lines. Human clinical trials carried out to date have addressed target cells naturally sequestered by means of anatomic compartments, reflecting the fact that adenoviral vectors administered via the systemic route exhibit profound hepatotropism, with more than 95% of vector localizing to the liver. Two significant consequences have derived from this tropism. First, ectopic expression of transgenes in the liver may elicit limiting hepatotoxicities (Lambright et al., 2001). Second, liver sequestration may limit effective (Zinn et al., 1998) levels of vector dose available for target cell infection. This fundamental barrier has thus restricted diseases to anatomic compartment applications and limited applications whereby vascular vector administration was suggested. Theoretically, the capacity to achieve true targeting would allow the exploitation of adenoviral vectors for a broader range of applications, potentially including systemic vascular administration contexts. To this end, characterization of the basis of adenoviral hepatic sequestration was attempted to direct strategies to "untarget" the liver. Zinn and co-workers demonstrated a significant contribution of CAR in their studies of the biodistribution of radiolabeled fiber knob (Zinn et al., 1998). Their assertion of a significant contribution of CAR to hepatic sequestration of adenoviral vectors is corroborated by studies demonstrating high levels of CAR mRNA in the liver. In further support of this model, several groups have employed retargeting complex-based approaches in the context of systemically administered adenoviral vectors and have shown mitigation of vector-mediated transgene expression in the liver of greater than 90% (Reynolds et al., 2000).

On this basis, approaches to ablate native tropism of serotype 5 adenoviruses have been endeavored via direct mutagenesis of the fiber knob. Roelvink and colleagues have successfully identified residues within loops of the fiber knob directly involved in the recognition of the native receptor CAR (Roelvink et al., 1999). Further, directed mutagenesis has allowed ablation of this recognition with abrogation of CAR-mediated gene transfer demonstrated in in vitro model systems. Direct analysis of these vectors, via systemic vascular administration, in murine models has demonstrated decrements in vector-mediated hepatic transgene expression paralleling results with the retargeting complex-based methods noted previously (Kirby et al., 1999). Further studies by this group have sought to obtain additional mitigation of liver sequestration via ablation of the RGD motif in the adenoviral penton base. These studies have appeared to demonstrate additional gain with respect to this goal with reduced liver transgene levels. In the aggregate, these studies appear to validate a major role of CAR in vector sequestration in the liver. Further, genetic capsid modifications seem to provide a valid means to mitigate this physiology.

Studies by Alemany and others, however, have failed to validate these findings (Leissner et al., 2000; Alemany and Curiel, 2001). Further, it has recently been recognized that additional factors are operative in determining target cell accessibility via a systemically administered adenoviral vector. A key role for innate immune clearance of adenoviruses has been demonstrated by several groups. On this basis, it is clear that simple ablation of CAR and integrin recognition will not be sufficient to wholly mitigate liver sequestration. Further, vascular barriers are clearly operative, as it has clearly been shown that no direct correlates can be demonstrated between target cell CAR levels and susceptibility to infection by systemically administered adenoviral vectors (Maillard et al., 1998; Fechner et al., 1999; Kuriyama et al., 2000).

These studies clearly establish that the current understanding of the basis of sequestration of systemically administered adenoviral vectors within the liver is incomplete. Further, the practical advancement in gene therapy approaches based on systemic vascular administration will be subservient to this further understanding, and to the development of technologies that practically accomplish the vectorologic goal of liver untargeting. Nonetheless, the potential gains are great. In this regard, Reynolds and colleagues have recently employed an approach to target the pulmonary vascular endothelial marker angiotensin-converting enzyme (ACE) (Reynolds et al., 2001). By virtue of a combination approach, based on retargeting complexes plus promoter-restricted transgene expression, target-to-liver ratios were augmented more than four orders of magnitude in the context of systemically administered adenoviral vectors. This study clearly establishes that dramatic targeting gains can be obtained in this most stringent delivery context. This was achieved via online technology, directly exploited via the conceptual approach of simple synergy of targeting approaches. Such improvements in targeted, cell-specific gene delivery via the systemic route will greatly advance the possibilities of endeavoring gene therapy for disseminated carcinoma. More generally it should be readily apparent from considerations raised in this chapter that many genetic capsid modification strategies effectively alter adenoviral tropism and that these alterations may be fully anticipated to allow direct therapeutic gain in gene therapy approaches; human trials will validate these gains as well as determine unanticipated limits of this technology. This latter possibility notwithstanding, it might rightly be argued that the dramatic potency enhancements demonstrated with the current HI loop RGD-4C vector already warrants its considerations as an alternative to unmodified adenoviral vectors in current clinical use.

CONCLUSION

The attempts to alter adenoviral tropism via genetic capsid modifications have realized direct practical benefit and have established the basis for further gene therapy advances. Studies to date have defined a diversity of capsid proteins that may be genetically modified. Further, in selected instances, targeting peptides incorporated within the adenoviral capsid can profoundly alter vector function. Most noteworthy in this regard, dramatic enhancements in adenoviral potency have been achieved that have accrued therapeutic benefits in model systems. These direct gains have been of a magnitude to warrant human testing of these new vectors species. The gains notwithstanding, the full benefits of practical gain deriving from the advanced generation adenoviral vectors have yet to be defined. Both toxic and immunologic reactions to adenovirues have been major factors limiting the full utilization of these vectors. Further, both phenomena are directly linked to adenoviral vector dose. The increased potency of infectivity-enhanced adenoviral vectors may allow the utilization of lower vector doses, with the attendant benefit of reduced toxic and immune sequelae impacting adenoviral vector use.

The direct challenges to realizing even fuller benefit from the current approaches are clear. Specifically, the ability to apply tropism-modified adenoviral vectors to a wider array of targets is subservient to the ability to achieve a higher level of selectivity for those targets. This capacity, in turn, is linked to the ability to incorporate optimal targeting motifs within the adenoviral capsid. Such a goal could potentially be achieved by two approaches: the realization of a means to configure functional single chain antibodies (scFv) into a defined adenoviral capsid protein site and the delineation of a process whereby phage-defined peptides function with fidelity within an adenoviral capsid locale. Both of these ambitious goals may be anticipated by the dramatic progress that has been made in direct engineering of the adenoviral capsid.

ACKNOWLEDGMENTS

This work is supported by grants from the United States Army Department of Defense DAMD17-00-1-0002, DAMD17-98-1-8571, National Institute of Health R01 CA83821, P50 CA83591, R01 CA86881, N01 C0-97110, the Lustgarten Foundation, LF043, and the CAP CURE Foundation to David T. Curiel.

REFERENCES

Alemany R, Curiel DT (2001): CAR-binding ablation does not change biodistribution and toxicity of adenoviral vectors. *Gene Ther* 8:1347–1353.

Arap W, Pasqualini R, Ruoslahti E (1998): Cancer treatment by targeted drug delivery to tumor vasculature in a mouse model. *Science* 279:377–380.

Barry MA, Dower WJ, Johnston SA (1996): Toward cell-targeting gene therapy vectors: selection of cell-binding peptides from random peptide-presenting phage libraries. *Nature Med* 2:299–305.

Bauerschmitz GJ, Lam JT, Kanerva A, Suzuki K, Nettelbeck DM, Dmitriev I, Krasnykh V, Mikheeva GV, Barnes MN, Alvarez RD, Dall P, Alemany R, Curiel DT, Hemminki A (2002): Treatment of ovarian cancer with a tropism modified oncolytic adenovirus. *Cancer Res.* In press.

Bergelson JM, Cunningham JA, Droguett G, Kurt-Jones EA, Krithivas A, Hong JS, Horwitz MS, Crowell RL, Finberg RW (1997): Isolation of a common receptor for Coxsackie B viruses and adenoviruses 2 and 5. *Science* 275:1320–1323.

Blackwell JL, Miller CR, Douglas JT, Li H, Peters GE, Carroll WR, Strong TV, Curiel DT (1999): Retargeting to EGFR enhances adenovirus infection efficiency of squamous cell carcinoma. *Arch Otolaryngol Head Neck Surg* 125:856–863.

Blackwell JL, Hui L, Krasnykh V, Carroll WR, Peters GE, Strong TV, Curiel DT (2000a): Improved molecular chemotherapy efficacy using an infectivity-enhanced adenovirus vector, submitted.

Blackwell JL, Li H, Navarro J, Dmitriev I, Krasnykh V, Richter CA, Shaw DR, Alwarez RD, Curiel DT, Strong TV (2000b): Using a tropism-modified adenoviral vector to circumvent inhibitory factors in ascites fluid. *Hum Gene Ther* 11:1657–1669.

Bouri K, Feero WG, Myerburg MM, Wickham TJ, Kovesdi I, Hoffman EP, Clemens PR (1999): Polylysine modification of adenoviral fiber protein enhances muscle cell transduction. *Hum Gene Ther* 10:1633–1640.

Chandler LA, Doukas J, Gonzalez AM, Hoganson DK, Gu DL, Ma C, Nesbit M, Crombleholme TM, Herlyn M, Sosnowski BA, Pierce GF (2000): FGF2-targeted adenovirus encoding platelet-derived growth factor-B enhances de novo tissue formation. *Mol Ther* 2:153–160.

Crawford-Miksza L, Schnurr DP (1996): Analysis of 15 adenovirus hexon proteins reveals the location and structure of seven hypervariable regions containing serotype-specific residues. *J Virol* 70:1836–1844.

Cripe TP, Dunphy EJ, Holub AD, Saini A, Vasi NH, Mahller YY, Collins MH, Snyder JD, Krasnykh V, Curiel DT, Wickham TJ, DeGregori J, Bergelson JM, Currier MA (2001): Fiber knob modifications overcome low, heterogeneous expression of the coxsackievirus-adenovirus receptor that limits adenovirus gene transfer and oncolysis for human rhabdomyosarcoma cells. *Cancer Res* 61:2953–2960.

Dmitriev I, Krasnykh V, Miller CR, Wang M, Kashentseva E, Mikheeva G, Belousova N, Curiel DT (1998): An adenovirus vector with genetically modified fibers demonstrates expanded tropism via utilization of a coxsackievirus and adenovirus receptor-independent cell entry mechanism. *J Virol* 72:9706–9713.

Dmitriev I, Kashentseva E, Seki T, Curiel DT (2001): Utilization of minor capsid polypeptides IX and IIIa for adenovirus targeting. *Mol Ther* 3:S167, 467.

Doorbar J, Winter G (1994): Isolation of a peptide anta[illegible] using phage display. *J Mol Biol* 244:361–369.

Douglas JT, Rogers BE, Rosenfeld ME, Michael SI, Feng M, Curiel DT (1996): Targeted gene delivery by tropism-modified adenoviral vectors. *Nat Biotechnol* 14:1574-1578.

Douglas JT, Miller CR, Kim M, Dmitriev I, Mikheeva G, Krasnykh V, Curiel DT (1999): A system for the propagation of adenoviral vectors with genetically modified receptor specificities. *Nat Biotechnol* 17:470–475.

Ebbinghaus C, Al-Jaibaji A, Operschall E, Schoffel A, Peter I, Greber UF, Hemmi S (2001): Functional and selective targeting of adenovirus to high-affinity Fcgamma receptor I-positive cells by using a bispecific hybrid adapter. *J Virol* 75:480–489.

Fechner H, Haack A, Wang H, Wang X, Eizema K, Pauschinger M, Schoemaker RG, van Veghel R, Houtsmuller AB, Schultheiss H-P, Lamers JMJ, Poller W (1999): Expression of Coxsackie adenovirus receptor and alphav-integrin does not correlate with adenovector targeting in vivo indicating anatomical vector barriers. *Gene Ther* 6:1520–1535.

Garcia-Castro J, Segovia JC, Garcia-Sanchez F, Lillo R, Gomez-Navarro J, Curiel DT, Bueren JA (2001): Selective transduction of murine myelomonocytic leukemia cells (WEHI-3B) with regular and RGD-adenoviral vectors. *Mol Ther* 3:70–77.

Goldman CK, Rogers BE, Douglas JT, Sosnowski BA, Ying W, Siegal GP, Baird A, Campain JA, Curiel DT (1997): Targeted gene delivery to Kaposi's sarcoma cells via the fibroblast growth factor receptor. *Cancer Res* 57:1447–1451.

Gomez-Navarro J, Curiel DT, Douglas JT (1999): Gene therapy for cancer. *Eur J Cancer* 35:2039–2057.

Gonzalez R, Vereecque R, Wickham TJ, Facon T, Hetuin D, Kovesdi I, Bauters F, Fenaux P, Quesnel B (1999a): Transduction of bone marrow cells by the AdZ.F(pK7) modified adenovirus demonstrates preferential gene transfer in myeloma cells. *Hum Gene Ther* 10:2709–2717.

Gonzalez R, Vereecque R, Wickham TJ, Vanrumbeke M, Kovesdi I, Bauters F, Fenaux P, Quesnel B (1999b): Increased gene transfer in acute myeloid leukemic cells by an adenovirus vector containing a modified fiber protein. *Gene Ther* 6:314–320.

Goodson RJ, Doyle MV, Kaufman SE, Rosenberg S (1994): High-affinity urokinase receptor antagonists identified with bacteriophage peptide display. *Proc Natl Acad Sci USA* 91:7129–7133.

Grill J, Van Beusechem VW, Van Der Valk P, Dirven CM, Leonhart A, Pherai DS, Haisma HJ, Pinedo HM, Curiel DT, Gerritsen WR (2001): Combined targeting of adenoviruses to integrins and epidermal growth factor receptors increases gene transfer into primary glioma cells and spheroids. *Clin Cancer Res* 7:641–650.

Gu DL, Gonzalez AM, Printz MA, Doukas J, Ying W, D'Andrea M, Hoganson DK, Curiel DT, Douglas JT, Sosnowski BA, Baird A, Aukerman SL, Pierce GF (1999): Fibroblast growth factor 2 retargeted adenovirus has redirected cellular tropism: evidence for reduced toxicity and enhanced antitumor activity in mice. *Cancer Res* 59:2608–2614.

Haisma H, Pinedo H, Rijswijk A, der Meulen-Muileman I, Sosnowski B, Ying W, Beusechem V, Tillman B, Gerritsen W, Curiel D (1999): Tumor-specific gene transfer via an adenoviral vector targeted to the pan-carcinoma antigen EpCAM. *Gene Ther* 6:1469–1474.

Haisma HJ, Grill J, Curiel DT, Hoogeland S, van Beusechem VW, Pinedo HM, Gerritsen WR (2000): Targeting of adenoviral vectors through a bispecific single-chain antibody. *Cancer Gene Ther* 7:901–904.

Harari OA, Wickham TJ, Stocker CJ, Kovesdi I, Segal DM, Huehns TY, Sarraf C, Haskard DO (1999): Targeting an adenoviral gene vector to cytokine-activated vascular endothelium via E-selectin. *Gene Ther* 6:801–807.

Hemminki A, Belousova N, Zinn KR, Liu B, Wang M, Chaudhuri TR, Rogers BE, Buchsbaum DJ, Siegal GP, Barnes MN, Gomez-Navarro J, Curiel DT, Alvarez RD (2001): An adenovirus with enhanced infectivity mediates molecular chemotherapy of ovarian cancer cells and allows imaging of gene expression. *Mol Ther* 4:223–231.

Hong JS, Engler JA (1996): Domains required for assembly of adenovirus type 2 fiber trimers. *J Virol* 70:7071–7078.

Hong SS, Galaup A, Peytavi R, Chazal N, Boulanger P (1999): Enhancement of adenovirus-mediated gene delivery by use of an oligopeptide with dual binding specificity. *Hum Gene Ther* 10:2577–2586.

Kasono K, Blackwell JL, Douglas JT, Dmitriev I, Strong TV, Reynolds P, Kropf DA, Carroll WR, Peters GE, Bucy RT, Curiel DT, Krasnykh V (1999): Selective gene delivery to head and neck cancer cells via an integrin targeted adenovirus vector. *Clin Cancer Res* 5:2571–2579.

Kelly FJ, Miller CR, Buchsbaum DJ, Gomez-Navarro J, Barnes MN, Alvarez RD, Curiel DT (2000): Selectivity of TAG-72-targeted adenovirus gene transfer to primary ovarian carcinoma cells versus autologous mesothelial cells in vitro. *Clin Cancer Res* 6:4323–4333.

Kirby I, Davison E, Beavil AJ, Soh CP, Wickham TJ, Roelvink PW, Kovesdi I, Sutton BJ, Santis G (1999): Mutations in the DG loop of adenovirus type 5 fiber knob protein abolish high-affinity binding to its cellular receptor CAR. *J Virol* 73:9508–9514.

Krasnykh VN, Mikheeva GV, Douglas JT, Curiel DT (1996): Generation of recombinant adenovirus vectors with modified fibers for altering viral tropism. *J Virol* 70:6839–6846.

Krasnykh V, Dmitriev I, Mikheeva G, Miller CR, Belousova N, Curiel DT (1998): Characterization of an adenovirus vector containing a heterologous peptide epitope in the HI loop of the fiber knob. *J Virol* 72:1844–1852.

Krasnykh V, Dmitriev I, Navarro JG, Belousova N, Kashentseva E, Xiang J, Douglas JT, Curiel DT (2000): Advanced generation adenoviral vectors possess augmented gene transfer efficiency based upon coxsackie adenovirus receptor-independent cellular entry capacity. *Cancer Res* 60:6784–6787.

Krasnykh V, Belousova N, Korokhov N, Mikheeva G, Curiel DT (2001): Genetic targeting of an adenovirus vector via replacement of the fiber protein with the phage T4 fibritin. *J Virol* 75:4176–4183.

Kuriyama N, Kuriyama H, Julin CM, Lamborn K, Israel MA (2000): Pretreatment with protease is a useful experimental strategy for enhancing adenovirus-mediated cancer gene therapy. *Hum Gene Ther* 11:2219–2230.

Lambright ES, Amin K, Wiewrodt R, Force SD, Lanuti M, Propert KJ, Litzky L, Kaiser LR, Albelda SM (2001): Inclusion of the herpes simplex thymidine kinase gene in a replicating adenovirus does not augment antitumor efficacy. *Gene Ther* 8:946–953.

Leissner P, Legrand V, Schlesinger Y, Spehner D, Weber J, Puvion-Dutilleul F, Mehtali M, Pavirani A (2000): Modification of the adenoviral tropism by genetic manipulations of the fiber. *Gene Med* S2:38, P30.

Magnusson MK, Hong SS, Boulanger P, Lindholm L (2001): Genetic retargeting of adenovirus: novel strategy employing "deknobbing" of the fiber. *J Virol* 75:7280–7289.

Maillard L, Ziol M, Tahlil O, Le Feuvre C, Feldman LJ, Branellec D, Bruneval P, Steg P (1998): Pre-treatment with elastase improves the efficiency of percutaneous adenovirus-mediated gene transfer to the arterial media. *Gene Ther* 5:1023–1030.

McDonald GA, Zhu G, Li Y, Kovesdi I, Wickham TJ, Sukhatme VP (1999): Efficient adenoviral gene transfer to kidney cortical vasculature using a fiber modified vector. *J Gene Med* 1:103–110.

Michael SI, Hong JS, Curiel DT, Engler JA (1995): Addition of a short peptide ligand to the adenovirus fiber protein. *Gene Ther* 2:660–668.

Miller CR, Buchsbaum DJ, Reynolds PN, Douglas JT, Gillespie GY, Mayo MS, Raben D, Curiel DT (1998): Differential susceptibility of primary and established human glioma cells to adenovirus infection: targeting via the epidermal growth factor receptor achieves fiber receptor-independent gene transfer. *Cancer Res* 58:5738–5748.

Nemerow GR (2000): Cell receptors involved in adenovirus entry. *Virology* 274:1–4.

Okegawa T, Li Y, Pong RC, Bergelson JM, Zhou J, Hsieh JT (2000): The dual impact of coxsackie and adenovirus receptor expression on human prostate cancer gene therapy. *Cancer Res* 60:5031–5036.

O'Neil KT, Hoess RH, Jackson SA, Ramachandran NS, Mousa SA, DeGrado WF (1992): Identification of novel peptide antagonists for GPIIb/IIIa from a conformationally constrained phage peptide library. *Proteins* 14:509–515.

Pasqualini R, Koivunen E, Ruoslahti E (1997): Alpha v integrins as receptors for tumor targeting by circulating ligands. *Nat Biotechnol* 15:542–546.

Pasqualini R, Koivunen E, Kain R, Lahdenranta J, Sakamoto M, Stryhn A, Ashmun RA, Shapiro LH, Arap W, Ruoslahti E (2000): Aminopeptidase N is a receptor for tumor-homing peptides and a target for inhibiting angiogenesis. *Cancer Res* 60:722–727.

Pereboev A, Pereboeva L, Curiel DT (2001): Phage display of adenovirus-5 fiber knob as a tool for specific ligand selection and validation. *J Virol* 75:7103–7113.

Printz MA, Gonzalez AM, Cunningham M, Gu D-L, Ong M, Pierce GF, Aukerman SL (2000): Fibroblast growth factor 2-retargeted adenoviral vectors exhibit a modified biolocalization pattern and display reduced toxicity relative to native adenoviral vectors. *Hum Gene Ther* 11:191–204.

Rancourt C, Rogers BE, Sosnowski BA, Wang M, Piche A, Pierce GF, Alvarez RD, Siegal GP, Douglas JT, Curiel DT (1998): Basic fibroblast growth factor enhancement of adenovirus-mediated delivery of the herpes simplex virus thymidine kinase gene results in augmented therapeutic benefit in a murine model of ovarian cancer. *Clin Cancer Res* 4:2455–2461.

Reynolds P, Dmitriev I, Curiel D (1999): Insertion of an RGD motif into the HI loop of adenovirus fiber protein alters the distribution of transgene expression of the systemically administered vector. *Gene Ther* 6:1336–1339.

Reynolds PN, Miller CR, Goldman CK, Doukas J, Sosnowski BA, Rogers BE, Gomez-Navarro J, Pierce GF, Curiel DT, Douglas JT (1998): Targeting adenoviral infection with basic fibroblast growth factor enhances gene delivery to vascular endothelial and smooth muscle cells. *Tumor Target* 3:156–168.

Reynolds PN, Zinn KR, Gavrilyuk VD, Balyasnikova IV, Rogers BE, Buchsbaum DJ, Wang MH, Miletich DJ, Grizzle WE, Douglas JT, Danilov SM, Curiel DT (2000): A targetable, injectable adenoviral vector for selective gene delivery to pulmonary endothelium in vivo. *Mol Ther* 2:562–578.

Reynolds PN, Nicklin SA, Kaliberova L, Boatman BG, Grizzle WE, Balyasnikova IV, Baker AH, Danilov SM, Curiel DT (2001): Combined transductional and transcriptional targeting improves the specificity of transgene expression in vivo. *Nat Biotechnol* 19:838–842.

Roelvink PW, Lizonova A, Lee JG, Li Y, Bergelson JM, Finberg RW, Brough DE, Kovesdi I, Wickham TJ (1998): The coxsackievirus-adenovirus receptor protein can function as a cellular attachment protein for adenovirus serotypes from subgroups A, C, D, E, and F. *J Virol* 72:7909–7915.

Roelvink PW, Mi Lee G, Einfeld DA, Kovesdi I, Wickham TJ (1999): Identification of a conserved receptor-binding site on the fiber proteins of CAR-recognizing adenoviridae. *Science* 286:1568–1571.

Rogers BE, Douglas JT, Ahlem C, Buchsbaum DJ, Frincke J, Curiel DT (1997): Use of a novel cross-linking method to modify adenoviral tropism. *Gene Ther* 4:1387–1392.

Rogers BE, Douglas JT, Sosnowski BA, Ying W, Pierce G, Buchsbaum DJ, Della-Manna D, Baird A, Curiel DT (1998): Enhanced in vivo gene delivery to human ovarian cancer xenografts utilizing a tropism-modified adenovirus vector. *Tumor Target* 3:25–31.

Russell SJ, Cosset FL (1999):Modifying the host range properties of retroviral vectors. *J Gene Med* 1:300–311.

Rux JJ, Burnett R (2000): Type-specific epitope locations revealed by X-ray crystallographic study of adenovirus type 5 hexon. *Mol Ther* 1:18–30.

Schneider H, Groves M, Muhle C, Reynolds PN, Knight A, Themis M, Carvajal J, Scaravilli F, Curiel DT, Fairweather NF, Coutelle C (2000): Retargeting of adenoviral vectors to neurons using the Hc fragment of tetanus toxin. *Gene Ther* 7:1584–1592.

Stevenson SC, Rollence M, Marshall-Neff J, McClelland A (1997): Selective targeting of human cells by a chimeric adenovirus vector containing a modified fiber protein. *J Virol* 71:4782–4790.

Suzuki K, Fueyo J, Krasnykh V, Reynolds PN, Curiel DT, Alemany R (2001): A conditionally replicative adenovirus with enhanced infectivity shows improved oncolytic potency. *Clin Cancer Res* 7:120–126.

Tillman BW, Gruijl TD, Bakker SA, Scheper RJ, Pinedo HM, Curiel TJ, Gerritsen WR, Curiel DT (1999): Maturation of dendritic cells accompanies high-efficiency gene transfer by a CD40-targeted adenoviral vector. *J Immunol* 162:6378–6383.

Tillman BW, Hayes TL, DeGruijl TD, Douglas JT, Curiel DT (2000): Adenoviral vectors targeted to CD40 enhance the efficacy of dendritic cell-based vaccination against human papillomavirus 16-induced tumor cells in a murine model. *Cancer Res* 60:5456–5463.

Tomko RP, Xu R, Philipson L (1997): HCAR and MCAR: the human and mouse cellular receptors for subgroup C adenoviruses and group B coxsackieviruses. *Proc Natl Acad Sci USA* 94:3352–3356.

Trepel M, Grifman M, Weitzman MD, Pasqualini R (2000): Molecular adaptors for vascular-targeted adenoviral gene delivery. *Hum Gene Ther* 11:1971–1981.

van Beusechem VW, van Rijswijk AL, van Es HH, Haisma HJ, Pinedo HM, Gerritsen WR (2000): Recombinant adenovirus vectors with knobless fibers for targeted gene transfer. *Gene Ther* 7:1940–1946.

Vanderkwaak TJ, Wang M, J Gm-N, Rancourt C, Dmitriev I, Krasnykh V, Barnes M, Siegal GP, Alvarez R, Curiel DT (1999): An advanced generation of adenoviral vectors selectively enhances gene transfer for ovarian cancer gene therapy approaches. *Gynecol Oncol* 74:227–234.

Vigne E, Mahfouz I, Dedieu JF, Brie A, Perricaudet M, Yeh P (1999): RGD inclusion in the hexon monomer provides adenovirus type 5-based vectors with a fiber knob-independent pathway for infection. *J Virol* 73:5156–5161.

Von Seggern DJ, Huang S, Fleck SK, Stevenson SC, Nemerow GR (2000): Adenovirus vector pseudotyping in fiber-expressing cell lines: improved transduction of Epstein-barr Virus-transformed B cells. *J Virol* 74:354–362.

Watkins SJ, Mesyanzhinov VV, Kurochkina LP, Hawkins RE (1997): The 'adenobody' approach to viral targeting: specific and enhanced adenoviral gene delivery. *Gene Ther* 4:1004–1012.

Wesseling JG, Bosma PJ, Krasnykh V, Kashentseva EA, Blackwell JL, Reynolds PN, Li H, Parameshwar M, Vickers SM, Jaffee EM, Huibregtse K, Curiel DT, Dmitriev I (2001): Improved gene transfer efficiency to primary and established human pancreatic carcinoma target cells via epidermal growth factor receptor and integrin-targeted adenoviral vectors. *Gene Ther* 8:969–976.

Wickham TJ, Carrion ME, Kovesdi I (1995): Targeting of adenovirus penton base to new receptors through replacement of its RGD motif with other receptor-specific peptide motifs. *Gene Ther* 2:750–756.

Wickham TJ, Roelvink PW, Brough DE, Kovesdi I (1996a): Adenovirus targeted to heparan-containing receptors increases its gene delivery efficiency to multiple cell types. *Nat Biotechnol* 14:1570–1573.

Wickham TJ, Segal DM, Roelvink PW, Carrion ME, Lizonova A, Lee GM, Kovesdi I (1996b): Targeted adenovirus gene transfer to endothelial and smooth muscle cells by using bispecific antibodies. *J Virol* 70:6831–6838.

Wickham TJ, Lee GM, Titus JA, Sconocchia G, Bakacs T, Kovesdi I, Segal DM (1997a): Targeted adenovirus-mediated gene delivery to T cells via CD3. *J Virol* 71:7663–7669.

Wickham TJ, Tzeng E, Shears LL, Roelvink PW, Li Y, Lee GM, Brough DE, Lizonova A, Kovesdi I (1997b): Increased in vitro and in vivo gene transfer by adenovirus vectors containing chimeric fiber proteins. *J Virol* 71:8221–8229.

Xia D, Henry LJ, Gerard RD, Deisenhofer J (1994): Crystal structure of the receptor-binding domain of adenovirus type 5 fiber protein at 1.7 A resolution. *Structure* 2:1259–1270.

Yoon SK, Mohr L, O'Riordan CR, Lachapelle A, Armentano D, Wands JR (2000): Targeting a recombinant adenovirus vector to HCC cells using a bifunctional Fab-antibody conjugate. *Biochem Biophys Res Commun* 272:497–504.

Zinn KR, Douglas JT, Smyth CA, Liu HG, Wu Q, Krasnykh VN, Mountz JD, Curiel DT, Mountz JM (1998): Imaging and tissue biodistribution of 99mTc-labeled adenovirus knob (serotype 5). *Gene Ther* 5:798–808.

Zinn KR, Chaudhuri TR, Buchsbaum DJ, Mountz JM, Rogers BE (2001): Simultaneous evaluation of dual gene transfer to adherent cells by gamma-ray imaging. *Nucl Med Biol* 28:135–144.

9

CONJUGATE-BASED TARGETING OF ADENO-ASSOCIATED VIRUS VECTORS

SELVARANGAN PONNAZHAGAN, PH.D., GIANDHAM MAHENDRA, PH.D., ALEKSANDR PEREBOEV, M.D., PH.D., DAVID T. CURIEL, M.D., AND JÜRGEN KLEINSCHMIDT, PH.D.

INTRODUCTION

Adeno-associated virus (AAV)-based vectors have recently gained popularity as potential vectors for gene therapy. The major advantages of using AAV are nonpathogenicity, long-term expression, and relatively low immunogenecity. Although AAV was believed to infect both dividing and nondividing cells, recent studies have clearly demonstrated variations in transduction among cell types. The identification of the host cell receptor, coreceptors, and other factors that facilitate AAV entry and intracellular processing, leading to transgene expression, indicates that deficiency in one or more of these molecules can limit a successful transduction. Thus, it is apparent that modifications, that will overcome these limitations will not only maximize the application of AAV vectors in gene therapy, but also lead to the development of targeted vectors for transgene delivery to specific cell types. This chapter presents the potential for the development of targeted AAV specifically by nongenetic approaches.

Vector Targeting for Therapeutic Gene Delivery, Edited by David T. Curiel and Joanne T. Douglas
ISBN 0-471-43479-5

LIMITATIONS OF AAV TRANSDUCTION

Based on several reports on the AAV life cycle, transduction of the vector primarily involves three major events. The first step is the binding of vector to molecules present on the host cell surface that provide anchorage. Recently, heparan sulfate glycoprotein has been identified as the cellular receptor for AAV (Summerford and Samulski, 1998). Subsequently, fibroblast growth factor 1 (FGF1) and integrin $\alpha V\beta 5$ have been reported as coreceptors involved in viral entry (Qing et al., 1999; Summerford et al., 1999). The second step following initial vector attachment is the internalization of virions, endosomal trafficking, and their rapid transport to the cell nucleus where the vector genome is released (Sanlioglu et al., 2000; Bartlett et al., 2000; Hansen et al., 2000). The third step in the transduction pathway is the conversion of the single-stranded vector genome into transcriptionally active double-stranded intermediates. Several factors have been reported that determine the efficacy of this step, such as involvement of host cell polymerases and other growth factors (Qing et al., 1997; Bartlett et al., 1998; Mah et al., 1998). Thus, it is evident that suboptimal vector transduction may be due to inadequacies in one or more of these events. Since different target cells in vivo possess different growth and metabolic characteristics, it is evident that optimal transduction may not occur equally in all of them. Hence, it is important to further understand the molecular events that regulate AAV transduction in each type of target to increase the transduction efficiency in these cells. Although the influence of intracellular events may vary depending on the cell type and its biological significance, the initial event of viral entry primarily depends on the availability of native receptors. Hence, it is conceivable that either the absence or low-level expression of receptor/coreceptor molecules in certain cell types, despite having optimal intracellular conditions for transgene expression, limit the potential application of recombinant AAV (rAAV) as a gene therapy vector. Thus, modifications to improve vector infectivity will broadly benefit transduction of rAAV to many cell types that either lack or have a low-level receptor expression.

INFECTION OF AAV VARIES WIDELY AMONG CELL TYPES

Although it was earlier believed that AAV infects both dividing and nondividing cells transcending a species barrier (Muzyczka, 1992), the initial report that certain human megakaryocytic leukemia cell lines were refractory to both wild-type (wt) and rAAV infection suggested the potential involvement of a cellular receptor for viral entry (Ponnazhagan et al., 1996; Mizukami et al., 1996). Following these observations, both a receptor and possible coreceptors have been identified (Summerford and Samulski 1998; Qing et al., 1999; Summerford et al., 1999). Studies with human primary bone marrow-derived CD34+ hematopoietic progenitor cells obtained from genetically distinct donors showed variations in AAV infection (Ponnazhagan et al., 1997a).

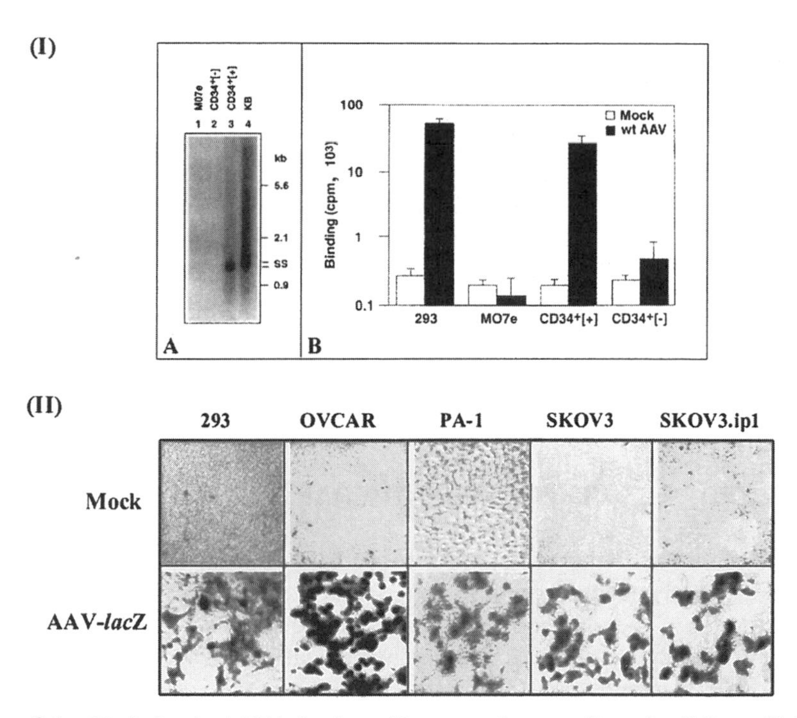

Figure 9.1 Variation in AAV infection of human primary cells and cell lines. (I) Variation in infection (A) and membrane binding (B) of wtAAV in $CD34^+$ cells from two donors who were either positive or negative for AAV infection. Whereas the internalized vector genome was determined by Southern blot analysis of low M_r DNA, membrane binding was determined by incubating ^{3}H-labeled wtAAV $CD34^+$ cells and counting the radioactivity after removing free virus. (II) Variation in transgene expression in human ovarian cancer cell lines following infection with rAAV encoding beta-galactosidase.

It was of interest to note in these studies the existence of a wide variation in transduction efficiencies, ranging from 0 to approximately 80%, based on the expression of the transgene. Further, binding studies with radiolabeled-AAV confirmed that the defect was in the binding affinity of the vector to cells that were negative for transgene expression (Fig. 9.1). Similar variations in the transduction efficiency were also reported in primitive progenitor cells from human cord blood (Fisher-Adams et al., 1996) and monocyte-derived dendritic cells (Ponnazhagan et al., 2001). In addition, we have also found variation in AAV transduction of human ovarian cancer cell lines (Fig. 9.1). While maximal expression of AAV transduction in vivo has been reported in skeletal muscle (Xiao et al., 1996; Monahan et al., 1998; Herzog et al., 1999; Song et al., 1998; Zhou et al., 1998; Murphy et al., 1997; Snyder et al., 1999) and brain (Kaplitt et al., 1994; During et al., 1998; Mandel et al., 1997; Xiao et al.,

1997), moderate levels of transduction have been reported in liver (Xiao et al., 1998; Snyder et al., 1999), lung (Afione et al., 1996; Flotte et al., 1993), eye (Grant et al., 1997; Jomary et al., 1997), heart (Su et al., 2000; Maeda et al., 1998), and hematopoietic cells (Ponnazhagan et al., 1997a,b; Hargrove et al., 1997; Schimmenti et al., 1998). Although it has been reported that transduction of AAV is higher in dividing cells, certain hematopoietic cells that undergo active cell division show poor transduction (Ponnazhagan et al., 1997a; Fisher-Adams et al., 1996). This phenomenon was found to correlate with the levels of expression of the receptor for AAV infection (Summerford and Samulski, 1998). Thus, it is evident that modification of vector tropism to target cells may overcome limitations in the infectivity of both nonpermissive cells and those with low-level AAV infection.

TARGETED-VECTORS FOR GENE THERAPY

The use of targeted viral vectors to achieve gene transfer to specific cell types holds several advantages over conventional, nontargeted vectors currently used in gene therapy. The resulting improvements in gene localization from targeted vectors are likely to reduce immunogenecity and toxicity associated with the vector dose, increase safety, and enable either systemic or target cell-specific administration of these vectors for multiple indications (Miller and Vile, 1995). Targeted vectors are designed specifically to transduce a particular cell type. To date, targeting of recombinant viral vectors has been attempted by genetic and conjugate-based methods. Whereas the principles of genetic targeting are typically based on modifications of the capsid structure to incorporate sequences of targeting ligands (Krasnykh et al., 2000; Peng and Russell, 1999), conjugate-based transduction is achieved by linking the unmodified recombinant vector to either a targeting ligand or antibody specific for cellular receptors via chemical or antibody bridges (Curiel, 1999).

SIGNIFICANCE OF THE DEVELOPMENT OF TARGETED-AAV

With the previously mentioned variations in infectivity of target cells by AAV due to differences in the level of receptor expression, it becomes important to develop strategies that would promote high-efficiency gene transfer. Two logical alternatives to achieve this are to enhance the expression of native receptor in the target cells or to modify the vector to achieve high-efficiency, cell-specific entry using alternate cellular pathways. The former approach is more complicated than it appears due to the nature of the AAV receptor and the possible requirement for coreceptors. For example, heparan sulfate proteoglycan (HSPG) encompasses a family of syndecans with complex molecular organization (Murdoch et al., 1992; Kallunki et al., 1992). Although HSPG is known to bind to AAV with high affinity, the nature of the binding and the affinity for different HSPG types are not well defined. Further, the molecular organization of a functional HSPG involves the appropriate localization of different domains in a unique spatial configuration. Thus, a better approach to enhance targeting of AAV to specific cell types is to make modifications to the vector to achieve transduction via alternate cellular receptors. Additionally, development of bispecific targeting reagents that will neutralize the binding of AAV to either the primary receptor or coreceptors should lead to the abrogation of transduction through the native entry pathway. Thus, the development of tropism-modified AAV through these methods will have a high likelihood to also abolish transduction of nontarget cells that express the AAV receptor and coreceptors.

In the last few years, attempts have been made to accomplish targeted transduction of rAAV through alternate cellular pathways with promise. The majority of these modifications have involved genetic incorporation of targeting ligands within the capsid protein. The conjugate-based vectors have been created by complexing rAAV to targeting ligands through chemical bridges to mediate viral entry through alternate cellular pathways.

ADVANTAGES AND LIMITATIONS IN THE DEVELOPMENT OF TARGETED AAV

Most of the current reports on the genetic modification of AAV capsids are based on hypothetical models derived from similarities with the capsid structures of other parvoviruses such as canine parvovirus (Girod et al., 1999; Rabinowitz et al., 1999; Wu et al., 2000; Grifman et al., 2001). Despite the identification of potential domains that are amenable to the inclusion of targeting ligands, the majority of mutations have resulted in vectors with low titer or poor infectivity. Hence, it is likely that the delineation of the crystallographic structure of the AAV capsid will lead to a better understanding of potential domains for genetic modifications. In addition, identification of serotypes of AAV, other than the more commonly used type 2, with differences in capsid protein sequence and the nature of cellular entry pathways underlines the need for better understanding of the three-dimensional structure to permit rational and efficient genetic modifications. Despite the potential of genetically modified AAV, an obvious technical limitation remains in the size of the targeting ligand that can be incorporated within the capsid structure without affecting assembly or infectivity.

Conjugate-based retargeting of AAV does not have a limitation on the size of targeting ligand that can be complexed to the vector capsid. Hence, a wide variety of molecules such as growth factors, monoclonal or single-chain antibodies against target cell receptors, and targeting peptides can be used as conjugates. However, major concerns in these types of vector modifications are the cumbersome steps involved in conjugation, the possible presence of unmodified vector after the modification steps, the in vivo stability of the complex, and potential host immunity against the components used in conjugation.

Although attempts so far to target AAV through a conjugate-based approach are limited to a single approach of targeting using a bispecific monoclonal antibody conjugate (Bartlett et al., 1999), information derived from other vectors and preliminary studies from our laboratory indicate that similar approaches may yield an efficacious outcome for the future development of targeted AAV.

TARGETING THROUGH BISPECIFIC MONOCLONAL ANTIBODY CONJUGATES

Bispecific antibody conjugates have been successfully used earlier in the targeting of recombinant adenoviral and retroviral vectors to cells that either lack or have low-level expression of the native receptors (Chu and Dornburg, 1997; Martin et al., 1998; Haisma et al., 2000). Similar to these works, recently Bartlett and co-workers demonstrated that infection of AAV through an alternate cellular receptor is possible using a bispecific monoclonal antibody conjugate (Bartlett et al., 1999) as shown schematically in Fig. 9.2. In these studies, a heterodimeric bispecific $F(ab'\gamma)_2$A20AP2 was prepared by linking half-cysteine residues on two monoclonal antibodies, A-20 (a monoclonal anti-

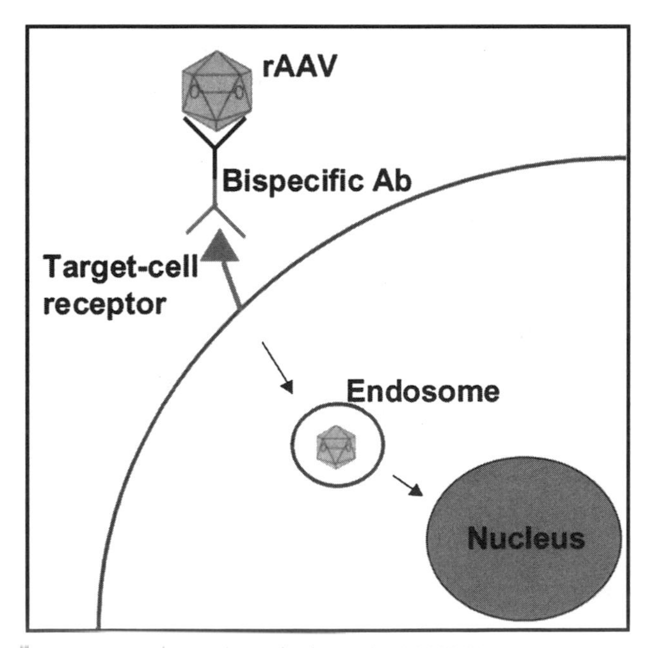

Figure 9.2 Strategy to achieve transduction of nonpermissive cells by rAAV through bispecific antibody. Infection of rAAV through an alternate cellular receptor is mediated by a bispecific monoclonal antibody conjugate with affinities to both the vector and a cell surface receptor.

body that recognizes intact AAV capsids) and AP-2 (a monoclonal IgG antibody that recognizes integrin $\alpha_{IIb}\beta_3$ (gpIIbIIIa; CD41) on human megakaryocytes and platelets [Handagama et al., 1993]). The Fab′ fragments were linked via thioether bonds using bismaleimidohexane, a bifunctional cross-linking reagent. The bispecific antibody was then subjected to reduction and alkylation to remove any untoward products, including F(ab′γ)$_2$ homodimers, which may have formed by oxidation or disulfide exchange. Further purification of the bispecific antibody through Sephacryl S-200 ensured removal of residual reactants.

Transduction of rAAV through alternate receptor following conjugation with the targeting bispecific antibody was demonstrated in DAMI, a human megakaryoblastoid cell line, and MO7e, a human megakaryocytic leukemia cell line, known to be resistant to AAV infection. Briefly, 2×10^6 infectious particles of a rAAV vector encoding β-galactosidase was preincubated with the bispecific antibody in a medium containing 20 mM HEPES at room temperature for 1 h following which the conjugate was used to infect DAMI or MO7e cells for 1 h at 37°C. The free virus was removed after infection and the cells were incubated for 24 h prior to determining the transgene activity. The results indicated that the bispecific A20AP2 F(ab′γ)$_2$ maximally increased AAV-βgal activity in $\alpha_{IIb}\beta_3$-positive DAMI and MO7e cells approximately 70-fold above the background. Further, preincubation of these cells with antibody to $\alpha_{IIb}\beta_3$ integrin abrogated this effect, indicating that the effect of retargeting is specific through $\alpha_{IIb}\beta_3$ integrin (Bartlett et al., 1999).

With increasing therapeutic application of monoclonal antibodies, further improvements, particularly in the production and purification and the in vivo stability of bispecific targeting antibodies may advance the usefulness of this strategy for future gene therapy applications.

RECOMBINANT ANTIBODIES AS EFFICACIOUS TARGETING MOIETIES

Some of the practical difficulties in the application of bispecific monoclonal antibody conjugates such as cross-linking and stability of the chemically linked antibodies can be overcome by using single-chain antibody (scFv) fragments genetically fused to targeting ligands or additional scFvs against a host cell receptor. Recent advances in genetic engineering have greatly facilitated the identification and conjugation of antibody fragments in recombinant systems. Novel multivalent and multispecific antibodies have been constructed by this method for a variety of biological and therapeutic applications (Huston et al., 1993). Such recombinant antibodies are created by polymerase chain reaction amplification of cDNAs encoding variable regions of immunoglobulin light and heavy chains obtained from B-lymphocytes and genetically fusing them with the addition of a flexible linker between the light and heavy chains for efficient folding (Bedzyk et al., 1990; Hoogenboom et al., 1991). Such an array of heterologous recombinant antibody genes is cloned within the genome of filamentous bacteriophages, which facilitates the selection of a specific scFv by biopanning with target antigens. Many candidate scFvs have so far been identified by this powerful method (Schultz et al., 2000). Furthermore, genetic coupling of a heterologous protein by fusion of such gene fragments generates new possibilities for immunotargeting. A variety of molecules have been successfully fused to scFvs and used for therapeutic targeting (Vitaliti et al., 2000; Loffler et al., 2000). Recently, the application of scFv-directed targeting using gene therapy vectors has been shown not only to be efficacious in both production and purification but also to be more stable than bispecific targeting conjugates produced by chemical cross-linking of two monoclonal antibodies (Pavlinkova et al., 1999). Such application of scFvs has been successfully performed mostly with adenoviral vectors (Watkins et al., 1997; Haisma et al., 2000) and to some extent with retroviral vectors (Benedict et al., 1999; Martin et al., 1998; Chu et al., 1997). Thus, the development of targeted AAV vectors by cloning scFvs against the AAV capsid may provide greater efficacy gains in the future.

In order to develop targeted AAV vectors by a high-affinity immunological method, we initiated cloning of an scFv against the AAV capsid by screening a phage-displayed antibody library as outlined in Fig. 9.3. In vitro-produced empty AAV capsids were used as the antigen against which to pan a human B-cell antibody library in the phage display system pSEX81 (GenBank accession #Y14584, kindly provided by Dr. Stefan Dübel, University of Heidelberg, Germany). Three rounds of panning resulted in a potentially positive clone with binding specificity to AAV-capsids (Mahendra et al., unpublished results). cDNA fusion of the variable heavy and light (vH and vL) domains from the two positive clonal populations were isolated from the pSEX vector and subcloned into bacterial expression vector pOPE under the control of T7 promoter. Sequences corresponding to the c-*myc* epitope and a 6-His tag were also included in-frame at the 3′ end of the scFv gene for identification and purification of the scFv, respectively. Following cloning, the scFv was expressed in *Escherichia coli* by induction with 20 μM IPTG and purified from soluble periplasmic proteins using a nickel-nitriloacetic acid (Ni-NTA) affinity column. The purified scFv was dialyzed against phosphate-buffered saline.

To determine the binding of the purified scFv to infectious AAV capsids, increasing concentrations of purified scFv (0–20 μg) were incubated at room temperature for 30 min with a constant amount (10^8 particles) of a rAAV encoding firefly luciferase. Following binding, the complex was purified through a Ni-NTA column. Eluted vector that retained binding to the scFv was dialyzed against cold phosphate-buffered saline (PBS) and concentrated using Centricon-100 filters to remove free scFv. 293 cells were infected with the purified vector at 37°C for 2 h, following which the cells were washed with PBS and grown for an additional 48 h. Luciferase activity was determined by lysing the cells 48 h postinfection. To determine if nonspecific binding of AAV-luc to the Ni-NTA column resulted in any transduction, equivalent amounts of

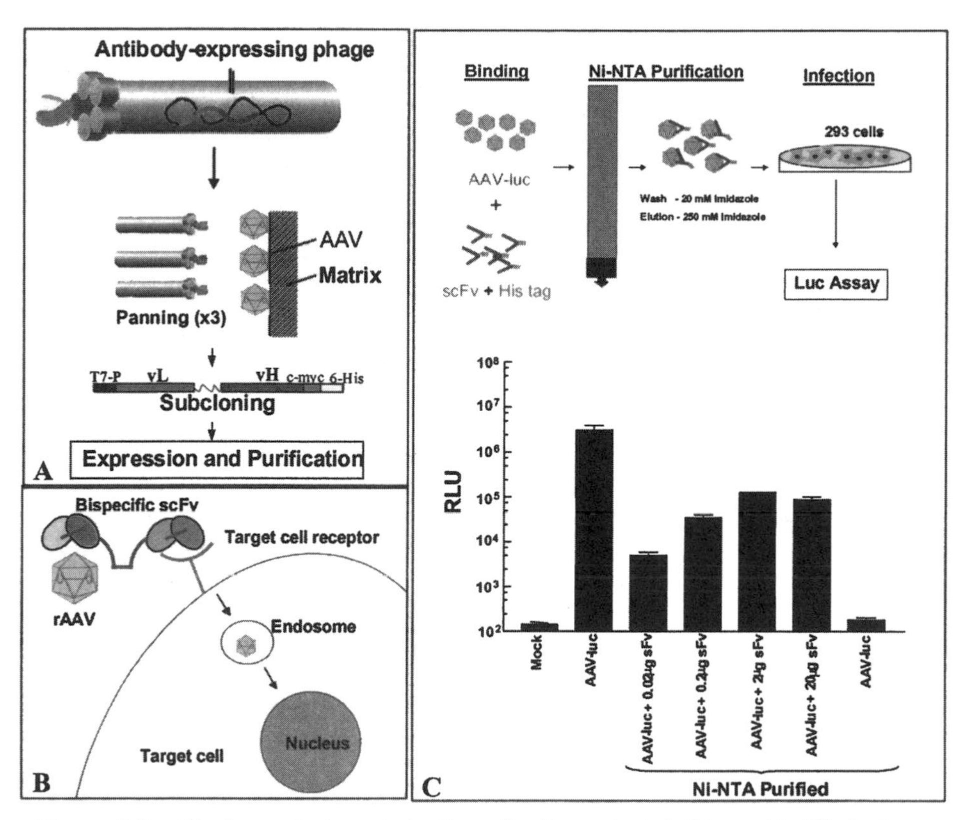

Figure 9.3 Cloning and characterization of scFv against AAV capsid. (*A*) A phage library expressing heterologous human antibodies is used to pan against AAV capsids fixed on a matrix. The cDNA obtained from a positive phage is subcloned in a prokaryotic expression vector for production and purification. (*B*) Schematic representation of targeting strategy through alternate cellular receptor using a bispecific scFv. (*C*) Specificity of binding of purified scFv to AAV capsids was confirmed by preincubating rAAV encoding luciferase with varying concentrations of scFv, purifying the complex through a Ni-NTA column and infecting 293 cells. Luciferase activity was determined 48 h later.

AAV-luc vector without prior binding to scFv were also subjected to Ni-NTA column purification under identical conditions.

Results, shown in Fig. 9.3C, indicated a dose-dependent increase in luciferase activity that correlated with increasing concentrations of the scFv up to 2 μg. Addition of scFv beyond this concentration did not result in further augmentation, which is likely due to saturation of binding. There was also no luciferase activity in cells transduced with the eluate of rAAV that was not preincubated with the scFv, suggesting the absence of nonspecific binding. When the transgene expression was compared between the scFv-conjugated vector and that of unmodified vector (10^8 particles of AAV-luc) not subjected to affinity purification, there was more than one-log higher luciferase activity in the latter, which may be due to either loss of scFv-conjugated vector during column purification or due to abrogation of transduction of the modified vector through the native entry pathway. Nonetheless, the data indicate that the cloned scFv possesses binding affinity for intact vector particles and will be useful in retargeting studies. The current focus is on developing a bispecific fusion protein using the scFv and a target cell-specific ligand or scFv.

POTENTIAL OF HIGH-AFFINITY TARGETING USING AVIDIN-BIOTIN INTERACTION

Streptavidin is a tetrameric protein of ~60 kDa which binds biotin with exceptional affinity ($K_d = 1 \times 10^{-15}$ M). The high affinity of biotin for streptavidin has made this pair of molecules very useful for many in vitro and in vivo applications (Wilbur et al., 1999). The majority of avidin–biotin in vivo applications, in a therapeutic context, are presently being used for tumor targeting (Wilbur et al., 1999; Saga et al., 1994; Nakaki, Takikawa, and Yamanaka, et al., 1997; Shi et al., 2000; Yao et al., 1998; Paganelli, Magnani, and Fazio, 1993). The development of pretargeting strategies using avidin–biotin interactions of therapeutic molecules or antibodies has not only resulted in increased efficacy gains but also greatly minimized the required dose of the therapeutic molecules due to the high-affinity binding of this system (Yao et al., 1998). Targeted immunotherapy of colon adenocarcinoma using a biotinylated anti-CEA monoclonal antibody and a biotinylated drug, neocarzinostatin, resulted in a five-fold increase in therapeutic efficacy (Nakaki, Takikawa, and Yamanaka, 1997). High-efficiency targeting of tumor cells by administration of a monoclonal antibody and a radiolabel using avidin–biotin interactions has been reported not only in preclinical studies but also in clinical trials (Sakahara and Saga, 1999). This method of tumor targeting has also resulted in high tumor to non-tumor targeting ratios in addition to reducing the background radioactivity of the directly labeled antibody (Sakahara and Saga, 1999). An avidin–biotin system has also been effectively utilized in the delivery of nerve growth factor to brain cells using transferrin receptors to overcome the blood–brain barrier (Li et al., 2000). Thus, the potential utility of avidin–biotin interaction, is evident from its efficacy, safety, and specificity of targeting.

In molecular conjugate vectors, targeting moieties have been attached to vectors mainly through electrostatic interactions or through bispecific monoclonal antibody conjugates. Although these methods have produced efficacious results in vitro, their realistic application in vivo is limited due to the lack of stability of the vector-conjugate complex. Such limitations in monoclonal antibody therapy for tumor patients have been overcome by using avidin–biotin interactions (Reilly, 1991). Thus, expanding the utility of the high-affinity, stable interaction of the avidin–biotin system to achieve modifications of vector tropism may result in greater efficacy gains. In this regard, the feasibility of this approach has been shown by Smith et al. who reported the high-efficiency transduction of c-Kit receptor-positive hematopoietic stem cells by biotinylated recombinant adenoviral vector, conjugated through an avidin bridge to a biotinylated stem cell factor (Smith et al., 1999).

In order to develop a targeting system using the avidin–biotin interactions, we constructed a genetic fusion of core-streptavidin (ST) and human epidermal growth factor (EGF) cDNA in a prokaryotic expression vector. The plasmid pSTE-2 containing the core-streptavidin gene was a kind gift of Dr. Stefan Dübel (University of Heidelberg, Germany). Coding sequences of human EGF were amplified by PCR with primers containing restriction sites and sequences of a 5-amino acid linker for in-frame cloning into vector pSTE-2 replacing the scFv gene at the 5′ end of the core-streptavidin. The 3′ end of the streptavidin gene also contained a 5-His tag for purification of the fusion protein through a nickel column.

E. coli was transformed with the recombinant plasmid, pEGF-ST, and expression of the fusion gene was achieved by induction with 20 μM IPTG. Purification of the EGF-streptavidin fusion protein was achieved using a Ni-NTA affinity column. Next, a rAAV vector encoding luciferase was biotinylated using 5 μg/ml N-hydroxysuccinimide ester water-soluble biotin. Following confirmation of vector biotinylation, we determined the ability of biotinylated-AAV, conjugated to an avidin-linked ligand, to transduce target cells through an alternate cellular receptor. To this end, 10^8 particles of rAAV-luc that was either unmodified or biotinylated were incubated with 1 μg of affinity-purified EGF-ST fusion protein, following which EGF receptor (EGFR)-positive SKOV3.ip1 cells and EGFR-negative MB-453 cells were independently transduced. Luciferase activity was determined 48 h posttransduction. Results, shown in Fig. 9.4, indicated a significant increase (>2 log) in transduction of EGF-streptavidin-conjugated rAAV only in EGFR-positive SKOV3.ip1 cells and not in EGFR-negative MB-453 cells, suggesting efficient retargeting of AAV by this method. The results of transduction with biotinylated vector alone also indicated that biotinylation of rAAV does not affect vector infection as compared to unmodified AAV-luc.

We also observed comparatively lower luciferase activity in both SKOV3.ip1 and MB-453 cells that were transduced with equal amounts of unmodified AAV-luc or biotinylated AAV-luc as well as with EGFR-targeted vector in MB-453 cells, which may be due to (1) the presence of vector particles that were not biotinylated, (2) the presence of unconjugated, biotinylated vector particles following incubation with the EGF-ST fusion protein, or (3) lack of abrogation of transduction of the modified vector through heparan sulfate glycoprotein receptor. Nonetheless, these data collectively indicate the possibility of utilizing the high-affinity biotin–avidin interaction in transduction of rAAV through alternate cellular pathways.

With the identification of potential domains in the AAV capsid that permit the genetic inclusion of targeting ligands without affecting the viral titer or

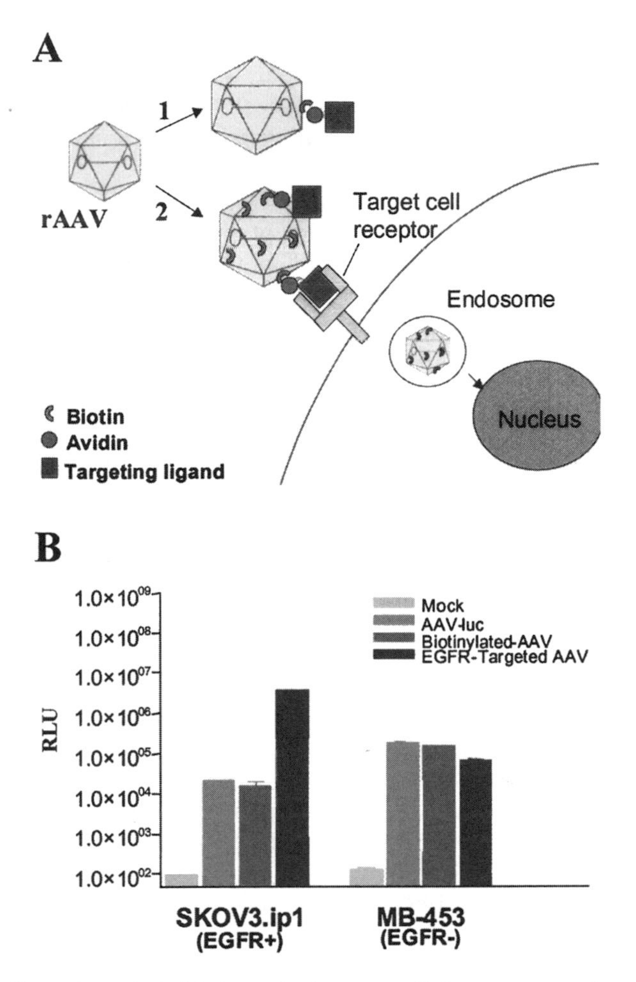

Figure 9.4 Targeting of rAAV through alternate cellular receptor using avidin–biotin bridges. (*A*) Schematic representation of modification of rAAV capsids to incorporate biotin either by genetic inclusion of transcarboxylase sequences (1) or by biochemical modification (2) followed by conjugation of an avidin-linked targeting ligand for infection through an alternate cellular receptor. (*B*) SKOV3.ip1 and MB-453 cells were either mock infected or infected with 10^8 particles of AAV-luc that was unmodified, or biotinylated alone, or biotinylated and conjugated to purified EGF-ST fusion protein, for 1 h at 37°C. Following infection, free virus was removed by washing with PBS and the cells incubated for 48 h. Luciferase activity was determined by lysing the cells and the activity expressed as relative light units (RLU) normalized to protein content of each cell lysate.

infectivity, modifications to genetically include the transcarboxylase sequence (Parrot and Barry, 2000), which is naturally biotinylated when expressed in cells grown in the presence of biotin, may further result in uniform biotinylation at a preferred site of the capsid without affecting the infectivity or intracellular trafficking of the modified vector.

CONJUGATE-BASED TARGETING OF AAV PLASMIDS

In addition to developing conjugate-based methods for retargeting of AAV vectors, the high-efficiency transfection of recombinant AAV plasmids, conjugated to cationic liposomes, has also been shown to be efficacious in gene transfer both in vitro and in vivo (Philip et al., 1994; Vieweg et al., 1995). Cationic liposomes spontaneously complex with plasmid DNA or RNA in solution and facilitate fusion of the complex with cells, resulting in efficient transfer of nucleic acids. Cationic liposome-DNA complexes have been used successfully to express transgenes in vivo after intravenous (Sakurai et al., 2001), intraperitoneal (Namiki, Takahashi, and Ohno, 1998), and intramuscular (Gregoriadis, Saffie, and de Souza, 1997) administration. Unlike viral vectors, liposomes are noninfectious and nonimmunogenic in vivo. Furthermore, since preformed cationic liposomes readily complex with DNA, there may not be significant limitations to the size of the transgene that can be delivered. Despite these advantages, liposomes have not been widely used because of low-efficiency gene transfer and inability to mediate long-term gene expression. Studies by independent groups have demonstrated that AAV plasmid DNA complexed to cationic liposomes not only facilitated enhanced transduction to both primary cells and established cell lines but also resulted in sustained expression of the transgene, which was significantly longer than cationic liposome-complexed transfection of plasmid DNA without AAV ITRs (Vieweg et al., 1995). Although the use of AAV plasmids rather than the vector differs in many ways, including the cellular entry pathway, intracellular trafficking, and, more importantly, limitations in chromosomal integration and long-term expression, the retention of the low M_r AAV genome in plasmid-transfected cells for up to 30 days and sustained expression of the transgene for that period indicates that this method of conjugated delivery may have potential application in the development of genetically modified tumor vaccine approaches (Vieweg et al., 1995).

CONCLUSIONS

Despite attempts to develop conjugate-based vectors for more than a decade, the potential application of such vectors in clinical gene therapy trials is highly limited because of concerns about the in vivo stability and possible toxicity and immunogenicity of the components used in conjugation. Thus, genetic modification of vector capsids remains the preferred means of development of targeted vectors. Nonetheless, the development of safe and efficacious conjugate-based targeting approaches for AAV transduction may still prove beneficial considering the small dimension of the AAV capsid, which highly restricts the size of targeting ligands which could be genetically incorporated. Further, the conjugate-based retargeting of AAV should be highly useful in ex vivo gene therapy applications, particularly for achieving high-efficiency transduction of hematopoietic stem cells, which are not easily transduced by the vector.

ACKNOWLEDGMENTS

We thank Dr. Arun Srivastava, Indiana University school of for his kind permission to include published findings. This work was supported by UAB Comprehensive Cancer Center American Cancer Society Institutional Research Grant 60-061-41, Career Development Award (NIH-SPORE grant in Ovarian Cancer [5 P50-CA8359]), National Institutes of Health R01 CA90850 and Muscular Dystrophy Association grants to S.P., National Institutes of Health P50 CA89019, R01 CA74242, R01 CA86881, N01 CO-97110, R01 CA68245, R03 CA90547, R01 HL67962, the Juvenile Diabetes Foundation 1-2000-23 to David T. Curiel, and the National Institutes of Health U19 DK57858 to Judy Thomas.

REFERENCES

Afione SA, Conrad CK, Kearns WG, Chunduru S, Adams R, Reynolds TC, Guggino WB, Cutting GR, Carter BJ, Flotte TR (1996): In vivo model of adeno-associated virus vector persistence and rescue. *J Virol* 70:3235–3241.

Bartlett JS, Kleinschmidt J, Boucher RC, Samulski RJ (1999): Targeted adeno-associated virus vector transduction of nonpermissive cells mediated by a bispecific F(ab′gamma)2 antibody. *Nat Biotechnol* 17:181–186.

Bartlett JS, Samulski RJ, McCown TJ (1998): Selective and rapid uptake of adeno-associated virus type 2 in brain. *Hum Gene Ther* 9:1181.

Bartlett JS, Wilcher R, Samulski RJ (2000): Infectious entry pathway of adeno-associated virus and adeno-associated virus vectors. *J Virol* 74:2777–2785.

Bedzyk WD, Weidner KM, Denzin LK, Johnson LS, Hardman KD, Pantoliano MW, Asel ED, Voss EW, Jr. (1990): Immunological and structural characterization of a high affinity anti-fluorescein single-chain antibody. *J Biol Chem* 265:18615–18620.

Benedict CA, Tun RY, Rubinstein DB, Guillaume T, Cannon PM, Anderson WF (1999): Targeting retroviral vectors to CD34-expressing cells: binding to CD34 does not catalyze virus-cell fusion. *Hum Gene Ther* 10:545–557.

Chu TH, Dornburg R (1997): Toward highly efficient cell-type-specific gene transfer with retroviral vectors displaying single-chain antibodies. *J Virol* 71:720–725.

Curiel DT (1999): Strategies to adapt adenoviral vectors for targeted delivery. *Ann NY Acad Sci* 886:158–171.

During MJ, Samulski RJ, Elsworth JD, Kaplitt MG, Leone P, Xiao X, Li J, Freese, A, Taylor JR, Roth RH, Sladek JR, O'Malley KL, Redmond DE (1998): In vivo expression of therapeutic human genes for dopamine production in the caudates of MPTP-treated monkeys using an AAV vector. *Gene Therapy* 5:820–827.

Fisher-Adams G, Wong KK Jr, Podsakoff G, Forman SJ, Chatterjee S (1996): Integration of adeno-associated virus vectors in CD34+ human hematopoietic progenitor cells after transduction. *Blood* 88:492–504.

Flotte TR, Afione SA, Conrad C, McGrath SA, Solow R, Oka H, Zeitlin PL, Guggino WB, Carter BJ (1993): Stable in vivo expression of the cystic fibrosis transmembrane conductance regulator with an adeno-associated virus vector. *Proc Natl Acad Sci USA* 90:10613–10617.

Girod A, Ried M, Wobus C, Lahm H, Leike K, Kleinschmidt J, Deleage G, Hallek M (1999): Genetic capsid modifications allow efficient re-targeting of adeno-associated virus type 2. *Nat Med* 5:1052–1056.

Grant CA, Ponnazhagan S, Wang XS, Srivastava A, Li T (1997): Evaluation of recombinant adeno-associated virus as a gene transfer vector for the retina. *Curr Eye Res* 16:949–956.

Gregoriadis G, Saffie R, de Souza JB (1997): Liposome-mediated DNA vaccination. *FEBS Lett* 402:107–110.

Grifman M, Trepel M, Speece P, Gilbert LB, Arap W, Pasqualini R, Weitzman M (2001): Incorporation of tumor-targeting peptides into recombinant adeno-associated virus capsids. *Mol Ther* 3:964–975.

Haisma HJ, Grill J, Curiel DT, Hoogeland S, van Beusechem VW, Pinedo HM, Gerritsen WR (2000): Targeting of adenoviral vectors through a bispecific single-chain antibody. *Cancer Gene Ther* 7:901–904.

Handagama P, Scarborough RM, Shuman MA, Bainton DF (1993): Endocytosis of fibrinogen into megakaryocyte and platelet α-granules is mediated by $\alpha_{IIb}\beta_3$ (glycoprotein IIb-IIIa). *Blood* 82:135–138.

Hargrove PW, Vanin EF, Kurtzman GJ, Nienhuis AW (1997): High-level globin gene expression mediated by a recombinant adeno-associated virus genome that contains the 3′ gamma globin gene regulatory element and integrates as tandem copies in erythroid cells. *Blood* 89:2167–2175.

Hansen J, Qing K, Kwon HJ, Mah C, Srivastava A (2000): Impaired intracellular trafficking of adeno-associated virus type 2 vectors limits efficient transduction of murine fibroblasts. *J Virol* 74:992–996.

Herzog RW, Yang EY, Couto LB, Hagstrom JN, Elwell D, Fields PA, Burton M, Bellinger DA, Read MS, Brinkhous KM, Podsakoff GM, Nichols TC, Kurtzman GJ, High KA. (1999): Long-term correction of canine hemophilia B by gene transfer of blood coagulation factor IX mediated by adeno-associated viral vector. *Nature Medicine* 5:56–63.

Hoogenboom HR, Griffiths AD, Johnson KS, Chiswell DJ, Hudson P, Winter G (1991): Multi-subunit proteins on the surface of filamentous phage: methodologies for displaying antibody (Fab) heavy and light chains. *Nucleic Acids Res* 19:4133–4137.

Huston JS, McCartney J, Tai MS, Mottola-Hartshorn C, Jin D, Warren F, Keck P, Oppermann H (1993): Medical applications of single-chain antibodies. *Int Rev Immunol* 10:195–217.

Jomary C, Vincent KA, Grist J, Neal MJ, Jones SE (1997): Rescue of photoreceptor function by AAV-mediated gene transfer in a mouse model of inherited retinal degeneration. *Gene Ther* 4:683–690.

Kallunki P, Tryggvason K (1992): Human basement membrane heparan sulfate proteoglycan core protein: a 467-kD protein containing multiple domain resembling elements of the low density lipoprotein receptor, laminin, neural cell adhesion molecules and epidermal growth factor. *J Cell Biol* 116:559–571.

Kaplitt MG, Leone P, Samulski RJ, Xiao X, Pfaff DW, O'Malley KL, During MJ (1994): Long-term gene expression and phenotypic correction using adeno-associated virus vectors in the mammalian brain. *Nat Genet* 8:148–154.

Krasnykh V, Douglas JT, van Beusechem VW (2000): Genetic targeting of adenoviral vectors *Mol Ther* 1:391–405.

Li XB, Liao GS, Shu YY, Tang SX (2000): Brain delivery of biotinylated NGF bounded to an avidin-transferrin conjugate. *J Nat Toxins* 9:73–83.

Loffler A, Kufer P, Lutterbuse R, Zettl F, Daniel PT, Schwenkenbecher JM, Riethmuller G, Dorken B, Bargou RC (2000): A recombinant bispecific single-chain antibody, CD19 × CD3, induces rapid and high lymphoma-directed cytotoxicity by unstimulated T lymphocytes. *Blood* 95:2098–2103.

Maeda Y, Ikeda U, Shimpo M, Ueno S, Ogasawara Y, Urabe M, Kume A, Takizawa T, Saito T, Colosi P, Kurtzman G, Shimada K, Ozawa K (1998): Efficient gene transfer into cardiac myocytes using adeno-associated virus (AAV) vectors. *J Mol Cell Cardiol* 30:1341–1348.

Mah C, Qing K, Khuntirat B, Ponnazhagan S, Wang XS, Kube DM, Yoder MC, Srivastava A (1998): Adeno-associated virus type 2-mediated gene transfer: role of epidermal growth factor receptor protein tyrosine kinase in transgene expression. *J Virol* 72:9835–9843.

Mandel RJ, Spratt SK, Snyder RO, Leff SE (1997): Midbrain injection of recombinant adeno-associated virus encoding rat glial cell line-derived neurotrophic factor protects nigral neurons in a progressive 6-hydroxydopamine-induced degeneration model of Parkinson's disease in rats. *Proc Natl Acad Sci USA* 94:14083–14088.

Martin F, Kupsch J, Takeuchi Y, Russell S, Cosset FL, Collins M (1998): Retroviral vector targeting to melanoma cells by single-chain antibody incorporation in envelope. *Hum Gene Ther* 9:737–746.

Miller N, Vile R (1995): Targeted vectors for gene therapy. *FASEB J* 9:190–199.

Mizukami H, Young NS, Brown KE (1996): Adeno-associated virus type 2 binds to a 150-kilodalton cell membrane glycoprotein. *Virology* 217:124–130.

Monahan PE, Samulski RJ, Tazelaar J, Xiao X, Nichols TC, Bellinger DA, Read MS, Walsh CE (1998): Direct intramuscular injection with recombinant AAV vectors results in sustained expression in a dog model of hemophilia. *Gene Ther* 5:40–49.

Murdoch AD, Dodge GR, Cohen I, Tuan RS, Iozzo RV (1992): Primary structure of the human heparan sulfate proteoglycan from basement membrane (HSPG2/Perlecan). A chimeric molecule with multiple domains homologous to the low-density lipoprotein receptor, laminin, neural cell adhesion molecules, and epidermal growth factor receptor. *J Biol Chem* 267:8544–8557.

Murphy JE, Zhou S, Giese K, Williams LT, Escobedo JA, Dwarki VJ (1997): Long-term correction of obesity and diabetes in genetically obese mice by a single intramuscular injection of recombinant adeno-associated virus encoding mouse leptin. *Proc Natl Acad Sci USA* 94:13921–13926.

Muzyczka N (1992): Use of adeno-associated virus as a general transduction vector for mammalian brain cells. *Curr Top Micobiol Immunol* 158:97–129.

Nakaki M, Takikawa H, Yamanaka M (1997): Targeting immunotherapy using the avidin-biotin system for a human colon adenocarcinoma in vitro. *J Int Med Res* 25:14–23.

Namiki Y, Takahashi T, Ohno T (1998): Gene transduction for disseminated intraperitoneal tumor using cationic liposomes containing non-histone chromatin proteins: cationic liposomal gene therapy of carcinomatosa. *Gene Therapy* 5:240–246.

Paganelli G, Magnani P, Fazio F (1993): Pretargeting of carcinomas with the avidin-biotin system. *Int J Biol Markers* 8:155–159.

Parrot MB, Barry MA (2000): Metabolic biotinylation of recombinant proteins in mammalian cells and in mice. *Mol Ther* 1:96–104.

Pavlinkova G, Beresford GW, Booth BJ, Batra SK, Colcher D (1999): Pharmacokinetics and biodistribution of engineered single-chain antibody constructs of MAb CC49 in colon carcinoma xenografts. *J Nucl Med* 40:1536–1546.

Peng KW, Russell SJ (1999): Viral vector targeting. *Curr Opin Biotechnol* 10:454–457.

Philip R, Brunette E, Kilinski L, Murugesh D, McNally MA, Ucar K, Rosenblatt J, Okarma TB, Lebkowski JS (1994): Efficient and sustained gene expression in primary T lymphocytes and primary and cultured tumor cells mediated by adeno-associated virus plasmid DNA complexed to cationic liposomes. *Mol Cell Biol* 14:2411–2418.

Ponnazhagan S, Mukherjee P, Wang X-S, Kurpad C, Qing K, Kube D, Mah C, Yoder M, Srour EF, Srivastava A (1997a): Adeno-associated virus 2-mediated transduction of primary human bone marrow derived CD34+ hematopoietic progenitor cells: Donor variation and correlation of expression with cellular differentiation. *J Virol* 71:8262–8267.

Ponnazhagan S, Yoder MC, Srivastava A (1997b): Adeno-associated virus 2-mediated transduction of murine repopulating hematopoietic stem cells and long-term expression of a human globin gene in vivo. *J Virol* 71:3098–3104.

Ponnazhagan S, Wang XS, Woody MJ, Luo F, Kang LY, Nallari ML, Munshi NC, Zhou SZ, Srivastava A (1996): Differential expression in human cells from the p6 promoter of human parvovirus B19 following plasmid transfection and recombinant adeno-associated virus 2 (AAV) infection: human megakaryocytic leukemia cells are non-permissive for AAV infection. *J Gen Virol* 77:1111–1122.

Ponnazhagan S, Mahendra G, Curiel DT, Shaw DR (2001): Adeno-associated virus-mediated transduction of human monocyte-derived dendritic cells: implications for ex vivo immunotherapy. *J Virol* 75:9493–9501.

Qing KY, Mah C, Hansen J, Zhou SZ, Dwarki VJ, Srivastava A (1999): Human fibroblast growth factor receptor 1 is a co-receptor for infection by adeno-associated virus 2. *Nature Med* 5:71–77.

Qing KY, Wang X-S, Kube DM, Ponnazhagan S, Bajpai A, Srivastava A (1997): Role of tyrosine phosphorylation of a cellular protein in adeno-associated virus 2-mediated transgene expression. *Proc Natl Acad Sci USA*, 94:10879.

Rabinowitz JE, Xiao W, Samulski RJ (1999): Insertional mutagenesis of AAV2 capsid and the production of recombinant virus. *Virology* 265:274–285.

Reilly RM (1991): Radioimmunotherapy of malignancies. *Clin Pharm* 10:359–375.

Saga T, Weinstein JN, Jeong JM, Heya T, Lee JT, Le N, Paik CH, Sung C, Neumann

RD (1994): Two-step targeting of experimental lung metastases with biotinylated antibody and radiolabeled streptavidin. *Cancer Res* 54:2160–2165.

Sakahara H, Saga T (1999): Avidin-biotin system for delivery of diagnostic agents. *Adv Drug Deliv Rev* 37:89–101.

Sakurai F, Nishioka T, Saito H, Baba T, Okuda A, Matsumoto O, Taga T, Yamashita F, Takakura Y, Hashida M (2001): Interaction between DNA-cationic liposome complexes and erythrocytes is an important factor in systemic gene transfer via the intravenous route in mice: the role of the neutral helper lipid. *Gene Therapy* 8:677–686.

Sanlioglu S, Benson PK, Yang J, Atkinson EM, Reynolds T, Engelhardt JF (2000): Endocytosis and nuclear trafficking of adeno-associated virus type 2 are controlled by rac1 and phosphatidylinositol-3 kinase activation. *J Virol* 74:9184–9196.

Schimmenti S, Boesen J, Claassen EA, Valerio D, Einerhand MP (1998): Long-term genetic modification of rhesus monkey hematopoietic cells following transplantation of adenoassociated virus vector-transduced CD34+ cells. *Hum Gene Ther* 9:2727–2734.

Schultz J, Lin Y, Sanderson J, Zuo Y, Stone D, Mallett R, Wilbert S, Axworthy D (2000): A tetravalent single-chain antibody-streptavidin fusion protein for pretargeted lymphoma therapy. *Cancer Res* 60:6663–6669.

Shi N, Boado RJ, Pardridge WM (2000): Antisense imaging of gene expression in the brain in vivo. *Proc Natl Acad Sci USA* 97:14709–14714.

Smith JS, Keller JR, Lohrey NC, McCauslin CS, Ortiz M, Cowan K, Spence SE (1999): Redirected infection of directly biotinylated recombinant adenovirus vectors through cell surface receptors and antigens. *Proc Natl Acad Sci USA* 96:8855–8860.

Snyder RO, Miao C, Meuse L, Tubb J, Donahue BA, Lin HF, Stafford DW, Patel S, Thompson AR, Nichols T, Read MS, Bellinger DA, Brinkhous KM, Kay MA (1999): Correction of hemophilia B in canine and murine models using recombinant adeno-associated viral vectors. *Nat Med* 5:64–70.

Song S, Morgan M, Ellis T, Poirier A, Chesnut K, Wang J, Brantly M, Muzyczka N, Byrne BJ, Atkinson M, Flotte TR (1998): Sustained secretion of human alpha-1-anti-
trypsin from murine muscle transduced with adeno-associated virus vectors. *Proc Natl Acad Sci USA* 95:14384–14388.

Su H, Lu R, Kan YW (2000): Adeno-associated viral vector-mediated vascular endothelial growth factor gene transfer induces neovascular formation in ischemic heart. *Proc Natl Acad Sci USA* 97:13801–13806.

Summerford C, Samulski RJ (1998): Membrane-associated heparan sulfate proteoglycan is a receptor for adeno-associated virus type 2 virions. *J Virol* 72:1438–1445.

Summerford C, Bartlett JS, Samulski RJ (1999): $\alpha V\beta 5$ integrin: A co-receptor for adeno-associated virus 2 infection. *Nature Med* 5:78.

Vieweg J, Boczkowski D, Roberson KM, Edwards DW, Philip M, Philip R, Rudoll T, Smith C, Robertson C, Gilboa E (1995): Efficient gene transfer with adeno-associated virus based plasmids complexed to cationic liposomes for gene therapy of human prostate cancer. *Cancer Res* 55:2366–2372.

Vitaliti A, Wittmer M, Steiner R, Wyder L, Neri D, Klemenz R (2000): Inhibition of tumor angiogenesis by a single-chain antibody directed against vascular endothelial growth factor. *Cancer Res* 60:4311–4314.

Watkins SJ, Mesyanzhinov VV, Kurochkina LP, Hawkins RE (1997): The 'adenobody' approach to viral targeting: specific and enhanced adenoviral gene delivery. *Gene Ther* 4:1004–1012.

Wilbur DS, Pathare PM, Hamlin DK, Stayton PS, To R, Klumb LA, Buhler KR, Vessella RL (1999): Development of new biotin/streptavidin reagents for pretargeting. *Biomol Eng* 16:113–118.

Wu P, Xiao W, Conlon T, Hughes J, Agbandje-McKenna M, Ferkol T, Flotte T, Muzyczka N (2000): Mutational analysis of the adeno-associated virus type 2 (AAV2) capsid gene and construction of AAV2 vectors with altered tropism. *J Virol* 74:8635–8647.

Xiao W, Berta SC, Lu MM, Moscioni AD, Tazelaar J, Wilson JM (1998): Adeno-associated virus as a vector for liver-directed gene therapy. *J Virol* 72:10222–10226.

Xiao X, Li J, Samulski RJ (1996): Efficient long-term gene transfer into muscle tissue of immunocompetent mice by adeno-associated virus vector. *J Virol* 70:8098–8108.

Xiao X, McCown TJ, Li J, Breese GR, Morrow AL, Samulski RJ (1997): Adeno-associated virus (AAV) vector antisense gene transfer in vivo decreases GABA(A) alpha1 containing receptors and increases inferior collicular seizure sensitivity. *Brain Res* 756:76–83.

Yao Z, Zhang M, Sakahara H, Saga T, Arano Y, Konishi J (1998): Avidin targeting of intraperitoneal tumor xenografts. *J Natl Cancer Inst* 90:25–29.

Zhou S, Murphy JE, Escobedo JA, Dwarki VJ (1998): Adeno-associated virus-mediated delivery of erythropoietin leads to sustained elevation of hematocrit in nonhuman primates. *Gene Ther* 5:665–670.

10

RECEPTOR TARGETING OF ADENO-ASSOCIATED VIRUS VECTORS

HILDEGARD BÜNING, PH.D., MARTIN ULRICH RIED, PH.D., AND MICHAEL HALLEK, M.D.

INTRODUCTION

The development of safe and efficient gene transfer vehicles is critical for the success of gene therapy. One of the most promising viral vectors is derived from the adeno-associated virus type 2 (AAV-2), a member of the parvovirus family. AAV-2 was discovered as a coinfecting agent during an adenovirus outbreak, without any apparent pathogenicity contributed by AAV-2 (Blacklow et al., 1968, 1971; Blacklow, 1988). Up to now, no human disease caused by AAV-2 has been detected. Moreover, AAV-2 seems to be protective against bovine papillomavirus and adenovirus mediated cellular transformation (Hermonat, 1989; Mayor, Houlditch, and Mumford, 1973; Khielf et al., 1991). AAV-2 does not induce strong cytotoxic effects and does not elicit a strong cellular immune response as commonly seen with other viral vectors (Carter and Samulski, 2000). Finally, AAV-2 has the unique potential to integrate site specifically into the q-arm of human chromosome 19 (Kotin et al., 1990; Samulski et al., 1991).

AAV-2 has a broad tissue tropism infecting such diverse organs as the brain, liver, muscle, lung, retina, and heart muscle. This makes AAV-2 attractive for

Vector Targeting for Therapeutic Gene Delivery, Edited by David T. Curiel and Joanne T. Douglas
ISBN 0-471-43479-5

in vitro gene transfer into various tissues (Carter and Samulski, 2000). AAV-2 vectors are now used for in vivo gene transfer [for details, see reviews by Monahan and Samulski (2000) and Tal (2000)]. However, these studies clearly demonstrate that clinically relevant gene expression levels can be reached only in the liver, unless vectors are administered directly into the target tissue or organ. Therefore, the studies emphasize the need to target AAV vectors in order to overcome this apparent limitation of a broad tissue tropism. Moreover, the targeting of AAV vectors would also enhance the safety and efficiency of AAV-mediated gene transfer in vivo. Therefore, increasing efforts are being undertaken to retarget AAV-2-based vectors to specific receptors and to generate selective, tissue- or organ-restricted vectors. So far, the studies summarized in this review show that it is possible—at least in principle—to target AAV-2 to a specific cell or organ. However, the targeting vectors still need to be optimized with regard to a further reduction of the wild-type AAV-2 tropism, or with regard to the infectious titer. With a rapidly increasing knowledge about the functional domains on the AAV-2 capsid which are involved in receptor binding and the subsequent steps of transmembrane and intracellular processing of the virion, it should be possible to create highly efficient AAV-2 targeting vectors.

GENOMIC ORGANIZATION OF AAV-2

AAV-2 is a single stranded, replication-deficient nonenveloped DNA virus (Rivadeneira et al., 1998) composed of an icosahedral protein capsid and a viral genome of 4680 nucleotides (Fig. 10.1). The AAV-2 genome encodes only two open reading frames (ORFs). It is flanked at both ends by the 145-bp inverted terminal repeat (ITR) sequences. The ITRs are required for encapsidation of the viral genome and seem to have enhancer and/or weak promoter activity. Moreover, the ITRs are one of the two viral elements necessary for the site-specific integration of wild-type AAV-2 and for the rescue of proviruses. The ITRs frame the two open reading frames *rep* and *cap*. The *rep* gene encodes four overlapping, multifunctional proteins (Rep78, Rep68, Rep52, and Rep40) controlled by two different promoters (Balagué et al., 1997). The large Rep proteins (Rep78 and its splice variant Rep68) are controlled by the p5 promoter and are necessary for viral DNA replication, transcriptional control, and site-specific integration. Rep52 and its splice variant Rep40 are known as small Rep proteins. They are transcribed from the p19 promoter and play an essential role in the accumulation of single-stranded progeny genomes used for packaging. The 3′ ORF *cap* accommodates the three capsid proteins VP1 (90 kDa), VP2 (72 kDa), and VP3 (60 kDa), which contribute to the 60 subunits of the AAV-2 viral capsid in a 1 : 1 : 20 ratio (Rabinowitz and Samulski, 2000). They are controlled by the p40 promoter, share the same stop codon, but have different initiation codons, resulting in progressively shorter proteins from VP1 to VP3. All three capsid proteins are necessary for the generation of infectious particles, although capsids are formed in the absence of VP1 (Hermonat et al.,

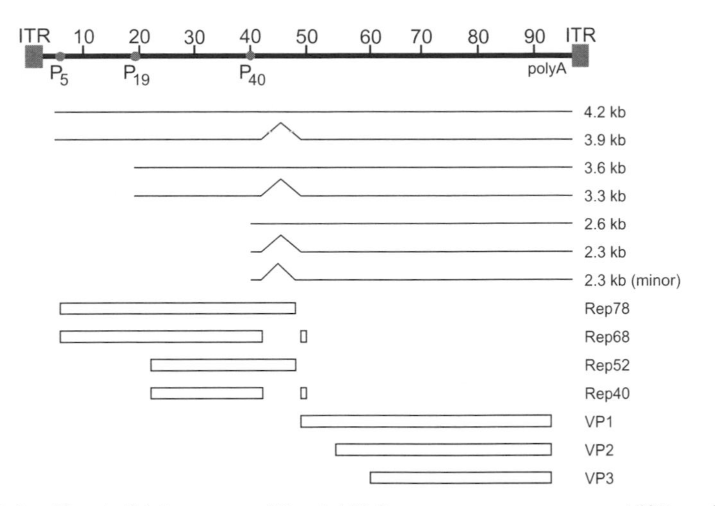

Figure 10.1 The AAV-2 genome. The AAV-2 genome encompasses 4680 nucleotides, divided into 100 map units. Indicated are the two inverted terminal repeats (ITRs), the three viral promoters at map positions 5 (p5), 19 (p19) and 40 (p40), and the common polyadenylation signal at map position 96. Large Rep proteins (Rep78 and Rep78) are under the control of the p5 promoter, whereas the small Rep proteins (Rep52 and Rep40) are driven by the p19 promoter, existing in spliced and unspliced isoforms. The *cap* genes encoding three different capsid proteins (VP1, VP2 and VP3) are under the control of the p40 promoter. (Thin lanes represent the mRNAs, open rectangles the proteins).

1984; Smuda and Carter, 1991; Tratschin, Miller, and Carter, 1984). The capsid assembly occurs in the nucleus (Wistuba et al., 1997, 1995). VP2 is responsible for the transport of VP3 into the nucleus (Hoque et al., 1999; Ruffing, Zentgraf, and Kleinschmidt, 1992). The encapsidation of the AAV-2 genome probably takes place in the nucleoplasm. Rep-tagged DNA seems to initiate packaging by interaction with capsid proteins (Dubielzig et al., 1999).

THREE-DIMENSIONAL STRUCTURE OF AAV-2 AND RELATED PARVOVIRUSES

In contrast to some related parvoviruses such as canine parvovirus (CPV) (Tsao et al., 1991), feline panleukopenia virus (Agbandje et al., 1993), minute virus of mice (Agbandje-McKenna et al., 1998), and the human parvovirus B19 (Agbandje et al., 1994), the three-dimensional structure of the AAV-2 capsid has not been determined. However, because the parvoviral capsids show structural homology, and antigenic epitopes are often located in homologous regions, it is expected that the AAV-2 capsid contains a core structure comprising an eight-stranded β-barrel motif. In other parvoviruses, the majority of the capsid surface is made up of large insertions lying between several strands of the β-barrel. The surface of the CPV capsid contains a hollow cylinder at the fivefold axis of symmetry, which is surrounded by a circular depression (canyon), a prominent protrusion at the threefold axis of symmetry (threefold spike), and a depression (dimple), spanning the twofold axis of symmetry (Chapman and Rossman, 1993; Tsao et al., 1991). As mentioned before, the highest homology is found in the β-barrel motif of various parvoviruses, while the surface regions show a higher variability and contribute to the specific host tropism of the viruses. Epitope mapping and mutagenesis studies of capsid proteins have allowed the determination of some critical antigenic, loop, and receptor binding regions of AAV-2 (Moskalenko et al., 2000; Rabinowitz et al., 1999; Wobus et al., 2000; Wu et al., 2000).

RECEPTORS MEDIATING THE CELLULAR UPTAKE OF AAV-2

Heparan sulfate proteoglycans (HSPGs) have been proposed to act as a primary or attachment receptor of AAV-2 (Summerford and Samulski, 1998). Since all adherent cells express glycosaminoglycans on their surface, this explains the broad tropism of the virus. Although no heparin binding motif has been identified so far, Wu et al. (2000) were able to map two regions involved in HSPG binding by alanine substitution and insertion of the hemagglutinin (HA) epitope YPVDVPDYA. These regions encompass amino-acid position 509 to 522 and 561 to 591 (in the VP3 region). Two types of coreceptors, $\alpha v\beta 5$ integrin and fibroblast growth factor receptor 1 (FGFR1) (Bartlett, Wilcher, and Samulski, 2000; Qing et al., 1999; Summerford, Bartlett, and Samulski, 1999), have also been identified. To date, their precise role in viral entry has not been determined in detail. After binding to the cell surface, AAV-2 becomes internalized by clathrin mediated endocytosis. Bartlett, Wilcher, and Samulski (2000) showed that AAV-2 is found inside endosomes and can penetrate into the cytosol, once it encounters a weakly acidic environment. Following its endosomal release, AAV-2 reaches the nuclear membrane and enters the nucleus probably via nuclear pore complexes. There AAV-2 enters its productive life cycle or, in the absence of a helper virus, is integrated at the AAVS1 site of chromosome 19 (Carter and Samulski, 2000).

AAV-SEROTYPES, OTHER THAN AAV-2

Most AAV vectors used are based on the AAV-2 serotype, since it was the first from which an infectious clone was available (Samulski et al., 1982). But since 50–96% of the population is seropositive for AAV-2, it is worth considering the use of other serotypes. In the meantime, five other serotypes have been cloned, sequenced, and analyzed with regard to their tropism (Chiorini et al., 1997, 1999; Muramatsu et al., 1996; Rutledge, Halbert, and Russell, 1998). With the exception of AAV-6, which has >99% amino-acid homology to AAV-1, all serotypes show a significantly different amino acid sequence of the capsid proteins, which is most prominent in VP3 (Rabinowitz and Samulski, 2000) and most obvious for AAV-4 and -5. It is interesting that neutralizing antibodies against AAV-4 or AAV-5 do not cross-react with other serotypes, whereas neutralizing antibodies against AAV-2 cross-react with AAV-1, AAV-3, and AAV-6 (for references see Rabinowitz and Samulski, 2000). Moreover, the occurrence of cross-reacting antibodies for AAV-1 and AAV-2 seems to depend on the route of adminstration (Xiao et al., 1999). Further studies will be useful to define the epitopes responsible for the generation of these neutralizing antibodies. This might then be used to create a second generation of "immune escape" AAV vectors that should not be eliminated by preexisting antibodies.

Despite their differences in VP3, AAV-1, -3, and -6 all bind to heparan sulfate proteoglycans, whereas AAV-4 and AAV-5 are insensitive to heparin competition (a soluble competitor for HSPG binding). AAV-4 requires α2-3 O-linked, whereas AAV-5 uses α2-3 N-linked sialic acid for cell binding (Kaludov et al., 2001). Interestingly, there are differences between HSPG binding serotypes with regard to the transduction efficiency to different cell types. For example, Xiao et al. (1999) found a better transduction of liver by AAV-2 as compared with AAV-1, while the opposite was seen in muscle. This points to differences in the viral entry, trafficking, and/or uncoating of the different AAV serotypes (Rabinowitz and Samulski, 2000). The identification of the responsible capsid domains will be of considerable interest and will help to understand the intracellular processing of AAV.

METHODS OF PACKAGING AND PURIFICATION OF AAV VECTORS

If *rep* and *cap* are provided in *trans* on a helper plasmid, 96% of the wild-type AAV genome can be removed and replaced by the transgene, since the only *cis* elements necessary for the generation of recombinant AAV are the ITRs (Carter and Samulski, 2000). The production and purification protocols to generate high-titer and highly purified viral preparations continue to undergo improvements (Summerford and Samulski, 1999; Hermens et al., 1999). But until now, in most of the cases rAAV is produced by transfection of the vector (contains the ITR-flanked transgene) and the helper plasmid (encodes Rep and Cap) into HeLa or 293 cells, followed by superinfection with adenovirus type 5 or by a triple transfection of vector-, helper- and an adenovirus helper plasmid in 293 cells (Girod et al., 1999; Grimm and Kleinschmidt, 1999; Xiao, Li, and Samulski, 1998; Ferrari et al., 1997). After harvesting, AAV is purified using iodixanol or CsCl gradient ultracentrifugation and/or chromatography (Chiorini et al., 1995; Hermens et al., 1999; Summerford and Samulski, 1999; Zolotukhin et al., 1999). After purification, *infectious* particle titers of AAV-2 of $>10^9$/ml can be easily reached, which is sufficient for most in vitro and in vivo experiments, at least in smaller rodents. However, when it comes to larger animals or human beings in clinical applications, it is desirable to enhance the target specificity of AAV vectors by receptor retargeting in order to reduce the amount of vector particles to be administered.

TARGETING OF AAV-2

In principle, at least, AAV-2 can be targeted to a specific cell in two different ways (Cosset and Russell, 1996):

1. The interaction between the viral vector and the target cell can be mediated by an associated molecule (e.g., a glycosid molecule or a bispecific antibody) which is bound at the viral surface and can interact with a cell surface molecule (Miller, 1996; Fig. 10.2B).
2. The cell-specific targeting of the vector can be mediated by a ligand directly inserted into the viral capsid (Walther and Stein, 1996; Fig. 10.2C).

For virus targeting by an intermediate molecule (Fig. 10.2B), it is not necessary to know the three-dimensional structure of the viral surface if high-affinity viral surface-binding molecules such as monoclonal antibodies are available. For this strategy, the stability of the interaction of the virus with the intermediate molecule, as well as the efficiency by which the complex is generated, are rate limiting. In addition, the intermediate molecules must bind to cell-specific receptors, which allow the uptake and correct intracellular processing of the virus. Finally, the process should allow a scale-up.

A combination of two important parameters is necessary for the successful generation of a targeting vector by direct modification of the capsid (Fig. 10.2C). The first parameter is a good choice of the insertion site to ensure that packaging of the mutant remains efficient and the inserted ligand is exposed

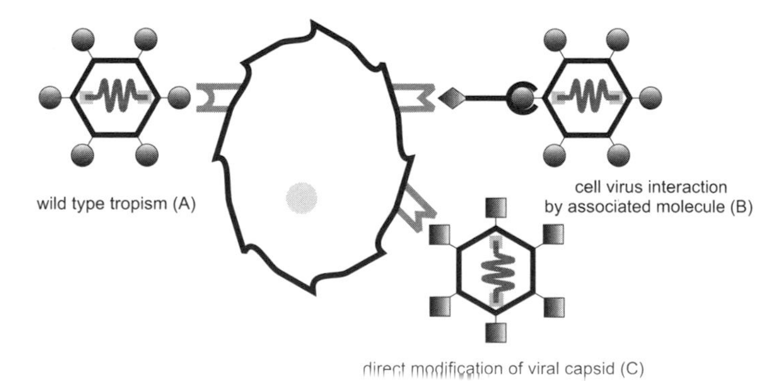

Figure 10.2 Two possibilities of targeting viral vectors. Viral vectors with wild type tropism (*A*) show a direct binding of structural capsid components to the cell surface receptor. In targeting vectors, the virus-cell interaction is mediated by a molecule associated with the capsid (*B*) or by a ligand directly inserted into the capsid (*C*).

on the cell surface. Unfortunately, the crystal structure of AAV is unknown. Therefore, two alternative strategies have been used to identify candidate positions for insertion of heterologous ligand: (1) sequence alignment between AAV-2 and other parvoviruses for which the X-ray crystal structure is known (Girod et al., 1999; Grifman et al., 2001); (2) a systematic, insertional mutagenesis of the whole AAV-2 capsid (Rabinowitz, Xiao, and Samulski, 1999; Wu et al., 2000). The second important parameter is the choice of the targeting peptide. It is not easy to predict the secondary structure of the ligand once it has been inserted into the AAV capsid region. Therefore, the ligand should be structure independent and not too large to avoid the destabilization of the whole capsid. Of course, the ligand should be cell type specific and be internalized in a way that allows an efficient transport and release of the viral DNA into the cell nucleus.

Both approaches and a combination thereof have been used to retarget AAV-2 (Yang et al., 1998; Girod et al., 1999; Bartlett et al., 1999; Wu et al., 2000; Grifman et al., 2001; Nicklin et al., 2001; Ried et al., 2002) and are described in the following paragraphs.

Targeting by Bispecific Antibodies

The feasibility of targeting AAV by a bispecific antibody mediating the interaction between virus and target cell (Fig. 10.2B) was first shown by Bartlett et al. (1999). The antibody used was generated by a chemical cross-link of the Fab′ arms of monoclonal antibodies against the $\alpha_{IIb}\beta_3$ integrin (AP-2 antibody) and the intact AAV-2 capsid (A20 antibody; Wistuba et al., 1995). The major ligand for $\alpha_{IIb}\beta_3$ is fibrinogen, which becomes internalized via endocytosis. Therefore, AAV-2 targeted to this integrin was expected to become internalized via receptor-mediated endocytosis, similar to wild-type virus. This targeting vector transduced MO7e and DAMI cells, which are not permissive for wild-type AAV-2 infection (70-fold above background). On cells not expressing the targeting receptor, a 90% reduction of AAV-2 transduction was seen. It remains to be determined whether this reduction was due to steric hindrance or some other mechanism. Another issue that remains to be solved is the stability of the virus-bispecific antibody complexes in vivo.

Targeting by Insertion of Single Chain Antibodies or Serpin Receptor Ligand at the N-Terminus of VP Proteins

The first attempt to alter the tropism of AAV-2 was described by Yang et al. (1998). They tried to insert a single-chain antibody against the human CD34 molecule, a cell surface molecule expressed on hematopoietic progenitor cells, at the 5′-ends of VP1, VP2, and VP3. Using a transcription and translation assay, they could express all the three different single-chain fragment variable region (scFv)-AAV-2 capsid fusion proteins. However, they failed to produce detectable rAAV-2 particles, when using either all three scFv-VP fusion proteins or one scFv-VP fusion with two other unmodified virion proteins. Therefore, they had to use all three wild-type AAV-2 capsid proteins for the packaging process in addition to one of the three single scFv-VP fusion proteins. With this procedure, intact viral particles could be generated, which were able to infect HeLa cells and showed an increased transduction of CD34-positive KG-1 cells. Although this approach demonstrated for the first time that targeting of AAV-2 by direct modification of the capsid is possible, only very low titers (1.9×10^2 transducing units/ml on KG-1) were achieved. Moreover, very heterogeneous viral preparations consisting of an unknown mixture of chimeric, targeting and wild-type AAV-2 particles were produced.

Wu et al. (2000) inserted the HA epitope YPVDVPDYA into the N-terminal regions of the VP1, VP2, and VP3 and the C-terminus of the *cap* ORF. They observed that the insertion of this and other epitopes at the N-termini of VP1 (VP1N) and VP3 (VP3N) and the C-terminus of the *cap* ORF (VPC) resulted either in no detectable particle (VP3N and VPC), or in a 2–3 log decrease of infectious and physical particle titers. In agreement with Yang et al. (1998), only the insertion at the N-terminus of VP2 was tolerated, but without the need to add the wild-type AAV-2 capsid proteins. Moreover, exchanging the HA epitope by the serpin receptor ligand KFNKPFVFLI (Ziady et al., 1997) resulted in a 15-fold higher infection of the lung epithelial cell line IB3 than by wild-type AAV-2. The fact that the N-terminal insertion of different peptides is tolerated in VP2 and allows targeting, albeit with low efficiency, probably reflects the exposure of the N-terminus of VP2 at the viral surface in analogy to CPV (Chapman and Rossmann, 1993; Weichert et al., 1998).

Targeting of rAAV-2 Vectors by Insertion of Ligands Inside the Viral Protein Sequences

The first successful demonstration that a genetic capsid modification can be used to retarget AAV-2 was described by Girod et al. (1999). A sequence alignment of AAV-2 and CPV identified 6 sites (amino-acid positions 261, 381, 447, 534, 573, 587) expected to accept the insertion of a ligand polypeptide without disruption of functions essential for the viral life cycle (Fig. 10.3A). At these positions, the sequence for the 14 amino-acid peptide L14 (QAGTFALRGDNPQG) was inserted into the capsid gene by PCR. The L14 peptide with the RGD motif of the laminin fragment P1 (Aumailley et al., 1990) was used for this approach, since it is the target for several cellular integrin receptors and can also serve as a viral receptor (Aumailley et al., 1990; White et al., 1993). Moreover, no specific secondary structure is required for recognition of the receptor (Aumailley et al., 1990). All the six mutants could be packaged with similar efficiency to wild-type AAV-2. All mutants showed an intact capsid structure by electron microscopy and in an A20 enzyme-linked immunosorbent assay (ELISA), which was used to identify and quantify intact AAV-2 viral capsids (Wistuba et al., 1997; Grimm et al., 1999). Using an ELISA, in which the virus mutants were bound to the plate by the A20 antibody, it was demonstrated that L14 was exposed at the viral surface when inserted in amino-acid

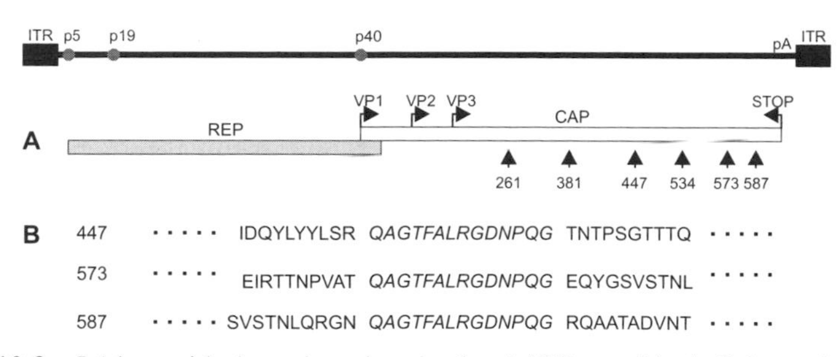

Figure 10.3 L14 peptide insertion sites in the AAV2 capsid. *A*. Schematic diagram of the two open reading frames *rep* and *cap*, the ITR structures, promoter start sites and poly A. *cap* encodes the three capsid protein VP1, VP2 and VP3. Sites of the L14 insertion described by Girod et al. (1999) are marked by arrows (numbers are the amino-acid positions N-terminal of the insertion). *B*. Sequence of L14 and flanking amino-acids. Only the insertions mutants which displayed the L14 peptide on the surfacing using an A20-coated ELISA plate are shown (Girod et al., 1999; numbering starts at the start codon of VP1).

positions 447, 573, and 587. In a cell binding assay insertions mutants I-447 and I-587 were able to bind B16F10 (mouse melanoma cell line) and RN22 (rat swannoma cells), which did not bind, and were not infected by, wild-type AAV-2. Finally, efficient transduction of B16F10 cells was observed using the AAV insertion mutant I-587 expressing Rep or β-galactosidase. Further investigations on insertions at the site I-587 with multiplication of the L14 sequences resulted in a decrease of packaging efficiency although the insertion of a 34 amino acid containing Z34C protein A domain of *Staphylococcus aureus* (see the following section) was well tolerated. These results show that the maximal length of the peptide tolerated at this position also depends on the sequence itself. The precise determinants of this tolerance are unknown.

The site I-587 was also successfully used for the insertion of an endothelial specific peptide isolated by phage display and allowed to generate an AAV-2 mutant able to infect endothelial cells such as human umbilical vein cells (HUVECs) and human saphenous vein endothelial cells (HSVECs) (Nicklin et al., 2001). In contrast to wild-type AAV-2 infection, the infection of the endothelial cells by the mutant could not be blocked by heparin, showing that the infection did not depend on HSPG. Moreover, heparin-binding studies showed that the mutant could not be efficiently retained in a heparin column, but was detected in the column wash, in contrast to the wild-type AAV-2 virus used as a control. The specificity of the binding shown by infection studies using different nonendothelial cell lines such as HepG2. Furthermore, the mutant seemed to follow an intracellular route different from wild-type AAV-2 since compounds such as bafilomycin A2 (an inhibitor of endosomal acidification) did not inhibit transduction. Taken together, all studies underline the potential to use the site 587 for the generation of AAV-2 targeting vectors (Fig. 10.2C) (Girod et al., 1999; Nicklin et al., 2001; Ried et al., 2002).

This result was confirmed recently by Grifman et al. (2001). They also used a similar alignment strategy to identify potential target sites of the AAV-2 capsid by expanding their alignment to parvoviruses other than CPV and to the other AAV serotypes (AAV 1, 3, 4, and 5). They found identical regions to Girod et al. (1999), and finally used sites 448 and 587 for their studies. They inserted the Myc epitope and a peptide with the sequence NGRAHA, identified by phage display, which is specific for CD13 (NGR receptor expressed on angiogenic vasculature and in many tumor cell lines). The insertion at 587 allowed cell-specific targeting to different cell lines tested (KS1767, a Kaposi's sarcoma cell line, and RD, a rhabdomyosarcoma cell line). Interestingly, deletion of the six amino acids (GNRQAA) at positions 586–591 resulted in the loss of heparin binding, whereas the insertion of the targeting peptide (NGRAHA) restored the heparin-binding ability. Taking into account that HSPG has a negative charge, the R residue at position 588 might have an essential role for HSPG binding. Another important residue in this region seems to be Q at 584, since a point mutation to S abrogated rAAV-2 transduction (Grifman et al., 2001).

Wu et al. (2000) constructed 93 mutants at 59 different positions in the AAV-2 capsid by site-directed mutagenesis. With these mutations they identified putative regions that are involved in the HSPG binding, as well as capsid domains that were exposed on the surface with the potential to tolerate the insertion of a ligand. These positions were 34 (in VP1), the N-terminus of VP2 (138), as well as 266, 328, 447, 522, 553, 591, 664 (in VP3). Although all the VP3 insertion could be immunoprecipitated by an antibody against the inserted HA epitope, only 266, 447, 591, and 664 were still infectious. For insertion mutant 522 this condition can be explained by loss of HSPG-binding ability, in agreement with Rabinowitz, Xiao, and Samulski (1999). For the other mutants, an explanation is lacking. The identification that amino acids 266, 447, and 591 at the surface of AAV, are able to tolerate a ligand insertion, confirmed again the results of Girod et al. (1999). I-261 expressed L14 at the cell surface as shown in an ELISA where the virus was bound directly to the plate (Girod et al., 1999). However, Wu et al. (2000) tested only position 34 in the VP1 sequence and the N-terminus of VP2 for targeting of AAV-2 to IB3 cells by substitution with the serpin receptor ligand (FVFLI for VP1 insertion; KFNKPFVFLI N-terminal insertion at VP2). It was shown that targeting and infection was possible with a 62-fold or 4-fold higher efficiency than wild-type or N-terminal VP-2 insertion mutants, respectively. Interestingly, both insertions, which were placed outside the potential HSPG binding regions were blocked by heparin, suggesting that the serpin-tagged mutants continued to use HSPG as the primary receptor and used the serpin receptor as the alternative (co-) receptor.

Generation of Universal Targeting Vectors by Combining Two Principles of Vector Targeting

Inspired by an earlier attempt for Sindbis virus (Ohno et al., 1997), we tried to use a general targeting vector using a truncated 34-amino acid peptide, Z34C, from protein A of *S. aureus* (Ried et al., 2002). Protein A recognizes and binds the Fc part of immunoglobulins (Ig), but not the variable Ig domain, which therefore remains able to bind the antigen. Z34C is derived of the protein A subunit B, which encompasses 56 amino acids and binds the Fc portion with a dissociation constant of about 10–50 nM (Sinha, Sengupta, and Ray, 1999). A 38-residue truncation of this domain, selected by phage display, was further truncated and stabilized by insertion of disulfide bonds and showed thereafter a dissociation constant of 20 nM. The insertion of Z34C at position 587 in the AAV-2 capsid (587Z34C) resulted in a 10-fold decrease of packaging efficiency in comparison to wild-type AAV-2. In contrast, the combination of an insertion with a nine amino-acid deletion (587Δ9Z34C) resulted in a packaging efficiency similar to wild-type AAV-2. Electron microscopy and A20-ELISA revealed a wild-type capsid morphology for both mutants, although three times more empty capsids were observed. Interestingly, the wild-type tropism of the Z34C insertion mutants decreased by 3–4 orders of magnitude, in agreement with Nicklin et al. (2001). The insertion of Z34C at 587 allowed the expression of the functional IgG-binding domain, as shown by binding studies of various antibodies. The capsid mutant 587Z34C bound antibodies more efficiently than 587Δ9Z34C, maybe because the binding domain was less accessible with the nine amino-acid deletion. In agreement, Grifman et al. (2001) showed that a substitution at residue 587 was less efficient than an insertion. Coupling 587Z34C virus with antibodies against CD29 (β1-integrin), CD117 (c-kit-receptor) or CXCR4 followed by infection of different hematopoietic cell lines resulted in specific, antibody-mediated transduction of the cell lines tested. No transduction could be detected without antibody, whereas the targeted infection could be blocked with soluble protein A or IgG molecules. In addition, no inhibition of transduction by the targeting vector was observed with heparin, demonstrating that the interaction of the 587Z24C mutants with the natural AAV-2 receptor HSPG was not essential for infection or transduction. Taken together, this targeting approach shows that a universal AAV targeting vector can be generated and loaded with different targeting molecules to mediate the transduction of specific cells. However, there remains much room for improvement, because the titers of these vectors are still too low.

UNSOLVED ISSUES

To efficiently retarget the AAV system it is necessary to understand the infectious biology of AAV better, including the virus–cell surface interactions, mechanisms of uptake, endosomal processing and release, nuclear transport, and mechanisms leading to gene expression. Moreover, a crystal structure would tremendously enhance our knowledge of the function of different capsid domains, whether they are exposed at the viral surface or not.

The identification of HSPG as the primary attachment receptor for AAV was an important achievement. However, no binding motif within the capsid

TABLE 10.1 Summary of Different Approaches for Targeting AAV to Specific Cell Lines

Author	Year	Method	Ligand	
				Sequence
Yang et al.	1998	N-terminal insertion at VP2	Single chain antibody against CD34	
Bartlett et al.	1999	Bispecific antibody	Fab: arm against $\alpha_{IIb}\beta_3$ integrin	
Girod et al.	1999	Genetic capsid modification at 587 (VP3 region)	L14 (RGD motif of the laminin fragment (P1)	QAGTFALRGDNPQG
Wu et al.	2000	N-terminal insertion at VP2	Serpin receptor ligand	KFNKPFVFLI
Wu et al.	2000	Genetic capsid modification at 34 (VP1 region)	Serpin receptor ligand	FVFLI
Grifman et al.	2001	Genetic capsid modification at 587 (VP3 region)	CD13 specific	NGRAHA
Nicklin et al.	2001	Genetic capsid modification at 587 (VP3 region)	Endothelial cell specific ligand	SIGYPLP
Ried et al.	submitted	Genetic capsid modification at 587 (VP3 region) plus antibody	Z34C domain of protein A plus antibodies against CD117, CD29 or CXCR4	

Notes: VP = viral capsid protein; tu = transducing units.
[a]Transducing units per ml for unmodified and mutant virus were determined and compared.
[b]Genomic titers of unmodified and mutant virus were used to transduce the target cell.
[c]Values calculated by the authors of this review from the data in the original paper.

Table 10.1 (*Continued*)

Target	Cell Line	Result			
		Titer of the modified AAV on target cell	Increase in target cell transduction in comparison to unmodified AAV	Titer of the modified AAV on wild-type permissive cells	Decrease in transduction of wild-type permissive cells in comparison to unmodified AAV
CD34	KG-1 (human acute myelogenous leukemia cell line)	1.9×10^2 tu/ml	>100-fold increase[a]	4.6×10^4 tu/ml	72.6% reduction[a, c]
Fibrinogen	MO7e, DAMI (human megakaryocyte cell lines)	titer?	70-fold increase	titer?	90% reduction
Integrin receptors	B16F10 (mouse) melanoma cell line)	5×10^4 tu/ml	10^4-fold increase[a]	6×10^5 tu/ml	98.8% reduction[a, c]
Serpin receptor	IB3 (lung epithelial cell line)	titer?	15-fold increase[a]	titer?	6-fold reduction[a]
Serpin receptor	IB3 (lung epithelial cell line)	titer?	62-fold increase[a]	titer?	6-fold reduction[a]
NGR receptor	KS 1767 (derived from Kaposi sarcoma), RD (rhabdomyosarcoma cell line)	titer?	10–20-fold increase[a]	titer?	94.4% reduction[a, c]
Unknown	HUVEC (human umbilical vein endothelial cells), HSVEC (human saphenous vein endothelial cells)	titer?	5.9-fold for HUVEC, 28.2-fold for HSVEC increase[b]	1×10^4 tu/ml	99.2% reduction[b, c]
CD117, CD29 or CXCR4	MO7e (human megakaryocyte cell lines), Jurkat (T cell leukemia cell line), Mec1 (B-CLL cell line	$1\text{–}3 \times 10^3$ tu/ml	>10^3-fold increase in specificity	2×10^4 tu/ml (without antibody)	99.9% reduction[a, b]

has been identified so far. This gap in our knowledge renders it difficult to specifically modify the natural viral tropism of AAV, despite some useful hints from the work of Rabinowitz et al. (1999) and Wu et al. (2000).

Understanding the intracellular processing of AAV targeting vectors will be most critical, because these AAV targeting vectors will eventually end up in a cellular compartment, from which they will never be released, or in which they will be processed in ways preventing nuclear processing or gene expression. Therefore, the success of creating AAV targeting vectors will ultimately depend on our ability to unveil the detailed mechanisms of AAV transport and processing. Some pieces of the puzzle are already known (Bartlett, Wilcher, and Samulski, 2000; Sanlioglu et al., 2000, Seisenberger et al., 2001), but the picture is not complete. In this regard, the attempts to target the AAV vector itself currently may help to uncover some important basic functions of AAV capsid proteins, as well as mechanisms of the infectious biology of AAV.

Another issue is the identification of the optimal ligand or targeting receptor. For the genetic modification strategy chosen by our group, the length and sequence of the ligand are very important, because the insertion of a peptide may result in profound alterations of the three-dimensional capsid structure. One theoretical possibility to overcome this problem is the combination of an insertion with a deletion, although in the absence of crystal structure of the AAV capsid, this can currently only be done by a trial-and-error approach. Another possibility is the insertion of a sequence that is able to form its own secondary structure, for example a loop closed by a cysteine bridge.

These difficulties do not exist when using an antibody or another bridging molecule to mediate the interaction between the viral surface and the target cell. However, these approaches will encounter problems like sufficient complex stability, potential of scaling up, and steric hindrance for virus uptake. To identify new ligands, phage display can be used. However, screening for a new ligand in the context of the AAV capsid itself (vector display), where a random peptide sequence is inserted into the capsid sequence and the viral pool generated is then screened on the target cell might even be more promising. Our group is currently performing these experiments to further improve the specific targeting of AAV vectors.

CONCLUSION

AAV is a very promising vector for human somatic gene therapy. However, its broad host range is a disadvantage for in vivo gene therapy, because a selective, tissue- or organ-restricted infection would be desired to increase the safety and efficiency for the gene transfer in vivo. Therefore, increasing efforts are being undertaken to target AAV-based vectors to specific receptors. The studies summarized in this review show that it is possible to target AAV-2 to a specific cell or organ (Table 10.1, pages 232–233). The most promising approach for this purpose is the genetic modification of the capsid protein. However, the current AAV targeting vectors leave much room for improvement with regard to fully eliminating the wild-type AAV-2 tropism or enhancing the infectious titer. Furthermore, a detailed understanding of the transmembrane and intracellular processing of wild-type AAV is necessary for specific modification of the tropism of this virus and and the creation of highly efficient AAV-2 targeting vectors.

ACKNOWLEDGMENTS

This work was supported by the Burda Foundation, the Deutsche Forschungsgemeinschaft (SFB 455), the Wilhelm Sander-Stiftung, and the Bayerische Forschungsstiftung. The authors thank all the members of the laboratory for many inspiring discussions and help during the work presented in part in this review.

REFERENCES

Agbandje M, McKenna R, Rossmann MG, Strassheim ML, Parrish CR (1993): Structure determination of feline panleukopenia virus empty particles. *Proteins* 16:155–171.

Agbandje M, Kajigaya S, McKenna R, Young NS, Rossmann MG (1994): The structure of human parvovirus B19 at 8A resolution. *Virology* 203:106–115.

Agbandje-McKenna M, Llamas-Saiz AL, Wang F, Tattersall P, Rossmann MG (1998): Functional implications of the structure of the murine parvovirus, minute virus of mice. *Structure* 6:1369–1381.

Aumailley M, Gerl M, Sonnenberg A, Deutzmann R, Timpl R (1990): Identification of the Arg-Gly-Asp sequence on laminin A chain as a latent cell binding site being exposed in fragment P1. *FEBS Lett* 262:82–86.

Balagué C, Kalla M, Zhang W-W (1997): Adeno-associated virus Rep78 protein and terminal repeats enhance integration of DNA sequences into cellular genome. *J Virol* 71:3299–3306.

Bartlett JS, Kleinschmidt J, Boucher RC, Samulski RJ (1999): Targeted adeno-associated virus vector transduction of nonpermissive cells mediated by bispecific F(ab′γ)2 antibody. *Nat Biotechnol* 17:181–186.

Bartlett JS, Wilcher R, Samulski RJ (2000): Infectious entry pathway of adeno-associated virus and adeno-associated virus vectors. *J Virol* 74:2777–2785.

Blacklow NR (1988): Adeno-associated viruses in humans. In *Parvoviruses and Human Disease*, Pattison J, ed., Boca Raton, FL: CRC Press pp. 165–174.

Blacklow NR, Hoggan MD, Kapikian AZ, Austin JB, Rowe WP (1968): Epidemiology of adeno-associated virus infection in a nursery population. *Am J Epidemiol* 89:368–378.

Blacklow NR, Hoggan MD, Sereno MS, Brandt CD, Kim HW, Parrott RH, Chanock

RM (1971): A seroepidemiologic study of adeno-associated virus infections in infants and children. *Am J Epidemiol* 94:359–366.

Chapman AS, Rossmann MG (1993): Structure, sequence and function correlation among parvoviruses. *Virology* 2:491–508.

Carter PJ, Samulski RJ (2000): Adeno-associated viral vectors as gene delivery vehicles. *Int J Mol Med* 6:17–27.

Chiorini J, Wendtner CM, Urcelay E, Safer B, Hallek M, Kotin R (1995): High-efficiency transfer of the T-cell costimulatory molecule B7-2 to lymphoid cells using high-titer recombinant adeno-associated virus vectors. *Hum Gene Ther* 6:1531–1541.

Chiorini JA, Yang L, Liu Y, Safer B, Kotin RM (1997): Cloning of adeno-associated virus type 4 (AAV4) and generation of recombinant AAV4 particles. *J Virol* 71:6823–6833.

Chiorini JA, Kim F, Yang L, Kotin RM (1999): Cloning and characterization of adeno-associated virus type 5. *J Virol* 73:1309–1319.

Cosset FL, Russell SJ (1996): Targeting retrovirus entry. *Gene Ther* 3:946–956.

Dubielzig R, King JA, Weger S, Kern A, Kleinschmidt JA (1999): Adeno-associated virus type 2 protein interactions: formation of pre-encapsidation complexes. *J Virol* 73:8989–8998.

Ferrari FK, Xiao X, McCarty D, Samulski RJ (1997): New developments in the generation of Ad-free, high-titer rAAV gene therapy vectors. *Nat Med* 3:1295–1297.

Girod A, Ried M, Wobus Ch, Lahm H, Leike K, Kleinschmidt J, Deleage G, Hallek M (1999): Genetic capsid modification allows efficient re-targeting of adeno-associated virus type 2. *Nat Med* 9:1052–1056.

Grifman M, Trepel M, Speece P, Gilbert LB, Arap W, Pasqualini R, Weitzman MD (2001): Incorporation of tumor-targeting peptides into recombinant adeno-associated virus capsid. *Mol Ther* 6:964–675.

Grimm D, Kleinschmidt JA (1999): Progress in adeno-associated virus type 2 vector production: promises and prospects for clinical use. *Hum Gene Ther* 10:2445–2450.

Grimm D, Kern A, Pawlita M, Ferrari FK, Samulski RJ, Kleinschmidt JA (1999): Titration of AAV-2 particles via a novel capsid ELISA: packaging of genomes can limit production of recombinant AAV-2. *Gene Ther* 6:1322–1330.

Hermens WTJMC, Ter Brake O, Dijkhuizen PA, Sonnemans MAF, Grimm D, Kleinschmidt JA, Verhaagen J (1999): Purification of recombinant adeno-associated virus by iodixanol gradient ultracentrifugation allows rapid and reproducible preparation of vector stocks for gene transfer in the nervous system. *Hum Gene Ther* 10:1885–1891.

Hermonant PL, Labow MA, Wright R, Berns KI, Muzyczka N (1984): Genetics of adeno-associated virus: isolation and preliminary characterization of adeno-associated virus type 2 mutants. *J Virol* 51:329–339.

Hermonat PL (1989): The adeno-associated virus Rep78 gene inhibits cellular transformation by bovine papilloma virus. *Virology* [illegible]

Hoque M, Ishizu K, Matsumoto A, Han SI, Arisaka F, Takayama M, Suzuki K, Kato K, Kanada T, Watanabe H, Handa H (1999): Nuclear transport of the major capsid protein is essential for adeno-associated virus capsid formation. *J Virol* 73:7912–7915.

Kaludov N, Brown KE, Walters RW, Zabner J, Chiorini JA (2001): Adeno-associated virus serotype 4 (AAV4) and AAV5 both require sialic acid binding for hemagglutination and efficient transduction but differ in sialic acid linkeage specificity. *J Virol* 75:6884–6893.

Khielf SN, Myers T, Carter BJ, Trempe JP (1991): Inhibition of cellular transformation by the adeno-associated virus rep gene. *Virology* 181:738–741.

Kotin RM, Siniscalco M, Samulski RJ, Zhu XD, Hunter L, Laughlin CA, McLaughlin S, Muzyczka N, Rocchi M, Berns KI (1990): Site-specific integration by adeno-associated virus. *Proc Natl Acad Sci USA* 87:2211–2215.

Mayor HD, Houlditch GS, Mumford DM (1973): Influence of adeno-associated virus on adenovirus-induced tumors in hamsters. *Nature* 241:44–46.

Monahan PE, Samulski RJ (2000): AAV vectors: is clinical success on the horizon?. *Gene Ther* 7:24–30.

Moskalenko M, Chen L, van Roey M, Donahue BA, Snyder RO, McArthur JG, Patel SD (2000): Epitope mapping of human anti-adeno-associated virus type 2 neutralizing antibodies: implications for gene therapy and virus structure. *J Virol* 74:1761–1766.

Muramatsu S, Mizukami H, Young NS, Brown KE (1996): Nucleotide sequencing and generation of an infectious clone of adeno-associated virus 3. *Virology* 221:108–217.

Miller AD (1996): Cell-surface receptors for retroviruses and implications for gene transfer. *Proc Natl Acad Sci USA* 93:11407–11413.

Nicklin AS, Büning H, Dishart KL, de Alwis M, Girod A, Hacker U, Thrasher AJ, Ali RR, Hallek M, Baker AH (2001): Genetic incorporation of the SIGYPLP peptide into adeno-associated virus-2 capsids directs efficient and selective gene transfer to human vascular endothelial cells. *Mol Ther* 4:174–181.

Ohno K, Sawai K, Lijima Y, Levin B, Meruelo D (1997): Cell-specific targeting of Sindbis virus vectors displaying IgG-binding domains of protein A. *Nat Biotechnol* 15:763–767.

Qing K, Mah C, Hansen J, Zhou S, Dwarki V, Scrivastava A (1999): Human fibroblast growth factor receptor 1 is a co-receptor for infection by adeno-associated virus 2. *Nat Med* 5:71–77.

Rabinowitz JE, Xiao W, Samulski RJ (1999): Insertional mutagenesis of AAV2 capsid and the production of recombinant Virus. *Virology* 265:274–285.

Rabinowitz JE, Samulski RJ (2000): Building a better vector: the manipulation of AAV virons. *Virology* 278:301–308.

Ried MU, Girod A, Leike K, Büning H, Hallek M (2002): Adeno-associated virus capoid displaying immunoglobulin binding-domains permit antibody-mediated vector retargeting to specific cell surface receptors. *J Virol* 76:4559–4566.

Rivadeneira ED, Popescu NC, Zimonjic DB, Chen GS, Nelson PJ, Ross MD, Dipaolo JA, Klotman ME (1998): Sites of recombinant adeno-associated virus integration. *Int J Oncol* 12:805–810.

Ruffing M, Zentgraf H, Kleinschmidt JA (1992): Assembly of viruslike particles by recombinant structural proteins of adeno-associated virus type 2 in insect cells. *J Virol* 66:6922–6930.

Rutledge EA, Halbert CL, Russell DW (1998): Infectious clones and vectors derived from adeno-associated virus (AAV) serotypes other than AAV type 2. *J Virol* 72:309–319.

Samulski RJ, Berns KJ, Tan M, Muzyczka N (1982): Cloning of adeno-associated virus into pBR322: rescue of intact virus from the recombinant plasmid in human cells. *Proc Nat Acad Sci USA* 79:2077–2081.

Samulski RJ, Zhu S, Xiao X, Brook JD, Housman DE, Epstein N, Hunter LA (1991): Targeted integration of adeno-associated virus (AAV) into human chromosome 19. *EMBO J* 10:3941–3950.

Sanlioglu S, Benson PK, Yang J, Atkinson EM, Reynolds T, Engelhardt JF (2000): Endocytosis and nuclear trafficking of Adeno-associated virus type 2 are controlled by Rac1 and Phosphatidylinositol-3 kinase activation. *J Virol* 74:9184–9196.

Seisenberger G, Ried M, Endreß T, Büning H, Hallek M, Braeuchle Ch. (2001): Real-time single molecule imaging of the infection pathway of an adeno-associated virus. *Science* 294:1929–1932.

Sinha P, Sengupta J, Ray PK (1999): Functional mimicry of Protein A of *Staphylococcus aureus* by a proteolytically cleaved fragment. *Biochem Biophys Res Com* 260:111–116.

Smuda JW, Carter BJ (1991): Adeno-associated viruses having nonsense mutations in the capsid genes: growth in mammalian cells containing an inducible amber suppressor. *Virology* 184:310–318.

Summerford C, Samulski RJ (1998): Membrane-associated heparan sulfate proteoglycan is a receptor for adeno-associated virus type 2 virions. *J Virol* 72:1438–1445.

Summerford C, Bartlett JS, Samulski RJ (1999): Avb5 integrin: a co-receptor for adeno-associated virus type 2 infection. *Nat Med* 5:78–82.

Summerford C, Samulski RJ (1999): Viral receptors and vector purification: new approaches for generating clinical-grade reagents. *Nat Med* 5:587–588.

Tal J (2000): Adeno-associated virus-based vectors in gene therapy. *J Biomed Sci* 7:279–291.

Tratschin JD, Miller IL, Carter BJ (1984): Genetic analysis of adeno-associated virus: properties of deletion mutants constructed in vitro and evidence for an adeno-associated virus replication function. *J Virol* 51:611–615.

Tsao J, Chapman MS, Agbandje M, Keller W, Smith K, Wu H, Luo M, Smith TJ, Rossmann MG, Compans RW (1991): The three-dimensional structure of canine parvovirus and its functional implications. *Science* 251:1456–1464.

Walther W, Stein U (1996): Cell type specific and inducible promotors for vectors in gene therapy as an approach for cell targeting. *J Mol Med* 74:379–392.

Weichert WS, Parker JSL, Wahid ATM, Chang SF, Meier W, Parrish CR (1998): Assaying for structural variation in the parvovirus capsid and its role in infection. *Virology* 250:106–117.

White JM (1993): Integrins as virus receptors. *Curr Biol* 3:596–599.

Wistuba A, Kern A, Weger S, Grimm D, Kleinschmidt JA (1997): Subcellular compartimentatization of adeno-associated virus type 2 assembly. *J Virol* 71:1341–1352.

Wistuba A, Weger S, Kern A, Kleinschmidt JA (1995): Intermediates of adeno-associated virus type 2 assembly: identification of soluble complexes containing Rep and Cap. *J Virol* 69:5311–5319.

Wobus C, Hugle-Dorr B, Girod A, Petersen G, Hallek M, Kleinschmidt JA (2000): Monoclonal antibodies against the adeno-associated virus type2 (AAV-2) capsid: epitope mapping and identification of capsid domains involved in AAV-2-cell interaction and neutralization of AAV-2 infection. *J Virol* 74:9281–9293.

Wu P, Xiao W, Conlon T, Hughes J, Agbandje-McKenna M, Ferkol T, Flotte T, Muzyczka N (2000): Mutational analysis of the adeno-associated virus type 2 (AAV2) capsid gene and construction of AAV2 vectors with altered tropism. *J Virol* 74:8635–8647.

Yang Q, Mamounas M, Gang Y, Kennedy S, Leaker B, Merson J, Wong-Staal F, Yu M, Barber JR (1998): Development of novel cell surface CD34-targeted recombinant adenoassociated virus vectors for gene therapy. *Hum Gene Ther* 9:1929–1937.

Xiao W, Chirmule N, Berta SC, McCullough B, Gao G, Wilson JM (1999): Gene therapy vectors based on adeno-associated virus type 1. *J Virol* 73:3993–4003.

Xiao X, Li J, Samulski RJ (1998): Production of high-titer recombinant adeno-associated virus vectors in the absence of helper adenovirus. *J Virol* 72:2224–2234.

Ziady AG, Perales JC, Ferkol T, Gerken T, Beegen H, Perlmutter DH, Davies PB (1997): Gene transfer into hepatoma cell lines via the serpin enzyme complex receptor. *Am J Physiol* 273:G545–G552.

Zolotukhin S, Byrne BJ, Mason E, Zolotukhin I, Potter M, Chesnut K, Summerford C, Samulski RJ, Muzyczka N (1999): Recombinant adeno-associated virus purification using novel methods improves infectious titer and yield. *Gene Ther* 6:973–985.

11

MECHANISMS OF RETROVIRAL PARTICLE MATURATION AND ATTACHMENT

ATSUSHI MIYANOHARA, PH.D. AND THEODORE FRIEDMANN, M.D.

INTRODUCTION

The eventual development of truly effective methods of in vivo gene delivery for gene therapy will require methods for delivering therapeutic genes to the correct target tissue specifically and efficiently, whether the gene delivery vector is a viral or a nonviral agent. A variety of viral vectors are available for such applications, including retroviral and lentiviral vectors. Retroviral vectors have been effective tools for developing many of our current concepts and techniques for models of gene therapy. They have many advantageous features, especially their ease of manipulation for vector construction and the generally stable transgene expression that results from provirus integration. They also have disadvantages, including their relatively low titers compared with several other viral vector systems, their instability in vivo, and their inability to transduce nonreplicating or slowly replicating cells such as neurons, hepatocytes, and others. Several of these deficiencies have been corrected partly by the development of methods to increase titers and alter vector tropism through methods for pseudotyping vectors with surrogate envelope components such as VSV-G protein. The development of the lentiviral vector system has also provided an efficient approach to gene transfer into postmitotic cells that exhibits

Vector Targeting for Therapeutic Gene Delivery, Edited by David T. Curiel and Joanne T. Douglas
ISBN 0-471-43479-5

many of the useful features of other retroviral systems and promises to become an important technique in future clinical studies. There are at least two principal difficulties in the use of these and all other viral vectors for therapeutic gene delivery. In the case of retroviruses, our present understanding of virus structure is insufficient to allow the design and production of truly targeted vectors through insertion of potentially cell-specific ligands into the viral capsid. In addition, there is a dearth of rigorous information on the pharmacokinetic and pharmacodynamic properties of these viruses; that is, the in vivo fate of vector particles, their interactions with the vascular endothelium as the first tissue barrier that they encounter, and the mechanisms that define their tissue uptake. In the case of systemic administration of retroviral vectors, as in the preclinical and clinical studies aimed at correction of the factor VIII deficiency of classical hemophilia, very little information is readily available on the distribution of systemically delivered virus particles and the exact identification of the tissue source of the factor VIII protein. In addition, even though the studies involving in vivo delivery of AAV vectors to muscle or liver for treatment of hemophilia B are extremely promising, variable patient response and treatment responses are likely to result from insufficient knowledge of the pharmacological properties of the vectors and variations in the mechanisms of tissue uptake and cell entry.

RETROVIRUS ASSEMBLY AND PSEUDOTYPING WITH VSV-G

Several classes of retroviruses and their vectors have been used to develop efficient gene transfer vectors, including vectors derived from Moloney murine leukemia virus (MLV) and human lentiviruses HIV-1 and HIV-2. The structure and life cycles of these viruses are well known and have been described extensively elsewhere (Yee, 1999). Knowledge of the structural and functional properties of the viruses has also been used in many ways to improve the gene transfer properties of vectors derived from them. We have been particularly interested to understand the mechanisms of virus assembly and attachment to target cells and to modify these agents to increase their titers, to improve their stability, and to learn how to specify their cell tropism and target them to defined cells and tissues in vivo.

Our approach to these and related questions has been to study the role of the virus-encoded envelope proteins in the viral properties and to develop methods to change the protein composition of the cell-derived membrane that viral particles acquire during the process of budding from a producer cell to produce particles ("pseudotyped" particles) with altered physiochemical, pharmacokinetic, and pharmacodynamic properties. In 1991 we reported that the virus-encoded envelope glycoprotein could be entirely replaced by the G protein of vesicular stomatitis virus, VSV-G, (Emi et al., 1991) to produce particles that were markedly more stable than native virus and that therefore could be concentrated to titers previously unattainable with retroviral vectors. We also found that such VSV-G-pseudotyped particles were able to infect a very broad spectrum of cells, many of which were previously not permissive for retrovirus infection, including insect, fish, and other nonmammalian eukaryotic cells (Burns et al., 1994; Lin et al., 1994; Burns et al., 1996). The expanded host range and cell tropism almost certainly were due largely to altered mechanisms by which the viral particles recognize and interact with receptors on the cell surface, but careful characterization of the infection process has been hampered to a great extent by the fact that the cell surface receptor for VSV-G has still not been identified rigorously. Evidence has been presented that phosphatidyl serine, phosphatidyl inositol, GM-3 ganglioside, and/or other components of the lipid bilayer, may serve as VSV-G receptors (Schlegel et al., 1983; Mastromarino et al., 1987). However, those studies have not been confirmed. Furthermore, the design of further modified VSV-G-pseudotyped vectors has been hampered by the fact that there is no tertiary structure available for VSV-G on which to base structural modifications.

Nevertheless, the development of VSV-G-pseudotyped virus particles has suggested a number of ways in which pseudotyped vectors could find useful application in gene therapy systems. One of the most important uses of VSV-G pseudotyping has been to permit the development of vectors from several kinds of lentiviruses, including human immunodeficiency virus type 1 (HIV-1) and HIV-2, feline and equine lentiviruses, and others (Naldini and Verma, 1999). These vectors have permitted efficient gene transfer into nonreplicating cells such as neurons and hepatocytes and suggested new approaches to serious diseases involving these tissues that were previously not suitable targets for gene therapy. For instance, a clinical study for AIDS gene therapy involving use of a VSV-G-pseudotyped HIV-based vector has recently been proposed and reviewed by local and federal regulatory and review agencies.

RETROVIRUS ATTACHMENT

The most efficient method of vector delivery would be one in which the vector is designed by suitable modification of the structure and composition of the virus particles to infect only the target cell type relevant to the disease phenotype for which a gene transfer-based therapy is being developed. The tropism of native viruses is defined by complex interactions of the virus components with primary and secondary receptors on the cell surface and is affected greatly by those mechanisms responsible for the initial attachment of virus to cell surface receptors.

The development of efficient methods for targeting retroviruses to specific cell and tissue in vivo will therefore require knowledge of the mechanisms by which the vectors recognize, attach to, and enter cells. With such knowledge in hand, it will become far more feasible than it is at present to design tissue-targeted vectors through the introduction of suitable ligands into the viral particles that recognize receptors on the cell surface and define the tropism and infectivity of vectors. This laboratory has devoted considerable effort to understanding the very earliest events of the virus–cell interaction and to identifying the cell surface components to which the virus particles first attach. It has generally been thought that the tropism of Moloney MLV-based retroviruses, as well as their initial attachment of virus to cells, is determined by the interaction of the viral envelope glycoprotein (env) with one of several classical retrovirus receptors. A number of such cell surface molecules have been identified (reviewed in Battini, Rasko, and Miller, 1999). We know now that the events of retroviral attachment and infection are more complex and involve additional cell surface molecules acting as coreceptors. Infection of susceptible cells with retroviral vectors is a complex, multistep process involving attachment of viral particles to target tissues and cells, interaction of the virus with, in many cases, more than one kind of cell surface receptor followed by membrane fusion events at the cell surface or in endosomes, and introduction of the virus into the cytoplasm for vector delivery to the nucleus. Although a number of cell surface proteins have been identified that are closely associated with uptake and infection by a number of viruses that have served as gene transfer vectors, including retroviruses, lentiviruses, adenoviruses, adeno-associated viruses, herpes simplex viruses, and others, it is becoming clear that alternative attachment and uptake pathways exist and that the originally described retroviral receptors are not entirely necessary or sufficient for virus attachment and/or infection. For instance, in the case of HIV, infection with HIV requires interaction of domains of the gp120 envelope protein with the CD4 receptor as well as interactions with chemokine receptors, especially CCR5 and CXCR4, which in turn induces membrane fusion and cell uptake. Additional interactions such as those of viral membrane components with cell surface polyanions such as heparan sulfate proteoglycans and possibly with other molecules have been reported (Tyagi et al., 2001; Saphire, Bobardt, and Gallay, 1999).

We have recently described new mechanisms for the initial attachment of MLV-based retroviral and lentiviral vector particles to cells, both in vitro and in vivo (Sharma, Miyanohara, and Friedmann, 2000). For these studies, we have used noninfectious retrovirus-like particles (VLPs) that are produced in large amounts in the absence of the viral envelope gene by retroviral packaging cells. The conventional interpretation of the lack of infectivity of these particles is that the VLPs are unable to attach to the cell surface retroviral receptors because they lack the virus-coded envelope glycoprotein. We have reported, however, that it is possible to convert such noninfectious particles to an infectious form in cell-free conditions in vitro by simple addition of the surrogate envelope protein VSV-G (Abe et al., 1998). More recently, we have found similar results with HIV-derived VLPs (Sharma, Miyanohara, and Friedmann, 2000). Moreover, we have found that the envelope-free particles from both MLV- and HIV-packaging cell lines attach very efficiently to cells. Very similar results have also been reported by Pizzato et al. (1999). Interestingly, we have also been able to show that such attached noninfectious particles can be made infectious by the addition of VSV-G after VLP attachment to the

TABLE 11.1 Infection of HT1080 Cells with HIV- and MoMLV-Based VLPs (cfu/ml)

Infection Method	HIV-GPR (LGFPRNL)	MLV-GPR (LZRNL)
GPR/VSV complex onto	1.2×10^5	6.9×10^5
HT1080 cells	1.4×10^5	4.0×10^5
GPR particles onto VSV-G-treated	7.8×10^4	3.8×10^5
HT1080 cells	9.0×10^4	9.8×10^4
GPR particles onto untreated cells	1.1×10^5	4.0×10^5
followed by VSV-G addition	8.2×10^4	2.4×10^5

cell (Table 11.1). These results demonstrate that MLV- and HIV-based retroviral particles are able to attach to cells by mechanisms completely independent of an interaction of the viral envelope proteins with their classical receptors (Fig. 11.1).

These studies indicate clearly that interaction between a viral envelope and a cell surface receptor is not at all necessary for, and may not be involved at all in, the initial virus binding to the cells but that a fusiogenic function provided by the native MLV or the surrogate VSV-G viral envelope is required for subsequent cell entry and infection. Based on these data, we have proposed that the earliest steps of retrovirus and retroviral vector attachment to target cells can occur by mechanisms entirely independent of any specific interactions between the retroviral envelope protein and traditional retroviral receptors (Sharma, Miyanohara, and Friedmann, 2000). Rather, we believe that other

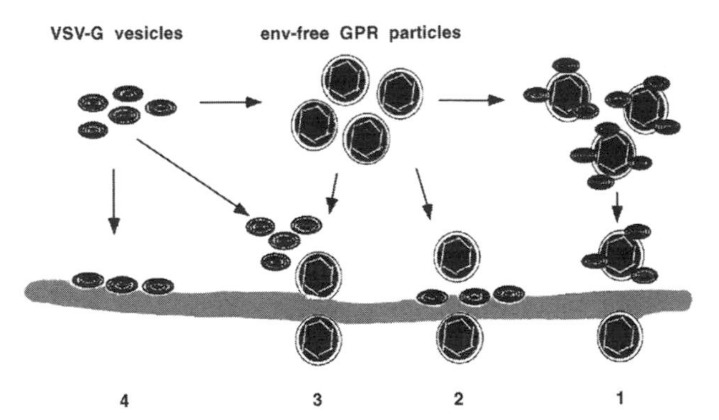

Figure 11.1 Mechanisms of retrovirus and lentivirus attachment to cells. Noninfectious envelope-free VLPs can be converted to the infectious form either by addition of VSV-G to the particles themselves (path 1), by prior treatment of target cells with VSV-G (path 2) or by attachment of VLPs to untreated cells followed by addition of VSV-G (path 3). The mechanisms involved in initial particle attachment to cells illustrated in path 3 obviously do not involve interaction of the retroviral receptor with a viral envelope protein and therefore point to the existence of an alternative cellular receptor for retrovirus and lentivirus particles.

components of the lipid bilayer serve as the first cellular receptors for this class of viruses and their vectors. Because glycosaminoglycans have also been identified as receptors in the attachment and infection of a variety of other viruses, including AAV and herpes viruses, and because of previous work implicating an interaction between components of the HIV virion with heparan sulfate, we hypothesize that glycosaminoglycans represent the initial attachment site for retroviral particles. Preliminary results in our laboratory have shown that heparin is able to compete effectively for cell surface receptor sites with envelope protein-free as well as with mature retroviral particles, supporting the hypothesis that heparan sulfate is at least one of the other coreceptors involved in retroviral attachment and infection.

RETROVIRUS TARGETING

As described elsewhere in this volume, a great deal of progress has been made in tissue-specific targeting of several classes of viruses, most notably adenoviruses. However, until now, progress toward targeting of retroviruses has been slower and more difficult, due at least partly to still unresolved uncertainties of the nature of the primary and secondary retroviral receptors themselves and of suitable sites in the virus capsid for insertion of retargeting ligand sequences. In light of our demonstration that the viral envelope protein is not required for initial cell attachment and the fact that all previous targeting attempts have focused on genetic modification of the envelope glycoprotein, we assume also that the difficulty that most investigators have had in targeting retroviruses may have been due to the fact that ligands have been chosen to drive an interaction with a secondary, rather than the primary, retroviral attachment receptor. Even with the insertion of a ligand to a specific cell surface protein receptor or other marker, continued nonspecific promiscuous attachment of retroviral vectors to cells through GAGs or adhesion molecules and adhesion receptors might readily interfere with the interaction of ligand-engineered envelope molecules with their receptors, making targeting through the latter mechanism more difficult.

Despite the uncertainty regarding the relative roles of primary and secondary retroviral receptors, several groups have reported successful retroviral vector targeting through the introduction of ligands into the envelope glycoproteins of several classes of retroviruses (Kasahara, Dozy, and Kan, 1994; Chu and Dornburg, 1997; Somia, Zoppe, and Verma, 1995; Hall et al., 2000). In a number of studies, Moloney-based retroviruses and spleen necrosis virus have been shown to display altered tropism and infectivity after the introduction into the envelope protein. Most recently, the insertion of the collagen-binding domain of the von Willebrand factor into the surface domain of the amphotropic retrovirus envelope protein has been reported to permit effective targeting in vivo to the exposed extracellular collagen matrix of tumors and to deliver a tumor-ablating cell cycle function to such tumors (Hall et al., 2000). That study has also been extended to a human clinical trial in patients with metastatic liver disease resulting from colon cancer. However, for each of those apparent successes, many more retroviral targeting studies have been hampered not only by inefficiency of targeting by envelope modification but also by the inability of the altered virus, once attached to a newly defined receptor, to enter the cell and produce an infection, presumably because of ineffective virus-cell fusion events. The reason that retargeting has proven to be more difficult with retroviruses than with other classes of viruses, most notably adenoviruses, is not entirely clear. But it seems self-evident that, since retroviruses are able to use several alternative and redundant mechanisms of cell attachment and entry, eventual precise targeting of such vectors would require a fuller understanding of the mechanisms involved in the very earliest attachment of viral particles to cells.

We have approached the issue of retroviral vector targeting by studies of subviral or quasi-viral particles containing the surrogate envelope protein VSV-G. The presumption underlying these studies is that targeting modifications for retroviruses will be most useful for VSV-G-pseudotyped MLV-based (Emi, Friedmann, and Yee, 1991; Burns et al., 1993; Yee et al., 1994; Yee, Friedmann, and Burns, 1994) and lentivirus-based vectors (Poeschla, Corbeau, and Wong-Staal, 1996; Naldini et al., 1996) and that such studies might be most easily carried out with noninfectious VLPs and even simpler gene transfer complexes containing VSV-G. Despite the paucity of the kind of tertiary structure information on VSV-G that would facilitate the identification of suitable sites for ligand insertion, we have made some progress in the preparation of simple gene complexes with VSV-G, most recently with complexes that consist only of plasmid and VSV-G. It has been known for some time that purified soluble VSV-G can be inserted into lipid bilayers of liposomes and lipid vesicles in cell-free systems in vitro (Petri and Wagner, 1979; Metsikko et al., 1986; Hug and Sleight, 1994). In those studies, VSV-G solubilized with detergents was found to be capable of insertion into liposomes and such VSV-G proteoliposomes were capable of inducing cell fusion in a pH-dependent fashion. We have extended this system to develop methods for the conversion of envelope-free, noninfectious particles to an infectious form in cell-free conditions in vitro (Abe et al., 1998; Sharma, Miyanohara, and Freidmann, 2000). We have previously demonstrated that VSV-G can be physically incorporated into lipofectin-DNA complexes to form fusogenic VSV-G-liposomes that demonstrate a markedly increased transfection efficiency that is, unlike native liposomes, not at all abrogated by serum (Abe, Miyanohara, and Freidmann, 1998). We have recently discovered that VSV-G vesicles prepared from conditioned medium of VSV-G expressing cells can associate with naked plasmid DNA in the absence of other viral proteins or other fusogenic factors and that such DNA-VSV-G complexes demonstrate a markedly enhanced transfection efficiency in a variety of recipient cells, even in the presence of serum (Miyanohara and Friedmann, in press) (Fig. 11.2 below).

We have not determined whether the enhanced gene-transferring activity of VSV-G-plasmid complexes results from enhanced attachment of the complex to target cells, from facilitated fusion and cell uptake, from enhanced release

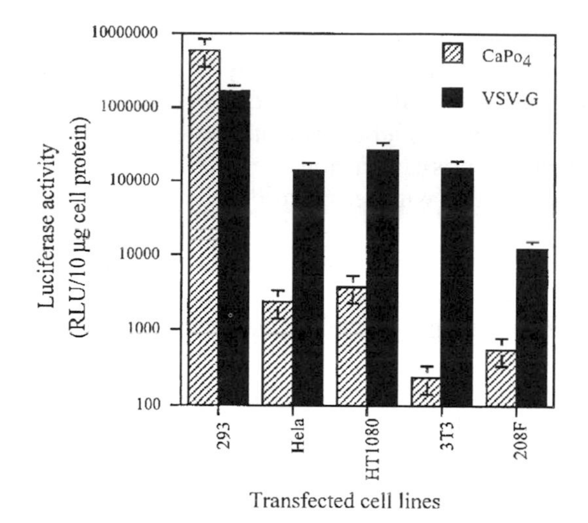

Figure 11.2 Comparison of DNA transfection mediated by calcium phosphate and VSV-G. Transfections were done with 12 well plates using 2 μg pCMVLuc DNA per well and luciferase activity was measured by established methods. The experiments were performed in triplicate and the data are means +/− standard deviations. Hatched bar: Calcium phosphate-mediated transfection; filled bar: DNA-VSV-G complex-mediated transfection.

from endosomes, or from other mechanisms. Whatever the mechanism, the end result is a method of carrying out gene transfer by naked DNA that may be more reminiscent of virus transduction than physical transfection-high efficiency and possibly involvement of specific cell surface receptors for attachment and/or uptake. Additional evidence for the quasi-viral nature of gene transfer by these complexes is suggested by the greatly increased susceptibility of several cell lines that are ordinarily relatively resistant to other transfection methods. For instance, rat 208F, mouse 3T3 cells, and human HT1080 cells are all known to be relatively refractory to lipofection- and calcium phosphate-mediated transfection. In the present studies, the VSV-G-plasmid complexes mediate quite efficient gene transfer into those cells. Interestingly, all three of these cell lines are readily susceptible to infection with VSV-G pseudotyped retroviral vectors, lending further support to the possibility that gene transfer with VSV-G-DNA complexes involves attachment and uptake of the complex through the VSV-G receptor and through the fusogenic properties of VSV-G. While it has been suggested that the VSV-G receptor may be phosphatidyl serine or another intrinsic component of the cell membrane, it has still not yet been unambiguously identified in any cell type. Characterization of the mechanisms of attachment and cell uptake of the complexes reported here may therefore illuminate and recapitulate the mechanisms involved in infection with VSV-G and/or VSV-G-pseudotyped vectors.

CONCLUSION

The development of optimally targeted retroviral vectors will be made increasingly feasible as our understanding of the mechanisms for cell attachment and uptake improves. The surrogate envelope protein VSV-G offers attractive opportunities for targeting pseudotyped retroviral vectors that take advantage of the properties of VSV-G to enhance production of high tier virus and to allow production of particularly useful vectors such as lentiviral vectors. Particularly enticing to us is the possibility of using immature, subviral particles and increasingly simple complexes of compacted genomes with VSV-G to serve as platforms for targeted gene delivery.

ACKNOWLEDGMENTS

This study was supported by grants from NIH (DK49023 and HL64730), from Center for AIDS Research (NIAID 2P30 AI 36214), and by funding from the UCSD School of Medicine for the UCSD Program in Human Gene Therapy. Portions of this material are derived from the manuscript, "The VSV-G envelope glycoprotein complexes with plasmid DNA and with genome- and envelope protein-free MLV retrovirus-like particles in cell-free conditions and enhances DNA transfection" by Okimoto, Friedmann and Miyanohara, in press in *Molecular Therapy*.

REFERENCES

Abe A, Miyanohara A, Friedmann T (1998): Enhanced gene transfer with fusogenic liposomes containing vesicular stomatitis virus G glycoprotein. *J Virol* 72:6159–6163.

Abe A, Chen S-T, Miyanohara A, Friedmann T (1998): In vitro cell-free conversion of noninfectious Moloney retrovirus particles to an infectious form by the addition of the vesicular stomatitis virus surrogate envelop G protein." *J Virol* 72:6356–6361.

Battini JL, Rasko JE, Miller AD (1999): A human cell-surface receptor for xenotropic and polytropic murine leukemia viruses: possible role in G protein-coupled signal transduction. *Proc Natl Acad Sci USA* 96:1385–1390.

Burns JC, Friedmann T, Driever W, Burrascano M, Yee JK (1993): Vesicular stomatitis virus G glycoprotein pseudotyped retroviral vectors: concentration to very high titer and efficient gene transfer into mammalian and non-mammalian cells. *Proc Natl Acad Sci USA* 90:8033–8037.

Burns J, Matsubara T, Lozinski G, Yee J-K, Friedmann T, Washabaugh CH, Tsonis PA (1994): Pantropic retroviral vector-mediated gene transfer, integration, and expression in newt limb cells. *Developmental Biol* 165:285–289.

Burns JC, McNeill L, Shimizu C, Matsubara T, Yee J-K, Friedmann T, Kurdi-Haidar

B, Maliwat E, Holt CE (1996): Retroviral gene transfer in Xenopus cell lines and embryos. *In vitro Cell Dev Biol-Animal* 32:78–84.

Chu T-HT, Dornburg R (1997): Toward highly efficient cell-type-specific gene transfer with retrovirual vectors displaying single chain antibodies. *J Virol* 71:720–725.

Emi N, Friedmann T, Yee JK (1991): Pseudotype formation of murine leukemia virus with the G protein of vesicular stomatitis virus. *J Virol* 65:1202–1207.

Hall FL, Liu L, Zhu NL, Stapfer M, Anderson WF, Beart RW, Gordon EM (2000): Molecular engineering of matrix-targeted retroviral vectors incorporating a surveillance function inherent in von Willebrand factor. *Hum Gene Ther*, 11:983–993.

Hug P, Sleight RG (1994): Fusogenic virosomes prepared by partitioning of vesicular stomatitis virus G protein into preformed vesicles. *J Biol Chem* 269:4050–4056.

Kasahara N, Dozy AM, Kan YW (1994): Tissue-specific targeting of retroviral vectors through ligand-receptor interaction. *Science* 266:1373–1376.

Lin S, Gaiano N, Culp P, Burns JC, Friedmann T, Yee J-K, Hopkins N (1994): Integration and germ line transmission of a pseudotyped retroviral vector in zebrafish. *Science* 265:666–669.

Mastromarino P, Conti C, Goldini P, Hauttencoeur B, Orsi N (1987): Characterization of membrane components of the erythrocyte involved in the attchment and fusion at acidic pH. *J Gen Virol* 68:2359–2369.

Metsikko K, van Meer G, Simmons K (1986): Reconstitution of the fusiogenic activity of vesicular stomatitis virus. *EMBO J* 5:3429–3435.

Naldini L, Verma IM (1999): Lentiviral Vectors. In *The Development of Human Gene Therapy*, Friedmann T, ed. New York: Cold Spring Harbor Laboratory Press, pp. 47–60.

Naldini L, Blomer U, Gallay P, Ory D, Mulligan R, Gage FH, Verma IM, Trono D (1996): In vivo gene delivery and stable transduction of nondividing cells by a lentiviral vector. *Science* 272:263–267.

Petri WAJ, Wagner RR (1979): Reconstitution into liposomes of the glycoprotein of vesicular stomatitis virus by detergent dialysis. *J Biochem* 254:4313–4316.

Pizzato M, Marlow SA, Blair ED, Takeuchi Y (1999): Initial binding of murine leukemia virus particles to cells does not require specific Env-receptor interaction. *J Virol* 73:8599–8611.

Poeschla E, Corbeau P, Wong-Staal F (1996): Development of HIV vectors for anti-HIV gene therapy. *Proc Natl Acad Sci USA* 93:11395–11399.

Saphire AC, Bobardt MD, Gallay PA (1999): Host cyclophilin A mediates HIV-1 attachment to target cells via heparans. *EMBO J* 18:6771–6785.

Schlegel R, Tralka TS, Willingham MC, Pastan I (1983): Inhibition of VSV binding and infectivity by phosphatidyl serine. *Cell* 639–646.

Sharma S, Miyanohara A, Friedmann T (2000): Separable mechanisms of attachment and cell uptake during retrovirus infection. *J Virol* 74:10790–10795.

Sharma S, Murai F, Miyanohara A, Friedmann T (1997): Non-infectious virus-like particles produced by Moloney murine leukemia virus-based retrovirus packaging cells deficient in viral envelope become infectious in the presence of lipofection reagents. *Proc Natl Acad Sci USA* 94:10803–10808.

Somia NV, Zoppe M, Verma IM (1995): Generation of targeted retroviral vectors by using single-chain variable fragment: an approach to *in vivo* gene delivery. *Proc Natl Acad Sci USA* 92:7570–7574.

Tyagi M, Rusnati M, Presta M, Giacca M (2001): Internalization of HIV-1 tat requires cell surface heparan sulfate proteoglycans. *J Biol Chem* 276:3254–3261.

Yee J-K (1999): Retroviral vectors. In *The Development of Human Gene Therapy*, Friedmann T, ed. New York: Cold Spring Harbor Laboratory Press, pp. 21–45.

Yee JK, Friedmann T, Burns JC (1994): Generation of high-titer pseudotyped retroviral vectors with very broad host range. *Methods Cell Biol* 43, Pt A:99–112.

Yee JK, Miyanohara A, LaPorte P, Bouic K, Burns JC, Friedmann T (1994): A general method for the generation of high-titer, pantropic retroviral vectors: highly efficient infection of primary hepatocytes 1994, *Proc Natl Acad Sci USA* 91:9564–9568.

12

TARGETING RETROVIRAL VECTORS USING MOLECULAR BRIDGES

JOHN A. T. YOUNG, PH.D.

INTRODUCTION

Approaches have been developed to target retrovirus vectors to specific cell types that do not involve modifying the viral envelope (Env) protein but instead employ molecular bridges to couple wild-type Env to specific cell surface receptors. To date, several types of molecular bridge have been described. The first is formed by cross-linked antibodies and ligands that link Env to a specific cell surface receptor. The second type is a chimeric bridge protein comprised of the functional domain of a retroviral receptor fused in-frame to either a ligand or a single-chain antibody. In this chapter, I review the progress made using each of these methods and provide a perspective on the future use of this viral targeting approach.

RETROVIRAL ENTRY

Retroviral entry is mediated by Env proteins, which consist of surface (SU) subunits involved in receptor binding and transmembrane (TM) subunits involved in the fusion of virus and cell membranes. These proteins exist as

Vector Targeting for Therapeutic Gene Delivery, Edited by David T. Curiel and Joanne T. Douglas
ISBN 0-471-43479-5

metastable trimers of heterodimers on the viral surface and for most retroviruses they are thought to undergo fusogenic conformational changes following specific interactions with cell surface receptors and/or coreceptors (Sommerfelt, 1999).

Retroviruses have evolved to use a variety of different cell surface proteins as their receptors. These include type-1 transmembrane proteins of the immunoglobulin-, low-density lipoprotein receptor-, and tumor necrosis factor receptor-protein families (Sommerfelt, 1999). Also, several multiple membrane-spanning proteins that function as chemokine receptors or as transporters of amino acids or ions have been exploited as retroviral receptors and coreceptors (Sommerfelt, 1999).

Studies of the human immunodeficiency virus (HIV-1) and avian sarcoma and leukosis viruses (ASLVs) have revealed that the receptor interaction is not only important for virus binding, but it also leads to structural and functional changes in Env necessary for the subsequent fusion of viral and cellular membranes. In the case of HIV-1, viral entry seems to occur by direct fusion of the viral membrane with the host cell plasma membrane following the sequential interaction of Env with the CD4 receptor and then with CC or CXC chemokine receptors (coreceptors) (Wyatt et al., 1998). CD4-binding induces structural changes in HIV-1 SU that lead to exposure of the chemokine receptor-binding site on the viral glycoprotein, thus promoting downstream steps of viral entry (Wyatt et al., 1998).

By contrast to HIV-1, ASLVs enter cells by a low pH-dependent mechanism that most likely involves virus trafficking to an acidic endosomal compartment where fusion occurs (Mothes et al., 2000). The TVA receptor for subgroup A ASLV (ASLV-A) triggers structural changes in the viral glycoprotein that alter the conformation of SU (Gilbert et al., 1995; Damico, Rong, and Bates, 1999), lead to the high-affinity binding of the viral glycoprotein to membranes (Hernandez et al., 1997; Damico, Crane, Bates, 1998), and prime the viral glycoprotein for low pH fusion activation (Mothes et al., 2000).

These two examples of retrovirus–receptor interactions clearly illustrate the fact that retroviral receptors are not only involved in virus binding but they also initiate changes in the viral glycoprotein necessary for membrane fusion. This postbinding function of retroviral receptors may explain why it has been so difficult to target retroviral vectors to novel cell surface receptors using recombinant ligand-Env fusion proteins (Zhao et al., 1999) (see also chapters 13 and 14). In this case, ligands also have to be accommodated in sites of Env where they do not interfere with protein biosynthesis and transport to the cell membrane for virion incorporation. In designing a retrovirus targeting approach it may therefore be desirable to leave the envelope glycoprotein untouched while taking advantage of molecular bridges that can couple virions to the target cell membrane. To date, several different types of molecular bridges have been described, namely cross-linked antibodies and ligands, as well as retroviral receptor-ligand bridge proteins.

MOLECULAR BRIDGES

Cross-Linked Antibodies and Ligands

The first types of molecular bridges that were employed for viral targeting were cross-linked antibodies that attached Env to specific cell surface proteins. This approach relied on binding an Env-specific antibody to virion surfaces and binding another antibody or a ligand to a specific cell surface marker (Fig. 12.1). These bound reagents were then cross-linked by another antibody (Goud et al., 1988) or by streptavidin (Roux et al., 1989; Etienne-Julan et al., 1992) (Fig. 12.1). This approach has been tested with ecotropic murine leukemia virus (MLV-E) vectors and with antibodies and ligands that bind to a variety of different cell surface markers including the transferrin receptor, major histocompatibility complex (MHC) class I and class II proteins, the epidermal growth factor receptor (EGFR), and the insulin receptor (Goud et al., 1988; Roux et al., 1989; Etienne-Julan et al., 1992) (Table 12.1).

Taken together, these studies revealed that retroviral targeting can be achieved through the use of Env-specific antibody-containing molecular bridges (Roux et al., 1989; Etienne-Julan et al., 1992). However, this method usually led to no viral infection or only to an extremely low level of targeted viral infection (Table 12.1).

A modification to this approach has involved using a fusion protein comprised of a cell type-specific ligand fused to protein A (Etienne-Julan et al.,

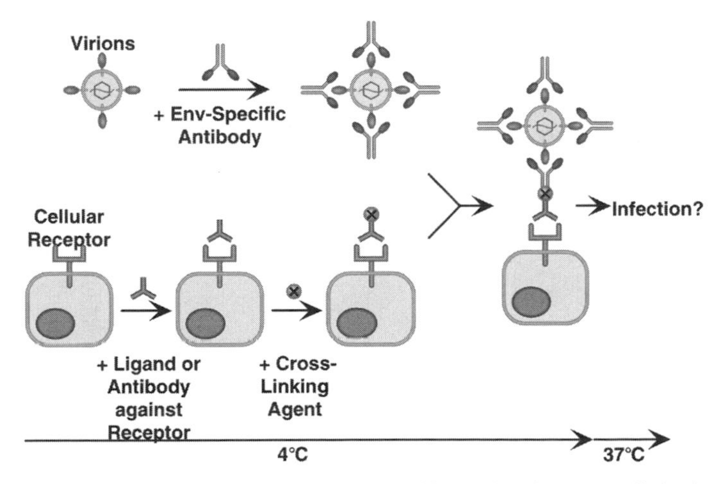

Figure 12.1 Retroviral targeting via Env-specific antibodies cross-linked to an antibody or to a ligand that binds to a specific cell surface receptor. The protocol shown here was taken from Goud et al., 1988; Roux et al., 1989; and Etienne-Julan et al., 1992.

TABLE 12.1 Efficiency of Retroviral Targeting Mediated by Cross-Linked Antibodies and Ligands

Virus-Specific Antibody	Cross-Linking Agent	Cellular Marker-Specific Antibody/Ligand	Cell Type	Efficiency (%)	Reference
anti-MLV SU	Secondary antibody	Antihuman transferrin receptor	Hep 2	0	Goud et al., 1988
Biotinylated anti-ecotropic MLV SU	Streptavidin	Biotinylated anti-MHC class I	HeLa	0.01–0.16	Roux et al., 1989[a]
		Biotinylated anti-MHC class II	HeLa	0.2–0.7	
		Biotinylated anti-EGFR	A431	1.4–2.4	Etienne-Julan et al., 1990[a]
		Biotinylated anti-transferrin receptor	HepG2	<0.002	
		Biotinylated EGF	A431	0.08–0.23	
		Biotinylated insulin	HeLa	0.016	
		Biotinylated asialofetuin	HepG2	<0.002	
None	Multivalent lectins	None	HeLa	<0.002	

[a]Numbers shown were obtained by comparing virus infection efficiency to that obtained with NIH-3T3 cells.

1992). This type of fusion protein should be capable of binding directly to a cell surface ligand receptor and to the Fc portion of an Env-specific antibody, thus eliminating the need for an additional cross-linking agent. However, this approach has so far failed to give rise to targeted viral entry (Etienne-Julan et al., 1992).

In summary, molecular bridges that contain Env-specific antibodies have, at least in some circumstances, been shown to lead to targeted retroviral infection. However, this approach has worked only at a low efficiency, presumably because the Env-specific antibodies fail to induce the postbinding changes in the viral glycoprotein that occur on contact with the natural viral receptor and that are required for membrane fusion.

Retroviral Receptor–Ligand Bridge Proteins

A different type of molecular bridge that has been described is a retroviral receptor–ligand bridge protein. These proteins consist of the extracellular domain of a retroviral receptor fused to a ligand or a single-chain antibody. These bridge proteins are bifunctional reagents that attach to specific cell surface receptors through their ligand moiety and to the viral glycoprotein through the retroviral receptor domain (Fig. 12.2). Importantly, these bridge proteins preserve the contact between the natural viral receptor and Env, thus increasing the likelihood that efficient targeted viral infection can be achieved.

To date, two distinct types of bridge protein have been described, based on

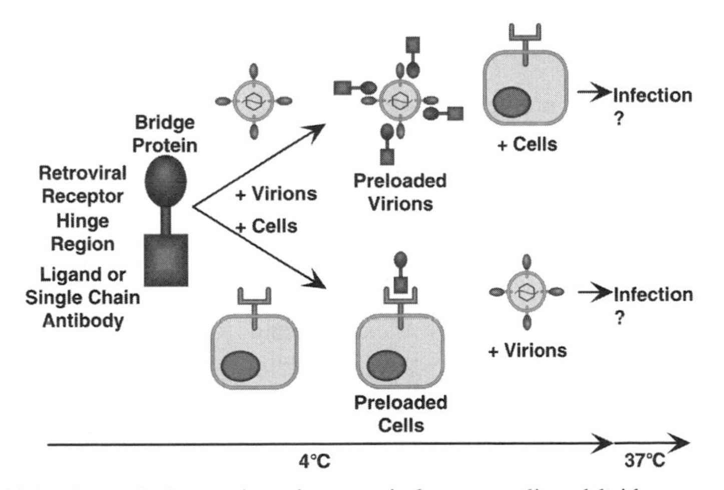

Figure 12.2 Retroviral targeting via retroviral receptor–ligand bridge proteins. The bridge protein consists of the functional domain of a retroviral receptor fused via a proline-rich hinge region to a ligand or to a single-chain antibody (Snitkovsky and Young, 1998; Boerger, Snitkovsky, and Young, 1999; Snitkovsky et al., 2000, 2001).

either the mCAT-1 receptor for MLV-E, or instead upon the TVA and TVB receptors for ASLVs.

Ecotropic MLV Receptor

The cellular receptor for MLV-E is mCAT-1, a cationic amino acid transporter with multiple membrane-spanning domains (reviewed in Sommerfelt, 1999). Determinants that are important for viral entry were mapped in mCAT-1 by exchanging amino acid residues between the murine receptor and its human homolog, which is not normally a viral receptor. These studies showed that it was possible to convert the human protein into a functional MLV-E receptor by replacing several amino acid residues within the predicted third extracellular domain (Albritton et al., 1993; Yoshimoto, Yoshimoto, and Meruelo, 1993).

Meruelo and colleagues have used this modified form of the human protein in an attempt to generate a bridge protein for retroviral targeting. Specifically they constructed TGFα-mH13, a soluble chimeric protein containing the first 50 amino acids of transforming growth factor-α (TGFα) fused in-frame to the third extracellular domain of the modified form of the human homolog of mCAT-1 (Ohno et al., 1995). They showed that TGFα-mH13, immobilized on a 96-well plate or attached to cell surface EGFR, was capable of binding to MLV-E (Ohno et al., 1995). However, they have not shown that this bridge protein can target viral infection via the EGFR (Ohno et al., 1995), probably because the third extracellular domain of the modified human protein cannot support viral entry when it is removed from its context within the multiple membrane-spanning receptor.

ASLV Receptors

ASLVs are divided into at least ten different viral subgroups based on receptor usage, host range, and sensitivity to neutralizing antibodies. To date, cellular receptors for four of these viral subgroups (ASLV-A, -B, -D, and -E) have been identified. TVA, the cellular receptor for ASLV-A, contains a single LDL-A module and is a member of the low-density lipoprotein receptor family (Bates, Young, and Varmus, 1993; Bates et al., 1998). TVB, the cellular receptor for subgroups B, D, and E ASLV, is a death receptor of the tumor necrosis factor receptor superfamily that is most closely related to the mammalian TRAIL receptors, DR4 and DR5 (Brojatsch et al., 1996; Smith et al., 1998; Adkins et al., 1997; Adkins, Brojatsch, and Young, 2000; Schulze-Osthoff et al., 1998).

ASLV receptors are especially well suited for a viral targeting strategy that employs retroviral receptor–ligand bridge proteins because they can render a variety of different mammalian cell types susceptible to viral infection. Furthermore, in contrast to most other retroviral receptors which, like mCAT-1, have multiple membrane-spanning domains, ASLV receptors are simple type-1 transmembrane proteins and therefore all of the functional determinants needed for virus interaction are contained within their single extracellular domains

(Sommerfelt, 1999). Therefore, the extracellular domains of these receptors can be produced as soluble forms that retain the ability to bind virus and stimulate entry. As discussed in the following section, ASLV receptor-ligand bridge proteins containing EGF, vascular endothelial growth factor (VEGF), heregulinβ1, or a single-chain antibody moiety (MR1) can support efficient and specific virus entry into cells that express the cognate target receptors.

EGF

EGF was chosen as the prototypical ligand to test the feasibility of targeting retroviral infection via ASLV receptor-ligand bridge proteins (Snitkovsky and Young, 1998; Boerger, Snitovsky, and Young, 1999), because EGF is a small ligand only 53 amino acids long and there was a concern that large bulky ligands might sterically interfere with virus entry (e.g., Buchholz et al., 1996). Also, the interaction between this ligand and its cellular receptor, EGFR, a tyrosine kinase receptor, was well characterized and several different receptor mutants were available that exhibited different cellular trafficking properties on binding ligand (DiFiore and Gill, 1999). More importantly, EGFR and related proteins that serve as heregulin receptors (erbB2, erbB3, and erbB4) are often overexpressed in human cancers (Yarden and Sliwkowski, 2001), making this type of receptor a clinically relevant target for a retroviral gene delivery approach.

Two synthetic genes were constructed encoding bridge proteins with the extracellular domains of either TVA (TVA-EGF) or TVB (TVB-EGF) fused in-frame to a proline-rich hinge region and to EGF (Snitkovsky and Young, 1998; Boerger, Snitkovsky, and Young, 1999). These chimeric proteins were shown to be capable of bridging ASLV SU to either wild-type, or kinase-deficient, forms of EGFR expressed on the surfaces of transfected mouse L cells (Snitkovsky and Young, 1998; Boerger, Snitkovsky, and Young, 1999). This binding was competed by EGF, confirming that the bridge protein was attached to the ligand-binding region of EGFR (Snitkovsky and Young, 1998; Boerger, Snitkovsky, and Young, 1999). Also like EGF, TVA-EGF was rapidly internalized into cells via the wild-type, but not the kinase-deficient EGFR, demonstrating that this bridge protein behaves in much the same way as the normal physiological ligand (Snitkovsky and Young, 1998).

The TVA-EGF and TVB-EGF bridge proteins were also shown to promote specific infection of cells when attached to cell surface EGFR prior to viral challenge. The level of targeted viral infection ranged from approximately 6% to 54% of that level obtained with control cells expressing transmembrane forms of these receptors (Snitkovsky and Young, 1998) (Table 12.2). Intriguingly, the level of TVA-EGF-dependent viral infection mediated by the kinase-deficient EGFR was approximately 2.5-fold higher than that level seen via the wild-type EGFR (Table 12.2) (Snitkovsky and Young, 1998). Although the reason for this difference between both forms of the receptor is not yet known, it may be related to the fact that these receptors are trafficked differently on

TABLE 12.2 Efficiency of Viral Targeting Mediated by ASLV Receptor-Containing Bridge Proteins

ASLV Receptor-Bridge Protein	Preloaded On To	Target Receptor	Efficiency of Infection[a] (%)		References
			−Bridge Protein	+Bridge Protein	
TVA-EGF	Cells	Control L cells (EGFR-minus)	0	0	Snitkovsky and Young, 1998
		L cells: WT-EGFR	0	6.60	
		L cells: ATP binding	0	16.5	
		mutant-EGFR	0	54.61	
TVB-EGF	Virions	Control L cells: (EGFR-minus)	0.04	0.03	Boerger, Snitkovsky, and Young, 1999
		L cells: WT-EGFR	0.01	28.12	
		L cells: ATP binding mutant-EGFR	0.06	137.50	
TVA-herβ1	Cells	Control NIH 3T3 cells (erbB2/3-minus)	0.07	0.05	Snitkovsky and Young, 2002
		NIH 3T3 cells: erbB2/erbB3	0.1	5.04	
TVA-VEGF		Control PAE cells (VEGFR-minus)	0.01	0.09	Snitkovsky et al., 2001
		PAE cells: VEGFR-2	0.02	8.72	
TVA-MR1		Control 293 cells (EGFRvIII-minus)	0.05	0.04	Snitkovsky et al., 2000
		293 cells: EGFRvIII	0.05	8.38	

Note: Data used to generate this table were derived from the references indicated.

[a]Efficiency of infection shown is compared relative to that seen with the same amount of virus added to matched cells that were engineered to express transmembrane forms of the TVA and the TVB receptors (100% infection control).

binding ligand. The ligand-bound form of the wild-type EGFR is rapidly mobilized out of cell surface caveolae, which are membrane microdomains rich in cholesterol and sphingolipids, and becomes incorporated into clathrin-coated pits where it is rapidly endocytosed. By contrast, the ligand-bound kinase-deficient receptor remains associated with caveolae for a longer time period and it is internalized more slowly (DiFiore and Gill, 1999).

The results obtained with TVA-EGF and TVB-EGF were encouraging, because they demonstrated that efficient and specific viral targeting can be achieved by adding ASLV receptor-ligand bridge proteins to cells before viral challenge. In the case of TVB-EGF, it has also been possible to achieve targeted viral infection by preloading pseudotyped MLV particles containing ASLV-B Env with TVB-EGF. These preloaded virions displayed a remarkable specificity for cells that expressed the EGFR (Boerger, Snitkovsky, and Young, 1999) (Table 12.2). Indeed, the level of targeted virus infection that was achieved with cells expressing the kinase-deficient EGFR was greater than that seen with control cells expressing a transmembrane form of TVB (Boerger, Snitkovsky, and Young, 1999) (Table 12.2). Also, as before, the level of infection that was observed with cells expressing the kinase-deficient EGFR exceeded that seen with cells expressing the wild-type protein (Boerger, Snitkovsky, and Young, 1999) (Table 12.2). It was also shown that TVB-EGF-loaded virions can be produced directly from viral vector packaging cells (that express the bridge protein along with all of the components of a viral vector) and that these virions retain their targeting specificity for cells that express the EGFR (Boerger, Snitkovsky, and Young, 1999).

Heregulin β1

Based on the results obtained with ASLV receptor-EGF bridge proteins, it was reasoned that replacing the EGF ligand with heregulin β1 should lead to targeted viral infection toward cells that express EGFR-related heregulin receptors (erbB2/erbB3/erbB4). Heregulin receptors are formed by erbB3 or erbB4 homodimers of or instead by erbB2/erbB3 and erbB2/erbB4 heterodimers (Stern, 2000).

This idea was tested by fusing the EGF-like domain of heregulin β1 to the extracellular domain of TVA, generating the TVA-herβ1 fusion protein (Snitkovsky and Young, 2002). This fusion protein bound specifically to transfected NIH 3T3 cells that express erbB2 and erbB3 and rendered these cells susceptible to infection by ASLV-A vectors (Snitkovsky and Young, 2002) (Table 12.2). Therefore, an ASLV receptor-heregulin β1 fusion protein can support targeted viral entry into cells that express heregulin receptors.

VEGF

VEGF was chosen as the next ligand to test because its receptors are expressed predominantly on endothelial cells. Therefore, any viral targeting strategy that employs VEGF has the potential for delivering genes specifically to endothelial cells and this may have clinical relevance, for example, in allowing for gene delivery into cells of the tumor vasculature. Several different VEGF receptors with cytoplasmic tyrosine kinase domains have been described: VEGFR-1 (flt-1), VEGFR-2 (KDR/flk-1), and neuropilins-1, and -2 (Neufeld et al., 1999). The nontyrosine kinase receptor VEGFR-3 (flt-4) also serves as a VEGF receptor (Neufeld et al., 1999).

VEGF was also of interest to test since it is a much more complex ligand than EGF. VEGF is a member of the cysteine-knot growth factor superfamily and it is expressed as an antiparallel disulfide-linked homodimer with two receptor-binding sites located at opposite ends of the molecule (Neufeld et al., 1999). Alternative splicing of a primary mRNA transcript gives rise to five distinct VEGF isoforms designated as VEGF121, VEGF145, VEGF165, VEGF189, and VEGF206 (Neufeld et al., 1999). Each of these VEGF isoforms (with the exception of VEGF121) binds extracellular membrane-associated heparan sulfate-containing proteoglycans (Neufeld et al., 1999). Since this feature would be expected to reduce the target cell specificity of infection via a VEGF-containing bridge protein, a modified form of this ligand (designated as VEGF110) lacking the C-terminal heparan-binding domain was used to construct the TVA-VEGF110 bridge protein (Snitkovsky et al., 2000).

TVA-VEGF110 was shown to bind to ASLV-A SU and to porcine aortic endothelial cells that had been transduced with a retroviral vector encoding VEGFR-2 (Snitkovsky et al., 2001). TVA-VEGF110 also supported efficient viral infection when it was added to these cells prior to viral challenge (Table 12.2). The level of targeted viral infection achieved was approximately 9% of that level seen with control cells that expressed a transmembrane form of the TVA receptor (Table 12.2) (Snitkovsky et al., 2001). Therefore, it is possible to efficiently target retroviral infection via VEGF receptors, using an ASLV receptor-VEGF bridge protein.

An ASLV Receptor Single Chain Antibody Bridge Protein

While targeted retroviral infection via ASLV receptor–ligand bridge proteins is useful, there are several potential limitations of this approach. For example, this approach limits the sites of virus attachment to the ligand-binding regions of cellular receptors and this may allow viral targeting in some, but not other, instances. Furthermore, by definition, the use of ligand-containing bridge proteins is limited to those receptors with known ligands. It was therefore of interest to test whether targeted retroviral infection could also be achieved using a bridge protein that contains a single-chain antibody moiety in place of the ligand. Single-chain antibodies have the advantage that they can be directed against a variety of sites on a target cellular receptor as well as against cellular factors for which there are no known ligands.

The single-chain antibody-containing bridge protein tested was TVA-MR1, which consists of the extracellular domain of TVA fused to the MR1 single-chain antibody (Snitkovsky et al., 2000). MR1 binds specifically to EGFRvIII, a mutant form of EGFR that is expressed on a variety of different human cancer cell types including those from glioblastoma, ovarian cancer, prostate cancer, lung cancer, and breast cancer (Huang et al., 1997). EGFRvIII results from a common deletion/rearrangement that occurs during amplification of the EGFR gene during tumor biogenesis. As a result, the protein exhibits an in-frame deletion of 267 amino acids from the extracellular domain and it is constitutively active in the absence of ligand binding (Huang et al., 1997). The MR1 antibody was raised against a synthetic peptide sequence that spans the two segments of the EGFR that are brought together as a consequence of this genetic rearrangement (Lorimer et al., 1996).

TVA-MR1 was shown to bind to ASLV-A SU and to human 293 cells that had been engineered to express a murine form of EGFRvIII (Snitkovsky et al., 2000). This binding was specifically blocked by a synthetic peptide containing the MR1 epitope but was unaffected by a scrambled version of this peptide

(Snitkovsky et al., 2000). Furthermore, the bridge protein did not bind to the parental human 293 cells, confirming that it binds specifically to EGFRvIII (Snitkovsky et al., 2000).

EGFRvIII-expressing cells preloaded with TVA-MR1 were very susceptible to infection by ASLV-A vectors (Snitkovsky et al., 2000). The level of targeted viral infection that was achieved was approximately 8–9% of that level obtained with cells that express a transmembrane form of TVA (Snitkovsky et al., 2000) (Table 12.2). TVA-MR1-dependent infection was specifically blocked when the bridge protein was incubated with the antibody epitope-containing peptide before it was added to cells, confirming that infection required the MR1-EGFRvIII contact (Snitkovsky et al., 2000).

CONCLUSIONS AND POTENTIAL CLINICAL APPLICATION

The studies described in this chapter have shown that retroviruses can be targeted to specific cell types by using molecular bridges. In those instances where an Env-specific antibody has been employed as a bridging component, targeted viral infection has been achieved but in general the levels of infection obtained are too low to be generally useful. By contrast, consistently high levels of targeted infection have been achieved through the use of ASLV receptor-ligand bridge proteins. Those bridge proteins containing the ligands EGF, VEGF, heregulin $\beta1$, or an EGFRvIII-specific single-chain antibody all functioned well to allow targeted viral infection when they were added to cells before viral challenge. In addition, in the case of TVB-EGF, it has been shown that targeted viral infection can be achieved with bridge protein-loaded virions. It is important to note that this method is not limited to the use of just ASLV vectors, because ASLV Env proteins can be incorporated onto other retroviral vectors including those based on MLV to create viral pseudotypes (e.g., see Boerger, Snitkovsky, and Young, 1999). It should also be possible to create viral pseudotypes with a lentiviral core and with ASLV Env proteins in order to target nondividing cell types for infection.

ASLV receptor-ligand bridge proteins have provided a versatile system for targeting retroviral infection to specific cell types. We have recently designated these proteins as "GATEs" (guided adaptors for targeted entry) (Snitkovsky and Young, in preparation). By changing the nature of the ligand, these reagents should be useful for targeting retroviral vectors to a variety of different cell types. By exchanging the viral receptor domain of this type of bridge protein, it should also be possible to target a number of other viral vectors to specific cell types. Indeed, other researchers have recently described a similar approach that has employed soluble coxsackie B and adenovirus receptor (CAR)-EGF/Fc fusion proteins to target adenoviral vectors to cells that express cognate receptors (Dmitriev et al., 2000; Ebbinghaus et al., 2001).

The results to date with soluble ALV receptor-ligand bridge proteins have all been obtained in vitro using bridge protein-loaded virions or instead using cultured cells that have been preloaded with the bridge protein prior to viral challenge. Although these results are encouraging, a number of significant hurdles still must be overcome before this can be considered a useful method for targeting retroviral vectors in in vivo gene therapy protocols. As with other retroviral vector targeting protocols, one of the most difficult challenges will be to generate viral vector stocks with sufficiently high titer and excellent bioavailability. The high titer problem may be overcome by subjecting virions to a selection protocol in order to identify those with altered Env proteins that can more efficiently mediate bridge protein-dependent viral entry. Nevertheless, the results obtained with ALV receptor-ligand bridge proteins represent an important first step toward the application of molecular bridges for cell-specific gene therapy using retroviral vectors.

ACKNOWLEDGMENTS

This work was supported by grant CA70810 from the National Cancer Institute and by grant DAMD 17-98-1-8488 from the Department of the Army.

REFERENCES

Adkins HB, Brojatsch J, Young JA (2000): Identification and characterization of a shared TNFR-related receptor for subgroup B, D, and E avian leukosis viruses reveal cysteine residues required specifically for subgroup E viral entry. *J Virol* 74:3572–3578.

Adkins HB, Brojatsch J, Naughton J, Rolls MM, Pesola JM, Young JA (1997): Identification of a cellular receptor for subgroup E avian leukosis virus. *Proc Natl Acad Sci USA* 94:11617–11622.

Albritton LM, Kim JW, Tseng L, Cunningham JM (1993): Envelope-binding domain in the cationic amino acid transporter determines the host range of ecotropic murine retroviruses. *J Virol* 64:2091–2096.

Bates P, Young JA, Varmus HE (1993): A receptor for subgroup A Rous sarcoma virus is related to the low density lipoprotein receptor. *Cell* 74:1043–1051.

Bates P, Rong L, Varmus HE, Young JA, Crittenden LB (1998): Genetic mapping of the cloned subgroup A avian sarcoma and leukosis virus receptor gene to the TVA locus. *J Virol* 72:2505–2508.

Boerger AL, Snitkovsky S, Young JA (1999): Retroviral vectors preloaded with a viral receptor-ligand bridge protein are targeted to specific cell types. *Proc Natl Acad Sci USA* 96:9867–9872.

Brojatsch J, Naughton J, Rolls MM, Zingler K, Young JA (1996): CAR1, a TNFR-related protein, is a cellular receptor for cytopathic avian leukosis-sarcoma viruses and mediates apoptosis. *Cell* 87:845–855.

Buchholz CJ, Schneider U, Devaux P, Gerlier D, Cattaneo R (1996): Cell entry by measles virus: long hybrid receptors uncouple binding from membrane fusion. *J*

Virol 70:3716–3723.

Damico RL, Crane J, Bates P (1998): Receptor-triggered membrane association of a model retroviral glycoprotein. *Proc Natl Acad Sci USA* 95(5):2580–2585.

Damico R, Rong LJ, Bates P (1999): Substitutions in the receptor-binding domain of the avian sarcoma and leukosis virus envelope uncouple receptor-triggered structural rearrangements in the surface and transmembrane subunits. *J Virol* 73:3087–3094.

Di Fiore PP, Gill GN (1999): Endocytosis and mitogenic signaling. *Curr Opin Cell Biol* 11:483–488.

Dmitriev I, Kashentseva E, Rogers BE, Krasnykh V, Curiel DT (2000): Ectodomain of coxsackievirus and adenovirus receptor genetically fused to epidermal growth factor mediates adenovirus targeting to epidermal growth factor receptor-positive cells. *J Virol* 74:6875–6884.

Ebbinghaus C, Al-Jaibaji A, Operschall E, Schoffel A, Peter I, Greber UR, Hemii S (2001): Functional and selective targeting of adenovirus to high-affinity Fcγreceptor I-positive cells by using a bispecific hybrid adapter. *J Virol* 75:480–489.

Etienne-Julan M, Roux P, Carillo S, Jeanteur P, Piechaczyk M (1992) The efficiency of cell targeting by recombinant retroviruses depends on the nature of the receptor and thc composition of thc artificial cell-virus linker. *J Gen Virol* 73:3251–3255.

Gilbert JM, Hernandez LD, Balliet JW, Bates P, White JM (1995): Receptor-induced conformational changes in the subgroup A avian leukosis and sarcoma virus envelope glycoprotein. *J Virol* 69:7410–7415.

Goud B, Legrain P, Buttin G (1988): Antibody-mediated binding of a murine ecotropic Moloney retroviral vector to human cells allows internalization but not the establishment of the proviral state. *Virology* 163:251–254.

Hernandez LD, Peters RJ, Delos SE, Young JAT, Agard DA, White JM (1997): Activation of a retroviral membrane fusion protein-soluble receptor-induced liposome binding of the ALSV envelope glycoprotein. *J Cell Biol* 139:1455–1464.

Huang HS, Nagane M, Klingbeil CK, Lin H, Nishikawa R, Ji XD, Huang CM, Gill GN, Wiley HS, Cavenee WK (1997): The enhanced tumorigenic activity of a mutant epidermal growth factor receptor common in human cancers is mediated by threshold levels of constitutive tyrosine phosphorylation and unattenuated signaling. *J Biol Chem* 272:2927–2935.

Lorimer IA, Keppler-Hafkemeyer A, Beers RA, Pegram CN, Bigner DD, Pastan I (1996): Recombinant immunotoxins specific for a mutant epidermal growth factor receptor: targeting with a single chain antibody variable domain isolated by phage display. *Proc Natl Acad Sci USA* 93:14815–14820.

Mothes W, Boerger AL, Narayan S, Cunningham JM, Young JAT (2000): Retroviral entry mediated by receptor priming and low pH triggering of an envelope glycoprotein. *Cell* 103:679–689.

Neufeld G, Cohen T, Gengrinovitch S, Poltorak Z (1999): Vascular endothelial growth factor (VEGF) and its receptors. *FASEB J* 13:9–22.

Ohno K, Brown GD, Meruelo D (1995): Cell targeting for gene delivery: use of fusion protein containing the modified human receptor for ecotropic murine leukemia virus. *Biochem Mol Med* 56:172–175.

Roux P, Jeanteur P, Piechaczyk M (1989) A versatile and potentially general approach to the targeting of specific cell types by retroviruses: application to the infection of human cells by means of major histocompatibility complex class I and class II antigens by mouse ecotropic murine leukemia virus-derived viruses. *Proc Natl Acad Sci USA* 86:9079–9083.

Schulze-Osthoff K, Ferrari D, Los M, Wesselborg S, Peter ME (1998): Apoptosis signaling by death receptors. *Eur J Biochem* 254:439–459.

Smith EJ, Brojatsch J, Naughton J, Young JA (1998): The CAR1 gene encoding a cellular receptor specific for subgroup B and D avian leukosis viruses maps to the chicken tvb locus. *J Virol* 72:3501–3503.

Snitkovsky S, Young JA (1998): Cell-specific viral targeting mediated by a soluble retroviral receptor-ligand fusion protein. *Proc Natl Acad Sci USA* 95:7063–7068.

Snitkovsky S, Young JA (2002): Targeting retroviral vector infection to cells that express heregulin receptors using a TVA-Heregulin Bridge protein. *Virology* 292:150–155.

Snitkovsky S, Niederman TM, Carter BS, Mulligan RC, Young JA (2000): A TVA-single-chain antibody fusion protein mediates specific targeting of a subgroup A avian leukosis virus vector to cells expressing a tumor-specific form of epidermal growth factor receptor. *J Virol* 74:9540–9545.

Snitkovsky S, Niederman TM, Mulligan RC, Young JA (2001): Targeting avian leukosis virus subgroup A vectors by using a TVA-VEGF bridge protein. *J Virol* 75:1571–1575.

Sommerfelt MA (1999): Retrovirus receptors. *J Gen Virol* 80:3049–3064.

Stern DF (2000): Tyrosine kinase signalling in breast cancer—ErbB family receptor tyrosine kinases. *Breast Cancer Res* 2:149–153.

Wyatt R, Kwong PD, Desjardins E, Sweet RW, Robinson J, Hendrickson WA, Sodroski JG (1998): The antigenic structure of the HIV gp120 envelope glycoprotein. *Nature* 393:705–711.

Yarden Y, Sliwkowski MX (2001): Untangling the ErbB signalling network. *Nat Rev Mol Cell Biol* 2:127–137.

Yoshimoto T, Yoshimoto E, Meruelo D (1993): Identification of amino acid residues critical for infection with ecotropic murine leukemia retrovirus. *J Virol* 6:1310–1314.

Zhao Y, Zhu LJ, Lee S, Li L, Chang E, Soong NW, Douer D, Anderson WF (1999): Identification of the block in targeted retroviral-mediated gene transfer. *Proc Natl Acad Sci USA* 96:4005–4010.

13

GENETIC TARGETING OF RETROVIRAL VECTORS

DAVID DINGLI, M.D. AND STEPHEN J. RUSSELL, M.D., PH.D.

INTRODUCTION

The goals of gene therapy are tissue-specific and high-level transgene expression in the targeted cells. Tissue-specific expression can be obtained either by specific uptake of the vector by the tissue of interest (transductional targeting) or by restricting expression to the target tissue (transcriptional targeting). Transductional targeting, the subject of this chapter, can be achieved by the direct injection of the vector into the organ, viscus, or cavity of interest or by regional perfusion (extrinsic targeting). However, not all organs are accessible for regional delivery approaches and direct injection may be undesirable. In such instances, intrinsic properties of the systemically administered vector can lead to specific transduction of the tissue of interest. This is intrinsic targeting and is the subject of this chapter. Retroviral vectors can integrate their reverse transcribed genome into the genome of the cell. This fact led many investigators to explore the feasibility of developing targetable and injectable vectors based on both the type C oncoretroviruses as well as lentiviruses. In this chapter we discuss the transductional targeting strategies that have been used to target retroviruses to specific tissues.

Vector Targeting for Therapeutic Gene Delivery, Edited by David T. Curiel and Joanne T. Douglas
ISBN 0-471-43479-5

TARGETING STRATEGIES

The goals of vector targeting for gene delivery are tissue-specific delivery and high-level transgene expression. The various targeting approaches can be categorized according to the nature of the molecular target (e.g., protease or receptor), the targeting element (e.g., soluble adaptor molecule or genetic modification of the viral envelope glycoprotein), or the nature of the host range modification (e.g., restriction or extension). Many laboratories have attempted to retarget retroviral vectors using a variety of strategies. Despite successful proof of principle, it has been the experience of many that this leads to very low transduction efficiency, so much so that the value of the targeting strategy is essentially lost. The first successful attempt at retargeting retrovirus entry was described in 1991. Investigators coated with lactose an ecotropic retrovirus coding for the β-galactosidase gene and claimed that this coated vector transduced hepatocytes specifically by binding to the asialoglycoprotein receptor. However, binding kinetics were slow and the vector had to be purified from unbound lactose (Neda, Wu, and Wu, 1991). Another approach has been the use of bifunctional cross-linkers that can bind both the retroviral membrane glycoprotein and a target molecule on the cell (tissue) of interest. In the first report of this approach, antibodies against the viral envelope glycoprotein of Moloney murine leukemia virus (MoMLV) (Mab 615) and against the human transferrin receptor (mAb 5E9) were cross-linked by an antimurine kappa light chain antibody. In this experiment, the retrovirus was internalized but the reporter gene (neomycin phosphotransferase) was not expressed. The authors concluded that binding and internalization were not sufficient to establish a proviral state (Goud, Legrain, and Buttin, 1988). Others have used biotinylated antibodies against both the viral membrane glycoprotein of MoMLV and a target cell antigen using streptavidin to cross-link the two (Roux, Jeanteur, and Piechaczyk, 1989; Etienne-Julan et al., 1992). However, transduction efficiency was poor and the experimental system required the sequential addition of antibodies and vector while keeping the cells at 0°C to prevent premature internalization of the bound antibody. It is unlikely that this strategy could be used for systemic targeting of vectors. More recently, Young and colleagues have retargeted avian leukosis virus (ALV) vectors to mammalian cells using a fusion protein composed of the ALV A or B receptor and epidermal growth factor (Snitkovsky and Young, 1998 and chapter 12). Pseudotyping vector particles with envelope glycoproteins from other viruses may expand (e.g., VSV-G) (Burns et al., 1993) or restrict (e.g., gp120) (Schnierle et al., 1996) the tropism of the vector. It is also possible to pseudotype retroviruses with receptor proteins in their envelope. Balliet and Bates (1998) expressed the RSV-A receptor, Tva in MLV-based vectors. The pseudotyped vector transduced cells expressing RSV-A on their membrane. The same authors also showed that incorporation of MCAT-1 as the envelope protein retargets vectors to cells expressing the ecotropic MLV envelope. CD4, CXCR4, and CCR5 have also been shown to pseudotype retroviral vectors and retarget them to cells infected with human immunodeficency virus (HIV) and expressing gp120 (Endres et al., 1997; Somia et al., 2000). However, the vector titers reported were low. Other investigators in the field have elected to target retroviral vectors using a genetic approach to modify the structure of the viral envelope glycoprotein. In this chapter, we discuss this approach emphasizing its strengths, limitations, and suggesting future directions that the field may take. We begin with a discussion of the structure and function of the retroviral envelope glycoprotein.

VIRAL ENVELOPE GLYCOPROTEINS: STRUCTURE AND FUNCTION

Retroviruses have a phospholipid envelope acquired during budding from the parent cell. It contains the viral membrane glycoprotein encoded by the *env* gene. The protein is synthesized as a precursor that is directed to the endoplasmic reticulum (ER) by its N-terminal signal sequence. The signal peptide is cleaved in the ER and high mannose carbohydrate chains are added on asparagine residues. The protein folds in the ER assisted by lumenal chaperone proteins. The folded protein is assembled into trimers and transported to the Golgi complex where O-linked glycosylation occurs and the N-linked carbohydrate chains are modified. A protease in the Golgi complex (furin) cleaves the protein into the surface (SU) and transmembrane (TM) components that are held together by noncovalent interactions as well as a labile disulfide bridge in some species. This cleavage step is essential for the protein to obtain its mature functional state that can trigger membrane fusion (Pinter et al., 1997). From the Golgi complex, the mature SU-TM protein is transported to the cell surface and is incorporated in the budding retrovirus particles. The exact details of how or whether the protein is concentrated at sites of virus budding are not known. Interaction between the cytoplasmic tail of the protein and the core of the virus is not essential since proteins without a cytoplasmic tail can be efficiently incorporated into budding viruses. The final maturation step of the protein occurs after its incorporation into a retroviral particle. Virus budding activates the viral protease that cleaves off the C-terminal 16 residues (known as the R-peptide) from the cytoplasmic tail of the TM protein (Rein et al., 1994). This final cleavage reaction is required for the protein to be able to trigger membrane fusion once it binds to its receptor.

Envelope glycoproteins of type C retroviruses have a modular design. (Fig. 13.1) The receptor-binding domain (RBD) of MLV is located in the N-terminal half of the SU (Heard and Danos, 1991). Two hypervariable regions within the RBD, VRA and VRB, are both involved in receptor binding and are thought to confer specificity for a particular receptor (Battini, Heard, and Danos, 1992). With the amphotropic envelope, VRA interacts with loops 4 and 5 of the PiT-2 receptor (a phosphate transporter that spans the membrane multiple times) while VRB binds to loop 2 (Tailor and Kabat, 1997). The RBD is attached via a proline-rich hinge to the C-terminal domain that interacts with the TM subunit leading to conformational changes and activation of the fusion process

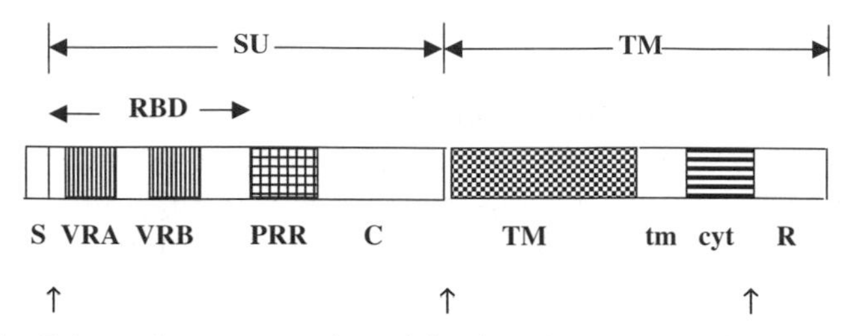

Figure 13.1 Schematic representation of the domain structure of the murine leukemia virus envelope glycoprotein. SU, surface subunit; TM, transmembrane subunit; S, signal peptide; RBD, receptor binding domain; VRA and VRB, variable regions A and B; PRR, proline rich region; C, SU carboxyl terminal domain; tm, TM anchor domain; cyt, cytoplasmic tail; R, R peptide. Vertical arrows indicate cleavage sites by specific enzymes during protein maturation.

(Pinter et al., 1997). Deletion of the RBD from the SU abolishes infection. However, infection can be restored to the truncated molecule if the RBD is provided in a soluble form, indicating that the RBD is not only required for virus attachment to the receptor but also for activation of the fusion process (Barnett, Davey, and Cunningham, 2001) Near the N-terminus of the RBD, there is a conserved peptide motif centered on a histidine residue. This motif is essential for triggering of postbinding events (Lavillette et al., 2000). The proline-rich region in the SU is important to maintain the structural integrity of the complex. Deletions and mutations in this part of the molecule destabilize the interaction between SU and TM resulting in shedding of the SU and loss of infectivity (Weimin Wu et al., 1998; Lavillette et al., 2000). The residues in the RBD that interact with the viral receptor form surface-exposed loops at the top of a β-sandwich (Fass et al., 1997). There are critical determinants of membrane fusogenicity in the RBD, PRR, C-terminal domain, TM fusion peptide, TM transmembrane domain as well as the cytoplasmic domain of TM (Bae, Kingsman, and Kingsman, 1997; Lavillette et al., 1998). Mutations at the N-terminus of SU that destroy fusion triggering can be compensated by other mutations in the C-terminus, providing further proof of the interaction between the two ends of the molecule to generate the functional SU component (Zavorotinskaya and Albritton, 1999).

The TM component can be divided into an extraviral domain, a membrane-spanning domain that anchors the complex into the viral membrane and a short C-terminal peptide that projects into the interior after the R peptide has been cleaved. In the extraviral domain, heptad repeat sequences have been identified that are thought to form trimeric leucine zippers (Fass, Harrison, and Kim, 1996). The fusion peptide, composed of hydrophobic amino acids is present close to the N-terminus of the TM (Jones and Risser, 1993; Zhu et al., 1998). In HIV, the ectodomain is composed of a trimer of identical helical hairpins in which a central trimeric coiled-coil composed of three N-terminal helices is wrapped in a layer of three antiparallel C-terminal helices (Jelesarov and Lu, 2001).

Receptor binding results in conformational changes in both SU and TM leading to activation of TM and its ability to trigger membrane fusion. Although the exact details of how fusion occurs are not known, it is thought that the mechanism of membrane fusion is similar to that of the influenza hemagglutinin glycoprotein. Triggering of membrane fusion by the influenza hemagglutinin has been extensively studied and probably, retroviral membrane fusion occurs in a similar fashion (White, 1992). It is thought that fusion starts with the dissociation of the SU and TM leading to conformational changes in the TM subunit exposing its 'fusion' peptide located in the N-terminus of the protein (Moore et al., 1990). In MLV, the last 16 residues in the cytoplasmic domain of the TM modulate fusogenicity (Rein et al., 1994).

Infection starts with attachment of the virus particle to its receptor on the target cell surface. Binding is specific and tight with a dissociation constant in the nanomolar range (Yu, Soong, and Anderson, 1995). A number of receptors have been identified and they all belong to a family of membrane transporters. As expected, they all have several membrane-spanning regions and are thought to have similar topologies. In mouse cells, the cationic amino acid transporter (CAT-1) is the receptor for the envelope glycoprotein gp70 present on the ecotropic Mo MLV (Wang et al., 1991a). The receptors for the gibbon ape leukemia virus (GALV) envelope and amphotropic 4070A MLV are the sodium dependent phosphate transporters PiT-1 (O'Hara et al., 1990) and PiT-2 (Miller, Edwards, and Miller, 1994) respectively. These are expressed in many tissues including the kidney, brain, heart, liver, muscle, and bone marrow, resulting in the wide host cell range of viruses with these envelopes. Both PiT-1 and PiT-2 are predicted to have 10 membrane-spanning domains with five extracellular loops (Kavanaugh and Kabat, 1996). The fact that all receptors for this group of retroviruses belong to the same class indicates that these receptors do not simply function to allow binding of the virus to the cell. Probably, they share common mechanisms involved in activation of the viral envelope glycoprotein to induce membrane fusion. There is increasing evidence that these receptors play active roles in triggering of the membrane fusion process (Weiss and Tailor, 1995; Siess, Kozak, and Kabat, 1996). Binding studies suggest that only a small subset of cell surface receptors bind virus productively, leading to functional infection. Therefore, not all receptors are equal and they are functionally diverse (Wang et al., 1991b; Battini et al., 1996). Retroviral receptors interact with the cytoskeleton and this is thought to be important for clustering, internalization, and intracellular transport of the viral particles. Disruption of actin microfilaments or cellular microtubules greatly decreases viral infection (Kizhatil and Albritton, 1997; Rodrigues and Heard, 1999). Thus the cell is an active participant in the early steps in virus entry. In the case of lentiviruses, a coreceptor belonging to the G-protein coupled chemokine receptor is required for infection.

Recently, Pizzato et al. (1999, 2001) demonstrated that both ecotropic and amphotropic retroviruses can be adsorbed to cells that do not display the cognate receptor on their plasma membrane. They studied virus–cell association and dissociation and found that the kinetics of these reactions are the same and independent of receptor expression. Interestingly, this nonspecific adhesion was observed only in adherent cells. Suspension cells exhibited poor nonspecific binding of vectors possibly because they lack a membrane component responsible for this initial interaction. Thus it seems that the initial virus–cell interaction is nonspecific, at least in adherent cell lines in tissue culture. The authors suggest that once the virus is nonspecifically bound, it can scan the membrane surface in two dimensions for the cognate receptor, leading to specific binding and activation of fusion. This may be advantageous because nonspecific binding allows the vector to search a surface for its receptor while unbound vectors have to search in three dimensions for the receptor. Similar observations with the envelope of vesicular stomatitis virus (VSV-G), a popular pseudotyping envelope, have been reported previously (Schlegel, Willingham, and Pastan, 1982). However, there is currently no direct evidence to support the idea that bound viruses can move across the surface of a cell to search for receptor sites. The chemical basis of this nonspecific adsorption is unclear; components of the extracellular matrix such as proteoglycans and glycosaminoglycans may be involved, although the evidence for this is inconclusive. These observations imply that the nonspecific adsorption may be important for retroviral infection and may have implications for targeting strategies. In particular, nonspecific binding might lead to significant vector wastage after systemic administration. It must be emphasized that these are in vitro observations (Pizzato et al., 1999) and have not been confirmed in vivo and such data cannot be extrapolated for in vivo biodistribution of vectors. It is possible that the nonspecific binding of vectors to cells is a reversible process and bound vector is in equilibrium with vector in the circulation. If this is the case, one would expect the in vivo biodistribution to be, in part, a function of blood flow. In addition, the vector has to be able to leave the circulation and bind to the targeted cells, presumably via its retargeted envelope. This subject is discussed further later in this chapter in the section on In Vivo Studies.

GENETIC MODIFICATION OF RETROVIRAL ENVELOPE GLYCOPROTEINS

Retroviral envelope glycoproteins are able to tolerate a variety of genetic modifications. Various investigators have described changes in the ecotropic MLV envelope glycoprotein that can only transduce mouse cells, the amphotropic MLV envelopes that can transduce human cells (4070A), and nonmurine envelopes such as the gibbon ape leukemia virus (GALV). In the case of the ecotropic envelope (Moloney murine leukemia virus, MoMLV), the aim has been to expand viral tropism to transduce a variety of human cells (host range extension). This targeting approach has been called direct targeting and has been actively investigated in many laboratories. With the 4070A, GALV and spleen necrosis virus (SNV) envelopes, the aim has been to restrict the ability of the vector to transduce multiple cell types, focusing instead on a target cell population (host range restriction). Glycoprotein modifications must be carefully engineered so as not to interfere with the proper folding of the protein, its functional maturation in the post ER compartment, and with its ability to trigger membrane fusion once it engages its receptor. Changes in specific regions in the receptor-binding domain (RBD), the polyproline hinge (PPR), the C-terminal domain of the SU, the fusion peptide, the R peptide, or the leucine zipper motif in the TM can enhance or inhibit fusion (see previous discussion).

Many peptides varying in size, glycosylation, oligomerization state, binding specificity, and receptor affinity have been displayed on surface glycoproteins. Foreign proteins and peptides have been displayed by insertion into surface exposed loops in the RBD, by substitution of these same peptide loops, by substitution of the N-terminal RBD or the entire SU glycoprotein, or as N-terminal extensions of the SU tethered to the underlying viral glycoprotein through cleavable or non-cleavable linkers (Russell, Hawkins, and Winter, 1993; Valesia-Wittmann et al., 1994; Cosset et al., 1995; Somia, Zoppe, and Verma, 1995; Schnierle et al., 1996; Hall et al., 1997; Konishi et al., 1998; Yajima et al., 1998; Benedict et al., 1999; Zhao et al., 1999; Lorimer and Lavictoire, 2000; Pizzato et al., 2001). Close to the extreme N-terminus, larger polypeptides such as growth factors and single-chain antibodies can be tolerated (Russell, Hawkins, and Winter, 1993; Chu et al., 1994; Kasahara, Dozy, and Kan, 1994; Chu and Dornburg, 1995; Han, Kasahara, and Kan, 1995; Matano et al., 1995; Jiang et al., 1998; Martin et al., 1998; Nguyen et al., 1998; Benedict et al., 1999, Engelstadter et al., 2000; Kuroki et al., 2000, Liu et al., 2000, Lorimer and Lavictoire, 2000; Pizzato et al., 2001). Protease cleavage sites can also be incorporated into linkers of N-terminally displayed peptides. These can be cleaved from the surface of the vector particles (Nilson et al., 1996; Morling et al., 1997; Peng et al., 1997, 1999). Others inserted peptides in the proline-rich hinge region of the SU component of the membrane glycoprotein. Weimin Wu et al. (1998) inserted a collagen-binding peptide, while Kayman et al. (1999) displayed a single-chain antibody in the same region.

With the amphotropic envelope (4070A), protein engineering is aimed to restrict the broad natural tropism of these envelope glycoproteins. Cytokines such as epidermal growth factor, stem cell factor, insulin-like growth factor 1 and interleukin 2 have been displayed at the N-terminus and shown to redirect binding to cells expressing the specific receptor (Morling et al., 1997; Peng et al., 1997; Fielding et al., 1998; Nguyen et al., 1998; Chadwick et al., 1999; Maurice et al., 1999). Protease-cleavable linkers have also been introduced into amphotropic vectors (Peng et al., 1997, 1998, 1999). A single-chain antibody against the high molecular weight melanoma-associated antigen (HMWMAA) displayed as an N-terminal extension also led to preferential transduction of melanoma cells (Martin et al., 1999). In a similar fashion, single-chain antibodies against T-cell markers (Engelstadter et al., 2000), CD34, Her2neu, and the transferrin receptor (Jiang et al., 1998), have been displayed on the envelope of SNV where they can confer new binding specificity. A summary of the displayed peptides is given in Table 13.1.

TABLE 13.1 Summary of the Ligands Displayed on Retroviral Envelopes for Retargeting

Insertion Site	Displayed Ligand	References
N-Terminal extension of truncated SU	Human erythropoietin	Kasahara, Dozy, and Kan, 1994
	ScFv directed against 2,4-dinitrophenol	Chu et al., 1994
	ScFv directed against human CEA-related surface protein B6.2	Chu and Dornburg, 1995
	Human CD4 surface domain	Matano et al., 1995
	α or β1 isoforms of human heregulin	Han, Kasahara, and Kan, 1995
	ScFv against human Her 2 neu	Jiang et al., 1998
	ScFv against human CD34	Jiang et al., 1998
	ScFv against human transferrin receptor	Jiang et al., 1998
	Human HGF	Nguyen et al., 1998
	ScFv against T cell-specific markers	Engelstadter et al., 2000
	vWF-derived collagen-binding domain	Hall et al., 2000
	Tumor vasculature targeting motifs	Liu et al., 2000
N-terminal extension of RBD	ScFv against hapten NIP	Russell, Hawkins, and Winter, 1993
	Integrin receptor binding peptide	Valsesia-Wittmann et al., 1994
	ScFv against human LDL-R	Somia, Zoppe, and Verma, 1995
	First 208 amino acids of amphotropic MoMLV surface protein	Cosset et al., 1995
	Human EGF	Cosset et al., 1995
	ScFv against human MHC class 1 antigens	Marin et al., 1996
	ScFv against CD3	Ager et al., 1996
	ScFv against colonic carcinoma cells	Ager et al., 1996
	Human EGF with factor Xa cleavable linker	Nilson et al., 1996
	Binding domain from HRG70	Schnierle et al., 1996
	19 amino acid peptide containing vWF derived collagen-binding domain	Hall et al., 1997
	Trimeric leucine zipper with factor Xa cleavable linker	Morling et al., 1997

TABLE 13.1 *(Continued)*

Insertion Site	Displayed Ligand	References
	Trimeric C-terminal domain of CD40L with factor Xa cleavable linker	Morling et al., 1997
	Human EGF with MMP cleavable linker	Peng et al., 1997
	Ram-1 binding peptide	Valsesia-Wittmann et al., 1997
	Human EGF with furinlike protease cleavable linker	Buchholz et al., 1998
	Human SCF	Fielding et al., 1998
	ScFv against CEA	Konishi et al., 1998
	ScFv against HMWMAA	Martin et al., 1998
	Ram-1 binding peptide coexpressed with hemagglutinin	Hatziioannou et al., 1998
	Human EGF with plasmin cleavable linker	Peng et al., 1998
	Murine SCF	Yajima et al., 1998
	ScFv against human Cd34	Benedict et al., 1999
	Human IGF-1	Chadwick et al., 1999
	ScFv against HMWMAA with MMP cleavable linker	Martin et al., 1999
	Human IL-2	Maurice et al., 1999
	Residues 116–261 of CD40L with MMP cleavable linker	Peng et al., 1999
	ScFv against human CD33	Zhao et al., 1999
	Human EGF with factor Xa cleavable linker	Fielding et al., 2000
	ScFv against CEA	Kuroki et al., 2000
	ScFv against α Folate Receptor (αFR)	Pizzato et al., 2001
Proline-rich region	16-amino acid peptide containing the collagen-binding domain	Weimin Wu et al., 1998
	ScFv against HIV-1 gp120	Kayman et al., 1999
	V1/V2 domain of HIV-1 gp120	Kayman et al., 1999
Disulfide bonded loop in N-terminal region of RBD	ScFv against mutant EGFRvIII	Lorimer and Lavictoire, 2000

Evaluating the Recombinant Retroviruses

Proving that the engineered envelope has been efficiently incorporated in the vector and is functionally active are essential steps in the evaluation of chimeric envelope glycoproteins. The protein may not be transported to the membrane and therefore not incorporated into budding viruses; it may be misfolded and retained in the ER and the displayed ligand may be lost by enzyme cleavage during maturation in the ER. The chimeric protein may be rapidly shed from the surface of the virus due to destabilization of the SU-TM linkage or may be unable to bind its cognate receptor or trigger the fusion process once it binds the receptor. Thus characterization of the chimeric envelope as well as the integrity of the vectors must be ascertained with each engineering attempt.

Polyclonal and monoclonal antibodies directed against common antigens in the SU as well as antibodies specific for the ligand incorporated in the SU are used to detect the chimeric proteins on immunoblots on proteins obtained from pelleted virus particles or cell lysates. Detection of the expressed protein will confirm that it can be processed by the cells, expressed on their membrane, and incorporated into virus particles. The presence and accessibility of any protease-cleavable signals incorporated into linker peptides can also be evaluated.

The engineered retroviral vectors should transduce the target cell with integration of the reverse transcribed genome into cellular genomic DNA. When assaying for reporter gene expression, one has to be aware of pseudotransduction with VSV-G pseudotyped vectors (Liu, Winther, and Kay, 1996; Gallardo et al., 1997). Retroviral vectors pseudotyped with VSV-G encoding β-galactosidase as the reporter gene were cultured with mouse hepatocytes under conditions not normally permissive to retroviral entry. In these experiments, high rates of reporter gene expression were detected and subsequently shown to be due to direct protein transfer rather than stable gene transfer (Liu, Winther, and Kay, 1996). This phenomenon can be ruled out by dilutions of the vector stock and looking at individual colonies, because pseudotransduction results in diffuse, low-level expression of the protein that does not increase with time.

Southern blotting after genomic DNA extraction confirms reverse transcription and integration of the retroviral genome. DNA is digested with restriction enzymes that give a predictable pattern of DNA products from within the integrated viral genome and the resulting blots are hybridized with virus-specific probes.

Virus–cell binding assays using cells displaying the targeted receptor with and without the ecotropic or amphotropic receptors must be done to prove the retargeting strategy. To confirm specificity it is important to demonstrate that binding can be competitively inhibited by the presence of free ligand or antibodies against the targeted receptor. Binding can be evaluated using different techniques, but flow cytometry after staining with fluorescent antibodies against the viral envelope is commonly used (Valsesia-Wittmann et al., 1996). However, using antibodies against SU does not discriminate between intact viral particles and free SU components bound to the cell. Thus, results from binding assays should be interpreted as evidence of retargeted SU binding to the targeted receptor and not as binding of viral particles to the targeted receptor (Pizzato et al., 1999).

A common problem with engineered retroviruses is low titer. All new constructs are titrated against suitable cell lines and reporter gene assays performed. Beta-galactosidase and green fluorescent protein are often used, the latter because of the ease of detection and quantification using flow cytometry. With antibiotic-resistance genes such as neomycin phosphotransferase or the hygromycin-resistance gene, one can select clones after plating cells in the selective medium and counting the number of surviving colonies. A common feature of the *Retrovirinae* is that they contain the enzyme reverse transcriptase (RT). Many commercially available RT assays can be used for normalization of viral titers. Thus, no single experiment can confirm retroviral integrity and a combination of techniques must be used to substantiate claims of retargeted vectors.

Despite many attempts at direct targeting with characterization of these vectors as discussed, transduction of the target cells has been poor to nonexistent. As an example, display of EGF on the ecotropic envelope of MLV allows binding to cells expressing EGFR but target cell transduction is very poor (Cosset et al., 1995). These results were unexpected but have led to a greater understanding of virus–cell interactions and have opened up new avenues for other targeting approaches.

Problems with Direct Retroviral Retargeting

Many reasons explain the problems encountered with direct retroviral targeting. One can postulate that the modified glycoprotein may not fold properly, have problems with assembly as a trimer, or fail to be transported to the cell surface. Of course, this will result in low viral titers, exhibited as poor target cell transduction. However, analysis of pelleted vector particles indicates that the chimeric envelopes are usually incorporated with reasonable efficiency. Binding studies show that these displayed peptides are usually functional and redirect attachment of the viruses to cells exhibiting the appropriate receptor. Thus, it is clear that in part, the problem is not one of binding but of postbinding events (Cosset et al., 1995; Zhao et al., 1999). The modified proteins may not be able to activate the fusion machinery possibly due to steric effects. These steric effects may be important at different levels of the viral entry process. The displayed polypeptide may interfere with the conformational changes that occur in both the envelope glycoprotein and the receptor that lead to activation of the fusion peptide in the TM subunit. One cannot predict a priori how modifications in the envelope glycoprotein will affect its function. However, some general rules are emerging that allow one to engineer the protein without interfering with the equally crucial postbinding events. Fusion of ligands at the extreme N-terminus of the protein rather than at the +7 position or inserting peptidic spacers between the displayed peptide and the underlying viral glycoprotein can overcome this problem. This may allow the vector to transduce the cell via its natural receptor, implying that the postbinding events leading to membrane fusion are intact (Ager et al., 1996; Valsesia-Wittmann et al., 1996, 1997). For membrane fusion to occur, the two membranes must be close together. When the vector is targeted by a polypeptide that binds to a receptor tyrosine kinase, it is possible that the phospholipid envelope of the bound vector is physically too far removed from that of the target cell for lipid mixing to occur, even if the fusion machinery is correctly triggered. Moreover, receptor dimerization may prevent SU depolymerization and may lead to endocytosis of the virus-receptor complex with acidification into endosomes and protease digestion that inactivates the vector. Thus while the virus may bind the cell with high specificity, productive infection would be impossible.

The insertion of spacer peptides between the envelope glycoprotein and the displayed ligand provides another effective strategy to overcome the postbinding block that prevents target cell transduction. Ager et al. showed that the transduction efficiency of ecotropic vectors with chimeric envelopes displaying an scFv strongly depended on the linkage position as well as the length of the spacer peptide. Viral titers were very low when the spacer peptide was only three amino acids long and fused to the SU at position +7. Fusion of the scFv to amino acid residue +1 via a heptapeptide led to a 10^5 increase in titer (Ager et al., 1996). In their constructs, the sequence AAAIEGR gave the best results when vectors were titrated on murine cells (NIH 3T3), but no targeting on human cells was achieved. Similar results were reported with both ecotropic and amphotropic envelopes (Valsesia-Wittmann et al., 1996, 1997). A series of glycine and serine residues in these linkers also prevents this postbinding block to viral entry (Hatziioannou et al., 1999). Presumably, these spacers prevent the steric hindrance created by the displayed ligand, allowing the conformational changes necessary that lead to activation of the membrane fusion apparatus.

Certain displayed polypeptide ligands abolish transduction even when the cell expresses the cognate receptor on its surface. As an example, amphotropic retroviral particles with envelope glycoproteins that display EGF are greatly impaired in their ability to transduce human cells expressing abundant EGF receptors (Cosset et al., 1995; Nilson et al., 1996). The EGF domain on the viral envelope directs the virus particle to bind with high affinity to the EGF receptor on the cell, but this prevents its productive interaction with the amphotropic receptor. Furthermore, the virus particle is treated as a natural ligand by the EGF receptor resulting in receptor dimerization and endocytosis via clathrin-coated pits and routing to lysosomes where the virus is inactivated or destroyed by the low pH. All tyrosine kinase receptors that have been targeted by ligand display exhibit viral sequestration. This receptor-mediated sequestration and rerouting can be prevented by the presence of soluble ligand (e.g., EGF) that blocks receptor sites (Fielding et al., 1998).

One can also use displayed polypeptides on engineered envelopes to inhibit binding of the viral envelope to its cognate receptor. This can be exploited as the basis for a useful targeting strategy when combined with protease cleavage (see following section). Morling et al. (1997) have engineered retroviral envelopes that display polypeptides on their N-terminus that tend to form trimers such as leucine zipper peptides or the C-terminal domain of CD40 ligand. These displayed peptides were linked to the envelope by protease-cleavable linkers. It is thought that when the viral envelope glycoprotein matures, these N-terminal peptides interact to form trimers that sterically prevent binding to the cognate receptor. In the absence of the protease, viral infectivity was poor but greatly increased, once the blocking ligand was cleaved off by the specific protease.

Thus there are many explanations for the low transduction efficiency of vectors having engineered SU glycoproteins. It appears that retroviruses have adapted to use specific receptors on cells not only because they allow binding of the viral particle to the cell (an essential first step) but also because they may activate membrane fusion and cell entry. All known receptors for the type C oncoretroviruses belong to the same class, suggesting that there are few receptor types that can trigger the conformational changes in the viral envelope glycoproteins necessary to trigger membrane fusion. This may limit our ability to efficiently retarget vectors using the approaches discussed so far.

Reconciliation of Vector Retargeting and High Transduction Efficiency

Vector retargeting and high transduction efficiency can be reconciled in the following four ways:

1. ***Chaperoning***. Investigators have attempted to overcome the block in retroviral entry by generating vectors that incorporate both the retargeted as well as the wild-type glycoprotein on their membrane. Thus, it has been reported that vectors expressing both the engineered envelope as well as the wild-type ecotropic envelope that does not have a receptor on human cells can successfully bind and transduce targeted human cells (Chu et al., 1994; Kasahara, Dozy, and Kan, 1994; Chu and Dornburg, 1995; Matano et al., 1995; Konishi et al., 1998; Martin et al., 1998). Such observations have also been reported with retargeted SNV vectors (Chu and Dornburg, 1995). The molecular mechanisms responsible for this enhanced transduction efficiency in the presence of wild-type envelope protein are unclear at present. In the case of SNV-derived chimeric vectors, it has been proposed that the presence of the wild-type envelope glycoprotein provides a helper function in membrane fusion by a low-affinity interaction with the D-type retroviral receptor present on human cells (Jiang et al., 1998). Co-incorporating a binding-defective hemagglutinin from influenza virus will also allow fusion to occur. Lin et al. (2001) displayed Flt-3 on MoMLV envelope and generated retroviral vectors that also displayed a binding defective influenza hemagglutinin (HAtmt)

on their phospholipid envelope. The retargeted vector could bind to cells expressing the Flt-3 receptor and binding was abrogated in the presence of soluble Flt-3 ligand. However, the titer of the vector generated was low (3.5×10^3 cfu/ml) and this approach must be optimized if it is to be of any therapeutic use.

2. ***Pseudotyping with engineered envelopes***. The influenza virus hemagglutinin (HA) can bind to different sialic acid-containing glycoproteins and trigger membrane fusion (Wiley and Skehel, 1987). The unbound HA can be activated by low pH as well as by partial denaturation or changes in temperature (Nobusawa and Nakajima, 1988). It appears that HA can tolerate the insertion of ligands without compromising its ability to induce membrane fusion. Hatziioannou et al. (1999) have engineered the HA of fowl plague virus to display polypeptides and single-chain antibodies. Retroviral vectors have been pseudotyped with chimeric influenza HA engineered to display on their N-terminus EGF, an scFv against MHC class 1 molecules and an antimelanoma antigen, and an IgG Fc-binding polypeptide. Retroviral vectors incorporated these engineered envelopes and transduced cells expressing the targeted receptor (e.g., EGFR, MHC class 1 antigens). If the HAs could be prevented from binding to their wild-type receptors, they might have the potential to provide a platform for retargeting many vectors using different displayed ligands. Retroviral incorporation of chimeric HA glycoprotein is of course a form of pseudotyping. Sawai and Meruelo (1998) pseudotyped MLV-based vectors with the Sindbis virus envelope engineered to display human chorionic gonadotrophin (hCG). The chimeric vector transduced choriocarcinoma cells expressing the hCG receptor but not receptor-negative cells. The same group engineered Sindbis envelope to display the IgG-binding domain of protein A. This vector can transduce cells in the presence of a monoclonal antibody directed against an antigen present on the target cell surface (Morizono et al., 2001). Retroviruses can also be pseudotyped with many other envelope glycoproteins including those derived from vesicular stomatitis virus (VSV-G) (Burns et al., 1993), semliki forest virus (SFV) (Suomalainen and Garoff, 1994), and lymphocytic choriomeningitis virus (LCMV) (Miletic et al., 1999). It may be possible to retarget by peptide display these envelopes as well and such studies are awaited.

3. ***Inverse targeting***. As discussed previously, engineering ligands to the N-terminus of the C-type retroviral envelope glycoprotein inhibits transduction of cells expressing the cognate receptor by receptor mediated sequestration (Cosset et al., 1995; Fielding et al., 1998; Chadwick et al., 1999). Studies have shown that the relative efficiencies of retroviral binding to the retargeted receptor and the natural receptor are dependent on the relative surface density of each receptor on the cell surface. Therefore, higher levels of the targeted receptor lead to sequestration, whereas

the presence of high amounts of the natural receptor lead to higher levels of productive infection (Chadwick et al., 1999).

This observation has been exploited to create inverse targeting. Fielding et al. (1998) generated retroviruses displaying either EGF or stem cell factor (SCF) on their envelope. Vectors displaying EGF could not transduce cell lines strongly expressing EGFR but they efficiently transduced hemopoietic cells strongly expressing KIT, the receptor for SCF. Conversely, SCF displaying vectors transduced epithelial cells (KIT-negative, EGFR-positive) but not hemopoietic cells expressing KIT. Thus, this approach has the potential for selective, in vitro transduction of hemopoietic cells, for example, with drug resistance genes such as the P-glycoprotein. Contaminating carcinoma cells expressing EGFR will not be transduced (Fielding et al., 1998).

4. ***Protease targeting***. The insertion of linker peptides between the displayed ligand and the viral envelope glycoprotein provides the opportunity to insert protease cleavable sequences that can be useful in tissue specific targeting. To this effect, sequences containing the cleavage signals recognized by plasmin (P-S-I-Q-Y-R/G-L), factor Xa protease (-I-E-G-R/), membrane-associated matrix metalloproteinases (P-L-G/L-W-A), and furin (R-X-(K/R)-R/) have been successfully incorporated and shown to be functional (Nilson et al., 1996; Morling et al., 1997; Peng et al., 1997, 1999; Buchholz et al., 1998). In a protease-targeted vector the displayed polypeptide blocks target cell transduction either by receptor-mediated sequestration or steric hindrance. Transduction will only occur if the displayed blocking polypeptide is cleaved off, allowing the underlying envelope glycoprotein to interact with its receptor and undergo the conformational changes necessary for membrane fusion and productive infection. High-affinity ligands were initially used to prove that this strategy works. Vectors displaying EGF linked to the envelope via a factor Xa-cleavable linker exhibited very poor transduction of cells expressing EGFR. However, these vectors became fully infectious when they were treated with factor Xa protease (Nilson et al., 1996).

   High-affinity ligands block transduction only in cells expressing the receptor for the displayed ligand. Therefore, Morling et al. (1997) designed vectors displaying homotrimeric leucine zipper peptides or globular domains that allow homotrimeric interactions similar to those found in the C-terminal domain of CD40 ligand or in tumor necrosis factor. These blocking domains block transduction in all cell types. It is thought that these polypeptides trimerize when the glycoprotein matures on the membrane surface and they form a "cap" at the tip of the envelope glycoprotein that blocks the receptor-binding site. Binding studies with these vectors showed that these polypeptides block the interaction between the viral envelope and the natural receptor on the cell. Treatment with factor Xa protease to cleave the displayed trimeric polypeptide restored full infectivity (Morling et al., 1997). Subsequently, blocking domains tethered to the SU glycoprotein through membrane-associated matrix metalloproteinases (MMPs) and plasmin-cleavable linkers were generated. These important enzymes trigger proteolysis of the matrix surrounding tumor cells, leading to tumor cell invasion, metastasis, and angiogenesis. Vectors that could be activated by MMP via cleavable linkers could discriminate between MMP-poor and MMP-rich cells in mixed populations. Similar results were obtained using the plasmin-activatable vectors (Peng et al., 1998). In addition, the same vectors could discriminate between MMP-rich and MMP-poor tumor xenografts in nude mice (Peng et al., 1999). Any protease with a known specific cleavage signal and which does not degrade the envelope glycoprotein itself can be used in this targeting approach. Once a specific cleavage signal is identified, it can be optimized using approaches discussed further below.

IN VIVO STUDIES

The retargeting studies just discussed were conducted in vitro. However, it is of interest to evaluate the distribution and expression of retargeted vectors in vivo. Systemically administered vectors face a very different environment from that in culture systems. Serum contains complement proteins and antibodies that have the potential to damage or inactivate retroviral vectors. For example, it was recently reported that human serum could inactivate VSV-G pseudotyped HIV-based vectors (DePolo et al., 2000). Moreover, blood flow, microvessel architecture (the presence of sinusoids and high endothelial venules), and the tissue distribution of targeted receptors, proteases, and their inhibitors are all expected to influence the biodistribution of targeted vectors administered into the bloodstream. These important issues can only be evaluated in animal models.

Peng et al. (1999) studied the in vivo activity of MMP-targeted vectors. Tumor xenografts from MMP-rich (HT1080) and MMP-poor (A431) cell lines were established in nude mice. Mice were injected intratumorally with EGF-displaying vectors, either noncleavable or MMP-cleavable, or with vectors bearing wild-type amphotropic envelopes. Three days later, the tumors were excised and tested for β-galactosidase activity. They demonstrated selective activation of the MMP-targeted vector in the MMP-rich HT1080 tumor, providing in vivo evidence of the feasibility of MMP-targeting. It seems that the presence of antiproteases did not hinder this targeted vector. However, systemic administration of these vectors was not reported.

In a recent study looking at the in vivo biodistribution of lentiviral vectors pseudotyped with the 4070A amphotropic envelope after intravenous administration in mice, Peng et al. (2001) showed that the highest levels of reporter gene expression were in the liver, spleen, heart, and skeletal muscle (in decreasing order). There was little expression in the lung, kidney, brain, ovaries, and bone marrow. The authors also looked at the in vivo biodistribution of VSV-G-pseudotyped vectors. Because the glycolipid receptor for VSV-G is ubiquitous, one would expect luciferase expression in many organs. Surprisingly, reporter gene expression was the same as with the amphotropic 4070A envelope vector. In the same experiment with an EGF-displaying amphotropic vector, the highest level of luciferase activity was found in the spleen and significantly less in the liver. There are very few EGFR-expressing cells in the spleen, whereas liver cells express EGFR at high levels (Mateo de Acosta et al., 1989). This provides in vivo evidence of receptor-mediated sequestration. In fact, liver expression was enhanced if the mice were perfused with soluble EGF before the targeted vector was injected. Thus it is clear that the main forces influencing vector biodistribution in vivo are blood flow and regional vascular anatomy and not cognate receptor expression.

Another approach to targeting by protein engineering has been described by Hall et al. (2000). They designed vectors displaying a collagen-binding peptide from von Willebrand factor (vWF) as an N-terminal extension of the SU of the 4070A envelope. Those vectors that also expressed the wild-type amphotropic envelope exhibited both high collagen binding affinity and high transduction efficiency with viral titers up to 10^7 CFU/ml. The same group also described an optimized construct with a single envelope glycoprotein having both extracellular matrix targeting as well as near wild-type amphotropic transduction efficiency. This vector was tested in two animal models in vivo. When neointimal injury was induced by balloon angioplasty in a rat carotid artery, intraarterial injection of the retargeted vector resulted in 20% of the neointimal cells expressing the reporter gene. Similarly, in a mouse model of liver metastases, intraportal infusion of the vector led to transduction of 1–3% of tumor cells (Hall et al., 2000). This approach has been called *tethered targeting* and is based on the mechanisms of wound healing and response to injury that are similar in all parts of the body. More recently, the same group reported that in nude mice with tumor xenografts, the intravenous injection of matrix-targeted vectors led to high tumor cell transduction (35%) as determined by β-galactosidase expression. The introduction of a cytocidal cyclin G1 in this targeted vector led to tumor regression with tumor cell necrosis and apoptosis (Gordon et al., 2001).

OPTIMIZATION OF GENETICALLY RETARGETED RETROVIRAL VECTORS

The initial choice of ligand displayed on the retroviral envelope for retroviral vector retargeting is usually guided by the literature. Often targeting is poor usually due to suboptimal linkage of the displayed peptide to the SU. Thus optimization of the displayed ligand is usually necessary.

Targeting cells without having a specific antigen or a monoclonal antibody can be accomplished using a combination of phage and retroviral display libraries. Engelstadter et al. (2000) identified retroviral vectors that can specifically transduce T cells using such a strategy. They immunized mice with human T cells and harvested their spleens. Total RNA was used to generate a cDNA library of single-chain antibody genes that was cloned into phagemids. This was expanded and screened for protein expression in *Escherichia coli.* Phage clones were selected for T-cell binding and their encoded single-chain variable fragments were cloned into the N-terminal region of the TM component of SNV vectors. These investigators initially generated 4.7×10^6 clones in their phagemid library. They selected 150 phagemids with significant T-cell binding and cloned all of the corresponding scFv genes into the SNV TM. Finally, one chimeric envelope (7A5) was identified that transduced T cells with high specificity. In a follow-up study, the same group demonstrated that it is possible to pseudotype MLV-based vectors with this chimeric envelope (SNV-7A5). This vector transduces T cells with high efficiency (Engelstadter et al., 2001).

Phage libraries must be grown in bacterial hosts that lack the enzymatic machinery necessary for the posttranslational modifications that occur to eukaryotic proteins. The result is often a misfolded or partially degraded protein. In contrast, replication-competent retroviruses can propagate in mammalian cells. As already discussed, many ligands have been displayed on the N-terminus of the SU component. These chimeric vectors can be subjected to selective pressure and viruses with optimized properties (binding and/or protease cleavage) have been isolated. Buchholz et al. (1998) generated a library of retroviruses displaying EGF linked to the N-terminus of SU by a randomized polypeptide linker. In this way, a retroviral library coding for 4×10^7 different linkers was generated. Viral supernatants were added to HT1080 cells that express high levels of the EGFR. Binding of viruses displaying the EGFR chimeric envelope on these cells resulted in virus sequestration and no productive infection. Thus, for HT1080 cells, productive infection can only occur if the EGF domain displayed on the chimeric envelope is cleaved off. The selected viruses were screened by polymerase chain reaction using virus-specific primers. The amplified DNA was sequenced and the linker sequences in all positive clones were found to be enriched in basic amino acids. Arginine/lysine motifs form the cleavage sites of the proprotein convertase family of proteases, including furin, an enzyme ubiquitously expressed in the endoplasmic reticulum. Immunoblots confirmed that the EGF domain was cleaved from the envelope by a ubiquitous intracellular protease. Thus, one can use the replicative potential of retroviruses in human cells together with recombinant DNA technology to generate large libraries that can be screened for the specific property of interest. In this way, optimized ligands for display on retroviral envelopes can be generated.

CONCLUSIONS AND FUTURE DIRECTIONS

The field of retroviral vector targeting has seen considerable progress in the last few years. The initial attempts at direct vector retargeting by envelope glycoprotein engineering have led to a better understanding of envelope–receptor interactions and membrane fusion. This has opened new targeting avenues that are giving meaningful results. Studies on the in vivo biodistribution of vectors have shown that the systemic delivery of specifically targeted retroviral vectors is feasible. However, there is an urgent need for additional work in this area to fully elucidate the forces that govern vector biodistribution in the in vivo setting. A better understanding of the interactions between retroviral envelope glycoproteins and their cognate receptors based on structural data is awaited. Nonspecific adsorption of vectors must be investigated more thoroughly in in vivo systems and its implications for systemic vector delivery defined. The advent of structural genomics should give a strong boost to the field by increasing our understanding of the rules that govern the three-dimensional (3D) structure of proteins and by describing the 3D structure of retroviral membrane glycoproteins and how they interact with their receptors. Finally, technology based on retroviral display libraries should help us to refine and optimize these targeting strategies.

REFERENCES

Ager S, Nilson BH, Morling FJ, Peng KW, Cosset FL, Russell SJ (1996): Retrovirus display of antibody fragments: interdomain spacing strongly influences vector infectivity. *Hum Gene Ther* 7:2157–2164.

Bae Y, Kingsman SM, Kingsman AJ (1997): Functional dissection of the Moloney murine leukemia virus envelope protein gp70. *J Virol* 71:2092–2099.

Balliet JW, Bates P (1998): Efficient infection mediated by viral receptors incorporated into retroviral particles. *J Virol* 72:671–676.

Barnett AL, Davey RA, Cunningham JM (2001): Modular organization of the Friend murine leukemia virus envelope protein underlies the mechanism of infection. *Proc Natl Acad Sci USA* 98:4113–4118.

Battini JL, Heard JM, Danos O (1992): Receptor choice determinants in the envelope glycoproteins of amphotropic, xenotropic, and polytropic murine leukemia viruses. *J Virol* 66:1468–1475.

Battini JL, Rodrigues P, Muller R, Danos O, Heard JM (1996): Receptor-binding properties of a purified fragment of the 4070A amphotropic murine leukemia virus envelope glycoprotein. *J Virol* 70:4387–4393.

Benedict CA, Tun RY, Rubinstein DB, Guillaume T, Cannon PM, Anderson WF (1999): Targeting retroviral vectors to CD34-expressing cells: binding to CD34 does not catalyze virus-cell fusion. *Hum Gene Ther* 10:545–557.

Buchholz CJ, Peng KW, Morling FJ, Zhang J, Cosset FL, Russell SJ (1998): In vivo selection of protease cleavage sites from retrovirus display libraries. *Nature Biotechnol* 16:951–954.

Burns JC, Friedmann T, Driever W, Burrascano M, Yee JK (1993): Vesicular stomatitis G glycoprotein pseudotyped retroviral vectors: concentration to a very high titer and efficient gene transfer into mammalian and nonmammalian cells. *Proc Natl Acad Sci USA* 90:8033–8037.

Chadwick MP, Morling FL, Cosset FL, Russell SJ (1999): Modification of retroviral tropism by display of IGF-1. *J Mol Biol* 285:485–494.

Chu TH, Dornburg R (1995): Retroviral vector particles displaying the antigen-binding site of an antibody enable cell-type-specific gene transfer. *J Virol* 69:2659–2663.

Chu TH, Martinez I, Shcay WC, Dornburg R (1994): Cell targeting with retroviral vector particles containing antibody-envelope fusion proteins. *Gene Ther* 1:292–299.

Cosset FL, Morling FJ, Takeuchi Y, Weiss RA, Collins MK, Russell SJ (1995): Retroviral retargeting by envelopes expressing an N-terminal binding domain. *J Virol* 69:6314–6322.

DePolo NJ, Reed JD, Sheridan PL, Townsend K, Sauter SL, Jolly DJ, Dubensky TW Jr (2000): VSV-G pseudotyped lentiviral particles produced in human cells are inactivated by human serum. *Mol Ther* 2:218–222.

Engelstadter M, Bobkova M, Baier M, Stitz J, Holtkamp N, Chu TH, Kurth R, Dornburg R, Buchholz CJ, Cichutek K (2000): Targeting human T cells by retroviral vectors displaying antibody domains selected from a phage display library. *Hum Gene Ther* 11:293–303.

Engelstadter M, Buchholz CJ, Bobkova M, Steidl S, Merget-Millitzer H, Willemsen RA, Stitz J, Cichutek K (2001): Targeted gene transfer to lymphocytes using murine leukemia virus vectors pseudotyped with spleen necrosis virus envelope glycoproteins. *Gene Ther* 9:1202–1206.

Endres MJ, Jaffer S, Haggarty B, Turner JD, Doranz BJ, O'Brien PJ, Kolson DL, Hoxie JA (1997): Targeting of HIV- and SIV-infected cells by CD4-chemokine receptor pseudotypes. *Science* 278:1462–1464.

Etienne-Julan M, Roux P, Carillo S, Jeanteur P, Piechaczyk M (1992): The efficiency of cell targeting by recombinant retroviruses depends on the nature of the receptor and the composition of the artificial cell-virus linker. *J Gen Virol* 73:3251–3255.

Fass D, Harrison SC, Kim PS (1996): Retrovirus envelope domain at 1.7 angstrom resolution. *Nat Struct Biol* 3:465–469.

Fass D, Davey RA, Hamson CA, Kim PS, Cunningham JM, Berger JM (1997): Structure of a murine leukemia virus receptor-binding glycoprotein at 2.0 angstrom resolution. *Science* 277:1662–1666.

Fielding AK, Maurice M, Morling FJ, Cosset FL, Russell SJ (1998): Inverse targeting of retroviral vectors: Selective gene transfer in a mixed population of hematopoietic and nonhematopoietic cells. *Blood* 91:1802–1809.

Fielding AK, Chapel-Fernandes S, Chadwick MP, Bullough FJ, Cosset FL, Russell SJ (2000): A hyperfusogenic gibbon ape leukemia envelope glycoprotein: targeting of a cytotoxic gene by ligand display. *Hum Gene Ther* 8:2183–2192.

Gallardo HF, Tan C, Ory D, Sadelian M (1997): Recombinant retroviruses pseudotyped with the vesicular stomatitis virus G glycoprotein mediate both stable gene transfer and pseudotransduction in human peripheral blood lymphocytes. *Blood* 90:952–957.

Gordon EM, Chen ZH, Liu L, Whitley M, Liu L, Wei D, Groshen S, Hinton DR, Anderson WF, Beart RW Jr, Hall FL (2001): Systemic administration of a matrix-targeted retroviral vector is efficacious for cancer gene therapy in mice. *Hum Gene Ther* 12:193–204.

Goud B, Legrain P, Buttin G (1988): Antibody-mediated binding of a murine ecotropic Moloney retroviral vector to human cells allows internalization but not the establishment of the proviral state. *Virology* 163:251–254.

Hall FL, Gordon EM, Wu L, Zhu NL, Skotzko MJ, Starnes VA, Anderson WF (1997): Targeting retroviral vectors to vascular lesions by genetic engineering of the MoMLV gp70 envelope protein. *Hum Gene Ther* 8:2183–2192.

Hall FL, Liu L, Zhu NL, Stapfer M, Anderson WF, Beart RW, Gordon EM (2000): Molecular engineering of matrix-targeted retroviral vectors incorporating a surveillance function inherent in von Willebrand factor. *Hum Gene Ther* 11:983–993.

Han X, Kasahara N, Kan YW (1995): Ligand-directed retroviral targeting of human breast cancer cells. *Proc Natl Acad Sci USA* 92:9747–9751.

Hatziionnou T, Valsesia-Wittmann S, Russell SJ, Cosset FL (1998): Incorporation of fowl plague virus hemagglutinin into murine leukemia virus particles and analysis of the infectivity of the pseudotyped retroviruses. *J Virol* 72:5313–5317.

Hatziioannou T, Delahaye E, Martin F, Russell SJ, Cosset FL (1999): Retroviral display of functional binding domains fused to the amino terminus of influenza hemagglutinin. *Hum Gene Ther* 10:1533–1544.

Heard JM, Danos O (1991): An amino-terminal fragment of the Friend murine leukemia virus envelope glycoprotein binds the ecotropic receptor. *J Virol* 65:4026–4032.

Jelesarov I, Lu M (2001): Thermodynamics of trimer-of-hairpins formation by the SIV gp41 envelope protein. *J Mol Biol* 307:637–656.

Jiang A, Chu TH, Nocken F, Cichutek K, Dornburg R (1998): Cell-type-specific gene transfer into human cells with retroviral vectors that display single-chain antibodies. *J Virol* 72:10148–10156.

Jones JS, Risser R (1993): Cell fusion induced by the murine leukemia virus envelope glycoprotein. *J Virol* 67:67–74.

Kasahara N, Dozy AM, Kan YW (1994): Tissue-specific targeting of retroviral vectors through ligand-receptor interactions. *Science* 266:1373–1376.

Kavanaugh MP, Kabat D (1996): Identification and characterization of a widely expressed phosphate transporter/retrovirus receptor family. *Kidney Int* 49:959–963.

Kayman SC, Park H, Saxon M, Pinter A (1999): The hypervariable domain of the murine leukemia virus surface protein tolerates large insertions and deletions, enabling development of a retroviral particle display system. *J Virol* 73:1802–1808.

Kizhatil K, Albritton LM (1997): Requirements for different components of the host cell cytoskeleton distinguish ecotropic murine leukemia virus entry via endocytosis from entry via surface fusion. *J Virol* 71:7145–7156.

Konishi H, Ochiya T, Chester KA, Begent RH, Muto T, Sugimura T, Terada M (1998): Targeting strategy for gene delivery to carcinoembryonic antigen-producing cancer cells by retrovirus displaying a single chain variable fragment antibody. *Hum Gene Ther* 9:235–248.

Kuroki M, Arakawa F, Khare PD, Kuroki M, Liao S, Matsumoto H, Abe H, Imakiire T (2000): Specific targeting strategies of cancer gene therapy using a single-chain variable fragment (scFv) with a high affinity for CEA. *Anticancer Res* 20:4067–4071.

Lavillette D, Maurice M, Roche C, Russell SJ, Sitbon M, Cosset FL (1998): A proline-rich motif downstream of the receptor binding domain modulates conformation and fusogenicity of murine retroviral envelopes. *J Virol* 72: 9955–9965.

Lavillette D, Ruggieri A, Russell SJ, Cosset FL (2000): Activation of a cell entry pathway common to type C mammalian retroviruses by soluble envelope fragments. *J Virol* 74: 295–304.

Lin AH, Kasahara N, Wu W, Stripecke R, Empig CL, Anderson WF, Cannon PM (2001): Receptor-specific targeting mediated by the coexpression of a targeted murine leukemia virus envelope protein and a binding-defective influenza hemagglutinin protein. *Hum Gene Ther* 12:323–332.

Liu ML, Winther BL, Kay MA (1996): Pseudotransduction of hepatocytes by using concentrated pseudotyped vesicular stomatitis virus G glycoprotein (VSV-G)-Moloney murine leukemia virus-derived retroviral vectors: Comparison of VSV-G and amphotropic vectors for hepatic gene transfer. *J Virol* 70:2497–2502.

Liu L, Anderson WF, Beart RW, Gordon EM, Hall FL (2000): Incorporation of tumor vasculature targeting motifs into moloney murine leukemia virus env escort proteins enhances retrovirus binding and transduction of human endothelial cells. *J Virol* 74:5320–5328.

Lorimer IA, Lavictoire SJ (2000): Targeting retrovirus to cancer cells expressing a mutant EGF receptor by insertion of a single chain antibody variable domain in the envelope glycoprotein receptor binding lobe. *J Immunol Meth* 237:147–157.

Marin M, Noel D, Valsesia-Wittman S, Brockly F, Etienne-Julan M, Russell S, Cosset FL, Piechaczyk M (2000): Targeted infection of human cells via major histocompatibility complex class I molecules by Moloney murine leukemia virus-derived viruses displaying single-chain antibody fragment-envelope fusion proteins. *J Virol* 70:2957–2962.

Martin F, Kupsch J, Takeuchi Y, Russell S, Cosset FL, Collins M (1998): Retroviral vector targeting to melanoma cells by single-chain antibody incorporation in envelope. *Hum Gene Ther* 9:737–746.

Martin F, Neil S, Kupsch J, Maurice M, Cosset F, Collins M (1999): Retrovirus targeting by tropism restriction to melanoma cells. *J Virol* 73:6923–6929.

Matano T, Odawara T, Iwamoto A, Yoshikura H (1995): Targeting infection of a retrovirus bearing a CD4-Env chimera into human cells expressing human immunodeficiency virus type 1. *J Gen Virol* 76:3165–3169.

Mateo de Acosta C, Justiz E, Skoog L, Lage A (1989): Biodistribution of radioactive epidermal growth factor in normal and tumor bearing mice. *Anticancer Res* 9:89–92.

Maurice M, Mazur S, Bullough FJ, Salvetti A, Collins MK, Russell SJ, Cosset FL (1999): Efficient gene delivery to quiescent interleukin-2 (IL-2)-dependent cells by murine leukemia virus-derived vectors harboring IL-2 chimeric envelope. *Blood* 94:401–410.

Miletic H, Bruns M, Tsiakas K, Vogt B, Rezai R, Baum C, Kuhlke K, Cosset FL, Ostertag W, Lother H, von Laer D (1999): Retroviral vectors pseudotyped with lymphocytic choriomeningitis virus. *J Virol* 73: 6114–6116.

Miller DG, Edwards RH, Miller AD (1994): Cloning of the cellular receptor for amphotropic murine retrovirus reveals homology to that for gibbon ape leukemia virus. *Proc Natl Acad Sci USA* 91:78–82.

Moore JP, McKeating JA, Weiss RA, Sattentau QJ (1990): Dissociation of gp120 from HIV-1 virions induced by soluble CD4. *Science* 250:1139–1142.

Morizono K, Bristol G, Xie YM, Kung SK, Chen IS (2001): Antibody directed targeting of retroviral vectors via cell-surface antigens. *J Virol* 75:8016–8020.

Morling FJ, Peng KW, Cosset FL, Russell SJ (1997): Masking of retroviral envelope functions by oligomerizating polypeptides. *Virology* 234:51–61.

Neda H, Wu CH, Wu GY (1991): Chemical modification of an ecotropic murine leukemia virus results in redirection of its target cell specificity. *J Biol Chem* 266:14143-14149.

Nguyen TH, Pages JC, Farge D, Briand P, Weber A (1998): Amphotropic retroviral vectors displaying hepatocyte growth factor-envelope fusion proteins improve transductional efficiency of primary hepatocytes. *Hum Gene Ther* 9:2469–2479.

Nilson BH, Morling FJ, Cosset FL, Russell SJ (1996): Targeting of retroviral vectors through protease-substrate interactions. *Gene Ther* 3:280–286.

Nobusawa E, Nakajima K (1988): Amino acid substitution at position 226 of the hemagglutinin molecule of influenza (H1N1) virus affects receptor binding activity but not fusion activity. *Virology* 167:8–14.

O'Hara B, Johann SV, Klinger HP, Blair DG, Rubinson H, Dunn KJ, Sass P, Vitek SM, Robins T (1990): Characterization of a human gene conferring sensitivity to infection with gibbon ape leukemia virus. *Cell Growth Differ* 1:119–127.

Peng KW, Morling FJ, Cosset FL, Murphy G, Russell SJ (1997): A gene delivery system activatable by disease-associated matrix metalloproteinases. *Hum Gene Ther* 8:729–738.

Peng KW, Morling FJ, Cosset FL, Russell SJ (1998): Retroviral gene delivery system activatable by plasmin. *Tumor Target* 3:112–120.

Peng KW, Vile R, Cosset FL, Russell SJ (1999): Selective transduction of protease-rich tumors by matrix-metalloproteinase-targeted retroviral vectors. *Gene Ther* 6:1552–1557.

Peng KW, Pham L, Ye H, Zufferey R, Trono D, Cosset FL, Russell SJ (2001): Organ distribution of gene expression after intravenous infusion of targeted and untargeted lentiviral vectors. *Gene Ther* 8:1456–1463.

Pinter A, Kopelman R, Li Z, Kayman SC, Sanders DA (1997): Localization of the labile disulfide bond between SU and TM of the murine leukemia virus envelope protein complex to a highly conserved CWLC motif in SU that resembles the active-site sequence of thiol-disulfide exchange enzymes. *J Virol* 71:8073–8077.

Pizzato M, Marlow SA, Blair ED, Takeuchi Y (1999): Initial binding of murine leukemia virus particles does not require specific Env-receptor interaction. *J Virol* 73:8599–8611.

Pizzato M, Blair ED, Fling M, Kopf J, Tomassetti A, Weiss RA, Takeuchi Y (2001): Evidence for nonspecific adsorption of targeted retrovirs vector particles to cells. *Gene Ther* 8:1088–1096.

Rein A, Mirro J, Haynes JG, Ernst SM, Nagashima K (1994): Function of the cytoplasmic domain of a retroviral transmembrane protein: p15E-p2E cleavage activates the membrane fusion capability of the murine leukemia virus Env protein. *J Virol* 68:1773–1781.

Rodrigues P, Heard JM (1999): Modulation of phosphate uptake and amphotropic murine leukemia virus entry by posttranslational modifications of PIT-2. *J Virol* 73:3789–3799.

Roux P, Jeanteur P, Piechaczyk M (1989): A versatile and potentially general approach to the targeting of specific cell types by retroviruses: Application to the infection of human cells by means of major histocompatibility complex class I and II antigens by mouse ecotropic murine leukemia virus-derived viruses. *Proc Natl Acad Sci USA* 86:9079–9083.

Russell SJ, Hawkins RE, Winter G (1993): Retroviral vectors displaying functional antibody fragments. *Nucleic Acids Res* 21:1081–1085.

Sawai K, Meruelo D (1998): Cell-specific transfection of choriocarcinoma cells by using Sindbis virus hCG expressing chimeric vector. *Biochem Biophys Res Commun* 248:315–323.

Schlegel R, Willingham MC, Pastan IH (1982): Saturable binding for vesicular stomatitis virus on the surface of Vero cells. *J Virol* 43:871–875.

Schnierle BS, Moritz D, Jeschke M, Groner B (1996): Expression of chimeric envelope proteins in helper cell lines and integration into Moloney murine leukemia virus particles. *Gene Ther* 3:334–342.

Siess DC, Kozak SL, Kabat D (1996): Exceptional fusogenicity of Chinese hamster ovary cells with murine retroviruses suggests roles for cellular factor(s) and receptor clusters in the membrane fusion process. *J Virol* 70:3432–3439.

Snitkovsky S, Young JA (1998): Cell-specific viral targeting mediated by a soluble retroviral receptor-ligand fusion protein. *Proc Natl Acad Sci USA* 95:7063–7068.

Somia NV, Zoppe M, Verma IM (1995): Generation of targeted retroviral vectors by using single-chain variable fragment: an approach to in vivo gene delivery. *Proc Natl Acad Sci USA* 92:7570–7574.

Somia NV, Miyoshi H, Schmitt MJ, Verma IM (2000): Retroviral vector targeting to human immunodeficiency virus type i-infected cells by receptor pseudotyping. *J Virol* 74:4420–4424.

Suomalainen M, Garoff H (1994): Incorporation of homologous and heterologous proteins into the envelope of Moloney murine leukemia virus. *J Virol* 68:4879–4889.

Tailor CS, Kabat D (1997): Variable regions A and B in the envelope glycoproteins of feline leukemia virus subgroup B and amphotropic murine leukemia virus interact with discrete receptor domains. *J Virol* 71:9383–9391.

Valsesia-Wittmann S, Drynda A, Deleage G, Aumailley M, Heard JM, Danos O, Verdier G, Cosset FL (1994): Modifications in the binding domain of avian retrovirus envelope protein to redirect the host range of retroviral vectors. *J Virol* 68:4609–4619.

Valsesia-Wittmann S, Morling FJ, Nilson BH, Takeuchi Y, Russell SJ, Cosset FL (1996): Improvement of retroviral retargeting by using amino acid spacers between an additional binding domain and the N-terminus of Moloney murine leukemia virus SU. *J Virol* 70:2059–2064.

Valsesia-Wittmann S, Morling FJ, Hatziioannou T, Russell SJ, Cosset FL (1997): Receptor co-operation in retrovirus entry: recruitment of an auxiliary entry mechanism after retargeted binding. *EMBO J* 16:1214–1223.

Wang H, Kavanagh MP, North RA, Kabat D (1991a): Cell surface receptor for ecotropic murine retroviruses is a basic amino acid transporter. *Nature* 352:729–731.

Wang H, Paul R, Burgeson RE, Keene DR, Kabat D (1991b): Plasma membrane receptors for ecotropic murine retroviruses require a limiting accessory factor. *J Virol* 65:6468–6477.

Weimin Wu B, Cannon PM, Gordon EM, Hall FL, Anderson WF (1998): Characterization of the proline-rich region of murine leukemia virus envelope protein. *J Virol* 72:5383–5391.

Weiss RA, Tailor CS (1995): Retrovirus receptors. *Cell* 82:531–533.

White JM (1992): Viral and cellular membrane fusion proteins. *Science* 258:917–924.

Wiley DC, Skehel JJ (1987): The structure and function of the hemagglutinin membrane glycoprotein of influenza virus. *Annu Rev Biochem* 56:365–394.

Yajima T, Kanda T, Yoshiike K, Kitamura Y (1998): Retroviral vector targeting human cells via c-Kit-stem cell factor interaction. *Hum Gene Ther* 9:779–787.

Yu H, Soong N, Anderson WF (1995): Binding kinetics of ecotropic (Moloney) murine leukemia retrovirus with NIH 3T3 cells. *J Virol* 69:6557–6562.

Zavorontinskaya T, Albritton LM (1999): Suppression of a fusion defect by second site mutations in the ecotropic murine leukemia virus surface protein. *J Virol* 73:5034–5042.

Zhao Y, Zhu L, Lee S, Li L, Chang E, Soong NW, Douer D, Anderson WF (1999): Identification of the block in targeted retroviral-mediated gene transfer. *Proc Nat Acad Sci USA* 96(7):4005–4010.

Zhu NL, Cannon PM, Chen D, Anderson WF (1998): Mutational analysis of the fusion peptide of Moloney murine leukemia virus transmembrane protein p15E. *J Virol* 72:1632–1639.

14

GENETIC ENGINEERING OF TARGETED RETROVIRAL VECTORS

ERLINDA M. GORDON, M.D., FREDERICK L. HALL, PH.D.,
ROBERT W. BEART, JR., M.D., AND W. FRENCH ANDERSON, M.D.

INTRODUCTION

Historical Perspectives

The pioneering clinical trials advancing the promise of human gene therapy employed retroviral vectors as the gene delivery vehicles of choice (Gordon and Anderson, 1994). These protocols primarily used an ex vivo approach, wherein target cells were removed from the patient, grown in vitro, transduced with a gene vector, and then engrafted into the patient (Rosenberg et al., 1990; Blaese et al., 1992). Devoid of viral genes, these replication-incompetent retroviral vectors accommodated up to ~8 kb of exogenous (i.e., therapeutic) DNA. Many clinical protocols using the ex vivo approach have been conducted in a bone marrow transplantation setting wherein retrovirally transduced autologous hematopoietic cells expressing a marker or therapeutic gene such as the multiple drug resistance gene, an adenosine deaminase cDNA (Dunbar et al., 1993), or an anti-human immunodeficiency virus (HIV) ribozyme gene, were infused intravenously into the recipient (Human Gene Therapy Protocols, National Institutes of Health (NIH)/Office of Biotechnology Activities, updated August 10, 2001). The retroviral producer cells themselves have been

Vector Targeting for Therapeutic Gene Delivery, Edited by David T. Curiel and Joanne T. Douglas
ISBN 0-471-43479-5

used in a number of brain tumor protocols consisting of direct intratumoral injection of herpes simplex virus thymidine kinase (HSVtk)-expressing producer cells followed by systemic infusion of ganciclovir (Oldfield et al., 1993). This particular clinical protocol was the first to advance to a Phase III clinical trial upon demonstration of tumor regression in some patients with malignant glioblastoma.

Subsequently, cell-based cancer immunotherapy regimens gained popularity, utilizing cultured irradiated tumor cells expressing a variety of cytokine transgenes such as tumor necrosis factor (Rosenberg, 1991), granulocyte macrophage colony stimulating factor (Chang, 1994), and interleukin 2 (Gansbacher et al., 1992; Bowman et al., 1998), which were then injected subcutaneously or intradermally to induce an antitumor immune response. More recently, autologous T cells or dendritic cells expressing major histocomptability complex (MHC) Class II antigens (e.g., B7, CD86) or tumor-specific antigens have been evaluated in clinical trials as potential cell-based vaccine therapies (Junghans, 1998; Bergsland, 1998). However, in spite of the initial demonstrations of safety and a number of anecdotal reports of tumor response, there has been no clear demonstration of an actual cure derived from any of these gene transfer regimens. Although safe and improved retroviral packaging systems have been developed to generate complement-resistant and highly infectious viral particles (Miller and Rosman, 1989; Pensiero et al., 1996; Kotani et al., 1994; Soneoka et al., 1995; Yang et al., 1999), the major limitation of retroviral vectors appears to be a logistical one, principal of which is the lack of tissue specificity (Anderson, 1998; Verma and Somia, 1997), and a resulting inability to deliver sufficient numbers of vector particles to target cells in vivo under physiological conditions. Clearly, "improved gene delivery methods were needed in order to give human tests a better chance of success" (Langreth and More, 1999).

The Quest for a Targeted Injectable Vector for Gene Delivery

At the turn of the century, gene therapy remains poised at the threshold of modernizing medicine. While holding great promise for the treatment of numerous diseases, the field of gene therapy has been disappointingly slow in the development of safe and efficient gene delivery systems (Anderson, 1998). Numerous attempts to target specific cell types have focused on modifying the receptor binding domain (SU) of the ecotropic Moloney murine leukemia virus (MuLV) envelope (Env) protein (Kasahara, Dozy, and Kan, 1994; MacKrell et al., 1996; Somia et al., 1995; Peng and Russell, 1999). While the selective binding of such modified vectors to specific cell types has been demonstrated to some extent in vitro, the resulting transduction efficiencies have been insufficient to be extended in vivo, thus precluding their further progress into clinical trials (Zhao et al., 1999). In recent years, we developed a targeting strategy that now enables gene therapy vectors to target diseased tissues within the body. Based on solid physiological principles of hematology and wound healing, we have engineered a series of targeted injectable vectors that accumulate at points in which the normal tissue architecture is disturbed. When introduced into the bloodstream, the guidance system performs a vital surveillance function, seeking out areas of injury, inflammation, and/or metastatic cancer. In this chapter, we present the evolution of various retroviral vector targeting strategies in our laboratories, with particular emphasis on the clinical potential of this lesion-seeking injectable vector system in the treatment of cancer, vascular lesions, and proliferative ocular diseases.

VECTOR TARGETING STRATEGIES

Targeting Cell Entry by Molecular Engineering of the MuLV Envelope Protein

The prototypical design of targeted MuLV envelope proteins involves genetic manipulations that insert a specific ligand or targeting motif into an endogenous or genetically engineered site within the primary structure of the ecotropic MuLV Env protein (Fig. 14.1; MacKrell et al., 1996; Hall et al., 1997; Wu et al., 1998) or within the 4070A (CAE) amphotropic Env protein to produce a chimeric Env (Morgan et al., 1993; Martin et al., 1998; Hall et al., 2000). This modified Env would exhibit an altered phenotype (Hall et al., 1997; Wu et al., 1998), enhanced viral binding to specific cells (Liu et al., 2000), or an extended tropism of the resulting vector (Masood et al., 2001). In many cases, however, the molecular modification of the Env protein prevents the obligatory conformational change that occurs in response to receptor binding and is required for viral fusion and core entry (Zhao et al., 1999). Consequently, severe loss of viral infectivity often occurs (Benedict et al., 1999; Morling et al., 1997; Cosset and Russell, 1996).

To circumvent this problem, a dual envelope configuration may be employed in which a wild-type 4070A amphotropic Env (CAE) is coexpressed and incorporated into retroviral particles along with the chimeric targeting Env (Zhu et al., 2001) to enhance amphotropic infectivity. To further refine this design, and to eliminate steric hindrances resulting from the rather large inoperative binding domain of the modified Env congeners, we developed a series of targetable "escort" Env proteins in which the entire receptor binding domain of the CEE+ Env—from the BstE II site to one of several unique (Avr II, Stu I, Pst I, or NgoM I) sites engineered into the proline-rich hinge region (Fig. 14.1; Wu et al., 1998)—was eliminated and replaced by either a tumor vasculature targeting motif (Liu et al., 2000), an IgG-binding domain (Masood et al., 2001), or a collagen-binding peptide (Fig. 14.1B; Hall et al., 2000). When a wild-type CAE Env was coexpressed with an engineered Env construct bearing a specific targeting motif, the vectors arrayed in either the dual envelope configuration (Hall et al., 2000) or the CAE plus escort Env protein configuration (Hall et al., 2000; Liu et al., 2000; Masood et al., 2001) exhibited the desired gain-of-function phenotype (e.g., enhanced viral binding to activated endothelial cells

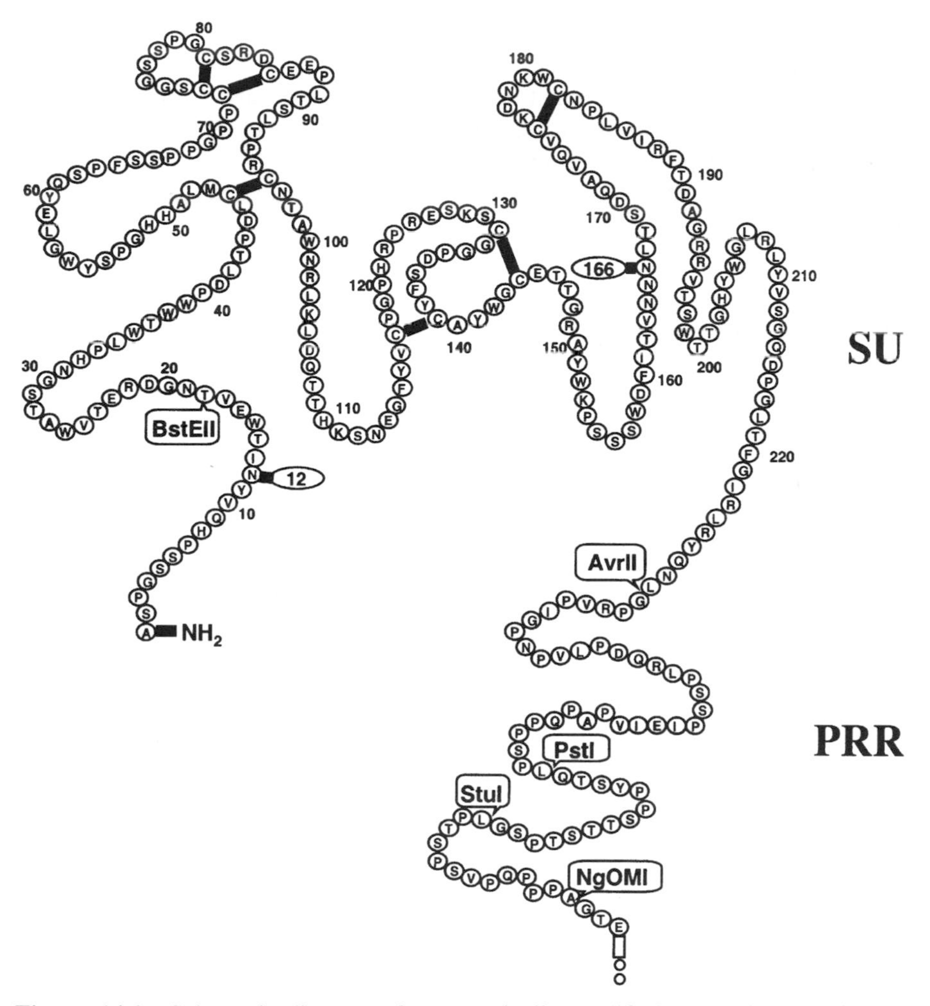

Figure 14.1 Schematic diagram of a strategically modified targeted ecotropic Env protein used to generate targeting constructs. The receptor-binding domain of the modified ecotropic Env protein illustrating a native BstE II cloning site (within the corresponding cDNA) between amino acids 18 and 19 of the mature surface (SU) protein, and a series of additional cloning sites engineered into the proline-rich (PRR) region of the CAE hinge which essentially replaces the ecotropic PRR region. Vectors displaying this chimeric CEE+/CAE Env protein exhibit ecotropic infectivity (Wu et al., 1998). The numbering and location of the N-linked glycosylation sites and disulfide bonds within the SU region are from MacKrell et al. (1996).

when compared to vectors bearing wild-type CEE or CAE Env), as well as high-titer amphotropic infectivity (Liu et al., 2000).

The utility of this approach is precedented in nature, as many types of wild-type viruses stably express dual envelope configurations, such as hemagglutinin or distinct attachment glycoproteins, in addition to membrane fusion proteins. Hence, these compound configurations may be utilized extensively in vector design to confer auxiliary targeting specificities. The escort Env protein configurations may be particularly suitable for insertion of large or bulky polypeptides, such as growth factors or molecular bridges (Masood et al., 2001) into retroviral vectors.

Ideally, a targeted Env protein could conceivably be engineered to exhibit both an integral gain-of-function and wild-type amphotropic infectivity within the context of a single Env polypeptide. One such design was recently achieved in the course of developing a lesion-targeted retroviral vector (Hall et al., 2000). The initial matrix-targeting constructs prepared as simple insertions into the ecotropic (CEE+) Env were remarkably successful, exhibiting both gain-of-function kinetics and ecotropic infectivity (Hall et al., 1997). However, initial attempts to translate the molecular engineering from the ecotropic (CEE+) to amphotropic (CAE) Env polypeptides were unsuccessful. In these experiments, we observed that direct transposition of the same targeting motif utilized in the ecotropic (CEE+) Env to a homologous site (BstE II) engineered by point mutation within the coding sequences of the amphotropic (CAE) Env severely compromised infectivity, presumably due to certain structural constraints that distinguish the amphotropic and ecotropic Env proteins. At this point, the assembly of targeted retroviral vectors bearing wild-type amphotropic Env and targeted escort proteins served as a useful initial technology (Hall et al., 2000; Liu et al., 2000).

Fortunately, nature often leaves clues for the molecular engineer. In the case of developing a targeted amphotropic Env, we noted a common feature between the amino (N) terminus of the amphotropic polypeptide and the collagen-binding domain (CBD) of von Willebrand factor (vWF). That is, the amino acid residues flanking the CBD of vWF form a Pro-His turn or kink (Pro-Arg in human vWF) just C-terminal to the minimal targeting domain, a structural peculiarity which is also present at amino acids 7 and 8 in the CAE Env. Hence, we engineered a structurally innocuous Pst I (Ala-Ala-Gly) cloning site into the coding sequences of the CAE Env at this position to generate CAEP (Fig. 14.2). CAEP now functions as a recipient construct into which various targeting motifs flanked by flexible, glycine-rich linkers could be inserted with minimal distortion of Env structure/function. As in the original targeted ecotropic Env design (Hall et al., 1997), a histidine residue was included in the N-terminal linker (Fig. 14.2; Hall et al., 2000) to promote an external conformation of the targeting peptide. The resulting constructs tolerated the insertion of a number of experimental targeting domains without compromising wild-type amphotropic infectivity (Hall et al., 2000, 2001).

The Classic Approach: Cell-Specific Targeting Strategies. Previous attempts to target retroviral vectors for cell-specific gene delivery include the insertion of auxiliary polypeptides, such as single-chain antibodies or peptide ligands, into retroviral envelope (Env) glycoproteins (Kasahara, Dozy, and

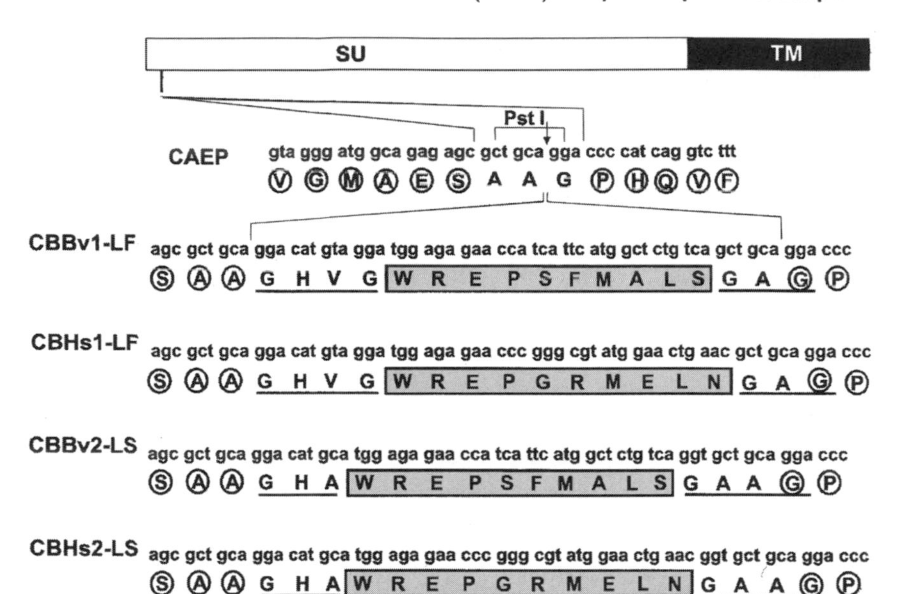

Figure 14.2 Design and engineering of the matrix-collagen-targeted amphotropic Env protein. Schematic diagram of the modified envelope structure and cloning strategy used to insert the WREPSFMALS (Bv) or the WREPGRMELN (Hs) collagen-binding decapeptides flanked by standard (LS) or special design (LF) linker residues into a unique Pst I site engineered within the N-terminal region of the CAE Env protein. The underlined residues correspond to oligonucleotides used to generate the respective cDNA inserts. The resultant vectors incorporating these Env congeners exhibit an integral gain-of-function while retaining wild-type amphotropic infectivity (Hall et al., 2000).

Kan, 1994; Cosset and Russell, 1996; Han et al.,1995). Alternatively, bifunctional cross-linkers that bridge the vector to a specific cell type have also been used (Yang et al., 1998). However, the performance of these targeted vectors has generally been impaired by loss of viral infectivity (Marin et al., 1996; Somia et al., 1995) due to postbinding block (Zhao et al., 1999), oligomerization (Morling et al., 1997), or sequestration into nonproductive pathways (Cosset et al., 1995).

More recently, an indirect immunologic approach, employing escort Env proteins, was evoked to target retroviral vectors in association with antibodies directed against endothelial cell-specific receptors (Masood et al., 2001). In this study, prokaryotic IgG-binding polypeptides were inserted by genetic engineering into the MuLV Env protein to target the vascular endothelial growth factor (VEGF) receptor, which is highly expressed in activated endothelial cells. Kaposi's sarcoma, KSY1, cells were used to model activated human endothelial cells that secrete high levels of VEGF, and express high levels of VEGF receptors, Flk-1/KDR and Flt-1. The modifications on the viral Env proteins included replacement of the entire receptor-binding region of the viral Env with either protein A or minimal (ZZ) IgG-binding domains. The truncated Env incorporating IgG-binding motifs (arrayed as escort proteins) provided the targeting function, while the coexpression of the wild type CAE Env protein enabled viral fusion and cell entry. An antihuman VEGF receptor (Flk-1/KDR) antibody served as a molecular bridge, directing the retroviral vector to the endothelial cell. Hence, the IgG-targeted vectors bound to the Flk-1/KDR antibody, which, in turn, bound to VEGF receptors expressed on Kaposi's sarcoma, KSY1, endothelial cells. The net effect of this configuration was increased viral fusion and infectivity of IgG-bound retroviral vectors when compared to nontargeted vectors bearing wild-type Env alone. Although IgG-binding polypeptides have been used to link monoclonal antibodies for targeting viruses to specific cell types (Ohno et al., 1997), these data provide the first proof of concept that IgG-binding vector/VEGF receptor antibody complexes may be used to enhance retroviral gene delivery to activated endothelial cells.

Another promising cell-targeting approach involves targeting adhesion receptors expressed on the surface of tumor-activated endothelial cells (Arap, Pasqualini, Ruoslahti, 1998). Selectively expressed adhesion receptors could also provide an advantageous locus for targeting gene therapy vectors to angiogenic tissues and/or tumor vasculature. Therefore, we engineered a series of Asn-Gly-Arg (NGR)-containing congeners of the presumptive cell-binding motif contained within the ninth Type III repeat of fibronectin, and displayed these tumor vasculature targeting motifs, TVTMs, within the context of MuLV Env escort proteins (Liu et al., 2000). Comparative studies of envelope incorporation into viral particles and evaluation of the cell-binding properties of the targeted vectors identified a subset of optimal TVTM congeners. Vectors displaying TVTM motifs as escort proteins significantly enhanced the transduction efficiency of both human KSY-1 and normal human umbilical vein endothelial (HUVE) cell cultures (Liu et al., 2000). To simplify the endothelial cell-targeted Env from a dual to a single Env configuration, selected TVTM inserts were cloned into the modified Env construct (CAEP) engineered to contain a unique Pst I restriction site near its N terminus (Hall et al., 2000). The resultant virions displaying TVTMs retained near wild-type infectivity yet significantly enhanced transduction efficiency in activated microvascular endothelial (SVR) cell cultures when compared to virions bearing the wild type (CAE) Env (Hall et al., 2001).

The Paradigm Shift: Targeting the Area of Pathology. Taken as a whole, the results of our previous work on vector targeting (Hall et al., 1997; Wu et al., 1998) support the general concept that strategically modifying retroviral envelopes to generate tissue or cell-specific targeted viral vectors is both feasible and advantageous (Anderson, 1995; Weiss and Tailor, 1995). However, in contrast to more standard approaches to engender cell-specific targeting by the display of polypeptide ligands (Kasahara, Dozy, and Kan, 1994; Valsesia-Wittman et al., 1994) or single-chain antibodies (Russell, 1993; Cosset et al., 1995; Somia, Zoppe, and Verma, 1995) on viral envelopes, we undertook the seemingly counterintuitive strategy of targeting the extracellular matrix (ECM) itself (Hall et al., 1997). We took advantage of the physiologic surveillance function present in a discrete collagen-binding domain of vWF and incorporated these coding sequences into the primary structure of the MuLV 4070A amphotropic Env. The resultant matrix-targeted vector exhibited a lesion-seeking feature, i.e., the ability to accumulate at sites of exposed collagen within the lesions created by growing tumors or vascular neointima. The hypothesis is that by binding the vector to the exposed collagen, the rapidly proliferating cancer cells or vascular neointimal cells in proximity will be more effectively transduced (Hall et al., 2000).

The vWF is a coagulation factor found in normal plasma that is synthesized by both platelets and endothelial cells (Ginsburg et al., 1989). The vWF performs a vital surveillance function by mediating platelet adhesion to sites of vascular injury (Ruggeri and Zimmerman, 1987). The recruitment of platelets into damaged vessels forms the initial hemostatic plug, which initiates a cascade of biochemical events that lead to the coagulation of blood and cessation of bleeding. Within the vWF molecule are distinct collagen-binding domains (Takaji et al., 1991; Tuan et al., 1996), which, together with glycoprotein Ib and glycoprotein IIb IIIa, promote the platelet-vessel wall interaction.

The relevant structural domains of vWF include a high-affinity collagen-binding domain (CBD) within the D2 domain of the mature polypeptide. Previous studies demonstrated that transposition of vWF-derived sequences into the ecotropic (CEE+) retroviral Env protein directs retroviral recruitment and accumulation at sites of exposed collagen in an animal model of vascular injury, enhancing both the transduction of reactive smooth muscle cells (Hall et al., 1997; Anderson, 1998) and the efficacy of gene therapy for vascular restenosis in rats (Gordon et al., 2001b; Xu et al., 2001). Recognizing that this ECM-targeting technology could have extensive clinical applications (Hollon, 2001), we generated a series of matrix-targeted amphotropic retroviral vectors for transduction of human cells.

Figure 14.2 shows diagrammatically the receptor-binding domain of the 4070A Env surface (SU) protein into which the collagen-binding polypeptides were inserted, by molecular cloning, into an engineered Pst I site (encoding Ala-Ala-Gly) between amino acids 7 and 8 of the mature CAE protein (Hall et al., 2000). Tandem synthetic oligonucleotides annealed into duplexes (as shown) were used to generate the respective cDNA inserts. Sequences encoding bovine, WREPSFMALS, (Bv), or human, WREPGRMELN, (Hs), vWF-derived collagen-binding domains were inserted into the Pst I cloning site. The inserts were flanked by either standard linkers (LS) or linkers modified to closely approximate native vWF (LF) flanking sequences. The resulting Env proteins were readily expressed in human 293T producer cells, were correctly processed, and were found to be incorporated equally to the wild-type CAE Env into retroviral particles. Each of the chimeric vectors bearing a collagen-binding motif exhibited significant collagen-binding affinity when compared to vectors bearing wild-type CEE or CAE Env (Hall et al., 2000).

In a rat carotid injury model of vascular restenosis, both intraarterial (by retrograde femoral artery catheterization) and intravenous (via femoral vein) injection of the matrix-targeted, but not the nontargeted, vector resulted in enhanced transduction of neointimal cells while sparing the normal noninjured contralateral artery (Fig. 14.3; Gordon et al., 2001). In a nude mouse model of liver metastasis, repeated portal vein injections of a matrix-targeted retroviral vector bearing a β-galactosidase gene resulted in enhanced transduction of metastatic tumor cells ($\geq 50\%$), in areas of active angiogenesis and notably, at points of overt tumor cell exit from the circulation (Fig. 14.4; Gordon et al., 2000).

In a human cancer xenograft in mice, a high level of transduction of tumor cells (~35%) was observed throughout the tumor nodules in animals treated

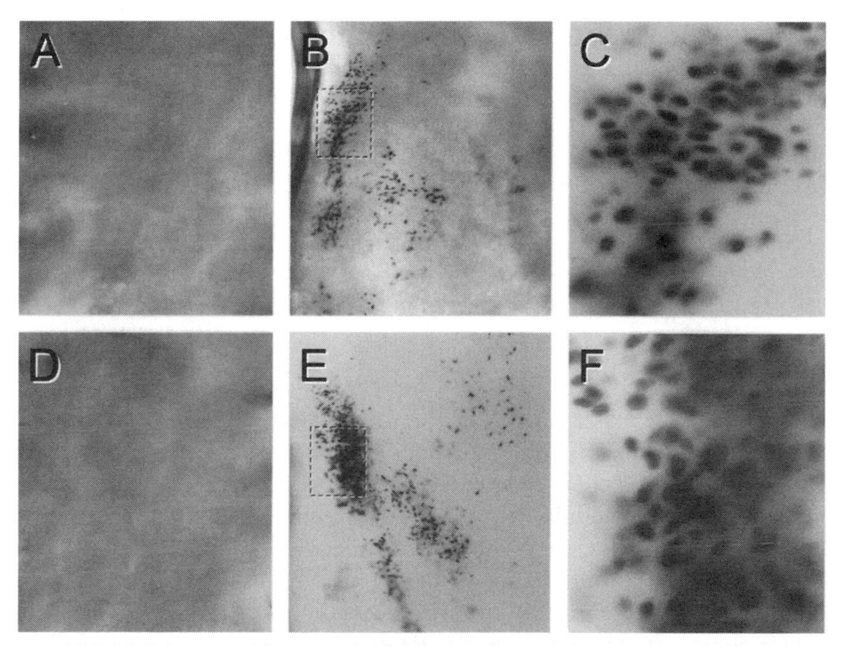

Figure 14.3 Lesion-specific transduction of balloon-injured carotid arteries in vivo using an IV administered matrix-targeted vector bearing a β-galactosidase gene. (*A*, *D*: ×400) X-gal staining for the presence of the β-galactosidase transgene shows no evidence of transduction in contralateral noninjured arterial segments of matrix-targeted vector-treated rats; (*B*, *E*: ×100; *C*, *F*: ×400). In contrast, X-gal staining demonstrates appreciable transduction of arterial segments obtained from matrix-targeted vector-treated rats.

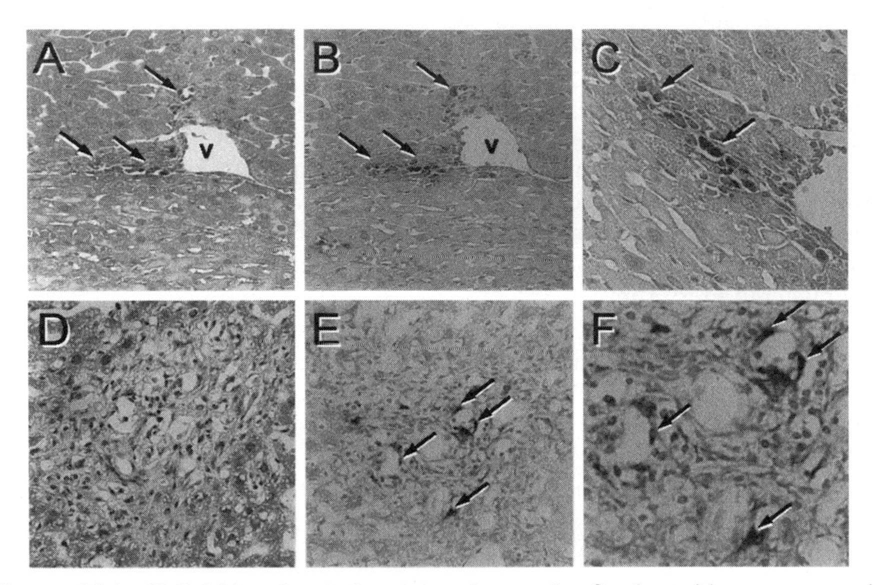

Figure 14.4 X-Gal histochemical staining detects the β-galactosidase transgene in metastatic tumor nodules. (A : ×40) Hematoxylin and eosin (H & E)-stained tissue sections of liver with metastatic nodule from a matrix-targeted β-galactosidase vector-treated mouse (B : ×40) X-gal-stained tissue section of A, counterstained with nuclear fast red stain; (C : ×400) higher magnification of B. Note: β-galactosidase-expressing tumor cells (cells with blue-staining nuclei) near a disrupted hepatic venule (v) are indicated by arrows. (D : ×40) H and E-stained tissue section of liver with metastatic nodule from a matrix-targeted β-galactosidase vector-treated mouse showing active angiogenesis within the tumor nodule; (E : ×40) X-gal-stained liver sections counterstained with nuclear fast red stain; (F : ×200) higher magnification of E. Note: β-galactosidase-expressing tumor stromal and endothelial cells are indicated by arrows.

with a matrix-targeted, but not the nontargeted, vector (Fig. 14.5; Gordon et al., 2001). Vector localization studies demonstrated the presence of immunoreactive vector particles within the tumor nodule at $t = 1$ hr after intravenous (IV) vector infusion, a phenomenon which was pronounced in angiogenic areas throughout the tumor nodule (Fig. 14.6B-C), while no immunoreactive vector particles were detected at $t = 24$ hr (Fig. 14.6A; Gordon et al., 2001a), indicating that viral entry into the tumor cells and/or physiologic clearance of extracellular virions had occurred. Further, biodistribution studies revealed weakly positive polymerase chain reactor (PCR) signals (<50 copies/μg DNA) in the liver and spleen, but not in testes, lung, brain, heart, kidney, and colon of animals treated with the matrix-targeted vector at a cumulative vector dose of 1.6×10^8 cfu per mouse (unpublished data). No histologic abnormality attributable to the vector was noted in nontarget organs. These findings indicate that the IV-infused matrix-targeted vector localized rapidly and preferentially in cancerous lesions as well as in the expected organs involved in viral clear-

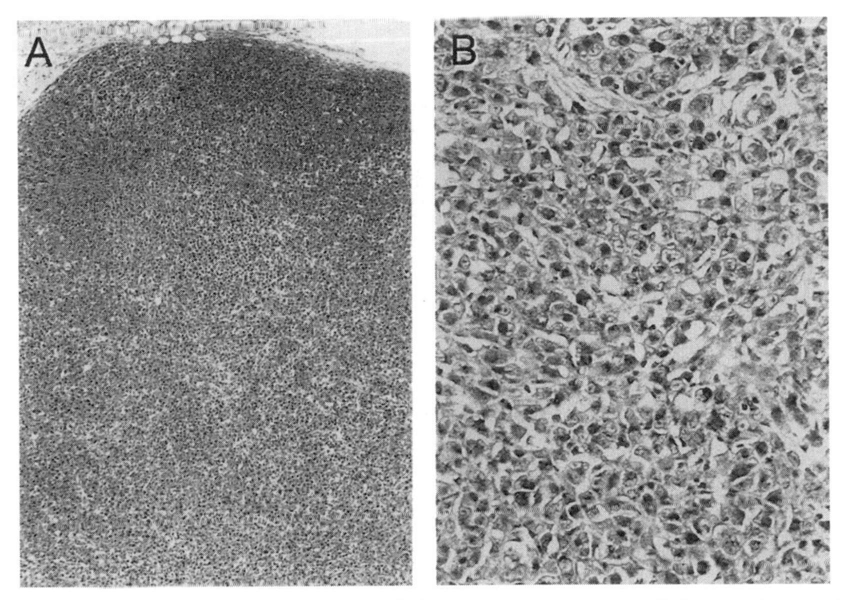

Figure 14.5 Immunohistochemical staining for the presence of the β-galactosidase transgene in a tumor nodule obtain from a matrix-targeted vector-treated animal. Tissue samples were harvested from tumor-bearing mice one day after the completion of a 10-day treatment cycle consisting of daily IV (tail vein) infusions of a matrix-targeted vector bearing a β-galactosidase gene. (*A*, ×40; *B*, ×200) A tumor nodule from a Mx-nBg vector-treated animal. β-galactosidase-expressing tumor cells and tumor endothelial cells are shown as reddish-brown nuclear stained cells, counterstained with methyl green.

ance (i.e., the reticuloendothelial system). Taken together, these data provide the first demonstration of site-specific vector localization and preferential transduction of tumor cells by intravenously administered matrix-targeted retroviral vectors. The in vivo studies using systemically administered matrix-targeted retrovial vectors are summarized in Table 14.1.

Targeting Genetic Elements Involved in Cell Cycle Control

The high degree of complexity and redundancy in growth factor signaling pathways has prompted the examination of conserved cell cycle control pathways as experimental targets in the design of novel cytostatic therapies (Braun-Dullaeus et al., 1998). While the retrovirus offers several advantages in terms of safety and efficacy by virtue of its innate ability to transduce proliferative cells (Gordon and Anderson, 1994), the overall effectiveness of antiproliferative gene therapy may be further enhanced by genetic targeting of convergent cell cycle control pathways (Gibbons and Dzau, 1996). Consequently, a number of novel gene therapy approaches to inhibit uncontrolled cell proliferation

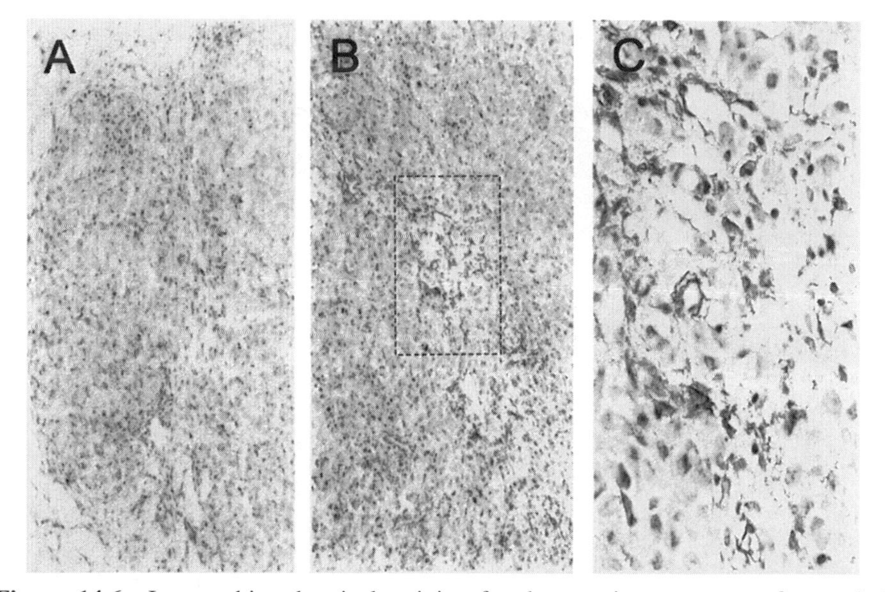

Figure 14.6 Immunohistochemical staining for the transient presence of retroviral particles (i.e., MuLV Env gp70 protein) in tumors of a matrix-targeted vector-treated mouse. (*A:* ×40) Staining for MuLV gp70 Env protein 24 h after IV injection of a matrix-targeted vector revealed no evidence for persistence of vector within the tumor nodule. (*B:* ×40) In contrast, positive staining for the Env protein was observed 1 h after IV infusion of a matrix-targeted vector. Using a monoclonal antirat antibody (83A25) directed against the MuLV gp70 Env protein followed by immunohistochemical staining, the presence of reddish-brown staining material indicates the accumulation of retroviral vector particles in areas of angiogenesis well within the tumor nodule (e.g., boxed area); (*C:* ×200) High magnification of the region boxed in B shows dispersion of the vector from the tumor vasculature.

have focused on specific cell cycle control elements, including oligodeoxynucleotides representing antisense constructs of cyclin-dependent protein kinase (CDK) subunits (Morishita et al., 1993), adenoviral vectors bearing CDK inhibitors (Chang et al., 1995), or vectors bearing constitutively active forms of the retinoblastoma (Rb) protein (Smith et al., 1997). Other studies have employed molecular decoy oligodeoxynucleotide strategies directed against the transcription factor E2F (Morishita et al., 1995), which regulates the induction of multiple cell cycle control genes. The reported efficacy of these experimental approaches supports the concept that cell cycle control elements that are selectively up-regulated in proliferative disorders would represent strategic therapeutic targets.

Recent studies in our laboratories have shown high level expression of the cyclin G1 protein, an inducible cell cycle control element (Wu et al., 1994), in cancerous lesions (Fig. 14.7) following balloon catheter injury in rodents (Fig. 14.8) and in nonhuman primates (Wu et al., 2000). Enforced expression

TABLE 14.1 In vivo Studies Using Systemically Administered Matrix-Targeted Retroviral Vectors

Author	Vector-Targeting Characteristics	Animal Model	Nature of Study	Route of Delivery
Hall et al., 1997	Insertion of collagen-binding motif in BstE II site of ecotropic (CEE+) *env*	Rat and rabbit carotid injury model of vascular restenosis	Vector localization and transduction	Common carotid artery installation with ligation
Hall et al., 2000	Insertion of collagen-binding motif in engineered Pst I site of 4070A amphotropic *env*	Rat carotid injury model of vascular restenosis; nude mouse model of liver metastasis	Vector transduction	Common carotid artery instillation with ligation; portal vein infusion
Gordon et al., 2000	Insertion of collagen-binding motif in engineered Pst I site of 4070A amphotropic *env*	Nude mouse model of liver metastasis	Vector transduction and efficacy	Portal vein infusion
Gordon et al., 2001	Insertion of collagen-binding motif in engineered Pst I site of 4070A amphotropic *env*	Subcutaneous tumor model in nude mouse	Vector localization, transduction, and efficacy	Tail vein infusion
Xu et al., 2001	Insertion of collagen-binding motif in engineered Pst I site of 4070A amphotropic *env*	Rat carotid injury model of vascular restenosis	Vector efficacy	Common carotid artery instillation with ligation
Gordon et al., 2001	Insertion of collagen-binding motif in engineered Pst I site of 4070A amphotropic *env*	Rat carotid injury model of vascular restenosis	Vector transduction and efficacy	Common carotid artery, femoral artery, and femoral vein infusion
Gordon et al., 2001	Insertion of collagen-binding motif in engineered Pst I site of 4070A amphotropic *env*	Liver metastasis and subcutaneous tumor models in nude mice	Biodistribution and vector toxicity	Portal vein and tail vein infusion

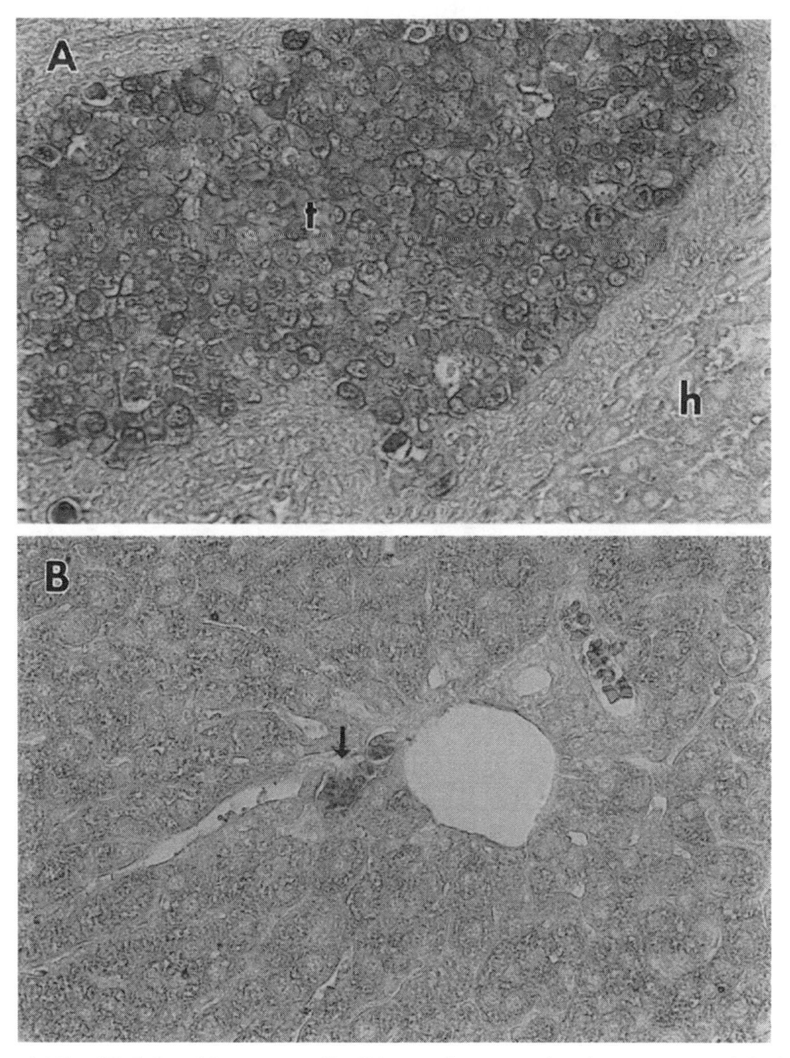

Figure 14.7 High-level human cyclin G1 protein expression in a tumor nodule from a nude mouse with experimental liver metastasis. (*A:* ×1000) Intense nuclear immunoreactivity for the human cyclin G1 protein (reddish-brown staining material) is observed within a large metastatic nodule in histologic sections of liver from a control vector-treated mouse; (*B:* ×1000) Intense nuclear immunoreactivity for the human cyclin G1 protein detects a small tumor nodule in a section of liver obtained from a matrix-targeted dnG1 vector-treated mouse.

of cyclin G1 in transfected cells in vitro accelerates the cell cycle and promotes clonal expansion (Smith et al., 1997), while blockade of cyclin G1 expression by antisense strategies induces cytostasis and cytolysis (Skotzko et al., 1995). Therefore, cyclin G1 appears to be a relevant and strategic target for thera-

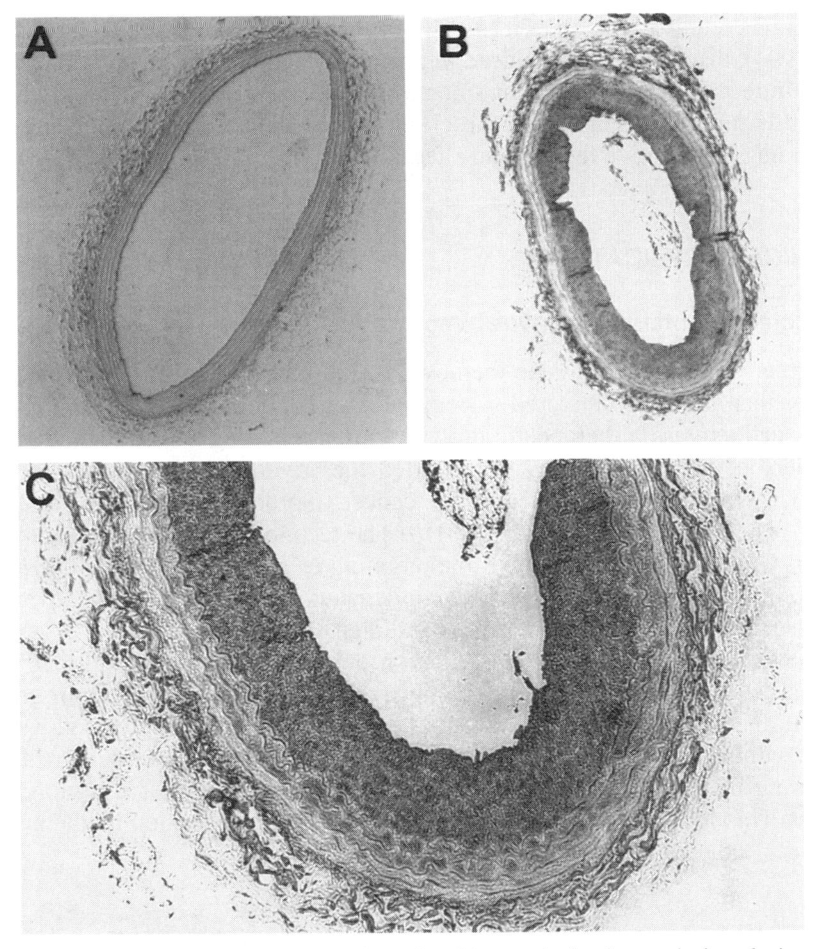

Figure 14.8 Enhanced expression of cyclin G1 protein in the neointima lesion of a balloon-injured rat artery. (*A:* ×40) A normal artery with negative staining for cyclin G1 protein; (*B:* ×40) In contrast, intense immunostaining for cyclin G1 protein is observed within the neointima lesion of a balloon-injured rat artery; (*C:* ×500) High magnification of *B* reveals intense nuclear immmunoreactivity for the cyclin G1 protein in activated smooth muscle cells in the medial layer, as well as in proliferative neointimal cells.

peutic intervention in cancer and other disorders involving uncontrolled cell proliferation.

To improve the therapeutic efficacy of the cyclin G1 knock-out constructs, a number of antisense cyclin G1 constructs were prepared in addition to a number of modified cyclin G1 expression constructs designed to function as potential dominant negative mutants (Xu et al., 2001). Among the cyclin G1 constructs tested, one particular mutant construct, a 630 bp fragment of cyclin G1 cDNA encoding amino acids 41 to 249 of the mature cyclin G1 protein (Wu et al., 1994), displayed a pronounced inhibitory effect on the growth and viability of cultured cancer and vascular smooth muscle cells (Xu et al., 2001). Therefore, the mutant cyclin G1 construct was selected for in vivo animal studies, providing an effective therapeutic alternative to previous antisense approaches.

CLINICAL APPLICATIONS

Targeted Injectable Retroviral Vectors for Somatic Gene Therapy

From its very inception, gene therapy has long held the ideal of an injectable vector that could be delivered systemically, target cells, transfer a therapeutic gene, and express a therapeutic protein without eliciting systemic side effects (Anderson, 1995). Recently, we reported the advent of a targeted injectable retroviral vector for gene therapy for cancer (Gordon et al., 2001b) and vascular restenosis (Gordon et al., 2001b). The technology employs a promising alternative approach that enables the targeting of retroviral vectors to *areas of pathology*, wherein connective tissue collagen is exposed, and takes advantage of tissue properties unique to tumors and their sites of metastases. The process of metastasis as well as tumor formation exposes collagen. Additionally, the formation of new blood vessels—angiogenesis—promotes rapid tumor growth and further susceptibility to retroviral transduction. The new blood vessels deep within the tumor are incompletely lined by normal endothelium, hence exposing the underlying connective tissue collagen to circulating blood elements. This feature makes solid tumors accessible to penetration by the matrix-collagen-targeted retroviral vectors.

Lesion-Targeted Injectable Retroviral Vectors for Cancer Gene Therapy

Two efficacy and safety studies using this matrix-targeted retroviral vector form the basis of an investigational new drug (IND) application that was submitted to the Food and Drug Administration in November, 2000. In one study, we assessed the antitumor effects of serial portal vein infusions of matrix-targeted vectors bearing a mutant cyclin G1 (dnG1) construct in a nude mouse model of liver metastasis. Here, a dramatic reduction in the size of tumor foci was observed in dnG1 vector-treated mice compared to those of control vector-treated animals (Fig. 14.9; Gordon et al., 2000). In another study, we demonstrated that intravenous infusion of a matrix-targeted retroviral vector distant to the tumor site enhanced gene delivery and efficacy of a cytocidal mutant cyclin G1 construct in human cancer xenografts in nude mice when compared to control vectors. In this case, the circulating vector necessarily transited the heart, passed through the microvasculature of the lungs, and returned to the heart before being distributed throughout the systemic circulation. It is important to

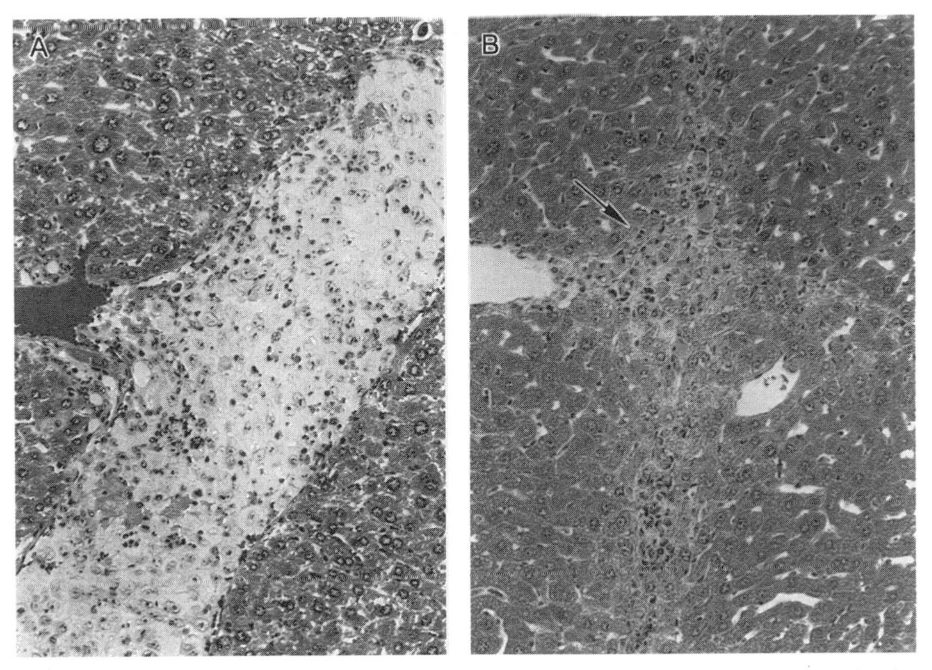

Figure 14.9 Reduction of tumor burden in the liver following repeated portal vein infusions of a matrix-targeted dnG1 vector in a nude mouse model of liver metastasis. (*A:* ×1000) H and E-stained tissue sections of liver from a control PBS-treated mouse reveals a large tumor nodule consisting of a heterogenous population of tumor cells, stromal cells, and endothelial cells; (*B:* ×1000) In contrast, H and E-stained section of liver obtained from a matrix-targeted dnG1 vector-treated mouse reveals a small tumor nodule consisting mainly of apoptotic tumor cells and hemosiderin-laden macrophages suggesting active clearance of degenerating tumor cells and tumor debris by the hepatic reticuloendothelial system.

note that the transduction efficiency and therapeutic efficacy observed under these formidable physiologic conditions is both unprecedented and important. Upon repeated IV injections of this targeted injectable vector, long-term efficacy studies showed sustained inhibition of tumor growth and/or complete tumor regression of animals treated with a matrix-targeted vector bearing a mutant cyclin G1 construct compared to those treated with phosphate-buffered saline (PBS) placebo, and a nontargeted vector (Figs. 14.10 and 14.11; Gordon et al., 2001a). The Kaplan Meier survival curve (representing the time to tumor quadrupling as the endpoint) of the animals is shown in Fig. 14.10. In a 6-week follow-up period, the tumors quadrupled in all placebo-treated mice, and in three of four mice treated with the nontargeted CAE-dnG1 vector. In contrast, none of the tumors quadrupled in mice treated with the Mx-dnG1 vector ($p = 0.004$; Tarone test for trend). In a 100-day follow-up period (Fig. 14.11), complete tumor resolution was achieved in 50% of Mx-dnG1 vector-treated

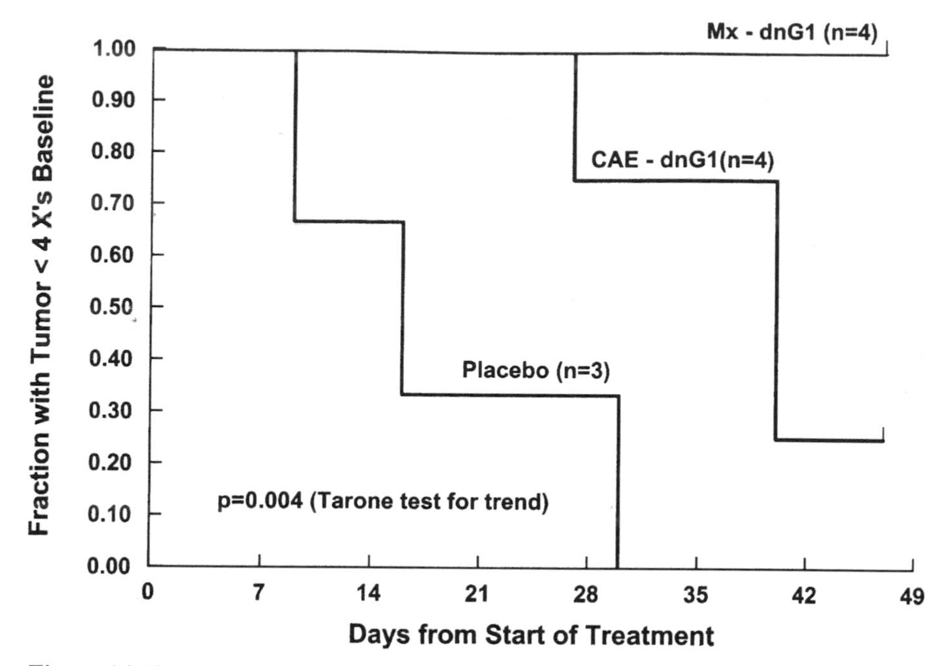

Figure 14.10 Kaplan-Meier survival studies demonstrate therapeutic efficacy in mice treated with matrix-targeted-dnG1 vector. The Kaplan-Meier survival curve (representing the time to tumor quadrupling as the endpoint) of animals is shown. The PBS placebo group, the nontargeted CAE-dnG1 group and the matrix-targeted Mx-dnG1 group are shown. The fraction surviving (representing animals with tumors that have not quadrupled; plotted on the vertical axis) is expressed as a function of time (days; plotted on the horizontal axis).

mice while tumor growth was inhibited in 50%. In contrast, comparative studies conducted with the PBS placebo-treated mice showed a rapid increase in tumor volumes to greater than five times their basal volumes within 6 weeks. One mouse was observed along with the Mx-dnG1 vector-treated mice until its tumor volume reached 1500 mm^3 after which it was euthanized for ethical reasons (Fig. 14.11).

A Clinical Protocol for Metastatic Cancer

A clinical protocol entitled "Tumor Site Specific Phase I Study Evaluating the Safety and Efficacy of Hepatic Arterial Infusion of a Matrix-targeted Retroviral Vector Bearing a Mutant Dominant Negative Cyclin G1 Construct (Mx-dnG1) as Treatment for Metastatic Colorectal Carcinoma" (Principal Investigator: Heinz-Josef Lenz, Co-Investigators: Robert W. Beart, Jr., W. French Anderson, Frederick L. Hall, and Erlinda M. Gordon) was presented to Recombinant DNA Advisory Committee (RAC) of the NIH for public discussion on

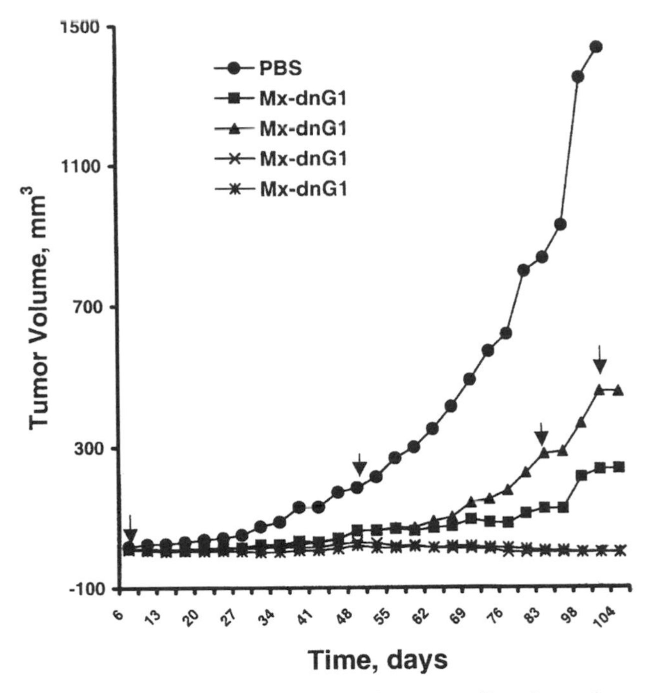

Figure 14.11 One hundred day long-term efficacy studies showed complete tumor regression ($n = 2$) or inhibition of tumor growth ($n = 2$) in matrix-targeted dnG1 vector-treated mice. Each mouse was treated with two 10–day treatment cycles (arrows). Each treatment cycle consisted of daily IV (tail vein) infusions of the matrix-targeted dnG1 retroviral vector. Two mice with partial responses were treated with two more cycles of Mx-dnG1 (arrows). Tumor volume (mm3), plotted on the vertical axis, is expressed as a function of time (days), which is plotted on the horizontal axis. Comparative studies were conducted with PBS placebo-treated control mice ($n = 4$) which were euthanized after their tumor volumes had increased to five times the baseline volume, except for one mouse which was observed along with the matrix-targeted dnG1 vector-treated mice until its tumor volume reached 1500 mm^3, after which it was euthanized.

December 13, 2000. The objectives of the study are: (1) to evaluate the dose-limiting toxicity (DLT) and maximum tolerated dose (MTD) of the Mx-dnG1 retroviral vector administered as hepatic arterial infusion; (2) to evaluate the pharmacodynamics of the Mx-dnG1 retroviral vector administered as hepatic arterial infusion; (3) to identify any objective tumor response to the Mx-dnG1 retroviral vector administered as hepatic artery infusion; and (4) to obtain preliminary data on molecular markers of tumor response. The population to be studied include male and female patients, >18 years old, with metastatic colorectal carcinoma. Nine to 15 patients (3 to 5 patients treated at each of four dose levels) will be evaluated. The gene transfer protocol involves hepatic arterial infusion via an IsoMed pump (Medtronics) of the Mx-dnG1 vector over 3 h daily, for a total of 5 days. The primary endpoint is clinical toxicity as defined by patient performance status, toxicity assessment score, hematologic, liver, and coagulation profile. The secondary endpoint is decrease in tumor size as detected by abdominal CT scan 3 weeks after completion of one treatment cycle.

Eight RAC members recommended that the protocol warranted public discussion. Dr. Stephen Russell, director of the Molecular Medicine Program at the Mayo Clinic Foundation, Rochester, Minnesota, served as ad hoc reviewer. Dr. Russell described the protocol as "a truly ground-breaking and elegant protocol." He stated that "if approved, it will be the first clinical gene therapy protocol employing vascular delivery of a retroviral vector. It will also be the first clinical use of a retroviral vector with a targeting modification in its envelope glycoprotein. It will be the first use of a retroviral vector produced by three plasmid co-transfection as opposed to the use of stable packaging cell lines. Finally, it will be the first clinical use of the dominant negative cyclin G1 therapeutic gene." The recommendations of the RAC are as follows:

- Characterization of the vector product should include determination of levels of trace contaminants such as SV40 T antigen DNA and protein, cellular DNA and debris, bovine serum, and characterization of the physicochemical properties of the vector to detect aggregation that may affect its biodistribution. The assay for detection of replication-competent retroviruses (RCR) should be defined as to the level of sensitivity. Passage of vector stocks on human cells would allow amplification of RCR as it is present and assaying on human cells as well as Mus dunni cells would allow maximal sensitivity.
- A large animal model (e.g., porcine or canine) should be employed for the analysis of any acute toxicities that may be associated with hepatic arterial administration of large volumes of the vector-containing supernatant that will be used in the clinical protocol.
- In regard to oversight, the establishment of a separate data and safety monitoring board would ensure that the trial received appropriate attention.
- To improve it comprehensibility, the informed consent document should be reorganized. In addition, potentially misleading terms such as "gene therapy" and "treatment" shoud be replaced with more neutral terms such as "gene transfer" and "infusion" or "intervention." A reference to a request for autopsy should also be included. Testing of sperm for retrovirus integration is an FDA requirement for retroviral vector protocols and should be included in the protocol and informed consent document.

Other Clinical Applications

Lesion-Targeted Injectable Retroviral Vectors for Vascular Restenosis. The clinical applications of this targeted injectable vector extend beyond cancer. For cardiovascular disorders, the pathologic lesions caused by catheter-based revascularization procedures for occlusive artery disease include disruption of the endothelium, exposure of extracellular matrix (ECM) proteins, and proliferation of vascular smooth muscle cells, which lead to neointima formation and restenosis (Schwartz, Hombs, and Topol, 1992; Meininger and Davis, 1992; Casscells, Engler, and Willerson, 1994; DeMeyer and Bult, 1997). The exposure of collagen at the site of balloon angioplasty/stent injury would promote site-specific accumulation of the matrix-targeted vector and enhancement of local effective vector concentration.

In support of this concept, both intraarterial and intravenous injections of a matrix-targeted vector in rats subjected to carotid balloon injury resulted in successful transduction of the injured artery without discernible transduction of the noninjured contralateral artery (Gordon et al., 2001b). Further, matrix targeting enabled the transfer of a cytocidal gene into balloon-injured rat arteries at levels that were sufficiently high to have impact on the neointima lesion. The comparative degrees of vascular neointima formation in animals that were treated with a PBS control, a targeted null vector (Mxnull), or a nontargeted antisense cyclin G1 vector (CAEaG1) were not significantly different among these groups (Table 14.2). In contrast, the matrix-targeted antisense cyclin G1 vector (MxaG1) induced a statistically significant inhibition of neointima formation when compared to the control groups and the nontargeted vector-treated group. Notably, the antiproliferative effect of antisense cyclin G1 was achieved at a lower vector dose than was used previously with a nontargeted antisense cyclin vector (Zhu et al., 1997). Further, treatment with the MxaG1 was not associated with significant thrombogenicity, inflammatory reaction, or necrosis of the arterial wall. These observations provide further evidence to support the potential safety and efficacy of this lesion-targeted vector system.

Lesion-Targeted Retroviral Vectors for Postexcimer Laser Corneal Haze. The use of photorefractive keratectomy (PRK) using excimer laser for correction of myopia, astigmatism, and hyperopia has gained increasing popularity in the ophthalmology community (Seiler and McDonnell, 1995; McDonnell, P. J., 2000). However, the postoperative phenomenon of partial loss of corneal transparency, known as corneal haze, limits the success of this procedure. Numerous studies indicate that corneal haze is mainly due to the uncontrolled proliferation of corneal keratocytes, which (1) produce extracellular matrix proteins and (2) migrate to the anterior stromal compartment of the cornea (Gaster et al., 1989; Fitzsimmons et al., 1992). Currently, there is no effective treatment to prevent these inappropriate wound healing responses. Therefore, experimental gene therapy emerges as a promising alternative.

Recently, we evaluated the efficacy and safety of topical eye drop applications of a matrix-targeted mutant cyclin G1 (Mx-dnG1) retroviral vector as preventive treatment of laser-induced corneal haze in rabbits (Behrens et al., 2001). Initial marking studies showed that the matrix-targeted vector induced a fourfold increase in transduction efficiency when compared to the nontargeted vector. The development of corneal haze in the rabbit model 2 weeks after laser

TABLE 14.2 Two Week Efficacy Studies Using Matrix-Targeted Versus Nontargeted Antisense Cyclin G1 Retroviral Vectors

Vector Name	Intima: Media Ratio	Percentage Inhibition (Significance Level)		
PBS Control $n = 7$	1.24 ± 0.12			
Mxnull $n = 7$	1.40 ± 0.14	vs. PBS	0	($p = 0.38$)
CAEaG1 $n = 6$	1.48 ± 0.19	vs. PBS	0	($p = 0.29$)
		vs. Mxnull	0	($p = 0.85$)
MxaG1 $n = 6$	0.85 ± 0.03	vs. PBS	31	($p = 0.02$)
		vs. Mxnull	39	($p = 0.03$)
		vs. CAEaG1	43	($p = 0.02$)

In vivo efficacy studies in a rat carotid injury model of vascular restenosis. Under general anesthesia (ketamine, 10 mg/kg; rompun, 5 mg/kg), in accordance with a protocol approved by the USC Institution Animal Care and Use Committee (IACUC, University of Southern California Keck School of Medicine, Los Angeles, CA, a 2F Intimax arterial embolectomy catheter (Applied Medical Resources Corp., Laguna Hills, CA, USA) was used to denude the carotid artery endothelium of Wistar rats (weights ranging from 375–425 gms) as previously described (Zhu et al., 1997). The catheter was inserted into the left external carotid artery which was ligated distally, and passed into the left common carotid artery (LCCA) which was transiently ligated proximally. The baloon was inflated to a volume equivalent to 7F on a French scale card, and passed 3 times along the length of the LCCA. In a two-week efficacy study, the internal carotid artery was transiently ligated just distal to the bifurcation, immediately after balloon injury. Then, 1×10^6 cfu of either the Mxnull, the MxaG1, the CAEaG1 or an equal volume of phosphate buffered saline (PBS) was instilled into the LCCA and held in place for 20 minutes. These animals were sacrificed 14 days after balloon injury and vector instillation. The intima to media ratio of arterial sections showing the greatest neointima was determined using an Optimas imaging analysis system (Optimas Corporation, Bothell, Washington, USA). The significance of differences among groups was evaluated by ANOVA. The tabulated results are expressed as arithmetic mean ± standard deviation.

surgery was inhibited by topical eye drop applications of the Mx-dnG1 vector and corneal ulcerations were not detected in the Mx-dnG1 vector-treated eyes. Moreover, biodistribution studies showed no evidence of vector dissemination in neighboring and distant nontarget organs. Based on these studies of safety and efficacy, a Phase I/II clinical trial is being planned as adjunctive treatment for superficial corneal opacity wherein ophthalmic instillation of the Mx-dnG1 retroviral vector will be given to patients after excimer laser phototherapeutic keratectomy.

CONCLUSIONS

In summary, we have developed a lesion-targeted vector system that unifies three targeting principles, namely: (1) an innate property of the retrovirus to integrate only in actively dividing cells, while sparing normal nondividing cells; (2) targeting the area of pathology for site-specific gene delivery and preferential transduction of tumors and vascular lesions; and (3) targeting genetic elements involved in cell cycle control for the treatment of cancer and other proliferative disorders. The initial clinical trial using this lesion-seeking vector involves patients with colorectal cancer metastatic to liver. Future studies could extend the utility of this targeted vector system for development of ancillary strategies, including the delivery of antiangiogenic constructs and/or immunomodulatory cytokine genes directly to cancerous lesions. The clinical applications of this enabling technology may extend to any disease wherein disruption of the vascular endothelium, ECM remodeling, and collagen deposition form the nexus for vector localization and concentration in vivo.

ACKNOWLEDGMENTS

Supported by grants from the American Heart Association Great Western States awarded to F.L.H. (1157-GI1) and E.M.G. (1156-GI1), and the Whittier Family Foundation, Pasadena California.

REFERENCES

Anderson WF (1995): Gene therapy. *Sci Am* 273:124–128.

Anderson WF (1998): Human gene therapy. *Nature* 392 (Suppl):25–30.

Arap W, Pasqualini R, Ruoslahti E (1998): Cancer treatment by targeted drug delivery to tumor vasculature in a mouse model. *Science* 279:377–380.

Benedict CA, Tun RY, Rubinstein DB, Guillaume T, Cannon PM, Anderson WF (1999): Targeting retroviral vectors to CD34-expressing cells: binding to CD34 does not catalyze virus-cell fusion. *Hum Gene Ther* 10:545–557.

Bergsland EK (1998): A Phase I/II study of hepatic infusion of autologous CC49-Zeta gene-modified T cells in patients with hepatic metastasis from colorectal cancer. Human Gene Therapy Protocols. NIH/Office of Biotechnology Activities. Updated August 10, 2001.

Blaese RM, Culver KW, Chang L, Anderson WF, Mullen C, Nienhuis A, Carter C, Dunbar C, Leitman S, Berger M (1992): Treatment of severe combined immunodeficiency disease (SCID) due to adenosine deaminase deficiency with CD34+ selected autologous peripheral blood cells transduced with a human ADA gene. Amendment to clinical research project, Project 90-C-195, January 10, 1992. *Human Gene Ther* 4:521–527.

Bowman LC, Grossman M, Rill D, Brown M, Zhong WY, Alexander B, Leimig T, Coustan-Smith E, Campana D, Jenkins J, Woods D, Brenner M (1998): Interleukin-2 gene modified allogeneic tumor cells for treatment of relapsed neuroblastoma. *Hum Gene Ther* 9:1303–1311.

Braun-Dullaeus RC, Mann MJ, Dzau VJ (1998): Cell cycle progression: New therapeutic target for vascular proliferative disease. *Circulation* 98:82–89.

Chang A (1994): Adoptive immunotherapy of cancer with activated lymph node cells primed in vivo with autologous tumor cells transduced with the GM-CSF gene. Protocol 9312–065. Human Gene Therapy Protocols. NIH/Office of Biotechnology Activities, updated August 10, 2001.

Casscells W, Engler D, Willerson JT (1994): Mechanisms of restenosis. *Tex Heart Inst J* 21:68–77.

Chang MW, Barr E, Seltzer J, Jiang YQ, Nabel GJ, Nabel EG, Parmacek MS, Leiden JM (1995): Cytostatic gene therapy for vascular proliferative disorders with a constitutively active form of the retinoblastoma gene product. *Science* 267:518–522.

Cosset FL, Morling FJ, Takeuchi U, Weiss RA, Collins MK, Russell SJ (1995): Retroviral targeting by envelopes expressing an N-terminal binding domain. *J Virol* 69:6314–6322.

Cosset FL, Russell SJ (1996): Targeting retrovirus entry. *Gene Ther* 3:946–956.

De Meyer GRY, Bult H (1997): Mechanisms of neointima formation-lessons from experimental models. *Vascul Med* 2:179–189.

Dunbar C, Chang L, Mullen C, Ramsey WJ, Carter C, Kohn D, Parkman R, Lenarsky C, Weinberg K, Culver KW, Anderson WF, Leitman S, Fleisher T, Klein H, Shearer G, Clerici M, McGarrity G, Bastian J, Hershfield MA (1993): Amendment to Clinical Research Project, Project 90–C-195. April 1, 1993. Treatment of severe combined immunodeficiency disease (SCID) due to adenosine deaminase deficiency with autologous lymphocytes transduced with a human ADA gene. *Hum Gene Ther* 10:477–488.

Fass D, Davey RA, Hamson CA, Kim PS, Cunningham JM, Berger JM (1997): Structure of a murine leukemia virus receptor-binding glycoprotein at 2.0 angstrom resolution. *Science* 277:1662–1666.

Fitzsimmons TD, Fagerholm P, Harfstrand A, Schenholm M (1992): Hyaluronic acid in the rabbit cornea after excimer laser superficial keratectomy. *Invest Ophthalmol Vis Sci* 33:3011–3016.

Gansbacher B, Motzer R, Houghton A (1992): Immunization with interleukin-2 secreting allogeneic HLA-A2–matched renal cell carcinoma cells in patients with advanced renal cell carcinoma. Human Gene Therapy Protocols. NIH/Office of Biotechnology Activities, updated August 10, 2001.

Gaster RN, Binder PS, Coalwell K, Berns M, McCord RC, Burstein NL (1989): Corneal surface ablation by 193 nm excimer laser and wound healing in rabbits. *Invest Ophthalmol Vis Sci* 30:90–98.

Gibbons GH, Dzau VJ (1996): Molecular therapies for vascular diseases. *Science* 272:689–693.

Ginsburg D, Konkle BA, Gill JC, Montgomery RR, Bockenstedt PL, Johnson TA, Yang AY (1989): Molecular basis of human von Willebrand disease: analysis of platelet von Willebrand mRNA. *Proc Natl Acad Sci USA* 86:3723–3727.

Gordon EM, Anderson WF (1994): Gene therapy using retroviral vectors. *Curr Opin Biotech* 5:611–616.

Gordon EM, Liu P, Chen ZH, Liu L, Whitley MD, Gee C, Groshen S, Hinton DR,

Beart RW, Hall FL (2000): Inhibition of metastatic tumor growth in nude mice by portal vein infusions of matrix-targeted retroviral vectors bearing a cytocidal cyclin G1 construct. *Cancer Res* 60:3343–3347.

Gordon EM, Liu PX, Chen ZH, Liu L, Whitley MD, Liu L, Wei D, Groshen S, Hinton DR, Anderson WF, Beart RW, Jr, Hall FL (2001a): Systemic administration of a matrix-targeted retroviral vector is efficacious for cancer gene therapy in mice. *Hum Gene Ther* 12:193–204.

Gordon EM, Zhu NL, Prescott MF, Chen ZH, Anderson WF, Hall FL (2001b): Lesion-targeted injectable vectors for vascular restenosis. *Hum Gene Ther* 12:1277–1287.

Hall FL, Gordon EM, Wu L, Zhu NL, Skotzko MJ, Starnes VA, Anderson WF (1997): Targeting retroviral vectors to vascular lesions by genetic engineering of the MoMuLV gp70 envelope protein. *Hum Gene Ther* 8:2183–2192.

Hall FL, Liu L, Zhu NL, Stapfer M, Anderson WF, Beart RW, Gordon EM (2000): Molecular engineering of matrix-targeted retroviral vectors incorporating a surveillance function inherent in von Willebrand factor. *Hum Gene Ther* 11:983–993.

Hall FL, Liu L, Wendler CB, Beart RW, Gordon EM (2001): Retroviral vectors bearing tumor vasculature targeting motifs enhance gene transfer in activated microvascular endothelial cell cultures, submitted.

Han L, Hofmann T, Chiang Y, Anderson WF (1995): Chimeric envelope glycoproteins constructed between amphotropic and xenotropic murine leukemia retroviruses. *Som Cell Mol Genet* 21:205–214.

Hollon T (2001): Gene therapy: taking it to the lesion. *The Scientist*, April 30, pp. 1, 16–17.

Junghans RP (1998): Phase I study of T cells modified with chimeric anti-CEA immunoglobulin T cell receptors (IgTCT) in adenocarcinoma. Human Gene Therapy Protocols. NIH/Office of Biotechnology Activities. Updated August, 2001.

Kasahara N, Dozy AM, Kan YW (1994): Tissue-specific targeting of retroviral vectors through ligand-receptor interactions. *Science* 266:1373–1376.

Kotani H, Newton PB, Zhang S, Chiang YL, Otto E, Weaver L, Blaese RM, Anderson WF, McGarrity GJ (1994): Improved methods of retroviral vector transduction and production for gene therapy. *Hum Gene Ther* 5:19–28.

Langreth R, More SD (1999): Researchers get dose of reality as logistics stymie gene therapy. *Wall Street Journal* :1.

Liu L, Anderson WF, Beart RW, Gordon EM, Hall FL (2000): Incorporation of tumor vasculature targeting motifs into Moloney murine leukemia virus env escort proteins enhances retrovirus binding and transduction of human endothelial cells. *J Virol* 74:5320–5328.

McDonnell PJ (2000): Emergence of refractive surgery. *Arch Ophthalmol* 118:1119–1120.

MacKrell AJ, Soong NW, Curtis CM, Anderson WF (1996): Identification of a subdomain in the Moloney murine leukemia virus envelope protein involved in receptor binding. *J Virol* 70:1768–1774.

Marin M, Noel D, Valsesia-Wittman S, Brockly F, Etienne-Julan M, Russell S, Cosset FL, Piechaczyk M (1996): Targeted infection of human cells via major histocompatibility complex class I molecules by Moloney murine leukemia virus-derived viruses displaying single-chain antibody fragment-envelope fusion proteins. *J Virol* 70:2957–2962.

Martin F, Kutsch J, Takuchi Y, Russell S, Cosset FL, Collins M (1998): Retroviral vector targeting to melanoma cells by single-chain antibody incorporation in envelope. *Hum Gene Ther* 9:737–746.

Masood R, Gordon EM, Whitley MD, Wu BW, Cannon P, Evans L, Anderson WF, Gill P, Hall FL (2001): Retroviral vectors bearing IgG-binding motifs for antibody-mediated targeting of vascular endothelial growth factor receptors. *Intl J Mol Med* 8:335–343.

Meininger GA, Davis MJ (1992): Cellular mechanisms involved in the vascular myogenic response. *Am J Physiol* 263:647–659.

Miller AD, Rosman GJ (1989): Improved retroviral vectors for gene transfer and expression. *Biotechniques* 7:984–990.

Morgan RA, Nussbaum O, Muenchau DD, Shu L, Couture L, Anderson WF (1993): Analysis of the functional and host range-determining regions of the murine ecotropic and amphotropic retrovirus envelope proteins. *J Virol* 67:4712–4721.

Morishita R, Gibbons GH, Ellison KE, Nakajima M, Zhang L, Kaneda Y, Ogihara T, Dzau VJ (1993): Single intraluminal delivery of antisense cdc2 kinase and proliferating-cell nuclear antigen oligonucleotides results in chronic inhibition of neointimal hyperplasia. *Proc Natl Acad Sci USA* 90:8474–8478.

Morishita R, Gibbons GH, Horiuchi M, Ellison KE, Nakajima M, Zhang L, Kaneda Y, Ogihara T, Dzau V (1995): A gene therapy strategy using a transcription factor decoy of the E2F binding site inhibits smooth muscle proliferation in vivo. *Proc Natl Acad Sci USA* 92:5855–5859.

Morling FJ, Peng KW, Cosset F-L, Russell SJ (1997): Masking of retroviral envelope functions by oligomerizing polypeptide adaptors. *Virology* 234:51–61.

Ohno K, Sawa IK, Iijima Y, Levin B, Meruelo D (1997): Cell-specific targeting of Sindbis virus vectors displaying IgG-binding domains of protein, A. *Nat Biotechnol* 15:763–767.

Oldfield EH, Ram Z, Culver KW, Blaese RM, DeVroom HL, Anderson WF (1993): Gene therapy for the treatment of brain tumors using intra-tumoral transduction with the thymidine kinase gene and intravenous ganciclovir. *Hum Gene Ther* 4:39–69.

Peng KW, Russell SJ (1999): Viral vector targeting. *Curr Opin Biotechnol* :454–457.

Pensiero MN, Wysocki CA, Nader K, Kikuchi GE (1996): Development of amphotropic murine retrovirus vectors resistant to inactivation by human serum. *Human Gene Ther* 7:1095–1101.

Rosenberg SA, Aebersold P, Cornetta K, Kasid A, Morgan RA, Moen R, Karson EM, Lotze MT, Yang, JC, Topalian SL (1990): Gene transfer into human-immunotherapy of patients with advanced melanoma, using tumor infiltrating lymphocytes modified by retroviral gene transduction. *N Engl J Med* 323:570–578.

Rosenberg SA (1991): Protocol 9110–011. Immunization of cancer patients using autologous cancer cells modified by insertion of the gene for interleukin2 (IL-2). Human Gene Therapy Protocols. NIH/Office of Biotechnology Activities, October 15, 1991.

Ruggeri ZM, Zimmerman TS (1987): von Willebrand factor and von Willebrand disease. *Blood* 70:895–904.

Russell SJ (1993): Retroviral vectors displaying functional antibody fragments. *Nucl Acid Res* 21:1081–1085.

Schwartz RS, Hombs DR, Topol J (1992): The restenosis paradigm revisited: an alternative proposal for cellular mechanisms. *Amer Coll Cardiol* 17:1284–1293.

Seiler T, McDonnell PJ (1995). Excimer laser photorefractive keratectomy. *Surg Ophthalmol* 40:89–118.

Skotzko MJ, Wu L, Anderson WF, Gordon EM, Hall FL (1995): Retroviral vector-mediated gene transfer of antisense cyclin G1 (*CYCG1*) inhibits proliferation of human osteogenic sarcoma cells. *Cancer Res* 55:5493–5498.

Smith ML, Kontny HU, Bortnick R, Fornace AJ, Jr. (1997): The p53-regulated cyclin G gene promotes cell growth: p53 downstream effectors cyclin g and Gadd45 exert different effects on cisplatin sensitivity. *Exp Cell Res* 230:61–68.

Somia NV, Zoppe M, Verma IM (1995): Generation of targeted retroviral vectors by using single-chain variable fragment: an approach to in vivo gene delivery. *Proc Natl Acad Sci USA* 92:7570–7574.

Soneoka Y, Cannon PM, Ramsdale EE, Griffiths JC, Romano G, Kingsman SM, Kingsman AJ (1995): A transient three-plasmid expression system for the production of high titer retroviral vectors. *Nucleic Acids Res* 23:628–633.

Takaji J, Fujisawa T, Sekiya F, Saito Y (1991): Collagen-binding domain within bovine propolypeptide of von Willebrand factor. *J Biol Chem* 266:5575–5579.

Tuan TL, Cheung L, Wu L, Yee A, Gabriel S, Han B, Morton L, Nimni ME, Hall FL (1996): Engineering, expression and renaturation of targeted TGF-beta fusion proteins. *Conn Tiss Res* 34:1–9.

Valsesia-Wittman S, Drynda A, Deleage G, Aumailley M, Heard JM, Danos O, Verdier G, Cosset FL (1994): Modification in the binding domain of avian retrovirus envelope protein to redirect the host range of retroviral vectors. *J Virol* 68:4609–4619.

Verma I, Somia N (1997): Gene therapy: promises, problems and prospects. *Nature* 389:239–242.

Weiss RA, Tailor CS (1995): Retrovirus receptors. *Cell* 82:531–533.

Wu L, Liu L, Yee A, Carbonaro-Hall D, Tolo VT, Hall FL (1994): Molecular cloning of the human CYCG1 gene encoding a G-type cyclin: overexpression in osteosarcoma cells. *Oncol Reports* 1:705–711.

Wu BW, Cannon PM, Gordon EM, Hall FL, Anderson WF (1998): Characterization of the proline-rich region of murine leukemia virus envelope protein. *J Virol* 72:5383–5391.

Wu K, Yee A, Zhu NL, Gordon EM, Hall FL (2000): Characterization of differential gene expression in monkey arterial neointima following balloon catheter injury. *Intl J Mol Med* 6:433–440.

Xu F, Prescott MF, Liu P, Chen ZH, Liau G, Gordon EM, Hall FL (2001): Long term inhibition of neointima formation in balloon-injured rat arteries by intraluminal instillation of a matrix-targeted retroviral vector bearing a cytocidal mutant cyclin G1 construct. *Intl J Mol Med* 8:19–30.

Yang Q, Mamounas M, Yu G, Kennedy S, Leaker B, Merson J, Wong-Staal F, Yu M, Barber JR (1998): Development of novel cell surface CD34–targeted recombinant adenoassociated virus vector for gene therapy. *Hum Gene Ther* 9:1929–1937.

Yang S, Delgado R, King SR, Woffendin C, Barker CS, Yang ZY, Xu L, Nolan GP, Nabel GJ (1999): Generation of retroviral vector for clinical studies using transient transfection. *Hum Gene Ther* 10:123–132.

Zhao Y, Zhu Y, Lee S, Li L, Chang E, Soong NW, Douer D, Anderson WF (1999): Identification of the block in targeted retroviral-mediated gene transfer. *Proc Natl Acad Sci USA* 96:4005–4010.

Zhu NL, Wu L, Liu PX, Gordon EM, Anderson WF, Starnes VA, Hall FL (1997): Down-regulation of cyclin G1 expression by retrovirus-mediated antisense gene transfer inhibits vascular smooth muscle cell proliferation and neointima formation. *Circulation* 96:628–635.

Zhu NL, Gordon EM, Liu L, Terramani TT, Anderson WF, Hall FL (2001): Collagen-targeted retroviral vectors displaying domain D2 of von Willebrand factor (vWF-D2) enhance gene transfer to human tissue explants. *IJPHO*, in press.

15

TARGETING MEASLES VIRUS ENTRY

ANTHEA L. HAMMOND, PH.D., RICHARD K. PLEMPER, PH.D., AND ROBERTO CATTANEO, PH.D.

MV AS A CYTOREDUCTIVE AGENT

Measles virus (MV) is a member of the Paramyxoviridae, a family of enveloped RNA viruses. It is part of the Mononegavirales group, which includes the other human pathogens mumps, human parainfluenza, respiratory syncytial, Ebola, and rabies viruses (listed according to their evolutionary proximity).

Infection with wild-type MV still causes the death of one million children yearly (Clements and Cutts, 1995). Most of this morbidity and mortality arises from secondary infections, susceptibility to which results from the profound immune suppression that accompanies infection (Borrow and Oldstone, 1995). Rarely, a lethal disease of the central nervous system called subacute sclerosing panencephalitis (SSPE) can develop 5–10 years after acute infection (Billeter and Cattaneo, 1991; Kristensson and Norrby, 1986).

However, measles can be effectively prevented by vaccination. The live attenuated vaccine strain MV-Edmonston (MV-Edm) has been licensed for more than 30 years and has proved to be one of the safest, most successful, and most cost-effective vaccines in use (Duclos and Ward, 1998). The vaccine virus is very rarely pathogenic; during 30 years of administration, very rare cases of vaccine-induced measles inclusion body encephalitis and thrombocytopenia have been reported (Norrby, 1995). Its use is recommended in all

Vector Targeting for Therapeutic Gene Delivery, Edited by David T. Curiel and Joanne T. Douglas
ISBN 0-471-43479-5

HIV-infected children other than those severely immunocompromised (Moss, Cutts, and Griffin, 1999; Palumbo et al., 1992).

Given this excellent safety record, MV-Edm represents an attractive choice for administration in a replication-competent form. Furthermore, its oncolytic properties make it particularly suitable for development as a cytoreductive agent designed to eliminate cancerous cells. There have been several reports of regression of Hodgkin's and non-Hodgkin's lymphoma following natural MV infection (Bluming and Ziegler, 1971; Taqi et al., 1981). In addition, clinical trials based on administration of the related Paramyxovirus mumps to 90 individuals with advanced and terminal cancer resulted in significant (although mostly short-lived) tumor regression in almost half of the patients, with minimal associated toxicity (Asada, 1974). And following promising results in preclinical oncolytic studies, the avian Paramyxovirus Newcastle Disease virus has recently entered phase I clinical trials (Lorence et al., 1994; Lorence, Rood, and Kelley, 1988).

The ability of MV to induce massive cell–cell fusion in infected cultures provides a further advantage for its use as a cytolytic agent. A single MV-infected cell can recruit many neighboring uninfected cells into a giant multinucleated syncytium, which ultimately dies. Fusion is mediated by the viral envelope proteins expressed at the surface of an infected cell, which bind viral receptors on surrounding uninfected cells and induce fusion with them. Other enveloped viruses induce syncytium formation in a similar manner (Klasse, Bron, and Marsh, 1998), leading to the term *fusogenic membrane glycoprotein* (FMG) to describe the class of viral envelope proteins that mediate this effect. Generating novel vectors that exploit this property is an attractive strategy for cytoreductive therapy. The delivery of FMGs in viral vectors designed to eliminate certain cell populations is one promising approach toward this goal (Bateman et al., 2000; Diaz et al., 2000; Fielding et al., 2000). Developing MV itself as a cytoreductive agent combines the cytotoxic potential of its own FMG with the opportunity for replication-competence offered by the safety profile of the vaccine strain. Indeed, recent work has demonstrated regression of human tumor xenografts in severe combined immunodeficiency (SCID) mice following intratumoral or systemic administration of MV-Edm (Grote et al., 2001; Peng et al., 2001). Retargeting the MV envelope glycoprotein to specific receptors offers the opportunity to restrict viral entry and thus cytotoxicity to desired cell populations, such as a specific tumor type.

This review describes the development to date of targeted recombinant MVs based on the vaccine strain. We focus on the methods used to generate recombinant MVs, the properties of the targeted viruses, and the limitations encountered thus far. We also discuss the feasibility of MV-based therapies and strategies to overcome potential difficulties.

MV REPLICATION AND PARTICLE STRUCTURE

The MV genome comprises 15,894 bases of nonsegmented, negative strand RNA. The RNA is tightly encapsidated by a helically arranged nucleocapsid

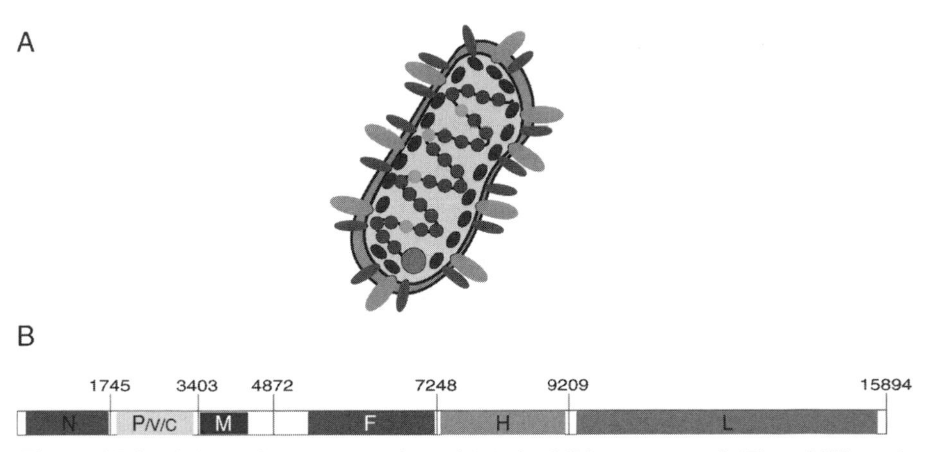

Figure 15.1 Schematic representation of (*A*) the MV genome and (*B*) an MV particle. The color code for proteins is indicated in the genome scheme. The lipid bilayer surrounding the particle is shown in blue. Figure also appears in Color Figure section.

(N) protein. Each molecule of N covers six bases, and therefore the genomes of MV and other Paramyxoviruses are efficiently replicated only when they encompass a number of nucleotides that is a multiple of six (Vulliemoz and Roux, 2001). A polymerase (L for large) and a polymerase cofactor (P for phosphoprotein) associate with the RNA and N to form a ribonucleoprotein (RNP) complex which is active in replication (Pringle, 1991). The genome comprises six viral genes in the following order (positive strand): 5′-N-P-M-F-H-L-3′ (Fig. 15.1B). The L polymerase transcribes and polyadenylates the six mRNAs sequentially; its lack of processivity results in a transcription gradient and therefore in a gradient of protein production, with N being most abundant and L least (Cattaneo et al., 1987).

The RNP complex is surrounded by the matrix protein (M) (Peeples, 1991), which bridges the viral envelope glycoproteins hemagglutinin (H) and fusion (F) with the RNP (Cathomen et al., 1998a; Morrison and Portner, 1991). H and F are the only two viral surface proteins and are postulated to interact with each other. H dimerizes via disulphide bonds at cysteines 139 and 154 (Plemper, Hammond, and Cattaneo, 2000) and is thought to exist at the cell surface as a tetramer. MV H is responsible for receptor binding and also provides support to the fusion function of F. F, which is postulated to form a trimer, is expressed as a precursor F_0 which is cleaved by furin in the trans-Golgi network to release F_1 and F_2. The N-terminus of F_1 contains the hydrophobic fusion peptide which, upon exposure, mediates membrane mixing (Lamb, 1993).

Two structural properties of the MV particle (Fig. 15.1A) appear favorable for its use as a vector. First, the RNP complex is helical and there is no size constraint on genome length as there is for icosahedral viruses. Secondly, MV particles are pleomorphic, again reflecting the relaxed constraints on their morphology. For more details on viral transcription and replication, see Figure 15.2 and the section MV Recovery System and Vectors.

THE NATURAL MV RECEPTORS AND VIRAL ENTRY

Two receptors have been identified for MV: SLAM (signaling lymphocytic activation molecule), a B- and T-cell specific glycoprotein (Hsu et al., 2001; Tatsuo et al., 2000b) and CD46, the ubiquitous regulator of complement activation (Dorig et al., 1993; Naniche et al., 1993). Vaccine strains of MV, including MV-Edm, seem able to use both CD46 and SLAM equivalently, while clinical isolates infect predominantly via SLAM, questioning the in vivo relevance of CD46 for the initial steps of viral replication in immune cells. Although this has yet to be studied in detail, initial reports showing that the majority of MVs isolated from throat swabs use just SLAM as a receptor (Ono et al., 2001a), and furthermore that other morbilliviruses use species-specific SLAMs for entry (Tatsuo, Ono, and Yanagi, 2001), support the theory that CD46 usage may be a consequence of in vitro adaptation (attenuation) of the virus. A number of wild-type MVs, however, do have the ability to infect efficiently cells via CD46 (Manchester et al., 2000).

Identifying the residues of MV H important for receptor interaction has so far only been achieved for CD46. MV hemadsorption, cell–cell fusion, and CD46 downregulation critically depend on amino acids 451 and 481 (Bartz et al., 1996; Lecouturier et al., 1996). A tyrosine at position 481 is strongly associated with efficient interaction of baculovirus-expressed MV H with CD46 in a cell-based binding assay (Hsu et al., 1998). Most vaccine strains carry this tyrosine, while the majority of wild-type strains analyzed possess an asparagine. The demonstration that H carrying asparagine at position 481 is not able to induce CD46 down-regulation or to bind CD46-positive cells supports this amino acid as being a critical determinant of CD46 recognition. A recombinant MV expressing a wild-type H with asparagine at position 481 can replicate in Vero cells (Johnston et al. 1999), however other studies have implicated amino acids 473–477 as required for efficient CD46 binding (Patterson et al., 1999). Our recent extensive mutational analysis of MV H has indicated that different binding sites for CD46 and SLAM exist; few residues are necessary to maintain CD46-dependent fusion and others SLAM-dependent fusion (S. Vongpunsawad et al., unpublished observations). The H protein also has a role in supporting fusion mediated by the F protein, and little is known regarding the regions of the molecule important for this function.

MV H is a major determinant of viral tropism, for both vaccine and wild-type strains (Tatsuo et al., 2000a). In infected humans, MV has been detected in monocytes (Esolen et al., 1993; Helin et al., 1999; Karp et al., 1996) and in T and B lymphocytes (Nakayama et al., 1995). Dendritic cells can be infected in vitro (Grosjean et al., 1997; Ohgimoto et al., 2001). In experimentally infected monkeys, which develop a disease similar to that of humans (McChesney et al.,

1997), prominent syncytia can be observed in both macrophage and lymphoid cells, similar to that seen in humans (Hall et al., 1971; McChesney et al., 1997; Sergiev, Ryazantseva, and Shroit, 1960). Mice transgenically expressing CD46 with human-like tissue specificity (Blixenkrone-Moller et al., 1998; Mrkic et al., 1998; Oldstone et al., 1999) have enabled MV dissemination and pathology to be studied in a small animal model. While MV replication in peripheral blood mononuclear cells (PBMCs) and lymphoid tissues was detected at a low level, generating a double transgenic CD46-positive, interferon type 1 receptor-negative mouse allowed detection of infected PBMCs for a longer time period (Roscic-Mrkic et al., 2001). In this model, macrophages appear to contribute significantly to MV dissemination systemically.

MV RECOVERY SYSTEM AND VECTORS

Recovery of infectious MV from cloned DNA was first achieved in 1995 (Radecke et al., 1995). In contrast to the systems previously available for the recovery of other negative strand RNA viruses (Lawson et al., 1995; Schnell, Mebatsion, and Conzelmann, 1994), the MV recovery system relies on a stable helper cell line expressing MV N and P proteins and T7 RNA polymerase (Fig. 15.2). This cell line is transfected with one plasmid encoding L (pEMC-La) and a second encoding the full-length MV antigenome (p(+)MV-NSe, Fig. 15.3), both driven by the T7 promoter. A positive strand MV antigenome is first generated by T7 polymerase and cotranscriptionally encapsidated by the endogenously expressed N and P proteins. The L protein, translated from an internal ribosome entry site (IRES) situated on a T7 RNA polymerase-driven transcript, transcribes these RNPs. MV N, P, and L-dependent replication then produces an encapsidated negative strand MV genome. From this, MV mRNA and new genomes are generated by the normal viral replication and transcription machinery, and infectious particles are assembled.

MV-BASED VIRUSES ENTERING CELLS THROUGH TARGETED RECEPTORS

Using the MV recovery system, engineering changes to the viral envelope glycoproteins is relatively straightforward. Both H and F genes can be altered individually in expression vectors (pCG-H and pCG-F, respectively) and the phenotype of the altered proteins assessed in transient transfection experiments. Transfer of the altered gene into the full-length cDNA is simple because unique restriction sites flank both genes (Fig. 15.3). Mirroring strategies to retarget retrovirus cell entry by ligand display on the SU component of the MLV or ALV envelope, ligand display on MV H protein has been attempted. So far, three recombinant MVs displaying targeting ligands on H (Fig. 15.3) have been

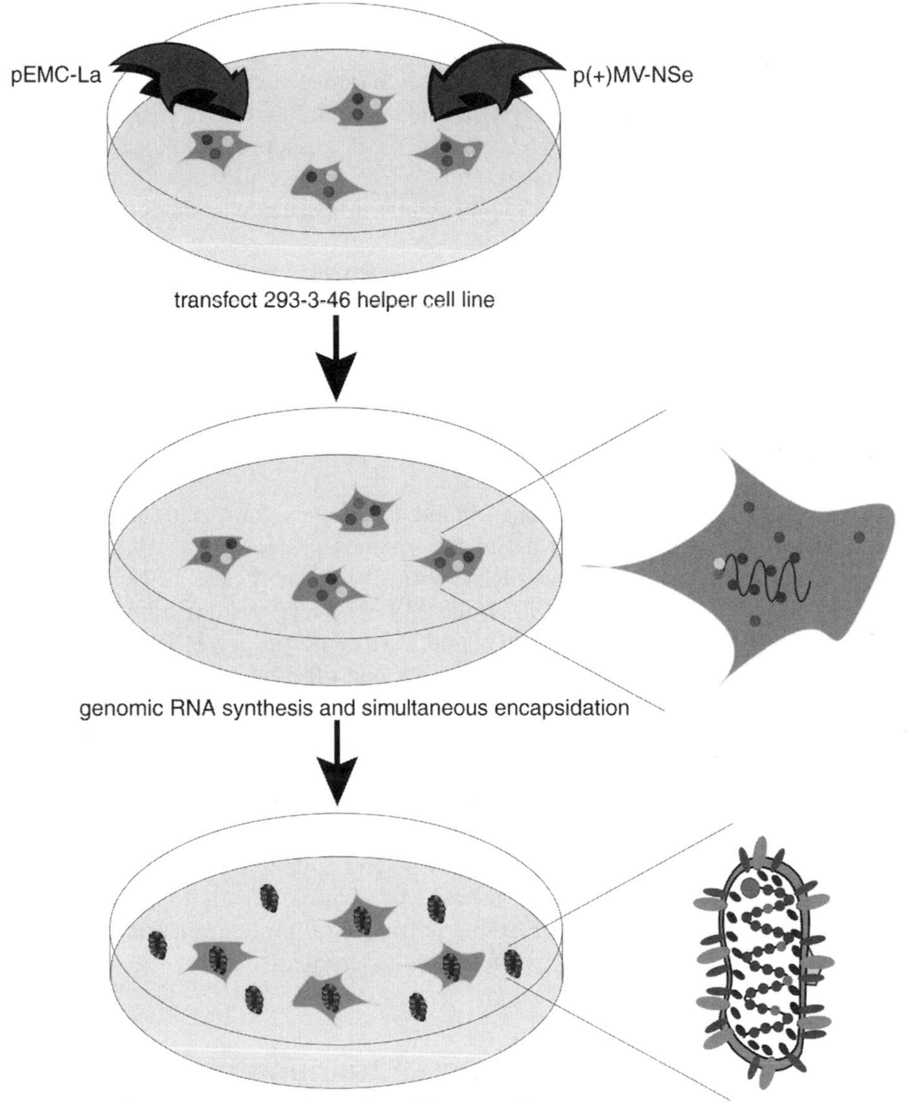

Figure 15.2 Recovery of MV from cloned DNA (Radecke et al., 1995). Plasmids encoding the full length MV antigenome (p(+)MV-NSe) and viral polymerase L (pEMC-La), both under the control of the T7 promoter are cotransfected in 293.3.46 cells constitutively expressing T7 polymerase (blue dot) and viral N (red dot) and P (green dot). T7 transcribes a positive strand MV antigenome that is encapsidated by N and P. From this template a negative strand genome is then transcribed by transiently expressed L protein. From this, viral mRNAs and new genomes are generated and infectious particles are produced and released. Overlay of the helper cells with highly fusogenic Vero cells facilitates detection of infectious centers. Figure also appears in Color Figure section.

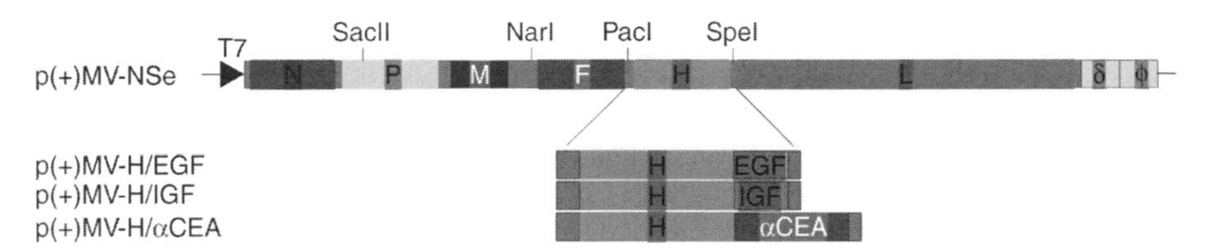

Figure 15.3 Genomic vectors for recovery of recombinant viruses. Plasmid p(+)MV-NSe specifies MV antigenomic RNA. Transcription with T7 RNA polymerase from a T7 promoter fused directly to the antigenome sequence yields RNA bearing the authentic nucleotides of the viral antigenomic 5′ terminus. The correct MV 3′ terminus is cleaved precisely by a hepatitis delta virus genomic ribozyme (δ) located immediately adjacent to the MV 3′-terminal nucleotides; the presence of T7 terminator (ϕ) increases the efficiency of this self-cleavage. Plasmid p(+)MV-NSe has been modified to introduce the unique restriction sites shown. Plasmids p(+)MV-H/EGF, p(+)MV-H/IGF, and p(+)MV-H/scAbCEA are all based on p(+)MV-NSe and contain chimeric MV H ORFs in which the EGF, IGF-1, and αCEA scAb ORFs, respectively, are fused to the 3′ terminus of H, separated by a sequence encoding a Factor Xa cleavage site. The chimeric H ORFs were generated in pCG expression vectors named pCG-H/EGF, pCG-H/IGF, and pCG-H/scAbCEA. Figure also appears in Color Figure section.

described which can infect cells through both CD46 and their targeted receptor (Hammond et al., 2001; Schneider et al., 2000).

The feasibility of ligand display on MV-Edm H was first assessed in transient expression. Epidermal growth factor (EGF) and insulin-like growth factor type 1 (IGF-1) were appended at the C-terminal, extracellular domain of H. In a further study, three versions of a single-chain antibody (scAb) specific for the tumor-expressed carcinoembryonic antigen (CEA) were appended to MV H. The three scAb forms carried different linker lengths of 0, 5, and 15 amino acids between V_H and V_L domains. In all protein hybrids, a factor Xa cleavable linker was inserted between H and the displayed specificity determinant.

When coexpressed with MV F protein in CD46-positive Vero cells, H-EGF and H carrying the long linker form of the scAb were able to induce syncytium formation to a level similar to that seen for unmodified H. H-IGF induced a slightly lower level of syncytia formation, and H carrying either the zero or short linker form of scAb failed to induce any cell–cell fusion. Since the length of linker between V_H and V_L constrains oligomerization of scAbs, it is expected that the zero and short forms of the scAb trimerized and dimerized, respectively, thus sterically inhibiting fusion. A steric or other effect may interfere with optimal fusion function in the case of the IGF display. The H-scAb proteins were also tested for their ability to mediate cell–cell fusion in human cells carrying both CD46 and CEA (HeLa-CEA), and in a mouse cell line lacking CD46 but expressing the targeted receptor, human CEA (MC38-CEA). Again, only the long linker form of H-scAb was active, with similar numbers of syncytia in HeLa-CEA cells as induced by unmodified H. Importantly, this chimeric H induced cell–cell fusion in MC38-CEA cells when coexpressed with F, while unmodified H did not. Characterization of this H-scAb protein revealed it had similar cell surface expression and similar potential for homo-oligomerization as unmodified H. Thus display of a 244 amino acid scAb did not impair processing or function of the underlying H protein.

The genes encoding H-EGF, H-IGF, and H displaying the long linker form of scAb were exchanged with the unmodified H gene in full length MV-Edm cDNA (p(+)MV-NSe, Fig. 15.3B). Viruses designated MV-H/EGF, MV-H/IGF, and MV-H/scAbCEA (formerly MV-HXL) respectively were recovered on Vero cells. In addition, the enhanced green fluorescent protein (eGFP) gene, was inserted upstream of N, yielding "green" versions of MV-H/EGF and MV-H/IGF.

In Vero cells, all three recombinant viruses replicate to levels similar to that of unmodified MV, demonstrating that the addition of targeting ligands of up to 250 amino acids does not abrogate entry via CD46. Significantly, all three viruses could efficiently infect rodent cells lacking the natural MV receptors but expressing the targeted receptor (Figs. 15.4 and 15.5). Using MV-H/scAbCEA, titers of approximately 10^5 pfu/ml were attained in the MC38-CEA cells.

The specificity of targeted MVs was demonstrated first by competing entry through the targeted receptor, and second by cleaving the displayed ligand by treatment of viral particles with Factor Xa. Competition of cell entry using soluble forms of EGF and IGF-1 ligands inhibited infectivity of MV-H/EGF and MV-H/IGF, respectively, by five- to sixfold on the relevant target cells. Competition with an anti-CEA mAb reduced the titer of MV-H/scAbCEA from 10^5 to 10^4 pfu/ml on MC38-CEA cells (Fig. 15.5). By cleaving the displayed ligand with Factor Xa, infectivity of all three viruses was reduced approximately 100-fold. Similar results have been obtained by displaying on H an scAb against CD38, a cell surface marker overexpressed in many myeloma cases (Peng et al., personal communication). Given the conserved nature of scAb structure, it is likely that the ability to display scAbs will be a general feature of MV H.

To date, three structurally distinct ligand types have been successfully displayed on the H protein of MV. None interfere with the ability of the underlying H protein to enter via CD46, and all can mediate entry through the desired target receptor. EGF and IGF-1 are about 60 and 90 amino acids in length; the scAb approaches 250 amino acids. The mechanism by which these ligands are able to induce fusion via a novel receptor is not yet clear. It has previously been suggested that MV, by analogy with other enveloped viruses such as influenza, undergoes a conformational change in its H and subsequently F protein upon receptor binding, which results in exposure of the hydrophobic fusion domain on F_1 and finally in membrane mixing (Fig. 15.6) (Lamb, 1993). That large specificity domains displayed on H can bind their targeted receptor and induce precisely the same sequence of conformational changes appears highly improbable. However, it may be possible that distinct conformations of MV H can lead to the formation of a reactive H intermediate, in turn inducing extrusion of the

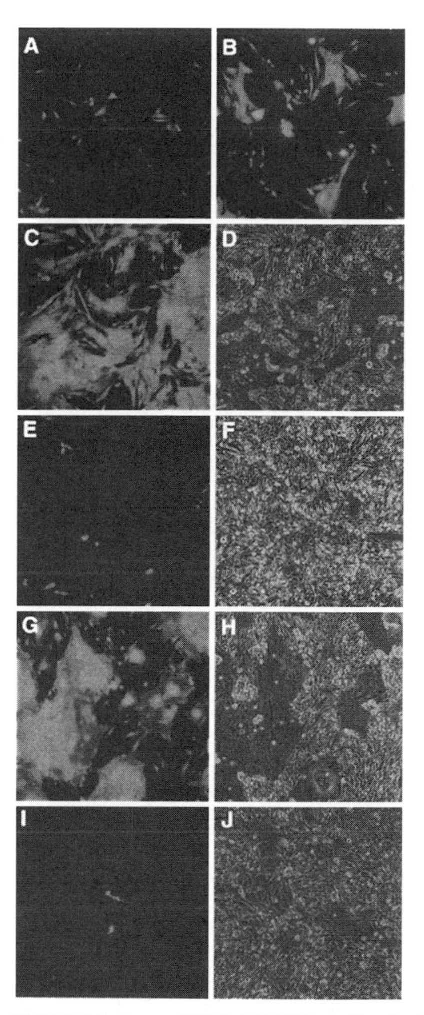

Figure 15.4 MV_{green}-H/EGF infects CHO-hEGFR cells. Infection of CHO-hEGFr cells (A to F), CHO-hEGFr.tr cells (*G* and *H*), and CHO cells (*I* and *J*) with MV_{green}-H/XhEGF (*A* to *D* and *G* to *J*) or MV (*E* and *F*). Cells were infected at a multiplicity of infection (MOI) of 1 and were monitored by fluorescent or phase-contrast microscopy 24 (*A*), 48 (*B*), or 72 (*C* to *J*) h after infection. (From Schneider et al., 2000, by permission from American Society for Microbiology.) Figure also appears in Color Figure section.

F fusion peptide and membrane mixing. Alternatively, a mechanism for MV entry that does not involve specific receptor-induced conformational changes in H and/or F may be proposed, as has previously been suggested to occur via an F receptor. To date, no evidence for a cell surface receptor for MV F has been described, thus this mechanism seems less probable.

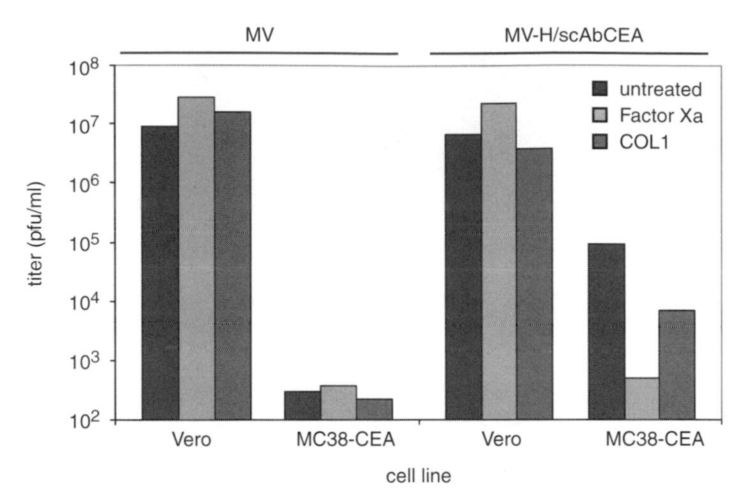

Figure 15.5 MV-H/scAbCEA specifically infects MC38-CEA cells. Vero cells (as control) and MC38-CEA cells were infected with MV or MV-H/scAbCEA (MOI 1), untreated or pretreated with 10 μg factor Xa protease. Also, cells were pretreated or not with 10 μg/ml anti-CEA COL1 antibody and then infected with MV or MV-H/scAbCEA (MOI 1). Virus released from the cells by freeze-thaw 72 h postinfection was quantified by $TCID_{50}$ titration on Vero cells. (From Hammond et al., 2001, by permission from American Society for Microbiology.)

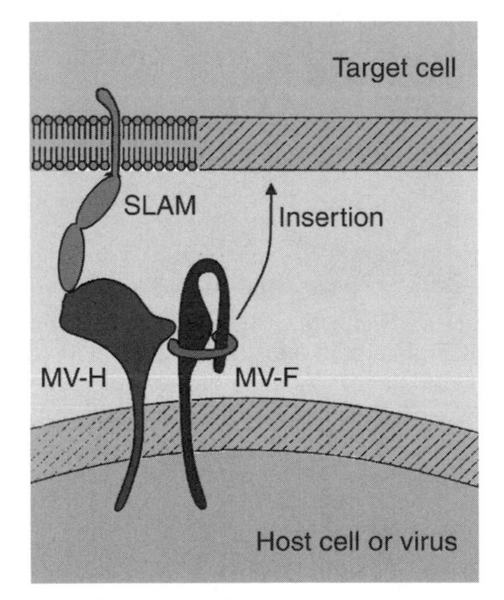

Figure 15.6 Model for MV entry. MV H protein is responsible for viral attachment to the V domain of the SLAM receptor (Ono et al., [illegible]), or to the [illegible] domain of CD46 (Buchholz et al., 1996), and also has a role in providing fusion support to F. This is hypothesized to occur via receptor binding inducing conformational changes in H and then in F, exposing the fusion peptide of F, which then mediates membrane mixing.

CONCLUSIONS

The ability to target MV entry by display on H of specificity domains, coupled with the demonstration of in vivo MV-induced regression of human tumor xenografts in a SCID mouse model (Grote et al., 2001; Peng et al., 2001) suggests MV has potential as a targeted cytoreductive agent, particularly in malignant tissue.

At present, the main obstacle to the development of truly retargeted MVs that maintain entry via a target receptor but which are no longer able to infect via natural MV receptors lies in reducing the interaction of H with CD46 and SLAM. Given that H not only functions as the viral attachment protein but also has an important role in fusion support for the F protein, mutants that retain fusion support while losing binding to one or both receptors should be sought. We have recently identified H mutants with reduced interaction with either CD46 or SLAM and are assessing these mutations in the context of recombinant virus; these data should give further insight into the nature of CD46 and SLAM binding sites. Another attractive approach for generating a truly retargeted MV is to develop a screening system based on applying a selective pressure, that is, the maintenance of fusion support, to a mutant library of targeted H molecules. Of note is the recent hypothesis that dissemination of MV in vivo may be primarily mediated by SLAM (Tatsuo et al., 2001). If this is the case, obtaining viruses with reduced interaction with SLAM but not necessarily CD46 may be sufficient to generate a targeted MV with minimal spread to nontargeted tissues.

More general concerns regarding the feasibility of targeted MV as a cytoreductive therapy must also be addressed. While tumor regression mediated by nontargeted MV-Edm has been demonstrated in vivo in a SCID mouse model of a human B cell lymphoma, assessing its antitumor activity in an immune competent mouse is necessary. These data will give insights into the cytoreductive properties of MV-Edm in the face of an active immune system and will also have important consequences for retargeted viruses. An immune competent syngeneic murine lymphoma model is currently being established in CD46 transgenic mice (A. K. Fielding, pers. communic.). On one hand, it is possible that the host immune system may destroy the virus before it has a chance to effect cytolysis. On the other hand, the immune response against MV may enhance its cytoreductive activity, recruiting antigen-presenting cells to the tumor, which may promote host antitumor activity. Preliminary data showing that passive transfer of MV antiserum in the SCID mouse model does not greatly affect tumor regression upon intratumoral virus injection suggests an intact immune system may not negate the antitumor response (Grote et al., 2001). Moreover, in some malignancies (for example multiple myeloma) levels of circulating neutralizing antibodies can be very low; in this case systemic delivery may be more effective, however, using a retargeted MV unable to infect nontumor cells via SLAM or CD46 would be a prerequisite in this immune-deficient scenario. Similarly, using a strictly retargeted MV, malignancies that can be accessed from the peritoneal cavity, such as ovarian cancer, may be potential therapeutic targets.

Conceivably, the efficacy of targeted MVs may be limited by the viral dose administered; indeed in SCID mice bearing human tumor xenografts, the degree of regression observed was highly dependent on the titer of MV-Edm administered (Grote et al., 2001). Generating targeted recombinants that replicate to higher titers may thus be a desirable goal. Mutations that enhance the cell–cell fusogenicity of the virus are conceivably also of value, enabling more extensive viral spread through the tumor mass. Previously, an MV lacking the tails of H and F was found to induce more extensive cell–cell fusion, although it was compromised in its titer (Cathomen et al., 1998b). Whether enhanced lateral spread or increased titer will be crucial for tumor regression, either from targeted or nontargeted MV-Edm, warrants testing. Mutants that result in both increased lateral spread and increased titer would, logically, be very desirable, and attempts to understand better the molecular basis for viral infectivity and to develop rapid mutagenesis and screening systems will both be of value toward this goal. We have recently developed a modified MV with these characteristics (Plemper et al., 2002), and are testing its efficacy in vivo.

In combination with modifications designed to improve specificity and efficacy, such as reduced interaction with natural receptors and enhanced cell–cell spread or viral release, recombinant MVs displaying targeting ligands may be applicable to a number of malignancies. In theory, any malignant cell type that expresses a tumor marker may be targeted by an MV displaying an scAb specific for the marker. Display of peptide ligands for cellular receptors has also proven feasible. The flexibility of MV entry through novel receptors will determine its breadth of applicability; to date the panel of receptors able to support entry of targeted MVs includes members of the immunoglobulin superfamily (carcinoembryonic antigen (CEA), CD38) and the tyrosine kinase receptor superfamily (EGF and IGF receptors). The natural receptors CD46 and SLAM belong to the regulators of complement activation family and the immunoglobulin superfamily, respectively. Whether only certain families of receptor can mediate targeted MV entry is currently unclear, however the promiscuity observed so far promises a versatile targeting platform. Furthermore, the conserved structure of scAbs suggests display of any chosen scAb on MV should be possible without interfering with H protein expression or viral assembly.

In conclusion, recombinant MVs displaying targeting ligands on the viral H protein present an attractive approach for developing novel targeted viruses. Strategies to ablate entry through the natural receptors while maintaining targeted entry may yield a retargeted monotropic virus. The cytolytic properties inherent to MV make it particularly suitable for cytoreductive therapy, and preliminary studies demonstrating its oncolytic activity in human tumor xenografts suggest that therapy of specific malignancies may be a promising application of targeted MVs.

ACKNOWLEDGMENTS

We thank the Mayo, Eisenberg and Siebens Foundations, and the German Research Foundation (DF6) for financial support, and S. Vongpunsawad, K.-W. Peng, S. J. Russell and A. Fielding for sharing unpublished information.

REFERENCES

Asada T (1974): Treatment of human cancer with mumps virus. *Cancer* 34:1907–1928.

Bartz R, Brinckmann U, Dunster LM, Rima B, Ter Meulen V, Schneider-Schaulies J (1996): Mapping amino acids of the measles virus hemagglutinin responsible for receptor (CD46) downregulation. *Virology* 224:334–337.

Bateman A, Bullough F, Murphy S, Emiliusen L, Lavillette D, Cosset FL, Cattaneo R, Russell SJ, Vile RG (2000): Fusogenic membrane glycoproteins as a novel class of genes for the local and immune-mediated control of tumor growth. *Cancer Res* 60:1492–1497.

Billeter MA, Cattaneo R (1991): Molecular biology of defective measles virus persisting in the human central nervous system. In *The Paramyxoviruses*, Kingsbury DW, ed. New York: Plenum Press, pp. 323–345.

Blixenkrone-Moller M, Bernard A, Bencsik A, Sixt N, Diamond LE, Logan JS, Wild TF (1998): Role of CD46 in measles virus infection in CD46 transgenic mice. *Virology* 249:238–248.

Bluming AZ, Ziegler JL (1971): Regression of Burkitt's lymphoma in association with measles infection. *Lancet* 2:105–106.

Borrow P, Oldstone MB (1995): Measles virus-mononuclear cell interactions. *Curr Topics Microbiol Immunol* 191:85–100.

Buchholz CJ, Schneider U, Devaux P, Gerlier D, Cattaneo R (1996): Cell entry by measles virus: long hybrid receptors uncouple binding from membrane fusion. *J Virol* 70:3716–3723.

Cathomen T, Mrkic B, Spehner D, Drillien R, Naef R, Pavlovic J, Aguzzi A, Billeter MA, Cattaneo R (1998a): A matrix-less measles virus is infectious and elicits extensive cell fusion: Consequences for propagation in the brain. *EMBO J* 17:3899–3908.

Cathomen T, Naim HY, Cattaneo R (1998b): Measles viruses with altered envelope protein cytoplasmic tails gain cell fusion competence. *J Virol* 72:1224–1234.

Cattaneo R, Rebmann G, Schmid A, Baczko K, ter Meulen V, Billeter MA (1987): Altered transcription of a defective measles virus genome derived from a diseased human brain. *EMBO J* 6:681–688.

Clements CJ, Cutts FT (1995): The epidemiology of measles: thirty years of vaccination. In *Measles Virus*, ter Meulen V, Billeter MA, eds. Berlin: Springer-Verlag, pp. 13–33.

Diaz RM, Bateman A, Emiliusen L, Fielding A, Trono D, Russell SJ, Vile RG (2000): A lentiviral vector expressing a fusogenic glycoprotein for cancer gene therapy. *Gene Ther* 7:1656–1663.

Dorig RE, Marcil A, Chopra A, Richardson CD (1993): The human CD46 molecule is a receptor for measles virus (Edmonston strain). *Cell* 75:295–305.

Duclos P, Ward BJ (1998): Measles vaccines: a review of adverse events. *Drug Safety* 19:435–454.

Esolen LM, Ward BJ, Moench TR, Griffin DE (1993): Infection of monocytes during measles. *J Infectious Dis* 168:47–52.

Fielding AK, Chapel-Fernandes S, Chadwick MP, Bullough FJ, Cosset FL, Russell SJ (2000): A hyperfusogenic gibbon ape leukemia envelope glycoprotein: targeting of a cytotoxic gene by ligand display. *Hum Gene Ther* 11:817–826.

Grosjean I, Caux C, Bella C, Berger I, Wild F, Banchereau J, Kaiserlian D (1997): Measles virus infects human dendritic cells and blocks their allostimulatory properties for CD4+ T cells. *J Exp Med* 186:801–812.

Grote D, Russell SJ, Cornu TI, Cattaneo R, Vile R, Poland GA, Fielding AK (2001): Live attenuated measles virus induces regression of human lymphoma xenografts in immunodeficient mice. *Blood* 97:3746–3754.

Hall WC, Kovatch RM, Herman PH, Fox JG (1971): Pathology of measles in rhesus monkeys. *Veterinary Pathol* 8:307–319.

Hammond AL, Plemper RK, Zhang J, Schneider U, Russell SJ, Cattaneo R (2001): Single-chain antibody displayed on a recombinant measles virus confers entry through the tumor-associated carcinoembryonic antigen. *J Virol* 75:2087–2096.

Helin E, Salmi AA, Vanharanta R, Vainionpaa R (1999): Measles virus replication in cells of myelomonocytic lineage is dependent on cellular differentiation stage. *Virology* 253:35–42.

Hsu EC, Iorio C, Sarangi F, Khine AA, Richardson CD (2001): CDw150(SLAM) is a receptor for a lymphotropic strain of measles virus and may account for the immunosuppressive properties of this virus. *Virology* 279:9–21.

Hsu EC, Sarangi F, Iorio C, Sidhu MS, Udem SA, Dillehay DL, Xu W, Rota PA, Bellini WJ, Richardson CD (1998): A single amino acid change in the hemagglutinin protein of measles virus determines its ability to bind CD46 and reveals another receptor on marmoset B cells. *J Virol* 72:2905–2916.

Johnston IC, ter Meulen V, Schneider-Schaulies J, Schneider-Schaulies S (1999): A recombinant measles vaccine virus expressing wild-type glycoproteins: consequences for viral spread and cell tropism. *J Virol* 73:6903–6915.

Karp CL, Wysocka M, Wahl LM, Ahearn JM, Cuomo PJ, Sherry B, Trinchieri G, Griffin DE (1996): Mechanism of suppression of cell-mediated immunity by measles virus. *Science* 273:228–231.

Klasse PJ, Bron R, Marsh M (1998): Mechanisms of enveloped virus entry into animal cells. *Adv Drug Delivery Rev* 34:65–91.

Kristensson K, Norrby E (1986): Persistence of RNA viruses in the central nervous system. *Ann Rev Microbiol* 40:159–184.

Lamb RA (1993): Paramyxovirus fusion: a hypothesis for changes. *Virology* 197:1–11.

Lawson ND, Stillman EA, Whitt MA, Rose JK (1995): Recombinant vesicular stomatitis viruses from DNA. *Proc Natl Acad Sci USA* 92:4477–4481.

Lecouturier V, Fayolle J, Caballero M, Carabana J, Celma ML, Fernandez-Munoz R, Wild TF, Buckland R (1996): Identification of two amino acids in the hemagglutinin glycoprotein of measles virus (MV) that govern hemadsorption, HeLa cell fusion, and CD46 downregulation: Phenotypic markers that differentiate vaccine and wild-type MV strains. *J Virol* 70:4200–4204.

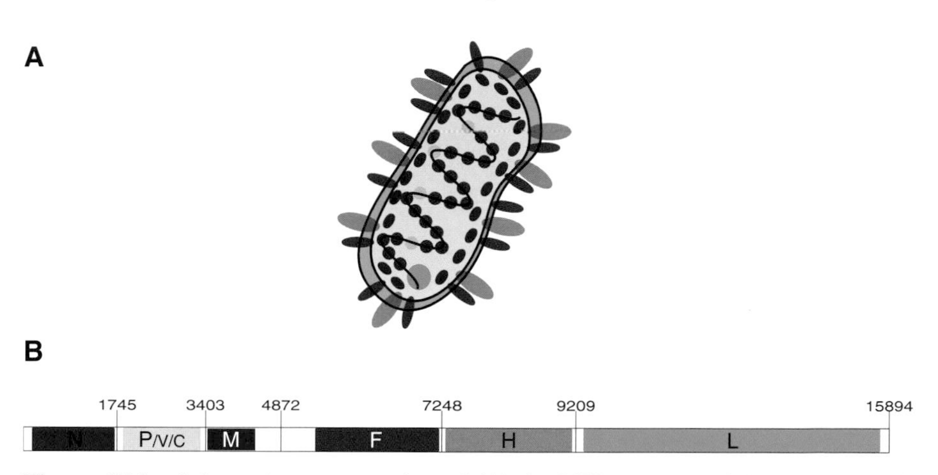

Figure 15.1 Schematic representation of (*A*) the MV genome and (*B*) an MV particle. The color code for proteins is indicated in the genome scheme. The lipid bilayer surrounding the particle is shown in blue.

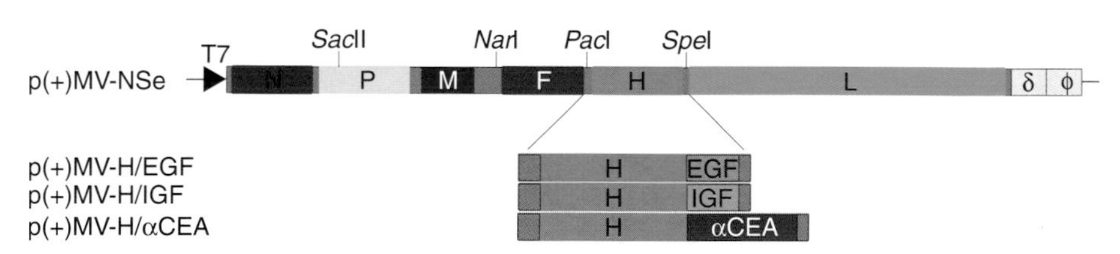

Figure 15.3 Genomic vectors for recovery of recombinant viruses. Plasmid p(+)MV-NSe specifies MV antigenomic RNA. Transcription with T7 RNA polymerase from a T7 promoter fused directly to the antigenome sequence yields RNA bearing the authentic nucleotides of the viral antigenomic 5′ terminus. The correct MV 3′ terminus is cleaved precisely by a hepatitis delta virus genomic ribozyme (δ) located immediately adjacent to the MV 3′-terminal nucleotides; the presence of T7 terminator (ϕ) increases the efficiency of this self-cleavage. Plasmid p(+)MV-NSe has been modified to introduce the unique restriction sites shown. Plasmids p(+)MV-H/EGF, p(+)MV-H/IGF, and p(+)MV-H/scAbCEA are all based on p(+)MV-NSe and contain chimeric MV H ORFs in which the EGF, IGF-1, and αCEA scAb ORFs, respectively, are fused to the 3′ terminus of H, separated by a sequence encoding a Factor Xa cleavage site. The chimeric H ORFs were generated in pCG expression vectors named pCG-H/EGF, pCG-H/IGF, and pCG-H/scAbCEA.

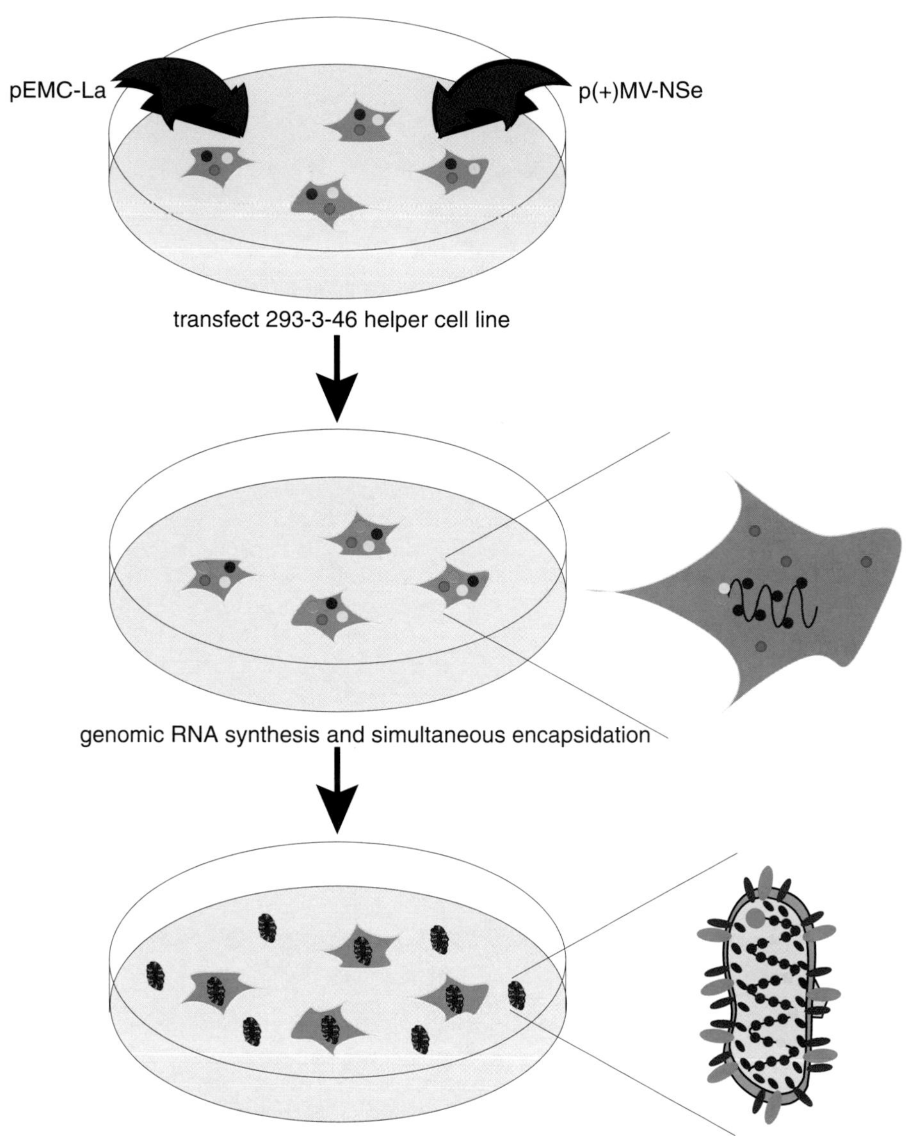

Figure 15.2 Recovery of MV from cloned DNA (Radecke et al., 1995). Plasmids encoding the full length MV antigenome (p(+)MV-NSe) and viral polymerase L (pEMC-La), both under the control of the T7 promoter are cotransfected in 293.3.46 cells constitutively expressing T7 polymerase (blue dot) and viral N (red dot) and P (green dot). T7 transcribes a positive strand MV antigenome that is encapsidated by N and P. From this template a negative strand genome is then transcribed by transiently expressed L protein. From this, viral mRNAs and new genomes are generated and infectious particles are produced and released. Overlay of the helper cells with highly fusogenic Vero cells facilitates detection of infectious centers.

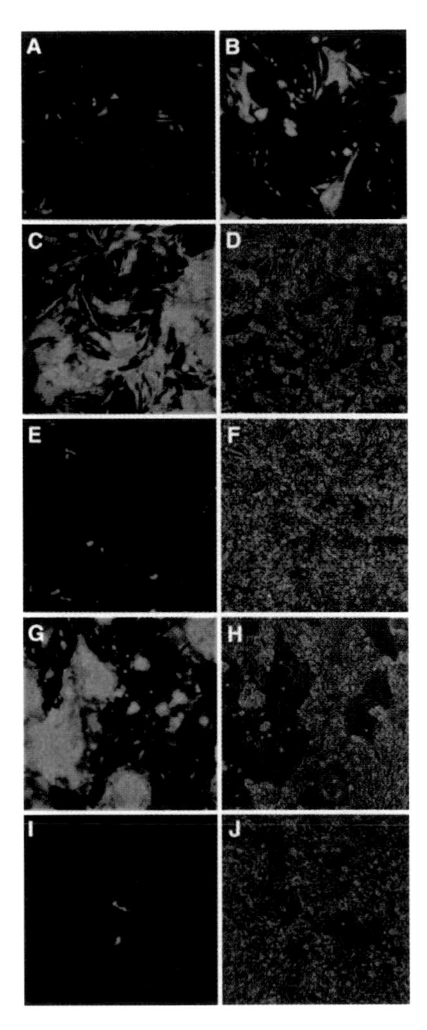

Figure 15.4 MV_{green}-H/EGF infects CHO-hEGFR cells. Infection of CHO-hEGFr cells (A to F), CHO-hEGFr.tr cells (*G* and *H*), and CHO cells (*I* and *J*) with MV_{green}-H/XhEGF (*A* to *D* and *G* to *J*) or MV (*E* and *F*). Cells were infected at a multiplicity of infection (MOI) of 1 and were monitored by fluorescent or phase-contrast microscopy 24 (*A*), 48 (*B*), or 72 (*C* to *J*) h after infection. (From Schneider et al., 2000, by permission from American Society for Microbiology.)

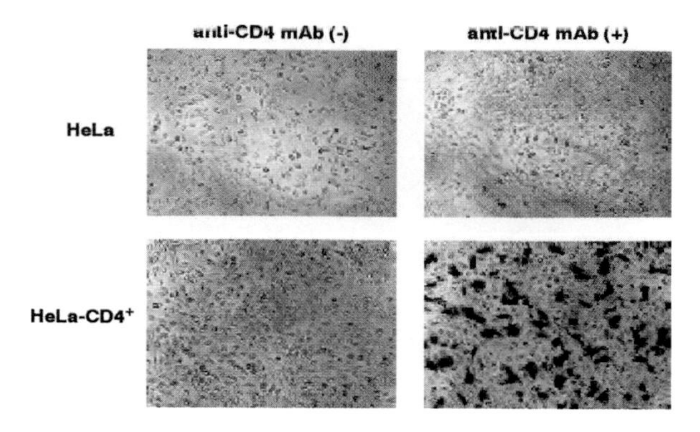

Figure 17.2 Infection of HeLa and HeLa-CD4^{+} cells with recombinant Sindbis virus derived from DH-BB-ZZ helper RNA that is transducing the bacterial *LacZ* gene. Viral supernatants (200 μl) were preincubated without or with anti-CD4 mAb (0.5 μg/ml) at room temperature for 1 h, and added to cells (2×10^5) in 6-well plates. After 1 h at room temperature, cells were washed with PBS and incubated in growth medium for 24 h. Viral infection was evaluated by X-Gal staining.

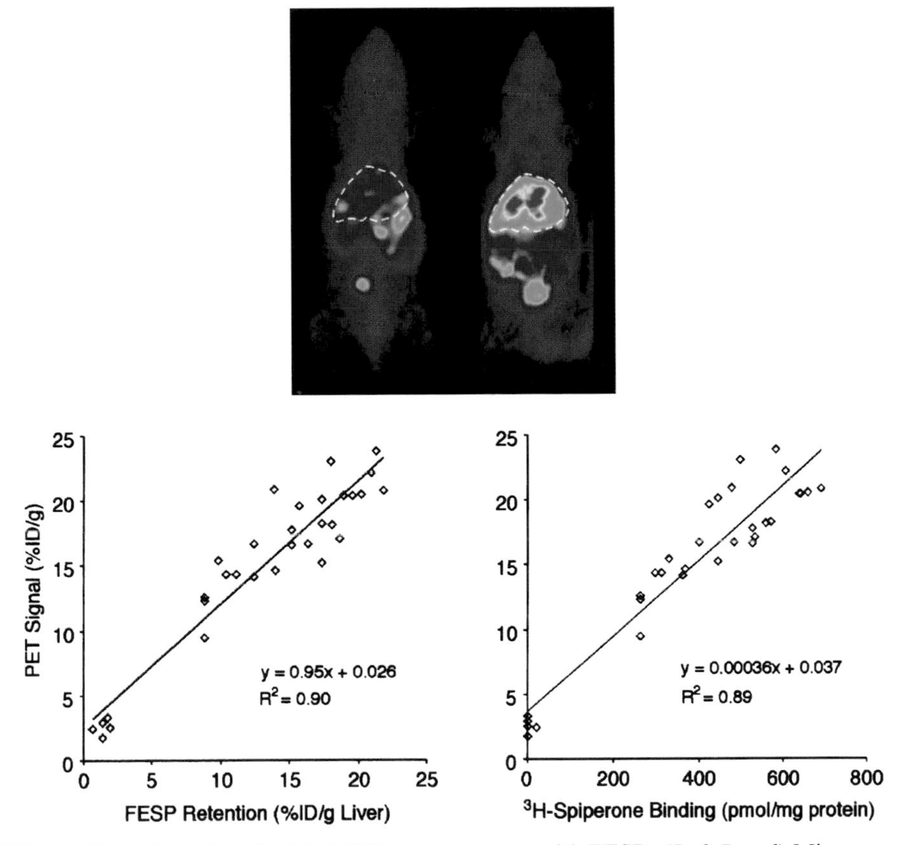

Figure 30.3 Imaging the D2R PET reporter gene with FESP. (*Left Panel*) Mice were injected intravenously with ad.βGal or ad.D2R (2×10^9 pfu). Two days later the mice were injected with FESP then imaged by microPET. Whole body coronal projections are shown, with the livers outlined in white. Red indicates the greatest intensity; purple the least. (*Right Panel*). Mice were injected with varying titers of ad.D2R ranging from 5×10^6 to 9×10^9 pfu, subjected to microPET analysis with FESP and then sacrificed. Liver samples were counted by well counting to determine the level of retained [^{18}F]-FESP and labeled metabolites. Additional liver samples were assayed for [^{3}H] spiperone binding. The graph on the left plots in vivo ^{18}F retention determined by region of interest analysis from the PET scan data versus the in vivo hepatic ^{18}F retention determined by well counting. The right graph shows in vivo ^{18}F retention determined by region of interest analysis from the PET scan data plotted versus the level of hepatic D2R receptor activity determined by [^{3}H] spiperone binding. (From Herschman et al. 2000, by permission.)

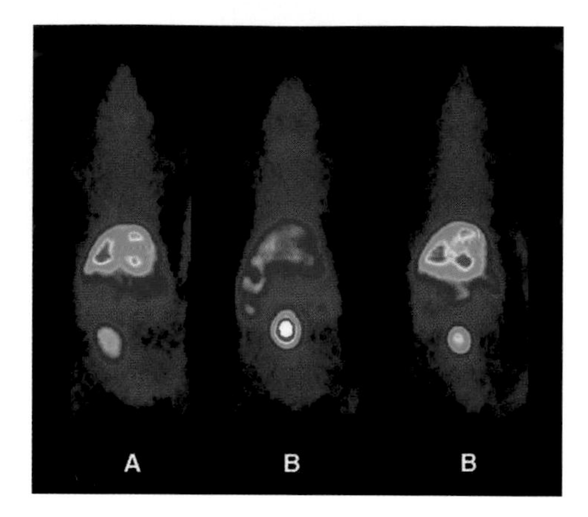

Figure 30.4 Pharmacologic demonstration of the specificity of the D2R/FESP PET reporter gene/PET reporter probe imaging system. Two mice were injected intravenously with ad.D2R. After 2 days, mouse B was injected intraperitoneally with (+)-butaclamol (2 mg/kg). After an additional hour, both mice were injected with FESP and subsequently imaged with microPET. Mouse B was injected a second time with FESP after another 3 days (5 days after virus injection) and again imaged with microPET. Summed whole body coronal images are shown. (From Herschman et al. 2000, by permission.)

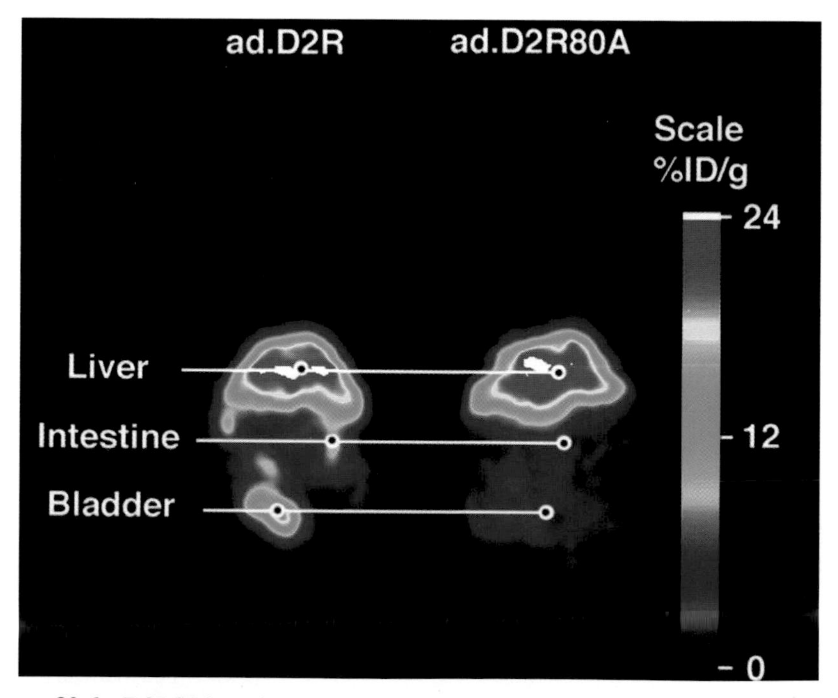

Figure 30.6 D2R80A and wild-type D2R are equally competent as PET reporter genes. Mice were injected intravenously with 2×10^9 pfu of ad.D2R or ad.D2R80A. Two days later, the mice were injected with FESP and imaged with microPET.

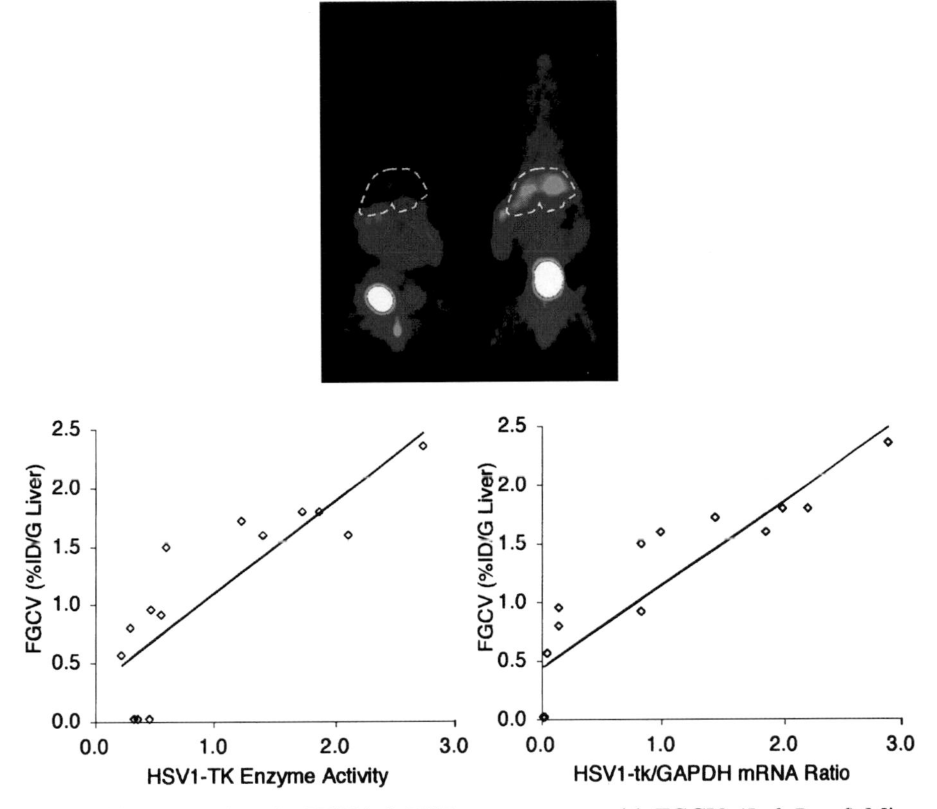

Figure 30.7 Imaging the HSV1-tk PET reporter gene with FGCV. (*Left Panel*) Mice were injected intravenously with 1×10^9 pfu of ad.βgal or ad.HSV1-tk. Two days later mice were injected with FGCV and imaged with microPET after 1 h. (*Right Panel*). Mice were injected intravenously with varying numbers of ad.HSV1-tk. After 2 days, the mice were injected with FGCV and sacrificed after 1 h. Liver samples were analyzed for FGCV retention (by well counting), HSV1-tk mRNA (by Northern blot), and HSV1-TK enzyme (by measuring conversion of [^{3}H]GCV to [^{3}H]GCV-P). (From Herschman et al., 2000, by permission.)

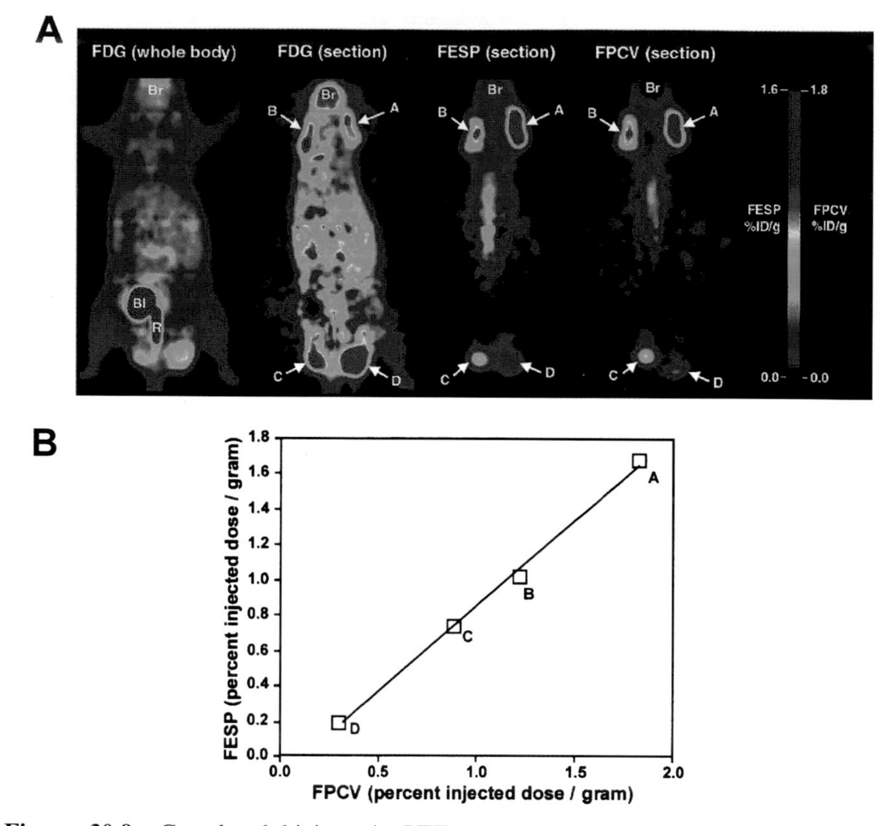

Figure 30.9 Correlated bicistronic PET reporter gene expression. (*A*) MicroPET imaging of xenograft tumors expressing differing levels of the pCMV-D2R-IRES-HSV1-sr39tk bicistronic plasmid. Three stably transfected C6 cell lines (A, B, and C) and C6 cells (D) were injected into four distinct sites on a nude mouse. Ten days later, the mouse was injected with FDG and analyzed by microPET. A microPET analysis was performed with FESP 24 h later. The following day a third microPET analysis was performed with FPCV. The FDG whole body image on the left averages all coronal (horizontal) planes. As a result, the tumors are not well visualized. The second FDG image, a set of coronal images that passes through all the tumors, shows FDG accumulation in each tumor. Br: brain; Bl: Bladder; R. Rectum. (*B*) Plot of FESP %ID/gm tissue versus FPCV %ID/gm of tissue. The data are obtained from regions of interest drawn on the images in panel A. The error bars represent the standard deviations for three regions of interest placed on the tumor images. The correlation between FESP and FPCV microPET signals is $r^2 = 0.99$. (From Yu et al., 2000, by permission.)

Lorence RM, Reichard KW, Katubig BB, Reyes HM, Phuangsab A, Mitchell BR, Cascino CJ, Walter RJ, Peeples ME (1994): Complete regression of human neuroblastoma xenografts in athymic mice after local Newcastle disease virus therapy. *J Nat Cancer Inst* 86:1228–1233.

Lorence RM, Rood PA, Kelley KW (1988): Newcastle disease virus as an antineoplastic agent: induction of tumor necrosis factor-alpha and augmentation of its cytotoxicity. *J Nat Cancer Inst* 80:1305–1312.

Manchester M, Eto DS, Valsamakis A, Liton PB, Fernandez-Munoz R, Rota PA, Bellini WJ, Forthal DN, Oldstone MB (2000): Clinical isolates of measles virus use CD46 as a cellular receptor. *J Virol* 74:3967–3974.

McChesney MB, Miller CJ, Rota PA, Zhu YD, Antipa L, Lerche NW, Ahmed R, Bellini WJ (1997): Experimental Measles. I. Pathogenesis in the normal and the immunized host. *Virology* 233:74–84.

Morrison T, Portner A (1991): Structure, function, and intracellular processing of the glycoproteins of paramyxoviridae. In *The Paramyxoviruses*, Kingsbury DW, ed. New York: Plenum Press, pp. 347–382.

Moss WJ, Cutts F, Griffin DE (1999): Implications of the human immunodeficiency virus epidemic for control and eradication of measles. *Clin Infect Dis* 29:106–112.

Mrkic B, Pavlovic J, Rulicke T, Volpe P, Buchholz CJ, Hourcade D, Atkinson JP, Aguzzi A, Cattaneo R (1998): Measles virus spread and pathogenesis in genetically modified mice. *J Virol* 72:7420–7427.

Nakayama T, Mori T, Yamaguchi S, Sonoda S, Asamura S, Yamashita R, Takeuchi Y, Urano T (1995): Detection of measles virus genome directly from clinical samples by reverse transcriptase-polymerase chain reaction and genetic variability. *Virus Res* 35:1–16.

Naniche D, Varior-Krishnan G, Cervoni F, Wild TF, Rossi B, Rabourdin-Combe C, Gerlier D (1993): Human membrane cofactor protein (CD46) acts as a cellular receptor for measles virus. *J Virol* 67:6025–6032.

Norrby E (1995): The paradigms of measles vaccinology. In *Measles Virus*, Fields BN, Knipe DM, et al., eds. New York: Raven Press, pp. 167–180.

Ohgimoto S, Ohgimoto K, Niewiesk S, Klagge IM, Pfeuffer J, Johnston IC, Schneider-Schaulies J, Weidmann A, ter Meulen V, Schneider-Schaulies S (2001): The haemagglutinin protein is an important determinant of measles virus tropism for dendritic cells in vitro. *J Gen Virol* 82:1835–1844.

Oldstone MB, Lewicki H, Thomas D, Tishon A, Dales S, Patterson J, Manchester M, Homann D, Naniche D, Holz A (1999): Measles virus infection in a transgenic model: virus-induced immunosuppression and central nervous system disease. *Cell* 98:629–640.

Ono N, Tatsuo H, Hidaka Y, Aoki T, Minagawa H, Yanagi Y (2001a): Measles viruses on throat swabs from measles patients use SLAM (CDw150), but not CD46, as a cellular receptor. *J Virol* 75:4399–4401.

Ono N, Tatsuo H, Tanaka K, Minagawa H, Yanagi Y (2001b): V domain of human SLAM (CDw150) is essential for its function as a measles virus receptor. *J Virol* 75:1594–1600.

Palumbo P, Hoyt L, Demasio K, Oleske J, Connor E (1992): Population-based study of measles and measles immunization in human immunodeficiency virus-infected children. *Ped Infect Dis J* 11:1008–1014.

Patterson JB, Scheiflinger F, Manchester M, Yilma T, Oldstone MB (1999): Structural and functional studies of the measles virus hemagglutinin: identification of a novel site required for CD46 interaction. *Virology* 256:142–151.

Peeples ME (1991): Paramyxovirus M proteins: Pulling it all together and taking it on the road. In *The Paramyxoviruses*, Kingsbury DW, ed. New York: Plenum Press, pp. 427–456.

Peng KW, Ahmann GJ, Pham L, Greipp PR, Cattaneo R, Russell SJ (2001): Systemic therapy of myeloma xenografts by an attenuated measles virus. *Blood* 98:2002–2007.

Plemper RK, Hammond AL, Cattaneo R (2000): Characterization of a region of measles virus hemagglutinin sufficient for its dimerization. *J Virol* 74:6485–6493.

Plemper RK, Hammond AL, Gerlier D, Fielding A, Cattaneo R (2002): Strength of envelope protein interaction modulates cytopathicity of measles virus. *J. Virol* 76:5051–5061.

Pringle CR (1991): The genetics of paramyxoviruses. In *The Paramyxoviruses*, Kingsbury DW, ed. New York: Plenum Press, pp. 1–39.

Radecke F, Spielhofer P, Schneider H, Kaelin K, Huber M, Dotsch C, Christiansen G, Billeter MA (1995): Rescue of measles viruses from cloned DNA. *EMBO J* 14:5773–5784.

Roscic-Mrkic B, Schwendener RA, Odermatt B, Zuniga A, Pavlovic J, Billeter MA, Cattaneo R (2001): Roles of macrophages in measles virus infection of genetically modified mice. *J Virol* 75:3343–3351.

Schneider U, Bullough F, Vongpunsawad S, Russell SJ, Cattaneo R (2000): Recombinant measles viruses efficiently entering cells through targeted receptors. *J Virol* 74:9928–9936.

Schnell MJ, Mebatsion T, Conzelmann KK (1994): Infectious rabies viruses from cloned cDNA. *Embo J* 13:4195–4203.

Sergiev PG, Ryazantseva NE, Shroit IG (1960): The dynamics of pathological processes in experimental measles in monkeys. *Acta Virologica* 4:265–273.

Taqi AM, Abdurrahman MB, Yakubu AM, Fleming AF (1981): Regression of Hodgkin's disease after measles. *Lancet* 1:1112.

Tatsuo H, Okuma K, Tanaka K, Ono N, Minagawa H, Takade A, Matsuura Y, Yanagi Y (2000a): Virus entry is a major determinant of cell tropism of Edmonston and wild-type strains of measles virus as revealed by vesicular stomatitis virus pseudotypes bearing their envelope proteins. *J Virol* 74:4139–4145.

Tatsuo H, Ono N, Tanaka K, Yanagi Y (2000b): SLAM (CDw150) is a cellular receptor for measles virus. *Nature* 406:893–897.

Tatsuo H, Ono N, Yanagi Y (2001): Morbilliviruses use signaling lymphocyte activation molecules (CD150) as cellular receptors. *J Virol* 75:5842–5850.

Vulliemoz D, Roux L (2001): "Rule of six": how does the Sendai virus RNA polymerase keep count? *J Virol* 75:4506–4518.

16

TARGETING OF POLIOVIRUS REPLICONS TO NEURONS IN THE CENTRAL NERVOUS SYSTEM

CASEY D. MORROW PH.D., MATHEW PALMER B.S., CHERYL A. JACKSON PH.D., LISA K. JOHANSEN PH.D., ANDREA BLEDSOE PH.D., DAVID D. ANSARDI PH.D., DONNA C. PORTER PH.D., CHARLES N. COBBS M.D., AND JEAN D. PEDUZZI PH.D.

POLIOVIRUS REPLICATION

Poliovirus is a member of the *Picornaviridae*, which consists of five genera: enterovirus, rhinovirus, cardiovirus, aphthovirus, and hepatovirus. Poliovirus is classified as an enterovirus because of the fecal/oral transmission cycle. For the most part, poliovirus infections are mild with clinical symptoms of mild gastritis and slight fever. In approximately one percent of infections, poliovirus reaches the central nervous system (CNS), where it replicates primarily in motor neurons of the anterior horn of the spinal cord. In some cases, this results in a flaccid paralysis (poliomyelitis), that can lead to permanent paralysis (Bodian, 1949; Sabin, 1956).

Poliovirus has been a topic of study for virologists, molecular biologists, and immunologists for over 30 years. The major features of the poliovirus life cycle have been elucidated. Poliovirus has a single-stranded genome of

Vector Targeting for Therapeutic Gene Delivery, Edited by David T. Curiel and Joanne T. Douglas
ISBN 0-471-43479-5

the plus sense (mRNA) that is about 7000 to 8000 base pairs in length (Kitamura et al., 1981). The virions are non-enveloped with an icosohedral shape of approximately 22–30 nm in diameter; the complete three-dimensional structure of the poliovirus is known (Hogle, Chow, and Filman, 1985). The replication cycle of poliovirus takes place in the cytoplasm; there are no DNA intermediates. Poliovirus infection begins with the interaction of the virion with a host cell surface molecule, CD155, which is a member of the immunoglobulin-like gene superfamily (Mendelsohn, Wimmer, and Racaniello, 1989). Although the normal function of CD155 is not known, similarities between this molecule and other cell adhesion-like molecules (nectins) suggest a common function (Lopez, Aoubala, and Jourdier, 1998; Satoh-Horikawa, Nakanishi, and Takahashi, 2000; Takahashi, Nakanishi, and Miyahara, 1999). CD155 is sufficient and necessary for infection of cells by poliovirus. Expression of this molecule in cells normally not susceptible to poliovirus infection, for example murine cells, renders these cells susceptible to poliovirus (Mendelsohn, Wimmer, and Racaniello, 1989; Ren et al., 1990; Ren and Racaniello, 1992a, 1992b). Following interaction with the receptor, the virus undergoes a conformational change in which the particle is converted to an altered (A) particle. Approximately 50–90% of the A particles are sloughed off the surface of a susceptible cell and can no longer interact with CD155 (Mandel, 1965; Putnak and Phillips, 1981; Rueckert, 1990). The particles that remain attached to the cell are internalized, perhaps by receptor-mediated endocytosis.

Following release of the viral RNA genome into the cytoplasm of the infected cell, translation of the entire genome produces a large, single polyprotein, which is processed in *cis* and *trans* by virally encoded proteases into the individual viral proteins. The proteolytic processing cascade of poliovirus is complex and many of the viral proteins serve dual functions in the poliovirus life cycle (Nicklin et al., 1986). The virus structural proteins are found at the amino terminal one-third of the long polyprotein (designated as P1). The remaining two thirds of the viral genome (P2 and P3) encode proteins required for the selective translation of viral genomes during the infection process and enzymes necessary to catalyze the synthesis of new viral nucleic acids (Rueckert, 1990). The infection process in vitro is rapid with most of the synthesis of new negative- and positive-sense RNAs complete 4 to 6 h postinfection; at approximately 6 h postinfection, assembly of new viruses occurs (Koch and Koch, 1985). The exact mechanism by which poliovirus RNA molecules are selectively encapsidated into virions is still not known. At approximately 8 to 10 h postinfection, host cell lysis occurs and virions are released into the surrounding medium.

POLIOVIRUS PATHOGENESIS

The natural host range for poliovirus is confined to humans and nonhuman primates (Sabin, 1956). In the host, poliovirus exhibits a restricted tissue tropism, due primarily to the restricted expression of CD155 (Freistadt, 1994; Freistadt,

Kaplan, and Racaniello, 1990; Hogle, Chow, and Filman, 1985; Mendelsohn et al., 1986). The translation and replication of the viral genome depends on cytoplasmic host cell proteins (Andino et al., 1993; Andino, Rieckhof, and Baltimore, 1990; Harris et al., 1994; Roehl et al., 1997; Rohll et al., 1994). Poliovirus shows a predilection for the infection of neurons, particularly motor neurons of the cervical and lumbar enlargements of the spinal cord; astrocytes, oliogdendrocytes, microglia, and inflammatory cells within the CNS are not generally infected by poliovirus (Bodian, 1949; Bodian and Horstmann, 1965; Jubelt et al., 1980). The infection of neurons results in dissolution of the cytoplasmic bodies responsible for neuronal protein synthesis (Nissl bodies), which leads to nuclear changes and shrinkage of the cytoplasm (chromatolysis). Changes in the neurons are accompanied by influx of inflammatory cells, including microglial, polymorphononuclear, and mononuclear phagocytes (Bodian, 1949). Further understanding of the pathogenesis of poliovirus has come from analysis of the gene expression of CD155 during development (Gromeier et al., 2000). In adult mice, immunofluorescence using a monoclonal antibody to CD155 revealed that, as would be expected, cells of the gut and neurons of the anterior horn in the spinal cord expressed the highest levels of CD155; even so, detection of the receptor on these cells using immunofluorescence is often very difficult. Recent studies have analyzed the expression of the CD155 promoter during development and in the adult (Gromeier et al., 2000). It was most active in early development, just prior to birth and was confined to the ventral midline structures in the developing CNS. Thereafter, the promoter was only minimally active, even in adult neurons. The results are consistent with the idea that the expression of CD155 in adults is very low, resulting in low levels (if any) of receptor expression on most cells (Gromeier et al., 2000).

A second feature of poliovirus cell tropism is the influence of the 5′ nontranslated region (5′NTR) of the viral genome. Poliovirus contains an unusually long 5′NTR, about 743 nucleotides that contain elements important for viral replication and translation. The 5′ terminal 108 nucleotides of the poliovirus genome are predicted to fold into a cloverleaf structure which has been shown to be essential for RNA replication (Andino et al., 1993; Andino, Reickhof, and Baltimore, 1990; Harris et al., 1994). A viral protein, 3CD, interacts with the cloverleaf structure (Andino et al., 1993). The 5′ RNA cloverleaf also contains signal(s) for viral translation. A host cell protein, poly (C) binding protein 2 (PCB2) has been shown to interact with the 5′ RNA stem-loop on a region opposite to 3CD. If PCB2 is bound to the cloverleaf, translation of the viral genome is favored; however, when 3CD is bound to the cloverleaf, translation is repressed. The differential binding of 3CD and PCB2 to the cloverleaf has been proposed to be a control switch between replication and translation of the viral genome (Andino, Reickhof, and Baltimore, 1990; Harris et al., 1994; Parsley et al., 1997). Additional cellular proteins such as the La autoantigen and the polypyrimidine track binding protein (PTB) interact with the poliovirus 5′ NTR and are essential for translation (Borman et al.,

1993; Hellen et al., 1994, 1993; Jang et al., 1990; Luz and Beck, 1990, 1991; Meerovitch, Pelletier, and Sonenberg, 1989; Svitkin et al., 1994). Differential expression of these proteins in cells can modulate the translation of poliovirus RNA and replication of the virus.

GENE EXPRESSION VECTORS BASED ON POLIOVIRUS

The complete nucleotide sequence of the poliovirus genome is known (Kitamura et al., 1981). The viral genome consists of a 5′ nontranslated region of approximately 743 nucleotides which contains elements required for RNA replication and translation (Kitamura et al., 1981). The RNA genome encodes a single long protein, which is subsequently processed by viral encoded proteases (Rueckert and Wimmer, 1984). The capsid proteins are encoded within a region of the genome, designated as P1, which is released from the translating polyprotein by the autocatalytic activity of a viral encoded protease, 2A, which immediately follows the P1 protein (Toyoda et al., 1986). Studies on the translation/replication of poliovirus were facilitated by the discovery that a cDNA of poliovirus can give rise to virus following transfection of cells (Racaniello and Baltimore, 1981). Subsequently, in vitro transcription of the cDNA resulted in RNA that upon transfection, gave rise to poliovirus. Previous studies from this laboratory and others have shown that foreign genes can be substituted within the P1 region and the in vitro transcribed RNA will maintain the capacity for self-amplification following introduction into cells (Ansardi et al., 1996; Ansardi, Porter, and Morrow, 1993; Porter et al., 1993, 1997; Porter, Ansardi, and Morrow, 1995) (Fig. 16.1). The foreign gene sequence must be inserted into the RNA genome such that the translational reading frame is maintained between the P2 and P3 region proteins in order to preserve the capacity of the RNA to undergo self-amplification. Studies on the requirements for cleavage by the 2A protease have revealed that the fourth upstream amino acid is important for efficient cleavage (Nicklin et al., 1987, 1986). Thus, foreign proteins expressed from the poliovirus replicon contain additional amino acids at the COOH terminus required for efficient 2A cleavage. Previous studies from this laboratory have found that the addition of approximately 8 to 10 amino acids of the P1 protein prior to the 2A cleavage results in efficient cleavage in most cases (Porter et al., 1993; Porter, Ansardi, and Morrow, 1995).

Recombinant poliovirus genomes that encode a variety of foreign genes substituted for all or part of P1 have been constructed. Previous studies from this laboratory have described the construction and characterization of replicons that encode proteins such as green fluorescent protein (GFP) (Bledsoe et al., 2000; Jackson et al., 2001) and luciferase (Ansardi et al., 2001; Porter et al., 1998), replicons that encode proteins that modulate the immune system and CNS, and replicons that can be used as recombinant vaccines. The replicon vectors can accommodate genes of up to 2500 base pairs without affecting the capacity of the replicon to be encapsidated (unpublished). All

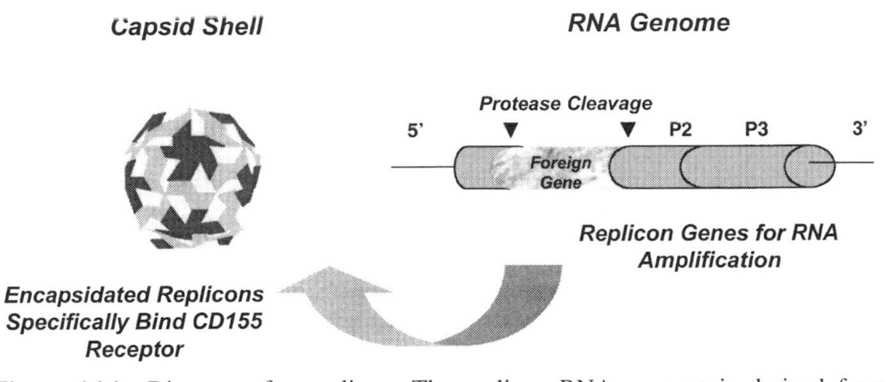

Figure 16.1 Diagram of a replicon. The replicon RNA genome is derived from the complete cDNA of poliovirus. The viral genome encoding the capsid (P1) has been substituted with foreign genes (up to 2500 base pairs). The remaining poliovirus genome, P2 and P3, encode proteins required for translation/replication of the replicon RNA. The replicon RNA can be generated from in vitro transcription of the replicon cDNA. Transfection of the RNA into cells results in the translation/replication of the replicon RNA and subsequent expression of the foreign gene. Protease cleavage sites (for 2A, at the amino terminus of P2) have been engineered into replicon genomes to facilitate release of the foreign proteins (black triangles).

To encapsidate the replicon RNAs, the P1 protein is provided in trans (usually from a recombinant vaccinia virus). The P1 protein is processed by viral proteases (provided by P2–P3 regions of the replicons), resulting in the encapsidated replicon into authentic poliovirions. The encapsidated replicons enter cells via the CD155 receptor like poliovirus.

of these replicons retain the features of self-amplification on introduction into the cytoplasm of target cells. To facilitate delivery of these replicons, we have developed a method to encapsidate the replicon RNA by providing the P1 protein in trans (Ansardi, Porter, and Morrow, 1993). For this, we have generated a recombinant vaccinia virus that expresses the P1 protein upon infection. Infection of cells with this vaccinia virus followed by transfection of replicon genomes results in the production of encapsidated replicons. Production of sufficient quantities of encapsidated replicons for experimental use can be achieved by passage of the encapsidated replicons in the presence of the recombinant vaccinia virus that encodes P1. Purification of encapsidated replicons can be achieved by using a variety of methods, although we have routinely used density gradient centrifugation to derive stocks of these encapsidated replicons (Ansardi, Porter, and Morrow, 1993). Preparations of encapsidated replicons are tested for vaccinia virus, and recombinant poliovirus by passage on HeLa cells, which are susceptible to both of these viruses (Bledsoe, Gillespie, and Morrow, 2000). Serial passage of replicons can be used to detect viral contamination, because replicons can only undergo one round of infection. Complete removal of the vaccinia virus is achieved because of the differential properties of poliovirus and vaccinia virus: size, susceptibility to detergents and salts.

To date, we have not observed recombination to generate infectious poliovirus. The reason for this is unclear, but could be related in part to the inability of the P1 mRNA to serve as a substrate for poliovirus RNA-dependent RNA-polymerase, which would be necessary for recombination to occur (Kirkegaard and Baltimore, 1986). Using this procedure, we have prepared preparations of up to 10^9 infectious units (representing nearly 10^{11} particles) that can be used for gene delivery both in vitro and in vivo.

GENE DELIVERY TO THE CNS

The ability to deliver and express foreign genes in cells of the CNS would benefit studies on basic cellular functions and would provide new avenues for therapeutics against neurological diseases and spinal cord injuries/trauma. Viral vectors have generated considerable attention for gene delivery to the CNS (see an excellent, comprehensive review by Hermens and Verhaagen, 1998). The first studies involved the use of vectors based on the DNA virus, herpes simplex virus (HSV). The natural tropism of HSV for neurons provided the impetus for the genetic manipulation of this large DNA virus for expression of foreign genes. These studies have relied on the development of attenuated HSV, in which the genes that encode proteins toxic to neurons have been eliminated or rendered temperature-sensitive. The foreign genes are cloned into the attenuated viruses, which retain their neuronal specificity. An alternative strategy has been the use of amplicon-based vectors in which foreign genes are encapsidated into HSV virions that are provided in trans by helper virus (usually an attenuated HSV). Although amplicon vectors provide a system for expression of large foreign genes with minimal HSV proteins, the drawbacks for this approach include the low titers of encapsidated amplicon and the absence of an effective means to eliminate the helper virus. Adenovirus-based vectors have also been used extensively for gene delivery to the CNS. Adenovirus-based vectors have many advantages, including the ability to generate large quantities of recombinant virus and the potential for specific targeting to different cell types of the CNS. Two potential drawbacks of both HSV and adenovirus for sustained gene delivery to the CNS involve the lack of integration of the foreign DNA (due to the episomal replication of these viruses in the nucleus) and the generation of a host cell immune response against the viral vector to eliminate transduced cells (Byrnes et al., 1995; Dewey et al., 1999). Two vector systems that appear, at least in the early going, to circumvent these problems, are based on adeno-associated virus (AAV) (Kaplitt and Makimura, 1997) and retroviruses (Hermens and Verhaagen, 1998). AAV are nonpathogenic and appear to not be very immunogenic. The intense interest in AAV as a gene transfer vector has resulted in the development of effective methods to generate large quantities of recombinant AAV that are virtually devoid of helper virus (AAV requires adenovirus gene products for replication). Expression of foreign genes in the CNS have been demonstrated using AAV vectors. Retrovirus-based vectors have also been tested for gene delivery in the CNS. In this case, the original vectors were based on murine leukemia virus (MuLV). The major drawback of systems based on MuLV is the inability of the virus to infect and integrate into postmitotic cells, such as neurons, which are the primary cell typc in the CNS. Recently, this limitation has been overcome through the development of vectors based on lentiviruses (e.g., human immunodeficiency virus HIV) (Blomer et al., 1997). Engineering of lentivirus vectors has resulted in the ability to successfully transduce the majority of the cells in the CNS, including postmitotic neurons. Recent studies have shown that lentivirus-based vectors are more effective than AAV at gene delivery within the CNS (Blomer et al., 1997).

RNA viruses have not been used as extensively as DNA viruses for gene delivery to the CNS. One reason is that because the RNA virus replication cycle does not include DNA intermediates (or integration into the host chromosome), the gene expression expected from such vectors would be transient. However, this concern is offset by the considerable potential of RNA viruses for high level foreign gene expression in the CNS due to the RNA genome amplification during replication. One RNA virus vector system is based on the alphavirus Sindbis virus (Gwag et al., 1998; Kerr et al., 2000). Previous studies have shown that this virus will effectively target neurons of the CNS, although a potential drawback of this vector is that infection leads to apoptosis of neurons. A second vector system has been recently reported based on measles virus, which has been used to express GFP in neurons (Duprex et al., 2000).

The development and use of poliovirus as a vector system for gene delivery to the CNS has many attractive features. It has been known for some time from analysis of post mortem tissue of the CNS of patients who had died from poliomyelitis that once poliovirus invades the CNS, it has a specific tropism for neurons, particularly motor neurons of the cervical and lumbar ventral horns of the spinal cord (Bodian, 1949). The destruction of these primary motor neurons resulted in the characteristic paralysis observed in poliomyelitis. Further understanding of the aspects of poliovirus pathogenesis was facilitated by the development of a mouse transgenic for the human poliovirus receptor (CD155) (Deatly et al., 1999, 1998; Ren et al., 1990; Ren and Racaniello, 1992). Following either intracerebral or intraspinal inoculation of wild-type poliovirus, examination of the tissues by in situ hybridization with RNA probes revealed that lesions in the spinal cord were most severe with inflammation and neuronal degeneration localized largely to the ventral horns. However, infected neuronal cells were detected in all areas of gray matter including the ventral horn, intermediate and interomediolateral columns, and dorsal horn. Although viral RNA was detected in the cytoplasm of neurons, both in the cell bodies and in their axonal and dendritic processes, viral RNA was not detected in vascular endothelial or glial cells. Viral replication was also not detected in the cerebral cortex, cerebellum, hippocampus, thalamus, hypothalamus, or olfactory bulb. If the virus was introduced to the animal via direct intracerebral inoculation, neurons in the brain were extensively infected and viral replication was detected in the cerebral cortex, pyramidal layer of the hippocampus, olfactory bulbs, thalamus, hypothalamus, and deep cerebellar nuclei. Although the route of inoculation may also determine which neurons become infected, it was clear that other resident cells of the CNS (astrocytes, oligodendrocytes, and microglia) were generally not susceptible to infection by poliovirus.

The exploitation of neurotropic viruses for gene delivery to the CNS has many potential applications for the treatment of neurological diseases and spinal cord injury/trauma. Two issues needed to be established for gene delivery to the CNS using poliovirus-based replicons. The first was to confirm that replicons maintained the tropism of poliovirus following introduction into the CNS. For these studies, we utilized replicons encoding luciferase (Bledsoe, Gillespie, and Morrow, 2000) or GFP (Bledsoe et al., 2000). Replicons were given by intraspinal inoculation or intrathecal delivery using a small catheter. Gene expression was monitored by direct enzymatic assay (luciferase) or by immunofluorescence using antibodies specific for GFP. In either case, we found that gene expression following delivery was evident as early as 6 h postinoculation, gene expression peaked by 12 to 24 h postinoculation, and the levels returned to background by 96 h post inoculation (Bledsoe, Gillespie, and Morrow, 2000; Bledsoe et al., 2000). This rapid gene expression correlated with the kinetics of the replication cycle for a single round of poliovirus infection in vivo. Analysis of the replicon-infected CNS revealed that the vast majority of infected cells were neurons, as judged by immunofluorescence using antibodies specific for neurons (Fig. 16.2). One of the more interesting aspects of the study was that expression of the foreign protein within the CNS occurred

Figure 16.2 Targeted gene expression in neurons by encapsidated replicons. Mice transgenic for CD155 were given replicons encoding GFP into the CNS by an intrathecal catheter. After 72 h, the animal was sacrificed and the spinal cord isolated; the samples were processed as previously described (Bledsoe, Gillespie, and Morrow, 2000; Bledsoe et al., 2000) and stained for GFP and NeuN (a neuronal marker).

A. Longitudinal frozen section through the lumbar enlargement in a poliovirus receptor (PVR) transgenic mouse that had received a single injection of replicons encoding GFP 72 h earlier. Anti-GFP staining using a biotinylated secondary antibody, Alexa 488 (green) fluorochrome. Excitation using an Argon laser reveals large, triangular profiles indicative of alpha motor neurons (white arrowhead). GFP expression is absent in the axonal tracts of the white matter (wm). Scale bar equals 40 μm. *B*. Identical section described in panel A. Section stained with anti-NeuN antibodies and visualized with a Krypton laser to excite the Alexa 594 (red) fluorochrome. Large triangular cells (motor neurons) are stained (white arrowhead). NeuN staining is confined to the gray matter of the spinal cord and is not found within the white matter. Scale bar equals 40 μm. *C*. Merged image of red and green channels of the confocal images described previously. Yellow fluorescence indicates the presence of anti-GFP fluorescence (green) coinciding with the NeuN staining (red) indicating neuronal identity (white arrowhead). GFP and NeuN expression are both absent in the white matter, wm. Scale bar equals 40 μm.

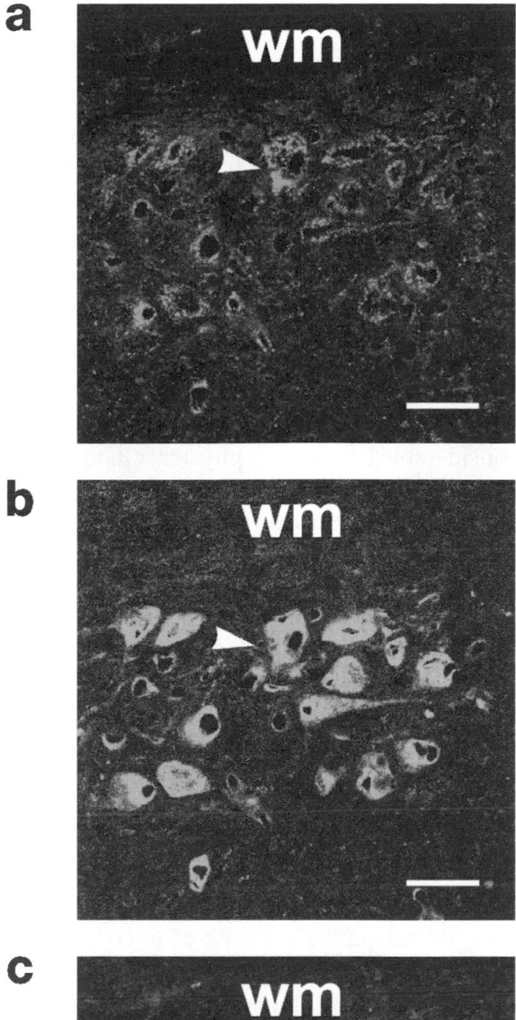
a
wm
b
wm

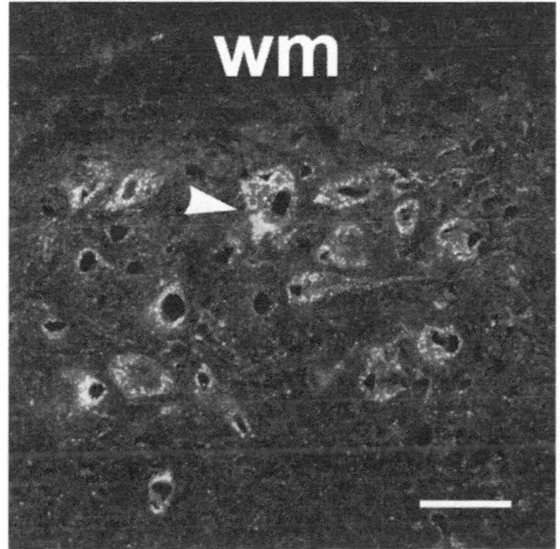
c
wm

at sites close to and far away from the site of inoculation. Most probably, replicons were able to access sites distal from the inoculation using transport via the cerebral spinal fluid (CSF). Consistent with what had been observed for poliovirus-infected CNS, we did not observe expression of the recombinant protein in other cells of the CNS, such as oligodendrocytes, astrocytes, or microglia. To exploit these features, we have analyzed the effects of replicons encoding small, biologically active molecules. Our first study utilized replicons encoding tumor necrosis factor α (TNF-α) to ascertain the effects of short, discrete, high-level expression of TNF-α in the CNS (Bledsoe et al., 2000). The results of our studies clearly demonstrate that biologically active TNF-α can be expressed within the CNS of animals leading to clear behavioral/physical defects and modulation of gene expression within the CNS.

A second consideration for the use of replicons for gene delivery to the neurons is safety. It was clear from previous studies that poliovirus infection of the CNS resulted in considerable behavioral/physical changes and neuronal destruction. The administration of wild-type poliovirus via the intraspinal/intrathecal routes directly into the CNS resulted in death of the animals by 96 h. In contrast, we have analyzed the effects of administration of replicons via intraspinal/intrathecal routes and have found no evidence of physical or behavioral deficits. In all cases tested thus far, we have found that animals given replicons encoding proteins such as GFP or luciferase all scored within the normal range using a system adapted from behavioral evaluation of rats that had undergone spinal cord injury (Jackson et al., 2001). There were no apparent behavioral or physical defects associated with administration of replicons in the CNS. Using a recently devised surgical procedure, we have been able to extend these studies to show that animals given up to 13 consecutive inoculations of replicons did not show any behavioral or physical changes from the control animals. Immunohistochemistry of the CNS tissue supported the conclusion that replicons did not induce CNS tissue damage. There was no infusion of inflammatory cells within the CNS of replicon-infected animals compared to control animals; there was no obvious loss of neurons from animals given replicons compared to control animals. Taken together, the results of these studies suggest that the delivery of replicons into the CNS results in the targeted expression of foreign proteins in neurons without discernable pathogenesis.

Poliovirus replicons then have a great potential in the treatment of spinal cord injury and other diseases or injuries of the CNS. Replicons overcome the difficulty of penetration and diffusion that limits the usefulness of growth factors and/or cytokines in the treatment of the brain and spinal cord. An added advantage is that replicons induce transient production (for a few days) of the protein encoded by the inserted RNA. For example, it would also be possible to have multiple injections of the replicons directed at the spinal cord or brain at particular times after injury or at particular stages in the disease process. Toward this goal, replicons have been constructed encoding cytokines or neurotrophic factors. A recent study suggests that administration of interleukin 10 (IL-10) shortly following spinal cord injury or trauma could have beneficial affects in the clinical outcome (Bethea et al., 1999). A replicon encoding IL-10 has been constructed and will be used for administration to animals that have received spinal cord injury or trauma via weight-drop. Replicons will be administered via intrathecal routes starting at approximately 24 h postinjury and given at 48 h intervals for up to 2 weeks. The recovery of these animals will be monitored using physical and behavioral tests. A second approach will use replicons encoding various neurotrophic factors such as neurotrophin 3 (NT-3), nerve growth factor (NGF), brain-derived neurotrophic factor (BDNF), and glial-derived neurotrophic factor (GDNF) which have been shown to have potential for improving the outcome in spinal cord injured animals (Horner and Gage, 2000). Replicons expressing individual neurotrophic factors or a combination of replicons expressing neurotrophic factors will be given to animals shortly after injury or given to animals that have a chronic spinal cord injury (paralysis). The recovery of the animals will again be assessed by physical and behavioral testing. The potential of replicons may be further extended by combining this treatment with infusion of neural stem cells obtained from the brain or bone marrow (Brazelton et al., 2000; Frisen et al., 1998; Gage, 2000; Horner and Gage, 2000).

NEW STRATEGIES FOR TARGETING OF REPLICONS

The tropism of poliovirus (and replicons) in the CNS is mainly a function of the expression of CD155 on neurons at levels that will allow infection of susceptible cells. Retargeting of replicons to non-CD155-expressing cells is possible, although difficult due to inherent features of the virus. During the natural infection process, the interaction of poliovirus (and replicons) with CD155 induces a conformational change, producing A particles. The change to an A particle is a prerequisite for infection, because the polio virion is not affected by conditions within the lysosome (low pH, enzymes) which are similar to the conditions in the gut. A recent study has addressed the issue of retargeting poliovirus using an IgG-CD155 fusion protein to induce A particle formation with cellular targeting to Fc receptor cells (Arita et al., 1999). In this case, poliovirus was shown to infect cells that were Fc receptor-positive and CD155-negative. A similar approach could be used for retargeting replicons.

A compounding issue for studies designed to retarget replicons though, is that replicon translation and replication can be influenced by intracellular proteins. This point is highlighted by recent studies in the laboratory where we have analyzed the effects of administration of replicons alone (nonencapsidated) into the CNS. In this case, we complexed the RNA with polycations (polyethylenimine, PEI) prior to direct administration into the CSF using an intrathecal catheter (Boussif et al., 1995; Fischer et al., 1999; Godbey, Wu, and Mikos, 1999; Poulain et al., 2000). We have analyzed the expression of the target foreign protein (in this case GFP) following administration and have found that, like the administration of encapsidated replicons, the expression was mainly confined to neurons; cells other than neurons could be transfected in vivo, but the proportions of labeled glial cells types were much lower than that for neurons. Because there would be no restriction for the replicon to enter other cells of the CNS such as astrocytes, oligodendrocytes, and microglia, we conclude from this experiment that the intracellular milieu of the neuron is also a determinant for replicon gene expression. To further highlight this result, we have analyzed the use of replicon:PEI complexes to be used for gene delivery into the CNS of rats. Again, we found the most exclusive expression of GFP in neurons, highlighting the tropism of replicons for neurons. Studies are underway to modify PEI with peptides known to interact with neuron-specific molecules (even CD155) to facilitate gene delivery by replicon : PEI conjugates. We have recently used phage display technology to identify peptides that can specifically interact with the CD155 receptor. These candidate peptides have been found which show specificity for the CD155 receptor. These peptides will be conjugated to PEI to enhance the specificity for cells expressing CD155. The ability to generate large quantities of conjugated peptide-PEI along with replicon RNA from in vitro transcription should facilitate the large-scale production required for clinical trials. Collectively, the results of our studies point to unique cellular targeting of poliovirus replicons to neurons that is dependent upon expression of a cellular receptor (CD155) and intracellular proteins. The exploitation of this tropism for gene delivery to CNS neurons using poliovirus-based vectors could provide new therapeutics for spinal cord injury and diseases.

ACKNOWLEDGMENTS

We thank Dee Martin for preparation of the manuscript. Lihua Feng is acknowledged for growing GFP replicons and Sylvia McPherson for construction of replicons (CFAR Molecular Biology Core, AI 27767). Histology was performed at the VSRC facility and we thank Pam Kontzen for guidance. Confocal imaging was carried out at UAB with help from Albert Tousson and Shawn Williams. Supported by grants from the NIH (CDM).

REFERENCES

Andino R, Rieckhof GE, Achacoso PL, Baltimore D (1993): Poliovirus RNA synthesis utilizes an RNP complex formed around the 5′-end of viral RNA. *EMBO J* 12:3587–3598.

Andino R, Rieckhof GE, Baltimore D (1990): A functional ribonucleoprotein complex forms around the 5′ end of poliovirus RNA. *Cell* 63:369–380.

Ansardi DC, Porter DC, Anderson MJ, Morrow CD (1996): Poliovirus assembly and encapsidation of genomic RNA. *Adv Virus Res* 46:1–68.

Ansardi DC, Porter DC, Jackson CA, Gillespie GY, Morrow CD (2001): RNA replicons derived from poliovirus are directly oncolytic for human tumor cells of diverse origins. *Cancer Res* 61:8470–8479.

Ansardi DC, Porter DC, Morrow CD (1993): Complementation of a poliovirus defective genome by a recombinant vaccinia virus which provides poliovirus P1 capsid precursor in *trans*. *J Virol* 67:3684–3690.

Arita M, Horie H, Arita M, Nomoto A (1999): Interaction of poliovirus with its receptor affords a high level of infectivity to the virion in poliovirus infections mediated by the Fc receptor. *J Virol* 73:1066–1074.

Bethea JR, Nagashima H, Acosta MC, Briceno C, Gomez F, Marcillo AE, Loor K, Green J, Dietrich WD (1999): Systemically administered interleukin-10 reduces tumor necrosis factor-alpha production and significantly improves functional recovery following traumatic spinal cord injury in rats. *J Neurotrauma* 16:851–863.

Bledsoe AW, Gillespie GY, Morrow CD (2000): Targeted foreign gene expression in spinal cord neurons using poliovirus replicons. *J Neurovirol* 6:95–105.

Bledsoe AW, Jackson CA, McPherson S, Morrow CD (2000): Cytokine production in motor neurons by poliovirus replicon vector gene delivery. *Nat Biotechnol* 18:964–969.

Blomer U, Naldini L, Kafri T, Trono D, Verma IM, Gage F (1997): Highly efficient and sustained gene transfer in adult neurons with a lentivirus vector. *J Virol* 71:6641–6649.

Bodian D (1949): Histopathological basis of clinical findings in poliomyelitis. *Am J Med* 563–578.

Bodian D, Horstmann DM (1965): Viral and rickettsial infections of man. In *Polioviruses*, Horstfall FL and Tamm I, eds. Philadelphia: Lippincott, p. 479–498.

Borman A, Howell MT, Patton JG, Jackson RJ (1993): The involvement of a spliceosome component in internal initiation of human rhinovirus RNA translation. *J Gen Virol* 74:1775–1788.

Boussif O, Lezoualc'h F, Zanta MA, Mergny MD, Scherman D, Demeneix B, Behr J-P (1995): A versatile vector for gene and oligonucleotide transfer into cells in culture and *in vivo:* Polyethylenimine. *Proc Natl Acad Sci USA* 92:7297–7301.

Brazelton TR, Rossi FMV, Keshet GI, Blau HM (2000): From marrow to brain: Expression of neuronal phenotypes in adult mice. *Science* 290:1775–1779.

Byrnes AP, Rusby JE, Wood MJA, Charlton HM (1995): Adenovirus gene transfer causes inflammation in the brain. *Neuroscience* 66:1015–1024.

Deatly AM, Coleman JW, McMullen G, McAuliffe JM, Jayarama V, Cupo A, Crowley JC, McWilliams T, Taffs RE (1999): Poliomyelitis in intraspinally inoculated poliovirus receptor transgenic mice. *Virology* 255:221–227.

Deatly AM, Taffs RE, McAuliffe JM, Nawoschik SP, Coleman JW, McMullen G, Weeks-Levy C, Johnson AJ, Racaniello VR (1998): Characterization of mouse lines transgenic with the human poliovirus receptor gene. *Microbial Pathogen* 25:43–54.

Dewey RA, Morrissey G, Cowsill CM, Stone D, Bolognani F, Dodd NJF, Southgate TD, Klatzmann D, Lassmann H, Castro MG, Lowenstein PR (1999): Chronic brain inflammation and persistent herpes simplex virus 1 thymidine kinase expression in survivors of syngeneic glioma treated by adenovirus-mediated gene therapy: Implications for clinical trials. *Nature Med* 5:1256–1263.

Duprex WP, McQuaid S, Roscic-Mrkic B, Cattaneo R, McCallister C, Rima BK (2000): In vitro and in vivo infection of neural cells by a recombinant measles virus expressing enhanced green fluorescent protein. *J Virol* 74:7972–7979.

Fischer D, Bieber T, Li Y, Elsasser HP, Kissel T (1999): A novel non-viral vector for DNA delivery based on low molecular weight, branched polyethylenimine: Effect of molecular weight on transfection effciency and cytotoxicity. *Pharm Res* 16:1273–1279.

Freistadt M (1994): Distribution of the poliovirus receptor in human tissue. In *Cellular receptors for animal viruses*, Wimmer E, ed. New York: Cold Springs Harbor Laboratory Press, pp. 445–461.

Freistadt MS, Kaplan G, Racaniello VR (1990): Heterogeneous expression of poliovirus receptor-related proteins in human cells and tissues. *Mol Cell Biol* 10:5700–5706.

Frisen J, Johansson CB, Lothian C, Lendahl U (1998): Central nervous system stem cells in the embryo and adult. *Cell Mol Life Sci* 54:935–945.

Gage FH (2000): Mammalian neural stem cells. *Science* 287:1433–1438.

Godbey WT, Wu KK, Mikos AG (1999): Poly(ethylenimine) and its role in gene delivery. *J Control Release* 60:149–160.

Gromeier M, Solecki D, Patel DD, Wimmer E (2000): Expression of the human poliovirus receptor/CD155 gene during development of the central nervous system: Implications for the pathogenesis of poliomyelitis. *Virology* 273:248–257.

Gwag BJ, Kim EY, Ryu BR, Won SJ, Ko HW, Oh YJ, Cho YG, Ha SJ, Sung YC (1998): A neuron-specific gene transfer by a recombinant defective Sindbis virus. *Brain Res Mol Brain Res* 63:53–61.

Harris KS, Xiang W, Alexander L, Lane WS, Paul AV, Wimmer E (1994): Interaction of poliovirus polypeptide 3CD^{pro} with the 5′ and 3′ termini of the poliovirus genome. *J Biol Chem* 269:27004–27014.

Hellen CUT, Pestova TV, Litterst M, Wimmer E (1994): The cellular polypeptide p57 (pyrimidine tract-binding protein) binds to multiple sites in the poliovirus 5′ non-translated region. *J Virol* 68:941–950.

Hellen CUT, Witherell GW, Schmid M, Shin SH, Pestova TV, Gil A, Wimmer E (1993): A cytoplasmic 57-kDa protein that is required for translation of picornavirus RNA by internal ribosomal entry is identical to the nuclear pyrimidine tract-binding protein. *Proc Natl Acad Sci USA* 90:7642–7646.

Hermens WTJMC, Verhaagen J (1998): Viral vectors, tools for gene transfer in the nervous system. *Prog Neurobiol* 55:399–432.

Hogle JM, Chow M, Filman DJ (1985): Three-dimensional structure of poliovirus at 2.9 angstroms resolution. *Science* 229:1358–1365.

Horner PJ, Gage FH (2000): Regenerating the damaged central nervous system. *Nature* 407:963–970.

Jackson CA, Cobbs C, Peduzzi JD, Novak M, Morrow CD (2001): Repetitive intrathecal injections of poliovirus replicons result in gene expression in neurons of the central nervous system without pathogenesis. *Hum Gene Ther* 12:1827–1841.

Jang SK, Pestova TV, Hellen CUT, Witherell W, Wimmer E (1990): Cap-independent translation of picornavirus RNA: structure and function of the internal ribosomal entry site. *Enzyme* 44:292–309.

Jubelt B, Gallez-Hawkins G, Narayan O, Johnson RT (1980): Pathogenesis of human poliovirus infection in mice. I. Clinical and pathological studies. *J Neuropathol Exp Neurol* 39:138–148.

Kaplitt MG, Makimura H (1997): Defective viral vectors as agents for gene transfer in the nervous system. *J Neurosci Methods* 71:125–132.

Kerr DA, Nery JP, Traystman RJ, Chau BN, Hardwick JM (2000): Survival motor neuron protein modulates neuron-specific apoptosis. *Proc Natl Acad Sci USA* 97:13312–13317.

Kirkegaard K, Baltimore D (1986): The mechanism of RNA recombination in poliovirus. *Cell* 47:433–443.

Kitamura N, Semler BL, Rothberg PG, Larsen GR, Adler CJ, Dorner AJ, Emini EA, Hanecak R, Lee JJ, van der Werf S, Anderson CW, Wimmer E (1981): Primary structure, gene organization and polypeptide expression of poliovirus RNA. *Nature (London)* 291:547–553.

Koch F, Koch G (1985): *The Molecular Biology of Poliovirus*. Vienna: Springer-Verlag.

Lopez M, Aoubala M, Jordier F (1998): The human poliovirus receptor related 2 protein is a new hematopoietic/endothelial homophilic adhesion molecule. *Blood* 92:4602–4611.

Luz N, Beck E (1990): A cellular 57 kDa protein binds to two regions of the internal translation initiation site of foot-and-mouth disease virus. *FEBS Lett* 269:311–314.

Luz N, Beck E (1991): Interaction of a cellular 57-kilodalton protein with the internal initiation site of foot-and-mouth disease virus. *J Virol* 65:6486–6494.

Mandel B (1965): The fate of the inoculum in HeLa cells infected with poliovirus. *Virology* 25:152–154.

Meerovitch K, Pelletier J, Sonenberg N (1989): A cellular protein that binds to the 5′-noncoding region of poliovirus RNA: implication for internal translation initiation. *Genes Dev* 3:1026–1034.

Mendelsohn C, Johnson B, Lionetti KA, Nobis P, Wimmer E, Racaniello VR (1986): Transformation of a human poliovirus receptor gene into mouse cells. *Proc Natl Acad Sci USA* 83:7845–7849.

Mendelsohn CL, Wimmer E, Racaniello VR (1989): Cellular receptor for poliovirus: Molecular cloning, nucleotide sequence, and expression of a new member of the immunoglobulin superfamily. *Cell* 56:855–865.

Nicklin MJH, Krausslich HG, Toyoda H, Dunn JJ, Wimmer E (1987): Poliovirus polypeptide precursors: Expression *in vitro* and processing by exogenous 3C and 2A proteinases. *Proc Natl Acad Sci USA* 84:4002–4006.

Nicklin MJH, Toyoda H, Murray MG, Wimmer E (1986): Proteolytic processing in the replication of polio and related viruses. *Bio/Technology* 4:33–42.

Parsley TB, Towner JS, Blyn LB, Ehrenfeld E, Semler BL (1997): Poly(rC) binding protein 2 forms a ternary complex with the 5′-terminal sequence of poliovirus RNA and the viral 3CD proteinase. *RNA* 3:1124–1134.

Porter DC, Ansardi DC, Choi WS, Morrow CD (1993): Encapsidation of genetically engineered poliovirus minireplicons which express human immunodeficiency virus type 1 Gag and Pol proteins upon infection. *J Virol* 67:3712–3719.

Porter DC, Ansardi DC, Morrow CD (1995): Encapsidation of poliovirus replicons encoding the complete human immunodeficiency virus type 1 *gag* gene by using a complementation system which provides the P1 capsid protein in*trans. J Virol* 69:1548–1555.

Porter DC, Ansardi DC, Wang J, McPherson S, Moldoveanu Z, Morrow CD (1998): Demonstration of the specificity of poliovirus encapsidation using a novel replicon which encodes enzymatically active firefly luciferase. *Virology* 243:1–11.

Porter DC, Wang J, Moldoveanu Z, McPherson S, Morrow CD (1997): Immunization of mice with poliovirus replicons expressing the C-fragment of tetanus toxin protects against lethal challenge with tetanus toxin. *Vaccine* 15:257–264.

Poulain L, Ziller C, Muller CD, Erbacher P, Bettinger T, Rodier JF, Behr JP (2000): Ovarian carcinoma cells are effectively transfected by polyethylenimine (PEI) derivatives. *Cancer Gene Ther* 7:644–652.

Putnak JR, Phillips BA (1981): Picornaviral structure and assembly. *Microbiol Rev* 45:287–315.

Racaniello VR, Baltimore D (1981): Molecular cloning of poliovirus DNA and determination of the complete nucleotide sequence of the viral genome. *Proc Natl Acad Sci (USA)* 78:4887–4891.

Ren R, Costantini F, Gorgacz EJ, Lee JJ, Racaniello VR (1990): Transgenic mice expressing a human poliovirus receptor: A new model for poliomyelitis. *Cell* 63:353–362.

Ren R, Racaniello VR (1992a) Human poliovirus receptor gene expression and poliovirus tissue tropism in transgenic mice. *J Virol* 66:296–304.

Ren R, Racaniello VR (1992b): Poliovirus spreads from muscle to the central nervous system by neural pathways. *J Infect Dis* 166:747–752.

Roehl HH, Parsley TB, Ho TV, Semler BL (1997): Processing of a cellular polypeptide by 3CD proteinase is required for poliovirus ribonucleoprotein complex formation. *J Virol* 71:578–585.

Rohll JB, Percy N, Ley R, Evans DJ, Almond JW, Barclay WS (1994): The 5′-untranslated regions of picornavirus RNAs contain independent functional domains essential for RNA replication and translation. *J Virol* 68:4384–4391.

Rueckert RR (1996): Picornaviridae: the viruses and their replication. In *Fields Virology*, 3rd ed. Fields BN, Knipe DM, Howley PM, eds. Philadelphia, Penn.: Lippincott-Raven Publishers, p. 609–654.

Rueckert RR, Wimmer E (1984): Systematic nomenclature of picornavirus proteins. *J Virol* 50:957–959.

Sabin AB (1956): Pathogenesis of poliomyelitis. Reappraisal in the light of new data. *Science* 123:1151–1157.

Satoh-Horikawa K, Nakanishi H, Takahashi K (2000): Nectin-3, a new member of immunoglobulin-like cell adhesion molecules that shows homophilic and heterophilic cell-cell adhesion activities. *J Biol Chem* 2000:10291–10299.

Svitkin YV, Meerovitch K, Lee HS, Dholakia JN, Kenan DJ, Agol VI, Sonenberg N (1994): Internal translation on poliovirus RNA: further characterization of La function in poliovirus translation in vitro. *J Virol* 68:1544–1550.

Takahashi K, Nakanishi H, Miyahara M (1999): Nectin/PRR: an immunoglobulin-like cell adhesion molecule recruited to cadherin-based adherens junctions through interaction with Afadin, a PDZ domain-containing protein. *J Cell Biol* 143:339–349.

Toyoda H, Nicklin MJH, Murray MG, Anderson CW, Dunn JJ, Studier FW, Wimmer E (1986): A second virus-encoded proteinase involved in proteolytic processing of poliovirus polyprotein. *Cell* 45:761–770.

17

GENERATION OF SAFE, TARGETABLE SINDBIS VECTORS THAT HAVE THE POTENTIAL FOR DIRECT IN VIVO GENE THERAPY

DANIEL MERUELO PH. D., BRANDI LEVIN B.A., AND CHRISTINE PAMPENO PH.D.

INTRODUCTION

Gene therapy for various inherited and acquired disorders would benefit greatly from the availability of a vector that has a high efficiency of gene expression and the ability to target cells in vivo. Of particular interest to our laboratory is cancer; therefore, although the vectors described in this chapter have potentially broad applicability, the focus of our discussion will be their utility for cancer therapy.

A number of transfection systems have been developed to deliver heterologous genes into in vivo tumors to investigate cancer gene therapy. A viral vector system that has numerous advantages is based on Sindbis virus. Sindbis virus has been studied extensively since its discovery in Egypt in 1953 (Hurlbut, 1953; Shah et al., 1960; Taylor and Hurlbut, 1953; Taylor et al., 1955). Gene transduction based on Sindbis virus, a member of the alphavirus

Vector Targeting for Therapeutic Gene Delivery, Edited by David T. Curiel and Joanne T. Douglas
ISBN 0-471-43479-5

genus, has been well studied (Altman-Hamamdzic et al., 1997; Bredenbeek et al., 1993; Gwag et al., 1998; Hariharan et al., 1998; Liljestrom and Garoff, 1991; Piper et al., 1994; Pugachev et al., 1995; Strauss and Strauss, 1994; Tsuji et al., 1998; Xiong et al., 1989). Sindbis vectors show extremely high efficiency of gene transfer. They are plus-strand RNA viruses that, through a process of amplification in the cytoplasm of infected cells, can express 10^5 active RNA species per cell within a few hours after infection. This level of RNA amplification allows for very high levels of expression of the transferred gene product, which would allow for prolonged expression were it not for the apoptotic nature of the virus (Balachandran et al., 2000; Jan, Chatterjee, and Griffin, 2000; Jan and Griffin, 1999; Levine et al., 1993). Based on recommendations for the handling of alphaviruses and other arboviruses in the laboratory (ACAV, 1980), Sindbis virus is considered fairly safe. Sindbis virus is endemic to many parts of the world, where it is associated with minimal, transient disease (McGill, 1995; Strauss and Strauss, 1994; Turrell, 1988). Whereas the majority of alphaviruses require Biosafety Level 3 practice and containment and/or vaccination, Level 2 containment is typically recommended for Sindbis virus. Level 2 practices and containment are assigned to viruses whose infection results either in no disease or in disease that is self-limited (ACAV, 1980; McGill, 1995; Turrell, 1988). Replication-incompetent Sindbis vectors derived from Sindbis viruses, such as those used in the current studies, can be considered even safer as their capacity to infect and replicate to cause viremia or disease is virtually nonexistent. The risk of these vectors is so low that Level 1 containment precautions are all that is recommended under some circumstances (ACGM, 2000). The capacity to infect and replicate to cause viremia can only be reacquired through recombination, which can be minimized and monitored. Sindbis vectors also avoid potential complications associated with chromosomal integration (Strauss and Strauss, 1994; Xiong et al., 1989). Recent methods have added substantial ease to engineering new Sindbis vector constructs capable of nonreplicative infection and further enhanced safety aspects of the vector (Strauss and Strauss, 1994). Because Sindbis virus is a blood-borne virus (Turrell, 1988), can cross the blood brain barrier (Altman-Hamamdzic et al., 1997), and has a sufficiently long half-life in blood (Byrnes and Griffin, 2000), vectors based on this virus are among the few available ones that are capable of migrating through the bloodstream to reach all cells of the body. This is a very important advantage.

A major goal of our research efforts has been to expand the utility of Sindbis vectors by rendering them directly targetable in vivo. Ultimately we envision a system in which the vector can enter the bloodstream by any of several possible routes and home directly to target cells located at one or more locations throughout the body. Such a targetable vector could be engineered to retain all the beneficial attributes of Sindbis vectors, such as safety and high expression capacity, while retaining other properties such as the apoptosis-inducing potential when these are desired. This chapter describes the status of our efforts to date in generating such vectors.

RETARGETING ALPHAVIRUSES

Undiminished Infectious Titers of Retargeted Alphaviruses

In trying to retarget alphaviruses, one must consider the chances that such efforts will affect virus infectivity. In contrast to other viruses, in which the mode of entry lends itself less well to retargeting efforts, alphaviruses are ideally suited to the introduction of modifications that permit retargeting. Extensive experimentation has established that the normal pathway of entry for alphaviruses involves endocytosis in clathrin-coated vesicles followed by transfer to endosomes (Doxsey et al., 1987; Helenius, 1984; Helenius et al., 1980; Helenius, Marsh, and White, 1982; Marsh, Bolzau, and Helenius, 1983; Marsh and Helenius, 1980; Marsh et al., 1982). There the low pH leads to the fusion domain in E1 being exposed and the virus envelope fusing with the endosomal membrane (Hoekstra and Kok, 1989; Kielian and Helenius, 1986; Kielian et al., 1990; Marsh, 1984; Marsh et al., 1982; Stegmann, Doms, and Helenius, 1989; Strauss and Strauss, 1994; White, 1990). Thus, Sindbis entry does not require fusion at the cell surface; mere binding to a cell receptor will lead to entry by receptor-mediated endocytosis. That is, E2, which is the Sindbis envelope molecule modified our laboratory (see following section), plays no role in fusion, which is solely driven by pH effects on E1 (Doxsey et al., 1987; Helenius, 1984; Helenius et al., 1980; Helenius, Marsh, and White, 1982; Marsh, Bolzau, and Helenius, 1983; Marsh and Helenius, 1980; Marsh et al., 1982; Strauss and Strauss, 1994). The Sindbis fusion element E1 is unchanged by our modifications. The presumption is that the fusion domain of E1, which is very hydrophobic, inserts into the target endosomal membrane and induces fusion of the two bilayers (Strauss and Strauss, 1994). This is why modifications to E2 that affect binding to cell surface ligands allowing successful retargeting of Sindbis virus and vectors have no impact on fusion and consequently do not diminish infectious titers of the vector.

Generation of Targetable Sindbis Vectors

To alter the tropism of Sindbis vectors to permit gene delivery to specific target cells, we ablated the endogenous viral tropism and introduced the desired targetable tropism (Ohno et al., 1997). To redirect the virus targeting (Fig. 17.1A), we incorporated into its envelope (E2) two (synthetic) binding domains (ZZ) of protein A. Staphylococcal protein A (SpA), the cell wall bound type-I Fc receptor, exhibits tight binding to many IgG, IgA, and IgM molecules at site(s) different from the antigen-combining site (Forsgren and Sjoquist, 1969; Ljungberg et al., 1993). The extracellular part of SpA contains a tandem repeat of five highly homologous IgG-binding domains (Moks et al., 1986; Sjodahl, 1977; Uhlen et al., 1984) designated (from the N terminus) E, D, A, B, and C, each of which includes about 58 amino acid residues. The Z domain is an

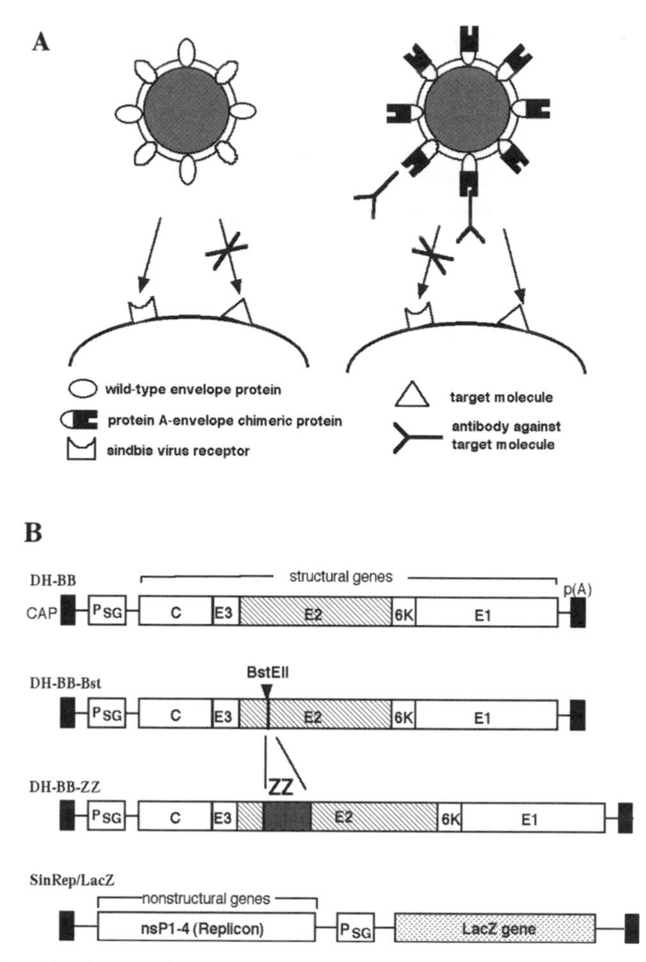

Figure 17.1 (*A*) Schematic strategy for retargeting a Sindbis virus vector. A wild-type Sindbis virus (left) binds to mammalian cells via its surface receptor, which is known to be highly conserved across species. A recombinant Sindbis virus displaying the IgG-binding domain of protein A (right) permits binding to a novel target molecule on the cell surface when used with a corresponding monoclonal antibody (mAb). (*B*) Schematic representation of recombinant helper constructs and a SinRep/LacZ expression vector. DH-BB is a parental helper plasmid that contains the genes for the structural proteins (capsid, E3, E2, 6K and E1) required for packaging of the Sindbis viral genome. DH-BB-Bst was constructed by introduction of a cloning site (BstEII) into the E2 glycoprotein between amino acids 71 and 74. The synthetic IgG-binding domain (ZZ) of protein A was inserted at BstEII in the DH-BB-Bst helper plasmid and DH-BB-ZZ was obtained. SinRep/LacZ, is a Sindbis virus-based expression vector that contains the packaging signal, nonstructural protein genes for replicating the RNA transcript and *LacZ* gene.

Abbreviations: P_{SG}, Sindbis viral subgenomic promoter; C, capsid; nsP1–4, non-structural protein genes 1–4; ZZ, synthetic IgG-binding domain of Protein A; p (A), polyadenylation signal.

engineered analog of the B domain originally developed as an affinity purification handle for fusion protein production (Nilsson et al., 1987; Uhlen et al., 1992). Domains E, D, A, B, C, and Z all bind the Fc domain of human IgG antibodies. The Z domain contains two amino acid substitutions relative to the B domain (Ala1->Val and Gly29->Ala). The B and Z domains exhibit identical binding affinities, on-rates, and off-rates in their interactions with Fc fragments of IgG antibodies (Jendeberg et al., 1995; Starovasnik et al., 1996). Medium-resolution nuclear magnetic resonance (NMR) solution structures of the B and Z domains reveal (Gouda et al., 1992; Jendeberg et al., 1996) an antiparallel three-helix bundle motif. A similar three-helical bundle structure has also been determined for the homologous E domain of SpA (Starovasnik et al., 1996). One important aspect of these protein A domains is their ability to acquire spontaneously the correct three-dimensional (3D) structure. The helix bundle of protein A is one of the smallest autonomously folding α-helical domains known (Boczko and Brooks, 1995; Kolinski and Skolnick, 1994; Olszewski et al., 1996a, 1996b). We reasoned that the ZZ-envelope chimeric virus vector, once successfully generated would need no further modification to target distinct cells. The targeting would simply be achieved by changing the monoclonal antibody (mAb) used.

The ZZ-envelope chimeric SV vector shows very low, if any, infectivity of baby hamster kidney (BHK) cells and all human cell lines tested. That is, its natural tropism has been ablated. Even more encouraging is the fact that, when used in conjunction with mAbs that react with cell surface antigens, the SpA-enveloped chimeric virus is able to transfer the *LacZ* gene into human cell lines with very high efficiency. Figure 17.2 demonstrates the exquisite specificity of

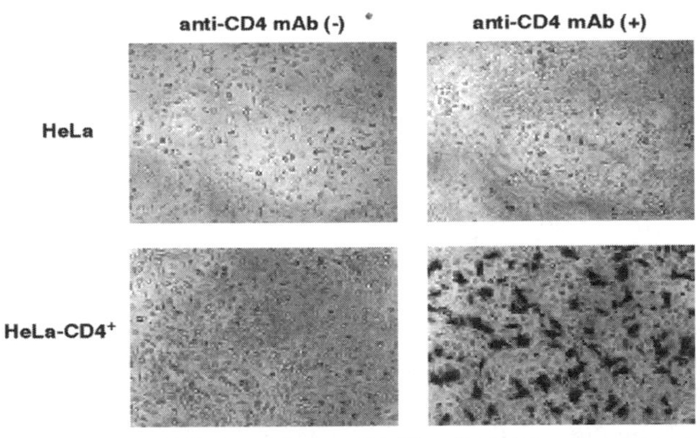

Figure 17.2 Infection of HeLa and HeLa-CD4$^+$ cells with recombinant Sindbis virus derived from DH-BB-ZZ helper RNA that is transducing the bacterial *LacZ* gene. Viral supernatants (200 μl) were preincubated without or with anti-CD4 mAb (0.5 μg/ml) at room temperature for 1 h, and added to cells (2×10^5) in 6-well plates. After 1 h at room temperature, cells were washed with PBS and incubated in growth medium for 24 h. Viral infection was evaluated by X-Gal staining. Figure also appears in Color Figure section.

the system. In this experiment, the SpA-envelope virus did not infect human HeLa or HeLa-CD4$^+$ cells. However, when virions were preincubated with an anti-CD4 mAb, the protein A-envelope chimeric virus could readily infect HeLa-CD4+ cells, but not HeLa cells lacking the transduced CD4 gene.

Similar results have been obtained with many other cell types by simply changing the antibody used in conjunction with the virus (Ohno et al., 1997). For example, to infect Daudi cells we can use the virus plus anti-HLADR antibodies and to infect HL-60 cells we can use virus plus anti-CD33 antibodies. As shown in Figure 17.3, when virus is used at a greater than 2 : 1 multiplicity of infection, virtually all cells in the population are infected and express the transduced gene (94% for Daudi and 97% for HL-60). The tropism of the recombinant virus depends on antigen–antibody interaction, because the SpA-envelope virus cannot infect targeted cells without the mAb and its cognate antigen on the cell surface. Thus, the tropism of the Sindbis virus vector can be altered to suit. These findings demonstrate that the vector system is versatile, highly efficient, and specific. Also, an added safety feature of these vectors is that they cannot infect cells in the absence of antibodies.

Fortunately, clinical experience gained over the past two decades has made selection of antibodies for human use a fairly sophisticated and successful endeavor. A number of antibodies are currently in clinical use, not only for imaging, but also for immunotherapy. The latter use indicates that their specificity is sufficiently high to preclude significant damage to organs or cells not intentionally targeted. In addition to the fact that the specificity and toxicity profiles of these antibodies are well known, many have been humanized to eliminate host immune responses. This group of antibodies would represent ideal targeting reagents for use in conjunction with our Sindbis vector.

We also have shown that it is possible to generate a directly targetable virus using our vector system. Altering the targeting element within the E2 region of Sindbis is straightforward. For example, a human chorionic gonadotropin (hCG)-expressing chimeric vector has been constructed (Sawai and Meruelo, 1998). Considering its size and the fact that hCG is dimeric, consisting of two dissimilar subunits, α and β, which are associated noncovalently, the fact that we were able to construct a Sindbis-hCG vector capable of targeting cells bearing LH/CG receptors attests to the versatility of the vector system. It also demonstrates the relative ease of engineering new constructs. Thus, besides the ease of substituting one antibody for another to alter targeting specificity, the Sindbis vectors can be modified to embody other determinants, which can target directly or indirectly, as desired.

Another advantage of the Sindbis vector system is that replication of alphaviruses occurs entirely in the cytoplasm of the infected cells as an RNA molecule, without a DNA intermediate. Consequently, levels of expression that can be achieved are generally much higher with alphavirus vectors than with many other vectors available. Furthermore, as there is no integration of the viral genome involved, insertional mutagenesis is not a factor when using Sindbis

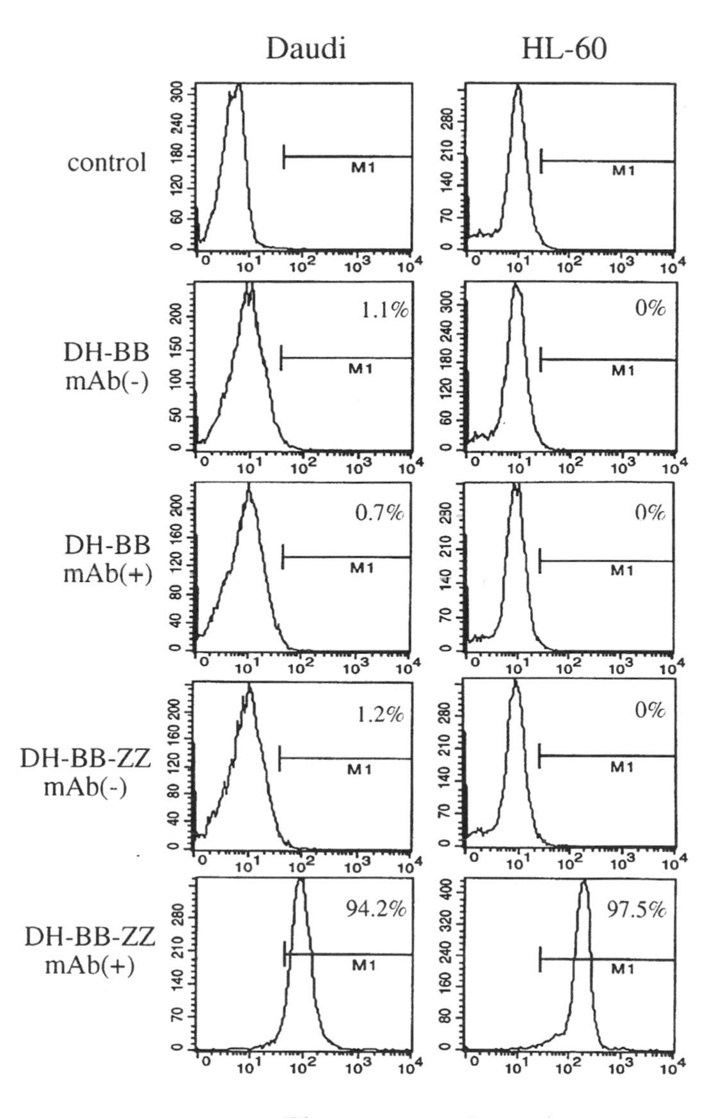

Figure 17.3 Antibody-dependent infectivities of recombinant Sindbis virus particles on Daudi and HL-60 suspension cells. Viral supernatants (500 μl) derived from DH-BB and DH-BB-ZZ transfected BHK cells were preincubated without or with 0.5 μg/ml of mAbs (anti-HLA-DR for Daudi and anti-CD33 for HL-60) at room temperature for 1 h, and added to cells (1×10^6) in 6-well plates. After 1 h incubation at room temperature, cells were washed with PBS and incubated in growth medium for 24 h. Control shows uninfected cells. Viral infection was evaluated by FACS-Gal analysis as described in Ohno et al., 1997. Positive percent of infected cells were shown in each panel.

virus or vectors. Finally, the maximum size of the insert is large enough to permit simultaneous delivery of two separate genes (Sawai et al., 1999).

RESULTS WITH TARGETABLE SINDBIS VECTORS

Expression of foreign gene products in vivo using Sindbis virus has been successfully demonstrated in a number of systems (Hariharan et al., 1998; Pugachev et al., 1995; Xiong et al., 1989). For example, a recombinant Sindbis vector has been used to sensitize cytotoxic T lymphocytes to heterologously expressed major histocompatability complex (MHC) class I-restricted antigens (Pugachev et al., 1995). In a number of recent unpublished studies we have shown that the vector can target specific cells in vivo.

In one experiment 10^7 human colon carcinoma cells, LS174T, were inoculated subcutaneously into 8–10 week old nude mice. When tumors grew to a visible size, animals received a single injection of Sindbis plus anti-carcinoembryonic antigen (CEA) antibodies or virus alone or antibody alone. The tumors were harvested 16–24 h later, teased into single cell suspensions, stained with a rhodamine-conjugated anti-CEA antibody and also exposed to fluorescein di-β-D-galactopyranoside (FDG), a substrate for the detection of β-galactosidase in single cells by flow cytometry assay. Tumors from mice receiving only antibody or only virus did not stain positively for β-galactosidase (0.11% and 0.03%, respectively). However, about 73% of tumor cells in mice injected with the virus-antibody complex were positive for β-galactosidase. In mice infected with the virus–antibody complex, a survey of other organs for β-galactosidase activity, revealed no activity over background except in the tumor cells.

GENERATION OF ALPHAVIRUS (SINDBIS) PACKAGING CELL LINES THAT PERMIT LARGE-SCALE PRODUCTION OF SINDBIS-BASED VECTORS

Overview

Application of viral vectors for gene therapy requires efficient means for their large-scale manufacture. A major obstacle to the widespread use of Sindbis vectors for gene therapy has been the inability to produce these vectors easily from packaging cell lines. To understand the issues involved in the case of Sindbis vectors, it is important to analyze how these vectors are produced.

At present recombinant Sindbis virus is generally produced by the electroporation of two RNA genomes that are synthesized in vitro. One genome, the replicon, directs the synthesis of the gene of interest along with the Sindbis RNA replicase (Fig. 17.4). This genome also contains a packaging signal allowing it to be incorporated into virion particles by the Sindbis structural proteins. The second, a helper genome [or, for greater safety, two split struc-

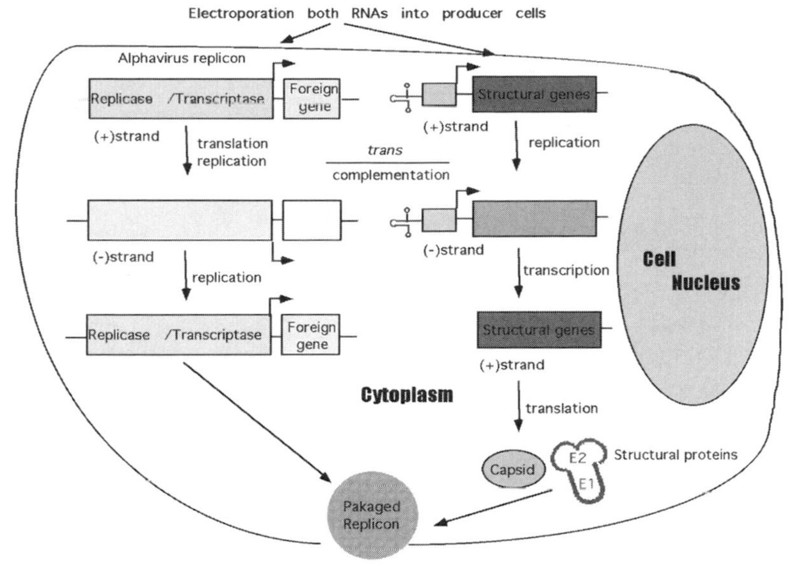

Figure 17.4 Packaging of replication-defective replicons by cotransfection of DHRNAs expressing the alphavirus structural proteins and the replicon. Defective-helper RNAs (DHRNAs) are designed to contain the *cis*-acting sequences required for replication as well as the subgenomic RNA promoter driving expression of the structural genes. Packaging of SIN replicons is achieved by efficient cotransfection of cells with both RNAs by electroporation. Replicase/transcriptase functions supplied by the vector RNA lead not only to its own amplification but also act in *trans* to allow replication and transcription of helper RNA. This results in synthesis of structural proteins that can package the replicon with > 10^7 infectious particles per milliliter being produced after only 16–24 h. This process has been described by Agapov et al. (1998).

tural genes helpers (Fig. 17.5)], directs the synthesis of viral structural proteins but lacks a packaging signal so that recombinant viral particles will undergo only one round of infection.

In addition to the need to prepare RNA in vitro and electroporate this RNA into cells, the lethality of Sindbis virus requires that this procedure be done anew each time virus is to be produced. In mammalian cells, apoptosis caused by Sindbis virus results in lethality to the infected cells, making it impossible to propagate Sindbis-virus infected cells. To overcome the lethality issue, one could construct an inducible system, which would allow propagation of the cells into high numbers before virus production is induced. While the producing cells would still die within days after induction, with an inducible system it would be possible to make large amounts of virus (such as would be needed for clinical applications). By simply preparing large amounts of uninduced cells and freezing aliquots, one could generate a master cell bank from which aliquots could be thawed as needed, grown to large numbers, and induced to produce virus.

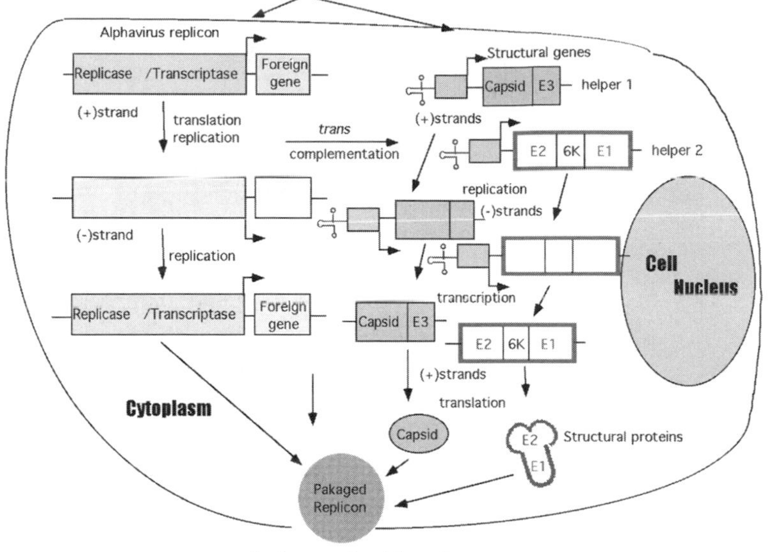

Figure 17.5 Cotransfection of two DHRNAs along with the expression replicon. One DHRNA (helper 1) expresses the capsid protein and a second DHRNA (helper 2) is designed for high-level expression of the virion glycoproteins. Helper 2 uses a deleted version of the capsid protein that is still able to function as a translational enhancer and an autoprotease but is defective for packaging. This approach is favored because it virtually eliminates the possibility that replication-competent viruses will arise by recombination. This process has been described by Agapov et al. (1998).

Polo et al. (1999) have developed a "stable" alphavirus packaging cell line that partially but not fully resolves the need to transfect RNA for vector production. Polo et al. (1999) describe the use of an inducible cell system that stably carries DNA for production of the viral structural proteins, but not the replicon and gene of interest. In this system, translation of the structural proteins (from helpers 1 and 2) is obtained only after synthesis of an authentic subgenomic mRNA by the vector-encoded replicase proteins (which are not encoded by the cells and must be electroporated).

Because the replicon is not present in the packaging cells, to obtain vector production, the system of Polo and colleagues (Polo et al., 1999) requires electroporation of all the cells with RNA encoding the Sindbis replicon and desired gene. To a large extent, this requirement imposes an unnecessary burden in the production of recombinant Sindbis vectors for therapeutic application. In particular, given the instability of RNA molecules, large scale production of in vitro Sindbis RNA genomes will be difficult and costly. Further, a complete master cell bank cannot be generated. As Polo et al. (1999) must have been aware of this drawback, it is assumed that they were not able to design their system in such a way that they could prevent the expression of replicon genes.

One reason for not being able to stably integrate the DNA encoding the Sindbis replicon is that production of genes encoded in the replicon, such as nsP2, is very toxic to mammalian cells.

Use of Mosquito Cells

One approach that may resolve the manufacturing issue is to use insect cells to produce Sindbis vectors. The rationale for the use of mosquito cells stems from the differing effects that Sindbis virus has after infection of mosquito cells and mammalian cells. Sindbis infection of most vertebrate cell lines results in massive cell death within 12 to 96 h. By contrast, infection of mosquito cells is often (but not always) accompanied by little if any cytopathology (Condreay and Brown, 1986; Karpf, Blake, and Brown, 1997; Karpf and Brown, 1998; Karpf et al., 1997; Luo and Brown, 1993; Miller and Brown, 1992). In mosquito cells the infection begins with an early acute phase, during which large amounts of virus are shed into the medium and some transient cellular changes occur (Condreay and Brown, 1986; Karpf, Blake, and Brown, 1997; Karpf and Brown, 1998; Karpf et al., 1997; Luo and Brown, 1993; Miller and Brown, 1992). This stage is followed by a prolonged persistent phase in which virus production is maintained at significantly lower levels and the cells continue to grow, divide, and produce virus through many passages (Condreay and Brown, 1986; Karpf, Blake, and Brown, 1997; Karpf and Brown, 1998; Karpf et al., 1997; Luo and Brown, 1993; Miller and Brown, 1992). As there is considerable variability in the degree of cytopathology, care must be used in the selection of mosquito cells for manufacturing purposes.

It appears that one reason why mosquito cell lines have not been thought of for the generation of packaging cell lines has to do with the frequently reported observation that, over time, titers of virus in mosquito cells drop significantly (Condreay and Brown, 1986; Karpf, Blake, and Brown, 1997; Karpf and Brown, 1998; Karpf et al., 1997; Luo and Brown, 1993; Miller and Brown, 1992). Studies with replication-competent virus (Karpf et al., 1997) have indicated that persistently infected cells such as C6/36 produced Sindbis virus at very low levels (less than10^3 plaque forming units (PFU)/ml/day) (Karpf et al., 1997). The decrease of virus production is thought to result from the release by mosquito cells of a polypeptide factor that downregulates virus production (Condreay and Brown, 1986, 1988; Karpf et al., 1997; Luo and Brown, 1993). One anti-Sindbis factor has been purified and found to be a hydrophobic polypeptide of 3200 Da.

However, the production of replication-incompetent virus is even more problematic than the drop in titers suggests. To understand why this may be so, one needs to examine how the persistent state is maintained for cells infected with replication-competent Sindbis virus. Upon primary infection of mosquito cells, vigorous viral replication occurs in all cells and a large amount of virus is shed into the medium (Condreay and Brown, 1986, 1988; Karpf et al., 1997; Luo and Brown, 1993). This is possible because the initial replication of the virus allows the infection to propagate from cell to cell. However, during the persistent infection only a fraction of the cells are actively replicating virus at any time. Thus, only a small percentage of the cells contain sufficient viral structural proteins to be detectable in an immunofluorescence assay and further, upon cloning of individual cells from the persistently infected population, only a fraction of the cell clones produce virus (Condreay and Brown, 1986, 1988; Karpf et al., 1997; Luo and Brown, 1993). Thus, it appears that upon viral infection of mosquito cells, all cells initially support viral replication but a large fraction of cells subsequently stop replicating virus (Condreay and Brown, 1986, 1988; Karpf et al., 1997; Luo and Brown, 1993). At first, such nonreproducing cells may remain resistant to superinfection, but ultimately the cells may become sensitive to reinfection by virus in the medium or by residual viral RNA in the cell, and in this way the culture remains persistently infected (Condreay and Brown, 1986, 1988; Karpf et al., 1997; Luo and Brown, 1993). Thus, at any point, the majority of cells in the culture have stopped replicating virus while others have just begun and the rest are somewhere in between.

A model in which a buildup of a viral *trans*-acting protease leads to the shutoff of minus-strand RNA synthesis followed by the decay of viral replicases producing plus-strand RNA has been postulated to explain this phenomenon (Condreay and Brown, 1986, 1988; Karpf et al., 1997; Luo and Brown, 1993). It has been shown that the uncleaved P123 is required for the production of minus-strand RNA templates and that if P123 is cleaved too rapidly by the protease present in nonstructural protein nsP2, minus-strand production does not take place (Condreay and Brown, 1986, 1988; Karpf, Blake, and Brown, 1997; Karpf et al., 1997; Luo and Brown, 1993; Miller and Brown, 1992; Strauss and Strauss, 1994). Because minus-strands are the first templates needed for RNA replication, this effectively means that no RNA replication can take place after infection and the appearance of sufficient *trans*-acting nsP2 protease to rapidly cleave P123. Existing minus-strands can continue to be used as a template to produce plus-strand RNA but these strands eventually disappear from the cytoplasm. This mechanism is thought to have evolved in order to regulate RNA replication such that minus-strand RNA is only produced early in infection but may also serve to exclude superinfecting virus.

If one envisions attempting to obtain a persistently producing vector packaging cell line with a replication-defective Sindbis vector, one might conclude that the situation is impossible, since the benefit of reinfection of the culture by virus emanating from the few producing cells is not possible. Hence as cells become permissive once again, presumably through elimination of viral RNA, they fail to meet replication-competent vectors from neighboring cells that would allow them to become reinfected. Under this circumstances it might be envisioned that vector titers would drop not only to very low levels but to zero.

Other complicating factors are the observations that different cell lines derived from mosquitoes exhibit wide variations in their ability to support viral replication in vitro, in their manner of virus production (e.g., secreted into the culture medium or produced internally), the extent of cytopathology associated with the acute phase of infection, and many other relevant parameters (Condreay and Brown, 1986, 1988; Karpf, Blake, and Brown, 1997; Karpf et al., 1997; Luo and Brown, 1993; Miller and Brown, 1992). Thus, any attempt to develop a packaging cell line from mosquito cells would appear a daunting problem that must overcome a number of intrinsic technical difficulties, as well as the notion that vector titers might be expected to drop to zero during the persistent phase of the infected state.

On examining this situation we realized that observations that have discouraged others might be overcome by a different approach. That is, a model in which a buildup of viral *trans*-acting protease leads to the shutoff of minus-strand RNA synthesis followed by the decay of viral replicases producing plus-strand RNA (Karpf et al., 1997), should be overcome by the use of plasmids driven by promoters that are constitutively expressed. In this approach, DNA is driving expression and thus minus-RNA strand degradation should have no impact on expression. Such packaging cells should not be subjected to the loss of expression, followed by "cure," followed by reinfection from the few producing cells in the culture.

In this strategy a superinfection block can actually increase safety. Any recombination event leading to the production of wild-type virus would under normal circumstances produce virus capable of, at least transiently, being expressed by the entire culture. Such an event would ultimately give rise to a persistently infected culture budding out wild-type virus that may infect patients receiving vectors made from such cultures. However, as cells expressing Sindbis vectors cannot be superinfected, cultures in which 100% of the cells are producing vectors would not be expected to support superinfection by wild-type recombinant viruses. Thus, while recombinations can still be expected to occur, and screening for such viruses must still be done, an additional safety factor is introduced by a block to superinfection. Like splitting genomes to reduce the likelihood that recombinants arise and propagate in the culture, every strategy that decreases the propagation of potential replication-competent virus provides an added margin of safety.

We are implementing this strategy to generate two general types of vector constructs: one that targets cells on the basis of protein A-antibody-cell surface antigen expression (or receptor-ligand interactions) and one that targets cells expressing the high-affinity laminin receptor (Strauss et al., 1994; Wang et al., 1992). Each of these vectors can be used to introduce any desired gene into selected cells. Common to both vectors is the SinRep5-X fragment, which encodes the Sindbis replicon and the desired gene (X) to be transfected (Fig. 17.6). This strand is packaging cell-positive so that it can be incorporated into progeny vectors made from the producing packaging cell line. The second strand differs depending on whether the vector is targetable via protein A(ZZ)-IgG interactions (or ligand-receptor interactions) with specific cells, or merely recognizes the high-affinity laminin receptor (HALR) present on cells to be infected (Fig. 17.7). The HALR-recognizing vector, known as

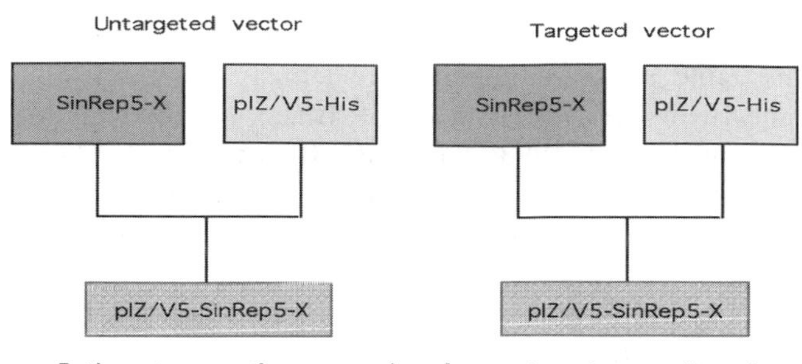

Figure 17.6 General construction of DNA Sindbis vectors capable of expression in mosquito cells: Replicon/ gene of interest strand.

DHBB/SinRep5-X, comprises the helper strand (DHBB) carrying the conventional Sindbis structural genes. This strand is packaging signal-negative so that its genome is not incorporated into the vectors produced, thereby rendering the progeny vectors replication incompetent. The ZZ-targetable vectors contain the ZZ-modified helper strand, DHBBZZ. It is also packaging cell-negative.

Our choice of mosquito cells has been the *Aedes albopictus* derived C6/36 cell line, which was obtained from the American type culture collection. This selection was based on numerous observations and experimental data. For example, as stated earlier, after infection not all mosquito cells are equally resistant to the cytopathic effects of Sindbis virus (Table 17.1). Extensive toxicity occurs with C7-10 (Table 17.1). Toxicity to C6/36 and u4.4 is minimal to zero (Table 17.1). Of these two, C6/36 shows the highest peak titers and budding of virus is clear, whereas u4.4 appears not to bud virus (Table 17.1). Again to minimize the chances that replication-competent virus could replicate in the culture, a superinfection block is desirable. Of the three lines, C7-10 does not show such a block. We also investigated a number of additional insect cell

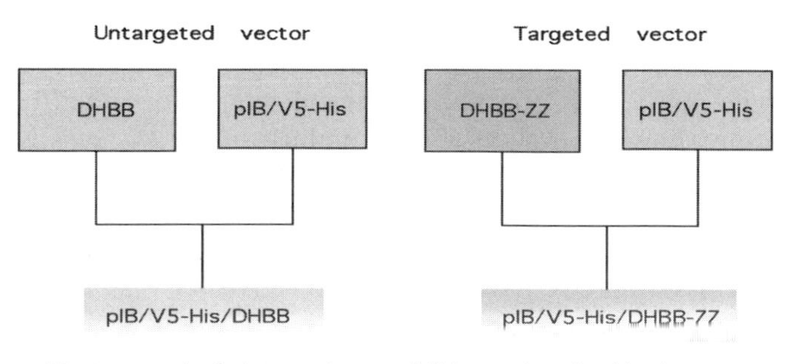

Figure 17.7 General construction of DNA Sindbis vectors capable of expression in mosquito cells: Helper strands.

TABLE 17.1 Effect of Sindbis Virus Replication in Various Mosquito-Derived Cell Lines

Cell Line	Percentage of Virus-Positive Cells Following Infection	Peak Titers	Cytotoxic Effects	Budding	Evidence of Vesicles Containing Virions Fusing Within Cells	Superinfection Possible
C7–10	100	2.3×10^9	Extensive and rapid in onset following infection	Yes	Yes	Yes
C6/36	100	2.3×10^9	Minimal	Yes	Yes	No
u4.4	100	3×10^8	None	No	Yes	No

Source: Data compiled from Boczko and Brooks, 1995; Doxsey et al., 1987; Helenius, 1984; Helenius et al., 1980; Helenius, Marsh, and White, 1982; Olszewski, Kolinski, and Skolnick, 1996a; Sawai and Meruelo, 1998.

lines, such as Schneider's Drosophila cell line 2 (ATCC #CRL-1963). Various considerations again tended to favor C6/36, such as extent of survival of the cells after transfection and selection with antibiotics such as blasticidin S, hygromycin B, and zeocin.

Preparation of Packaging Cell Lines

For generating the C6/36 packaging cell line, the two plasmids used to generate the Sindbis DHBB or DHBBZZ vectors were separately cloned into the pIB/V5-His vector. The packaging cell line is constructed by taking advantage of the pIZ/V5-His vector, which is specifically designed to express gene sequences in insect cells; that is, the OpIE2 promoter provides high-level, constitutive expression of the gene of interest in insect cells. The Sindbis replicase and subgenomic promoter are excised from plasmid SinRep5-X (the source of X, i.e., the gene incorporated into the replicase fragment of the Sindbis vector, varies), pIZ/V5 is cut and the fragments are ligated in the correct orientation. Zeocin-resistance allows selection of the clones so that untransfected cells die. The vector helper plasmid (DHBB or DHBBZZ) of the virus is separately cloned into the vector pIB/V5-His, which carries a selectable marker for blasticidin S. After a stable clone has been generated with the pIZ/V5SinRep5-X plasmid, the clone is transfected with pIB/V5-His/DHBB or pIB/V5-His/DHBBZZ and the cells (Fig. 17.8) subjected to double selection using both zeocin and blasticidin S. It is important to realize that only cells expressing the drug resistance gene will survive in the culture. Hence, after a short period of selection, 100% of the cells must be expressing the selectable marker. Whereas for most constructs the expression of the drug resistance gene is not necessarily associated with expression of the viral encoded gene, in at least some instances this association is obligatory. For example, in some constructs we use the hygromycin-GFP fusion gene as the selectable

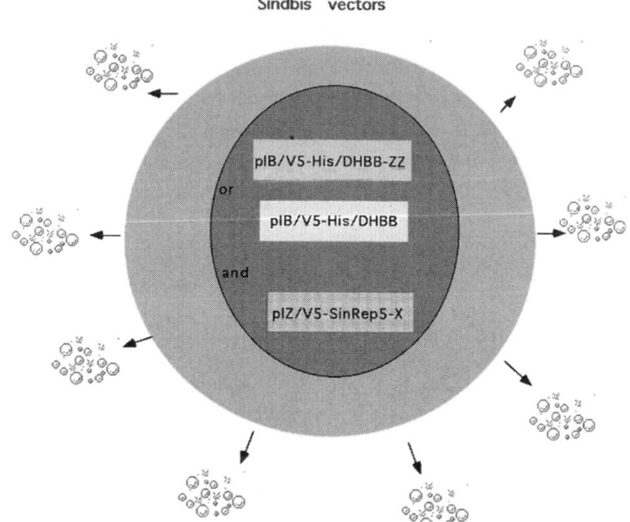

Figure 17.8 Schematic of packaging cell line based on insect cells.

marker, replacing the β-galactosidase reporter gene for X. When these cultures are transfected, it is obligatory for the survival of cells for the virally encoded hygromycin-GFP fusion gene to be expressed. The added advantage is that these cells should express GFP, allowing further selection among cells (e.g., one can select for cells expressing the highest amount of GFP using fluorescence-activated cell sorting).

A separate factor that has to be considered in the development of a packaging cell line is the use of antibiotics for selection of cells in which the Sindbis construct (e.g., pIZ/V5-His or pIB/V5-His) is expressed. Lacal et al. (1980) have shown that several antibiotics including anthelmycin, blasticidin S, destomycin A, gougerotin, hygromycin B, and edeine complex, known to powerfully block translation in cell-free systems, do not inhibit protein synthesis in intact mouse L and 3T6 cells nor in hamster BHK 21 cells, due to failure to cross the cell plasma membrane. However, after viral infection, these antibiotics exhibit a marked blockade of translation related to the permeability changes induced by viral infection. The inhibition of protein synthesis by hygromycin B in virus-infected BHK cells was studied by Lacal et al. (1980) over the time course of infection with the alphavirus Semliki forest virus. These authors observed that the entry of hygromycin B into virus-infected cells paralleled the inhibition of cellular protein synthesis; that is, the cells became permeable to this antibiotic at the time the shut-off of host translation occurred. Simultaneously, a marked inhibition of viral RNA synthesis by hygromycin B resulted, likely as a consequence of the inhibition of the viral replicase synthesis. As a consequence, a major reduction in the virus yield was obtained (74% reduction with blasticidin S and 76% reduction with hygromycin B) after treatment of virus-infected cells with several antibiotics. Therefore, to increase vector yield it might be necessary to release cells from the antibiotic selection pressure some time before production of virus batches, particularly those that might be dedicated for clinical applications. It is unlikely that this release will lead to loss of the vector-producing plasmid over a brief interval of at most several weeks.

Using the foregoing procedures we have generated a number of C6/36-based packaging cell lines, one of which has now been maintained in the laboratory for approximately a year. For this particular line, the pIB/V5-His/DHBB and pIZ/V5-His/SinRep5-LacZ plasmids were utilized. Cells are kept under both blasticidin S (50 μg/ml) and zeocin (600 μg/ml) selection.

Supernatants collected at various points from packaging lines generated by the methods described here have demonstrated infectivity both in vitro and in vivo by a variety of assays. For example, in one recent experiment media were removed from the cell culture and tested at various dilutions on BHK cells encoding a reporter luciferase gene (BHK-luc) that is activated only by Sindbis virus/vector infection (Olivo, Frolov, and Schlesinger, 1994). Typical titers of Sindbis vectors produced from BHK cells by the standard procedure involving in vitro RNA production and electroporation are shown in Fig. 17.9. Titers of virus produced from transfected C6/36 titers and assayed in BHK-luc cells are shown in Figure 17.10. Titers of the continuously producing insect cells are comparable to titers obtained from BHK cells (Fig. 17.9).

In a second experiment we examined the ability of the established viral infection to interfere with superinfection by the same virus (homologous interference). Riedel and Brown (1979) have shown that *A. albopinctus* cells persis-

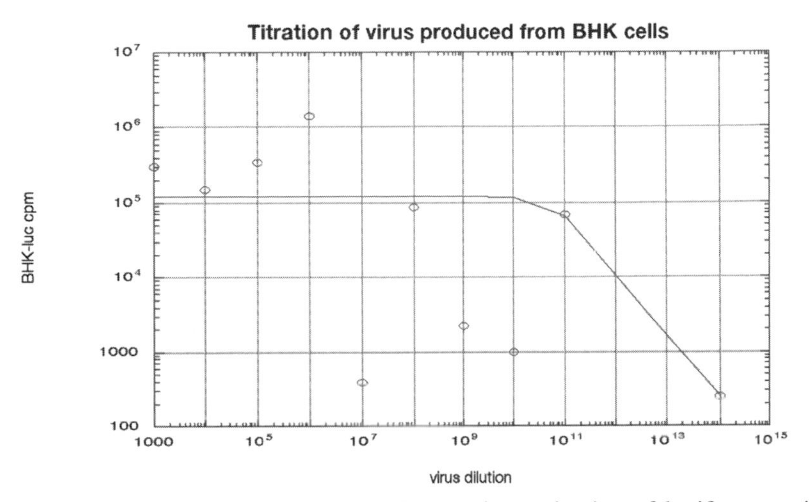

Figure 17.9 Viral titers were determined by testing activation of luciferase activity in BHK-luc cells as described in Olivo, Frolov, and Schlesinger, (1994).

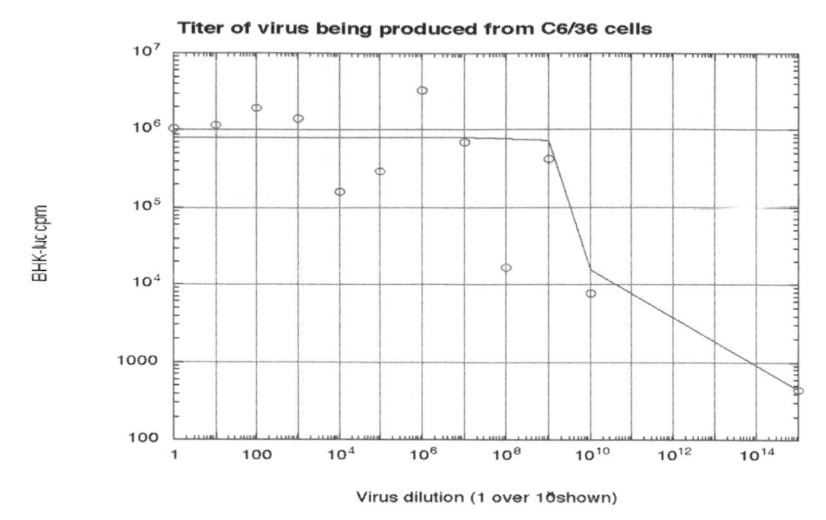

Figure 17.10 Viral titers were determined as described in the caption, Figure 17.9.

tently infected with Sindbis virus released an antiviral agent into the surrounding medium. The antiviral agent inhibits sindbis virus production in cultured insect cells. Typically, these cultures produce virus at levels representative of the persistent phase of infection; the high levels of virus production characteristic of the acute phase of infection are not found.

In our experiment the packaging cells producing virus for many months were infected with Sindbis vector and compared to C6/36 cells that had never been infected with Sindbis or made to produce Sindbis virus by transfection of the Sindbis vector plasmids. As shown in Figure 17.11, a dramatic difference

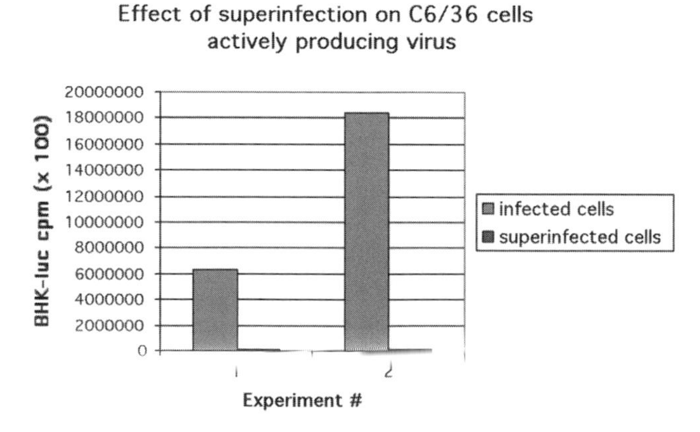

Figure 17.11 Viral titers were determined as described in the caption of Figure 17.9. Numbers shown are 5x actual readouts to adjust for dilution.

was observed between the effects of Sindbis infection on naïve C6/36 and the packaging C6/36 cultures.

CONCLUSION

We believe that our efforts, built on knowledge generated by many investigators, many of whom we have not been able to cite in this brief chapter, have allowed our laboratory to advance the day when clinical application of Sindbis vectors can become a reality. We know now that it is possible to cause complete regression with Sindbis vectors of some xenografted tumors grown in severe combined immunodeficiency (SCID) mice (unpublished data), and are engaged in extending these studies to a variety of other tumor types. Also, the issue of manufacturing, which has been problematic, if not resolved fully has been advanced to the point where generation of permanent cell lines may become possible in the not too distant future. All in all, we believe that Sindbis vectors appear to offer great promise for use in human therapy.

REFERENCES

ACAV (1980): Laboratory safety for arboviruses and certain other viruses of vertebrates. The Subcommittee on Arbovirus Laboratory Safety of the American Committee on Arthropod-Borne Viruses. *Am J Trop Med Hyg* 29:1359–1381.

ACGM (2000): Guidance on commonly used viral vectors (United Kingdom: Health and Safety Commission's Advisory Committe on Genetic Modification), pp. Part 2–Annex III.

Agapov EV, Frolov I, Lindenbach BD, Pragai, BM, Schlesinger S, Rice CM (1998): Noncytopathic sindbis virus RNA vectors for heterologous gene expression. *Proc Natl Acad Sci USA* 95:12989–12994.

Altman-Hamamdzic S, Groseclose C, Ma JX, Hamamdzic D, Vrindavanam NS, Middaugh LD, Parratto NP, Sallee FR (1997): Expression of beta-galactosidase in mouse brain: utilization of a novel nonreplicative Sindbis virus vector as a neuronal gene delivery system. *Gene Ther* 4:815–822.

Balachandran S, Roberts PC, Kipperman T, Bhalla KN, Compans RW, Archer DR, Barber GN (2000): Alpha/beta interferons potentiate virus-induced apoptosis through activation of the FADD/Caspase-8 death signaling pathway. *J Virol* 74:1513–1523.

Boczko EM, Brooks CL, 3rd (1995): First-principles calculation of the folding free energy of a three-helix bundle protein. *Science* 269:393–396.

Bredenbeek PJ, Frolov I, Rice CM, Schlesinger S (1993): Sindbis virus expression vectors: packaging of RNA replicons by using defective helper RNAs. *J Virol* 67:6439–6446.

Byrnes AP, Griffin DE (2000): Large-plaque mutants of Sindbis virus show reduced binding to heparan sulfate, heightened viremia, and slower clearance from the circulation. *J Virol* 74:644–651.

Condreay LD, Brown DT (1988): Suppression of RNA synthesis by a specific antiviral activity in Sindbis virus-infected Aedes albopictus cells. *J Virol* 62:346–348.

Condreay LD, Brown DT (1986): Exclusion of superinfecting homologous virus by Sindbis virus-infected *Aedes albopictus* (mosquito) cells. *J Virol* 58:81–86.

Doxsey SJ, Brodsky FM, Blank GS, Helenius A (1987): Inhibition of endocytosis by anti-clathrin antibodies. *Cell* 50:453–463.

Forsgren A, Sjoquist J (1969): Protein A from Staphylococcus aureus. VII. Physicochemical and immunological characterization. *Acta Pathol Microbiol Scand* 75:466–480.

Gouda H, Torigoe H, Saito A, Sato M, Arata Y, Shimada I (1992): Three dimensional solution structure of the B domain of staphylococcal protein A: comparisons of the solution and crystal structures. *Biochemistry* 31:9665–9672.

Gwag BJ, Kim EY, Ryu BR, Won SJ, Ko HW, Oh YJ, Cho YG, Ha SJ, Sung YC (1998): A neuron-specific gene transfer by a recombinant defective Sindbis virus. *Brain Res Mol Brain Res* 63:53–61.

Hariharan MJ, Driver DA, Townsend K, Brumm D, Polo JM, Belli BA, Catton DJ, Hsu D, Mittelstaedt D, McCormack JE, Karavodin L, Dubensky TW, Jr., Chang SM, Banks TA (1998): DNA immunization against herpes simplex virus: enhanced efficacy using a Sindbis virus-based vector. *J Virol* 72:950–958.

Helenius A (1984): Semliki Forest virus penetration from endosomes: a morphological study. *Biol Cell* 51:181–185.

Helenius A, Marsh M, White J (1982): Inhibition of Semliki forest virus penetration by lysosomotropic weak bases. *J Gen Virol* 58:47–61.

Helenius A, Kartenbeck J, Simons K, Fries E (1980): On the entry of Semliki forest virus into BHK-21 cells. *J Cell Biol* 84:404–420.

Hoekstra D, Kok JW (1989): Entry mechanisms of enveloped viruses. Implications for fusion of intracellular membranes. *Biosci Rep* 9:273–305.

Hurlbut HS (1953): The experimental transmission of coxsackie-like viruses by mosquitoes. *J Egypt Med Assoc* 36:495–498.

Jan JT, Chatterjee S, Griffin DE (2000): Sindbis virus entry into cells triggers apoptosis by activating sphingomyelinase, leading to the release of ceramide. *J Virol* 74:6425–6432.

Jan JT, Griffin DE (1999): Induction of apoptosis by Sindbis virus occurs at cell entry and does not require virus replication. *J Virol* 73:10296–10302.

Jendeberg L, Tashiro M, Tejero R, Lyons BA, Uhlen M, Montelione GT, Nilsson B (1996): The mechanism of binding staphylococcal protein A to immunoglobin G does not involve helix unwinding. *Biochemistry* 35:22–31.

Jendeberg L, Persson B, Andersson R, Karlsson R, Uhlen M, Nilsson B (1995): Kinetic analysis of the interaction between protein A domain variants and human Fc using plasmon resonance detection. *J Mol Recognition* 8:270–278.

Karpf AR, Brown DT (1998): Comparison of Sindbis virus-induced pathology in mosquito and vertebrate cell cultures. *Virology* 240:193–201.

Karpf AR, Blake JM, Brown DT (1997): Characterization of the infection of *Aedes albopictus* cell clones by Sindbis virus. *Virus Res* 50:1–13.

Karpf AR, Lenches E, Strauss EG, Strauss JH, Brown DT (1997): Superinfection exclusion of alphaviruses in three mosquito cell lines persistently infected with Sindbis virus. *J Virol* 71:7119–7123.

Kielian M, Jungerwirth S, Sayad KU, DeCandido S (1990): Biosynthesis, maturation, and acid activation of the Semliki Forest virus fusion protein. *J Virol* 64:4614–4624.

Kielian M, Helenius A (1986): Entry of alpha viruses. In *The Togaviridae and Flaviviridae*, Schlesinger S, Schlesinger MJ, eds. New York: Plenum Publishing, pp. 91–119.

Kolinski A, Skolnick J (1994): Monte Carlo simulations of protein folding. II. Application to protein A, ROP, and crambin. *Proteins* 18:353–366.

Lacal JC, Vazquez D, Fernandez-Sousa JM, Carrasco L (1980): Antibiotics that specifically block translation in virus-infected cells. *J Antibiot (Tokyo)* 33:441–446.

Levine B, Huang Q, Isaacs JT, Reed JC, Griffin DE, Hardwick JM (1993): Conversion of lytic to persistent alphavirus infection by the bcl-2 cellular oncogene. *Nature* 361:739–742.

Liljestrom P, Garoff H (1991): A new generation of animal cell expression vectors based on the Semliki Forest virus replicon. *Biotechnology (NY)* 9:1356–1361.

Ljungberg UK, Jansson B, Niss U, Nilsson R, Sandberg BE, Nilsson B (1993): The interaction between different domains of staphylococcal protein A and human polyclonal IgG, IgA, IgM and F(ab′)2: separation of affinity from specificity. *Mol Immunol* 30:1279–1285.

Luo T, Brown DT (1993): Purification and characterization of a Sindbis virus induced peptide which stimulates its own production and blocks virus RNA synthesis. *Virology* 194:44–49.

Marsh M (1984): The entry of enveloped viruses into cells by endocytosis. *Biochem J* 218:1–10.

Marsh M, Bolzau E, Helenius A (1983): Penetration of Semliki Forest virus from acidic prelysosomal vacuoles. *Cell* 32:931–940.

Marsh M, Wellsteed J, Kern H, Harms E, Helenius A (1982): Monensin inhibits Semliki Forest virus penetration into culture cells. *Proc Natl Acad Sci USA* 79:5297–5301.

Marsh M, Helenius A (1980): Adsorptive endocytosis of Semliki Forest virus. *J Mol Biol* 142:439–454.

McGill PE (1995): Viral infections: alpha-viral arthropathy. *Baillieres Clin Rheumatol* 9:145–150.

Miller ML, Brown DT (1992): Morphogenesis of Sindbis virus in three subclones of *Aedes albopictus* (mosquito) cells. *J Virol* 66:4180–4190.

Moks T, Abrahmsen L, Nilsson B, Hellman U, Sjoquist J, Uhlen M (1986): Staphylococcal protein A consists of five IgG-binding domains. *Eur J Biochem* 156:637–643.

Nilsson B, Moks T, Jansson B, Abrahmsen L, Elmblad A, Holmgren E, Henrichson C, Jones TA, Uhlen M (1987): A synthetic IgG-binding domain based on staphylococcal protein A. *Protein Eng* 1:107–113.

Ohno K, Sawai K, Iijima Y, Levin B, Meruelo D (1997): Cell-specific targeting of Sindbis virus vectors displaying IgG-binding domains of protein A. *Nat Biotechnol* 15:763–767.

Olivo PD, Frolov I, Schlesinger S (1994): A cell line that expresses a reporter gene in response to infection by Sindbis virus: a prototype for detection of positive strand RNA viruses. *Virology* 198:381–384.

Olszewski KA, Kolinski A, Skolnick J (1996a): Does a backwardly read protein sequence have a unique native state? *Protein Eng* 9:5–14.

Olszewski KA, Kolinski A, Skolnick J (1996b): Folding simulations and computer redesign of protein A three-helix bundle motifs. *Proteins* 25:286–299.

Piper RC, Slot JW, Li G, Stahl PD, James DE (1994): Recombinant Sindbis virus as an expression system for cell biology. *Methods Cell Biol* 43:55–78.

Polo JM, Belli BA, Driver DA, Frolov I, Sherrill S, Hariharan MJ, Townsend K, Perri S, Mento SJ, Jolly DJ, Chang SM, Schlesinger S, Dubensky TW, Jr. (1999): Stable alphavirus packaging cell lines for Sindbis virus and Semliki Forest virus-derived vectors. *Proc Natl Acad Sci USA* 96:4598–4603.

Pugachev KV, Mason PW, Shope RE, Frey TK (1995): Double-subgenomic Sindbis virus recombinants expressing immunogenic proteins of Japanese encephalitis virus induce significant protection in mice against lethal JEV infection. *Virology* 212:587–594.

Riedel B, Brown DT (1979): Novel antiviral activity found in the media of Sindbis virus-persistently infected mosquito (*Aedes albopictus*) cell cultures. *J Virol* 29:51 60.

Sawai K, Ikeda H, Ishizu A, Meruelo D (1999): Reducing cytotoxicity induced by Sindbis viral vectors. *Mol Genet Metab* 67:36–42.

Sawai K, Meruelo D (1998): Cell-specific transfection of choriocarcinoma cells by using Sindbis virus hCG expressing chimeric vector. *Biochem Biophys Res Commun* 248:315–323.

Shah KV, Johnson HN, Rao TR, Rajagopalan PK, Lamba BS (1960): Isolation of five strains of Sindbis virus in India. *Indian J Med* 48:300–308.

Sjodahl J (1977): Structural studies on the four repetitive Fc-binding regions in protein A from Staphylococcus aureus. *Eur J Biochem* 78:471–479.

Starovasnik MA, Skelton NJ, O'Connell MP, Kelley RF, Reilly D, Fairbrother WJ (1996): Solution structure of the E-domain of staphylococcal protein A. *Biochemistry* 35:15558–15569.

Stegmann T, Doms RW, Helenius A (1989): Protein-mediated membrane fusion. *Annu Rev Biophys Biophys Chem* 18:187–211.

Strauss JH, Strauss EG (1994): The alphaviruses: gene expression, replication, and evolution. *Microbiol Rev* 58:491–562.

Strauss JH, Wang KS, Schmaljohn AL, Kuhn RJ, Strauss EG (1994): Host-cell receptors for Sindbis virus. *Arch Virol Suppl* 9:473–484.

Taylor RM, Hurlbut HS, Work TH, Kingsbury JR, Frothingham TE (1955): Sindbis Virus: A newly recognized arthropod-transmitted virus. *Am J Trop Med Hyg* 4:844–846

Taylor RM, Hurlbut HS (1953): Isolation of coxsackie-like viruses from mosquitoes. *J Egypt Med Assoc* 36:489–494.

Tsuji M, Bergmann CC, Takita-Sonoda Y, Murata K, Rodrigues EG, Nussenzweig RS, Zavala F (1998): Recombinant Sindbis viruses expressing a cytotoxic T-lymphocyte epitope of a malaria parasite or of influenza virus elicit protection against the corresponding pathogen in mice. *J Virol* 72:6907–6910.

Turrell MJ (1988): Horizontal and vertical transmission of viruses by insect and tick vectors. In *The Arboviruses: Epidimiology and Ecology*, Monath TP, ed. Boca Raton, Fla.: CRC Press, pp. 127–152.

Uhlen M, Forsberg G, Moks T, Hartmanis M, Nilsson B (1992): Fusion proteins in biotechnology. *Curr Opin Biotechnol* 3:363–369.

Uhlen M, Guss B, Nilsson B, Gatenbeck S, Philipson L, Lindberg M (1984): Complete sequence of the staphylococcal gene encoding protein A. A gene evolved through multiple duplications. *J Biol Chem* 259:1695–1702.

Wang KS, Kuhn RJ, Strauss EG, Ou S, Strauss JH (1992): High-affinity laminin receptor is a receptor for Sindbis virus in mammalian cells. *J Virol* 66:4992–5001.

White JM (1990): Viral and cellular membrane fusion proteins. *Annu Rev Physiol* 52:675–697.

Xiong C, Levis R, Shen P, Schlesinger S, Rice CM, Huang HV (1989): Sindbis virus: An efficient, broad host range vector for gene expression in animal cells. *Science* 243:1188–1191.

18

REDIRECTING THE TROPISM OF HSV-1 FOR GENE THERAPY APPLICATIONS

QING BAI PH.D., EDWARD A. BURTON M.D., PH.D., WILLIAM F. GOINS PH.D., AND JOSEPH C. GLORIOSO PH.D.

INTRODUCTION

Herpes simplex virus (HSV) is an enveloped neurotropic DNA virus (Fig. 18.1; reviewed in Roizman and Sears, 1996). The mature virion consists of the following components:

- A trilaminar lipid envelope in which are embedded viral proteins and glycoproteins responsible for cellular entry (Rajcani and Vojvodova, 1998; Spear, 1993a; 1993b; Stevens and Spear, 1997; and see the following)
- A matrix of proteins, the tegument, which forms a layer between the envelope and the underlying capsid. Functions of the tegument proteins include: induction of viral gene expression (Batterson and Roizman 1983; Campbell, Palfreyman, and Preston, 1984; Mackem and Roizman, 1982); shutoff of host protein synthesis immediately following infection (Kwong and Frenkel, 1987, 1989; Kwong, Kruper, and Frenkel, 1988; Read and Frenkel, 1983); and virion assembly functions
- An icosadeltahedral capsid, typical of the herpesvirus family (Homa and Brown, 1997; Newcomb et al., 1999)

Vector Targeting for Therapeutic Gene Delivery, Edited by David T. Curiel and Joanne T. Douglas
ISBN 0-471-43479-5

- A core of toroidal dsDNA (Furlong, Swift, and Roizman, 1972; Homa and Brown, 1997; Puvion Dutilleul, Pichard, and Leduc, 1985)

Replication-defective HSV particles can be generated by carrying out a series of genetic modifications to the wild-type virus, which destroy the tightly regulated viral lytic gene expression cascade (Burton et al., 2001a; DeLuca, McCarthy, and Schaffer, 1985; Krisky et al., 1998b; Samaniego, Neiderhiser, and DeLuca, 1998). The resulting vectors have many favorable properties for multiple gene transfer applications:

- Nonpathogenicity (Krisky et al., 1998b; Samaniego, Neiderhiser, and DeLuca, 1998).
- Large capacity for the insertion of therapeutic or experimental transgene sequences (Akkaraju et al., 1999; Krisky et al., 1998a).
- Easy production of large quantities.
- Wild-type helper virus does not contaminate replication-defective vector stocks, which are grown in complementing cell lines.
- High infectivity (Moriuchi et al., 2000).
- Broad host cell range (Goins et al., 1999, 2001; Gomez Navarro et al., 2000; Goss et al., 2001; Huard et al., 1998; Marconi et al., 2000; Oligino et al., 1999).
- Natural latent life cycle in which the virus remains within nuclei of neuronal and nonneuronal cells for the lifetime of the host, without disturbing host cell metabolism.
- Latency promoter system (Chen et al., 1995; Dobson et al., 1989; Goins et al., 1994; Zwaagstra et al., 1989), which allows long-term expression of transgenes (Goins et al., 1999; Lachmann and Efstathiou, 1997; Wolfe et al., 2001).

Uses of HSV Vectors

Clinical applications of HSV vectors are still in the early stages of development and are limited to the use of attenuated, conditionally replicating vectors in malignant glioma patients (Markert et al., 2000; Rampling et al., 2000). However, there is a wealth of potential clinical uses for this vector system. We have recently published detailed reviews covering experimental applications of HSV vectors in animal models of disease (Burton et al., 2001a, 2001b). The following points represent a selection of likely disease targets that may be particularly amenable to intervention using HSV vectors, because of unique features in the viral life cycle:

- Peripheral nervous system (PNS)—The latent behavior of the vector in its natural host tissue (Goins et al., 1994), and the nontoxicity of disabled vectors (Krisky et al., 1998b), have been exploited to allow prolonged transgene delivery in experimental models of peripheral neuropathy (Goins et al., 1999, 2001) and chronic pain (Goss et al., 2001; Wilson et al., 1999).

- Central nervous system (CNS)—Defective virus is able to transduce CNS neurons safely and drive expression of transgenes in the CNS (Krisky et al., 1998b; Marconi et al., 1999). Genes have been delivered that encode proteins performing neuroprotective functions (Natsume et al., 2001; Yamada et al., 2001, 1999) in addition to using the meninges as a depot site for the local synthesis of anti-inflammatory cytokines that are secreted into the cerebrospinal (CSF) fluid (Martino et al., 2000a, 2000b).
- Malignancy—The ability to produce high titers of pure viral preparations easily enables transduction of large numbers of cells within a tumor at high multiplicity. In addition, owing to the large cloning capacity of HSV vectors, multiple transgenes may be delivered simultaneously (Krisky et al., 1998a). This is of particular interest in the situation arising within a tumor, where there are heterogeneous cell populations bearing distinct mutational burdens. We have shown that simultaneous, coordinated, multiple genetic and nongenetic treatments are more effective in experimental tumor eradication than single agents (Marconi et al., 2000; Moriuchi et al., 1998; Niranjan et al., 2000).
- Secreted proteins—The infective cycle that ensues when a replication-defective HSV vector enters cells of nonneural tissue resembles that occurring during wild-type latency in neurons. We have shown that long-term transgene expression within nonneural tissue is possible, allowing the use of joints and ligaments as depot sites for the synthesis of circulating proteins (Wolfe et al., 2001) or locally acting secreted proteins (Oligino et al., 1999). This raises the possibility of novel therapeutic strategies for hitherto intractable problems, such as the common form of hemophilia in which the cDNA encoding the missing gene product is large, and the issue of sustained gene expression has been difficult to resolve.
- Stem cells—Recent data show that bone marrow-derived stem cells bear the receptors for HSV cell entry (Wechuck et al., in prep.) and that these cells may be transduced using replication-defective vectors (Gomez Navarro et al., 2000). Transduction of these cell populations with nonintegrating HSV vectors would allow transient expression of multiple transgenes, without effecting genomic modification. This route might prove useful to drive stem cells into particular pathways of differentiation, but to allow their normal function once committed to particular lineages.

Advantages of Targeting

The broad host cell range of the unmodified vector is advantageous when developing a system with sufficient flexibility to allow use in multiple circumstances. However, in some situations, it may be desirable or necessary to restrict the expression of therapeutic transgenes to predefined subpopulations of cells. One way of achieving this goal might be through redirecting the tropism of HSV vectors; conceivably, this could be combined with targeted transcription to ensure a strictly defined spatial or temporal expression pattern. The following paragraphs outline some advantages that may be gained by a successful targeted tropism strategy:

- CNS—The ability to target cell entry to particular subsets of neurons might allow the development of invaluable experimental reagents for neuroscience applications, in addition to raising the exciting prospect of selective modification of the physiology of predefined neural pathways. A simple example might be selective gene replacement in diseases characterized by null mutations; restoration of normal function of a neural network might only follow reestablishment of the wild-type expression pattern, which could not be achieved without one or more targeting strategies. In addition, targeting might allow the transduction of neural elements relatively noninvasively by means of the CSF pathways. Inoculation of the CSF with vector bearing wild-type cell entry determinants invariably results in transduction of meningeal tissue, for which HSV displays avid tropism. Removal of this tropism might allow passage of vector into the neuropil and result in targeted transduction of neural populations throughout the length of the neuraxis.
- PNS—Targeting HSV entry to nerve endings either in skin (sensory) or muscle end plates (motor) would allow a larger effective dose to be delivered to the targeted neurons, by avoiding the sink effect of transduction of surrounding epithelial and supporting tissue. This would confer the additional advantage of avoiding the use of larger, more immunogenic, doses of vector, which might result in local tissue damage at the site of inoculation or immune clearance of transduced cells. In addition, in anterior horn cell diseases, it may be possible to target expression of neurotrophic agents to motor neurons, avoiding their expression in sensory neurons with accompanying problems (e.g., enhancement of pain transmission).
- Cancer—The technology to target entry of the vector into cancer cells would allow the selective delivery of toxic gene products into malignant cells, giving rise to selective tumor lysis while sparing normal neighboring tissue. Alternatively, it may be possible to introduce cytoprotective or radio-resistance genes selectively into normal glia or neurons, while allowing the destruction of infiltrating glial tumor tissue.
- Stem cells—Targeting vector tropism to stem cell populations in vivo may obviate the need for complex procedures to isolate and purify these cells ex vivo. Conceivably, this may be advantageous in avoiding the removal of stem cells from their native environment, allowing the retention of potentially useful properties such as multipotency.

In recent years, fundamental steps have been taken toward unraveling the basic mechanisms by which HSV enters cells. In the following paragraphs, we review published literature that helps elucidate the processes governing the entry of the HSV nucleocapsid and tegument proteins into cells, and discuss ways in which we have started to manipulate these mechanisms with the ultimate goal of dictating the infectious targets of recombinant virions.

THE MECHANISMS BY WHICH HSV-1 ENTERS CELLS

HSV-1 has evolved a complex series of virion components, with complementary or redundant functions, that allow the mature particle to deliver its DNA and protein payload directly into the cytoplasm of infected cells (Fig. 18.1) (Morgan, Rose, and Mednis, 1968; Spear 1993a). This transplasmalemmal route of entry distinguishes HSV-1 from many other types of virus, which bind to internalizing receptors and enter the cytoplasm from within the endosome compartment. The complexity of the HSV-1 cell entry process inevitably means that difficulty is encountered in attempting to alter the end result of the entry cascade by manipulating its components. A rational approach to engineering the entry of this virus requires a detailed understanding of the processes presiding over HSV cytopenetration during normal circumstances.

Early electron microscopic studies suggested that HSV virions enter cells by a two-step process: initial cell attachment is followed by fusion of the viral envelope with the cell membrane (Morgan, Rose, and Mednis, 1968). The latter

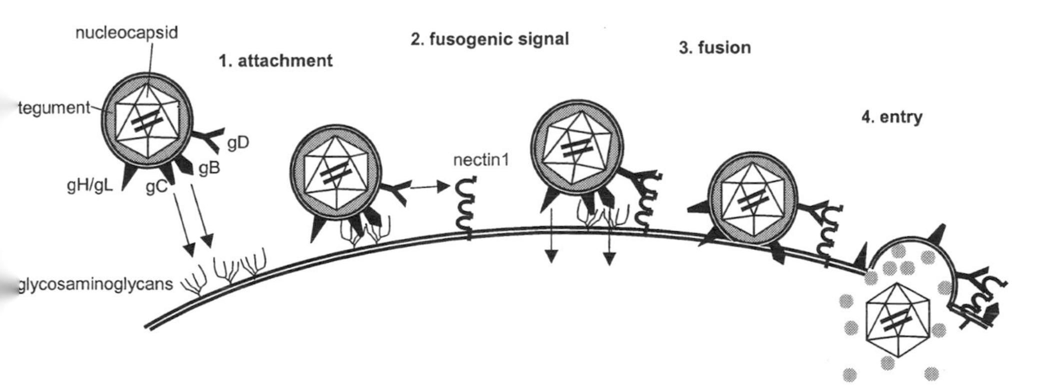

Figure 18.1 Mechanism of cell entry of wild type HSV-1. The important structural features and surface determinants of the mature virion are depicted schematically. Initial attachment occurs by binding of virion gB and gC to cell surface glycosaminoglycans. This allows viral gD to interact with a secondary receptor; HveC is shown here for illustrative purposes. The resulting fusogenic signal allows gB and gH/L to effect virion-cell membrane fusion, with consequent entry of the viral nucleocapsid and tegument proteins into the cell.

step results in the appearance of viral surface components in the plasmalemma (Para, Baucke, and Spear, 1980). Release of the viral nucleocapsid and tegument proteins into the cytoplasm ensues. For the purposes of this discussion, it is convenient to consider these steps separately, although in reality the entry cascade proceeds rapidly and seamlessly from one phase to another.

HSV Virion Surface Components

Eighty-four viral genes are contained within the 152 kb HSV-1 genome (McGeoch et al., 1988, 1986, 1985). Of these, approximately one-fifth encode virion surface and envelope components:

- There are eleven viral glycoproteins, designated gB, gC, gD, gE, gG, gH, gI, gJ, gK, gL, gM [reviewed in Roizman and Sears (1996) and Rajcani and Vojvodova, 1998]. With the exception of gK, all of these are embedded in the trilaminar envelope of the mature virion and therefore are available at the surface of the particle to interact with appropriate cellular components.
- The viral envelope also contains two (U_L20, U_L34) and possibly other (U_L24, U_L45, $U_L49.5$) nonglycosylated intrinsic membrane proteins [reviewed in Roizman and Sears (1996) and Stevens and Spear (1997)].

Of the glycoproteins, gB (Cai, Gu, and Person, 1988), gD (Fuller and Spear 1987; Highlander et al., 1987; Ligas and Johnson, 1988), gH (Desai, Schaffer, and Minson, 1988; Forrester et al., 1992) and gL (Hutchinson et al., 1992; Roop, Hutchinson, and Johnson, 1993) are essential to enable viral replication in cell culture, whereas the other glycoproteins are dispensable. Of the intrinsic envelope proteins, only that encoded by the U_L34 gene is essential for replication in cell culture (Barker and Roizman, 1992; MacLean et al., 1991; Visalli and Brandt, 1991; Ye and Roizman, 2000); it appears to have an important role in the intracellular trafficking of virions, rather than a role in cell entry per se.

The functions of many of the viral surface components, especially those whose presence is not requisite for viral replication, remain obscure. In the following discussion, details of structure and function are limited to those viral components with at least one defined role in the entry process that might be manipulated to modify viral tropism.

Cell Attachment

The first phase in the entry cascade consists of viral attachment (adsorption) to cells (Morgan, Rose, and Mednis, 1968). The initial attachment is a dynamic process that depends on the binding of viral components to glycosaminoglycan (GAG) moieties of cell surface proteoglycans (Shieh et al., 1992; Spear et al., 1992). Specific GAGs shown to be involved in the initial HSV-1 binding event include heparan sulfate (HS) (Gruenheid et al., 1993), chondroitin sulfate (Banfield et al., 1995b), and dermatan sulfate (Williams and Straus, 1997). All consist of long stretches of sulfated, and therefore negatively charged, carbohydrate chains. Their necessity in the initial virion–cell attachment process is exemplified by the following data specifically pertaining to heparan sulfate:

- There is absence of detectable adsorption of wild-type virus onto the surface of cell lines that either do not express HS genetically (Shieh et al., 1992), or have been enzymatically treated to remove HS from the cell surface (WuDunn and Spear, 1989).
- Heparin, which shares structural and chemical features with heparan sulfate, is able to reduce HSV cell binding by competing with cell surface HS for binding sites on HSV (WuDunn and Spear, 1989).

The virion surface glycoprotein gC has been implicated in the initial adsorption event (Herold et al., 1991, 1994) by the following lines of evidence:

- Mutant particles devoid of gC show approximately 65% reduced attachment to cells (Laquerre et al., 1998b).
- Anti-gC neutralizing antibodies inhibit cell attachment of the virus (Fuller and Spear, 1985), whereas neutralizing antibodies to other virion surface components impair fusion without adversely affecting attachment (Fuller and Spear, 1985; Highlander et al., 1987, 1988).
- Glycoprotein C is able to bind to heparin on an affinity column (Herold et al., 1991).

The presence of a minimal set of sulfated residues in the heparan sulfate chain is essential for gC binding (Feyzi et al., 1997). HS binding has been localized to a discrete area of gC by monoclonal antibodies, mutagenesis, and peptide competition (Trybala et al., 1994). Critical amino acid residues in this region include a positively charged arginine-rich domain (Mardberg et al., 2001).

Interestingly, gC-null mutant viruses retain a degree of infectivity and cell binding. The binding is further reduced by deletion of gB from a gC-null virus, or of HS from the cells (Herold et al., 1994). This implies that gB also plays a role in the initial cell attachment step to HS, although isolated gB-null mutants are more impaired for cell entry than attachment (Cai, Gu, and Person, 1988). Purified gB binds to a heparin affinity column (Herold et al., 1991) and plasmon resonance biosensor experiments show that the gB ectodomain binds heparan and dermatan sulfates (Williams and Straus, 1997). There is a lysine-rich domain within gB (poly-K or simply pK); the high positive charge of this region is thought to favor binding to heparan moieties by electrostatic attraction. Indeed, chemically synthesized polylysine inhibits HSV cell attachment by competing with viral receptors for cellular HS (Langeland et al., 1988).

Deletion of the pK region results in a mutant form of gB with impaired binding to HS (Laquerre et al., 1998b). This mutant is expressed on the cell surface and is able to rescue a gB-deleted virus, resulting in the production of particles that show approximately 20% reduction in cell attachment, but which retain cell entry functions (Laquerre et al., 1998b). Introduction of the pK^- gB mutant into a gC-null background results in a mutant virus with approximately 80% reduction in cell binding, but which retains fusion activity (Laquerre et al., 1998b). The residual cell attachment and entry observed following removal of gC and gB-pK, or removal of all cell surface GAGs (Banfield et al., 1995a), is attributable to interactions between gD and cellular receptors and is considered in the next section. The relevance to the present discussion is that (1) a triple manipulation (gC, gB, gD) may be necessary to completely abolish virion binding to nontargeted cells, and (2) the gB : pK^-, gC^- background may be a useful starting point from which targeted cell binding could be accomplished, with preservation of virion–cell fusion functions.

Receptor Recognition

Following the initial adsorption event, a secondary binding event occurs between viral gD and a cellular receptor. This was initially suggested by three observations:

1. Inactivated virus antagonizes wild-type virus entry, provided the inactivated virions contain gD (Johnson and Ligas, 1988).
2. Soluble gD competes with HSV-1 for a cellular binding site and prevents viral entry but not adsorption (Johnson, Burke, and Gregory, 1990).
3. Expression of gD on the cell surface antagonizes transplasmalemmal viral entry by competing for cellular receptors but does not prevent attachment, resulting in viral endocytosis (Campadelli Fiume et al., 1988).

Subsequent work identified two structurally unrelated HSV-1 secondary receptors that bind gD following gC/gB/HS-dependent adsorption, resulting in commitment to virion-cell fusion.

1. Herpesvirus entry mediator A (HVEM/HveA). HveA is a member of the tumor necrosis factor α/nerve growth factor (TNFα/NGF) receptor superfamily. The HveA cDNA was initially isolated from a HeLa cell expression library by virtue of its ability to confer susceptibility to HSV infection when expressed in infection-resistant CHO-K1 cells (Montgomery et al., 1996). Further studies showed that gD binds directly to HveA (Whitbeck et al., 1997), that truncated soluble HveA blocks HSV infection by competing with cellular receptors for gD binding (Whitbeck et al., 2001), and that some anti-gD

monoclonal antibodies capable of blocking HSV infection also inhibit binding of gD to HveA (Nicola et al., 1998). Published data imply that HveA has a limited expression pattern in vivo and is probably not the major entry mediator during natural infection of mucosae and neurons (Montgomery et al., 1996).

2. Herpesvirus entry mediator C (HveC)/poliovirus receptor related protein-1 (PRR-1)/nectin-1. Following the demonstration that certain HSV-2 strains and HSV-1 mutants can enter cells by interaction with poliovirus receptor-related protein 2 (PRR-2 or herpes virus entry mediator B—HveB), the closely related PRR-1 became a candidate receptor to mediate entry of other HSV strains. Initially, it was shown that PRR-1 could mediate entry of HSV-1 into CHO-K1 cells; the receptor was thus redesignated herpes virus entry mediator C (HveC) (Geraghty et al., 1998). Soluble HveC can antagonize viral entry into susceptible cells by competing for binding to viral gD (Geraghty et al., 1998). Subsequent studies showed that viral gD could directly bind to HveC and that anti-gD antibodies that block infection can prevent gD-HveC binding (Krummenacher et al., 1998).

HveC is a type 1 transmembrane glycoprotein and a member of the immunoglobulin (Ig) superfamily (Geraghty et al., 1998). Like other members of the Ig family, the ectodomain of HveC contains immunoglobulin homology domains, an amino (N)-terminal variable (V)-homology domain followed by two C2-like domains. Pairs of cysteine residues, which form disulfide bonds in the mature protein, delimit each of these domains, resulting in the adoption of the characteristic V- and constant (C)-domain loops. HveC is a protein isoform resulting from alternative splicing of the primary transcript from the nectin1 gene (Cocchi et al., 1998; Geraghty et al., 1998; Lopez et al., 2001). The gene gives rise to three splice variants, which differ in their 3′ exons. The resulting proteins have identical ectodomains, but differ in the nature or presence of their transmembrane and endodomains. The nomenclature is confusing—the prototype receptor, HveC, strictly refers to nectin1β (Geraghty et al., 1998). Nectin1δ does not contain a transmembrane domain and is secreted into the extracellular space (Lopez et al., 2001). All three isoforms have been shown to mediate HSV-1 cell entry. The physiological cellular function of nectin1 is poorly characterized at present. Nectin1 null mutations in humans result in a type of ectodermal dysplasia syndrome (Suzuki et al., 2000).

Role of gD in Receptor Recognition Further genetic mapping studies have shown that the important domain for gD binding lies within the V-domain of nectin1/HveC (Krummenacher et al., 1999). Indeed, cell surface expression of the V-domain is both necessary and sufficient to mediate entry of HSV-1 into nonsusceptible cells (Krummenacher et al., 1999). The requirements for the remainder of the protein bearing the nectin1 V-domain are extremely lax. We recently showed that the V-domain alone in solution, independent of any cell association, was sufficient to trigger entry of HSV into nonsusceptible cells

(Bai et al., in prep.). It appears, therefore, that the important secondary signal for cell entry passes from cell to virus, rather than vice versa. Interaction of gD with its cognate receptor triggers the remainder of the entry cascade, resulting in virus–cell fusion.

Thus, gD can bind to two structurally unrelated cellular receptors (Krummenacher et al., 1998), with the same end result (activation of viral–cell fusion). This intriguing observation raises the complex question of how two presumably different binding events are able to instigate similar secondary consequences. Genetic and structural studies of viral gD have started to address this question.

Scanning insertional mutagenesis has defined four important functional domains of gD in which mutations destroy the receptor function without impairing the structure or trafficking of the glycoprotein (Chiang, Cohen, and Eisenberg, 1994). These regions are located at the N-terminal (region I: amino acids 27–43), middle (region II: amino acids 126–161) and carboxy (C)-terminal (region III: amino acids 225–246, region IV: amino acids 277–310) areas of the gD ectodomain. We have undertaken a detailed and systematic mutagenesis study of regions I (Bai et al., in prep.) and III–IV (AliShah et al., in prep.) to try to elucidate which amino acid residues are responsible for the binding and entry functions of gD. These data demonstrate that the receptor-binding and cell entry functions of gD are dissociable. The binding sites for HveA and HveC are overlapping, but rely on different residues at the N-terminus. In addition, a different (but overlapping) subset of residues for each secondary receptor is responsible for triggering viral entry following gD binding. The nature of protein conformational changes of the wild-type and mutant gD molecules induced by receptor binding may become apparent once the crystal structures of the receptor/gD complexes are fully available; the crystal structure of truncated HveA ectodomain bound to truncated gD ectodomain was recently published (Carfi et al., 2001). These fascinating preliminary data provide structural support for our functional mutagenesis studies. The consequences of these data for gD-mediated viral targeting are discussed in Engineerng Receptor-Recognizing Glycoprotein D (p. 390).

Virus–Cell Fusion

Following the binding of gD to its cognate cell surface receptor, the viral envelope fuses with the cell membrane, allowing release of the virion contents into the cell (Morgan, Rose, and Mednis, 1968). It is unclear whether the gD-receptor complex is directly involved in effecting this step, or whether it performs a purely regulatory role. Three other HSV glycoproteins have been implicated in this step of the entry cascade: gB, gH, and gL. The latter two glycoproteins form a heterodimer in the virion, referred to as gH/L (Hutchinson et al., 1992; Roop, Hutchinson, and Johnson, 1993; Westra et al., 1997). The evidence for the central role of these components in envelope–plasmalemmal fusion may be summarized thus:

- Temperature-sensitive (ts) gB (Gage et al., 1992; Haffey and Spear, 1980; Manservigi, Spear, and Buchan, 1977; Sarmiento, Haffey, and Spear, 1979) and gH (Desai, Schaffer, and Minson, 1988) mutants and gB-null (Cai, Gu, and Person, 1988), gH-null (Forrester et al., 1992), and gL-null (Roop, Hutchinson, and Johnson, 1993) viruses can bind to cells, but cannot enter.
- Anti-gB (Highlander et al., 1988; Navarro, Paz, and Pereira, 1992), anti-gH (Fuller, Santos, and Spear, 1989; Gompels and Minson, 1986) blocking antibodies allow adsorption but not fusion of the virus to the cell.
- Various rate-of-entry mutants in the gB gene have been described (Bzik et al., 1984; DeLuca et al., 1982; Highlander et al., 1989; Saharkhiz Langroodi and Holland, 1997).
- The expression of gB, gH, and gL in cells with gD and HveA/C is sufficient to cause membrane fusion and polykaryote formation (Manservigi, Spear, and Buchan, 1977; Noble et al., 1983; Pertel et al., 2001; Shieh and Spear 1994; Terry Allison et al., 1998); this implies that gB, gH, gL may form a minimal set of fusogenic determinants in the presence of the regulatory gD-HveA/C signal.

The exact biophysical nature of the events resulting in envelope–plasmalemma fusion remains obscure. Indeed, this step has proved difficult to study in the context of the virus, and it is debatable whether the cell–cell fusion that occurs in the presence of the HSV fusogenic components is an adequate model of virion–cell fusion. With relevance to the present discussion, the absence of cell-specific determinants in the final part of the viral–cell fusion cascade implies that this process would not be an appropriate target for modifications aimed at altering the cell range of the vector.

STRATEGIES FOR MODIFYING THE TROPISM OF HSV

It follows from the previous section that strategies aiming to modify the entry of HSV into cells might target any of the earlier steps in the virion–cell entry cascade. We have started to address the optimal way of redirecting the tropism of HSV-1 by altering viral components. Two broad strategies may be envisaged: manipulation of the transplasmalemmal entry cascade and the transendosomal entry pathway.

Manipulation of the Transplasmalemmal Entry Cascade

We have made targeted modifications to viral glycoproteins to try to modify the subset of cellular receptors that HSV will bind to, with a view to targeting viral entry through effecting virion fusion at the plasmalemma in predefined cell subpopulations.

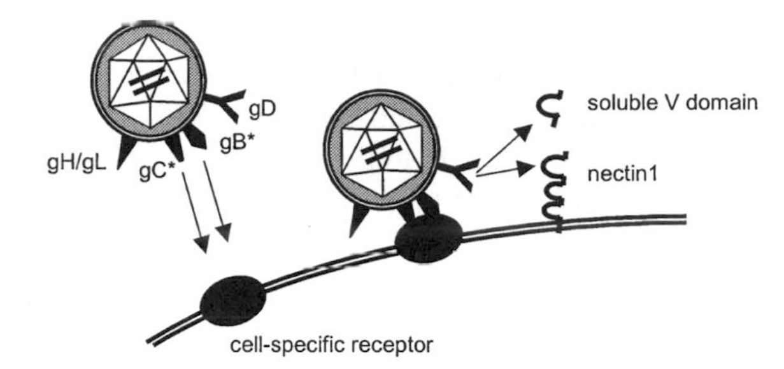

Figure 18.2 Targeted tropism by engineering cell attachment. Attachment of the virus to the cell occurs through modified gB and gC (denoted gB^* and gC^*). These modifications result in loss of GAG binding, but acquisition of a new binding specificity to a cellular receptor. Following targeted adsorption, entry proceeds through interaction of viral gD with a cellular gD receptor, or by incubation of the virus with soluble truncated HveC.

Targeted adsorption The first step in the entry cascade is the binding of viral gB/gC to HS. We hypothesized that modification of gC/gB, or their replacement with a ligand for another receptor would target the adsorption of virions to predetermined cell subtypes (Fig. 18.2).

To examine this hypothesis more closely, we generated engineered virions that harbored the $gBpK^-$ mutation, and in which gC was replaced by gC-erythropoietin (EPO) fusion proteins (Laquerre et al., 1998a) (Fig. 18.3). Initial studies confirmed that replacement of different parts of the N-terminal ectodomain of gC with EPO was possible, and that the fusion proteins were expressed at the cell surface in transient assays. Two of the three fusion proteins retained the ability to bind to a GST-soluble EPO receptor (EREx) by far-Western analysis.

Engineered viruses were generated by recombining the chimeric genes encoding gC-EPO into the gC locus of a vector containing the $gBpK^-$ mutation and a marker gene at the gC locus. The desired recombinants were screened by reporter assay and verified by Southern blot analysis. Immunoprecipitation analysis of supernatant-derived virus showed that the chimeric gC-EPO proteins were present within the virions, and their localization within the viral envelope was confirmed by complement-dependent neutralization assay (Laquerre et al., 1998a).

The gC-EPO recombinant viruses did not show appreciable binding to HS, but did bind to EREx immobilized on an affinity column (Laquerre et al., 1998a). Using cells that are not susceptible for HSV entry but express the EPO receptor (FD-EPO) (Nagata, Nishida, and Todokoro, 1997), it was possible to demonstrate binding of virus to the EPO receptor expressed on the cell surface. Furthermore, the virus was able to stimulate growth of EPO-dependent cells by binding to the receptor.

It was necessary to use cells that are not permissive for HSV entry in these experiments, to distinguish virion binding via gC-EPO to EPO receptor from

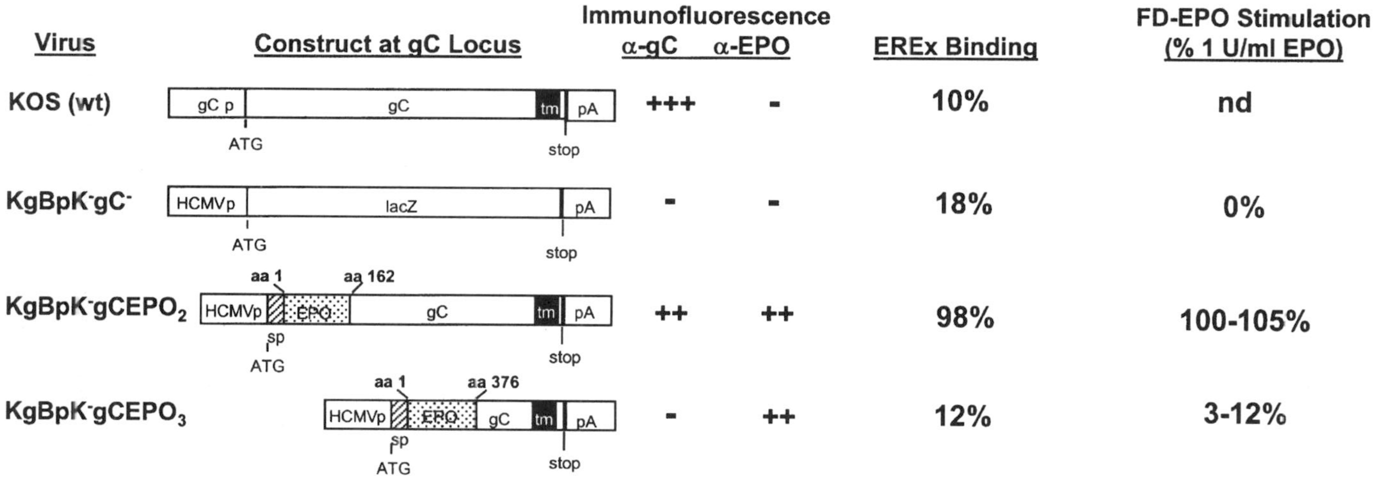

Virus	Construct at gC Locus	Immunofluorescence α-gC	Immunofluorescence α-EPO	EREx Binding	FD-EPO Stimulation (% 1 U/ml EPO)
KOS (wt)	gC p, ATG, gC, tm, stop, pA	+++	-	10%	nd
KgBpK⁻gC⁻	HCMVp, ATG, lacZ, stop, pA	-	-	18%	0%
KgBpK⁻gCEPO$_2$	HCMVp, ATG, sp, aa 1, EPO, aa 162, gC, tm, stop, pA	++	++	98%	100-105%
KgBpK⁻gCEPO$_3$	HCMVp, ATG, sp, aa 1, EPO, aa 376, gC, tm, stop, pA	-	++	12%	3-12%

Figure 18.3 Chimeric erythropoietin/gC ligands. Constructs incorporating segments of gC- and EPO-encoding sequences were generated and recombined into the gC locus of a parent vector. The chimeric proteins resulting from expression of the constructs in cells were assayed for cell surface expression of gC or EPO epitopes by immunofluorescence, and their ability to bind to the erythropoietin receptor (EREx—see text) and stimulate the receptor (FD-EPO stimulation).

a binding event between gD and HveA or HveC. Thus it was thus not possible to demonstrate transplasmalemmal viral entry or plaque formation in these experiments. However, it remains possible that, in the presence of a cognate cellular gD receptor, retargeting viral attachment to a new cell surface receptor would give rise to targeted transplasmalemmal cell entry. We are currently evaluating whether gC^{-}EPO; $gBpK^{-}$ vectors preferentially enter cells expressing the EPO receptor in the context of HveA or HveC cell surface expression. An alternative means to achieve targeted cell entry, through modification of cell adsorption, utilizes soluble truncated HveC as a trigger for attached virions to enter cells, independent of the presence of a cell surface gD receptor. We have shown that the V-domain of HveC alone is adequate to stimulate entry of virus into non-susceptible cells by interacting with virion gD in solution (Bai et al., in prep.); it remains possible that the entry of gC^{-}EPO; $gBpK^{-}$ particles into the new target cells could follow the transplasmalemmal pathway if viral gD was bound by soluble HveC to generate a fusogenic signal. This may confer an additional advantage, preventing low-grade nontargeted cell entry by blocking the potential interaction between virion gD and Hve A or HveC expressed on the surface of nontargeted cells.

One of the consequences of EPO signaling is internalization of the receptor and its subsequent degradation in the endosome compartment. Interestingly, gC-EPO virus was also internalized into the endosome compartment of the EPO receptor-positive cells, providing further evidence that a specific and avid binding event had occurred between the recombinant virus and its redefined cognate cellular receptor. This observation prompted us to examine the trans-endosomal entry pathway in more detail as discussed later in this chapter. It is unclear whether entry to the endosome compartment would result from binding between the virion and cell surface EPO receptor in the presence of a gD receptor.

Engineering Receptor-Recognizing Glycoprotein D An alternative to engineering the cell attachment receptors gC and gB might be to modify the viral attachment/entry glycoprotein gD (Fig. 18.4A). If gD could be modified in a way that (1) prevented its binding to HveA and HveC, (2) retained its cell entry signaling function, and (3) generated a new binding specificity, it might be possible to limit HSV entry to cells in which the appropriate receptor was expressed. Thus, the virion could bind to all GAG-expressing cells, but the fusogenic signal to the virion would only be generated in the context of the target cell expressing the new gD-binding receptor.

Such modifications to gD would be complex and would demand an intimate knowledge of crucial amino acid residues involved in HveA and HveC binding, and in the subsequent events that trigger viral entry functions. Our [illegible] investigations to define the dissociable binding and entry functions of amino acid residues within the functional domains of gD are described earlier in this chapter. The receptor-binding domains are complex conformational epitopes. This complicates the rational design of engineered gD proteins, which would

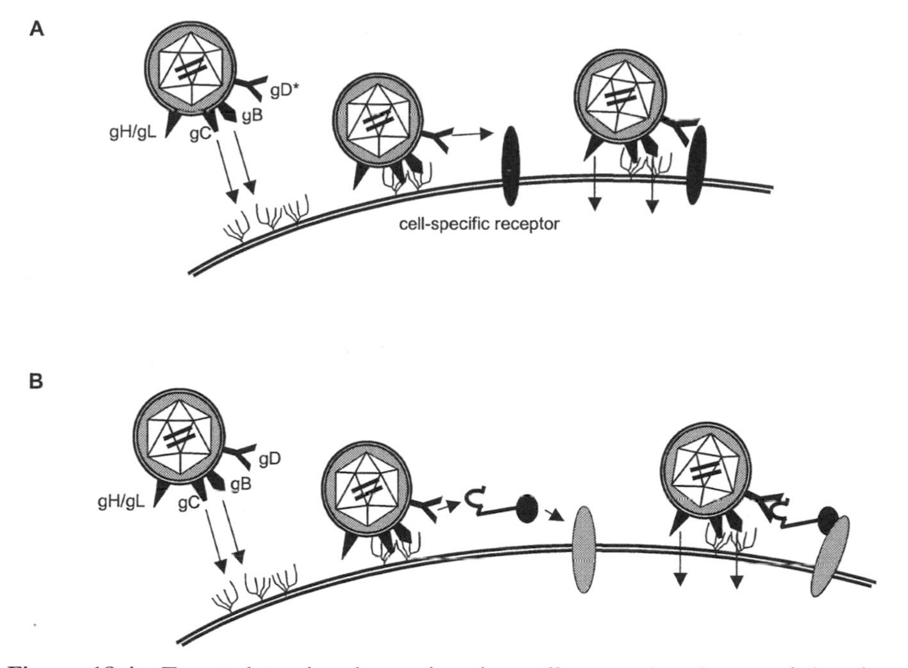

Figure 18.4 Targeted tropism by engineering cell entry. Attachment of the virus occurs through gB/C binding to HS moieties. Targeted entry is achieved by one of two mechanisms: *A*. Genetic modifications to gD alter its binding specificity, allowing the cell entry cascade to be triggered when engineered gD (denoted gD*) interacts with a cell-specific receptor. *B*. Binding of wild-type gD to a heterologous receptor occurs via a bispecific adapter molecule intermediary. Binding of the adapter to virion gD prevents cell entry through cell surface gD receptors.

have to present a specific ligand epitope in an appropriate conformation to the cellular receptor while destroying the wild-type HveA/C receptor-binding activities. Importantly, this would have to be achieved without disrupting the complex secondary or tertiary structure responsible for maintaining appropriate features of the discontinuous regions of gD involved in the signaling consequences of receptor binding. Furthermore, binding of the ligand to its new target would have to induce a similar series of changes in the chimeric protein to those induced by binding of the wild-type gD to its native cognate receptors in order to trigger the remainder of the virion–cell fusion process. This latter point might not be as insurmountable as it appears—recall that the gD-signaled fusion cascade proceeds following binding to either of two structurally diverse cellular receptors; it follows that the structural and biophysical features of the receptor-binding events that trigger the fusogenic signal might not require interactions with particularly stringent parameters. Experiments aiming to identify domains that can be replaced to allow new ligand-specific gD activation are ongoing.

Use of Soluble Receptors and Adapters The use of the HveC V-domain as a soluble trigger to gD has been described earlier in this chapter. Targeting using this method would depend upon soluble gD being able to compete with cellular gD, preventing non HS-dependent binding to the cell (Fig. 18.4B). We have demonstrated that the truncated form of the HveC receptor is able to antagonize viral entry into susceptible cells, implying that occupancy of virion gD by a soluble HveC V domain, in the context of a gC^-, $gBpK^-$ virus, is adequate to prevent nontargeted virion adsorption (Bai et al., in prep.).

A logical development of this idea is to use an HS-binding deficient virus, but to engineer cell-specific attachment to take place via a soluble HveC adapter. In this scheme, the chimeric adapter molecule provides both the cell-specific binding and the gD trigger signal. The rationale is that the isolated V-domain can both trigger adsorption-dependent entry into nonsusceptible cells and antagonize entry into permissive cells by competing with cell-surface HveC. Thus an adapter molecule might contain the HveC V-domain at its N-terminal and a ligand for a specific cellular receptor at its C-terminal. It follows that use of such an adapter would cause both cell-specific attachment and triggering of virus–cell fusion, while preventing nontargeted cell binding by occupying viral gD. We are currently evaluating several such adapter molecules, including chimeric adapters expressed on the surface of the virion. One potential advantage of soluble adapters is that subsequent retargeting of the vector to a new receptor would not necessitate the generation of novel recombinant virions; the techniques for synthesizing new adapter proteins are relatively straightforward and less time-consuming than those involved in the isolation of novel recombinant viruses.

The Transendosomal Entry Pathway

Our initial studies on modified adsorption prompted us to consider targeting HSV entry through the transendosomal pathway. We have studied two types of mechanisms for effecting targeted endocytosis of virions: redirecting cell adsorption to an internalized receptor (Fig. 18.5) and pseudotyping HSV.

Redirecting Cell Adsorption to an Internalized Cellular Receptor The studies examining modification of adsorptive targets of HSV by genetically modifying gC and gB are described previously (Laquerre et al., 1998a). The prerequisite properties of a virus developed for targeted entry through the trans-endosomal pathway by this technique would be: (1) deletion of GAG-binding domains of gB and gC to prevent nonspecific binding to nontarget cells; (2) deletion of gD (or the HveA and HveC-binding activity of gD) to prevent non-specific entry to unintended cells thorough gD HveA/C interactions; and (3) virion surface expression of a ligand for a cell-specific determinant that is internalized upon receptor occupancy. The initial experiments on EPO-targeting demonstrated the feasibility of this approach.

Problems remain with this strategy, however. The virus enters the endo-

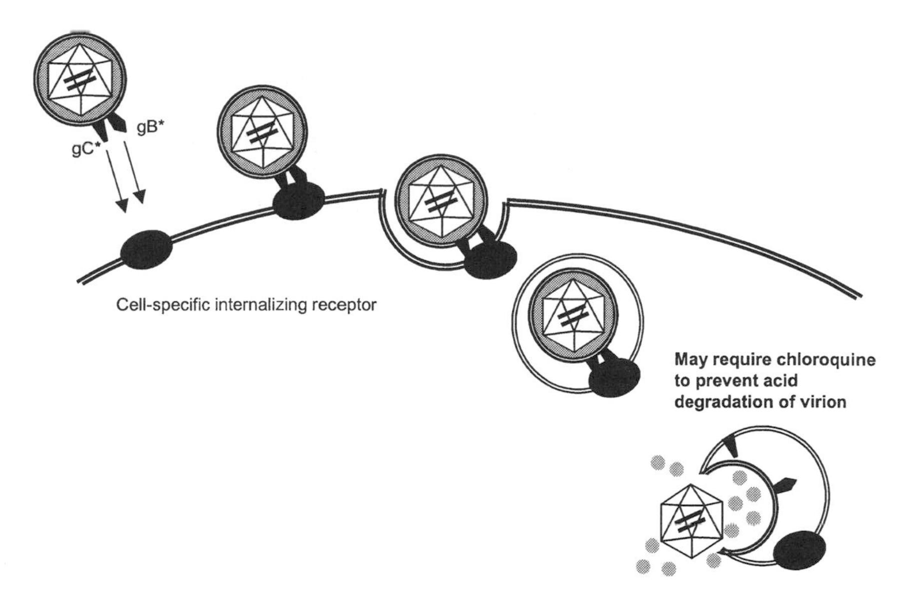

Figure 18.5 Targeted tropism through the endosomal pathway. Attachment of the virus is engineered as in Figure 18.2, but an internalizing receptor is targeted. The virion is endocytosed and released from the endosomal compartment into the cytoplasm. The release mechanism is complex and may involve pseudotyping HSV with surface components from viruses that normally use this route of cell entry or using agents to prevent acid degradation of the virion.

some compartment but still requires a complex series of processes to trigger fusion within the endosomal membrane. HSV fusion occurs at the cell membrane in the presence of physiological pH (Rajcani and Vojvodova, 1998; Spear, 1993a). The endosomal compartment is characterized by progressive acidification as the endosome matures. Many other viruses take advantage of the acidification to activate a pH-dependent fusion mechanism. The pH-independent HSV entry mechanism, however, fails under these conditions, and the virion becomes degraded. The problem is further compounded by the requirement that gD is removed from the viral surface to enable targeted endocytosis, in that the primary fusogenic trigger is then missing from the virion. These adverse considerations led us to consider an alternative possibility.

Pseudotyping HSV Using Surface Determinants from Another Virus

The host range of a virus may be modified using the surface receptors of another virus. This approach has been successfully used to broaden the host range of human immunodeficiency virus (HIV) vectors, using the G-protein of vesicular stomatitis virus (VSV) as a surface receptor (Emi, Friedmann, and Yee, 1991; Naldini et al., 1996). We reasoned that the G-spike glycoprotein from the surface of VSV might be able to replace the complex glycoprotein array expressed on the surface of the HSV virion and provide proof of concept that pseudotyping HSV was possible. Moreover, use of a receptor that had evolved to effect viral entry through the transendosomal entry pathway might allow HSV to deliver its contents to the cytoplasm via this route.

To test this hypothesis, we attempted to pseudotype HSV using VSV/HSV fusion proteins. A mutant virus was utilized in which the short unique segment of the HSV genome U_S3–8 (encoding gD in addition to nonessential glycoproteins gE, gG, gJ, and gI) was deleted. The gD-null phenotype of this vector was transiently rescued using a series of plasmid expression cassettes encoding chimeric VSV-G/HSV glycoprotein B and D fusion proteins (Anderson et al., 2000). Chimeras containing the ectodomain of VSV-G linked to either the C-terminal or transmembrane domain of either gD or gB were generated. A number of different constructs were tested to facilitate the identification of chimeras that were effectively packaged into the virion, because it was previously suggested that the structural requirements for incorporation of glycoproteins into the envelope of the mature virus particle were stringent with respect to transmembrane or endodomains (Gilbert et al., 1994; Huff et al., 1988; Raviprakash et al., 1990; Skoff and Holland, 1993). It was demonstrated that VSV-G chimeras containing the transmembrane domain of gD, or a truncated gB transmembrane domain, were incorporated into the viral envelope efficiently, and that the wild-type VSV-G protein was incorporated rather less efficiently (Anderson et al., 2000). The latter, however, was able to partially rescue the gD-deficient phenotype, whereas the chimeric proteins were nonfunctional. Neutralizing anti-VSV-G antibodies blocked the partial VSV-G mediated rescue (Anderson et al., 2000).

The poor efficiency of phenotypic rescue of the U_S3–8 null virus by VSV-G may be attributable to either inefficient incorporation of the foreign viral glycoprotein or acid degradation of HSV in the endosome compartment. We have studied this issue further in order to elucidate and exploit the mechanisms contributing to partial rescue. A recombinant virus was generated in which the VSV-G expression cassette was incorporated into the genome of the U_S3–8 deleted virus (the resulting vector is null for gD, gE, gG, gJ, gI, but expresses VSV-G). The recombinant particle enters cells possessing the VSV receptor. However, an abortive infection ensues, culminating in endosomal degradation of the virion at low pH, similar to that observed when the gC-EPO expressing recombinant enters cells (Goins et al., in prep.). The use of lysomotropic agents that raise endosomal pH, such as chloroquine (Nash and Buchmeier, 1997), enables release of viral contents from the endosome into the cytoplasm, resulting in plaque formation.

An acidic environment is required for the activation of VSV-G fusion functions (Fredericksen and Whitt 1996, 1998). It follows that, in the presence of lysomotropic drugs, fusion of the VSV-G expressing, U_S3–8 null, recombinant with the endosomal membrane must be mediated by HSV fusion glycoproteins (gB or gH/gL). It thus appears likely that, in this situation, VSV-G functions as a ligand for an internalized receptor, rather than a pH-dependent mediator of viral envelope fusion with the endosome membrane. We are currently examining the mechanism in more detail. In addition, we are seeking to identify other agents that may allow HSV contents to enter the cytoplasm following targeting of the virion to internalized cellular receptors.

CONCLUSIONS

We have only recently gained access to sufficient information about HSV entry into cells to start investigating ways of modifying this complex process. In some other types of vectors, cell entry is controlled by one or two glycoproteins, and simply modifying cell binding is adequate to direct targeted tropism. In contrast, the cell entry mechanisms of HSV are sufficiently complex that a series of complementary and carefully engineered manipulations will be necessary to efficiently redirect cell entry to cellular populations of interest. However, the many advantages of the HSV vector system encourage us to believe that the effort expended in this considerable intellectual puzzle may eventually give rise to reagents of enormous utility and significance. Furthermore, the detailed genetic manipulation and resulting engineering of protein–protein interactions necessary to achieve our goal will teach us much about more fundamental biological issues.

REFERENCES

Akkaraju GR, Huard J, Hoffman EP, Goins WF, Pruchnic R, Watkins SC, Cohen JB, Glorioso JC (1999): Herpes simplex virus vector-mediated dystrophin gene transfer and expression in MDX mouse skeletal muscle. *J Gene Med* 1:280–289.

Anderson DB, Laquerre S, Goins WF, Cohen JB, Glorioso JC (2000): Pseudotyping of glycoprotein D-deficient herpes simplex virus type 1 with vesicular stomatitis virus glycoprotein G enables mutant virus attachment and entry. *J Virol* 74:2481–2487.

Banfield BW, Leduc Y, Esford L, Schubert K, Tufaro F (1995a): Sequential isolation of proteoglycan synthesis mutants by using herpes simplex virus as a selective agent: evidence for a proteoglycan-independent virus entry pathway. *J Virol* 69:3290–3298.

Banfield, BW, Leduc Y, Esford L, Visalli RJ, Brandt CR, Tufaro F (1995b): Evidence for an interaction of herpes simplex virus with chondroitin sulfate proteoglycans during infection. *Virology* 208:531–539.

Barker DE, Roizman B (1992): The unique sequence of the herpes simplex virus 1 L component contains an additional translated open reading frame designated UL49.5. *J Virol* 66:562–566.

Batterson W, Roizman B (1983): Characterization of the herpes simplex virion-associated factor responsible for the induction of alpha genes. *J Virol* 46:371–377.

Burton EA, Huang S, Goins WF, Glorioso JC (2001a): Use of the herpes simplex virus genome to construct gene therapy vectors. In Viral Vectors for Gene Therapy: Methods and Protocols, Machida C, ed. Totowa, NJ: Humana Press, in press.

Burton EA, Wechuck JB, Wendell SK, Goins WF, Fink DJ, Glorioso JC (2001b): Multiple applications for replication-defective herpes simplex virus vectors. *Stem Cells* 19:358–377.

Bzik DJ, Fox BA, DeLuca NA, Person S (1984): Nucleotide sequence of a region of the herpes simplex virus type 1 gB glycoprotein gene: Mutations affecting rate of virus entry and cell fusion. *Virology* 137:185–190.

Cai WH, Gu B, Person S (1988): Role of glycoprotein B of herpes simplex virus type 1 in viral entry and cell fusion. *J Virol* 62:2596–2604.

Campadelli Fiume G, Arsenakis M, Farabegoli F, Roizman B (1988): Entry of herpes simplex virus 1 in BJ cells that constitutively express viral glycoprotein D is by endocytosis and results in degradation of the virus. *J Virol* 62:159–167.

Campbell ME, Palfreyman JW, Preston CM (1984): Identification of herpes simplex virus DNA sequences which encode a trans-acting polypeptide responsible for stimulation of immediate early transcription. *J Mol Biol* 180:1–19.

Carfi A, Willis SH, Whitbeck JC, Krummenacher C, Cohen GH, Eisenberg RJ, Wiley DC (2001): Herpes simplex virus glycoprotein D bound to the human receptor HveA. *Mol Cell* 8:169–179.

Chen X, Schmidt MC, Goins WF, Glorioso JC (1995): Two herpes simplex virus type 1 latency-active promoters differ in their contributions to latency-associated transcript expression during lytic and latent infections. *J Virol* 69:7899–7908.

Chiang HY, Cohen GH, Eisenberg RJ (1994): Identification of functional regions of herpes simplex virus glycoprotein gD by using linker-insertion mutagenesis. *J Virol* 68:2529–2543.

Cocchi F, Menotti L, Mirandola P, Lopez M, Campadelli Fiume G (1998): The ectodomain of a novel member of the immunoglobulin subfamily related to the poliovirus receptor has the attributes of a bona fide receptor for herpes simplex virus types 1 and 2 in human cells. *J Virol* 72:9992–10002.

DeLuca N, Bzik DJ, Bond VC, Person S, Snipes W (1982): Nucleotide sequences of herpes simplex virus type 1 (HSV-1) affecting virus entry, cell fusion, and production of glycoprotein gb (VP7). *Virology* 122:411–423.

DeLuca NA, McCarthy AM, Schaffer PA (1985): Isolation and characterization of deletion mutants of herpes simplex virus type 1 in the gene encoding immediate-early regulatory protein ICP4. *J Virol* 56:558–570.

Desai PJ, Schaffer PA, Minson AC (1988): Excretion of non-infectious virus particles lacking glycoprotein H by a temperature-sensitive mutant of herpes simplex virus type 1: Evidence that gH is essential for virion infectivity. *J Gen Virol* 69:1147–1156.

Dobson AT, Sederati F, Devi Rao G, Flanagan WM, Farrell MJ, Stevens JG, Wagner EK, Feldman LT (1989): Identification of the latency-associated transcript promoter by expression of rabbit beta-globin mRNA in mouse sensory nerve ganglia latently infected with a recombinant herpes simplex virus. *J Virol* 63:3844–3851.

Emi N, Friedmann T, Yee JK (1991): Pseudotype formation of murine leukemia virus with the G protein of vesicular stomatitis virus. *J Virol* 65:1202–1207.

Feyzi E, Trybala E, Bergström T, Lindahl U, Spillmann D (1997): Structural requirement of heparan sulfate for interaction with herpes simplex virus type 1 virions and isolated glycoprotein C. *J Biol Chem* 272:24850–24857.

Forrester A, Farrell H, Wilkinson G, Kaye J, Davis Poynter N, Minson T (1992): Construction and properties of a mutant of herpes simplex virus type 1 with glycoprotein H coding sequences deleted. *J Virol* 66:341–348.

Fredericksen BL, Whitt MA (1996): Mutations at two conserved acidic amino acids in the glycoprotein of vesicular stomatitis virus affect pH-dependent conformational changes and reduce the pH threshold for membrane fusion. *Virology* 217:49–57.

Fredericksen BL, Whitt MA (1998): Attenuation of recombinant vesicular stomatitis viruses encoding mutant glycoproteins demonstrate a critical role for maintaining a high pH threshold for membrane fusion in viral fitness. *Virology* 240:349–358.

Fuller AO, Spear PG (1985): Specificities of monoclonal and polyclonal antibodies that inhibit adsorption of herpes simplex virus to cells and lack of inhibition by potent neutralizing antibodies. *J Virol* 55:475–482.

Fuller AO, Spear PG (1987): Anti-glycoprotein D antibodies that permit adsorption but block infection by herpes simplex virus 1 prevent virion-cell fusion at the cell surface. *Proc Natl Acad Sci USA* 84:5454–5458.

Fuller AO, Santos RE, Spear PG (1989): Neutralizing antibodies specific for glycoprotein H of herpes simplex virus permit viral attachment to cells but prevent penetration. *J Virol* 63:3435–3443.

Furlong D, Swift H, Roizman B (1972): Arrangement of herpesvirus deoxyribonucleic acid in the core. *J Virol* 10:1071–1074.

Gage PJ, Sauer B, Levine M, Glorioso JC (1992): A cell-free recombination system for site-specific integration of multigenic shuttle plasmids into the herpes simplex virus type 1 genome. *J Virol* 66:5509–5515.

Geraghty RJ, Krummenacher C, Cohen GH, Eisenberg RJ, Spear PG (1998): Entry of alphaherpesviruses mediated by poliovirus receptor-related protein 1 and poliovirus receptor. *Science* 280:1618–1620.

Gilbert R, Ghosh K, Rasile L, Ghosh HP (1994): Membrane anchoring domain of herpes simplex virus glycoprotein gB is sufficient for nuclear envelope localization. *J Virol* 68:2272–2285.

Goins WF, Sternberg LR, Croen KD, Krause PR, Hendricks RL, Fink DJ, Straus SE, Levine M, Glorioso JC (1994): A novel latency-active promoter is contained within the herpes simplex virus type 1 UL flanking repeats. *J Virol* 68:2239–2252.

Goins WF, Lee KA, Cavalcoli JD, O'Malley ME, DeKosky ST, Fink DJ, Glorioso JC (1999): Herpes simplex virus type 1 vector-mediated expression of nerve growth factor protects dorsal root ganglion neurons from peroxide toxicity. *J Virol* 73:519–532.

Goins WF, Yoshimura N, Ozawa H, Yokoyama T, Phelan M, Bennet N, de Groat WC, Glorioso JC, Chancellor MB (2001): Herpes simplex virus vector-mediated nerve growth factor expression in bladder and afferent neurons: potential treatment for diabetic bladder dysfunction. *J Urology* 165:1748–1754.

Gomez Navarro J, Contreras JL, Arafat W, Jiang XL, Krisky D, Oligino T, Marconi P, Hubbard B, Glorioso JC, Curiel DT, Thomas JM (2000): Genetically modified CD34+ cells as cellular vehicles for gene delivery into areas of angiogenesis in a rhesus model. *Gene Ther* 7:43–52.

Gompels U, Minson A (1986): The properties and sequence of glycoprotein H of herpes simplex virus type 1. *Virology* 153:230–247.

Goss JR, Mata M, Goins WF, Wu HH, Glorioso JC, Fink DJ (2001): Antinociceptive effect of a genomic herpes simplex virus-based vector expressing human proenkephalin in rat dorsal root ganglion. *Gene Ther* 8:551–556.

Gruenheid S, Gatzke L, Meadows H, Tufaro F (1993): Herpes simplex virus infection and propagation in a mouse L cell mutant lacking heparan sulfate proteoglycans. *J Virol* 67:93–100.

Haffey ML, Spear PG (1980): Alterations in glycoprotein gB specified by mutants and their partial revertants in herpes simplex virus type 1 and relationship to other mutant phenotypes. *J Virol* 35:114–128.

Herold BC, WuDunn D, Soltys N, Spear PG (1991): Glycoprotein C of herpes simplex virus type 1 plays a principal role in the adsorption of virus to cells and in infectivity. *J Virol* 65:1090–1098.

Herold BC, Visalli RJ, Susmarski N, Brandt CR, Spear PG (1994): Glycoprotein C-independent binding of herpes simplex virus to cells requires cell surface heparan sulphate and glycoprotein B. *J Gen Virol* 75:1211–1222.

Highlander SL, Sutherland SL, Gage PJ, Johnson DC, Levine M, Glorioso JC (1987): Neutralizing monoclonal antibodies specific for herpes simplex virus glycoprotein D inhibit virus penetration. *J Virol* 61:3356–3364.

Highlander SL, Cai WH, Person S, Levine M, Glorioso JC (1988): Monoclonal antibodies define a domain on herpes simplex virus glycoprotein B involved in virus penetration. *J Virol* 62:1881–1888.

Highlander SL, Dorney DJ, Gage PJ, Holland TC, Cai W, Person S, Levine M, and Glorioso JC (1989): Identification of mar mutations in herpes simplex virus type 1 glycoprotein B which alter antigenic structure and function in virus penetration. *J Virol* 63:730–738.

Homa FL, Brown JC (1997): Capsid assembly and DNA packaging in herpes simplex virus. *Rev Med Virol* 7:107–122.

Huard J, Goins WF, Akkaraju GR, Krisky D, Oligino T, Marconi P, Glorioso JC (1998): Gene transfer to muscle and spinal cord using herpes simplex virus-based vectors. In *Stem Cell Biology and Gene Therapy*, Quesenberry PJ, Stein GS, Forget B, Weissman S, eds. New York: John Wiley, pp 179–200.

Huff V, Cai W, Glorioso JC, Levine M (1988): The carboxy-terminal 41 amino acids of herpes simplex virus type 1 glycoprotein B are not essential for production of infectious virus particles. *J Virol* 62:4403–4406.

Hutchinson L, Browne H, Wargent V, Davis Poynter N, Primorac S, Goldsmith K, Minson AC, Johnson DC (1992): A novel herpes simplex virus glycoprotein, gL, forms a complex with glycoprotein H (gH) and affects normal folding and surface expression of gH. *J Virol* 66:2240–2250.

Johnson DC, Ligas MW (1988): Herpes simplex viruses lacking glycoprotein D are unable to inhibit virus penetration: Quantitative evidence for virus-specific cell surface receptors. *J Virol* 62:4605–4612.

Johnson DC, Burke RL, Gregory T (1990): Soluble forms of herpes simplex virus glycoprotein D bind to a limited number of cell surface receptors and inhibit virus entry into cells. *J Virol* 64:2569–2576.

Krisky DM, Marconi PC, Oligino TJ, Rouse RJ, Fink DJ, Cohen JB, Watkins SC, Glorioso JC (1998a): Development of herpes simplex virus replication-defective multigene vectors for combination gene therapy applications. *Gene Ther* 5:1517–1530.

Krisky DM, Wolfe D, Goins WF, Marconi PC, Ramakrishnan R, Mata M, Rouse RJ, Fink DJ, Glorioso JC (1998b): Deletion of multiple immediate-early genes from herpes simplex virus reduces cytotoxicity and permits long-term gene expression in neurons. *Gene Ther* 5:1593–1603.

Krummenacher C, Nicola AV, Whitbeck JC, Lou H, Hou W, Lambris JD, Geraghty, RJ Spear PG, Cohen GH, Eisenberg RJ (1998): Herpes simplex virus glycoprotein D can bind to poliovirus receptor-related protein 1 or herpesvirus entry mediator, two structurally unrelated mediators of virus entry. *J Virol* 72:7064–7074.

Krummenacher C, Rux AH, Whitbeck JC, Ponce de Leon M, Lou H, Baribaud I, Hou W, Zou C, Geraghty RJ, Spear PG, Eisenberg RJ, Cohen GH (1999): The first immunoglobulin-like domain of HveC is sufficient to bind herpes simplex virus gD with full affinity, while the third domain is involved in oligomerization of HveC. *J Virol* 73:8127–8137.

Kwong AD, Frenkel N (1987): Herpes simplex virus-infected cells contain a function(s) that destabilizes both host and viral mRNAs. *Proc Natl Acad Sci USA* 84:1926–1930.

Kwong AD, Frenkel N (1989): The herpes simplex virus virion host shutoff function. *J Virol* 63:4834–4839.

Kwong AD, Kruper JA, Frenkel N (1988): Herpes simplex virus virion host shutoff function. *J Virol* 62:912–921.

Lachmann RH, Efstathiou S (1997): Utilization of the herpes simplex virus type 1 latency-associated regulatory region to drive stable reporter gene expression in the nervous system. *J Virol* 71:3197–3207.

Langeland N, Moore LJ, Holmsen H, Haarr L (1988): Interaction of polylysine with the cellular receptor for herpes simplex virus type 1. *J Gen Virol* 69:1137–1145.

Laquerre S, Anderson DB, Stolz DB, Glorioso JC (1998a): Recombinant herpes simplex virus type 1 engineered for targeted binding to erythropoietin receptor-bearing cells. *J Virol* 72:9683–9697.

Laquerre S, Argnani R, Anderson DB, Zucchini S, Manservigi R, Glorioso JC (1998b): Heparan sulfate proteoglycan binding by herpes simplex virus type 1 glycoproteins B and C, which differ in their contributions to virus attachment, penetration, and cell-to-cell spread. *J Virol* 72:6119–6130.

Ligas MW, Johnson DC (1988): A herpes simplex virus mutant in which glycoprotein D sequences are replaced by beta-galactosidase sequences binds to but is unable to penetrate into cells. *J Virol* 62:1486–1494.

Lopez M, Cocchi F, Avitabile E, Leclerc A, Adelaide J, Campadelli Fiume G, Dubreuil P (2001): Novel, soluble isoform of the herpes simplex virus (hsv) receptor nectin1 (or prr1-higr-hvec) modulates positively and negatively susceptibility to hsv infection. *J Virol* 75:5684–5691.

Mackem S, Roizman B (1982): Structural features of the herpes simplex virus alpha gene 4, 0, and 27 promoter-regulatory sequences which confer alpha regulation on chimeric thymidine kinase genes. *J Virol* 44:939–949.

MacLean AR, ul Fareed M, Robertson L, Harland J, Brown SM (1991): Herpes simplex virus type 1 deletion variants 1714 and 1716 pinpoint neurovirulence-related sequences in Glasgow strain 17+ between immediate early gene 1 and the 'a' sequence. *J Gen Virol* 72:631–639.

Manservigi R, Spear PG, Buchan A (1977): Cell fusion induced by herpes simplex virus is promoted and suppressed by different viral glycoproteins. *Proc Natl Acad Sci USA* 74:3913–3917.

Marconi P, Simonato M, Zucchini S, Bregola G, Argnani R, Krisky D, Glorioso JC, Manservigi R (1999): Replication-defective herpes simplex virus vectors for neurotrophic factor gene transfer *in vitro* and *in vivo. Gene Ther* 6:904–912.

Marconi P, Tamura M, Moriuchi S, Krisky DM, Niranjan A, Goins WF, Cohen JB, Glorioso JC (2000): Connexin 43-enhanced suicide gene therapy using herpesviral vectors. *Mol Ther* 1:71–81.

Mardberg K, Trybala E, Glorioso JC, Bergstrom T (2001): Mutational analysis of the major heparan sulfate-binding domain of herpes simplex virus type 1 glycoprotein C. *J Gen Virol* 82:1941–1950.

Markert JM, Medlock MD, Rabkin SD, Gillespie GY, Todo T, Hunter WD, Palmer CA, Feigenbaum F, Tornatore C, Tufaro F, Martuza RL (2000): Conditionally replicating herpes simplex virus mutant, G207 for the treatment of malignant glioma: results of a phase I trial. *Gene Ther* 7:867–874.

Martino G, Poliani PL, Furlan R, Marconi P, Glorioso JC, Adorini L, Comi G (2000a): Cytokine therapy in immune-mediated demyelinating diseases of the central nervous system: A novel gene therapy approach. *J Neuroimmunol* 107:184–190.

Martino G, Poliani PL, Marconi PC, Comi G, Furlan R (2000b): Cytokine gene therapy of autoimmune demyelination revisited using herpes simplex virus type-1-derived vectors. *Gene Ther* 7:1087–1093.

McGeoch DJ, Dolan A, Donald S, Rixon FJ (1985): Sequence determination and genetic content of the short unique region in the genome of herpes simplex virus type 1. *J Mol Biol* 181:1–13.

McGeoch DJ, Dolan A, Donald S, Brauer DH (1986): Complete DNA sequence of the short repeat region in the genome of herpes simplex virus type 1. *Nucleic Acids Res* 14:1727–1745.

McGeoch DJ, Dalrymple MA, Davison AJ, Dolan A, Frame MC, McNab D, Perry LJ, Scott JE, Taylor P (1988): The complete DNA sequence of the long unique region in the genome of herpes simplex virus type 1. *J Gen Virol* 69:1531–1574.

Montgomery RI, Warner MS, Lum BJ, Spear PG (1996): Herpes simplex virus-1 entry into cells mediated by a novel member of the TNF/NGF receptor family. *Cell* 87:427–436.

Morgan C, Rose HM, Mednis B (1968): Electron microscopy of herpes simplex virus. I. Entry. *J Virol* 2:507–116.

Moriuchi S, Oligino T, Krisky D, Marconi P, Fink D, Cohen J, Glorioso JC (1998): Enhanced tumor cell killing in the presence of ganciclovir by herpes simplex virus type 1 vector-directed coexpression of human tumor necrosis factor-alpha and herpes simplex virus thymidine kinase. *Cancer Res* 58:5731–5737.

Moriuchi S, Krisky DM, Marconi PC, Tamura M, Shimizu K, Yoshimine T, Cohen JB, Glorioso JC (2000): HSV vector cytotoxicity is inversely correlated with effective TK/GCV suicide gene therapy of rat gliosarcoma. *Gene Ther* 7:1483–1490.

Nagata Y, Nishida E, Todokoro K (1997): Activation of JNK signaling pathway by erythropoietin, thrombopoietin, and interleukin-3. *Blood* 89:2664–2669.

Naldini L, Blomer U, Gallay P, Ory D, Mulligan R, Gage FH, Verma IM, Trono D (1996): *In vivo* gene delivery and stable transduction of nondividing cells by a lentiviral vector. *Science* 272:263–267.

Nash TC, Buchmeier MJ (1997): Entry of mouse hepatitis virus into cells by endosomal and nonendosomal pathways. *Virology* 233:1–8.

Natsume A, Mata M, Goss J, Huang S, Wolfe D, Oligino T, Glorioso JC, Fink D (2001): Bcl-2 and GDNF delivered by HSV-mediated gene transfer act additively to protect dopaminergic neurons from 6-OHDA-induced degeneration. *Experimental Neurology* 169:231–238.

Navarro D, Paz P, Pereira L (1992): Domains of herpes simplex virus I glycoprotein B that function in virus penetration, cell-to-cell spread, and cell fusion. *Virology* 186:99–112.

Newcomb WW, Homa FL, Thomsen DR, Trus BL, Cheng N, Steven A, Booy F, Brown JC (1999): Assembly of the herpes simplex virus procapsid from purified components and identification of small complexes containing the major capsid and scaffolding proteins. *J Virol* 73:4239–4250.

Nicola AV, Ponce de Leon M, Xu R, Hou W, Whitbeck JC, Krummenacher C, Montgomery RI, Spear PG, Eisenberg RJ, Cohen GH (1998): Monoclonal antibodies to distinct sites on herpes simplex virus (HSV) glycoprotein D block HSV binding to HVEM. *J Virol* 72:3595–3601.

Niranjan A, Moriuchi S, Lunsford LD, Kondziolka D, Flickinger JC, Fellows W, Rajendiran S, Tamura M, Cohen JB, Glorioso JC (2000): Effective treatment of experimental glioblastoma by HSV vector-mediated TNFalpha and HSV-tk gene transfer in combination with radiosurgery and ganciclovir administration. *Mol Ther* 2:114–120.

Noble AG, Lee GT, Sprague R, Parish ML, Spear PG (1983): Anti-gD monoclonal antibodies inhibit cell fusion induced by herpes simplex virus type 1. *Virology* 129:218–224.

Oligino T, Ghivizzani S, Wolfe D, Lechman E, Krisky D, Mi Z, Evans C, Robbins P, Glorioso J (1999): Intra-articular delivery of a herpes simplex virus IL-1Ra gene vector reduces inflammation in a rabbit model of arthritis. *Gene Ther* 6:1713–1720.

Para MF, Baucke RB, Spear PG (1980): Immunoglobulin G(Fc)-binding receptors on virions of herpes simplex virus type 1 and transfer of these receptors to the cell surface by infection. *J Virol* 34:512–520.

Pertel PE, Fridberg A, Parish ML, Spear PG (2001): Cell fusion induced by herpes simplex virus glycoproteins gB, gD, and gH-gL requires a gD receptor but not necessarily heparan sulfate. *Virology* 279:313–324.

Puvion Dutilleul F, Pichard E, Leduc EH (1985): Influence of embedding media on DNA structure in herpes simplex virus type 1. *Biol Cell* 54:195–198.

Rajcani J, Vojvodova A (1998): The role of herpes simplex virus glycoproteins in the virus replication cycle. *Acta Virol* 42:103–118.

Rampling R, Cruickshank G, Papanastassiou V, Nicoll J, Hadley D, Brennan D, Petty R, MacLean A, Harland J, McKie E, Mabbs R, Brown M (2000): Toxicity evaluation of replication-competent herpes simplex virus (ICP 34.5 null mutant 1716) in patients with recurrent malignant glioma. *Gene Ther* 7:859–866.

Raviprakash K, Rasile L, Ghosh K, Ghosh HP (1990): Shortened cytoplasmic domain affects intracellular transport but not nuclear localization of a viral glycoprotein. *J Biol Chem* 265:1777–1782.

Read GS, N Frenkel (1983): Herpes simplex virus mutants defective in the virion-associated shutoff of host polypeptide synthesis and exhibiting abnormal synthesis of alpha (immediate early) viral polypeptides. *J Virol* 46:498–512.

Roizman B, Sears AE (1996): Herpes simplex viruses and their replication. In *Fields Virology*, Fields BN, Knipe DM, Howley PM, eds. Philadelphia: Lippincott-Raven, pp. 2231–2295.

Roop C, Hutchinson L, Johnson DC (1993): A mutant herpes simplex virus type 1 unable to express glycoprotein L cannot enter cells, and its particles lack glycoprotein H. *J Virol* 67:2285–2297.

Saharkhiz Langroodi A, Holland TC (1997): Identification of the fusion-from-without determinants of herpes simplex virus type 1 glycoprotein B. *Virology* 227:153–159.

Samaniego LA, Neiderhiser L, DeLuca NA (1998): Persistence and expression of the herpes simplex virus genome in the absence of immediate-early proteins. *J Virol* 72:3307–3320.

Sarmiento M, Haffey M, Spear PG (1979): Membrane proteins specified by herpes simplex viruses. III. Role of glycoprotein VP7(B2) in virion infectivity. *J Virol* 29:1149–1158.

Shieh MT, Spear PG (1994): Herpesvirus-induced cell fusion that is dependent on cell surface heparan sulfate or soluble heparin. *J Virol* 68:1224–1228.

Shieh MT, WuDunn D, Montgomery RI, Esko JD, Spear PG (1992): Cell surface receptors for herpes simplex virus are heparan sulfate proteoglycans. *J Cell Biol* 116:1273–1281.

Skoff AM, Holland TC (1993): The effect of cytoplasmic domain mutations on membrane anchoring and glycoprotein processing of herpes simplex virus type 1 glycoprotein C. *Virology* 196:804–816.

Spear PG (1993a): Entry of alphaherpesviruses into cells. Seminars in *Virology* 4:167–180.

Spear PG (1993b): Membrane fusion induced by herpes simplex virus. In *Viral Fusion Mechanisms*, Bentz J eds. Boca Raton: CRC Press, pp. 201–232.

Spear PG, Shieh MT, Herold BC, WuDunn D, Koshy TI (1992): Heparan sulfate glycosaminoglycans as primary cell surface receptors for herpes simplex virus. *Adv Exp Med Biol* 313:341–353.

Steven AC, Spear PG (1997): Herpesvirus capsid assembly and envelopment. In *Structural Biology of Viruses*, Chiu W, Burnett R, Garcea R, eds. New York: Oxford University Press, pp. 512–533.

Suzuki K, Hu D, Bustos T, Zlotogora J, Richieri Costa A, Helms JA, Spritz RA (2000): Mutations of PVRL1, encoding a cell-cell adhesion molecule/herpesvirus receptor, in cleft lip/palate-ectodermal dysplasia. *Nat Genet* 25:427–430.

Terry Allison T, Montgomery RI, Whitbeck JC, Xu R, Cohen GH, Eisenberg RJ, Spear PG (1998): HveA (herpesvirus entry mediator A), a coreceptor for herpes simplex virus entry, also participates in virus-induced cell fusion. *J Virol* 72:5802–5810.

Trybala E, Bergstrom T, Svennerholm B, Jeansson S, Glorioso JC, Olofsson S (1994): Localization of a functional site on herpes simplex virus type 1 glycoprotein C involved in binding to cell surface heparan sulphate. *J Gen Virol* 75:743–752.

Visalli RJ, Brandt CR (1991): The HSV-1 UL45 gene product is not required for growth in Vero cells. *Virology* 185:419–423.

Westra DF, Glazenburg KL, Harmsen MC, Tiran A, Jan Scheffer A, Welling GW, Hauw The T, Welling Wester S (1997): Glycoprotein H of herpes simplex virus type 1 requires glycoprotein L for transport to the surfaces of insect cells. *J Virol* 71:2285–2291.

Whitbeck JC, Peng C, Lou H, Xu R, Willis SH, Ponce de Leon M, Peng T, Nicola AV, Montgomery RI, Warner MS, Soulika AM, Spruce LA, Moore WT, Lambris JD, Spear PG, Cohen GH, Eisenberg RJ (1997): Glycoprotein D of herpes simplex virus (HSV) binds directly to HVEM, a member of the tumor necrosis factor receptor superfamily and a mediator of HSV entry. *J Virol* 71:6083–6093.

Whitbeck JC, Connolly SA, Willis SH, Hou W, Krummenacher C, Ponce De Leon M, Lou H, Baribaud I, Eisenberg RJ, Cohen GH (2001): Localization of the gD-binding region of the human herpes simplex virus receptor, hveA. *J Virol* 75:171–180.

Williams RK, Straus SE (1997): Specificity and affinity of binding of herpes simplex virus type 2 glycoprotein B to glycosaminoglycans. *J Virol* 71:1375–1380.

Wilson SP, Yeomans DC, Bender MA, Lu Y, Goins WF, Glorioso JC (1999): Anti-hyperalgesic effects of infection with a preproenkephalin-encoding herpes virus. *Proc Natl Acad Sci USA* 96:3211–3216.

Wolfe D, Goins WF, Kaplan TJ, Capuano SV, Fradette J, Murphey-Corb M, Robbins PD, Cohen JB, Glorioso JC (2001): Herpesvirus-mediated systemic delivery of nerve growth factor. *Molecular Therapy* 3:61–69.

WuDunn D, Spear PG (1989): Initial interaction of herpes simplex virus with cells is binding to heparan sulfate. *J Virol* 63:52–58.

Yamada M, Natsume A, Mata M, Oligino T, Goss J, Glorioso J, Fink DJ (2001): Herpes simplex virus vector-mediated expression of Bcl-2 protects spinal motor neurons from degeneration following root avulsion. *Exp Neurol* 168:225–230.

Yamada M, Oligino T, Mata M, Goss JR, Glorioso JC, Fink DJ (1999): Herpes simplex virus vector-mediated expression of Bcl-2 prevents 6-hydroxydopamine-induced degeneration of neurons in the substantia nigra in vivo. *Proc. Natl. Acad. Sci. USA* 96:4078–4083.

Ye GJ, Roizman B (2000): The essential protein encoded by the UL31 gene of herpes simplex virus 1 depends for its stability on the presence of UL34 protein. *Proc Natl Acad Sci USA* 97:11002–11007.

Zwaagstra J, Ghiasi H, Nesburn AB, Wechsler SL (1989): In vitro promoter activity associated with the latency-associated transcript gene of herpes simplex virus type 1. *J Gen Virol* 70:2163–2169.

19

ENGINEERING TARGETED BACTERIOPHAGE AS EVOLVABLE VECTORS FOR THERAPEUTIC GENE DELIVERY

DAVID LAROCCA, PH.D., AND ANDREW BAIRD, PH.D.

INTRODUCTION

A fundamental challenge for gene therapy is the development of gene delivery vectors that target genes to the appropriate cells in the body for maximal therapeutic effect while minimizing vector toxicity (Hodgson, 1995; Kay, Liu, and Hoogerbrugge, 1997; Verma and Somia, 1997). One approach that takes advantage of the high transduction efficiency of animal viruses is to modify the tropism of a viral vector by attaching or engineering a targeting ligand onto the viral capsid or envelope (Goldman et al., 1997; Gu et al., 1999; McDonald et al., 1999; Rogers et al., 1998). Alternatively, synthetic vectors can be assembled by combining DNA with a condensing agent, a genetic or chemically linked targeting ligand, and other elements needed for intracellular trafficking (Buschle et al., 1995; Medina-Kauwe, Kasahara, and Kedes, 2001; Plank et al., 1994; Sosnowski et al., 1996). Bacteriophage (phage) vectors are simple genetic packages that offer an attractive alternative to viral and syn-

Vector Targeting for Therapeutic Gene Delivery, Edited by David T. Curiel and Joanne T. Douglas
ISBN 0-471-43479-5

thetic approaches. Phage vector targeting is simplified compared to conventional vectors because (1) there is no need to ablate native tropism and (2) the relatively simple phage coat structure can tolerate fusion to a wide variety of targeting ligands including peptides, growth factor ligands, and antibodies without disrupting assembly. Moreover, phage display technology allows for the directed evolution of the phage coat and thus there is the potential to evolve phage for greater efficiency and specificity. Finally, phage vector production is potentially more cost-effective because high titer stocks can be prepared from bacterial supernatants. These characteristics led researchers to explore the use of phage as a gene therapy vector. Here we review strategies for adapting phage vectors for targeted gene delivery using both rational design and directed evolution.

BACKGROUND ON BACTERIOPHAGE

Twort and d'Herelle independently reported the discovery of bacteriophage in the 1920s and described them as filterable particles that infect and lyse bacteria. Phage were initially envisioned as a natural means of combating bacterial infections. This work was quickly overshadowed by the discovery of penicillin in the 1940s. However, once again there is a strong interest in the use of therapeutic phage to treat human disease because of the appearance of antibiotic-resistant strains (Lederberg, 1996). Because of the simple genetic structure of bacteriophages, research on phage genetics laid the groundwork for much of modern molecular biology. The application of phage for drug discovery began in 1985 when George Smith demonstrated that foreign proteins could be displayed on the surface of filamentous phage by genetic fusion of a foreign gene with a phage coat protein gene (Smith, 1985). This linkage of phenotype and genotype forms the basis of phage display technology whereby affinity-selected proteins displayed on phage are identified by sequencing their encoding DNA "tags" within the phage genome. Nonlytic filamentous phage are by far the most common tool for the display and directed evolution of binding ligands and are the first phage to be adapted for targeted mammalian cell gene delivery (Larocca et al., 1998). This chapter focuses primarily on filamentous phage, although it may be possible to adapt other kinds of phages (e.g., lambda, T4) for gene delivery.

Phage Structure and Function

Filamentous phages are a group of long cylindrical bacterial viruses that contain a circular single-stranded DNA genome (Fig. 19.1). The most extensively studied Ff class (f1, fd, and M13) infects *Escherichia coli* through the tip of the F pilus. Unlike lytic phage, filamentous phage do not kill their host but rather are produced by extrusion from the bacterial cell wall. Continuous production of phage from a growing culture of infected bacteria results in accu-

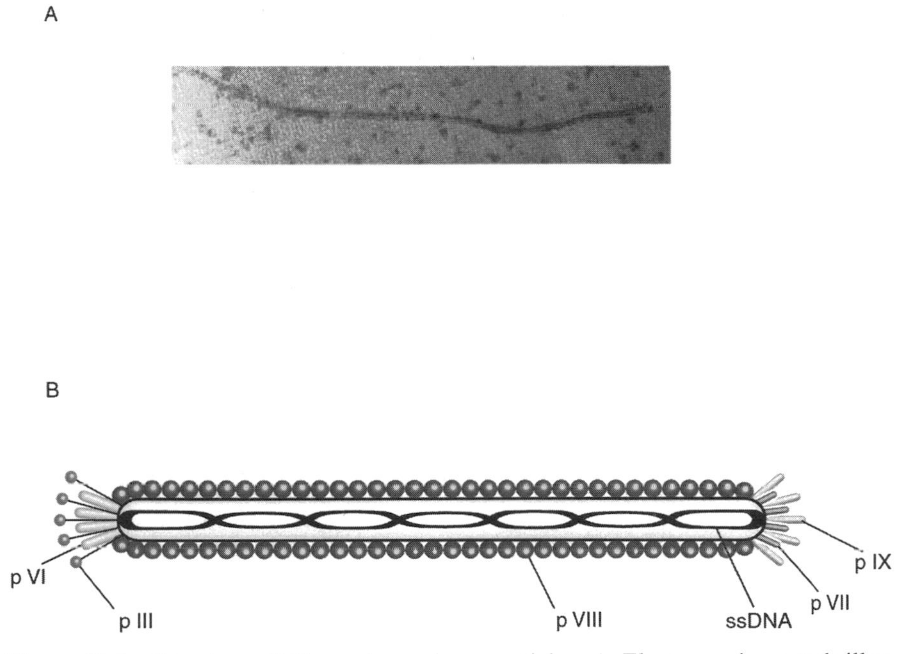

Figure 19.1 Structure of filamentous phage particles. *A*. Electron micrograph illustrating long filamentous structure of phage particle. *B*. Schematic diagram of phage coat.

mulation of 10^{11}–10^{12} particles per milliliter. The somewhat flexible filaments are about 6.5 nm in diameter and 930 nm in length and consist of a closed circular single-stranded DNA genome surrounded by a protein sheath of approximately 2700 copies of the major coat protein, pVIII (Fig. 19.1). The simple 6.4 kb genome contains 11 genes and an intergenic region involved in replication and packaging. At one end of the filament there are about five copies each of the minor coat proteins, pVII and pIX. At the other end, there are about five copies each of additional minor coat proteins, pIII and pVI. Coat proteins pIII and pIX extend out from the ends of the particle, while pVI and pVII are closer to the genomic DNA in the pVIII sheath.

The pIII protein is required for infection and formation of stable phage particles. It consists of three domains each separated by a flexible glycine-rich linker region. The bipartite amino (N) terminal domain forms a knoblike structure at the tip of the phage particles. The first domain, N1, is responsible for insertion of the phage coat proteins into the cell membrane at the F-pilus and translocation of the DNA genome into the cell. The second domain, N2, is needed for binding to the F-pilus. The third or carboxy (C) terminal domain (CT) of pIII is needed for stable formation and release of the phage particle from the bacterial cell wall. See Webster (Webster, 2001) for an in-depth description of filamentous phage biology.

Phage Display

Phage display originated with the creation of the first fusion phage in 1985, which displayed a fragment of Eco RI endonuclease (Smith, 1985). Smith recognized the potential of the flexible structure of pIII at the tip of filamentous phage to tolerate genetic fusion to foreign proteins. In this study, he also demonstrated that an antibody against Eco RI endonuclease could be used to affinity-select the Eco RI fusion phage from a 1000-fold excess of wild-type phage (Smith, 1985). Within a few years, several studies demonstrated the power of phage display for selecting and evolving proteins from large combinatorial libraries of peptides and antibodies displayed on phage (Barbas et al., 1991; Marks et al.; 1991; Scott and Smith, 1990; Smith and Scott, 1993).

A remarkably wide range of proteins has been displayed on pIII, and most of them retain at least some activity (Smith and Petrenko, 1997). For example, biologically active growth factors, cytokines, and peptide hormones have been displayed (Bass, Greene, and Wells, 1990; Buchli, Wu, and Ciardelli, 1997; Gram et al., 1993; Rousch et al., 1998; Saggio, Glaoguen, and Laufer, 1995; Szardenings et al., 1997). Various forms of antibody fragments, including Fabs and single-chain antibodies (scFvs) have also been displayed (Hoogenboom et al., 1998). In addition to pIII, the major coat protein, pVIII, has been used for displaying a variety of peptides, proteins, and antibodies. Small peptides (six to eight amino acids) can be displayed on all copies of pVIII in the phage coat to create phage particles that are "landscaped" with altered peptide sequences (Petrenko et al., 1999). Larger proteins including antibodies can be displayed with the inclusion of wild-type pVIII in the phage coat so that phage assembly is not disrupted (Greenwood, Willis, and Perham, 1991; Markland et al., 1991). To a lesser extent other minor coat proteins pVI, pVII, and pIX (Gao et al., 1999; Jespers et al., 1995) have been used to anchor displayed proteins and several C-terminal fusions have recently been made to pIII and pVIII (Fuh et al., 2000; Fuh and Sidhu, 2000) suggesting greater flexibility in genetically modifying phage particles than originally anticipated.

Directed Evolution Using Phage Display Libraries

Phage display technology makes use of the principles of Darwinian evolution (diversity, selection, and amplification) to isolate and evolve proteins with desirable properties such as affinity for a target molecule or cell. Diversity is generated genetically by insertion of synthetic DNA-encoding random peptides or DNA-encoding antibody repertoires from naïve or immunized animals, or randomly mutated DNA encoding a synthetic or naturally occurring ligand. Selection from phage display libraries is performed by repeated rounds of "panning" against a desired target linked to a solid phase. Following incubation of the library with the target, unbound phages are removed by extensive washing. Because of their stable structure, the bound phage particles can be recovered under harsh conditions such as low pH, chaotropic agents, or protease cleavage. Amplification of selected phage is then accomplished by propagation in host bacteria. The phage are then purified and used as input phage for the next round of panning. This process is reiterated until the complexity of the library is sufficiently reduced to allow characterization of individual clones by DNA sequencing.

Phage display libraries were initially selected by affinity for a simple target, such as a purified protein, but more recently, methods have been developed for selection against complex targets such as live cells (De Kruif et al., 1995; Figini et al., 1998; Hoogenboom et al., 1999; Pereira et al., 1997; Szardenings et al., 1997) or organ vasculature (Pasqualini, 1999; Pasqualini and Ruoslahti, 1996a). Although technically more challenging, selection against complex targets can identify cell-specific binding peptides without prior knowledge of the targeted cell surface receptor. Selection can also be based on a function such as cellular internalization or transcytosis (Barry, Dower, and Johnston, 1996; Ivanenkov and Menon, 2000; Poul et al., 2000) or, as discussed in the following section, gene delivery (Kassner et al., 1999). Once it has been identified, the binding peptide, ligand, or antibody can be further evolved by random mutagenesis or gene shuffling (Stemmer, 1994), followed by additional cycles of selection and amplification. With the adaptation of phage for gene delivery, it becomes possible to directly apply phage display technology toward the directed evolution of gene delivery vectors.

RATIONALE FOR TARGETED PHAGE GENE DELIVERY

Conferred Tropism

Bacteriophage vectors have many desirable properties of both viral and nonviral gene transfer vectors without the significant drawbacks (Table 19.1). Phage vectors, like synthetic vectors, are simple genetic particles with no intrinsic tropism for mammalian cells. Therefore, they have a much lower potential for toxicity from expressed viral proteins, replication-competent virus and/or uptake of vector into untargeted tissues, compared to animal viral vectors.

Table 19.1 Comparison of Phage Vectors with Animal and Synthetic Vectors

Property	Animal Virus	Synthetic Vector	Phage
Intrinsic safety	Low	High	High
Simplicity	Low	High	High
Cost Efficiency	Low	High	High
Reproducibility	High	Low	High
Transduction Efficiency	High	Low	Low
Genetic Targeting	Limited	Limited	Favorable
Directed Evolution	Limited	Limited	Favorable

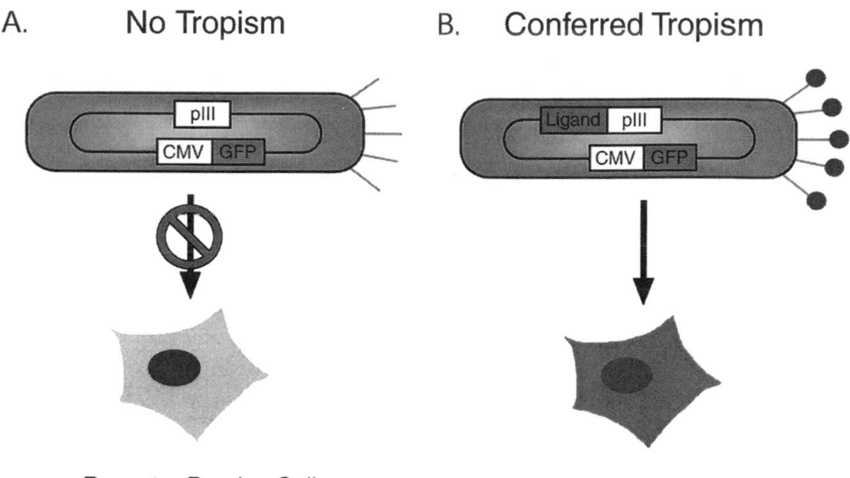

Figure 19.2 Conferred cell tropism. *A.* Filamentous bacteriophage particles have no intrinsic tropism for mammalian cells but can be genetically modified for targeted mammalian cell transduction. *B.* Modified phage particles that carry a reporter gene, GFP, are targeted to cell surface receptors by fusing a ligand to the phage coat protein, pIII. Gene delivery and subsequent transgene expression occurs only when the appropriate ligand is displayed on the phage coat and its receptor is on target cells.

Because phage lack intrinsic tropism for mammalian cells, phage vectors that contain a transgene, such as green fluorescent protein (GFP) or β-galactosidase and the appropriate transcriptional control elements, do not transduce cells unless introduced by nonspecific chemical transfection (Yokoyama-Kobayashi and Kato, 1993, 1994). However, specific tropism can be engineered into similarly modified phage by displaying a ligand on the phage surface, thus enabling targeted gene delivery and subsequent transduction (Fig. 19.2). Accordingly, the displayed ligand confers tropism for cells bearing the appropriate receptor. Thus far, the efficiency of gene delivery by animal viral vectors, which have evolved mechanisms for host cell entry, trafficking, and replication exceeds phage or synthetic vectors. However, the genetic flexibility and potential for directed phage evolution suggest that it will be possible to engineer modified phage for highly efficient targeted gene delivery.

Simplified Targeting

A variety of targeting ligands have been used successfully to target phage particles to mammalian cell surface receptors (Table 19.2). The initial studies that demonstrated transduction by ligand-targeted phage used phagemid particles that were targeted noncovalently with an avidin–biotin bridge to basic fibroblast growth factor (FGF2) (Larocca et al., 1998). These studies demonstrated

TABLE 19.2 Targeting Phage Vectors to Cells via Cell Surface Receptors

		Targeting Ligand	Receptor	Reference
In vitro	Binding/ internalization	RGD peptide	Integrins	Hart et al., 1994
		Anti-HER2	HER2	Becerril, Poul, and Marks, 1999
		Selected peptides	Unknown	Barry, Dower, and Johnston, 1996; Ivanenkov, Felici, and Menon, 1999a
	Transduction	FGF2	FGFR	Larocca et al., 1998
		EGF	EGFR	Larocca et al., 1999; Kassner et al., 1999
		Anti-HER2	HER2	Poul and Marks, 1999
		Adenovirus penton base	$\alpha v\beta 3$, $\alpha v\beta 5$ integrins	Di Giovine et al., 2001
In vivo	Binding/ internalization	Selected peptides	Endothelial peptidases	Pasqualini and Ruoslahti, 1996a
		Selected peptides	Unknown	Samoylova and Smith, 1999
	Transduction	EGF	EGFR	Burg et al., 2002

that ligand-targeted phage particles could deliver a transgene (GFP) to mammalian cells using a receptor-mediated pathway. Subsequent studies showed that phage-mediated transduction of various mammalian cell types could be obtained using genetically targeted phage (Kassner et al., 1999; Larocca et al., 2001, 1999). Phage display vectors were engineered to display the targeting ligand (FGF2/EGF), as a pIII fusion protein to target GFP to cell lines bearing the appropriate receptors. Transduction by phage was ligand-, dose-, and time-dependent and specific for the targeting ligand's cognate receptor (Kassner et al., 1999; Larocca et al., 1999). These results have been confirmed independently by Poul and co-workers (Poul and Marks, 1999) using antibody-targeted phage to target HER2 receptor-bearing breast tumor cells and by Di Giovine and colleagues (Di Giovine et al., 2001) using adenoviral penton base protein to target phage particles to HeLa cells. In these studies, background transduction by untargeted phages was low, confirming that ablation of native targeting is not needed to engineer phage particles for targeted gene delivery. Thus far, phage have been targeted using genetic fusions to the pIII gene, but targeting with proteins displayed on pVIII as well as other minor coat proteins should be feasible. Potentially any peptide, antibody, or natural ligand that can be displayed on phage could be used to target phage for gene delivery including those cell-type or organ-specific proteins or peptides that have been selected from large highly diverse phage display libraries.

Vector Evolution

By engineering phage for gene delivery to mammalian cells using known ligands, it becomes feasible to apply the power of phage display technology to genetically select novel phage-displayed ligands capable of targeted gene delivery to a desired cell type or tissue. Genetic selection of ligand-display phage has been demonstrated using a method termed *LIVE* (ligand identification via expression) (Kassner et al., 1999). LIVE selection is based on the recovery of ligand-targeted phage from genetically transduced target cells (Fig. 19.3). The desired target cells are transfected with a phage display library constructed in a vector that carries a transgene expression cassette. The library consists

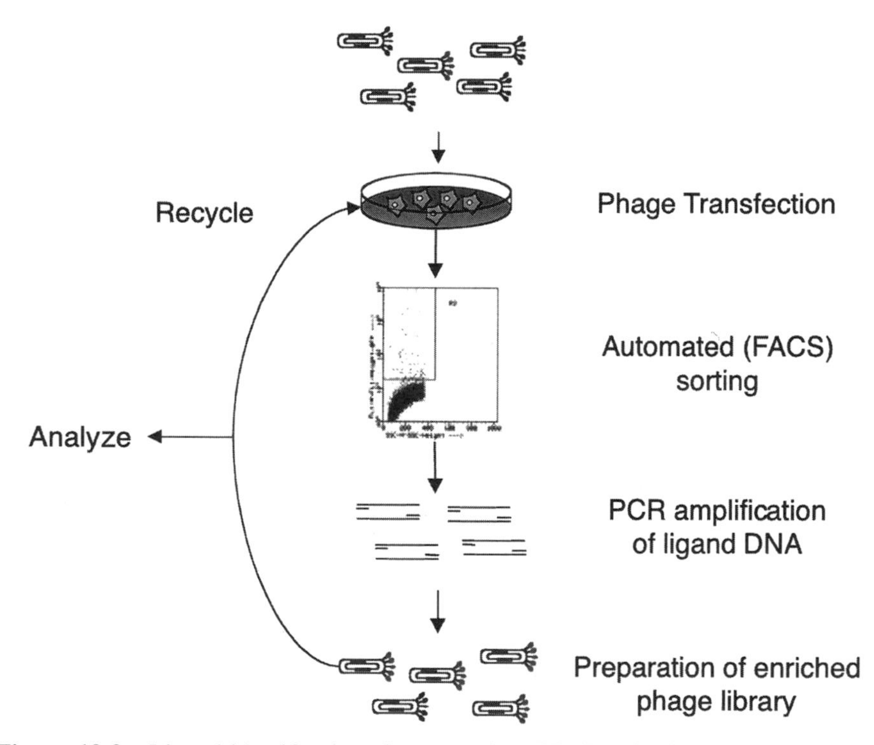

Figure 19.3 Ligand identification via expression (LIVE) selection strategy. A phage display library that carries a reporter gene (i.e., GFP) is contacted with target cells. Phage particles that display ligands enter cells and deliver the reporter gene. Automated FACS sorting is used to isolate GFP-positive cells from which the internalized phage particles are reconstituted using PCR amplification. The reconstituted phage particles are used as input phages for the next cycle of enrichment. When the complexity of the library is sufficiently reduced, the ligands are characterized by DNA sequencing.

of a highly diverse population of random peptides, antibodies, or natural ligands (cDNAs) displayed on the phage coat. Cells that internalize phage are isolated by expression of either a reporter gene (e.g., GFP, β-galactosidase, luciferase) or a selectable drug resistance gene (e.g., neomycin phosphotransferase). For example, autofluorescent GFP-positive cells can be isolated using a fluorescence-activated cell sorting (FACS). Alternatively, drug resistant cells expressing the *neo* gene can be isolated by growth in the antibiotic, G418. The ligand-encoding sequences are recovered from the selected cells using the polymerase chain reaction (PCR) and the appropriate primers for amplification. The amplified DNA fragments are then subcloned back into the phage vector from which phage particles are prepared for the next round of selection. Each cycle of transfection, transduction, and phage recovery enriches for ligand-display phage that are capable to binding, internalizing, and transducing the target cells. Kassner and co-workers demonstrated that EGF-targeted phage can be enriched over one million-fold after three to four rounds of LIVE selection against EGF-receptor bearing cells (Kassner et al., 1999). A significant advantage of using the same vector for both selection and gene transfer is that it ensures that the selected ligand is active in the structural context of the gene transfer vehicle. This is not always the case when transferring a targeting ligand from the selection vehicle (i.e., phage) to the gene transfer vehicle (an animal virus or molecular conjugate).

The ability to produce large diverse libraries of vector from which targeted phage gene transfer vectors can be selected genetically opens up the possibility of evolving phage for other desirable characteristics such as improved efficiency, serum stability, and resistance to host immune response. The relatively short generation time and the comparatively simple genetics of phage make directed evolution with phage vectors more practical than with animal viral vectors. Phage libraries can typically be constructed with 10^7–10^{10} individual members and conveniently prepared from bacterial cultures. It has not been practical to generate comparable libraries in animal viruses, such as retroviruses and adenoviruses. Moreover, modification of viral envelopes and capsids by proteins larger than small peptides typically has adverse effects on assembly or infectivity. Thus, a particularly attractive advantage of using phage vectors for gene delivery is the ability to direct the evolution of the vector. Accordingly, unique phage vectors could be tailored for different gene therapy targets or applications.

Safety

Although further studies are needed to determine the safety of using bacterial viruses for human gene therapy, the concept of using a bacterial virus for mammalian cell gene delivery has several intrinsic advantages with regard to safety. Modified phage could be regarded as nonproductive animal viral vectors that can be produced with the convenience of a phage vector. The lack of intrinsic tropism for mammalian cells should minimize the risk of internalization by nontargeted tissues. Moreover, phages that are internalized are not likely to synthesize phage proteins in a eukaryotic cell. Even if small amounts of phage protein were made in the foreign milieu of the eukaryotic cell, phage morphogenesis and assembly is even more unlikely. Therefore, phage transgene delivery and expression might be accomplished with minimal risk of productive infection or of a cellular immune response. The use of phagemid vectors that carry only a phage origin of replication but can be packaged into phage particles by infection with a helper phage that supplies the necessary structural and morphogenic genes would reduce the likelihood of productive infection even further. Recent studies describe mammalian cell transduction by EGF-displaying phagemids that in addition to lacking nearly all phage structural genes are noninfectious in bacteria because they also lack full-length pIII (Larocca et al., 2001). The ability to engineer such highly targeted gene transfer phages that are not only cell specific but also have no tropism for bacteria could be an important additional safety consideration.

ADAPTATION OF PHAGE FOR TARGETED GENE DELIVERY

Mechanism of Phage Gene Delivery

The mechanism of phage vector-mediated transduction of mammalian cells is not completely understood, however, targeted phage particles are likely to pass through thresholds that are similar to the early stages of viral infection (Fig. 19. 4). The initial step is cell surface recognition and binding. For many animal viruses this is a two-step process involving both a low-affinity surface scanning interaction and high-affinity receptor binding. In the case of adenoviral particles, binding is mediated by a high affinity receptor (the coxsackievirus and adenovirus receptor, CAR) and internalization by an RGD-integrin interaction (Wickham et al., 1993). Thus far, phage vectors have been targeted with one ligand displayed on pIII. However, it might be beneficial to engineer phage particles that mimic the dual receptor-targeting approach of animal viruses by expressing one ligand on the major coat protein and another one on a minor coat protein. Following binding and internalization into an endosomal compartment, the phage particle must then escape the endosome and traffic to the nucleus. At some point either before or after nuclear entry, the phage particles need to be uncoated. Finally, the single-stranded genome is presumably converted to double-stranded DNA for transcription of transgene mRNA and its subsequent translation in the cytoplasm. Further understanding of these processes will be important for engineering improved phage vector targeting and transduction efficiency.

Binding and Internalization

A variety of cell surface recognition molecules have been used to confer specific targeting to phage particles via cell-specific ligand-receptor binding (Table

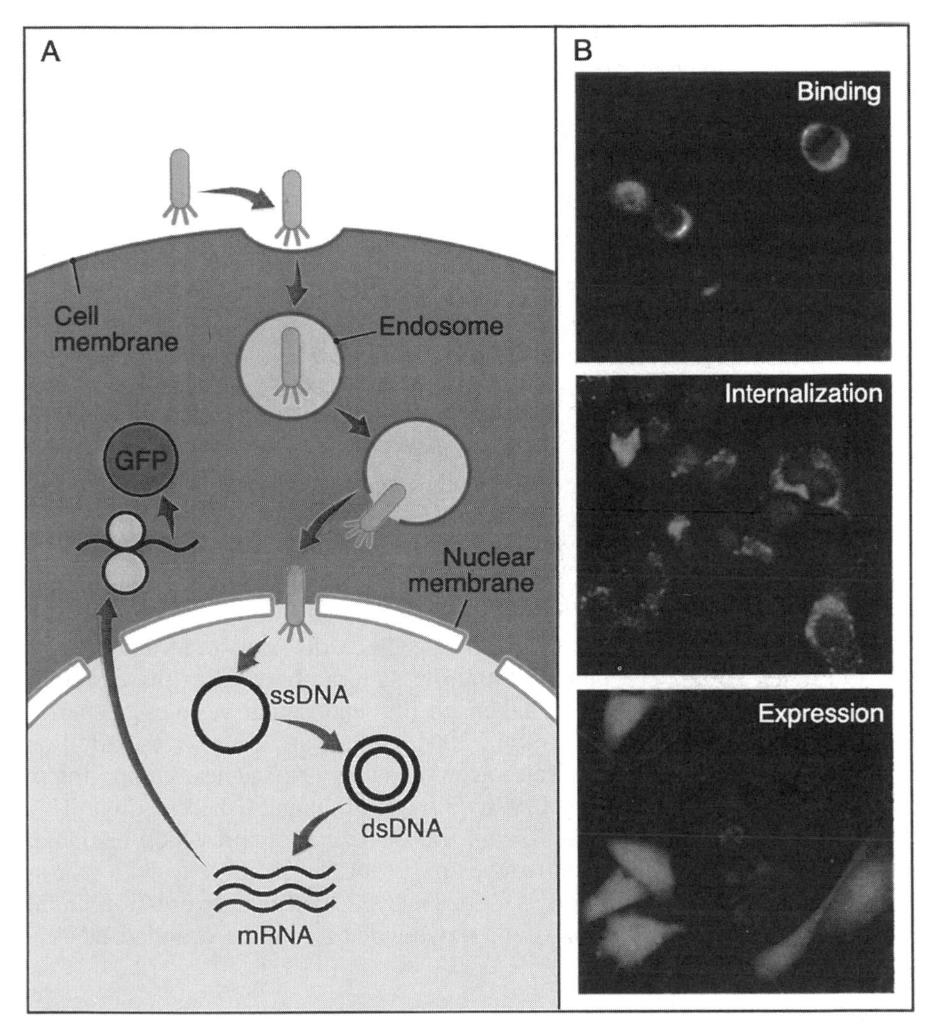

Figure 19.4 Schematic model of phage-mediated gene transfer. *A*. Ligand-targeted phage bind to cell surface receptors and are internalized by receptor-mediated endocytosis. Phage particles escape the endosome and enter the nucleus. The phage genome is unpacked and converted to double-stranded DNA by host DNA repair system. The transgene (GFP) mRNA is transcribed under the control of a mammalian promoter (CMV) and transported to the cytoplasm for translation resulting in accumulation of GFP protein. *B*. Visualization of EGF-phagemid particle binding, internalization, and GFP transgene expression. Human prostate carcinoma cells were incubated for 2 h with multivalent EGF-phagemid particles, fixed, and stained with an antiphage antibody and Texas Red secondary antibody. Internalized phage are stained by removing bound phage with low pH buffer, fixing, and permeabilizing the cells before staining (Larocca et al., 2001). Nuclei are stained with DAPI. Autofluorescent GFP-expressing cells are visualized using an epifluorescent microscope and FITC filter set. Magnification = 400×.

19.2). Hart and colleagues used an RGD-containing peptide that recognizes cell surface integrins to target phage to cells and demonstrated that the phages were internalized (Hart et al., 1994). RGD sequences are used by certain bacteria and viruses for cell invasion (Bergelson et al., 1992; Isberg, 1991; Wickham et al., 1993) and therefore it was reasonable to assume this interaction could be used to deliver macromolecules such as DNA or phage particles. Indeed, the adenoviral RGD-containing penton base protein was recently used to target phage into HeLa cells (Di Giovine et al., 2001). Barry and co-workers also demonstrated internalization of targeted phage particles using cell-specific peptides that were selected from phage display libraries by recovering phage particles from cell lysates after acid washing to remove externally bound phage particles (Barry, Dower, and Johnston, 1996). Ivanenkov and colleagues have obtained similar results by selecting internalizing peptides displayed on pVIII (Ivanenkov et al., 1999a).

Internalization of ligand-targeted phage particles appears to be very efficient in cells that express sufficient amounts of receptor. For example, nearly all cells expressing receptors for EGF or HER2 internalize appropriately targeted phage particles (Becerril, Poul, and Marks, 1999; Larocca et al., 2001). Importantly, little or no background staining is observed when cells are treated with equivalent doses of untargeted control phage (Becerril, Poul, and Marks, 1999; Larocca et al., 2001). The punctate staining pattern observed in these studies indicates that phage particles are taken up into endosomal vesicles (Becerril, Poul, and Marks, 1999; Larocca et al., 2001). At about 24 h after the addition of targeted phage, the phage particles accumulate in a perinuclear compartment (Ivanenkov, Felici, and Menon, 1999b; Burg et al., unpublished results). This could account for a delay in significant transgene expression which begins at about 48 h after transfection and reaches maximum levels at 72 to 96 h (Kassner et al., 1999; Poul and Marks, 1999). Delayed transgene expression could also be the result of conversion of single-stranded to double-stranded DNA.

Transgene Expression

Although early studies demonstrated receptor-mediated internalization of targeted phage, the potential of phage to transduce cells was not appreciated. It seemed unlikely that phage particles could survive the hostile endosomal environment, traffic to the nucleus, uncoat, and express a transgene in a eukaryotic cell. Remarkably, initial studies with FGF2-targeting showed that phage-mediated transduction could be obtained through a cell surface receptor-mediated pathway (Larocca et al., 1998). Since these studies, several investigators have demonstrated transgene delivery to mammalian cells using genetically desplayed FGF2, EGF, anti-HER2, adenoviral penton base targeted phage and selected peptides (Di Giovine et al., 2001; Kassner et al., 1999; Larocca et al., 1999; Poul and Marks, 1999; Burg, unpublished). The efficiency of targeted phage transduction under normal conditions is from 1% to 4%, however, targeted phage are likely to be more efficient than ligand- or antibody-targeted

DNA-complexes (Buschle et al., 1995; Sosnowski et al., 1996) because much less DNA is needed to obtain a similar transduction efficiency, although direct comparison of targeted phage with targeted synthetic vectors is needed to confirm this. Recently optimization of ligand display and genotoxic treatment of target cells has been used to obtain phage-mediated transduction efficiencies as high as 45% (Burg et al., 2002; Larocca et al., 2001) (see following section). It is therefore reasonable to predict that genetic adaptations of phage vectors that facilitate the processes leading to transgene expression will result in highly efficient phage vectors suitable for gene therapy.

IMPROVING PHAGE-MEDIATED GENE TRANSFER

Several approaches could be taken to increase phage-mediated transduction. Assuming phage binding and internalization can be maximized by designing more efficient ligand-display phage, further improvements in transduction will likely result from increasing the efficiency of postinternalization events. Accordingly, phage transduction might be improved by modifying phage particles for efficient escape from the endosome and translocation into the nucleus. Other targets for phage improvement include increasing conversion of the single-stranded genome to double-stranded DNA and subsequent phage DNA replication, integration, and transgene expression. Improvements in phage vector efficiency are likely to come from a combination of rationally designed modifications and directed evolution.

Rational Design

Recent studies show that optimization of the number of EGF molecules/particle (valency) to three to five copies/particles increases the transduction efficiency of targeted phage to about 10% of transfected cells (Larocca et al., 2001). Phage vectors do not always display foreign proteins at full valency because of degradation or mistranslation of the fusion protein. However, stable multivalent display of EGF on phagemid particles was obtained by fusing EGF to the C-terminal domain of pIII and rescuing with a gene III-deleted helper phage. The valency of EGF in this construct exceeded that of either a phage or phagemid vector where EGF was fused to full-length pIII (Larocca et al., 2001). Phagemid rescue with wild-type helper phage results in monovalent display (one or fewer EGF molecules/phage particle) because the helper phage provides wild-type pIII, which is incorporated into phage particles along with the EGF-pIII fusion protein. However, the same vector produces multivalent particles when rescued with a gene III deleted helper phage such as the one developed by Rakonjac and colleagues (Rakonjac, Jovanovic, and Model, 1997), because virtually all the pIII protein available is fused to EGF (Fig. 19.5). Comparison of monovalent to multivalent phagemid particles prepared from the same vector shows that increased valency directly correlates with

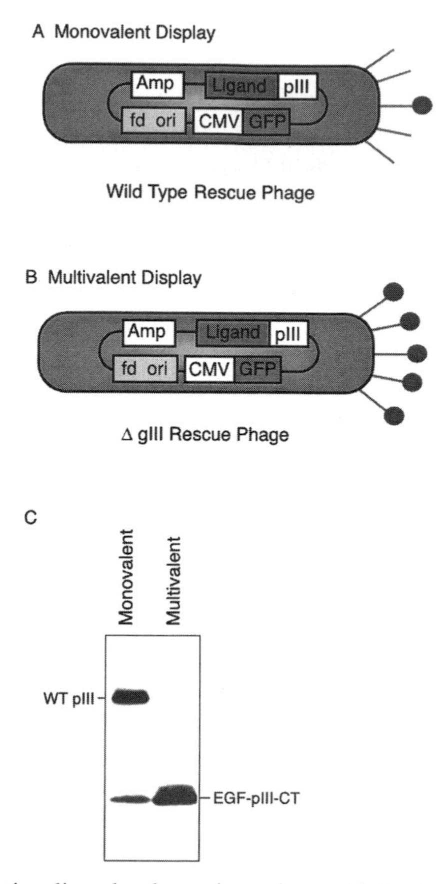

Figure 19.5 Regulating ligand valency in a phagemid display system. The valency of ligand display in a phagemid vector is determined by the rescue phage. *A*. Monovalent display particles are prepared by rescuing with wild-type R408 helper phage. A combination of wild-type pIII and ligand pIII fusion are displayed on the phage tip. *B*. Multivalent phagemid particles are prepared from the same phagemid vector by rescuing with the gIII deleted R408d3 helper phage. Virtually all the pIII displayed on the phage is the ligand pIII fusion protein. *C*. A Western blot of purified monovalent and multivalent phagemid particles probed with anti-pIII antibody. Monovalent phage contain a mixture of wild-type pIII (upper band) and EGF-pIII-CT fusion protein (lower band). In this case, the EGF-pIII fusion protein is smaller than wild-type pIII because it is fused to the C-terminal (CT) domain. Multivalent phage contain predominantly EGF-pIII-CT fusion protein (lower band).

both increased phage internalization and GFP transgene transduction efficiency (Larocca et al., 2001). Additional studies have demonstrated similar findings using phage particles targeted with antibodies or peptides (Poul and Marks, 1999; Ivanenkov, Felici, and Menon, 1999b). Increased transduction by multivalent phage particles could be the result of increased receptor dimerization or clustering and subsequent internalization caused by interaction with ligand dimers presented on the phage coat. Indeed, peptides are often more active than the equivalent free peptide when displayed multivalently on phage or covalently linked to force dimerization (Ballinger et al., 1999; Wrighton et al., 1997). However, increased phage particle avidity and the effects of receptor dimerization on downstream endosomal trafficking could also contribute to increased transduction (Marsh et al., 1995).

The observation that even monovalent EGF phagemid particles are more efficient than the longer EGF-phage particles suggests that particle size can affect transduction efficiency (Larocca et al., 2001). Accordingly, compact phagemid particles might internalize and traffic more efficiently than larger phage particles. Phagemid vectors have additional advantages because production yields are higher and they can incorporate larger foreign DNA inserts while maintaining a relatively small particle size. The use of a phagemid also minimizes the amount of prokaryotic sequences in the vector that could potentially interfere with transgene expression. Therefore, phagemids may be preferred over phage vectors for gene delivery.

Another approach to improving phage-mediated transduction is to add protein sequences that facilitate postinternalization events such as endosomal escape, nuclear translocation, and phage particle uncoating and replication. For example, endosomolytic sequences from animal viruses have been used to increase transduction by synthetic vectors (Plank et al., 1994; Sosnowski et al., 1996). Phage vectors could be engineered using a similar approach by displaying both targeting and endosomolytic peptides on separate phage coat proteins. Following release from the endosome, the phage particles must translocate to the nucleus. In replicating cells, nuclear transfer may occur during replication when the nuclear membrane is compromised. However, it might be advantageous to direct import of the phage particle into the nucleus, particularly for transfer of genes into nonreplicating cells. Nuclear localization signal (NLS) sequences from SV40, for example, have been linked to nontargeted synthetic vectors to promote nuclear import (Zanta, Belguise-Valladier, and Behr, 1999; Zhang et al., 1999; Bremner, Seymour, and Pouton, 2001). It is interesting to note that NLS sequences have recently been shown also to promote nuclear import of genes delivered through a receptor-mediated pathway (Chan and Jans, 2001). Thus, it would be reasonable to incorporate NLS sequences into phage particles to promote nuclear uptake.

Following nuclear import, single-stranded phage DNA presumably must be converted to transcriptionally active double-stranded DNA for transgene expression to occur. Transduction by recombinant adeno-associated virus (AAV) vectors, which are also single-stranded DNA, has been shown to be limited by leading-strand synthesis (Fisher et al., 1996). Accordingly, AAV transduction is enhanced by genotoxic treatments (UV, topoisomerase I inhibitors, heat shock) (Alexander, Russell, and Miller, 1994; Ferrari et al., 1996; Russell, Alexander, and Miller, 1995), which presumably act by inducing host cell DNA repair proteins that in turn increase DNA strand conversion. These genotoxic treatments also significantly improve transduction by filamentous phage (Burg et al., 2002). For example, treatment of target cells with a topoisomerase I inhibitor increases transduction by EGF-targeted phage to as high as 45% in certain human carcinoma cell lines (Burg et al., 2002). The mechanism of increased transduction is likely to involve DNA strand conversion, although other explanations are possible given the pleiotropic affects of the treatments. In the absence of DNA-damaging agents, AAV strand conversion and transduction can be enhanced genetically by the addition of the adenoviral E1 orf-6 gene (Fisher et al., 1996). Taken together, these data suggest that targeted phage transduction could be increased by engineering the phage genome to include genes or DNA elements that facilitate DNA strand conversion.

Another approach to engineering more efficient phage transduction is to create phage–viral hybrid vectors. Accordingly, phage could be modified to include viral or chromosomal DNA replication elements, such as an EBV replication origin for selective replication in EBV-infected cells and integration elements from retroviruses or AAV for genome integration. Thus, the fate of the vector DNA (episomal vs. integration) could be determined by the choice of viral DNA elements.

Directed Evolution

Although rational design is an effective way of engineering vector improvements, the choice of accessory elements to be inserted must be determined largely by the tedious and time-consuming process of trial and error. Phage vector development would, therefore, be accelerated by taking advantage of the ability to apply directed evolution to the development of phage vectors. It is reasonable to assume that reiterative transfection and genetic selection (i.e., LIVE) from highly diverse phage libraries will yield phage with improved transduction efficiency. LIVE and other selection strategies could be applied to direct phage evolution toward particles with desirable properties for targeted therapeutic gene transfer. For example, phage could be selected for increased efficiency, prolonged serum half-life, decreased immunogenicity, selective tissue targeting, improved intracellular trafficking, stable episomal replication, or integration (Fig. 19.6). Indeed, Merril and co-workers have demonstrated the feasibility of directed evolution of phage in vivo using repeated rounds of phage library injection and recovery of surviving phage to select long-lived λ phage mutants that escape serum clearance (Merril et al., 1996). Sokoloff and colleagues used a similar strategy to identify peptides that protect circulating T7 phage from clearance by complement (Sokoloff et al., 2000). Negative selection is another approach that might be used to identify phage with low immunogenic potential. In this case, immunoreactive phages are repeatedly removed from the phage population by adsorption with immune serum (Jenne et al., 1998).

Pasqualini and Rouslahti have demonstrated that affinity selection from phage display libraries can be applied in vivo to identify peptides that target

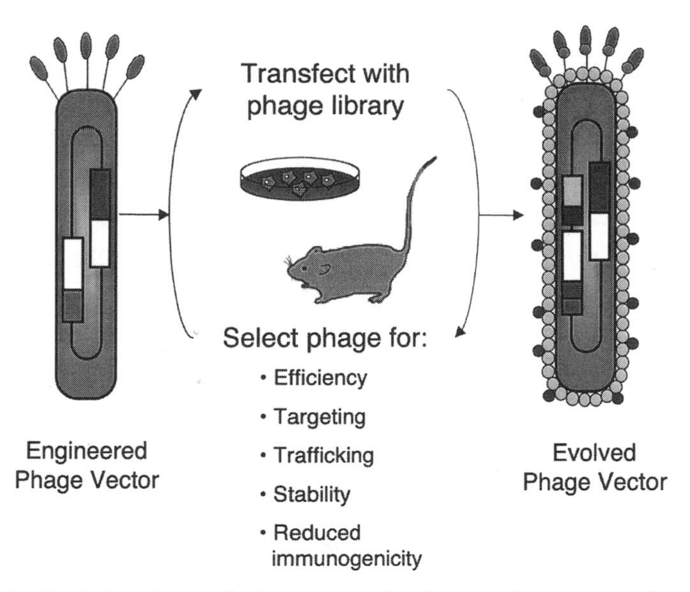

Figure 19.6 Evolving phage display vectors for therapeutic gene transfer. The principles of directed evolution used in phage display could be applied to phage vectors to tailor them for gene delivery to specific target cells in vivo. Reiterative transfection and genetic selection are used to engineer phages with favorable efficiency, stability, targeting, immunogenicity, and so forth for in vivo gene transfer.

specific organ vasculature such as brain, kidney, and xenografted tumors (Arap, Pasqualini, and Ruoslahti, 1998; Pasqualini, 1999; Pasqualini and Ruoslahti, 1996a, 1996b). Peptide-displaying phage were selected in vivo by repeated cycles of phage library injection into mice and recovery of phage from targeted organs. This technique has led to the discovery of organ- and tumor-targeting phages that recognize unique molecular addresses present in organ vasculature. Phage selection in vivo may also be useful for identifying targeted phage that bind and internalize in a target tissue. Samoylova and Smith (1999) have selected phage capable of internalization into heart and skeletal muscle, indicating the potential of long filamentous particles to pass through vasculature into tissues. The observation that systematically administered phage can penetrate tissues suggests the possibility of using genetic selection to identify improved transducing phage in vivo. These data suggest that both in vitro and in vivo selection could be applied to diverse libraries of phage variants to evolve phage for therapeutic gene delivery.

Alternative Phage Vectors

In addition to filamentous phage, other genetic packages may be useful for targeted gene delivery. For example, alternative display systems have been developed using λ (Sternberg and Hoess, 1995; Maruyama, Maruyama, and

Brenner, 1994) and T4 phages (Efimov, Napluev, and Mesyanzhinou, 1995; Jiang, Abu-Shilbayeh, and Rao, 1997). Strategies for the display of foreign proteins of λ include display on the D protein (Sternberg and Hoess, 1995) and the tail fiber (Maruyama, Maruyama, and Brenner, 1994). Provided they can be appropriately targeted to mammalian cells, these double-stranded DNA phages might be advantageous because they would not require strand conversion for transgene expression. Indeed, significant transduction of mammalian cells is obtained using chemical transfection with λ phage (Okayama and Berg, 1985), indicating that λ phage particles can become uncoated and express their double-stranded DNA in mammalian cells. Dunn obtained low levels of transduction using RGD displaying λ phage (Dunn, 1996). Eguchi and co-workers have recently obtained transduction levels as high as 30% in COS cells transfected with λ phage targeted with the human immunodeficiency virus (HIV) Tat protein transduction domain (Eguchi et al., 2001). However, the data suggest that gene delivery by Tat-phage involves caveolae formation and is not receptor-mediated. Thus, further studies are needed to determine if double-stranded DNA phage systems will be applicable to targeted gene delivery in mammalian cells.

TARGETED PHAGE GENE THERAPY

Targeting Phage In Vivo

The development of phage vectors for targeted gene delivery is still in its early stages compared to viral or even nonviral vectors. Clearly, further research on the biodistribution, half-life, and host response to systemically delivered phages is needed to develop phage vectors for human gene therapy. Assuming that phage vector efficiency can be optimized by rational design or directed evolution, targeted phage vectors could provide an economical means of therapeutic gene delivery in a variety of applications such as gene replacement, genetic vaccines, therapeutic angiogenesis, tissue engineering, and cancer gene therapy. In the near term, phage vectors may be particularly useful for delivery of immunomodulatory or prodrug genes. For example, EGF receptor-targeted herpes simplex viral thymidine kinase (HSV-TK) gene delivery has been shown to be effective in a mouse tumor model at relatively low transduction efficiency because HSV-TK sensitizes both transduced cells and nearby "bystander" cells to the drug ganciclovir (Chen et al., 1998). It is also likely that nuclear entry by phage would be facilitated in replicating tumor cells and that the leaky vasculature of tumors would be accessible to systemically delivered phage particles. Indeed, Burg and co-workers have reported GFP transgene delivery and expression in human prostate carcinoma cells by EGF-targeted phagemid particles in a mouse xenograft tumor model (Burg et al., 2002). GFP expression was observed when phage was delivered either intratumorally or systemically and camptothecin was used in these experiments to enhance phage-mediated transduction. Accordingly, combination therapy with genotoxic chemotherapy agents, like camptothecin, and therapeutic phage may prove to be useful for treating solid tumors.

Targeting Phage Ex Vivo

Phage vectors may also be of use in ex vivo gene therapy where cells are removed from the body, transfected in vitro with a therapeutic gene, and then delivered to the patient. This strategy overcomes many concerns associated with systemic gene delivery because transfection occurs outside the body. For example, phage vectors could be used to deliver therapeutic genes to primary cell cultures that are otherwise difficult to transfect by identifying and targeting cell-surface receptors. Along these lines, Burg and co-workers have reported successful gene delivery by EGF-phage to primary cell cultures prepared from rat olfactory bulb, striatum, cortex, and dorsal root ganglia (Burg et al., 2001). Thus, targeted phage vectors offer an alternative to chemical transfection and viral gene delivery for ex vivo gene delivery. By targeting phage to specific cell surface receptors, it may be possible to deliver genes to specific subtypes of cells within a primary cell population.

CONCLUSIONS

For more than a decade, phage display vectors have been recognized as valuable tools for the discovery and directed evolution of protein-binding pairs. It is only recently that they have also been established as vectors for receptor-targeted gene delivery to mammalian cells. Already, phage transduction rates from 10% to 45% have been observed, in vitro, suggesting the potential to improve vector efficiency considerably to levels that rival animal viral vectors. The lack of native tropism and the ability to target cells with a phage-displayed ligand should allow for the development of highly specific cell-targeting phage for gene delivery. The elucidation of the mechanisms involved in phage gene transfer and the fate of systemically delivered phage will help in the rational design of improved phage vectors. However, the ability to evolve phage vectors for gene delivery opens up new possibilities for the development of gene therapy vectors. Accordingly it is now feasible to use genetic selection from vector display libraries to evolve safe and efficient gene delivery vectors that are custom tailored for specific indications. For example, phage vector libraries could be selected for phage that deliver genes to different target organs by selected routes of administration such as oral or airway delivery. In addition, selection of transducing phage from large libraries of phage variants will be useful for identifying functional proteins for cell targeting and intracellular trafficking that could be useful for other gene delivery systems as well as phage.

REFERENCES

Alexander IE, Russell DW, Miller, AD (1994): DNA-Damaging agents greatly increase the transduction of nondividing cells by adeno-associated virus vectors. *J Virol* 68:8282–8287.

Arap W, Pasqualini R, Ruoslahti E, (1998): Cancer treatment by targeted drug delivery to tumor vasculaturc in a mousc modcl. *Science* 279:377–380.

Ballinger MD, Shyamala V, Forrest LD, Deuter-Reinhard M, Doyle LV, Wang JX, Panganiban-Lustan L, Stratton JR, Apell G, Winter JA, Doyle MV, Rosenberg S, Kavanaugh WM (1999): Semirational design of a potent, artificial agonist of fibroblast growth factor receptors. *Nat Biotechnol* 17:1199–1204.

Barbas CF 3rd, Kang AS, Lerner RA, Benkovic SJ (1991): Assembly of combinatorial antibody libraries on phage surfaces: The gene III site. *Proc Natl Acad Sci USA* 88:7978–7982.

Barry MA, Dower WJ, Johnston SA, (1996): Toward cell-targeting gene therapy vectors: Selection of cell-binding peptides from random peptide-presenting phage libraries. *Nat Med* 2:299–305.

Bass S, Greene R, Wells JA (1990): Hormone phage: An enrichment method for variant proteins with altered binding properties. *Proteins* 8:309–314.

Becerril B, Poul MA, Marks JD (1999): Toward selection of internalizing antibodies from phage libraries. *Biochem Biophys Res Commun* 255:386–393.

Bergelson JM, Shepley MP, Chan BM, Hemler ME, Finberg RW (1992): Identification of the integrin VLA-2 as a receptor for echovirus 1. *Science* 255:1718–1720.

Bremner KH, Seymour LW, Pouton CW (2001): Harnessing nuclear localization pathways for transgene delivery. *Curr Opin Mol Ther* 3:170–177.

Buchli PJ, Wu Z, Ciardelli TL (1997): The functional display of interleukin-2 on filamentous phage. *Arch Biochem Biophys* 339:79–84.

Burg MA, Jensen-Pergakes K, Ravey P, Gonzalez AM, Baird A, Larocca D. (2001): Receptor-targeted gene delivery to carcinoma cells using filamentous phage. *Mol Ther* 3:S179 (abstract).

Burg MA, Jensen-Pergakes K, Gonzalez AM, Ravey P, Baird A, Larocca D. (2002): Enhanced phagemid particle gene transfer in camptothecin-treated carcinoma cells. *Cancer Res* 62:977–981.

Buschle M, Cotten M, Kirlappos H, Mechtler K, Schaffner G, Zauner W, Birnstiel ML, Wagner E (1995): Receptor-mediated gene transfer into human T lymphocytes via binding of DNA/CD3 antibody particles to the CD3 T cell receptor complex. *Hum Gene Ther* 6:753–761.

Chan CK, Jans DA (2001): Enhancement of MSH receptor- and GAL4-mediated gene transfer by switching the nuclear import pathway. *Gene Ther* 8:166–171.

Chen J, Gamou S, Takayanagi A, Ohtake Y, Ohtsubo M, Shimizu N (1998): Targeted in vivo delivery of therapeutic gene into experimental squamous cell carcinomas using anti epidermal growth factor receptor antibody: immunogene approach. *Hum Gene Ther* 9:2673–2681.

De Kruif J, Terstappen L, Boel E, Logtenberg T (1995): Rapid selection of cell subpopulation-specific human monoclonal antibodies from a synthetic phage antibody library. *Proc Natl Acad Sci USA* 92:3938–3942.

Di Giovine M, Salone B, Martina Y, Amati V, Zambruno G, Cundari E, Failla CM, Saggio I (2001): Binding properties, cell delivery, and gene transfer of adenoviral penton base displaying bacteriophage. *Virology* 282:102–112.

Dunn IS (1996): Mammalian cell binding and transfection mediated by surface-modified bacteriophage lambda. *Biochimie* 78:856–861.

Efimov VP, Nepluev IV, Mesyanzhinov VV (1995): Bacteriophage T4 as a surface display vector. *Virus Genes* 10:173–177.

Eguchi A, Akuta T, Okuyama H, Senda T, Yokoi, Inokuchi H, Fujita S, Hayakawa T, Takeda K, Hasegawa M, Nakanishi M (2001): Protein transduction domain of HIV-1 tat protein promotes efficient delivery of DNA into mammalian cells. *J Biol Chem* 276:26204–26210.

Ferrari FK, Samulski T, Shenk T, Samulski RJ (1996): Second-strand synthesis is a rate-limiting step for efficient transduction by recombinant adeno-associated virus vectors. *J Virol* 70:3227–3234.

Figini M, Obici L, Mezzanzanica D, Griffiths A, Colnaghi MI, Winter G, Canevari S (1998): Panning phage antibody libraries on cells: isolation of human Fab fragments against ovarian carcinoma using guided selection. *Cancer Res* 58:991–996.

Fisher KJ, Gao G, Weitzman MD, DeMatteo R, Burda JF, Wilson JM (1996): Transduction with recombinant adeno-associated virus for gene therapy is limited by leading-strand synthesis. *J Virol* 70:520–532.

Fuh G, Sidhu SS (2000): Efficient phage display of polypeptides fused to the carboxy-terminus of the M13 gene-3 minor coat protein. *FEBS Lett* 480:231–234.

Fuh G, Pisabarro MT, Li Y, Quan C, Lasky LA, Sidhu SS (2000): Analysis of PDZ domain-ligand interactions using carboxyl-terminal phage display. *J Biol Chem* 275:21486–21491.

Gao C, Mao S, Lo CH, Wirsching P, Lerner RA, Janda KD (1999): Making artificial antibodies: A format for phage display of combinatorial heterodimeric arrays. *Proc Natl Acad Sci USA* 96:6025–6030.

Goldman CK, Rogers BE, Douglas JT, Sosnowski BA, Ying W, Siegal GP, Baird A, Campain JA, Curiel DT (1997): Targeted gene delivery to Kaposi's sarcoma cells via the fibroblast growth factor receptor. *Cancer Res* 57:1447–1451.

Gram H, Strittmayer U, Lorenz M, Glück D, Zenke G (1993): Phage display as a rapid gene expression system: Production of bioactive cytokine-phage and generation of neutralizing monoclonal antibodies. *J Immunol Methods* 161:169–176.

Greenwood J, Willis AE, Perham RN (1991): Multiple display of foreign peptides on a filamentous bacteriophage. *J Mol Biol* 220:821–827.

Gu D, Gonzalez AM, Printz MA, Doukas J, Ying W, D'Andrea M, Hoganson DK, Curiel DT, Douglas JT, Sosnowski BA, Baird A, Aukerman SL, Pierce GF (1999): Fibroblast growth factor 2 retargeted adenovirus has redirected cellular tropism: Evidence for reduced toxicity and enhanced antitumor activity in mice. *Cancer Res* 59:2608–2614.

Hart SL, Knight AM, Harbottle RP, Mistry A, Hunger H-D, Cutler DF, Williamson R, Coutelle C (1994): Cell binding and internalization by filamentous phage displaying a cyclic Arg-Gly-Asp-containing peptide. *J Biol Chem* 269 (17):12468–12474.

Hodgson CP (1995): The vector void in gene therapy. *Bio/Technology* 13:222–225.

Hoogenboom HR, de Bruine Ap, Hufton SE, Hoet RM, Arends JW, Roovers RC

(1998): Antibody phage display technology and its applications. *Immunotechnology* 4(1):1–20.

Hoogenboom HR, Lutgerink JT, Pelsers MMAL, Roush MJMJ, Coote J, van Neer N. de Bruine A, van Nieuwenhoven FA, Glatz JFC, Arends JW (1999): Selection-dominant and nonaccessible epitopes on cell-surface receptors revealed by cell-panning with a large phage antibody library. *European J Biochem* 260:774–784.

Isberg RR (1991): Discrimination between intracellular uptake and surface adhesion of bacterial pathogens. *Science* 252:934–938.

Ivanenkov VV, Felici F, Menon AG (1999a): Targeted delivery of multiivalent phage display vectors into mammalian cells. *Biochim Biophys Acta* 1448:463–472.

Ivanenkov VV, Felici F, Menon AG (1999b): Uptake and intracellular fate of phage display vectors in mammalian cells. *Biochim Biophys Acta* 1448:450–462.

Ivanenkov VV, Menon AG (2000): Peptide-mediated transcytosis of phage display vectors in MDCK cells. *Biochem Biophys Res Commun* 276:251–257.

Jenne S, Brepoels K, Collen D, Jespers L (1998): High resolution mapping of the B cell epitopes of staphylokinase in humans using negative selection of a phage-displayed antigen library. *J Immunol* 161:3161–3168.

Jespers LS, Messens JH, De Keyser A, Eeckhout D, Van den Brande I, Gansemans YG, Lauwereys MJU, Vasuk GP, Stanssens PE (1995): Surface expression and ligand-based selection of cDNAs fused to filamentous phage gene VI. *Biotechnology (NY)* 13:378–382.

Jiang J, Abu-Shilbayeh L, Rao VB (1997): Display of a PorA peptide from Neisseria meningitidis on the bacteriophage T4 capsid surface. *Infect Immun* 65:4770–4777.

Kassner PD, Burg MA, Baird A, Larocca D (1999): Genetic selection of phage engineered for receptor-mediated gene transfer to mammalian cells. *Biochem Biophys Res Commun* 264:921–928.

Kay MA, Liu D, Hoogerbrugge PM (1997): Gene therapy. *Proc Natl Acad Sci USA* 94:12744–12746.

Larocca D, Witte A, Johnson W, Pierce GF, Baird A (1998): Targeting bacteriophage to mammalian cell surface receptors for gene delivery. *Hum Gene Ther* 9:2393–2399.

Larocca D, Kassner P, Witte A, Ladner R, Pierce GF, Barid A (1999): Gene transfer to mammalian cells using genetically targeted filamentous bacteriophage. *FASEB J* 13:727–734.

Larocca D, Jensen-Pergakes K, Burg M, Baird A (2001): Receptor-targeted gene delivery using multivalent phagemid particles. *Mol Ther* 3:476–484.

Lederberg J (1996): Smaller fleas . . . ad infinitum: therapeutic bacteriophage redux. *Proc Natl Acad Sci USA* 93:3167–3168.

Markland W, Roberts BL, Sexena MJ, Guterman SK, Ladner RC (1991): Design, construction and function of a multicopy display vector using fusions to the major coat protein of bacteriophage M13. *Gene* 109:13–19.

Marks JD, Hoogerboom HR, Bonnert TP, McCafferty J, Griffiths AD, Winter G (1991): By-passing immunization. Human antibodies from V-gene libraries displayed on phage. *J Mol Biol* 222:581-597.

Marsh EW, Leopold PL, Jones NL, Maxfield FR (1995): Oligomerized transferrin receptors are selectively retained by a lumenal sorting signal in a long-lived endocytic recycling compartment. *J Cell Biol* 129:1509–1522.

Maruyama IN, Maruyama HI, Brenner S (19940): λ foo: a λ phage vector for the expression of foreign proteins. *Proc Natl Acad Sci USA* 91:8273–8277.

McDonald GA, Zhu G, Li Y, Kovesdi I, Wickham TJ, Sukhatme VP (1999): Efficient adenoviral gene transfer to kidney cortical vasculature utilizing a fiber modified vector. *J Gene Med* 1:103–110.

Medina-Kauwe LK, Kasahara N, Kedes L (2001): 3PO, a novel nonviral gene delivery system using engineered Ad5 penton proteins. *Gene Ther* 8:795–803.

Merril CR, Biswas B, Carlton R, Jensen NC, Creed GJ, Zullo S, Adhya S (1996): Long-circulating bacteriophage as antibacterial agents. *Proc Natl Acad Sci USA* 93:3188–3192.

Okayama H, Berg P (1985): Bacteriophage lambda vector for transducing a cDNA clone library into mammalian cells. *Mol Cell Biol* 5:1136–1142.

Pasqualini R (1999): Vascular targeting with phage peptide libraries. *Q J Nucl Med* 43:159–162.

Pasqualini R, Ruoslahti E (1996a): Organ targeting in vivo using phage display peptide libraries. *Nature* 380:364–366.

Pasqualini R., Ruoslahti E. (1996b): Tissue targeting with phage peptide libraries. *Mol Psychiatry* 1:421–422.

Pereira S, Maruyama H, Siegel D, Van Belle P, Elder D, Curtis P, Herlyn D (1997): A model system for detection and isolation of a tumor cell surface antigen using antibody phage display. *J Immunol Methods* 203(1):11–24.

Petrenko VA, Smith GP, Gong X, Quinn T (1999): A library of organic landscapes on filamentous phage. *Protein Eng* 9:797–801.

Plank C, Oberhauser B, Mechtler K, Koch C, Wagner E (1994): The influence of endosome-disruptive peptides on gene transfer using synthetic virus-like gene transfer systems. *J Biol Chem* 269:12918–12924.

Poul M, Marks JD (1999): Targeted gene delivery to mammalian cells by filamentous bacteriophage. *J Mol Biol* 288:203–211.

Poul MA, Becerril B, Nielsen UB, Morisson P, Marks JD (2000): Selection of tumor-specific internalizing human antibodies from phage libraries. *J Mol Biol* 301:1149–1161.

Rakonjac J, Jovanovic G, Model P (1997): Filamentous phage infection-mediated gene expression: Construction and propagation of the *gIII* deletion mutant helper phage R408d3. *Gene* 198:99–103.

Rogers BE, Douglas JT, Sosnowski BA, Ying W, Pierce GF, Buchsbaum DJ, Della Manna D, Baird A, Curiel DT (1998): Enhanced in vivo gene delivery to human ovarian cancer xenografts utilizing a tropism-modified adenovirus vector. *Tumor Targeting* 3:25–31.

Rousch M, Lutgerink JT, Coote J, de Bruine A, Arends JW, Hoogenboom HR (1998): Somatostatin displayed on filamentous phage as a receptor-specific agonist. *Br J Pharmacol* 125:5–16.

Russell DW, Alexander IE, Miller AD (1995): DNA synthesis and topoisomerase inhibitors increase transduction by adeno-associated virus vectors. *Proc Natl Acad Sci USA* 92:5719–5723.

Saggio I, Gloaguen I, Laufer R (1995): Functional phage display of ciliary neurotrophic factor. *Gene* 152:35–39.

Samoylova TI, Smith BF (1999): Elucidation of muscle-binding peptides by phage display screening. *Muscle Nerve* 22:460–466.

Scott JK, Smith GP (1990): Searching for peptide ligands with an epitope library. *Science* 249:386–390.

Smith GP (1985): Filamentous fusion phage: Novel expression vectors that display cloned antigens on the virion surface. *Science* 228:1315–1317.

Smith, GP, Petrenko VA (1997): Phage Display. *Chem Rev* 97:391–410.

Smith GP, Scott JK (1993): Libraries of peptides and proteins displayed on filamentous phage. *Methods Enzymol* 217:228–257.

Sokoloff AV, Bock I, Zhang G, Sebestyen MG, Wolff JA (2000): The interactions of peptides with the innate immune system studied with use of T7 phage peptide display. *Mol Ther* 2:131–139.

Sosnowski BA, Gonzalez AM, Chandler LA, Buechler YJ, Pierce GF, Baird A (1996): Targeting DNA to cells with basic fibroblast growth factor (FGF2). *J Biol Chem* 271:33647–33653.

Stemmer WPC (1994): DNA shuffling by random fragmentation and reassembly: In vitro recombination for molecular evolution. *Proc Natl Acad Sci USA* 91:10747–10751.

Sternberg N, Hoess RH (1995): Display of peptides and proteins on the surface of bacteriophage lambda. *Proc Natl Acad Sci USA* 92:1609–1613.

Szardenings M, Tornroth S, Mutulis F, Muceniece R, Keinanen K, Kuusinen A, Wikberg JE (1997): Phage display selection on whole cells yields a peptide specific for melanocortin receptor 1. *J Biol Chem* 272:27943–27948.

Verma IM, Somia N (1997): Gene therapy—promises, problems and prospects. *Nature* 389:239–241.

Webster R (2001): *Filamentous Phage Biology: Phage Display: A Laboratory Manual.* Cold Spring Harbor, NY: Cold Spring Harbor Laboratory Press.

Wickham TJ, Mathias P, Cheresh DA, Nemerow GR (1993): Integrins alpha v beta 3 and alpha v beta 5 promote adenovirus internalization but not virus attachment. *Cell* 73:309–319.

Wrighton NC, Balasubramanian P, Barbone FP, Kashyap AK, Farrell FX, Jolliffe LK, Barrett RW, Dower WJ (1997): Increased potency of an erythropoietin peptide mimetic through covalent dimerization. *Nat Biotechnol* 15:1261–1265.

Yokoyama-Kobayashi M, Kato S (1993): Recombinant f1 phage particles can transfect monkey COS-7 cells by DEAE dextran method. *Biochem Biophys Res Commun* 192:935–939.

Yokoyama-Kobayashi M, Kato S (1994): Recombinant f1 phage-mediated transfection of mammalian cells using lipopolyamine technique. *Anal Biochem* 223:130–134.

Zanta MA, Belguise-Valladier P, Behr JP (1999): Gene delivery: A single nuclear localization signal peptide is sufficient to carry DNA to the cell nucleus. *Proc Natl Acad Sci USA* 96:91–96.

Zhang F, Andreassen P, Fender P, Geissler E, Hernandez JF, Chroboczek J (1999). A transfecting peptide derived from adenovirus fiber protein. *Gene Ther* 6:171–181.

20

TARGETING BACTERIOPHAGE VECTORS

ISABELLA SAGGIO, PH.D.

INTRODUCTION

Viruses represent the most efficient system offered by nature to encapsidate, protect, and deliver DNA into cells. This fact has induced many scientists to exploit eukaryotic viruses as vectors for gene transfer in mammalian cells. In this chapter we present innovative delivery systems, based on viruses, but of prokaryotic origin. These should provide several theoretical advantages as compared to traditional gene transfer vectors.

The physical complexity of eukaryotic viruses has frequently represented an obstacle for the modification of capsid proteins for the production of targeted vectors. It has been shown that the manipulation of the retroviral envelope can alter its fusogenic properties, therefore lessening viral infectivity (Russell and Cosset, 1999). The complex structure of the trimeric fiber of adenovirus has allowed only restrained genetic modifications, *i.e.*, small peptides inserted in defined regions of the fiber knob, limiting the available pattern of targeting ligands (Dmitriev et al., 1998, Roelvink et al., 1998). Furthermore, complete ablation of natural vector tropism, essential for selective targeting, is particularly complex in the case of eukaryotic viruses. The adenoviral infectious pathway, for example, sustains a residual infectivity of viruses ablated in their primary receptor-binding capacity, through the penton base—integrin interaction (Roelvink et al., 1998). Conversely, phage-based vectors are expected to

Vector Targeting for Therapeutic Gene Delivery, Edited by David T. Curiel and Joanne T. Douglas
ISBN 0-471-43479-5

be highly targetable: They cannot enter and transduce mammalian cells unless specific, selected ligands are displayed on the capsids. A further advantage is the fact that the manipulation of phage capsid proteins and the molecular maturation of exogenous phage-displayed peptides have been extensively and successfully performed, as detailed further in this chapter.

Another advantage of prokaryotic systems, as compared to eukaryotic, is the low cost of large-scale preparation procedures, an aspect that is particularly relevant for clinical applications.

A third advantageous characteristic of phage-based systems is the prokaryotic nature of the vector backbone, which reduces the chances of complementation or recombination events between vector and host DNA sequences. By contrast, it has been described that eukaryotic virus-based preparations can be contaminated with helper virus, or can contain replication-competent particles, originating from homologous recombination/complementation between *cis* and *trans* eukaryotic viral sequences (Lochmüller, et al., 1994; Mehir et al., 1996).

Another problematic aspect, largely documented for eukaryotic viruses, is vector immunogenicity and/or toxicity. It has been shown that *de novo* synthesis of viral proteins is, at least in part, responsible for toxicity and transient transgene expression mediated by E1-deleted adenoviral vectors, as a consequence of the activation of major histocompatibility complex class I-restricted cytotoxic T lymphocytes (Yang, Ertl, and Wilson, 1994; Zoltick et al., 2001). To alleviate this problem, a new generation of adenoviral vectors that are ablated of most viral sequences has been produced. Though these gutless constructs are, effectively, less immunogenic, they require a helper virus for particle amplification, resulting in a risk of producing contaminated stocks (Sandig et al., 2000). Conversely, in vivo injection of recombinant phages should induce a limited immunological response, consequent only to input proteins, even when, to allow phage replication, a helper virus is used for amplification procedures.

Taken together, the potential advantages of phage-based systems have encouraged several research groups to develop bacteriophage chimeras, engineered to transfer genes to mammalian cells. In the following sections we detail the specific characteristics of the M13 and lambda viruses, of display systems based on these viruses, and the construction and biological characteristics of transducing vectors derived therefrom.

USAGE OF BACTERIOPHAGE FOR DISPLAY OF PROTEINS AND PEPTIDES

Filamentous Phage: Structure and Biology

Filamentous bacteriophage M13, f1 and fd belong to a larger family of bacteriophages infecting different bacterial species, and have been extensively studied and utilized in molecular biology for many years. M13, f1, and fd phage infect *Escherichia coli* male strains because their receptor on the bacterial surface is the tip of the sex pilus encoded by the F episome. Phage particles include a single-stranded circular genome. The complete nucleotide sequence of M13, f1, and fd phage has been determined and many vectors deriving from these phage have been described to allow cloning and sequencing of DNA inserts. The genomic DNA is about 6400 nucleotides long and codes for ten proteins: three are required for phage DNA synthesis (the products of genes II, V, and X); two, encoded by genes I and IV, serve phage morphogenesis; and five are virion structural proteins (those encoded by genes III, VI, VII, VIII, and IX). All the genes are very close, when not overlapping, in the genome. The viral particle, about a micron in length and about 6–7 nm in diameter, consists of the DNA molecule encapsidated in a protein coat constituted of about 2700 copies per virion of the major coat protein pVIII. At one end of the viral particle there are three to five copies each of pIII and pVI (406 and 112 amino acids long, respectively), that are involved in host–cell binding and in the termination of the assembly process. At the other end are three to five copies each of pVII and pIX (33 and 32 amino acids long, respectively), that are required for initiation of assembly and for maintenance of virion stability. The structure of the M13 phage is shown schematically in Fig. 20.1.

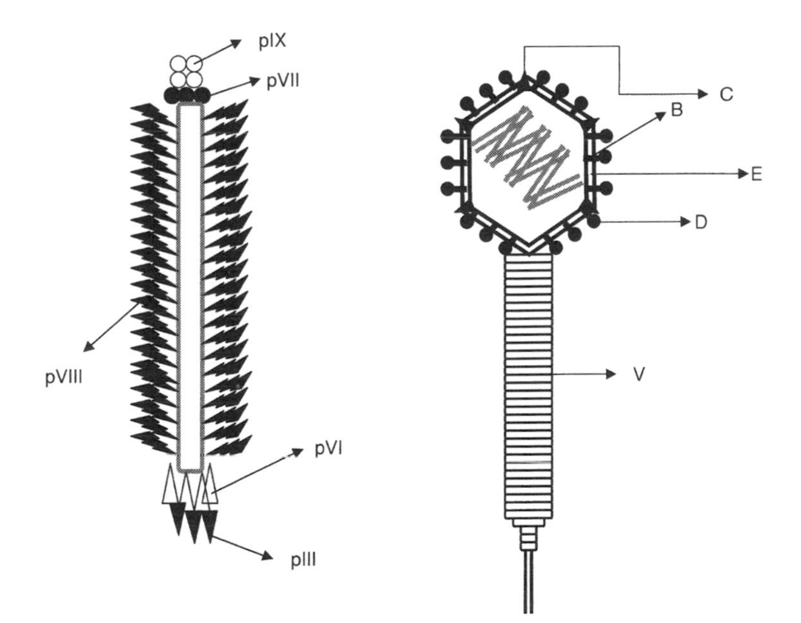

Figure 20.1 Schematic representation of M13 and lambda phage. DNA molecules are depicted in blue. The principal capsid proteins are pointed out by arrows; red color indicates capsidic peptides utilized to host exogenous sequences (see details in the text).

The filamentous phage infectious cycle begins with injection of single-stranded DNA injection into the cell. The complementary (–) strand is immediately synthesized by host enzymes. Following the production of a pool of double-stranded genomic DNA, phage proteins are synthesized and viral progeny are produced. Phage assembly and production takes place at the membrane level and occurs without the lysis or death of the host cell (Felici et al., 1995; Campbell, 1996; Rasched and Oberer, 1986, and references therein). The major phenotypic effect of phage infection and production is a slowing down of the bacterial growth rate, which causes the formation of plaques.

Filamentous Phage: Peptide Display

Three structural proteins have been exploited for peptide display on filamentous capsids: the most common is pIII, followed by pVIII, and rarely pVI. pVI has been used for at least one library. Given the large number of potential binding sites per particle, pVIII display has been exploited as a source of affinity matrices or vaccine candidates. pIII is a 42 kD protein that serves as the scaffold for almost all phage display libraries. Recombinant phage can be constructed so that all copies of a phage contain the insert (3–5 for pIII, or 2700–3000 for pVIII), or, alternatively, only a fraction of the structural proteins can contain the insert. The latter can be constructed by the inclusion of a second gene encoding the protein either in the viral genome itself, or through the construction of DNA plasmids called *phagemids*, which code for recombinant phage proteins that are expressed and subsequently inserted into the inner bacterial membrane, followed by incorporation into viruslike particles during morphogenesis (Felici et al., 1995).

Protein display on filamentous phage has a broad range of applications, which include drug and target discovery, protein evolution, and rational drug design. An impressive list of ligands have been displayed on the surface of filamentous phage, including peptides, antibody fragments, enzymes, cDNA libraries, growth factors, receptors, and protein scaffolds. In all cases the direct link between the ligand genotype and phenotype allows the enrichment of specific phage by, for example, selection on an immobilized target. Phage that display a relevant ligand will be retained, whereas nonadherent phage will be washed away. Bound phage are then recovered from the surface, reinfected into bacteria and grown for further enrichment and eventually for analysis of binding.

In 1990 Smith and Scott described the construction of a pIII-based-examer peptide library, detailing its potential for epitope mapping (Scott and Smith, 1990). In the same year Devlin and co-workers presented data concerning the construction and screening of a 15-amino acid peptide library displayed as a fusion to the pIII protein (Devlin, Panganiban, and Devlin, 1990). In 1991 Felici and collaborators described a multivalent pVIII-based peptide library and its use for selection of antibody ligands (Felici et al., 1991). In the years since then, many groups have been involved in phage peptide library construction and screening with many different ligands such as antibodies, patients' sera, receptors, and enzymes (Folgori et al., 1994; Felici et al., 1995, and references therein). Peptide libraries have also been exploited for the identification of peptide ligands for nonproteinaceous molecules, such as streptavidin, biotin (Saggio and Laufer, 1993), and concanavalin A (Scott et al., 1992).

Further applications of peptide libraries have been exploited for in vivo selection of cell ligands. Pasqualini and co-workers have elegantly proved the possibility of recovering tissue-specific recombinant phage. Following intravenous injection of peptide libraries into mice and subsequent phage rescue from different organs, these investigators could identify peptides able to mediate phage localization to brain and kidney blood vessels (Pasqualini and Ruoslahti, 1996). Phage have also been used for display of proteins and protein domains/scaffolds. Antibody fragment-display has represented a major breakthrough in the phage display field. Fab and single-chain antibody (scFv) fragments, that is V_L and V_H joined by means of a polypeptide tether, have been fused to the pVIII or to the pIII proteins (reviewed in Rader and Barbas, 1997 and references therein). Recombinant antibodies can be isolated from repertoires of antibody genes of immunized or nonimmunized donors, from libraries of germline genes, or from antibody gene repertoire synthetic libraries.

Although antibodies and peptides dominate phage display, several groups have worked on the expression of cytokines, growth factors, and hormones. This field was opened up by the pioneering work of Lowman and Wells on human growth hormone (Lowman and Wells, 1993). These scientists described the possibility of exploiting the phage system for in vitro evolution of hormone-binding properties. An affinity selection of a hormone library displayed on phage allowed the selection of variants with 400-fold higher receptor affinity as compared to wild-type ligand. With analogous experiments we could identify ciliary neurotrophic factor variants characterized by receptor superagonist properties (Saggio et al., 1995).

Filamentous bacteriophages modified to target selected receptors have been recently proposed for gene delivery purposes. Eukaryotic viral vectors are the most frequently used system for gene transfer in mammalian cells for gene therapy applications. In comparison to these, phage present two major advantages: the easier manipulation of the capsidic proteins and the absence of natural tropism for mammalian cells, allowing targeted transduction. The idea of using phage as gene delivery vectors was anticipated in the work of Hart and co-workers that described cell internalization of filamentous phage displaying a cyclic, RGD-containing, integrin-binding peptide (Hart et al., 1994). Phage internalization was also analyzed by Becerril and collaborators who showed that scFv-displaying phage were endocytosable (Becerril, Poul, and Marks, 1999), and by Ivanenkov and colleagues who described internalization of phage multivalently displaying cell-specific peptides (Ivanenkov, Felici, and Menon, 1999). Proof-of-concept of the potential of filamentous phage as a gene delivery vector was obtained by Larocca and collaborators. They first exploited a nongenetic modification of phage particles to transduce the green fluorescent protein in mammalian cells through the receptors for basic fibroblast growth factor (FGF2) or epidermal growth factor (EGF) (Larocca et al., 1998). Subsequently, they presented data indicating that genetic-based multivalent display of a receptor ligand insured up to 10% efficiency of reporter gene transduction in mammalian cells (Larocca et al., 1999, 2001). Contemporary to Larocca's publications were the data from Poul and Marks who showed transduction of ErbB2-expressing cells using phage displaying an anti-ErbB2 scFv, isolated by selection of a phage-antibody library on breast tumor cells and recovery of infectious phage from within the cell (Poul and Marks, 1999). In all cases it has been proposed that the gene transfer characteristics of recombinant phage could be improved by direct in vivo selection of better transducers from a library of ligand variants displayed on the phage.

Lambda Phage: Structure and Biology

Since bacteriophage lambda was first used as a cloning vehicle in the early 1970s, more than 400 different vectors have been described. The genome of bacteriophage lambda is a double stranded DNA molecule, 45,802 bp in length. The DNA is carried in the particle as a linear molecule with single-stranded cohesive termini. Soon after entering a host bacterium, the cohesive termini associate by base-pairing to form a circular molecule with two staggered nicks, which are rapidly sealed by host enzymes to generate a closed circular DNA molecule that serves as the template for transcription during the early, uncommitted phase of infection. During lytic growth, the circular DNA directs the synthesis of about 30 proteins required for its replication, the assembly of bacteriophage particles, and cell lysis. In its lysogenic state, bacteriophage lambda DNA is integrated into the bacterial genome, is replicated as part of the bacterial chromosome, and is thus transmitted to progeny bacteria like a chromosomal gene. During establishment and maintenance of lysogeny, only a small number of lambda genes are expressed.

Lambda bacteriophage genes are functionally organized in genetic clusters. The left-hand region of the genome includes genes whose products are used to package the viral DNA into bacteriophage heads and to assemble infectious virions from filled heads and preformed tails. The central region codes for functions involved in gene regulation, establishment and maintenance of lysogeny, and genetic recombination. Many genes of the central region are not essential for lytic growth and can be sacrificed during construction of bacteriophage lambda vectors to make room for segments of foreign DNA. The right-hand region contains essential genes used in replication of bacteriophage lambda and lysis of infected bacteria.

If a lytic phase is undertaken, tandem polymers of bacteriophage genomes are produced as well as prohead particles in which DNA molecules are packaged. This process is performed in the presence of the protein FI. The head is locked in place around the DNA in the presence of the proteins FII and D. At this stage, the head has a diameter of about 60 nm. The noncontractile tail shaft of bacteriophage lambda, consisting of 32 rings of V protein forming a hollow tube 9 nm in diameter, is then connected to mature lambda heads (Campbell, 1996, and references therein). The structure of a lambda phage is shown schematically in Fig. 20.1.

The substantial difference between filamentous and lytic phage life cycles resides in the fact that lytic phage particle assembly takes place intracellularly.

Lambda Phage: Peptide Display

Recently a number of display systems employing lytic phage have been described. Display methods have been developed based on fusion to the C-terminus of the lambda phage tail protein pV, and both to the N- and C-terminus of capsid protein D. In 1995 Sternberg and Hoess described the production and characterization of recombinant lambda phage displaying, as fusion to the N-terminus of lambda capsid protein D, the eight residue peptide hormone AII, or the 65 amino acid B1 IgG-binding domain from group G *Streptococcus* (Sternberg and Hoess, 1995). These authors showed the possibility of both *cis* and *trans* genetic strategies for assembly of fusion proteins to phage capsids. In the same year, Dunn described the assembly of functional bacteriophage lambda virions incorporating C-terminal peptide or protein fusion with the major tail protein, V (Dunn, 1995). In 1996, Mikawa and collaborators published data on recombinant lambda virions containing D proteins engineered in their N- or C-termini (Mikawa, Maruyama, and Brenner, 1996). In 2000, Santi and co-workers described the construction and characterization of two lambda surface-displayed cDNA expression libraries derived from human brain and mouse embryo. These libraries were affinity-purified on different ligands and allowed rapid selection of specific clones (Santi et al., 2000). The general strategy for selection is essentially the same as that for filamentous phage, in that it involves incubation of the lambda phage-displayed peptide repertoire with immobilized antigen. Phage displaying a binding ligand on their surface are retained and the nonbinding phage are washed away. Adherent phage are then recovered by the addition of bacteria and then infected cells are used to generate phage for a second round of selection.

Recombinant lambda phage has been proposed as a gene delivery vector for mammalian cells. In this perspective, as compared to filamentous phage, lambda presents some important advantages. These include the double-stranded nature of the genome, the potentially high cloning capacity, and, for specific applications, the intracellular location of particle assembly. In 1996 Dunn described bacteriophage lambda-mediated gene delivery in mammalian cells (Dunn, 1996). In his experiments, Dunn used virions whose tail tube major subunit (V) proteins were modified with a cyclizable RGD peptide. Recombinant phage were able to selectively transfect COS cells with a reporter gene. More recent work was published by Eguchi and collaborators on phage displaying a peptide derived from human immunodeficiency virus (HIV) Tat. (Eguchi et al., 2001). In this case the peptide was fused to the head protein D. These authors showed that Tat peptide-phage was able to transduce mammalian cells with an efficiency comparable to that of commercially available lipid-based transfection systems (DOTAP).

Taken together, the data presented in these studies show that both filamentous and lambda phage have been successfully modified to functionally display a number of exogenous proteins, and that, when these proteins are cell receptor ligands, phage can be endocytosed and reporter gene expression can be obtained. The easy manipulation of phage and the possibility of improving transduction properties with selection protocols make them a promising tool in the field of gene transfer vectorology.

ADENOVIRUS PENTON BASE

Adenovirus: Structure and Biology

Adenovirus is a nonenveloped, 35.9 kb linear double-stranded DNA virus. The protein coat comprises 240-exons and 12 pentons forming 252 capsomers in an icosahedral structure. Each penton is composed of a penton base and a fiber. The viral genome is inside of the capsid and has a terminal peptide (TP) covalently linked at the 5′ end. It carries two groups of genes, the early and the late transcription units. The early genes consist of E1A, E1B, E2, E3, E4, and two delayed units, IX and IVa2. The late genes consist of L1 to L5. Structural analysis of human adenovirus type 2 has been achieved by X-ray crystallography combined with image reconstruction of intact particles from cryo-electron micrographs. The seven proteins that assemble into the capsid form a particle of 70–80 nm excluding the fiber protein. This protrudes from the 12 vertices and is a highly extended molecule, 37 nm in length in adenovirus type 2 (Shenk, 1996).

The adenovirus enters the cell by high affinity fiber binding to the coxsackievirus adenovirus receptor protein (CAR) (K_d = 7.9 nM for type 5 fiber knob–CAR interaction) (Bergelson et al., 1997; Kirby et al., 2001), and internalizes via $\alpha v\beta 3$ and $\alpha v\beta 5$ integrins (Wickham et al., 1993). In this way the adenovirus reaches the endosomal pathway and avoids lysosomal degradation. Inside the endosome, a stepwise disassembly program takes place allowing the adenovirus to release its genome into the nucleus. During this process, the pH of the endosome decreases, leading to the release of the fiber from the virion and the dissociation of the penton base. The resulting rupture of the endosome allows the DNA to escape from inside the degraded capsid and to enter the nucleus (Wickham et al., 1994). As early as 60 min after infection, the adenovirus begins to transcribe its genome in the host cell (Greber et al., 1993). To complete the life cycle after infection the virus has to replicate its DNA and assemble the virion within the host cell.

In recent years the adenoviruses have become vectors of choice for a number of gene transfer studies because of their (1) in vitro and in vivo stability, (2) high titers, (3) high cloning capacity, and (4) ability to efficiently infect both dividing and nondividing cells (Trapnell, 1993; Kochaneck et al., 1996).

Penton Base and Its Interaction with Integrin Receptors

In studies carried out more than 30 years ago, Everett and Ginsberg reported the discovery of a soluble factor encoded by adenovirus that caused cells to detach from glass or plastic surfaces in vitro. This cell-detaching activity was later shown to be mediated by the virus penton base. Since cultured cells adhere to vitronectin through the recognition of its RGD sequence by cell surface αv integrins, the presence of an RGD peptide sequence in the penton base, together with its cell-detaching activity, suggested that it may also interact with cell surface integrins. Work by Wickham and co-workers (Wickham et al., 1993), confirmed that adenoviruses exploit the interaction between penton base and αvβ3 and αvβ5 integrins for viral entry into cells. In fact, (1) antibodies directed against these receptors, or soluble penton base, block viral internalization in αv integrin-expressing cells, (2) infection is inhibited by soluble RGD peptides; and (3) viral entry is inefficient in αv negative cells. Direct binding of adenovirus to soluble recombinant αvβ5 was described by Mathias and co-workers in 1998 (Mathias, Galleno, and Nemerow, 1998). These authors showed as well that not only type 2 adenovirus uses this integrin subtype for viral infection, but also serotypes 3, 4, 5, and 37 (Mathias et al., 1994). In addition to αvβ3 and αvβ5, other integrins have been implicated, directly or indirectly, in adenovirus infection. Huang and co-workers have described that adenovirus binding to hematopoietic cells can be blocked by anti-αMβ2 antibodies (Huang et al., 1996), while Davison and collaborators have shown that the TS2/16 β1integrin-activating antibody renders melanoma cells more susceptible to adenoviral infection (Davison et al., 1997). Croyle and collaborators have published in vitro and in vivo data suggesting that integrin α6β1 plays a role in adenoviral infection of the intestinal epithelium (Croyle et al., 1998). Recent data from Li and co-workers demonstrated that integrin αvβ1 also serves as an adenoviral coreceptor (Li et al., 2001a). These authors showed that function-blocking antibodies directed against β1, but not β3, β5 or α5, integrin subunits, block adenoviral infection and viral endocytosis.

Although adenovirus tropism is stringently dependent on the high-affinity binding of the fiber capsid protein to CAR, the virus–integrin interaction mediates the key infection step of particle internalization. This takes place through the formation of clathrin-coated endocytic vesicles, followed by vesicle permeabilization. This process appears to be preferentially mediated by integrin $\alpha v\beta 5$ (Wickham, et al., 1994). Data consistent with the adenovirus exploitation of the clathrin-coated pit pathway were presented by Wang and co-workers, who described significant impairment of type 5 adenoviral infection in HeLa cells overexpressing a dynamin dominant-negative mutant (Wang et al., 1998). Adenovirus internalization by αv integrins requires activation of phosphoinositide-3-OH kinase (PI3K), a downstream effector of the focal adhesion kinase (FAK), a cell-signaling molecule associated with integrin-mediated cellular processes (Li et al., 1998a). In contrast, adenoviral entry was shown to be independent of the αv integrin-mediated cell motility pathways, i.e., the ERK1/ERK2 MAP kinase pathway and myosin light chain kinase signaling. It has also been demonstrated that members of the Rho family of GTPases act downstream of P13K to promote viral endocytosis (Li et al., 1998b; Rauma et al., 1999). p130^{CAS}, a molecule playing a critical role in the organization of the actin cytoskeleton, has also been proposed to be a key component in the signaling complex that regulates adenoviral internalization (Li et al., 2000).

Structure–function relationships of soluble recombinant penton base have been extensively studied by Boulanger and co-workers. These authors have produced several mutants of this adenoviral capsid protein using the baculovirus expression system. Structure–function analysis has shown that discrete protein domains are involved in the functional properties of the molecule. Pentamerization of the penton base was found to be dependent on three amino acid positions, Trp119, Tyr553, and Lys556 of type 2 penton base. The fiber-penton base interaction was affected by mutations in positions Arg254, Cys432, Trp439, Arg340, and in the 547-556 stretch. Interaction of the penton base with the cell was confirmed for the previously identified integrin-binding motifs RGD^{340} and LDV^{287} and was functionally and/or topologically related to other discrete regions which include positions Trp119, Trp165, Cys246, Cys432, and Trp439 of adenovirus type 2 penton base (Karayan et al., 1997).

Recently, Hong and colleagues have shown that baculovirus-expressed penton base, both as a monomer and as a pentamer, is able to recapitulate the entire entry pathway of an adenovirion, i.e., enter the cell through the endocytic pathway, promote self-vesicular escape, and target the nucleus, crossing the nuclear membrane through the nuclear pores (Fig. 20.2) (Hong et al., 1999).

Recent data have described a nonviral system using engineered adenovirus type 5 penton base proteins. Medina-Kauwe and co-workers produced a penton base fusion protein containing a polylysine sequence complexed with DNA. The delivery efficiency of this system was tested in CAR-positive cells in vitro. It was found that the transduction mediated by about 10^9 complexes was equivalent to that mediated by 5×10^6 adenoviral particles (Medina-Kauwe, Kasahara, and Kedes, 2001).

Taken together, these results show that the penton base, alone or in combination with other molecules, can be used as a gene delivery vector.

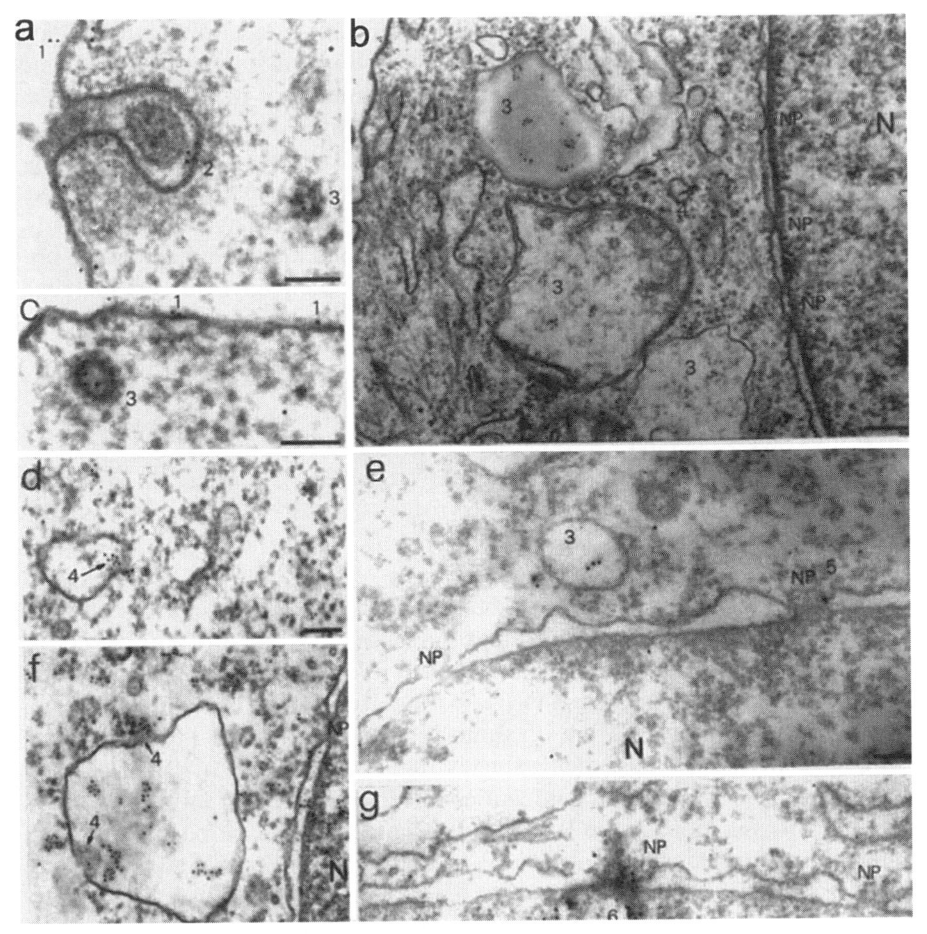

Figure 20.2 Immunogold l abeling and electron microscopy analysis of penton base interaction with HeLa cells. Hypothetical steps of the penton base entry pathway are arbitrarily numbered as follows: (1) attachment to plasma membrane, (2) endocytosis, (3) intravesicular step, (4) vesicular escape and entry into the cytoplasm, (5) docking at the nuclear pore, and (6) translocation across the nuclear pore. Note that clathrin-coated vesicles in panels *c* and *a* contain gold grain-labeled penton base. Arrows in panels *b*, *d*, and *e* point to regions of discontinuities in the endosomal membrane. N = nucleus, C = cytoplasm, NP = nuclear pore. Bar represents 200 nm in panel *b*, and 100 nm in all other panels.

INTEGRIN PROPERTIES

Integrins as Viral Receptors

The term *integrin* was applied in 1987 to describe a family of structurally, immunochemically and functionally related cell surface heterodimeric receptors, that integrated the extracellular matrix with the intracellular cytoskeleton to mediate cell migration and adhesion. The 3 original subunits ($\beta 1$, $\beta 2$, and $\beta 3$) have been expanded to at least 8, and the number of α subunits stands at 17. These subunits interact noncovalently in a restricted manner to form more than 20 family members. Table 20.1 summarizes the major extracellular ligands of integrins. The list, undoubtedly incomplete, includes a large number of extracellular matrix proteins (bone matrix proteins, collagens, fibronectins, fibrinogen, laminins, vitronectin, and von Willebrand factor), reflecting the primary function of integrins in cell adhesion to extracellular matrices. Included in the list are many microorganisms that utilize integrins to gain entry into cells (Plow et al., 2000; Humphries, 2000).

An interesting case study of microorganism/integrin interaction is that of foot-and-mouth disease virus (FMDV), a highly contagious virus (Ferguson, Donnelly, and Anderson, 2001). The seven FMDV serotypes are members of the *Aphthovirus* genus of the family of *Picornaviridae*. Picornaviruses are small, nonenveloped, single-stranded, positive-sense RNA viruses that cause many important diseases of humans and animals. The virus capsid is made of 60 copies each of four virus-encoded proteins, VP1 to VP4. The VP1 includes, in an exposed conformationally flexible loop, a highly conserved RGD tripeptide. This feature, together with direct evidence, has implicated integrins as FMDV coreceptors (Jackson et al., 1997). An interesting aspect of the biology of receptor recognition by these viruses is the observation that strains adapted for growth in cultured cell lines acquire high affinity for the heparan sulfate proteoglycans. By contrast, field isolates of FMDV tend to use RGD-dependent integrins as receptors for infection. Receptor usage variations have been correlated to mutations in the RGD peptide or at neighboring position thought to be critical for integrin interaction (Baranowski, Riuz-Yarabo, and Domingo, 2001). To date, three integrin subtypes have been implicated as FMDV receptor candidates, $\alpha v\beta 3$, $\alpha v\beta 6$, and $\alpha 5\beta 1$. Jackson and co-workers have suggested that $\alpha v\beta 6$ could be an alternative ligand in cell lines that do not express significant levels of $\alpha v\beta 3$ (Jackson et al., 2000).

As indicated in Table 20.1, numerous other viruses exploit integrin interaction in the infection cycle: $\alpha 2\beta 1$ and $\alpha 4\beta 1$ are used by SA11 rotavirus (Hewish, Takada, and Coulson, 2000), $\alpha v\beta 3$ and $\beta 1$ are receptors of the human parechovirus1 (Triantafilou et al., 2000), and $\alpha v\beta 5$ has been described as a coreceptor in adeno-associated virus type 2 infection (Summerford, Bartlett, and Samulski, 1999).

Another interesting case of microorganism–integrin interaction is invasin, a *Yersinia pseudotuberculosis* surface protein that allows bacterial internalization. This 986-amino acid long protein binds at least four different integrins containing the $\beta 1$ chain. Tran Van Nhieu and Isberg published elegant data showing that bacterial internalization mediated by $\beta 1$ chain integrins is determined by ligand affinity and receptor density (Tran Van Nhieu and Isberg, 1993).

TABLE 20.1 Integrins, Ligands, and Recognition Sequences

Integrin	Natural Ligand	Recognition Sequence	Oher Ligands
$\alpha 10\beta 1$	Collagens		
$\alpha 11\beta 1$	Collagens		
$\alpha 1\beta 1$	Collagens	R . . . D	
	Laminins		
$\alpha 2\beta 1$	Collagens	RGD?	Echovirus-1
	Laminins	YYGDLR	Rotavirus
	$\alpha 3\beta 1$	FYFDLR	
	Tenascin	DGEA	
	Fibronectin		
	ACAM		
$\alpha 3\beta 1$	Laminins	RGD?	Invasin
	Collagens		
	$\alpha 3\beta 1$, $\alpha 2\beta 1$		
	Fibronectin		
	Epiligrin		
	Thrombospondin		
$\alpha 4\beta 1$	Fibronectin	LDV	Invasin
	VCAMs	IDS	Rotavirus
		RGD?	
		IDA	
$\alpha 4\beta 7$	Fibronectin	LDV	
	ACAM	IDS	
	VCAMs	LDT	
$\alpha 5\beta 1$	Fibronectin	KQAGDV?	Invasin
$\alpha 10\beta 1$	Denatured collagen fibrinogen	LDV?	
		IDS?	
		RGD	
$\alpha 6\beta 1$	Laminins,		Adenovirus?
	Sperm fertilin		Invasin
$\alpha 6\beta 4$	Laminins		
$\alpha 7\beta 1$	Laminins	RGD?	
$\alpha 8\beta 1$	Fibronectin	RGD	
	Tenascin-C		
$\alpha 9\beta 1$	Tenascin-C	AEIDGIEL	
$\alpha E\beta 7$	E-cadherin		
$\alpha Ib\beta 3$	Collagens		
$\alpha IIb\beta 3$	Fibrinogen	KQAGDV	*Borrelia burgdoferi*
	Collagens	RGD	
	Denatured collagen fibronectin		
	Vitronectin		
	Von Willebrand factor		
	Deorsin		
	Disintegrins		
	Plasminogen thrombospondin		
$\alpha L\beta 2$	ICAMs	L	Adenovirus?
		IET	
$\alpha M\beta 2$	Factor X	RLD	Adenovirus?
	ICAMs	IDS	*Canadida albicans*
	Fibrinogen	RGD?	

TABLE 20.1 ***(Continued)***

Integrin	Natural Ligand	Recognition Sequence	Oher Ligands
$\alpha v\beta 1$	Fibronectin	RGD	
	Vitrogens		
$\alpha v\beta 3$	Vitronectin	RLD	Adenovirus 2, 12, 9, 3, 5
	Fibrinogen	KRLDGS	HIV-Tat protein
	Osteopontin	RGD	
	Von Willebrand factor		
	Bone sialoprotein		
	Denatured collagen tenascin-C		
	Disintegrins		
	Fibronectin		
	Laminins		
	Thrombospondin		
$\alpha v\beta 5$	Vitronectin	RGD	Adenovirus 2, 12, 9, 3, 5, 4, 19, 37
	Bone sialoprotein fibronectin		HIV-Tat protein
$\alpha v\beta 6$	Tenascin-C	RGD	
	Fibronectin	DLXXL	
$\alpha v\beta 8$	Fibronectin	RGD	
$\alpha x\beta 2$	Fibrinogen	GPR	

Another microorganism-derived protein that interacts with integrin receptors is the Tat protein of HIV type 1. This transactivating factor contains a highly conserved RGD tripeptide, and has been shown to be able to bind integrin receptors on the cell surface. This protein and peptides derived therefrom have also been shown to be able to penetrate, and to mediate penetration of exogenous molecules, through cell membranes (Watson and Edwards, 1999).

Taken together these data show that, although a survey of different virus groups illustrates that a number of different cell receptors are exploited for viral infection, integrins are targeted by microorganisms with high frequency. The reasons behind this phenomenon could be functional or geographical. In fact, it could be speculated that integrins could be functionally prone to particle internalization, thanks to their intimate connections with the cyoskeleton. A simpler, hypothesis is that viruses tend to target highly diffused molecules to statistically improve the success of infection.

Integrins as Targetable Molecules

Numerous studies have exploited integrins as targeting molecules. The PubMed web site (http://www.ncbi.nlm.nih.gov/PubMed/) lists about a hundred papers published within the past [illegible] to [illegible] years only on the subject "integrin & gene delivery." A major study has been performed by Hart and co-workers on synthetic integrin-targeted vectors. This group published data describing the use of RGD peptides complexed with DNA through a polylysine moiety for transfection of human cells (Hart et al., 1997). In 2000, Jenkins and co-workers proposed the use of a new class of integrin-directed vectors for in vivo gene delivery. These consisted of a cationic liposome, an integrin-binding peptide derived from *Yersinia*, with a polylysine tail and a plasmid DNA, which combined electrostatically to form a complex whose in vivo transfection efficiency was similar to that of an adenoviral vector and greater that that of the cationic liposome DOTAP (Jenkins et al., 2000).

Several research studies have shown the possibility of inserting integrin-binding peptides into viral capsid proteins to create new tropisms. Incorporation of the RGD-4C peptide into the HI loop of the fiber protein (Dmitriev et al., 1998) or in the hexon adenoviral protein (Vigne et al., 1999), significantly improved transfection efficiency in CAR-negative cells (see chapter 8 of this book).

As described in previous sections, the incorporation of integrin-binding peptides into filamentous or lambda phage capsid proteins has allowed particle internalization in mammalian cells (Dunn, 1996; Hart et al., 1994).

In summary, the success of integrin-binding moieties in mediating cell internalization in numerous viral and nonviral systems has made them a useful tool for the construction of gene delivery vectors.

PHAGE PENTON BASE CHIMERAS

Taking into account the advantages of phage as gene delivery vectors and the specific biological properties of the adenoviral penton base, we have thought to construct phage chimeras, based on a filamentous or phage lambda backbone, displaying full-length adenoviral penton base or a peptide derived therefrom to target genes into integrin-expressing cells.

Penton Base Display on Filamentous Phage

To create recombinant phage monovalently displaying the penton base protein, the full-length adenovirus type 2 penton base gene or its central domain (penton base 286–393) was inserted into a phagemid containing a truncated version of the filamentous phage capsid protein pIII (Hogenboom et al., 1991). Three parameters were considered for the choice of the 286–393 fragment: (1) size of the insert, to limit interference with phage functionality, (2) inclusion of the integrin-binding motif, RGD, (3) structural conformation of the insert. To this regard, we took into account published data on cryo-electron micrograph visualization of adenoviral (Stewart et al., 1997), and performed structure-prediction analysis of the penton base sequence. Taken together, the data indicated that the 286–393 amino acidic stretch would include the RGD integrin-binding motif surrounded by α-helices, expected to give a structural conformation to the exposed loop. To produce recombinant vectors, bacteria transformed with phagemids were infected with a helper phage. Quantification of a recombinant pIII/phage preparation indicated that higher amounts were present in the phage stock displaying the penton base fragment 286–393, as compared to the full length penton base-expressing particles: 1/23 and 1/118 recombinant pIII/total pIII, respectively. These results can be correlated with the difficulty of displaying large inserts—the full length penton base in adenovirus 2 is 571 amino acid long—without affecting phage assembly and infectious properties.

As shown in Figure 20.3, both penton base-displaying phage and particles displaying the 286–392 stretch retained the properties of the adenoviral capsid protein. In fact, phage were able to bind "classical" adenoviral receptors, *i.e.*, integrins $\alpha v\beta 3$, $\alpha v\beta 5$, and also $\alpha 3\beta 1$ and $\alpha 5\beta 1$ subtypes.

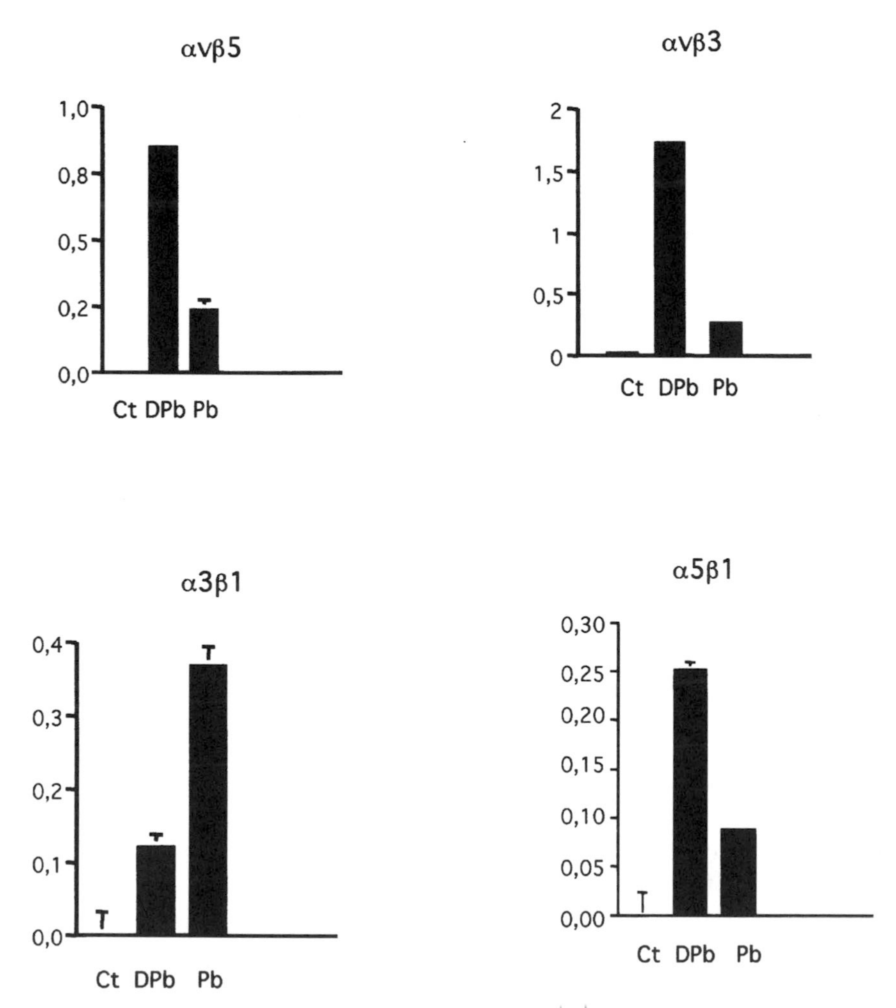

Figure 20.3 Interaction of recombinant phage with immobilized integrins. Immobilized integrins were incubated with penton base (pp) phage (4×10^{12}) (Pb), ΔPb-phage (1×10^{12} particles/well) (DPb), or control phage (4×10^{12} particles/well) (Ct). Ligand–ligand interaction was revealed with an anti-M13 monoclonal antibody. Data are presented as average values of optical density from duplicate measurements.

Cell Binding and Internalization of Penton Base Phage

As described in previous sections, the penton base protein, both as a monomer and as a pentamer is able to promote self-internalization in mammalian cells (Fig. 20.2) (Hong et al., 1999). By electron microscopy and immunofluorescence we could observe that these penton base properties were retained by recombinant phage. In fact, HeLa cells incubated with recombinant phage particles at 4°C, showed filamentous phagelike structures along the cell surface when analyzed by electron microscopy, (Fig. 20.4). Immunofluorescence stud-

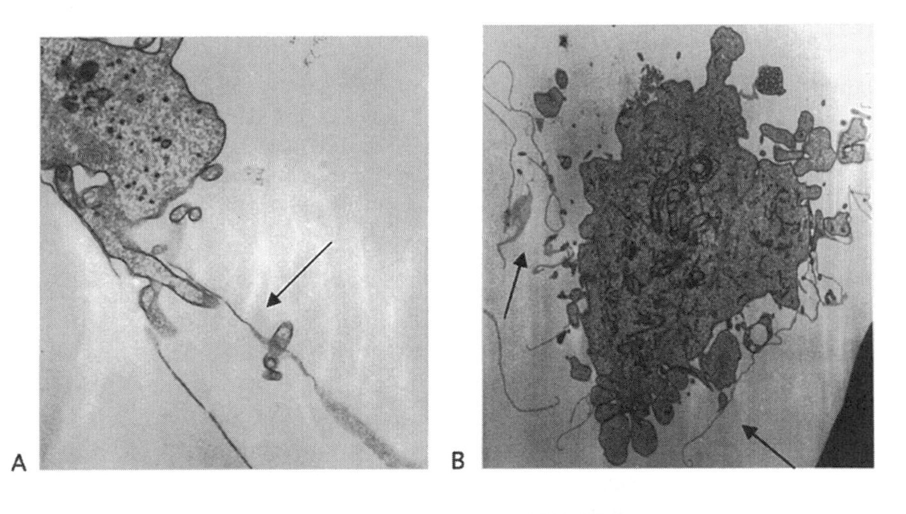

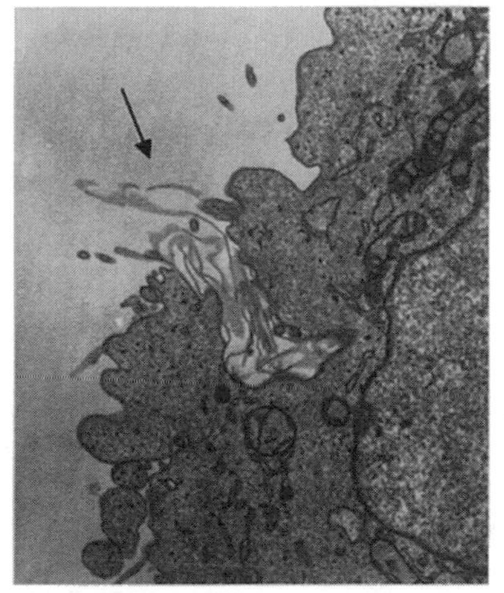

Figure 20.4 Binding of recombinant phage to mammalian cells. HeLa cells were incubated at 4°C with 3×10^{12} and 9×10^{12} particles of ΔPb-phage (*B*, *C*) and Pb-phage (*A*), respectively. After incubation, cells were processed for electron microscopy analysis. Original magnification: *A*, 15500×; *B*, 5200×; *C*, 11500×. Arrows indicate filamentous phagelike structures.

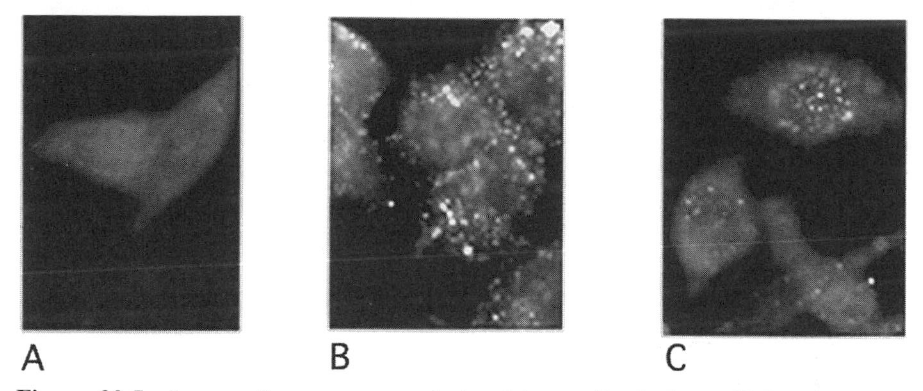

Figure 20.5 Immunofluorescence analysis of internalized phage. HeLa cells were incubated with phage particles: for control phage (*A*) and ΔPb-phage (*B*) 3×10^{12} were used, for Pb-phage (*C*) 9×10^{12}. Incubation time was 1 h at 4°C, followed by 1 h at 37°C. Antibodies for phage detection were added following cell permeabilization. Cells were examined using a fluorescence microscope with a 40× objective.

ies on cells incubated at 37°C, proved that recombinant phage could be detected intracellularly using anti-M13 monoclonal antibodies (Fig. 20.5).

It is interesting to note that, as described for adenovirus (Li et al., 1998a), penton base phage-mediated endocytosis was found to be sensitive to specific kinase inhibitors. Phage internalization was in fact inhibited when infection was performed in the presence of wortmannin, an inhibitor of PI3K kinase, but its endocytosis was unaltered when the same experiment was performed in the presence of the myosin light chain kinase inhibitor, ML7-hydrochloride. These data suggest that recombinant phage activate the same kinase pathway as adenovirus.

Phage-Mediated Transduction of Mammalian Cells

To create gene delivery vectors for mammalian cells, a eukaryotic green fluorescent protein (GFP) expression cassette was inserted into the modified phagemids. Engineered phage were used to infect cells differing for integrin expression, that is HeLa cells that are weakly positive for αv and strongly positive for β1 subtypes, Cs-1/β3 cells displaying high quantities of β1 subtypes, or Cs-1 cells that are β1- and αv-negative (Di Giovine et al., 2001). Consistently with integrin expression, penton base phage were able to transduce GFP in HeLa cells and Cs-1/β3, but not in Cs-1, with a peak of 4% efficiency in HeLa cells. Specificity of transduction was revealed in competition experiments. As shown in Fig. 20.6, the GRGDSP integrin-specific peptide could inhibit phage transduction by 90%, while only partial inhibition was obtained with the GRGESP control peptide. These data reconfirm (1) the use of recombinant phage as gene transfer vectors for mammalian cells, (2) the exportability of penton base properties to a different viral backbone, (3) the use of integrins as targetable molecules.

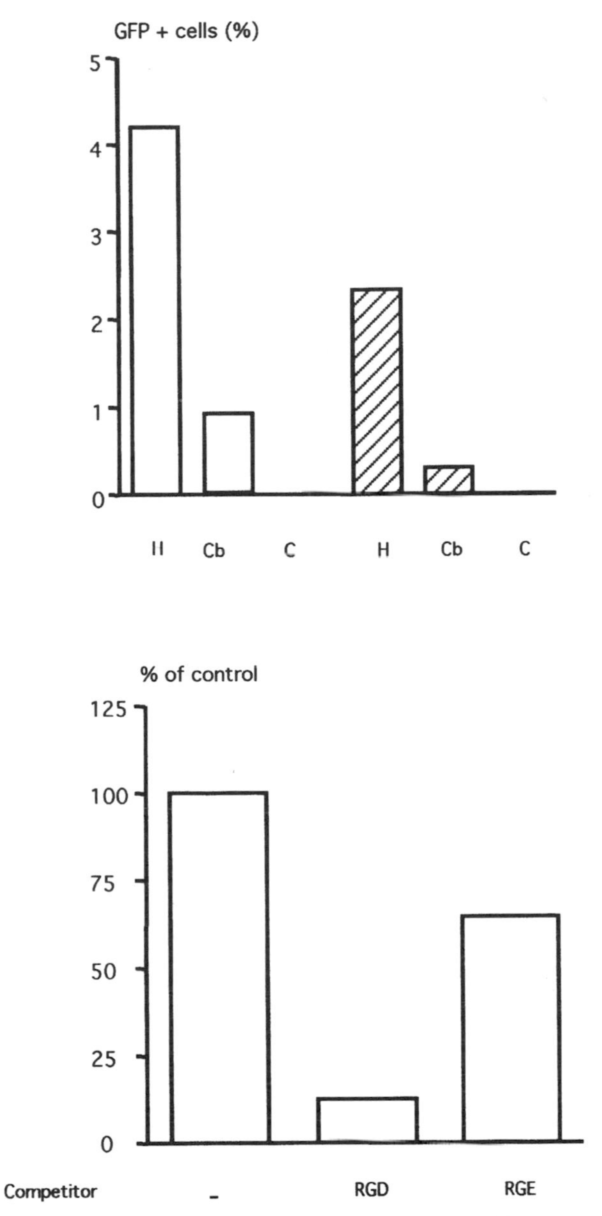

Figure 20.6 Phage-mediated transduction in mammalian cells. HeLa (H), Cs-1/β3 (Cb), and Cs-1(C) cells were incubated with 2×10^{13} particles of Pb-GFP-phage or of ΔPb-GFP-phage. After 72 h, cells were analyzed with FACS. White and striped bars correspond to ΔPb-GFP phage and Pb-GFP-phage infection, respectively. For competition, HeLa cells were preincubated 1 h at 4°C with GRGDSP (RGD) or GRGESP (RGE) peptides 4.86 μM, and then infected with 2×10^{13} particles of ΔPb-GFP-phage.

EVOLUTION OF THE CHIMERIC CONCEPT

Affinity Selection on Integrin Receptors

A major advantage of the phage system is the possibility of evolving functional properties of displayed proteins by panning libraries of variants on specific substrates. The same concept can be applied for evolution of penton base properties in terms of receptor affinity and selectivity for targeting purposes. For proof of concept, we performed micropanning experiments with recombinant and control phage on various cell lines and recovered bound or internalized phage particles. As shown in Table 20.2, it was possible to selectively enrich for recombinant phage on human cell lines, both when the internalized and the bound fractions were analyzed.

The results just described indicate that recombinant phage can be affinity-selected on integrin-expressing cells, that phage can also be recovered intracellularly, and that enrichment yields vary in cell lines differentially expressing integrin receptors. This information suggests the possibility of improving phage penton base properties for specific targeting applications.

Lambda Phage: A New Backbone for Penton Base

Although the penton base/filamentous phage chimera was found to be effective for gene delivery, its efficacy was limited to a maximum of 4% at the conditions tested. A way to improve delivery could be to substitute the filamentous phage backbone with the lambda one. Doing so could offer specific advantages for gene transfer purposes, that is, multivalency, double-stranded genome, the possibility of hosting large peptide inserts, and high genome capacity (Table 20.3). Encouraged by these theoretical possibilities, we constructed recombinant lambda phage displaying the penton base fragment 286–393 as a fusion to the N-terminus of the capsidic protein D. We used a *trans* strategy, that is, we constructed a recombinant phagemid including the penton base-derived peptide fused to the D-encoding gene; this was used to transform bacteria infected with a lambda phage containing the D protein under the control of an amber stop codon (Santi et al., 2000). Lambda phage produced with this

TABLE 20.2 Differences in Phage Enrichment Yields In Vivo

	Binding		Internal	
	Pb	ΔPb	Pb	ΔPb
HeLa	8 ± 3	15 ± 6	9 ± 3	40 + 16
CS-1	1	1	1	1
CS-1/β3	4 ± 1	77 ± 19	2 ± 1	12 ± 3

Note: Cells/well were plated and incubated with 10^{12} phage particles. Results are expressed as fold of enrichment with respect to control phage.

TABLE 20.3 Phage Properties as Gene Delivery Vectors for Mammalian Cells: λ versus M13

	λ Phage	M13 Phage
Genome size (bp)	46000	6400
Particle diameter (nm)	60	6–7
Genome nature	Double-stranded	Single-stranded
Max insert size (amino acids)	1023 (β-galactosidase)	571 (penton base)
Max copy number	400 (D protein)	3000 (pVIII protein
Library display (publications)	1	>100
Particle assembly	Intracellular	Membrane

system are expected to display up to 400 copies of recombinant D/phage. Enzyme-linked immunosorbent assay (ELISA) experiments performed on the integrin receptor $\alpha v\beta 3$ showed that comparable binding results can be observed when using 10^8–10^9 plague-forming units (pfu) of recombinant lambda phage, and 10^{12} recombinant filamentous particles. This significant improvement in affinity/avidity observed for the lambda backbone as compared to the M13 one, is probably related to multivalent display of the penton base peptide.

Taking into account the recombinant lambda phage avidity/affinity properties observed in vitro, and lambda biological properties (see Table 20.3), it is predicted that these viruses could represent a promising vector for targetable gene transduction.

CONCLUSIONS

In this chapter, we have described how it is possible to manipulate filamentous or lambda phage to display exogenous proteins and also how these systems can be used to functionally transfer reporter genes in mammalian cells. Although the results obtained with phage-based vectors are encouraging, a long road has still to be run along to obtain phage-based vectors suitable for gene therapy applications.

A main point to be addressed is efficiency. We have reported that, upon infection with recombinant phage-targeting integrin receptors, only 4% of HeLa cells can be transduced efficiently (Di Giovine et al., 2001). Data from Larocca and collaborators indicate that, at a multiplicity of infection (MOI) of 10^5, only 10% of the cells are transduced through the EGF receptor (Larocca et al., 2001). In the same order of magnitude are the results of Poul and Marks that have described an efficiency of transduction of 4%, after infection of 5×10^4 cells with 10^{12} phage particles directed toward ErbB2 molecules (Poul and Marks, 1999). Although these levels of reporter expression are comparable to those obtained with nonviral vectors, they are definitely much lower than those reported for eukaryotic viral systems (Kremer and Perricaudet, 1995; Robbins, Tahara, and Ghivizzani, 1998).

The process that allows exogenous gene expression in mammalian cells is a multistep pathway that includes vector binding to cell membrane, vector internalization, vector disassembly, DNA entry into the nucleus, double stranded DNA synthesis, DNA transcription, and RNA translation. It is possible that one or more of these steps limit phage-mediated transduction. The experiments of Poul and Marks, of Larocca and collaborators, and also our preliminary results on lambda phage, indicate that multivalent display of receptor ligands reduces the MOI/efficiency ratio. Further amelioration of phage vectors could be obtained through experiments of in vivo selection of high performance cell-binding, internalizing, and transducing chimeras from phage-displayed libraries of ligand variants.

Another cause for phage inefficiency could be the intracellular fate of viral particles. It is likely that internalized phage are directed to the lysosomal pathway and, consequently, mostly degraded. To favor safe delivery of transgene DNA to the nucleus, two approaches can be envisaged. Escape sequences, such those described for influenza virus (Wiley and Skehel, 1987) and adenoviruses (Cotten et al., 1994, 1993), could be exploited to promote active viral exit from the endosomes. This approach has been successfully used for the improvement of nonviral gene transfer system (Wagner, 1999). Another possibility could be to promote DNA migration to the nucleus, by means of shuttle molecules. Dean and collaborators have shown that a specific segment of the SV40 enhancer can, for example, significantly facilitate DNA transport into the nucleus, enhancing transgene expression levels (Li, et al., 2001b; Dean, et al., 1999).

For in vivo applications, another matter to be considered for phage-based vectors is their toxicity and/or immunogenicity. It can be foreseen that capsid proteins will be immunogenic, possibly toxic. To limit the antigenicity phage-based vectors will definitely need to be improved, both in terms of efficacy and of expression stability. To obtain long-term transduction, phage genomes could be engineered with specialized integration sequences derived from eukaryotic viruses, such those present in adeno-associated and retroviral genomes. It is noteworthy that chimeras between adenovirus and retrovirus (Bilbao, et al. 1997), and between baculovirus and adeno-associated virus (Palombo, et al., 1998) have been already successfully tested. In conclusion, it could be speculated that a suitable gene transfer vector could be obtained combining the best from different worlds: the world of eukaryotic viruses, that of bacteriophage, and, that of human fantasy.

ACKNOWLEDGMENTS

I am grateful to B. Salone, Y. Martina, S. Piersanti, G. DiZenzo, M. Di Giovine, G. Zambruno, V. Amati, and C. Failla who contributed to some of the work that I described. I am especially grateful to P. Boulanger who gave permission to use EM images presented in Figure 20.2. I am also thankful to L. Cordier for critical reading of the manuscript. I acknowledge the Istituto Pasteur Fondazione Cenci Bolognetti, the CNR, Progetto Finalizzato Biotecnologie, and the Consorzio Interuniversitario Biotecnologie for support.

REFERENCES

Baranowski E, Riuz-Yarabo C, Domingo E (2001): Evolution of cell recognition by viruses. *Science* 292:1102–1105.

Becerril B, Poul M, Marks, J (1999): Toward selection of internalizing antibodies from phage libraries. *Biochem Biophys Res Comm* 255:386–393.

Bergelson JM, Cunningham JA, Droguett G, Kurt-Jones EA, Krithivas A, Hong JS, Horwitz MS, Crowell RL, Finberg RW (1997): Isolation of a common receptor for Coxsackie B viruses and Adenoviruses 2 and 5. *Science* 275:1320–1323.

Bilbao G, Feng M, Rancourt C, Jackson WJ, Curiel D (1997): Adenoviral/retroviral vector chimeras: a novel strategy to achieve high-efficiency strable transduction in vivo. *FASEB J* 11:624–634.

Campbell A (1996): Bacteriophages. In Fields B, Knipe D, Howley P, eds. *Virology*. Philadelphia: Lippincott Williams & Wilkins.

Cotten M, Wagner E, Zatloukal K, Birnstiel M (1993): Chicken adenovirus (CELO virus) particles augment receptor-mediated DNA delivery to mammalian cells and yield exceptional levels of stable transformants. *J Virol* 67:3777–3785.

Cotten M, Saltik M, Kursa M, Wagner E, Maass G, Birnstiel M (1994): Psoralen treatment of adenovirus particles eliminates virus replication and transcription while maintaining the endosomolytic activity of the virus capsid. *Virology* 205:254–261.

Croyle M, Stone M, Amidon G, Roessler B (1998): In vitro and in vivo assessment of adenovirus 41 as a vector for gene delivery to the intestine. *Gene Ther* 5:645–654.

Davison E, Diaz R, Hart I, Santis G, Marshall J (1997): Integrin $\alpha 5\beta 1$-mediated Adenovirus infection is enhanced by the integrin-activating antibody TS2/16. *J Virol* 71:6204–6207.

Dean D, Dean B, Muller S, Smith L (1999): Sequence requirements for plasmid nuclear import. *Exp Cell Res* 253:713–722.

Devlin J, Panganiban L, Devlin P (1990): Random peptide libraries: A source of specific protein binding molecules. *Science* 249:404–406.

Di Giovine M, Salone B, Martina Y, Amati V, Zambruno G, Cundari E, Failla CM, Saggio I (2001): Binding properties, cell delivery and gene transfer of Adenoviral penton base displaying bacteriophage. *Virology* 282:102–112.

Dmitriev I, Krasnykh V, Miller CMW, Kashentseva E, Mikheeva G, Belousova N, Curiel D (1998): An adenovirus vector with genetically modified fibers demonstrates expanded tropism via utilization of a coxsackievirus and adenovirus receptor-independent cell entry mechanism. *J Virol* 72:9706–9713.

Dunn I (1995): Assembly of functional bacteriophage lambda virions incorporating C-terminal peptide or protein fusions with the major tain protein. *J Mol Biol* 248:497–506.

Dunn I (1996): Mammalian cell binding and transfection mediated by surface-modified bacteriophage lambda. *Biochimie* 78:856–871.

Eguchi A, Akuta H, Senda T, Yokoi H, Inokuchi H, Fujita S, Hayakawa T, Takeda K, Hasegawa M, Nakanishi M (2001): Protein transduction domain of HIV-1 Tat protein promotes efficient delivery of DNA into mammalian cells. *J Biol Chem* 276:26204–26210.

Felici F, Castagnoli L, Musacchio A, Jappelli R, Cesareni G (1991): Selection of antibody ligands from a large library of oligopeptides expressed on a multivalent exposition vector. *J Mol Biol* 222:301–310.

Felici F, Luzzago A, Monaci P, Nicosia A, Sollazzo M, Traboni C (1995): Peptide and protein display on the surface of filamentous bacteriophage. *Biotechnol Ann Rev* 1:149–183.

Ferguson N, Donnelly C, Anderson R (2001): The foot-and-mouth epidemic in Great Britain: pattern of spread and impact of interventions. *Science* 292:1155–1160.

Folgori A, Tafi R, Meola A, Felici F, Galfré G, Cortese R, Monaci P, Nicosia A (1994): A general strategy to identify mimotopes of pathological antigens using only random peptide libraries and human sera. *EMBO J* 13:2236–2243.

Greber U, Willets M, Webster P, Helenius A (1993): Stepwise dimantling of Adenovirus 2 during entry into cells. *Cell* 75:477–486.

Hart S, Knight A, Harbottle R, Mistry A, Hunger H, Cutler D, Williamson R, Coutelle C (1994): Cell binding and internalization by filamentous-phage displaying a cyclic Arg-Gly-Asp containing peptide. *J Biol Chem* 269:12468–12474.

Hart S, Collins L, Gustafsson K, Fabre J (1997): Integrin-mediated transfection with peptides containing arginine-glycine-aspartic acid domains. *Gene Ther* 4:1225–1230.

Hewish M, Takada Y, Coulson B (2000): Integrins alpha2beta1 and alpha4beta1 can mediated rotavirus attachment and entry into cells. *J Virol* 74:228–236.

Hogenboom HR, Griffiths AD, Johnson KS, Chiswell DJ, Hudson P, Winter G (1991): Multi-subunit proteins on the surface of filamentous phage: methodologies for displaying antibody (Fab) heavy and light chains. *Nucleic Acid Res* 19:4133–4137.

Hong S, Gay B, Karayan L, Dabauvalle M, Boulanger P (1999): Cellular uptake and nuclear delivery of recombinant adenovirus penton base. *Virology* 262:163–177.

Huang S, Kamata T, Takada Y, Ruggeri Z, Nemrow G (1996): Adenovirus interaction with distinct integrins mediates separate events in cell entry and gene delivery to hematopoietic cells. *J Virol* 70:4502–4508.

Humphries M (2000): *Integrin Structure.* Biochemical Society, pp. 311–340.

Ivanenkov V, Felici F, Menon A (1999): Targeted delivery of multivalent phage display vectors into mammalian cells. *Biochim Biophys Acta* 1448:463–472.

Jackson TY, Sharma A, Ghazaleh R A, Blakemore WE, Ellard FM, Simmons DL, Newman JWI, Stuart DI, King MQ (1997): Arginine-glycine-aspartic acid-specific binding by foot-and-mouth disease viruses to the purified integrin $\alpha v\beta 3$ in vitro. *J Virol* 71:8357–8361.

Jackson T, Sheppard D, Denyer M, Blakemore W, King A (2000): The epithelial integrin $\alpha v\beta 6$ is a receptor for foot-and-mouth disease virus. *J Virol* 74:4949–4956.

Jenkins R, Herrick S, Meng Q, Kinnon C, Laurent G, McAnulty R, Hart S (2000):

An integrin-targeted non-viral vector for pulmonary gene therapy. *Gene Ther* 7:393–400.

Karayan L, Hong SS, Gay B, Tournier J, Angeac AD, Boulanger P (1997): Structural and functional determinants in Adenovirus Type 2 penton base recombinant protein. *J Virol* 71:8678–8679.

Kirby I, Lord R, Davison E, Wickham T, Roelvink P, Kovesdi I, Sutton B, Santis G (2001): Adenovirus type 9 fiber knob binds to the Coxsackie B virus-Adenovirus receptor (CAR) with lower affinity than fiber knobs of other CAR-binding Adenovirus serotypes. *J Virol* 75:7210–7214.

Kochaneck S, Clemens PR, Mitani K, Chen HH, Chan S, Caskey CT (1996): A new adenoviral vector: replacement of all viral coding sequences with 28 kb of DNA independently expressing both full length dystrophin and beta-galactosidase. *Proc Natl Acad Sci USA* 93:5731–5736.

Kremer E, Perricaudet M (1995): Adenovirus and adeno-associated virus mediated gene transfer. *Brit Med Bull* 51:31–44.

Larocca D, Witte A, Johnoson W, Pierce GF, Baird A (1998): Targeting bacteriophage to mammalian cell surface receptors for gene delivery. *Hum Gene Ther* 9:2393–2399.

Larocca D, Kassner P, Witte A, Ladner R, Pierce G, Baird A (1999): Gene transfer to mammalian cells using genetically targeted filamentous bacteriophage. *FASEB J* 6:727–734.

Larocca D, Jensen-Pergakes K, Burg M, Baird A (2001): Receptor-targeted gene delivery using multivalent phagemid particles. *Molecular Ther* 3:476–484.

Li E, Stupack D, Klemke R, Cheresh D, Nemerow G (1998a): Adenovirus endocytosis via ?v integrins requires phosphoinositide-3–OH kinase. *J Virol* 72:2055–2061.

Li E, Stupak D, Bokoch G, Nemerow G (1998b): Adenovirus endocytosis requires actin cytoskeleton reorganization mediated by the Rho family GTPases. *J Virol* 72:8806–8812.

Li E, Stupack D, Brown S, Klemke R, Shlaepfer D, Nemerow G (2000): Association of p130CAS with phosphatidylinositol3–OH kinase mediates Adenovirus cell entry. *J Biol Chem* 275:14729–14735.

Li E, Brown S, Stupack D, Puente X, Cheresh D, Nemerow G (2001a): Integrin $\alpha v\beta 1$ is an Adenovirus coreceptor. *J Virol* 75:5405–5409.

Li S, MacLaughlin F, Fewell J, Gondo M, Wang J, Nicol F, Dean D, Smith L (2001b): Muscle-specific enhancement of gene expression by incorporation of SV40 enhancer in the expression plasmid. *Gene Ther* 8:494–497.

Lochmüller H, Jani H, Huard J, Prescott S, Simoneau M, Massie B, Karpati GG (1994): Emergence of early region 1–containing replication-competent adenovirus in stocks of replication-defective adenovirus recombinants (delta E1 + delta E3) during multiple passages in 293 cells. *Hum Gene Ther* 5:1485–1492.

Lowman HB, Wells JA (1993): Affinity maturation of human growth hormone by monovalent phage display. *J Mol Biol* 234:564–578.

Mathias P, Galleno M, Nemerow G (1998): Interactions of soluble recombinant integrin $\alpha v\beta 5$ with human Adenovirus. *J Virol* 72:8669–8675.

Mathias P, Wickham T, Moore M, Nemerow G (1994): Multiple adenovirus serotypes use α-v integrins for infection. *J Virol* 68:6811–6814.

Medina-Kauwe LK, Kasahara N, Kedes L (2001): 3PO, a novel nonviral gene delivery system using engineered Ad5 penton proteins. *Gene Ther* 8:795–803.

Mehir K, Armentano D, Cardoza L, Choquette T, Berthelette P, White G, Couture L, Everton M, Keegan J, Martin J, Pratt D, Smith M, Smith A, Wadsworth S (1996): Molecular characterization of replication-competent variants of Adenovirus vectors and genome modifications to prevent their occurrence. *J Virol* 70:8459–8467.

Mikawa Y, Maruyama I, Brenner S (1996): Surface display of proteins on bacteriophage lambda heads. *J Mol Biol* 262:21–30.

Palombo F, Monciotti A, Recchia A, Cortese R, Ciliberto G, La Monica N (1998): Site-specific integration in mammalian cells mediated by a new hybrid baculo-adeno-associated virus vector. *J Virol* 72:5025–5034.

Pasqualini R, Ruoslahti E (1996): Organ targeting in vivo using phage display peptide libraries. *Science* 380:364–366.

Plow E, Haas T, Zhang L, Loftus J, Smith J (2000): Ligand binding to integrins. *J Biol Chem* 275:21785–21788.

Poul M, Marks J (1999): Targeted gene delivery to mammalian cells by filamentous phage. *J Mol Biol* 288:203–211.

Rader C, Barbas C III (1997): Phage display of combinatorial antibody libraries. *Curr Opin Biotechnol* 8:503–508.

Rasched I, Oberer E (1986): Ff coliphages: Structural and functional relationships. *Microbiol Rev* 50:401–427.

Rauma T, Tuukkanen J, Bergelson J, Denning G, Hautala T (1999): rab5 GTPase regulates Adeonvirus endocytosis. *J Virol* 73:9664–9668.

Robbins P, Tahara H, Ghivizzani S (1998): Viral vectors for gene therapy. *Trends Biotechnol* 16:35–40.

Roelvink PW, Lizonova A, Lee JGM, Li Y, Bergelson JM, Finberg RW, Brough DE, Kovesdi I, Wickham TJ (1998): The Cocksackievirus-Adenovirus receptor protein can function as a cellular attachment protein for adenovirus serotypes from subgroups A,C, D, E, and F. *J Virol* 72:7909–7915.

Russell S, Cosset F (1999): Modifying host range properties of retroviral vectors. *J Gene Med* 1:300–311.

Saggio I, Laufer R (1993): Biotin binders selected from a random peptide library expressed on phage. *Biochem J* 293:613–616.

Saggio I, Gloaguen I, Poiana G, Laufer R (1995): CNTF variants with increased biological potency and receptor selectivity define a functional site of receptor interaction. *EMBO J* 14:3045–3054.

Sandig V, Youil R, Bett A, Franlin L, Oshima M, Maione D, Wang F, Metzker M, Savino R, Caskey C (2000): Optimization of the helper-dependent adenovirus system for production and potency in vivo. *Proc Natl Acad Sci USA* 97:1002–1007.

Santi E, Capone S, Mennuni C, Lahm A, Tramontano A, Luzzago A, Nicosia A (2000): Bacteriophage lambda display of complex cDNA libraries: a new approach to functional genomics. *J Mol Biol* 296:497–508.

Scott J, Smith G (1990): Searching for peptide ligands with an epitope library. *Science* 249:386–390.

Scott J, Loganathan D, Easley R, Gong X, Goldstein I (1992): A family of concanavalin

A-binding peptides from a hexapeptide epitope library. *Proc Natl Acad Sci USA* 89:5398–5402.

Shenk T (1996): Adenoviridae: The viruses and their replication. In Fields B, Knipe D, Howley P, eds. *Virology*. Philadelphia; Lippincott Williams & Wilkins.

Sternberg N, Hoess R (1995): Display of peptides and proteins on the surface of bacteriophage λ. *Proc Natl Acad Sci USA* 92:1609–1613.

Stewart P, Chiu C, Huang S, Muir T, Zhao Y, Chait B, Mathias P, Nemerow G (1997): Cryo-EM visualization of an exposed RGD epitope on adenovirus that escapes antibody neutralization. *EMBO J* 16:1189–1198.

Summerford C, Bartlett J, Samulski R (1999): AlphaVbeta5 integrin: a co-receptor for adeno-associated virus type 2. *Nat Med* 5:78–82.

Tran Van Nhieu G, Isberg R (1993): Bacterial internalization mediated by β1 chain integrins is determined by ligand affinity and receptor density. *EMBO J* 12:1887–1895.

Trapnell BC (1993): Adenoviral vectors for gene transfer. *Adv Drug Deliv Rev* 12:185–199.

Triantafilou K, Triantafilou M, Takada Y, Fernandez N (2000): Human parechovirus 1 utilizes integrin alphavbeta3 and alphavbeta1 as receptors. *J Virol* 74:5856–5862.

Vigne E, Mahfouz I, Dedieu J, Brie A, Perricaudet M, Yeh P, (1999): RGD inclusion in the hexon monomer provides adenovirus type 5–based vectors with a fiber knob-independent pathway of infection. *J Virol* 73:5156–5161.

Wagner E (1999): Application of membrane-active peptides for non viral gene delivery. *Adv Drug Deliv Rev* 38:279–289.

Wang K, Huang SA, Kapoor-Munshi A, Nemerow G (1998): Adenovirus internalization and infection require dynamin. *J Virol* 72:3455–3458.

Watson K, Edwards R (1999): HIV-1-transactivating (Tat) protein: both a target and a tool in therapeutic approaches. *Biochem Pharmacol* 58:1521–1528.

Wickham TJ, Mathias P, Cheresh DA, Nemerow GR (1993): Integrins $\alpha v\beta 3$ and $\alpha v\beta 5$ promote adenovirus internalization but not virus attachment. *Cell* 73:309–319.

Wickham T, Filardo J, Cheresh D, Nemerow G (1994): Integrin $\alpha v\beta 5$ selectively promotes adenovirus mediated cell membrane permealization. *J Cell Biol* 127:257–264.

Wiley D, Skehel J (1987): The structure and function of the hemagglutinin membrane glycoprotein of influenza virus. *Ann Rev Biochem* 56:365–394.

Yang Y, Ertl HC, Wilson JM (1994): MHC class I-restricted cytotoxic T lymphocytes to viral antigens destroy hepatocytes in mice infected with E1–deleted recombinant adenoviruses. *Immunity* 1:433–442.

Zoltick P, Chirmule N, Schnell M, Gao G, Hughes J, Wilson JM (2001): Biology on E1–deleted Adenovirus vectors in nonhuman primate muscle. *J Virol* 75:5222–5229.

PART III

TRANSCRIPTIONAL TARGETING

21

TUMOR/TISSUE-SELECTIVE PROMOTERS

MARIO FERNANDEZ, PH.D. AND NICK LEMOINE, PH.D.

INTRODUCTION

Cancer is a disease that it is becoming increasingly common in modern society, and conventional therapeutic approaches are inadequate to manage the problem completely. In the normal situation, the cells of the human body have the capacity to duplicate when required to maintain homeostasis. However, when the control mechanisms that maintain the cell cycle are lost, the affected cell is able to keep dividing to form a tumor mass, which may then spread to distant organs. This characteristic makes cancer one of the most difficult diseases to treat, because it can appear in almost any tissue. With conventional cancer therapy (chemotherapy and radiotherapy) the lack of specificity may cause nondesired side effects, which in turn restrict the dose that can be safely administered, and finally compromise the outcome of the treatment. The introduction of gene therapy for cancer has the potential for a great improvement, because it aims to eliminate the origin of the disease. However, one of the main questions in the design of an effective gene therapy is how to target the desired cells without causing any damage to normal cells. This issue has attracted a remarkable effort by a great number of investigators. Targeting can be achieved at two levels: at the point of delivery by modifying or using the natural tropism of the vector, and at the transcriptional level by restricting expression of the therapeutic gene to selected cells. With respect to the first approach, many investi-

Vector Targeting for Therapeutic Gene Delivery, Edited by David T. Curiel and Joanne T. Douglas
ISBN 0-471-43479-5

gators have exploited delivery targeting through receptor-mediated interactions with membrane molecules (Barsov, Payne, and Hughes, 2001; Etienne-Julan et al., 1992; Pereboev, Pereboeva, and Curiel, 2001; Suzuki et al., 2001; Vyas, Singh, and Sihorkar, 2001; Wesseling et al., 2001 and chapters 1–20 of this book). The final goal with this targeting is to be able to administer safely and systemically the modified viruses to cancer cells. However, more specific and restricted targets must be found for this approach to be fully effective at the clinic (Galanis, Vile, and Russell, 2001; Wickham, 2000).

Over the past decade enormous strides have been made in understanding the molecular mechanisms that govern the transcriptional events in a normal cell and those that convert a normal cell into a malignant one. Transcription in eukaryotic cells is the process by which the information contained in a DNA sequence (e.g., a gene) is transcribed into messenger RNA. This process is highly regulated and involves the enzymatic activity of RNA polymerase II together with several other proteins called transcription factors. The initiation of transcription occurs when the RNA polymerase II binds a specific region in a position 5′ to the ATG start codon of the gene to be transcribed; this region is called the promoter. In eukaryotic cells, the promoter region is characterized by the presence of four consensus sequences that can be found in almost all the promoters studied: a TATA box normally located at position −30 with respect to the start of the sequence from which the transcription takes place; a CAAT box situated at position −80; GC boxes placed at several positions; and finally an octamer box also placed at different positions depending of the promoter. All these sequences increase the rate of transcription of the corresponding gene and are necessary for the correct positioning of the RNA polymerase II complex over the ATG sequence. However, additional sequences can affect the rate of transcription either positively (called enhancers) or negatively (called repressors), and in addition, thousands of genes are also transcriptionally regulated by environmental signals (stress, hypoxia, hormones, etc.) through the presence of sequences named response elements. All these types of sequences present in the transcriptional regulatory elements of a gene are able to bind specific proteins that help the RNA polymerase II to initiate the transcription and synthesize protein.

It is now recognized that dysregulated transcription is a major contributor to the activation of several classes of genes involved in tumorigenesis. When the expression of a gene is modified in a malignant cell, its regulatory sequences might be exploited for targeting the expression of a therapeutic construct. With this goal in mind, much research effort has been applied to the study of promoters (Tables 21.1 and 21.2) that are tissue- or tumor-specific (Harrington, Linardakis, and Vile, 2000; Miller and Whelan, 1997; Nettelbeck, Jerome, and Muller, 2000). But from a practical point of view, reality is not so "ideal." In some cases, the promoters lose their tissue specificity in the context of the delivery system, and in other cases the activity of the promoter is too weak to be therapeutically useful. For all of these reasons, transcriptional targeting is becoming an issue of great interest as improvements over existing designs are being developed, as will be discussed in this chapter.

TABLE 21.1 Tissue-Specific Promoters Used for Gene Therapy

Promoter	Specificity	Reporter/ Therapeutic Gene	Tested In Vivo	Reference
Tyrosinase	Melanocyte	+/+	Yes	Siders, Halloran and Fenton, 1998; Vile and Hart, 1993a
Prostate-specific antigen (PSA)	Prostate cells	+/+	Yes	Martiniello-Wilks et al., 1998
Kallikrein 2	Prostate cells	+/+	Yes	Yu, Sakamoto, and Henderson, 1999
DF3/MUC1	Epithelial cells	+/+	Yes	Chen et al., 1995
Mouse amylase	Pancreas	+/−	Yes	Dematteo et al., 1997
Neuron-specific enolase	Neuron cells	−/+	Yes	Morelli et al., 1999
Albumin	Liver	+/+	Yes	Miyatake et al., 1999
Muscle creatinine kinase	Skeletal muscle cells	+/−	Yes	Larochelle, et al., 1997
Thyroglobulin	Thyroid gland	−/+	No	Shimura, et al., 2001
Surfactant protein B	Bronchial and alveolar cells	−/+	Yes	Doronin, et al., 2001
Murine preproendothelin-1	Angiogenic endothelial cells	+/−	Yes	Varda-Bloom et al., 2001

TABLE 21.2 Tumor-Specific Promoters Used for Gene Therapy

Promoter	Specificity	Reporter/ Therapeutic Gene	Tested In Vivo	Reference
Carcinoembryonic antigen (CEA)	Several adenocarcinomas	+/+	Yes	Konishi et al., 1999; Takeuchi et al., 2000
Alphafetoprotein (AFP)	Liver cancer	+/+	Yes	Hirano et al., 2001
HER2/ERBB2	Breast, pancreas, and gastrointestinal cancer	−/+	Yes	Harris et al., 1994
Telomerase	Almost any kind of cancer	+/+	Yes	Majumdar et al., 2001
Hypoxia-responsive element (HRE)	Solid tumors	+/-	No	Boast et al., 1999
Grp78	Solid tumors	−/+	Yes	Chen et al., 2000a
Hexoquinase II	Solid tumors	+/−	No	Katabi et al., 1999
Osteocalcin	Bone tumors	+/+	Yes	Cheon et al., 1997
MidKine (MK)	Ovarian cancer	+/+	Yes	Casado et al., 2001
Cyclooxygenase-2	Ovarian cancer	+/+	Yes	Casado et al., 2001

TISSUE-SPECIFIC PROMOTERS

Tyrosinase

Melanogenesis is tightly regulated at the transcriptional level. This control is exerted through the tyrosinase gene promoter and the promoters of the genes of the tyrosinase-related proteins TRP-1 and TRP-2 (Korner and Pawelek, 1982; Mason, 1948). The expression of melanin is mainly restricted to pigment cells called melanocytes in the epidermis although its expression has been detected also in cells of non-melanocyte origin (Bouchard et al., 1989; Guida et al., 2000). The study of the 5′ flanking regions of the genes for tyrosinase, TRP-1 and TRP-2 proteins in mouse and human has revealed different elements that make the expression of these genes so restricted (Jackson et al., 1991; Kluppel et al., 1991; Ruppert et al., 1988; Shibata et al., 1992a, 1992b). A series of enhancer elements has also been described (Diaz et al., 1998; Shibata et al., 1992a). For this reason the tyrosinase promoter has been proposed as a good candidate to use in gene therapy. The first studies made with the tyrosinase and TRP-1 promoters demonstrated tight regulation of the β-galactosidase reporter gene in cells from melanocytic origin in comparison with a broad range of other cancer cell lines (Vile and Hart, 1993a). Thereafter, this promoter has been used in vitro and in vivo to drive transcription of the thymidine kinase (TK) gene (Siders, Halloran, and Fenton, 1998; Vile and Hart, 1993b). Another group had shown interesting results using interleukin-2 (IL-2), IL-4, or macrophage-colony-stimulating factor (M-CSF) (Vile and Hart, 1994) and a modified version of the receptor for tumor necrosis factor alpha (TNFα) (Bazzoni and Regalia, 2001) under the control of the mouse tyrosinase promoter. In the latter case, the mouse promoter showed a very restricted pattern of expression, thus reducing the potential risk of side effects. Only very low levels of the reporter gene were found in cells of endothelial, lung, and liver origin, which are primary targets for the adenovirus used as vector of the therapeutic gene. Also, it has been used to drive the cytosine deaminase (CD) cassette (Cao et al., 1999c) and *Escherichia coli* purine nucleoside phosphorylase (PNP) (Hughes et al., 1995; Park et al., 1999) as therapeutic genes. The PNP enzyme/prodrug system needs to be tightly regulated because of its potency. The use of the tyrosinase promoter with this system has proved to be a very good candidate for restricting expression only to melanoma cells, but has also demonstrated the importance of using specific enhancer sequences when the expression from the promoter is weak. In every case the specificity and the activity of the tyrosinase promoter have proved to be very reliable in these different constructs, as well as relatively strong when enhancer sequences are added [about 40% of the cytomegalovirus (CMV) promoter in transient reporter assays] and also effective in vitro and in vivo.

Prostate-Specific Antigen and Human Kallikrein 2

Prostate cancer is one of the most common malignant diseases worldwide and its incidence has been increasing since the introduction of screening tests to detect it (Barry 2001; Hsing, Tsao, and Devesa, 2000). Presently, the best marker of this disease is an abnormal level of the prostate-specific antigen (PSA). This protein belongs to a family of serine proteases called kallikreins, which is composed of three members termed hK1, hK2, and hK3 (PSA) (Clements 1989, 1994; Young et al., 1996). The detection of PSA has been traditionally used as a marker of this type of cancer, but recently studies have shown that the presence of hK2 is also a good marker of the disease and correlates positively with cancer grade and progression (Darson et al., 1997; Slawin et al., 2000). Both proteins are expressed in normal and transformed prostate epithelium (Young, Andrews, and Tindall, 1995), and regulated by androgens (Young, Andrew, and Tindall, 1995). Its 5′ flanking regions have been intensively studied (Mitchell et al., 2000; Murtha, Tindall, and Young, 1993; Riegman et al., 1991, 1989; Yu et al., 1999b) and regions identified that are responsive to androgen, and also several consensus binding sites for common transcription factors (Yu et al., 1999b). The PSA promoter has been used to drive several different therapeutic genes. In one of the first studies, Lee and colleagues used antisense sequences against DNA polymerase-α and topoisomerase II α expressed under the control of this promoter, demonstrating a cytotoxic effect restricted to cell lines of prostate origin (Lee et al., 1996). In another study, PNP and TK genes (Martiniello-Wilks et al., 1998) were inserted using the same adenoviral vector and tested in vivo. The same constructs were first compared in vitro, and it was shown that the PNP system was much better at causing cell death than the TK system (Lockett et al., 1997). But surprisingly, both therapeutic genes reduced tumor growth with respect to controls by 75% and 80% respectively, and also extended the survival of the treated mice (Martiniello-Wilks et al., 1998). Also, p53 (Lee et al., 2000), diphtheria toxin A (DT-A) (Pang, 2000), and CD genes (Uchida et al., 2001) have been assayed in vitro and in vivo with good results. In all these cases, the specificity of the promoter has been proved, although the transcriptional activity without an enhancer is weak (Segawa et al., 1998) with respect to strong promoters like CMV. However, in the presence of appropriate enhancers, its activity increases significantly (Uchida et al., 2001). Of interest are the studies currently ongoing with the adenovirus CN706 in which the E1A protein is driven by the PSA enhancer-promoter (Chen et al., 2000b; Rodriguez et al., 1997; Simons et al., 2000). Preliminary results show that this adenovirus is well tolerated and has tumoricidal activity.

With respect to hK2, the use of the promoter is in an early stage of research but it is also demonstrating good targeting specificity to prostate tissue in vitro and in vivo. Although the activity of the promoter is weak, the use of enhancers has greatly increased the expression of reporter genes while the promoter retains its specificity, making it a promising candidate to exploit in gene therapy design (Xie et al., 2001). Also noteworthy are the results obtained with the attenuated replication-competent adenovirus CV764 (Yu et al., 1999b). This virus is similar to the CN706 virus but in this case a further level of transcriptional control is introduced by placing the E1B protein under the control of the hK2 promoter. This adenovirus has proved to be more specific for PSA-positive cells than the original design and it is now being improved for use in gene therapy for prostate cancer (Yu et al., 1999a).

DF3/MUC1 Antigen

The human MUC1 antigen is a polymorphic epithelial mucin expressed mainly at the apical surface of epithelial cells lining the glands or ducts of the stomach, pancreas, lungs, trachea, kidney, uterus, and salivary and mammary glands (Girling et al., 1989; Zotter et al., 1988). This protein belongs to a family of eight members, which share common characteristics such as high molecular weight and the presence of amino acid residues that allow O-glycosylation of the proteins (Gendler et al., 1987; Siddiqui et al., 1988). MUC1 is overexpressed in human breast and other carcinomas (Abe and Kufe 1987; Friedmanm, Hayes, and Kufe, 1986) due to transcriptional up regulation (Abe and Kufe 1990; Graham et al., 2001). This has attracted attention to the sequences of the MUC1 gene that regulate its cell type-specific transcription (Abe and Kufe 1993; Graham et al., 2001; Kovarik et al., 1996, 1993; Morris and Taylor-Papadimitriou 2001; Zaretsky et al., 1999). An important study demonstrated the selectivity that an enhancer sequence in the MUC1 promoter (Abe and Kufe, 1993) confers to a minimal generic promoter (Manome et al., 1994). Positive and negative MUC1 human cancer cell lines were transfected with a construct containing the TK gene under the control of its own minimal promoter. The MUC1-positive cells showed a marked sensitivity (0% survival) to high concentrations of ganciclovir (10 μg/ml GCV) in comparison with cells that do not express MUC1 (70% survival). Also this construction has been validated in vivo using an adenovirus as delivery system for the same therapeutic cassette (Chen et al., 1995). In addition, the same group has engineered an adenovirus in which the E1A protein is driven by the MUC1 promoter (Kurihara et al., 2000). This modified adenovirus was injected into tumor xenografts of MUC1-positive cells causing inhibition of tumor growth restricted to these positive cells, which indicates the safety of this promoter. Moreover, it was shown that this construction was able to accommodate additionally the TNF-α gene under the control of a CMV promoter, increasing the therapeutic index. The MUC1 promoter has been also used in the context of adenoviral vectors to transfer genes for receptors to tumor cells so that they are able to bind radiolabeled peptides with good results (Stackhouse et al., 1999). Finally, this promoter has been introduced into retroviral vectors to drive the TK gene in vitro (Ring et al., 1997). In this work, the MUC1 enhancer region was combined with the erbB2 promoter that enhanced the sensitivity of MUC1-positive cells to the toxic action of GCV. This result shows that it is possible to achieve a tighter specificity to a particular type of cancer knowing at the molecular level which genes are upregulated and using this knowledge to design better vectors.

Other Tissue-Specific Promoters

A great effort is being made to understand the global gene transcriptional profiles that characterize various tissues and disease states. So the number of available promoters specific for a determined tissue is growing constantly. In the pancreas, the activity of the mouse amylase promoter has been analyzed (Dematteo et al., 1997). In this work it has been shown that the amylase promoter restricted the expression of a reporter gene to pancreatic acinar cells and also retained the capacity to be inducible in an adenoviral vector. Also the rat neuron-specific enolase promoter is being studied with promising results (Andersen et al., 1993; Navarro et al., 1999; Peel et al., 1997). In one of these studies, the promoter driving the Fas ligand gene in an adenoviral vector shows a very restricted pattern of expression in vitro, and more importantly in vivo with no toxicity in comparison with the same construction using the CMV promoter (Ambar et al., 1999; Morelli et al., 1999). Also the albumin promoter has been studied as a tool to direct therapeutic genes to the liver used in xenograft mouse models with success (Cao et al., 1997; Hafenrichter et al., 1994; Miyatake et al., 1997; Miyatake et al., 1999).

TUMOR-SPECIFIC PROMOTERS

Carcinoembryonic Antigen

The carcinoembryonic antigen (CEA) was first described in 1965 by Gold and Freedman (Gold and Freedman, 1965). It was found to be associated with cancer and it is a useful marker of the diagnosis and prognosis of the disease (Graham et al., 1998). CEA is a highly glycosylated protein that belongs to a family of 12 genes of which only 7 appear to be expressed (Hammarstrom, 1999). The CEA protein is expressed during the early stages of development and thereafter during adult life is restricted to colon, stomach, and some other organs (Nap et al., 1992, 1988; Prall et al., 1996). However, it is overexpressed in tumors of epithelial origin such as colorectal and gastric carcinomas, or lung adenocarcinoma (Robbins et al., 1993; Shi, Tacha, and Itzkowitz, 1994; Thompson et al., 1993). The promoter of the CEA gene has been isolated and analyzed extensively (Schrewe et al., 1990; Willcocks and Craig 1990). It is not a classic promoter in the sense that it does not have TATA or CAAT boxes (Schrewe et al., 1990), but the organization of the promoter has been elucidated (Hauck and Stanners, 1995). It has been shown that its transcriptional activity is similar to the activity of a strong promoter such as SV40 (Cao et al., 1998). In terms of gene therapy, the CEA promoter has been widely applied in several types of cancer. In adenocarcinoma of the lung it has been assayed in combination with the TK gene in vitro in colon and lung cancer cells, and in vivo with tumor xenografts in mouse models (Konishi et al., 1999; Osaki et al., 1994). Moreover, it has been tested using pancreatic carcinoma cell lines in vitro as in vivo (DiMaio et al., 1994; Ohashi et al., 1998). Altogether, these studies have shown the feasibility of the CEA promoter to direct a suicide gene therapy with the TK gene to CEA-expressing cancer cells. Recently, one group used a dominant negative H-ras mutant as a therapeutic gene driven by the CEA promoter (Takeuchi et al., 2000). This construction was able to inhibit the growth of pancreatic tumor cells in nude mice without damaging the liver.

The suitability of the CEA promoter in gastric cancer has also been tested with different therapeutic genes. An adenovirus containing the CEA promoter driving the TK gene led to a reduction of tumor growth and an efficient bystander killing effect in vivo in mouse models (Tanaka et al., 1997). Moreover, its usefulness has been demonstrated driving CD as a therapeutic gene in gastric cancer cells in vitro (Lan et al., 1996). More recently, the same group has shown, using the same construction, that in a mouse model with intraperitoneally disseminated MKN45 tumors transduction with the CD gene combined with treatment with the prodrug 5-fluorocytosine (5FC) increased the survival of the mice (Lan et al., 1997). Interestingly, the CEA promoter has also been used to develop a transgenic mouse model of gastric carcinoma (Thompson et al., 2000) using the SV40 T antigen. Finally, similar constructions have been already tested for colorectal carcinoma (Brand et al., 1998; Cao et al., 1999a, 1999b). A recent study has used a retroviral vector that contains the CEA promoter driving the CD gene (Humphreys et al., 2001). This construction has been shown to sensitize rat K12 colonic adenocarcinoma cells to the effects of the prodrug 5FC, but also that in vivo administration via the hepatic artery of the retrovirus followed by the injection of the prodrug was able to suppress the previously established liver metastases.

Alpha-Fetoprotein

Together with CEA, the promoter of the α-fetoprotein (AFP) gene is one of the most extensively used in gene therapy. The protein is most abundant during the early stages of development and its expression is transcriptionally downregulated after birth (Peyton, Ramesh, and Spear, 2000). But in cancer, AFP is upregulated by mechanisms that are not fully understood (Spear, 1999). This profile has attracted effort in the identification of the sequences that govern AFP transcription, and improvement of its weak transcriptional activity (Watanabe, Saito, and Tamaoki, 1987). The promoter has been used with therapeutic genes (TK and CD genes) (Hirano et al., 2001; Kanai et al., 1997) with excellent results. It was shown that in vivo the therapeutic genes were able to inhibit the growth of hepatic tumors even when the vector was administered via the portal vein of the mice (Hirano et al., 2001). Also, replication-selective adenoviruses have been constructed with this promoter (Hallenbeck et al., 1999; Ohashi et al., 2001). In this case, essential genes for the life cycle of the virus, including E1A and E1B, were placed under the transcriptional control of the AFP promoter, resulting in viruses that were only able to replicate in cells that express the AFP protein (e.g., tumor cells). In each of these studies it was shown that this approach is able to reduce the tumor growth without affecting surrounding cells, and in one of the studies the systemic administration of the adenoviral construction was demonstrated to be nontoxic for the host animals (Ohashi et al., 2001).

HER2/ERBB2

HER2 is a proto-oncogene that encodes a transmembrane protein implicated in the control of cell proliferation (De Placido et al., 1998). This gene is commonly overexpressed in carcinomas of the breast, pancreas, and gastrointestinal tract (Ring et al., 1996). Its selectivity has been demonstrated using the CD gene against a panel of positive and negative HER2 expressing cell lines (Harris et al., 1994) and this cassette has been already used in a phase I clinical trial for patients with cutaneous metastases of HER2-positive breast cancer (Pandha et al., 1999).

Telomerase Promoter

As the knowledge of how eukaryotic cells replicate their DNA has grown, new candidates for exploitation in gene therapy have appeared. A complex called telomerase, located at the telomere sequences, is responsible for compensating against the incomplete replication of chromosome ends (Greider and Blackburn, 1985). This complex is composed of a protein called hTERT (human telomerase reverse transcriptase) with the polymerase activity and an RNA named hTR (human telomerase RNA) that acts like a primer (Greider and Blackburn, 1989). What has attracted a lot of interest in the study of this complex is the fact that its activity is present in almost 95% of cancer cells examined as well as in highly proliferative cells like germ-line cells and hematopoietic cells but not in human somatic cells (Poole, Andrews, and Tollefsbol, 2001). This characteristic makes the telomerase components themselves very promising candidates to use as targets of different therapeutic approaches against cancer, but also the sequences that regulate their expression (Elenitoba-Johnson, 2001; Meyerson, 2000). The promoters of the hTR and hTERT genes have been cloned and analyzed (Cong, Wen, and Bacchetti, 1999; Feng et al., 1995). Several studies have shown that the rate-limiting component of the telomerase activity is the hTERT enzyme rather than the hTR, because the mRNA for the hTR subunit is found in both telomerase activity-positive and -negative tissues, while the mRNA for the hTERT protein is only found in immortal cells (Hahn and Meyerson, 2001). The hTERT promoter has been used in combination with different pro-apoptotic genes like Bax and caspases 6 and 8 (Gu et al., 2000; Komata et al., 2001; Koga et al., 2000). In the case of the Bax gene it is noteworthy that the activity of the hTERT promoter was comparable to that of the strong CMV promoter and also that the good selectivity for tumor cells in vivo of this promoter prevented the hepatotoxicity observed in a previous study in which the Bax gene was under the control of a nonspecific promoter (Gu et al., 2000). Also the hTERT promoter has been used in combination with the TK gene confirming the high selectivity for targeting cancer cells in vivo in comparison with the same suicide gene under the control of the nonspecific CMV promoter (Majumdar et al., 2001).

ENHANCERS AND INSULATORS

Like many other therapeutic approaches, the use of tissue- or tumor-specific promoters for gene therapy is not exempt from problems. Probably the two most important are the weakness of transcriptional activity shown by some promoters studied and the loss of specificity in the context of viral vectors. It has been demonstrated that sequences from adenoviral and retroviral vectors alter the properties of some promoters and change their specificity, or even silence them. For example, it has been shown that in the context of a retroviral vector the liver albumin promoter loses its specificity and is able to direct high levels of expression of a reporter gene in fibroblast cells. However, the liver-specific promoter of the alpha-1 antitrypsin gene is not affected by the retroviral sequences in the same vector (Wu et al., 1996). Also, in the context of an adenoviral vector the c-erbB2 promoter can lose its specificity, affected by *cis*-acting sequences from the viral backbone (Ring et al., 1996). A similar observation has been made using the human ventricular/slow muscle myosin light chain 1 promoter (Shi, Wang, and Worton, et al., 1997). The most likely explanation appears to be the presence of enhancer sequences within the adenoviral genome. One of the best characterized is the activity of the E1A enhancer. Since most of the designs involved the insertion of the therapeutic gene and specific promoter in this region, it is not surprising that this enhancer could affect the final activity of the specific promoter. To overcome this limitation, one strategy available is the cloning of the promoter of interest near the right inverted terminal repeat of adenovirus 5 (Ad5) vectors, which diminishes the interference of the viral E1A enhancer region with the promoter used (Rubinchik et al., 2001). Another approach is the use of insulators, which are sequences naturally occurring in the DNA genome of invertebrate and vertebrate species. The chicken β- and γ-globin locus control regions have been used in the context of retrovirus and adenoviruses, giving good results (Emery et al., 2000; Steinwaerder and Lieber, 2000). Moreover, it has been possible to restore the selectivity of the ERBB2 promoter in the context of an Ad5 virus using the bovine growth hormone transcriptional stop signal (Vassaux, Hurst, Lemoine, 1999). With respect to the suboptimal activity of promoters, one option is to add enhancer sequences in order to improve transcription, but further novel approaches are emerging such as the use of artificial promoters (see Chapter 22) (Nettelbeck, Jerome, and Muller, 1998).

FUTURE AND CLINICAL APPLICATION

The range of potential promoters and enhancers that can be exploited for application in gene therapy for cancer continues to increase, and will no doubt be expanded by the results of expression profiling studies of different tumor and tissue types. Although the initial studies used various suicide genes under selective transcription control in nonreplicating vectors, the future is likely to involve application in replication-competent viruses, which have already shown promise with conditional promoters driving early gene expression. A great attraction of selective transcriptional regulatory elements is the ability to restrict therapeutic gene expression (and replication of vectors) to reduce potential toxicity to noncancer tissues. This is likely to be increasingly important in clinical applications where safety is paramount. And thus there is a growing need to take all the excellent results being obtained in model systems to clinical application where the efficacy can be really tested. As our knowledge of transcriptional targeting increases, more accurate vectors are being designed. Indeed, several transcriptionally regulated vectors have been already used in clinical trials with different types of cancer such as prostate (Doehn and Jocham, 2001; Herman et al., 1999; Koeneman et al., 2000; Shalev et al., 2000) and breast cancer (Pandha et al., 1999). The initial results obtained are encouraging but also showed that although transcriptional control of the therapeutic gene is an issue of great importance, other issues, such as efficiency of transduction, need to be addressed. With all the studies actually ongoing to optimize the activity and selectivity of promoter elements and also the new approaches using artificial promoters, we may expect a great progress in gene therapy for cancer.

REFERENCES

Abe M, Kufe DW (1987): Identification of a family of high molecular weight tumor-associated glycoproteins. *J Immunol* 139:257–261.

Abe M, Kufe D (1990): Transcriptional regulation of DF3 gene expression in human MCF-7 breast carcinoma cells. *J Cell Physiol* 143:226–231.

Abe M, Kufe D (1993): Characterization of cis-acting elements regulating transcription of the human DF3 breast carcinoma-associated antigen (MUC1) gene. *Proc Natl Acad Sci USA* 90:282–286.

Ambar BB, Frei K, Malipiero U, Morelli AE, Castro MG, Lowenstein PR, Fontana A (1999): Treatment of experimental glioma by administration of adenoviral vectors expressing Fas ligand. *Hum Gene Ther* 10:1641–1648.

Andersen, JK, Frim DM, Isacson O, Breakefield XO (1993) :Herpesvirus-mediated gene delivery into the rat brain: specificity and efficiency of the neuron-specific enolase promoter. *Cell Mol Neurobiol* 13:503–515.

Barry, MJ (2001): Prostate-specific-antigen testing for early diagnosis of prostate cancer. *N Engl J Med* 344:1373–1377.

Barsov, EV, Payne WS, Hughes SH (2001): Adaptation of chimeric retroviruses in vitro and in vivo: isolation of avian retroviral vectors with extended host range. *J Virol* 75:4973–4983.

Bazzoni, F, Regalia E (2001): Triggering of antitumor activity through melanoma-specific transduction of a constitutively active tumor necrosis factor (TNF) R1 chimeric receptor in the absence of TNF-alpha. *Cancer Res* 61:1050–1057.

Boast K, Binley K, Iqball S, Price T, Spearman H, Kingsman S. Kingsman A, Naylor S (1999): Characterization of physiologically regulated vectors for the treatment of ischemic disease. *Hum Gene Ther* 10:2197-2208.

Bouchard B, Fuller B, Vijayasaradhi S, Houghton A (1989): Induction of pigmentation in mouse fibroblasts by expression of human tyrosinase cDNA. *J Exp Med* 169:2029–2042.

Brand K, Loser P, Arnold W, Bartels T, Strauss M (1998): Tumor cell-specific transgene expression prevents liver toxicity of the adeno-HSVtk/GCV approach. *Gene Ther* 5:1363–1371.

Cao G, Kuriyama S, Du P, Sakamoto T, Kong X, Masui K, Qi Z (1997): Complete regression of established murine hepatocellular carcinoma by in vivo tumor necrosis factor alpha gene transfer. *Gastroenterology* 112:501–510.

Cao G, Kuriyama S, Gao J, Mitoro A, Cui L, Nakatani T, Zhang X, Kikukawa M, Pan X, Fukui H, Qi Z (1998): Comparison of carcinoembryonic antigen promoter regions isolated from human colorectal carcinoma and normal adjacent mucosa to induce strong tumor-selective gene expression. *Int J Cancer* 78:242–247.

Cao G, Kuriyama S, Cui L, Nagao S, Pan X, Toyokawa Y, Zhang X, Nishiwaki I, Qi Z (1999a): Analysis of the human carcinoembryonic antigen promoter core region in colorectal carcinoma-selective cytosine deaminase gene therapy. *Cancer Gene Ther* 6:572–580.

Cao G, Kuriyama S, Gao J, Kikukawa M, Cui L, Nakatani T, Zhang X, Tsujinoue H, Pan X, Fukui H, Qi Z (1999b): Effective and safe gene therapy for colorectal carcinoma using the cytosine deaminase gene directed by the carcinoembryonic antigen promoter. *Gene Ther* 6:83–90.

Cao G, Zhang X, He X, Chen Q, Qi Z (1999c): A safe, effective in vivo gene therapy for melanoma using tyrosinase promoter-driven cytosine deaminase gene. *In Vivo* 13:181–187.

Casado E, Gomez-Navarro J, Yamamoto M, Adachi Y, Collidge CJ, Arafat WO, Barker SD, Wang MH, Mahasreshti PJ, Hemminki A, Gonzalez-Baron M, Barnes MN, Pustilnik TB, Siegal GP, Alvarez RD, Curiel DT (2001): Strategies to accomplish targeted expression of transgenes in ovarian cancer for molecular therapeutic applications, *Clin Cancer Res* 7:2496-2504.

Chen L, Chen D, Manome Y, Dong Y, Fine HA, Kufe DW (1995): Breast cancer selective gene expression and therapy mediated by recombinant adenoviruses containing the DF3/MUC1 promoter. *J Clin Invest* 96:2775–2782.

Chen X, Zhang D. Dennert G, Hung G, Lee AS (2000a): Eradication of murine mammary adenocarcinoma through HSVtk expression directed by the glucose-starvation inducible grp78 promoter. *Breast Cancer Res Treat* 59:81-90.

Chen Y, Yu DC, Charlton D, Henderson DR (2000b): Pre-existent adenovirus antibody inhibits systemic toxicity and antitumor activity of CN706 in the nude mouse LNCaP xenograft model. implications and proposals for human therapy. *Hum Gene Ther* 11:1553–1567.

Cheon J, Ko SC, Gardner TA, Shirakawa T. Gotoh A. Kao C, Chung LW (1997): Chemogene therapy: osteocalcin promoter-based suicide gene therapy in combination with methotrexate in a murine osteosarcoma model. *Cancer Gene Ther* 4:359-365.

Clements JA (1989): The glandular kallikrein family of enzymes: tissue-specific expression and hormonal regulation. *Endocr Rev* 10:393 419.

Clements JA (1994): The human kallikrein gene family: a diversity of expression and function. *Mol Cell Endocrinol* 99:C1–6.

Cong YS, Wen J, Bacchetti S (1999): The human telomerase catalytic subunit hTERT: organization of the gene and characterization of the promoter. *Hum Mol Genet* 8:137–142.

Darson MF, Pacelli A, Roche P, Rittenhouse HG, Wolfert RL, Young CY, Klee GG, Tindall DJ, Bostwick DG (1997): Human glandular kallikrein 2 (hK2) expression in prostatic intraepithelial neoplasia and adenocarcinoma: a novel prostate cancer marker. *Urology* 49:857–862.

Dematteo RP, McClane SJ, Fisher K, Yeh H, Chu G, Burke C, Raper SE (1997): Engineering tissue-specific expression of a recombinant adenovirus: selective transgene transcription in the pancreas using the amylase promoter. *J Surg Res* 72:155-161.

De Placido S, Carlomagno C, De Laurentiis M, Bianco AR (1998): c-erbB2 expression predicts tamoxifen efficacy in breast cancer patients. *Breast Cancer Res Treat* 52:55–64.

Diaz RM, Eisen T, Hart IR, Vile RG (1998): Exchange of viral promoter enhancer elements with heterologous regulatory sequences generates targeted hybrid long terminal repeat vectors for gene therapy of melanoma. *J Virol* 72:789–795.

DiMaio JM, Clary BM, Via DF, Coveney E, Pappas TN, Lyerly HK (1994): Directed enzyme pro-drug gene therapy for pancreatic cancer in vivo. *Surgery* 116:205–213.

Doehn C, Jocham D (2001): Technology evaluation: CV-787, Calydon Inc. *Curr Opin Mol Ther* 3:204-210.

Doronin K, Kuppuswamy M, Toth K, Tollefson AE, Krajcsi P, Krougliak V, Wold WS (2001): Tissue-specific, tumor-selective, replication-competent adenovirus vector for cancer gene therapy. *J Virol* 75:3314-3324.

Elenitoba-Johnson KS (2001): Complex regulation of telomerase activity: implications for cancer therapy. *Am J Pathol* 159:405–410.

Emery DW, Yannaki E, Tubb J, Stamatoyannopoulos G (2000): A chromatin insulator protects retrovirus vectors from chromosomal position effects. *PNAS* 97:9150–9155.

Etienne-Julan M, Roux P, Bourquard P, Carillo S, Jeanteur P, Piechaczyk M (1992): Cell targeting by murine recombinant retroviruses. *Bone Marrow Transplant* 9:139–142.

Feng J, Funk WD, Wang SS, Weinrich SL, Avilion AA, Chiu CP, Adams RR, Chang E, Allsopp RC, Yu J et al. (1995): The RNA component of human telomerase. *Science* 269:1236–1241.

Friedman EL, Hayes DF, Kufe DW (1986): Reactivity of monoclonal antibody DF3 with a high molecular weight antigen expressed in human ovarian carcinomas. *Cancer Res* 46:5189–5194.

Galanis E, Vile R, Russell SJ (2001): Delivery systems intended for in vivo gene therapy of cancer: targeting and replication competent viral vectors. *Crit Rev Oncol Hematol* 38:177–192.

Gendler SJ, Burchell JM, Duhig T, Lamport D, White R, Parker M, Taylor-Papadimitriou J (1987): Cloning of partial cDNA encoding differentiation and tumor-associated mucin glycoproteins expressed by human mammary epithelium. *Proc Natl Acad Sci USA* 84:6060–6064.

Girling A, Bartkova J, Burchell J, Gendler S, Gillett C, Taylor-Papadimitriou J (1989): A core protein epitope of the polymorphic epithelial mucin detected by the monoclonal antibody SM-3 is selectively exposed in a range of primary carcinomas. *Int J Cancer* 43:1072–1076.

Gold P, Freedman SO (1965): Specific carcinoembryonic antigens of the human digestive system. *J Exp Med* 122:467–481.

Graham RA, Wang S, Catalano PJ, Haller DG (1998): Postsurgical surveillance of colon cancer: preliminary cost analysis of physician examination, carcinoembryonic antigen testing, chest x-ray, and colonoscopy. *Ann Surg* 228:59–63.

Graham RA, Morris JR, Cohen EP, Taylor-Papadimitriou J (2001): Up-regulation of MUC1 in mammary tumors generated in a double-transgenic mouse expressing human MUC1 cDNA, under the control of 1.4-kb 5' MUC1 promoter sequence and the middle T oncogene, expressed from the MMTV promoter. *Int J Cancer* 92:382–387.

Greider CW, Blackburn EH (1985): Identification of a specific telomere terminal transferase activity in Tetrahymena extracts. *Cell* 43:405–413.

Greider CW, Blackburn EH (1989): A telomeric sequence in the RNA of Tetrahymena telomerase required for telomere repeat synthesis. *Nature* 337:331–337.

Gu J, Kagawa S, Takakura M, Kyo S, Inoue M, Roth JA, Fang B (2000): Tumor-specific transgene expression from the human telomerase reverse transcriptase promoter enables targeting of the therapeutic effects of the Bax gene to cancers. *Cancer Res* 60:5359–5364.

Guida G, Gallone A, Maida I, Boffoli D, Cicero R (2000): Tyrosinase gene expression in the Kupffer cells of *Rana esculenta L. Pigment Cell Res* 13:431–435.

Hafenrichter DG, Ponder KP, Rettinger SD, Kennedy SC, Wu X, Saylors RS, Flye MW (1994): Liver-directed gene therapy: evaluation of liver specific promoter elements. *J Surg Res* 56:510–517.

Hahn WC, Meyerson M (2001): Telomerase activation, cellular immortalization and cancer. *Ann Med* 33:123–129.

Hallenbeck PL, Chang YN, Hay C, Golightly D, Stewart D, Lin J, Phipps S, Chiang YL (1999): A novel tumor-specific replication-restricted adenoviral vector for gene therapy of hepatocellular carcinoma. *Hum Gene Ther* 10:1721–1733.

Hammarstrom S (1999): The carcinoembryonic antigen (CEA) family: structures, suggested functions and expression in normal and malignant tissues. *Semin Cancer Biol* 9:67–81.

Harrington KJ, Linardakis E, Vile RG (2000): Transcriptional control: an essential component of cancer gene therapy strategies? *Adv Drug Deliv Rev* 44:167–184.

Harris JD, Gutierrez AA, Hurst HC, Sikora K, Lemoine NR (1994): Gene therapy for cancer using tumour-specific prodrug activation. *Gene Ther* 1:170–175.

Hauck W, Stanners CP (1995): Transcriptional regulation of the carcinoembryonic antigen gene. *J Biol Chem* 270:3602-3610.

Herman JR, Adler HL, Aguilar-Cordova E, Rojas-Martinez A, Woo S. Timme TL,

Wheeler TM, Thompson TC, Scardino PT (1999): In situ gene therapy for adenocarcinoma of the prostate: a phase I clinical trial. *Hum Gene Ther* 10:1239-1249.

Hirano T, Kaneko S, Kaneda Y, Saito I, Tamaoki T, Furuyama J, Kobayashi K, Ueki T, Fujimoto J (2001): HVJ-liposome-mediated transfection of HSVtk gene driven by AFP promoter inhibits hepatic tumor growth of hepatocellular carcinoma in SCID mice. *Gene Ther* 8:80–83.

Hsing AW, Tsao L, Devesa SS (2000): International trends and patterns of prostate cancer incidence and mortality. *Int J Cancer* 85:60–67.

Hughes BW, Wells AH, Bebok Z, Gadi VK, Garver RI, Parker WB, Sorscher EJ (1995): Bystander killing of melanoma cells using the human tyrosinase promoter to express the Escherichia coli purine nucleoside phosphorylase gene. *Cancer Res* 55:3339–3345.

Humphreys MJ, Ghaneh P, Greenhalf W, Campbell F, Clayton TM, Everett P, Huber BE, Richards CA, Ford MJ, Neoptolemos JP (2001): Hepatic intra-arterial delivery of a retroviral vector expressing the cytosine deaminase gene, controlled by the CEA promoter and intraperitoneal treatment with 5-fluorocytosine suppresses growth of colorectal liver metastases. *Gene Ther* 8:1241–1247.

Jackson IJ, Chambers DM, Budd PS, Johnson R (1991): The tyrosinase-related protein-1 gene has a structure and promoter sequence very different from tyrosinase. *Nucleic Acids Res* 19:3799–3804.

Kanai F, Lan KH, Shiratori Y, Tanaka T, Ohashi M, Okudaira T, Yoshida Y, Wakimoto H, Hamada H, Nakabayashi H, Tamaoki T, Omata M (1997): In vivo gene therapy for alpha-fetoprotein-producing hepatocellular carcinoma by adenovirus-mediated transfer of cytosine deaminase gene. *Cancer Res* 57:461–465.

Katabi MM, Chan HL, Karp SE, Batist G (1999): Hexokinase type II: a novel tumor-specific promoter for gene-targeted therapy differentially expressed and regulated in human cancer cells. *Hum Gene Ther* 10:155-164.

Kluppel M, Beermann F, Ruppert S, Schmid E, Hummler E, Schutz G (1991): The mouse tyrosinase promoter is sufficient for expression in melanocytes and in the pigmented epithelium of the retina. *Proc Natl Acad Sci USA* 88:3777–3781.

Koeneman KS, Kao C, Ko SC, Yang L, Wada Y, Kallmes DF, Gillenwater JY, Zhau HE, Chung LW, Gardner TA (2000): Osteocalcin-directed gene therapy for prostate-cancer bone metastasis. *World J Urol* 18:102-110.

Koga S, Hirohata S, Kondo Y, Komata T, Takakura M, Inoue M, Kyo S, Kondo S (2000): A novel telomerase-specific gene therapy: gene transfer of caspase-8 utilizing the human telomerase catalytic subunit gene promoter. *Hum Gene Ther* 11:1397–1406.

Komata T, Kondo Y, Kanzawa T, Hirohata S, Koga S, Sumiyoshi H, Srinivasula SM, Barna BP, Germano IM, Takakura M, Inoue M, Alnemri ES, Shay JW, Kyo S, Kondo S (2001): Treatment of malignant glioma cells with the transfer of constitutively active caspase-6 using the human telomerase catalytic subunit (human telomerase reverse transcriptase) gene promoter. *Cancer Res* 61:5796–5802.

Konishi F, Maeda H, Yamanishi Y, Hiyama K, Ishioka S, Yamakido M (1999): Transcriptionally targeted in vivo gene therapy for carcinoembrionic antigen-producing adenocarcinoma. *Hiroshima J Med Sci* 48:79–89.

Korner A, Pawelek J (1982): Mammalian tyrosinase catalyzes three reactions in the biosynthesis of melanin. *Science* 217:1163–1165.

Kovarik A, Peat N, Wilson D, Gendler SJ, Taylor-Papadimitriou J (1993): Analysis of the tissue-specific promoter of the MUC1 gene. *J Biol Chem* 268:9917–9926.

Kovarik A, Lu PJ, Peat N, Morris J, Taylor-Papadimitriou J (1996): Two GC boxes (Sp1 sites) are involved in regulation of the activity of the epithelium-specific MUC1 promoter. *J Biol Chem* 271:18140–18147.

Kurihara T, Brough DE, Kovesdi I, Kufe DW (2000): Selectivity of a replication-competent adenovirus for human breast carcinoma cells expressing the MUC1 antigen. *J Clin Invest* 106:763–771.

Lan KH, Kanai F, Shiratori Y, Okabe S, Yoshida Y, Wakimoto H, Hamada H, Tanaka T, Ohashi M, Omata M (1996): Tumor-specific gene expression in carcinoembryonic antigen-producing gastric cancer cells using adenovirus vectors. *Gastroenterology* 111:1241–1251.

Lan KH, Kanai F, Shiratori Y, Ohashi M, Tanaka T, Okudaira T, Yoshida Y, Hamada H, Omata M (1997): In vivo selective gene expression and therapy mediated by adenoviral vectors for human carcinoembryonic antigen-producing gastric carcinoma. *Cancer Res* 57:4279-4284.

Lee, CH, Liu M, Sie KL, Lee MS (1996): Prostate-specific antigen promoter driven gene therapy targeting DNA polymerase-alpha and topoisomerase II alpha in prostate cancer. *Anticancer Res* 16:1805–1811.

Lee SE, Jin RJ, Lee SG, Yoon SJ, Park MS, Heo DS, Choi H (2000): Development of a new plasmid vector with PSA-promoter and enhancer expressing tissue-specificity in prostate carcinoma cell lines. *Anticancer Res* 20:417–422.

Lockett LJ, Molloy PL, Russell PJ, Both GW (1997): Relative efficiency of tumor cell killing in vitro by two enzyme-prodrug systems delivered by identical adenovirus vectors. *Clin Cancer Res* 3:2075–2080.

Majumdar AS, Hughes DE, Lichtsteiner SP, Wang Z, Lebkowski JS, Vasserot AP (2001): The telomerase reverse transcriptase promoter drives efficacious tumor suicide gene therapy while preventing hepatotoxicity encountered with constitutive promoters. *Gene Ther* 8:568–578.

Manome Y, Abe M, Hagen MF, Fine HA, Kufe DW (1994): Enhancer sequences of the DF3 gene regulate expression of the herpes simplex virus thymidine kinase gene and confer sensitivity of human breast cancer cells to ganciclovir. *Cancer Res* 54:5408–5413.

Martiniello-Wilks R, Garcia-Aragon J, Daja MM, Russell P, Both GW, Molloy PL, Lockett LJ, Russell PJ (1998): In vivo gene therapy for prostate cancer: preclinical evaluation of two different enzyme-directed prodrug therapy systems delivered by identical adenovirus vectors. *Hum Gene Ther* 9:1617–1626.

Mason H (1948): The chemistry of melanin: III. Mechanism of the oxidation of droxyphenylalanine by tirosinase. *J Biochem Chem* 172:83–99

Meyerson M (2000): Role of telomerase in normal and cancer cells. *J Clin Oncol* 18:2626–[illegible]

Miller N, Whelan J (1997): Progress in transcriptionally targeted and regulatable vectors for genetic therapy. *Hum Gene Ther* 8:803–815.

Mitchell SH, Murtha PE, Zhang S, Zhu W, Young CY (2000): An androgen response element mediates LNCaP cell dependent androgen induction of the hK2 gene. *Mol Cell Endocrinol* 168:89–99.

Miyatake S, Iyer A, Martuza RL, Rabkin SD (1997): Transcriptional targeting of herpes simplex virus for cell-specific replication. *J Virol* 71:5124–5132.

Miyatake SI, Tani S, Feigenbaum F, Sundaresan P, Toda H, Narumi O, Kikuchi H, Hashimoto N, Hangai M, Martuza RL, Rabkin SD (1999): Hepatoma-specific antitumor activity of an albumin enhancer/promoter regulated herpes simplex virus in vivo. *Gene Ther* 6:564–572.

Morelli AE, Larregina AT, Smith-Arica J, Dewey RA, Southgate TD, Ambar B, Fontana A, Castro MG, Lowenstein PR (1999): Neuronal and glial cell type-specific promoters within adenovirus recombinants restrict the expression of the apoptosis-inducing molecule Fas ligand to predetermined brain cell types, and abolish peripheral liver toxicity. *J Gen Virol* 80:571–583.

Morris JR, Taylor-Papadimitriou J (2001): The Sp1 transcription factor regulates cell type-specific transcription of MUC1. *DNA Cell Biol* 20:133–139.

Murtha P, Tindall DJ, Young CY (1993): Androgen induction of a human prostate-specific kallikrein, hKLK2: characterization of an androgen response element in the 5' promoter region of the gene. *Biochemistry* 32:6459–6464.

Nap M, Mollgard K, Burtin P, Fleuren GJ (1988): Immunohistochemistry of carcino-embryonic antigen in the embryo, fetus and adult. *Tumour Biol* 9:145–153.

Nap M, Hammarstrom ML, Bormer O, Hammarstrom S, Wagener C, Handt S, Schreyer M, Mach JP, Buchegger F, von Kleist S et al. (1992): Specificity and affinity of monoclonal antibodies against carcinoembryonic antigen. *Cancer Res* 52:2329–2339.

Navarro V, Millecamps S, Geoffroy MC, Robert JJ, Valin A, Mallet J, Gal La Salle GL (1999): Efficient gene transfer and long-term expression in neurons using a recombinant adenovirus with a neuron-specific promoter. *Gene Ther* 6:1884–1892.

Nettelbeck DM, Jerome V, Muller R (1998): A strategy for enhancing the transcriptional activity of weak cell type-specific promoters. *Gene Ther* 5:1656–1664.

Nettelbeck DM, Jerome V, Muller R (2000): Gene therapy: designer promoters for tumour targeting. *Trends Genet* 16:174–181.

Ohashi M, Kanai F, Tanaka T, Lan KH, Shiratori Y, Komatsu Y, Kawabe T, Yoshida H, Hamada H, Omata M (1998): In vivo adenovirus-mediated prodrug gene therapy for carcinoembryonic antigen-producing pancreatic cancer. *Jpn J Cancer Res* 89:457–462.

Ohashi M, Kanai F, Tateishi K, Taniguchi H, Marignani PA, Yoshida Y, Shiratori Y, Hamada H, Omata M (2001): Target gene therapy for alpha-fetoprotein-producing hepatocellular carcinoma by E1B55k-attenuated adenovirus. *Biochem Biophys Res Commun* 282:529–535.

Osaki T, Tanio Y, Tachibana I, Hosoe S, Kumagai T, Kawase I, Oikawa S, Kishimoto T (1994): Gene therapy for carcinoembryonic antigen-producing human lung cancer cells by cell type-specific expression of herpes simplex virus thymidine kinase gene. *Cancer Res* 54:5258–5261.

Pandha HS, Martin LA, Rigg A, Hurst HC, Stamp GW, Sikora K, Lemoine NR (1999): Genetic prodrug activation therapy for breast cancer: A phase I clinical trial of erbB-2-directed suicide gene expression. *J Clin Oncol* 17:2180–2189.

Pang S (2000): Targeting and eradicating cancer cells by a prostate-specific vector carrying the diphtheria toxin A gene. *Cancer Gene Ther* 7:991–996.

Park BJ, Brown CK, Hu Y, Alexander HR, Horti J, Raje S, Figg WD, Bartlett DL (1999): Augmentation of melanoma-specific gene expression using a tandem melanocyte-specific enhancer results in increased cytotoxicity of the purine nucleoside phosphorylase gene in melanoma. *Hum Gene Ther* 10:889–898.

Peel AL, Zolotukhin S, Schrimsher GW, Muzyczka N, Reier PJ (1997): Efficient transduction of green fluorescent protein in spinal cord neurons using adeno-associated virus vectors containing cell type-specific promoters. *Gene Ther* 4:16–24.

Pereboev A, Pereboeva L, Curiel DT (2001): Phage display of adenovirus type 5 fiber knob as a tool for specific ligand selection and validation. *J Virol* 75:7107–7113.

Peyton DK, Ramesh T, Spear BT (2000): Position-dependent activity of alpha-fetoprotein enhancer element III in the adult liver is due to negative regulation. *PNAS* 97:10890–10894

Poole JC, Andrews LG, Tollefsbol TO (2001): Activity, function, and gene regulation of the catalytic subunit of telomerase (hTERT). *Gene* 269:1–12.

Prall F, Nollau P, Neumaier M, Haubeck HD, Drzeniek Z, Helmchen U, Loning T, Wagener C (1996): CD66a (BGP), an adhesion molecule of the carcinoembryonic antigen family, is expressed in epithelium, endothelium, and myeloid cells in a wide range of normal human tissues. *J Histochem Cytochem* 44:35–41.

Riegman PH, Vlietstra RJ, van der Korput JA, Romijn JC, Trapman J (1989): Characterization of the prostate-specific antigen gene: a novel human kallikrein-like gene. *Biochem Biophys Res Commun* 159:95–102.

Riegman PH, Vlietstra RJ, van der Korput JA, Brinkmann AO, Trapman J (1991): The promoter of the prostate-specific antigen gene contains a functional androgen responsive element. *Mol Endocrinol* 5:1921–1930.

Ring, CJ, Harris JD, Hurst HC, Lemoine NR (1996): Suicide gene expression induced in tumour cells transduced with recombinant adenoviral, retroviral and plasmid vectors containing the ERBB2 promoter. *Gene Ther* 3:1094–1103.

Ring CJ, Blouin P, Martin LA, Hurst HC, Lemoine NR (1997): Use of transcriptional regulatory elements of the MUC1 and ERBB2 genes to drive tumour-selective expression of a prodrug activating enzyme. *Gene Ther* 4:1045–1052.

Robbins PF, Eggensperger D, Qi CF, Schlom J (1993): Definition of the expression of the human carcinoembryonic antigen and non-specific cross-reacting antigen in human breast and lung carcinomas. *Int J Cancer* 53:892–897.

Rodriguez R, Schuur ER, Lim HY, Henderson GA, Simons JW, Henderson DR (1997): Prostate attenuated replication competent adenovirus (ARCA) CN706: a selective cytotoxic for prostate-specific antigen-positive prostate cancer cells. *Cancer Res* 57:2559–2563.

Rubinchik S, Lowe S, Jia Z, Norris J, Dong J (2001): Creation of a new transgene cloning site near the right ITR of Ad5 results in reduced enhancer interference with tissue-specific and regulatable promoters. *Gene Ther* 8:247–253.

Ruppert S, Muller G, Kwon B, Schutz G (1988): Multiple transcripts of the mouse tyrosinase gene are generated by alternative splicing. *Embo J* 7:2715–2722.

Schrewe H, Thompson J, Bona M, Hefta LJ, Maruya A, Hassauer M, Shively JE, von Kleist S, Zimmermann W (1990): Cloning of the complete gene for carcinoembryonic antigen: analysis of its promoter indicates a region conveying cell type-specific expression. *Mol Cell Biol* 10:2738–2748.

Segawa T, Takebayashi H, Kakehi Y, Yoshida O, Narumiya S, Kakizuka A (1998): Prostate-specific amplification of expanded polyglutamine expression: a novel approach for cancer gene therapy. *Cancer Res* 58:2282–2287.

Shavle M, Kadmon D, Teh BS, Butler EB, Aguilar-Cordova E, Thompson TC, Herman JR, Adler HL, Scardino PT, Miles BJ (2000): Suicide gene therapy toxicity after multiple and repeat injections in patients with localized prostate cancer. *J Urol* 163:1747-1750.

Shi ZR, Tacha D, Itzkowitz SH (1994): Monoclonal antibody COL-1 reacts with restricted epitopes on carcinoembryonic antigen: an immunohistochemical study. *J Histochem Cytochem* 42:1215–1219.

Shi Q, Wang Y, Worton R (1997): Modulation of the specificity and activity of a cellular promoter in an adenoviral vector. *Hum Gene Ther* 8:403–410.

Shibata K, Muraosa Y, Tomita Y, Tagami H, Shibahara S (1992a): Identification of a cis-acting element that enhances the pigment cell-specific expression of the human tyrosinase gene. *J Biol Chem* 267:20584–20588.

Shibata K, Takeda K, Tomita Y, Tagami H, Shibahara S (1992b): Downstream region of the human tyrosinase-related protein gene enhances its promoter activity. *Biochem Biophys Res Commun* 184:568–575.

Shimura H, Suzuki H, Miyazaki A, Furuya F, Ohta K, Haraguchi K, Endo T, Onaya T (2001): Transcriptional activation of the thyroglobulin promoter directing suicide gene expression by thyroid transcription factor-1 thyroid cancer cells. *Cancer Res* 61:3640-3646.

Siddiqui J, Abe M, Hayes D, Shani E, Yunis E, Kufe D (1988): Isolation and sequencing of a cDNA coding for the human DF3 breast carcinoma-associated antigen. *Proc Natl Acad Sci USA* 85:2320–2323.

Siders WM, Halloran PJ, Fenton RG (1998): Melanoma-specific cytotoxicity induced by a tyrosinase promoter-enhancer/herpes simplex virus thymidine kinase adenovirus. *Cancer Gene Ther* 5:281–291.

Simons JW, Mikhak B, Van der Poel HG, DeMarzo AM, Rodriguez R, Goemann MA, Nelson WG, Li S, Detorie N, Hampert UM, Ramakrishna N, DeWeese TL (2000): Molecular and clinical activity of CN706, a PSA-selective oncolytic Ad5 vector in a phase I trial in locally recurrent prostate cancer following radiation therapy. *Mol Ther* 1:S144–S146.

Slawin KM, Shariat SF, Nguyen C, Leventis AK, Song W, Kattan MW, Young CY, Tindall DJ, Wheeler TM (2000): Detection of metastatic prostate cancer using a splice variant-specific reverse transcriptase-polymerase chain reaction assay for human glandular kallikrein. *Cancer Res* 60:7142–7148.

Spear BT (1999): Alpha-fetoprotein gene regulation: lessons from transgenic mice. *Semin Cancer Biol* 9:109–116.

Stackhouse MA, Buchsbaum DJ, Kancharla SR, Grizzle WE, Grimes C, Laffoon K, Pederson LC, Curiel DT (1999): Specific membrane receptor gene expression targeted with radiolabeled peptide employing the erbB-2 and DF3 promoter elements in adenoviral vectors. *Cancer Gene Ther* 6:209–219.

Steinwaerder DS, Lieber A (2000): Insulation from viral transcriptional regulatory elements improves inducible transgene expression from adenovirus vectors in vitro and in vivo. *Gene Ther* 7:556–567.

Suzuki K, Fueyo J, Krasnykh V, Reynolds PN, Curiel DT, Alemany R (2001): A conditionally replicative adenovirus with enhanced infectivity shows improved oncolytic potency. *Clin Cancer Res* 7:120–126.

Takeuchi M, Shichinohe T, Senmaru N, Miyamoto M, Fujita H, Takimoto M, Kondo S, Katoh H, Kuzumaki N (2000): The dominant negative H-ras mutant, N116Y, suppresses growth of metastatic human pancreatic cancer cells in the liver of nude mice. *Gene Ther* 7:518–526.

Tanaka T, Kanai F, Lan KH, Ohashi M, Shiratori Y, Yoshida Y, Hamada H, Omata M (1997): Adenovirus-mediated gene therapy of gastric carcinoma using cancer-specific gene expression in vivo. *Biochem Biophys Res Commun* 231:775–779.

Thompson J, Mossinger S, Reichardt V, Engels U, Beauchemin N, Kommoss F, von Kleist S, Zimmermann W (1993): A polymerase-chain-reaction assay for the specific identification of transcripts encoded by individual carcinoembryonic antigen (CEA)-gene-family members. *Int J Cancer* 55:311–319.

Thompson J, Epting T, Schwarzkopf G, Singhofen A, Eades-Perner AM, van Der Putten H, Zimmermann W (2000): A transgenic mouse line that develops early-onset invasive gastric carcinoma provides a model for carcinoembryonic antigen-targeted tumor therapy. *Int J Cancer* 86:863–869.

Uchida A, O'Keefe DS, Bacich DJ, Molloy PL, Heston WD (2001): In vivo suicide gene therapy model using a newly discovered prostate-specific membrane antigen promoter/enhancer: a potential alternative approach to androgen deprivation therapy. *Urology* 58:132–139.

Varda-Bloom N, Shaish A, Gonen A, Levanon K, Greenbereger S, Ferber S, Levkovitz H, Castel D, Goldberg I, Afek A, Kopolovitc Y, Harats D (2001): Tissue-specific gene therapy directed to tumor angiogenesis. *Gene Ther* 8:819–827.

Vassaux G, Hurst HC, Lemoine NR (1999): Insulation of a conditionally expressed transgene in an adenoviral vector. *Gene Ther* 6:1192–1197.

Vile R, Hart I (1993a): In vitro and in vivo targeting of gene expression to melanoma cells. *Cancer Res* 53:962–967

Vile, RG, Hart IR (1993b): Use of tissue-specific expression of the herpes simplex virus thymidine kinase gene to inhibit growth of established murine melanomas following direct intratumoral injection of DNA. *Cancer Res* 53:3860–3864.

Vile RG, Hart IR (1994): Targeting of cytokine gene expression to malignant melanoma cells using tissue specific promoter sequences. *Ann Oncol* 5 (Suppl 4):59–65.

Vyas SP, Singh A, Sihorkar V (2001): Ligand-receptor-mediated drug delivery: an emerging paradigm in cellular drug targeting. *Crit Rev Ther Drug Carrier Syst* 18:1–76.

Watanabe K, Saito A, Tamaoki T (1987): Cell-specific enhancer activity in a far upstream region of the human alpha-fetoprotein gene. *J Biol Chem* 262:4812–4818.

Wesseling JG, Bosma PJ, Krasnykh V, Kashentseva EA, Blackwell JL, Reynolds PN, Li H, Parameshwar M, Vickers SM, Jaffee EM, Huibregtse K, Curiel DT, Dmitriev I (2001): Improved gene transfer efficiency to primary and established human pancreatic carcinoma target cells via epidermal growth factor receptor and integrin-targeted adenoviral vectors. *Gene Ther* 8:969–976.

Wickham TJ (2000): Targeting adenovirus. *Gene Ther* 7:110–114.

Willcocks TC, Craig IW (1990): Characterization of the genomic organization of human carcinoembryonic antigen (CEA): comparison with other family members and sequence analysis of 5' controlling region. *Genomics* 8:492–500.

Wu X, Holschen J, Kennedy SC, Ponder KP (1996): Retroviral vector sequences may interact with some internal promoters and influence expression. *Hum Gene Ther* 7:159–171.

Xie X, Zhao X, Liu Y, Young CY, Tindall DJ, Slawin KM, Spencer DM (2001): Robust prostate-specific expression for targeted gene therapy based on the human kallikrein 2 promoter. *Hum Gene Ther* 12:549–561.

Young CY, Andrews PE, Tindall DJ (1995): Expression and androgenic regulation of human prostate-specific kallikreins. *J Androl* 16:97–99.

Young CY, Seay T, Hogen K, Charlesworth MC, Roche PC, Klee GG, Tindall DJ (1996): Prostate-specific human kallikrein (hK2) as a novel marker for prostate cancer. *Prostate Suppl* 7:17–24.

Yu DC, Chen Y, Seng M, Dilley J, Henderson DR (1999a): The addition of adenovirus type 5 region E3 enables calydon virus 787 to eliminate distant prostate tumor xenografts. *Cancer Res* 59:4200–4203.

Yu DC, Sakamoto GT, Henderson DR (1999b): Identification of the transcriptional regulatory sequences of human kallikrein 2 and their use in the construction of calydon virus 764, an attenuated replication competent adenovirus for prostate cancer therapy. *Cancer Res* 59:1498–1504.

Zaretsky JZ, Sarid R, Aylon Y, Mittelman LA, Wreschner DH, Keydar I (1999): Analysis of the promoter of the MUC1 gene overexpressed in breast cancer. *FEBS Lett* 461:189–195.

Zotter S, Hageman PC, Lossnitzer A, Mooi WJ, Hilgers J (1988): Tissue and distribution of human polymorphic epithelial mucin. *Cancer Rev* 11:12:55–101.

22

PROMOTER OPTIMIZATION AND ARTIFICIAL PROMOTERS FOR TRANSCRIPTIONAL TARGETING IN GENE THERAPY

DIRK M. NETTELBECK, PH.D., DAVID T. CURIEL, M.D., AND ROLF MÜLLER, PH.D.

INTRODUCTION

Selectivity of intervention is essential to most therapeutic strategies. In gene therapy, optimization of gene transfer vectors, including vector targeting, is crucial for the development of efficient clinical protocols (Anderson, 1998; Nabel, 1999). Considering the diversity of diseases, it is evident that any single gene delivery system cannot be universally applicable. Individually optimized vectors are therefore required. Transcriptional targeting facilitates spatially controlled, inducible, or physiologically regulated therapy by utilizing regulatory DNA sequences—promoters, enhancers, and/or silencers—to drive targeted expression of the therapeutic gene. Furthermore, a specific therapeutic strategy for treatment of malignancy, viral oncolysis, is based on the replacement of viral regulatory elements by selective promoters to control the expression of essential viral genes with the goal of obtaining tumor-specific condi-

Vector Targeting for Therapeutic Gene Delivery, Edited by David T. Curiel and Joanne T. Douglas
ISBN 0-471-43479-5

tionally replicating viruses. Of note, transcriptional targeting is the only feasible strategy for certain applications, such as radiation-induced gene therapy, which link genetic and conventional therapeutic approaches. Furthermore, selective or inducible promoters can be combined with transductional targeting to form a multilevel targeting approach. This combination may result in increased selectivity (1) by superimposing two levels of selectivity for the same target or (2) by combining different targeting modalities, for example tissue targeting with proliferation-dependent gene expression (dual specificity).

DESIGNER PROMOTERS

Key to the realization of the full potential of transcriptional targeting is the availability of highly specific and effective promoters for a wide spectrum of diseases. In addition, the promoter of choice must be functional in the heterologous context of a given gene therapy vector (vector compatibility). Thus, size restrictions and potential interference between vector sequences and promoter activity are of potential concern. Few promoters that fulfill these requirements have been described to date. Furthermore, for some applications, suitable natural promoters might not exist. Therefore, recent research efforts have focused on the development of "designer promoters" for a specific disease or vector context. In this regard, the following criteria have to be considered: (1) an adequate level of promoter activity in the target cell (ON status), comparable to the activity of strong, but not selective promoters used in standard nontargeted gene therapy studies (like the cytomegalovirus [CMV] promoter/enhancer), (2) negligible promoter activity in nontarget tissues (OFF status), that is minimal "leakiness," and (3) size limitations imposed by the vector of choice.

In general, the necessary activity and tightness of a candidate promoter depend on the therapeutic gene of choice, that is, on the potency of its encoded product on target versus nontarget cells. The goal is to achieve the highest possible therapeutic index for the specific combination of selective promoter and therapeutic gene. Technical strategies to achieve this aim include (1) the optimization of native promoters, (2) the construction of synthetic promoters, and (3) the engineering of vector backbone DNA sequences in order to retain the expression profile of incorporated promoters.

Fortunately, progress in elucidating basic questions of molecular cell biology provides gene therapists with the necessary tools for the development of targeted vectors. For example, the determination of viral cell binding and cell entry mechanisms facilitates the development of transductionally targeted vectors. Likewise, manipulation of promoter activity and size as well as the design of synthetic promoters are based on the characterization of molecules and mechanisms of transcriptional control. Specifically, the modular structure of both *cis*-acting (promoters, enhancers, and silencers) as well as *trans*-acting (transcription factors) regulators represents the basis for promoter optimization and design. The number, diversity, orientation, and placement of regulatory elements within a candidate promoter are critical parameters determining its performance. Their precise combination must be determined by detailed structure–function analyses. Similarly, recombinant transcription factors can be engineered by fusing natural or artificial domains that mediate DNA-binding, transactivation, or ligand-dependent regulation, thus deriving designer molecules able to regulate gene expression in the desired fashion.

The strategies for promoter optimization and development of artificial promoters described in this chapter represent a scenario for the translation of progress in basic biological research into therapeutic approaches of modern molecular medicine.

PROMOTER OPTIMIZATION: DELETION AND MULTIMERIZATION

Cell-specific or otherwise selective promoters are frequently inefficient activators of transcription or are comprised of extended DNA sequences too large for incorporation into gene transfer vectors. Naturally occurring or disease-associated activating point mutations within promoters have been exploited to increase transgene expression. However, activating mutations have been found only in a few cases, such as the α-fetoprotein (*AFP*) and multiple drug resistence (*mdr*) promoters (Ishikawa et al., 1999; Stein, Walther, and Shoemaker, 1996), and the levels of activation were minimal. Clearly, a more potent strategy is to optimize promoters by recombinant DNA technology. The simplest approach for expression optimization and simultaneous size reduction is to eliminate promoter sequences that do not obviously contribute to its transcriptional strength and/or selectivity. In addition, the multimerization of important regulatory elements can be advantageous. Both strategies have been successfully applied to several promoters as illustrated in the following paragraphs.

An exhaustive promoter analysis and optimization study for the regulatory region of the carcinoembryonic antigen (*CEA*) gene, which is transcriptionally upregulated in adenocarcinomas, has been performed by Richards, Austin, and Huber, 1995. An upstream enhancer element and a tetramer of a tumor-specific promoter element fused to the core promoter generated a highly active and selective promoter. Simultaneously, the promoter size was reduced considerably. The fusion of distant enhancer elements to a core promoter is a strategy successfully applied to several transcriptional regulatory regions, such as the hepatocellular carcinoma-specific *AFP* gene (Wills et al., 1995; Mawatari et al., 1998), the prostate specific genes for prostate-specific antigen (*PSA*) (Pang et al., 1997; Latham et al., 2000) and human kallikrein 2 (*hK2*) (Xie et al., 2001), and the melanocyte-specific human and murine tyrosinase genes (Siders, Halloran, and Fenton, 1996, 1998; Park et al., 1999).

For both tyrosinase constructs an increase in activity and a size compatible with most gene therapy application have been achieved by combining an enhancer dimer with the core promoters (Siders, Halloran, and Fenton, 1996). For the human construct, the size was reduced from 2.7 kb to 650 bp, with a five-fold increase in activity. Melanocyte-specific cytotoxicity was shown for these promoter constructs in combination with the herpes simplex virus thymidine kinase (HSV-tk)/gancyclovir (GCV) and purine nucleoside phosphorylase (PNP)/6-methyl purine phosphorylase (6-MPDR) prodrug activation systems in both adenoviral and liposomal vectors (Siders et al., 1998; Park et al., 1999).

Of note, optimized albumin, *AFP*, *PSA*, and *hK2* promoters have been successfully applied for the transcriptional targeting of herpes simplex virus (HSV) or adenovirus (Ad) replication to tumor cells (Miyatake et al., 1997, 1999; Rodriguez et al., 1997; Yu, Sakamoto, and Henderson, 1999; Hallenbeck et al., 1999). In this scenario, expression of the essential immediate early viral genes *ICP4* (HSV) or *E1* (Ad) is driven by a tumor/tissue-specific promoter.

Transcriptional regulatory sequences located within the transcribed sequence have also been utilized for transcriptional targeting. For example, in the endothelium-specific and angiogenesis associated flk-1 gene, a tissue-specific enhancer was located within the first intron. Accordingly, transcriptional targeting to tumor endothelium was achieved with a construct containing the flk-1 promoter and a 3′-located intron (Kappel et al., 1999; Heidenreich, Kappel, and Breier, 2000).

CHIMERIC PROMOTERS

Synthetic promoters with optimized activity and/or specificity can also be engineered by linking transcriptional control elements derived from different genes, yielding chimeric promoters. Different strategies have been pursued in this context: (1) combining defined elements of different genes with the same specificity, (2) combining specific cellular promoters with viral enhancers to increase transcriptional activity, (3) combining transcriptional control elements mediating different types of specificity, (4) utilizing disease-specific transcription factor binding elements fused to heterologous core promoters, and (5) utilizing specific repressor elements in combination with ubiquitously active promoters.

Combining Defined Elements of Different Genes of the Same (Tissue–) Specificity

Chimeric promoters for enhanced liver- or hepatocellular carcinoma (HCC)-specific gene expression have been derived by fusing a tetramer of the apolipoprotein E enhancer to the human alpha-antitrypsin (*hAAT*) promoter (Okuyama et al., 1996), or by linking the *AFP* enhancer to the albumin promoter (Su et al., 1996, 1997). For the latter construct, the IC_{50} value for HSV-tk/GCV cell killing in vitro was inversely correlated to the level of AFP (but not albumin) expression of HCC cells.

A particularly sophisticated approach for generating chimeric promoters is the random combination of (tissue-) specific elements followed by the isolation of the most active and specific construct. Li and co-workers assembled 5 to

20 muscle-specificity-mediating elements in random order in such a way that they were exposed on the same side of the DNA double helix (Li et al., 1999). This synthetic enhancer library was cloned upstream of a minimal alpha–actin promoter driving the luciferase reporter gene and subsequently analyzed for gene expression. The most promising constructs showed specific activity that was stronger than the CMV promoter/enhancer in differentiating muscle cells in vitro and after intramuscular injection in mice.

Activity Enhancement by Strong Viral Enhancers

This strategy has been pursued by cloning the CMV promoter/enhancer upstream of a 600 bp *PSA* promoter, resulting in a prostate-specific increase in promoter activity (Pang et al., 1995). However, this strategy is not universally applicable, as it depends on the presence of a tissue-specific repressor element to ablate the activity of the viral enhancer in nontarget tissues. Thus, linking the SV40 enhancer to the vascular smooth muscle cell alpha actin promoter (Keogh et al., 1999) or to the endothelium-specific von Willebrand factor (vWF) promoter (Nettelbeck, Jérôme, and Müller, 1998) resulted in activation, but loss of specificity.

In a different approach, fusion of muscle-specific elements to the CMV promoter/enhancer or partial replacement of the CMV promoter/enhancer with these elements, resulted in chimeric promoters with enhanced activity both in muscle cells and after intramuscular injection in vivo (Barnhart et al., 1998). However, these constructs showed considerable activity in nontarget cells. Interestingly, this study did not show a clear correlation between promoter activity in vivo and in vitro, thus underscoring the importance of in vivo promoter analysis in the screening process.

Combining Transcriptional Control Elements Mediating Different Types of Specificity

This strategy has been investigated in two studies, establishing combined tissue- or tumor-specific and inducible transcriptional control. In the first study, the human beta-interferon promoter was linked to the B-cell specific IgHC enhancer, resulting in B-cell-specific, virus-inducible gene expression (Engelhardt et al., 1990). The second study combined a 300 bp *AFP* promoter with the *VEGF* gene-derived hypoxia-responsive element for the targeting of liver tumors. This construct resulted in hypoxia-induced, HCC-specific and, relative to the *AFP* promoter alone, increased reporter gene expression and therapeutic gene mediated toxicity (Ido et al., 2001).

Utilizing Disease-Specific Transcription Factor Binding Elements in Combination with Heterologous Core Promoters.

To utilize disease-specific transcription factors for transcriptional targeting in gene therapy, the corresponding DNA elements, or multiple copies thereof, are

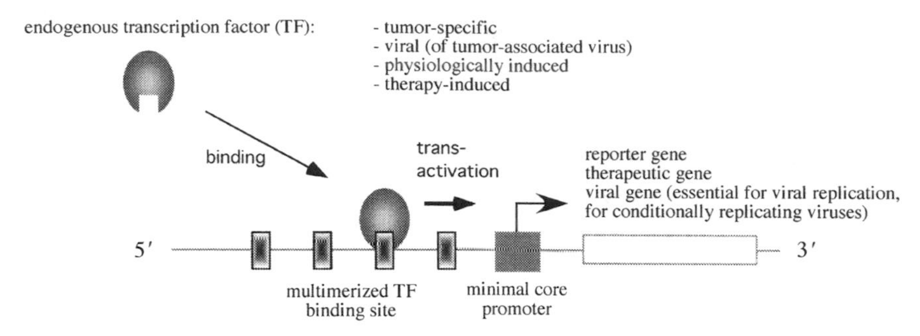

Figure 22.1 Outline of a synthetic promoter activated by disease-specific transcription factors. A synthetic promoter induced by disease-specific transcription factors consists of multiple copies of the corresponding transcription factor binding site and a minimal core promoter. Binding of the transcription factors to their binding sites in target cells activates transcription from the core promoter, which shows no activity per se. Tumor specific transcription factors can result from genomic translocations that generate fusion molecules with novel properties. The Epstein-Barr virus nuclear antigen 1 (EBNA-1) is an example for a tumor-associated viral transcription factor. Other transcription factors are induced by tumor-associated hypoxia or conventional therapeutic regimens like irradiation. Synthetic transcription factor responsive promoters have been utilized to drive expression of therapeutic genes (gene therapy) or to control expression of essential viral genes to trigger selective viral replication (virotherapy).

cloned within a synthetic construct upstream of a core promoter (Fig. 22.1). The core promoter is the locale for binding of the transcription initiation complex, but this is insufficient for appreciable transcriptional activity. Efficient gene expression from the core promoter depends on the additional binding of transcriptional activators to their cognate elements. Like the core promoter, the upstream elements are normally not capable of mediating efficient transcription on their own.

An example of this strategy is the exploitation of the tumor-specific chimeric transcription factor PAX3-FKHR, a fusion protein of the paired box gene 3 (PAX3) DNA-binding domain and the forkhead in rhabdomyosarcoma (FKHR) transactivation domain that results from a genomic rearrangement found in alveolar rhabdomyosarcoma (ARMS). Since expression of the parental transcription factor PAX3 is restricted to prenatal development, the tumor-specific chimeric transcription factor can be applied to target gene expression to alveolar rhabdomyosarcoma. For this purpose, an artificial promoter consisting of repeated PAX3 binding sites in front of a heterologous minimal adenoviral *E1B* promoter was engineered and shown to be strictly PAX3-FKHR-dependent (Massuda et al., 1997). This sophisticated approach might find broader application because translocations that generate chimeric transcription factors are frequently found in human cancers.

In a similar approach, viral transcription factors have been utilized for the transcriptional targeting of transgenes to virus-associated tumors. Specifically, multimerized binding sites for the Epstein-Barr Virus (EBV) encoded transcription factors EBNA-1 or -2, cloned upstream of a minimal core promoter, have been applied to the targeting of Burkitt's lymphoma and other EBV-associated tumors (Judde et al., 1996; Gutierrez et al., 1996; Franken et al., 1996). Because viruses have been associated with several human malignancies, this approach may be applicable to a variety of cancers. An interesting aspect of both approaches described previously is the repressor activity of the PAX3 and EBNA-1 binding elements in the absence of the corresponding transcription factors, which leads to tight transcriptional control and remarkable inducibility.

In a similar scenario, artificial promoters responsive to physiologically activated transcription factors have been engineered. For the targeting of hypoxic tumors, hypoxia-responsive elements (HREs) from different genes have been cloned upstream of the minimal SV40 promoter and analyzed for induction by oxygen deprivation. A trimer of the phosphoglycerate kinase gene HRE was shown to mediate a >100-fold activation in the presence of 0.1% oxygen, resulting in stronger gene expression than the CMV promoter/enhancer (Boast et al., 1999; Binley et al., 1999). Another study demonstrated that the identity of the core promoter is critical in such a setting. Hence, a pentamer of the VEGF-derived HRE mediated hypoxia-dependent gene expression in combination with the CMV or adenovirus *E1B* minimal promoters, but not with the elongation factor 1 α (EF-1 α) core promoter (Shibata, Giaccia, and Brown, 2000). Clearly, an advantage of these artificial hypoxia-dependent promoters is their applicability to a broad range of solid tumors. In addition, increased specificity was achieved by combining the HRE repeats with a tumor-specific core promoter (see foregoing text).

A further scenario is the utilization of multimerized binding sites for therapy-induced transcription factors. The goal of this concept is to combine gene therapy with conventional therapies, for example, the temporally and/or spatially restricted expression of proteins, which sensitize cells to the inducing conventional therapeutic regimen. In this regard, the heat shock factor 1 (HSF-1) transcription factor was utilized for hyperthermia-induced gene expression by fusing multiple HSF-1 binding sites, termed heat shock elements (HSE), upstream of the heat shock protein 70b promoter. This synthetic promoter resulted in increased heat-inductivity of reporter gene expression compared to the hsp70b promoter alone (Brade et al., 2000). Another therapy-inducible synthetic promoter consists of multiple radiation responsive elements upstream of the minimal CMV promoter. Radiation inducibilty of this synthetic promoter has been shown to be superior to the wild-type radiation-responsive egr-1 promoter (Marples et al., 2000).

A different concept is the utilization of disease-specific transcription factors to target viral oncolysis. Recently, an adenovirus with colon cancer restricted replication competence was generated by replacing the promoters of the essential viral genes E1B and E2 with multimerized binding sites for the transcription factor Tcf4 (Brunori et al., 2001). This transcription factor is constitutively activated by mutations in the beta-catenin and adenomatous polyposi coli (*APC*) genes, a hallmark of colon cancer.

Utilizing Specific Repressor Elements in Combination with Ubiquitously Active Enhancers or Promoters

Repressor elements that mediate repression selectively in nontarget tissues represent a potentially interesting option to achieve tissue-specific gene expression. Their utilization for gene therapy purposes is feasible, if combined with ubiquitously active transcriptional control elements. Millecamps and co-workers achieved targeted gene expression in the context of adenoviral vectors with a promoter construct harboring multiple copies of the neuron-restrictive silencer factor (NRSF) binding sites upstream of the ubiquitous phosphoglycerate kinase promoter (Millecamps et al., 1999). Since NRSF is expressed only in nonneuronal cells, this artificial promoter is specifically active in neurons.

RECOMBINANT TRANSCRIPTIONAL ACTIVATORS: INDUCIBILITY, AMPLIFICATION, AND DUAL SPECIFICITY

Transcriptional control is mediated by *cis*-acting regulatory elements and *trans*-acting transcription factors that bind to these elements. Artificial transcriptional control systems can therefore also be based on engineered designer transcription factors or recombinant transcriptional activators (RTAs, Fig. 22.2A). Transcription factors are composed of multiple functionally independent domains (modules), including a DNA-binding domain (DBD) that directs the protein to specific binding sites, and a transactivation domain (TAD) that controls the transcription rate through interactions with other transcription factors and the transcription machinery. RTAs can be engineered by combining

Figure 22.2 Enhancing promoter activity by use of recombinant transcriptional activators (RTAs). (*A*) RTAs are synthetic fusion proteins containing a potent transactivation domain (for example of the HSV-VP16 protein) fused to a DNA-binding domain (for example of the yeast Gal4 or the bacterial LexA protein). In addition, a ligand-binding domain (for example of a steroid hormone receptor) can be incorporated to trigger ligand-dependent RTA activity. The RTA is encoded by a corresponding fusion gene. Binding of the RTA to its binding site results in activation of gene expression. For amplification of specific promoter activity (*B*) The RTA is expressed from the (tissue-) specific but weak promoter. The RTA binds to synthetic binding sites upstream of a second promoter and activates transcription and gene expression via its strong transactivation domain. A further level of specificity can be achieved if the second promoter is also specific and thus activated in target cells only (*C*). (*B* See Segawa et al., 1998; *C* see Nettelbeck, Jérôme and Müller, 1998).

modules that exhibit the desired characteristics. Frequently, a DBD of non-mammalian origin—which does not bind to mammalian sequences, thus preventing adverse effects on endogenous gene expression—is combined with a TAD of choice (for example a strong activator). In addition, ligand-binding domains (LBDs), derived, for example, from steroid hormone receptors, can be incorporated into the RTA, resulting in ligand-controlled activity of the RTA

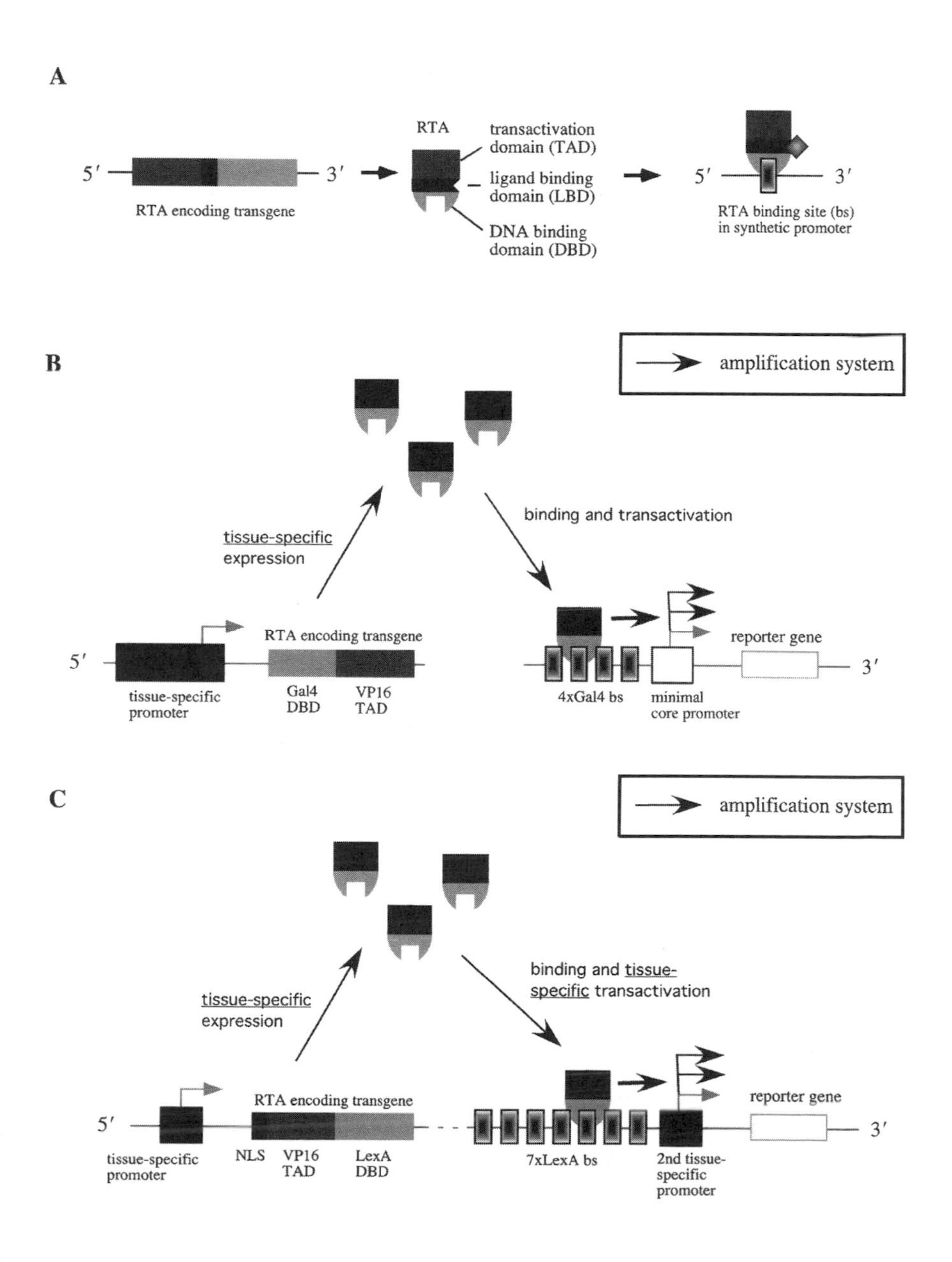

and RTA-dependent promoters. Artificial promoters regulated by RTAs have also been shown to be functional in vivo (Oligino et al., 1996; Fang et al., 1998). They have been developed for different purposes, including pharmacologically regulated gene expression, amplification of promoter activity, and combining tissue-specific and cell cycle regulated gene expression.

Pharmacologically Regulated Gene Expression

Several RTA-driven artificial promoter systems have been developed to implement inducible gene expression (for references see Harvey and Caskey, 1998). A common feature of these systems is that the gene of interest is expressed under control of a synthetic RTA responsive promoter. The RTA itself is expressed constitutively, whereas DNA-binding of the RTA, and thus transgene expression, is drug dependent. One drug-inducible promoter system, termed Tet-OFF, utilizes the bacterial tet repressor fused to the HSV VP16 transactivation domain (tTA). This RTA binds to the tet operator DNA element in absence of tetracycline or its analogue doxocycline. A reverse phenotype mutant of the tet repressor fused to the VP16 TAD (rtTA) binds to the tet operator in presence of doxocycline, thus establishing the Tet-ON system. A further variation of the tet system is the fusion of the tetR to the transcription repression domain of the transcription factor KRAB. In this approach, gene expression depends on dissociation of the RTA from the tet operator.

A further group of pharmacologically controlled synthetic promoters is represented by the hormone-inducible promoter systems. Here, the RTA consists of the yeast Gal4 DBD fused to a hormone receptor and the VP16 TAD or the KRAB repression domain. DNA binding of the RTA, and consequently gene expression or repression, are induced, for instance, by the progesterone analogue RU486. This component, but not endogenous progesterone, binds to a truncated progesterone receptor module of the RTA.

The rapamycin inducible or dimerizer system is triggered by an RTA consisting of two subunits, a DBD and a TAD fusion protein. These fusion proteins dimerize in the presence of rapamycin, thus reconstituting a functional RTA capable of inducing expression of the gene of interest via a corresponding DNA element.

Of note, spatial targeting of inducible transcriptional control is feasible by expressing the drug responsive RTA from an appropriate tissue-specific promoter (Smith-Arica et al., 2000).

Amplification of Promoter Activity

A further application of RTAs is the amplification of the activity of specific but weak promoters in order to obtain gene expression appropriate for therapeutic purposes. In this scenario, the weak promoter drives expression of a strong RTA (harboring, for example, the powerful HSV VP16 transactivation domain). In a second construct, an RTA-responsive promoter drives expression of the gene of interest (Fig. 22.2B and C). This strategy is more generally applicable than the deletion/multimerization approach for enhancing promoter activity, since the latter depends on the presence of defined promoter elements. However, as even low-level expression of the powerful RTA in nontarget cells would result in loss of specificity, the "tightness" of the applied tissue-specific promoter is critical for the development of an RTA-driven promoter amplification system. Segawa et al. used the PSA promoter to drive expression of a Gal4-VP16 fusion protein. This RTA then activates transgene expression through Gal4 binding sites placed upstream of a minimal tk promoter, resulting in enhanced transgene expression (Fig. 22.2B; Segawa et al., 1998).

In a different system, the specific promoter drives expression of both the gene encoding the RTA, here a LexA-DBD/VP16-TAD fusion protein, and the transgene. Incorporation of multimeric LexA binding sites upstream of the specific but weak promoter results in specific amplification of gene expression. With this approach both the endothelium-specific vWF and the gastrointestinal-specific sucrase isomaltase promoters were enhanced >100-fold in target cells. No increase in transgene expression over the activity of promoterless constructs was observed in control cells, resulting in high levels (>1000-fold) of specificity (Nettelbeck, Jérôme, and Müller, 1998). The tightness of this approach was achieved by two mechanisms conveying specificity: (1) specific expression of the RTA and (2) restriction of RTA-mediated transactivation to target cells by the specific promoter (Fig. 22.2C).

Dual-Specific Transcriptional Control Combining Tissue-Specific and Cell Cycle-Regulated Gene Expression

Unrestrained proliferation is a hallmark of cancer. Thus, anticancer therapeutics frequently target proliferating cells. Limitations of this strategy are imposed by (1) the resistance of tumor cells not proliferating at the time of therapy and (2) a dose-limiting toxicity due to adverse effects on healthy proliferative tissues, such as the bone marrow and the epithelium of digestive organs. In gene therapy, transcriptional targeting strategies can potentially circumvent these drawbacks by constructing proliferation-dependent vectors rather than effectors by utilizing cell cycle regulated promoters (Nettelbeck et al., 1998). To target nonproliferating tumor cells, effector systems with a local bystander effect on nonproliferating neighboring cells can be utilized. Furthermore, the dual-specificity strategy combining tissue-specific and cell cycle-regulated transcription control can abrogate toxicity to proliferating nontarget tissue.

Two generally applicable dual-specificity designer promoter systems have been described. The first approach (Fig. 22.3A) is based on a heterodimeric RTA, which drives expression of the therapeutic gene. Dual-specificity results from tissue-specific expression of the first RTA subunit, here by the optimized tyrosinase promoter described previously, and proliferation-specific expression

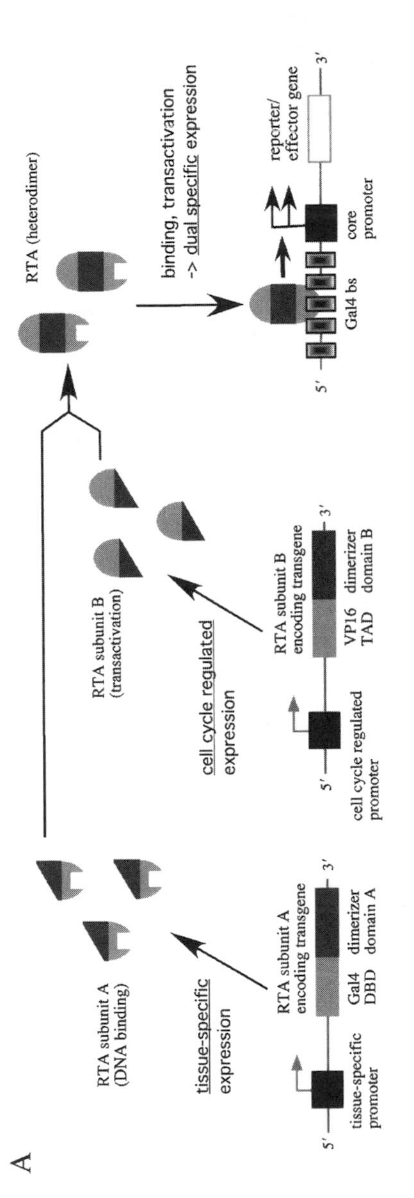

A
RTA subunit A (DNA binding)
tissue-specific expression
RTA subunit A encoding transgene
5′
3′
tissue-specific promoter
Gal4 DBD
dimerizer domain A
RTA subunit B (transactivation)
cell cycle regulated expression
RTA subunit B encoding transgene
cell cycle regulated promoter
VP16 TAD
dimerizer domain B
RTA (heterodimer)
binding, transactivation -> dual specific expression
Gal4 bs
core promoter
reporter/ effector gene

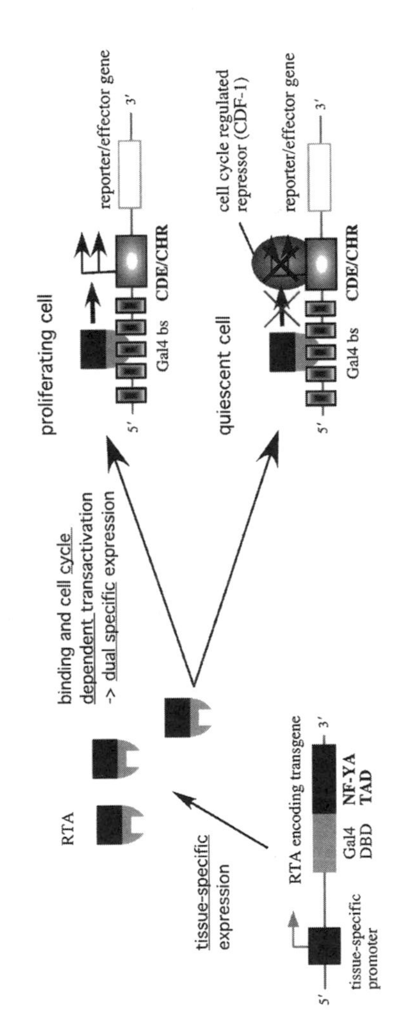

B
RTA
tissue-specific expression
RTA encoding transgene
5′
3′
tissue-specific promoter
Gal4 DBD
NF-YA TAD
binding and cell cycle dependent transactivation -> dual specific expression
proliferating cell
Gal4 bs
CDE/CHR
reporter/effector gene
quiescent cell
cell cycle regulated repressor (CDF-1)

of the second subunit, here by the *cyclin A* promoter (Jérôme and Müller, 1998, 2001). The second dual specificity system (Fig. 22.3B) is based on the mechanism of cell cycle dependent transcriptional repression of the *cyclin A* promoter. Here, the RTA, a fusion protein of the Gal4-DBD and the TAD of the transcription factor NF-Y, is expressed under control of the tissue-specific optimized tyrosinase promoter. In the target tissue, the RTA subsequently binds to multimerized Gal4 binding sites replacing the upstream regulatory sequence of the *cyclin A* promoter. Cell cycle regulated gene expression is mediated by a transcriptional repressor, which binds to the *cyclin A* core promoter specifically in non-proliferating cells. This repressor specifically inhibits transactivation by the NF-Y transactivation domain. For this system, a cell cycle specificity similar to that of the wild-type *cyclin A* promoter, but restricted to melanoma cells, has been achieved (Nettelbeck, Jérôme, and Müller, 1999).

The described RTA-dependent promoter systems involve the expression of nonhuman proteins, which are potentially immunogenic in patients. For clinical gene therapy it may be advantageous to avoid an immune response to the transduced cells. To this end, it has been reported that the viral VP16 domain can be replaced without loss of activity with the TAD of human p65, a subunit of the transcription factor NFκB (Rivera et al., 1996). On the other hand, RTAs with a DBD of a human transcription factor could bind to their cognate DNA sites in the genome of transduced cells potentially triggering detrimental interferences with endogenous gene expression. Recently, human zink-finger derived DBDs with altered DNA sequence specificities have been engineered by exchanging only a few amino acids. These DBDs combine both lower immunogenicity than viral DBDs with high specificity for the transgenic target sequence—an 18 bp sequence of choice—relative to genomic sites (Xu et al., 2001).

Figure 22.3 Dual specificity promoter systems: combining tissue specificity and cell cycle regulation by utilization of RTAs. (*A*) A DNA-binding subunit expressed from a tissue-specific promoter and a transactivating subunit expressed from a cell cycle-regulated promoter interact to form a heterodimeric RTA via fused dimerization domains. Expression of the functional, heterodimeric RTA is therefore restricted to proliferating cells of a certain tissue type. The RTA binds to Gal4 binding sites (bs) of a synthetic reporter/effector construct and activates gene expression from the downstream core promoter (Jérôme and Müller, 1998, 2001) (*B*) An RTA that consists of the Gal4 DNA-binding domain fused to the NF-YA transactivation domain is expressed from a tissue-specific promoter and binds constitutively to Gal4 binding sites of a synthetic reporter/effector construct. The transcriptional repressor CDF-1 binds to the downstream CDE/CHR element specifically in the G_0/G_1 phases of the cell cycle and inhibits transactivation by NF-YA. As a consequence, expression of the transgene is restricted to proliferating cells of a specific type (from Nettelbeck, Jérôme and Müller 1999).

THE CRE/LOX SYSTEM

The bacteriophage P1-derived Cre/lox system consists of the Cre recombinase and a pair of loxP DNA elements. The Cre recombinase mediates site-specific excisional deletion of a DNA sequence that is flanked by a pair of loxP sequences resulting in a residual loxP site left in the original DNA sequence and the excised DNA circularized at the loxP site (Fig. 22.4). For transcriptional targeting in gene therapy, the Cre/lox system is utilized for these conceptual strategies (see Fig. 22.4): (1) as an ON switch for the amplification and prolongation of promoter activity (gene activation strategy) or (2) as an OFF switch to selectively delete a target gene (gene inactivation strategy).

The Cre/lox Gene Activation Strategy

The Cre/lox ON switch (Fig. 22.4A) has been described in several recent gene therapy studies. In this scenario, the reporter or therapeutic gene of interest separated from a strong (and constitutive) promoter by a transcriptional stop cassette, that is a marker gene or a polyA transcription termination sequence, flanked by a pair of loxP sites. In a second construct, the specific promoter of choice drives expression of the Cre recombinase. Activity of the specific promoter in target cells results in expression of Cre and subsequent excision of the stop cassette in the reporter/therapeutic construct leading to strong and constitutive expression of the gene of interest. By means of this strategy, thyroglobulin or CEA promoter driven HSV-tk/GCV therapies of thyroid carcinoma and adenocarcinoma, respectively, have been improved significantly without loss of specificity (Nagayama et al., 1999; Kijima et al., 1999). In a different context, the Cre/lox ON system switches the transient activity of a radiation-activated promoter into enhanced and persistent expression of the gene of interest (Scott et al., 2000).

Figure 22.4 Switching genes ON and OFF with the Cre/lox system. (*A*) In the Cre/lox gene activation strategy a transcription stop cassette flanked by loxP sites is deleted by Cre recombinase selectively in target cells where the specific promoter that drives Cre expression is active. This strategy results in strong and prolonged expression of the gene of interest. In cells not targeted by the specific promoter, therapeutic gene expression is prevented by the transcription termination signal, the stop cassette, which is located between constitutive promoter and gene of interest. (*B*) In contrast, in the Cre/lox gene inactivation strategy the gene of interest is excised by Cre recombinase. This strategy has been recently applied for the development of conditionally replicative adenoviruses. Here, Cre expression is controlled by a p53 responsive promoter resulting in excision of the essential E1A gene from the virus genome in healthy cells (p53 wild-type), but not in p53 mutant tumor cells. Thus, the virus specifically replicates in tumor cells that lack functional p53 (Nagayama et al., 2001).

The Cre/lox Gene Inactivation Strategy

The Cre/lox gene inactivation strategy (Fig. 22.4B) has been applied to the transcriptional targeting of lytic adenovirus replication to p53 mutant tumor cells. p53, a transcriptional activator, is mutated, deleted, or inactivated in roughly 50% of human tumors. Flanking the essential viral E1A gene by a pair of loxP sites and regulating Cre expression with a p53-responsive pro-

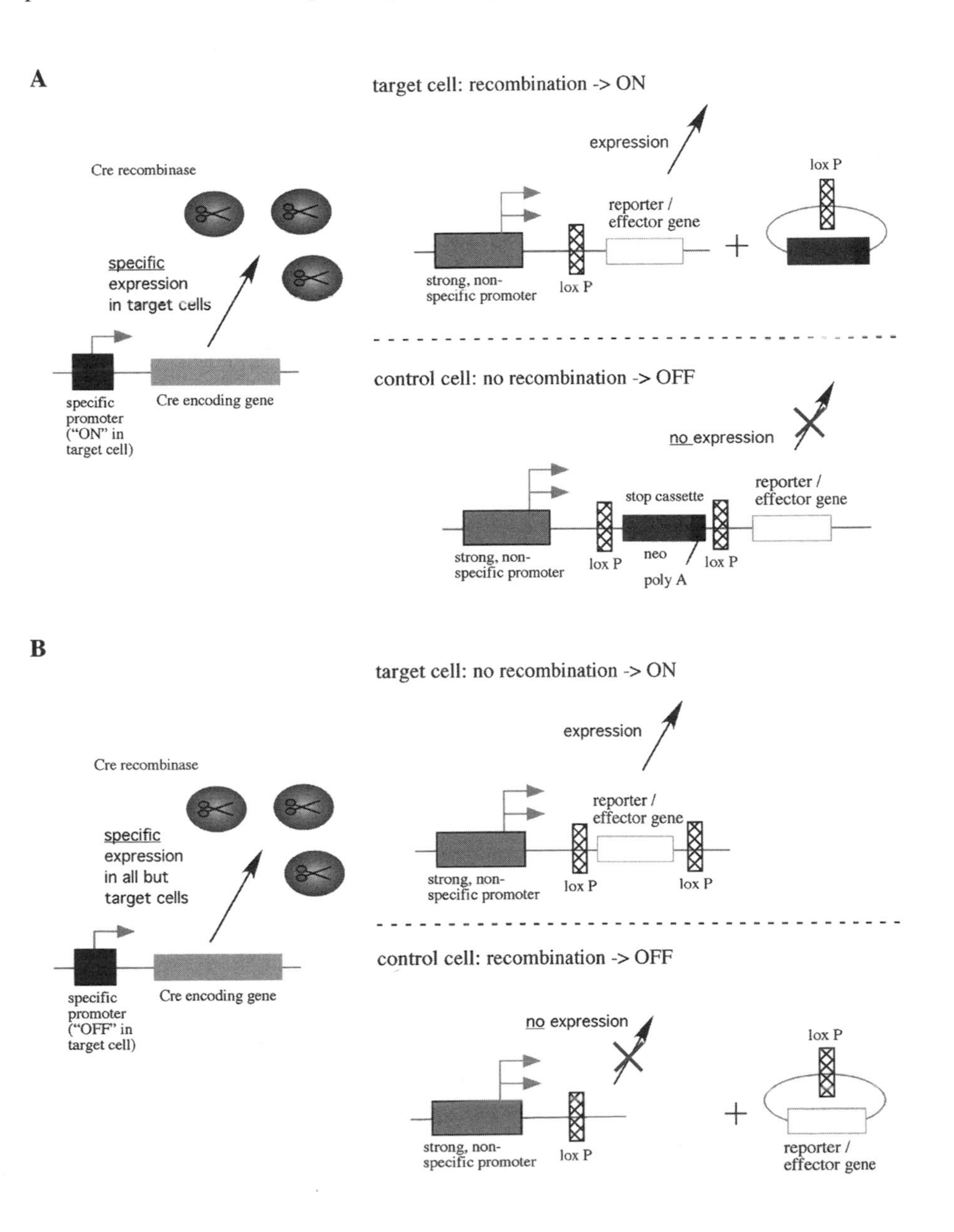

moter results in deletion of the viral gene and loss of viral replication capacity in nonmalignant, p53 wild-type cells. In contrast, in tumor cells without functional p53, the virus stays intact and retains its replicative potency, resulting in oncolysis (Nagayama et al., 2001).

NONPROMOTER ELEMENTS FOR OPTIMIZED TRANSCRIPTIONAL TARGETING AND PROMOTER FIDELITY IN VIRAL VECTOR GENOMES

Noncoding DNA Sequences Other Than Promoters, Enhancers, and Silencers

Noncoding DNA sequences other than promoters, enhancers, and silencers have shown benefit for gene therapeutic purposes. For instance, gene expression can be increased by DNA elements mediating posttranscriptional events, as exemplified by the incorporation of an *intron* into the nontranslated region of the therapeutic gene. Increased gene expression results from splicing of the intron-containing pre-mRNA, as splicing can promote mRNA export and translation. Nevertheless, the activity of such elements depends on the candidate gene (Schambach et al., 2000) and has to be reevaluated in the context of a given promoter.

Virus-derived internal ribosome entry site (IRES) sequences are other elements of interest for transcriptional targeting in gene therapy. These sequences mediate cap-independent internal translation initiation. By this means, a single specific promoter can drive expression of two transgenes via a bicistronic IRES containing mRNA (Harries et al., 2000). However, the activity of IRES sequences depends on the fused cDNA and in some instances it can be advantageous to apply two promoters rather than an IRES sequence (Nettelbeck, Jérôme, and Müller, 1998).

Noncoding sequences not triggering transcriptional control can also mediate (or improve) specific gene expression. This possibility was revealed for an element located in the 3′ nontranslated sequence of hypoxia-dependent genes that regulates RNA stability in a hypoxia-dependent manner (Damert et al., 1997; Boast et al., 1999; Shibata, Giaccia, and Brown, 2000).

Promoter Fidelity in the Context of the Vector of Choice

Crucial for the application of transcriptional targeting in gene therapy is the fidelity of promoters in the context of the vector of choice (vector compatibility of a candidate promoter). Viral vectors contain sequences with transcriptional regulatory activity, which can be detrimental to the fidelity of a candidate-specific promoter, as reported for adenoviral and retroviral vectors (Vile et al., 1994, 1995; Ring et al., 1996; Shi, Wang, and Worton, 1997). Whereas some promoters might not be influenced by viral sequences, others lose their specificity in the vector context.

A potentially detrimental adenoviral regulatory element is the E1A enhancer, which cannot be deleted, as it overlaps with the packaging signal. It is located in proximity to the conventional site of transgene incorporation. In this context, promoter interference has been avoided by two strategies: engineering of the adenovirus vector backbone or incorporation of insulator elements. The first strategy has been realized by the creation of transgene incorporation sites less prone to promoter interference in the virus genome (Rubinchik et al., 2001). Insulator elements are defined as DNA sequences that do not influence promoter activity per se, but block promoter activation by enhancers when placed between them. Insulator activity in the context of adenoviral vectors has been shown for a 1.2 kb element derived from the chicken beta-globin locus (Steinwaerder and Lieber, 2000) and for the bovine growth hormone polyadenylation signal (Vassaux, Hurst, and Lemoine, 1999).

In retroviral vectors, enhancer elements within the long terminal repeat (LTR) represent potential interfering sequences. Whereas some promoters have been shown to lose specificity in an internal position within the retroviral genome, other reports demonstrate specific transcriptional control by replacing the viral LTR enhancer with these or other promoters, for example the muscle creatine kinase enhancer (Ferrari et al., 1995), the tyrosinase promoter/enhancer (Vile et al., 1995; Diaz et al., 1998), the human preproendothelin promoter (Jager, Zhao, and Porter, 1999), or a synthetic hypoxia-responsive promoter (Boast et al., 1999). Transcriptional targeting in retroviral gene transfer faces a further level of promoter deregulation resulting from the chromosomal integration of the vector genome and interference by neighboring chromosomal elements. In this regard, flanking the transgene by matrix or scaffold attachment regions (MARs or SARs) or by a locus control region (LCR) can result in position-independent transgene activity after genomic integration (McKnight et al., 1992; Kalos and Fournier, 1995; Dang, Auten, and Plavec, 2000; Kowolik, Hu, and Yee, 2001).

CONCLUSIONS

There is no doubt that vector optimization and vector specificity are essential for translating gene therapy concepts into successful clinical regimens. To this end, transcriptional targeting is a powerful strategy for mediating pharmacologically, physiologically, or spatially regulated gene therapy. Furthermore, the conditionality of viral replication required by the recent therapeutic approach of viral oncolysis can be achieved by transcriptional targeting. Preclinical studies have proven the feasibility of transcriptional targeting for mediating effective and selective gene therapy and viral oncolysis, resulting in significantly reduced toxic side effects. Clinical studies applying transcriptional targeting are currently underway.

Based on the rapid progress in research on transcriptional regulation in recent years, a plethora of novel strategies for the optimization of promoter activity and for the development of artificial promoters has been successfully established. These strategies resulted in reduced promoter leakiness and in increased promoter activity and specificity in different vector systems. The utilization of this transcriptional targeting science, in combination with novel technologies to determine disease signatures, might promote the derivation of a new generation of targeted vectors and their application in molecular medicine.

ACKNOWLEDGMENTS

We are grateful to Joel Glasgow for critical reading of the manuscript. Research in the authors' laboratories was supported by grants from the National Cancer Institute R01 CA74242, R01 CA86881, N01 CO-97110, R01 CA68245, P50 CA89019, R01 HL50255, and from the Juvenile Diabetes Foundation 1-2000-23 to DTC; and from the Deutsche Forschungsgemeinschaft and the Dr. Mildred Scheel Stiftung to RM. DMN is supported by a postdoctoral fellowship from the Deutsche Forschungsgemeinschaft.

REFERENCES

Anderson WF (1998): Human gene therapy. *Nature* 392: 25–30.

Barnhart KM, Hartikka J, Manthorpe M, Norman J, Hobart P (1998): Enhancer and promoter chimeras in plasmids designed for intramuscular injection: a comparative *in vivo* and *in vitro* study. *Hum Gene Ther* 9:2545–2553.

Binley K, Iqball S, Kingsman A, Kingsman S, Naylor S (1999): An adenoviral vector regulated by hypoxia for the treatment of ischaemic disease and cancer. *Gene Ther* 6:1721–1727.

Boast K, Binley K, Iqball S, Price T, Spearman H, Kingsman S, Kingsman A, Naylor S (1999): Characterization of physiologically regulated vectors for the treatment of ischemic disease. *Hum Gene Ther* 10:2197–2208.

Brade AM, Ngo D, Szmitko P, Li PX, Liu FF, Klamut HJ (2000): Heat-directed gene targeting of adenoviral vectors to tumor cells. *Cancer Gene Ther* 7:1566–1574.

Brunori M, Malerba M, Kashiwazaki H, Iggo R (2001): Replicating adenoviruses that target tumors with constitutive activation of the wnt signaling pathway. *J Virol* 75:2857–2865.

Damert A, Machein M, Breier G, Fujita MQ, Hanahan D, Risau W, Plate KH (1997): Up-regulation of vascular endothelial growth factor expression in a rat glioma is conferred by two distinct hypoxia-driven mechanisms. *Cancer Res* 57:3860–3864

Dang Q, Auten J, Plavec I (2000): Human beta interferon scaffold attachment region inhibits de novo methylation and confers long-term, copy number-dependent expression to a retroviral vector. *J Virol* 74:2671–2678

Diaz RM, Eisen T, Hart IR, Vile RG (1998): Exchange of viral promoter/enhancer elements with heterologous regulatory sequences generates targeted hybrid long terminal repeat vectors for gene therapy of melanoma. *J Virol* 72:789–795.

Engelhardt JF, Kellum MJ, Bisat F, Pitha PM (1990): Retrovirus vector-targeted inducible expression of human beta-interferon gene to B-cells. *Virology* 178:419–428.

Fang B, Ji L, Bouvet M, Roth JA (1998): Evaluation of GAL4/TATA *in vivo*. Induction of transgene expression by adenovirally mediated gene codelivery. *J Biol Chem* 273:4972–4975.

Ferrari G, Salvatori G, Rossi C, Cossu G, Mavilio F (1995): A retroviral vector containing a muscle-specific enhancer drives gene expression only in differentiated muscle fibers. *Hum Gene Ther* 6:733–742.

Franken M, Estabrooks A, Cavacini L, Sherburne B, Wang F, Scadden DT (1996): Epstein-Barr virus-driven gene therapy for EBV-related lymphomas. *Nat Med* 2:1379–1382.

Gutierrez MI, Judde JG, Magrath IT, Bhatia KG (1996): Switching viral latency to viral lysis: a novel therapeutic approach for Epstein-Barr virus-associated neoplasia. *Cancer Res* 56:969–972.

Hallenbeck PL, Chang YN, Hay C, Golightly D, Stewart D, Lin J, Phipps S, Chiang YL (1999): A novel tumor-specific replication-restricted adenoviral vector for gene therapy of hepatocellular carcinoma. *Hum Gene Ther* 10:1721–1733.

Harries M, Phillipps N, Anderson R, Prentice G, Collins M (2000): Comparison of bicistronic retroviral vectors containing internal ribosome entry sites (IRES) using expression of human interleukin-12 (IL-12) as a readout. *J Gene Med* 2:243–249.

Harvey DM, Caskey CT (1998): Inducible control of gene expression: prospects for gene therapy. *Curr Opin Chem Biol* 2:512–518.

Heidenreich R, Kappel A, Breier G (2000): Tumor endothelium-specific transgene expression directed by vascular endothelial growth factor receptor-2 (Flk-1) promoter/enhancer sequences. *Cancer Res* 60:6142–6147.

Ido A, Uto H, Moriuchi A, Nagata K, Onaga Y, Onaga M, Hori T, Hirono S, Hayashi K, Tamaoki T, Tsubouchi H (2001): Gene therapy targeting for hepatocellular carcinoma: selective and enhanced suicide gene expression regulated by a hypoxia-inducible enhancer linked to a human alpha-fetoprotein promoter. *Cancer Res* 61:3016–3021.

Ishikawa H, Nakata K, Mawatari F, Ueki T, Tsuruta S, Ido A, Nakao K, Kato Y, Ishii N, Eguchi K (1999): Utilization of variant-type of human a-fetoprotein promoter in gene therapy targeting hepatocellular carcinoma. *Gene Ther* 6:465–470.

Jager U, Zhao Y, Porter CD (1999): Endothelial cell-specific transcriptional targeting from a hybrid long terminal repeat retrovirus vector containing human prepro-endothelin-1 promoter sequences. *J Virol* 73:9702–9709.

Jérôme V, Müller R (1998): Tissue-specific, cell cycle-regulated chimeric transcription factors for the targeting of gene expression to tumor cells *Hum Gene Ther* 9:2653–2659.

Jérôme V, Müller R (2001): A synthetic leucine zipper-based dimerization system for combining multiple promoter specificities. *Gene Ther* 8:725–729.

Judde JG, Spangler G, Magrath I, Bhatia K (1996): Use of Epstein-Barr virus nuclear antigen-1 in targeted therapy of EBV- associated neoplasia *Hum Gene Ther* 7:647–653.

Kalos M, Fournier RE (1995): Position-independent transgene expression mediated by boundary elements from the apolipoprotein B chromatin domain . *Mol Cell Biol* 15:198–207.

Kappel A, Ronicke V, Damert A, Flamme I, Risau W, Breier G (1999): Identification of vascular endothelial growth factor (VEGF) receptor-2 (Flk-1) promoter/enhancer sequences sufficient for angioblast and endothelial cell-specific transcription in transgenic mice. *Blood* 93:4284–4292.

Keogh MC, Chen D, Schmitt JF, Dennehy U, Kakkar VV, Lemoine NR (1999): Design of a muscle cell-specific expression vector utilising human vascular smooth muscle alpha-actin regulatory elements. *Gene Ther* 6:616–628.

Kijima T, Osaki T, Nishino K, Kumagai T, Funakoshi T, Goto H, Tachibana I, Tanio Y, Kishimoto T (1999): Application of the Cre recombinase/loxP system further enhances antitumor effects in cell type-specific gene therapy against carcinoembryonic antigen-producing cancer. *Cancer Res* 59:4906–4911.

Kowolik CM, Hu J, Yee JK (2001): Locus control region of the human cd2 gene in a lentivirus vector confers position-independent transgene expression. *J Virol* 75:4641–4648.

Latham JP, Searle PF, Mautner V, James ND (2000): Prostate-specific antigen promoter/enhancer driven gene therapy for prostate cancer: construction and testing of a tissue-specific adenovirus vector. *Cancer Res* 60:334–341.

Li X, Eastman EM, Schwartz RJ, Draghia AR (1999): Synthetic muscle promoters: activities exceeding naturally occurring regulatory sequences. *Nat Biotechnol* 17:241–245.

Marples B, Scott SD, Hendry JH, Embleton MJ, Lashford LS, Margison GP (2000): Development of synthetic promoters for radiation-mediated gene therapy. *Gene Ther* 7:511–517.

Massuda ES, Dunphy EJ, Redman RA, Schreiber JJ, Nauta LE, Barr FG, Maxwell IH, Cripe TP (1997): Regulated expression of the diphtheria toxin A chain by a tumor-specific chimeric transcription factor results in selective toxicity for alveolar rhabdomyosarcoma cells. *Proc Natl Acad Sci USA* 94:14701–14706.

Mawatari F, Tsuruta S, Ido A, Ueki T, Nakao K, Kato Y, Tamaoki T, Ishii N, Nakata K (1998): Retrovirus-mediated gene therapy for hepatocellular carcinoma: selective and enhanced suicide gene expression regulated by human alpha-fetoprotein enhancer directly linked to its promoter. *Cancer Gene Ther* 5:301–306.

McKnight RA, Shamay A, Sankaran L, Wall RJ, Hennighausen L (1992): Matrix-attachment regions can impart position-independent regulation of a tissue-specific gene in transgenic mice. *Proc Natl Acad Sci USA* 89:6943–6947.

Millecamps S, Kiefer H, Navarro V, Geoffroy MC, Robert JJ, Finiels F, Mallet J, Barkats M (1999): Neuron-restrictive silencer elements mediate neuron specificity of adenoviral gene expression. *Nat Biotechnol* 17:865–869.

Miyatake S, Iyer A, Martuza RL, Rabkin SD (1997): Transcriptional targeting of herpes simplex virus for cell-specific replication. *J Virol* 71:5124–5132.

Miyatake SI, Tani S, Feigenbaum F, Sundaresan P, Toda H, Narumi O, Kikuchi H, Hashimoto N, Hangai M, Martuza RL, Rabkin SD (1999): Hepatoma-specific antitumor activity of an albumin enhancer/promoter regulated herpes simplex virus *in vivo*. *Gene Ther* 6:564–572.

Nabel GJ (1999): Development of optimized vectors for gene therapy. *Proc Natl Acad Sci USA* 96:324–326.

Nagayama Y, Nishihara E, Iitaka M, Namba H, Yamashita S, Niwa M (1999): Enhanced efficacy of transcriptionally targeted suicide gene/prodrug therapy for thyroid carcinoma with the Cre-loxP system. *Cancer Res* 59:3049–3052.

Nagayama Y, Nishihara E, Namba H, Yokoi H, Hasegawa M, Mizuguchi H, Hayakawa T, Hamada H, Yamashita S, Niwa M (2001): Targeting the replication of adenovirus to p53-defective thyroid carcinoma with a p53-regulated Cre/loxP system. *Cancer Gene Ther* 8:36–44.

Nettelbeck DM, Jérôme V, Müller R (1998): A strategy for enhancing the transcriptional activity of weak cell type-specific promoters. *Gene Ther* 5:1656–1664.

Nettelbeck DM, Jérôme V, Müller R (1999): A dual specificity promoter system combining cell cycle-regulated and tissue-specific transcriptional control. *Gene Ther* 6:1276–1281.

Nettelbeck DM, Zwicker J, Lucibello FC, Gross C, Liu N, Brüsselbach S, Müller R (1998): Cell cycle regulated promoters for the targeting of tumor endothelium. *Adv Exp Med Biol* 451:437–440.

Okuyama T, Huber RM, Bowling W, Pearline R, Kennedy SC, Flye MW, Ponder KP (1996): Liver-directed gene therapy: a retroviral vector with a complete LTR and the ApoE enhancer-alpha 1-antitrypsin promoter dramatically increases expression of human alpha 1-antitrypsin *in vivo. Hum Gene Ther* 7:637–645.

Oligino T, Poliani PL, Marconi P, Bender MA, Schmidt MC, Fink DJ, Glorioso JC (1996): *In vivo* transgene activation from an HSV-based gene therapy vector by GAL4:vp16. *Gene Ther* 3:892–899.

Pang S, Taneja S, Dardashti K, Cohan P, Kaboo R, Sokoloff M, Tso CL, Dekernion JB, Belldegrun AS (1995): Prostate tissue specificity of the prostate-specific antigen promoter isolated from a patient with prostate cancer. *Hum Gene Ther* 6:1417–1426.

Pang S, Dannull J, Kaboo R, Xie Y, Tso CL, Michel K, deKernion JB, Belldegrun AS (1997): Identification of a positive regulatory element responsible for tissue- specific expression of prostate-specific antigen. *Cancer Res* 57:495–499.

Park BJ, Brown CK, Hu Y, Alexander HR, Horti J, Raje S, Figg WD, Bartlett DL (1999): Augmentation of melanoma-specific gene expression using a tandem melanocyte-specific enhancer results in increased cytotoxicity of the purine nucleoside phosphorylase gene in melanoma *Hum Gene Ther* 10:889–898.

Richards CA, Austin EA, Huber BE (1995): Transcriptional regulatory sequences of carcinoembryonic antigen: identification and use with cytosine deaminase for tumor-specific gene therapy. *Hum Gene Ther* 6:881–893.

Ring CJ, Harris JD, Hurst HC, Lemoine NR (1996): Suicide gene expression induced in tumour cells transduced with recombinant adenoviral, retroviral and plasmid vectors containing the ERBB2 promoter. *Gene Ther* 3:1094–1103.

Rivera VM, Clackson T, Natesan S, Pollock R, Amara JF, Keenan T, Magari SR, Phillips T, Courage NL, Cerasoli FJ, Holt DA, Gilman M (1996): A humanized system for pharmacologic control of gene expression [see comments]. *Nat Med* 2:1028–1032.

Rodriguez R, Schuur ER, Lim HY, Henderson GA, Simons JW, Henderson DR (1997): Prostate attenuated replication competent adenovirus (ARCA) CN706: a selective cytotoxic for prostate-specific antigen-positive prostate cancer cells. *Cancer Res* 57:2559–2563.

Rubinchik S, Lowe S, Jia Z, Norris J, Dong J (2001): Creation of a new transgene cloning site near the right ITR of Ad5 results in reduced enhancer interference with tissue-specific and regulatable promoters. *Gene Ther* 8:247–253.

Schambach A, Wodrich H, Hildinger M, Bohne J, Krausslich HG, Baum C (2000): Context dependence of different modules for posttranscriptional enhancement of gene expression from retroviral vectors. *Mol Ther* 2:435–445.

Scott SD, Marples B, Hendry JH, Lashford LS, Embleton MJ, Hunter RD, Howell A, Margison GP (2000): A radiation-controlled molecular switch for use in gene therapy of cancer. *Gene Ther* 7:1121–1125.

Segawa T, Takebayashi H, Kakehi Y, Yoshida O, Narumiya S, Kakizuka A (1998): Prostate-specific amplification of expanded polyglutamine expression: a novel approach for cancer gene therapy. *Cancer Res* 58:2282–2287.

Shi Q, Wang Y, Worton R (1997): Modulation of the specificity and activity of a cellular promoter in an adenoviral vector *Hum Gene Ther* 8:403–410.

Shibata T, Giaccia AJ, Brown JM (2000): Development of a hypoxia-responsive vector for tumor-specific gene therapy. *Gene Ther* 7:493–498.

Siders WM, Halloran PJ, Fenton RG (1996): Transcriptional targeting of recombinant adenoviruses to human and murine melanoma cells. *Cancer Res* 56:5638–5646.

Siders WM, Halloran PJ, Fenton RG (1998): Melanoma-specific cytotoxicity induced by a tyrosinase promoter-enhancer/herpes simplex virus thymidine kinase adenovirus. *Cancer Gene Ther* 5:281–291.

Smith-Arica JR, Morelli AE, Larregina AT, Smith J, Lowenstein PR, Castro MG (2000): Cell-type-specific and regulatable transgenesis in the adult brain: adenovirus-encoded combined transcriptional targeting and inducible transgene expression. *Mol Ther* 2:579–587.

Stein U, Walther W, Shoemaker RH (1996): Vincristine induction of mutant and wild-type human multidrug- resistance promoters is cell-type-specific and dose-dependent. *J Cancer Res Clin Oncol* 122:275–282.

Steinwaerder DS, Lieber A (2000): Insulation from viral transcriptional regulatory elements improves inducible transgene expression from adenovirus vectors *in vitro* and *in vivo. Gene Ther* 7:556–567.

Su H, Chang JC, Xu SM, Kan YW (1996): Selective killing of AFP-positive hepatocellular carcinoma cells by adeno-associated virus transfer of the herpes simplex virus thymidine kinase gene. *Hum Gene Ther* 7:463–470.

Su H, Lu R, Chang JC, Kan YW (1997): Tissue-specific expression of herpes simplex virus thymidine kinase gene delivered by adeno-associated virus inhibits the growth of human hepatocellular carcinoma in athymic mice. *Proc Natl Acad Sci USA* 94:13891–13896.

Vassaux G, Hurst H, Lemoine N (1999): Insulation of a conditionally expressed transgene in an adenoviral vector. *Gene Ther* 6:1192–1197.

Vile RG, Nelson JA, Castleden S, Chong H, Hart IR (1994): Systemic gene therapy of murine melanoma using tissue specific expression of the HSVtk gene involves an immune component. *Cancer Res* 54:6228–6234.

Vile RG, Diaz RM, Miller N, Mitchell S, Tuszyanski A, Russell SJ (1995): Tissue-specific gene expression from Mo-MLV retroviral vectors with hybrid LTRs containing the murine tyrosinase enhancer/promoter. *Virology* 214:307–313.

Wills KN, Huang WM, Harris MP, Machemer T, Maneval DC, Gregory RJ (1995): Gene therapy for hepatocellular carcinoma: chemosensitivity conferred by adenovirus-mediated transfer of the HSV-1 thymidine kinase gene. *Cancer Gene Ther* 2:191–197.

Xie X, Zhao X, Liu Y, Young CY, Tindall DJ, Slawin KM, Spencer DM (2001): Robust prostate-specific expression for targeted gene therapy based on the human kallikrein 2 promoter. *Hum Gene Ther* 12:549–561.

Xu L, Zerby D, Huang Y, Ji H, Nyanguile OF, de los Angeles JE, Kadan MJ (2001): A versatile framework for the design of ligand-dependent, transgene-specific transcription factors. *Mol Ther* 3:262–273.

Yu DC, Sakamoto GT, Henderson DR (1999): Identification of the transcriptional regulatory sequences of human kallikrein 2 and their use in the construction of calydon virus 764, an attenuated replication competent adenovirus for prostate cancer therapy. *Cancer Res* 59:1498–1504.

23

PHYSIOLOGICAL TARGETING

KATIE BINLEY, PH.D.

INTRODUCTION

The ability to deliver a transgene to the correct disease location and achieve expression at a relevant therapeutic level is important for the development of gene delivery systems with a high therapeutic index. A number of different promoter systems show great promise for restricting expression of the transgene to the appropriate target tissue. Tissue-specific promoters restrict expression spatially in a cell-specific manner whereas pharmacologically responsive promoters are controlled temporally through the administration of chemical repressors or activators. This chapter focuses on the development of physiologically responsive promoters that combine both spatial and temporal control of gene expression.

Many physiological changes occur in tissues at the onset of disease. For example, decreases in oxygen and nutrient availability are common features of ischemic diseases where blood vessels leading to the tissue are abnormal or become occluded. Cells respond to these adverse conditions by activating a myriad of genes that increase the chances of survival. Many of these genes are regulated by control mechanisms that are ubiquitous to most cell types.

The transcriptional responses to these conditions can be harnessed to provide disease-related control of gene expression. In diseased tissue this expression may be spatially and temporally controlled by the presence of the abnormal condition. If this environment were important to the pathology of the disease state, gene expression would be terminated only when the disease resolves. If

Vector Targeting for Therapeutic Gene Delivery, Edited by David T. Curiel and Joanne T. Douglas
ISBN 0-471-43479-5

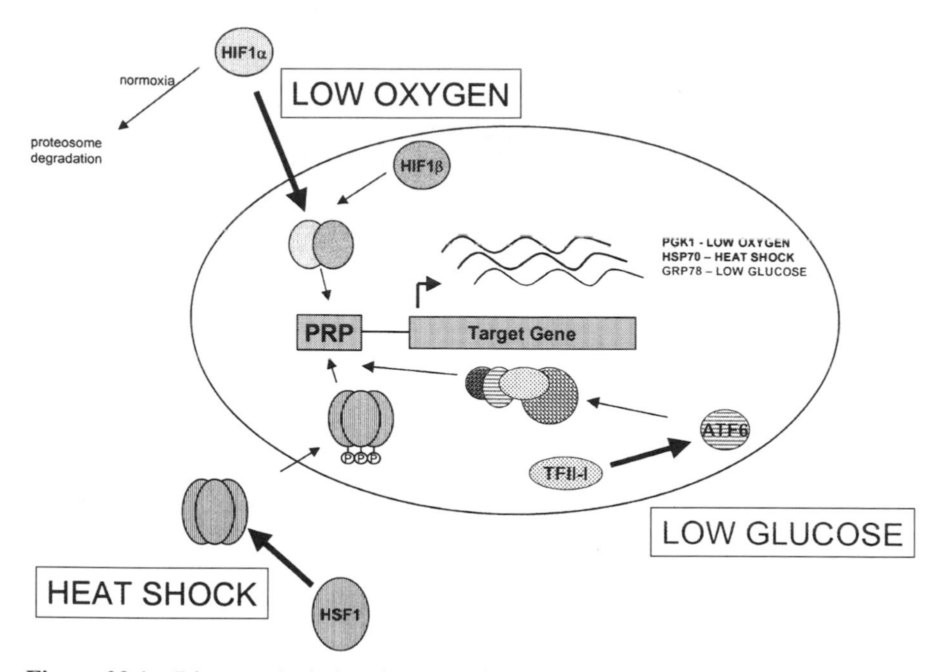

Figure 23.1 Diagram depicting the transcriptional responses within the cell to different physiological stresses: low oxygen (hypoxia), heat shock, and low glucose. PRP-physiologically regulated promoter.

not, the gene expression would be temporally fixed to the duration of the environmental change. Therefore, physiological promoters offer the potential for controlling therapeutic gene expression using the host cell's normal transcriptional response to a physiological condition. Although tissue-specific promoters also utilize the target cell's transcriptional machinery, physiological regulation provides a unifying system for many disease targets where a particular environmental condition is a common feature. In addition, the activity of physiological promoters is often reflected by the severity of the disease condition. Therefore, the extent, duration, and spatial expression of the transgene can be simultaneously regulated.

In this chapter, the development of gene therapy modalities targeted to three key physiological conditions are discussed (Fig. 23.1). These are (1) low oxygen, (2) low glucose, and (3) heat shock.

HYPOXIA-MEDIATED GENE REGULATION

Cells that constitute a tumor are exposed to severe physiological stresses. Solid tumors in particular show relatively aberrant vascularization resulting in a physiology ranging from intermittent to absent perfusion, causing dramatic reductions in essential nutrients including oxygen. Solid tumors therefore provide a good model for examining the potential of hypoxia (low oxygen)-regulated promoters for gene therapy applications. Direct oxygen tension measurements in patient tumors show a median range of 1.3–3.9% O_2 with 0.33% O_2 being commonly measured in solid tumors, whereas concentrations in normal tissues are significantly higher, 3.1–8.7% O_2 (Vaupel, 1993). In addition to cancer, hypoxia plays a fundamental role in the pathophysiology of common causes of mortality including ischemic heart disease, stroke, chronic lung disease, and congestive heart failure. Generally, organisms respond to hypoxia via a range of homeostatic mechanisms that aim to increase respiration and blood flow. Systemic hypoxia may occur in the context of chronic lung disease or congenital heart disease. This induces the expression of the hormone erythropoietin (EPO) to simulate red blood cell production to ultimately increase the oxygen-carrying capacity of the blood. Local hypoxia/ischemia occurs in tumors and can induce the expression of vascular endothelial growth factor (VEGF), a major regulator of angiogenesis, thereby increasing oxygen delivery.

HIF-1/HRE Signaling Pathways

At the cellular level, molecular adaptation to hypoxia brings about an increase in the synthesis of glucose transporters and glycolytic enzymes that enable the cell to maintain adenosine triphosphate (ATP) production in the presence of low oxygen. The *cis*-acting regulatory elements that control the expression of hypoxically regulated genes have been well studied leading to the identification of a consensus HIF-1 (hypoxia inducible factor 1) DNA binding site (HBS) within the hypoxia response element, HRE (Table 23.1).

The HBS acts as a binding site for a heterodimeric transcription factor, HIF-1, which consists of two subunits, HIF1α and HIF1β (Wang and Semenza, 1993).

TABLE 23.1 Compilation of Consensus Enhancer Sequences that Bind *trans*-Activating Factors Known To Be Stabilized/Modulated under Physiological Stress

Enhancer	Consensus $5' \Rightarrow 3'$	Trans-Activating Factors	References
HIF-1 Binding site (HBS)	RCGTG	HIF-1	Wenger and Gassmann (1997)
Heat shock element	nGAAn triple repeat	HSF-1	Williams and Morimoto (1990)
Endoplasmic reticulum stress response element (ESRE)	(CCAAT)N9 (CCACG)	ERSF/TFII-I, NF-Y, YY1, ATF-6	Parker et al. (2001)

The β subunit (also known as the aryl hydrocarbon nuclear translocator, ARNT) is a common subunit for several transcription factors. HIF1α is unique to HIF-1 and is the oxygen-regulated subunit. Other oxygen regulated subunits, HIF2α (EPAS-1) and HIF3α are also able to form heterodimers with HIF1β and bind to the HBS (Tian, McKnight, and Russell, 1997; Gu et al., 1998). The oxygen-regulated subunits are not functionally redundant; they display unique patterns of developmental and spatial expression in different tissues (Jain et al., 1998).

The activation of HIF-1 is an important defense mechanism triggered by a deficiency of molecular oxygen. The ensuing adaptation renders the cell more capable of surviving and functioning in hypoxic/adverse conditions. HIF1α is labile in normoxic conditions with a half-life of less than 15 min (Wang et al., 1995, Huang et al., 1996). In the presence of oxygen, HIF1α is targeted for destruction by an E3 ubiquitin ligase containing the von Hippel Lindau tumor suppressor protein (pVHL) (Cockman et al., 2000; Maxwell et al., 1999). The interaction between HIF1α and pVHL is regulated via oxygen-dependent hydroxylation of a HIF1α proline residue (Ivan et al., 2001; Jaakkola et al., 2001). Immediately upon a decrease in oxygen tension, HIF1α is stabilized and translocated to the nucleus. There it forms a heterodimer with HIF1β before binding to the HBS present in either the 5′ or 3′ regions of hypoxically regulated genes (Fig. 23.1).

Optimal stabilization and DNA-binding activity of the HIF-1 transcription factor increases exponentially with decreasing oxygen tensions (Fig. 23.2). The

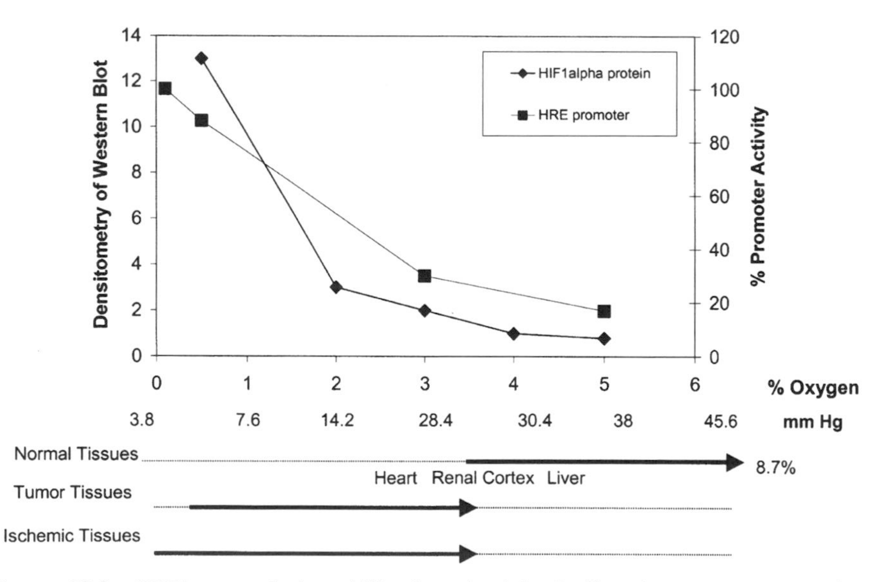

Figure 23.2 HIF1α protein is stabilized at physiologically relevant oxygen tensions that correlate with the activity of a synthetic hypoxia regulated, HRE, promoter which in this case is three copies of the *PGK*1 HRE fused to the minimal SV40 promoter (based on Boast et al., 1999).

TABLE 23.2 Examples of Genes Where Regulation of Expression in Hypoxia Occurs via HIF-1 Independent Mechanisms

Gene(s)	*Cis*-Acting Element	*Trans*-Activating Factor	Reference
Proinflammatory cytokines e.g., IL8, TNFα	NFκB binding	NFκB	Bowie and O'Neill (2000)
Metallothionein I	Antioxidant response element (ARE)	AP1 (?)	Waleh et al., (1998)
Metallothionein II	Metal response element (MRE)	MTF-1	Murphy et al., (1999)
Early growth response-1	Serum response elements and ets binding sites EBS/SRE	Elk1/SRF (?)	Yan et al., (1999)

maximal response occurs at 0.5% oxygen and reoxygenation results in rapid degradation of HIF1α (Jiang et al., 1996). The possession of an oxygen-regulated HIF factor is ubiquitous to all tissue types studied so far, making the HIF-1/HRE promoter system an attractive option for regulating a therapeutic transgene in diseased tissue where hypoxia is a feature.

The HIF-1/HRE system is believed to be the main mediator of hypoxic gene regulation. However, the expression of some hypoxically regulated genes occurs via mechanisms independent of HIF-1. Examples of some of the genes and the *trans*- and *cis*-acting factors involved in their regulation are shown in Table 23.2.

Using Hypoxia to Drive Heterologous Gene Expression

Endogenous full-length hypoxia-regulated promoters give rise to only a few-fold induction in response to low oxygen. The effectiveness of the native HREs present in these promoters is influenced by other regulatory elements within the gene that affect both basal and induced expression. Although the murine phosphoglycerate kinase 1 (*PGK*1) promoter gives a 2–3-fold induction of gene expression in response to hypoxia, a smaller fragment (−270 to −523) gives a 15-fold induction (Firth et al., 1994). Similarly, the full-length *EPO* promoter gives a 3.6-fold hypoxic induction compared to 6-fold induction with a 53 bp fragment (Blanchard et al., 1992). Therefore, in isolation the HRE shows a significant improvement in the induction ratio compared to the full-length promoter.

Further improvements to the hypoxic response have been made through the development of synthetic promoters where HREs are fused to heterologous promoters (Firth et al., 1994; Boast et al., 1999; Shibata et al., 1998; Shibata, Giaccia, and Brown, 2000). The context and spacing of the HIF-1 DNA-binding sites (HBS) is critical for the hypoxic response. This is exemplified by the diversity in the response to hypoxia between HREs isolated from different hypoxically regulated promoters (Boast et al., 1999). In addition to the HBS itself, ancillary sequences within the HRE are known to be important for the hypoxic response (Kimura et al., 2001).

The criteria for selecting the "best" hypoxic promoter include low basal expression and high hypoxia inducibility. Low basal expression is important for gene therapy applications such that there is minimal inappropriate gene expression in surrounding nontarget tissue. High inducible expression is important so that the therapeutic protein is expressed above a threshold level under which it may be ineffective.

Multimerizing the HRE monomeric unit markedly improved the specificity of the hypoxia responsive promoters (Fig. 23.3). For example, three copies of the murine *PGK*1 HRE (3×*PGK*1) were better than two, giving 18-fold versus 11-fold induction ratios respectively (Firth et al., 1994). However, the hypoxic induction ratio does not increase indefinitely with increasing numbers of HREs such that 10 multimeric copies of the *VEGF* HRE were not significantly better than 5 (Shibata et al., 2000).

The potency, that is the absolute expression level under hypoxia, has been improved through changing the minimal heterologous promoter to which the multimerized HREs are fused. Replacing the minimal thymidine kinase (*TK*) promoter with the minimal SV40 promoter resulted in an increase in the induction ratio from 18-fold (Firth et al., 1994) to 146-fold (Boast et al. 1999) where promoter activity was inversely proportional to the oxygen concentration (Fig. 23.2). Although the difference in induction ratios is likely in part due to the

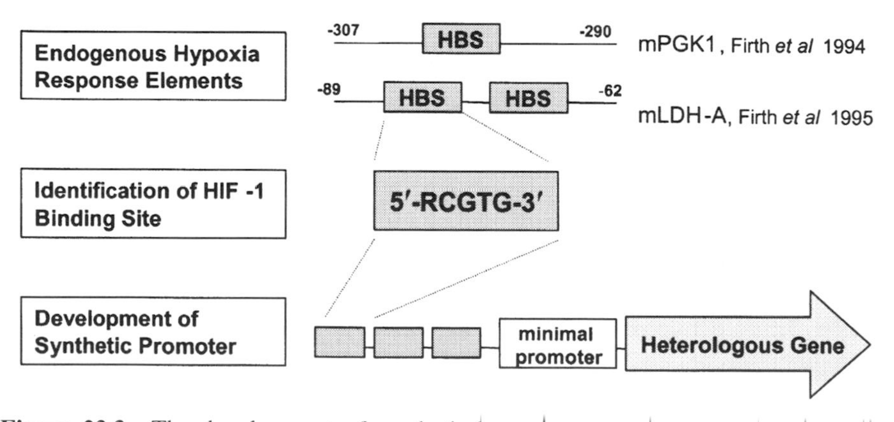

Figure 23.3 The development of synthetic hypoxia responsive promoters from the initial identification of the HIF-1 DNA binding consensus (HBS) to the construction an optimized synthetic promoter.

different cell lines used in these two studies, the induced expression levels with the 3×*PGK*1/minimal SV40 promoter configuration were especially high. In fact, induced expression levels matched that of the same reporter gene driven by the strong human CMV/IE promoter. In similar studies, five copies of the *VEGF* HRE in combination with a minimal E1b promoter gave only modest induced expression levels whereas in combination with the minimal CMV/IE promoter the induced expression levels also matched that of the CMV/IE promoter (Shibata, Giaccia, and Brown, 2000).

Plasmids harboring HRE-driven expression cassettes have been examined for their efficacy in *in vivo* applications. The 3×*PGK*1 HRE fused to the 9-27 (interferon-inducible gene) promoter driving a cytosine deaminase 2 (*CD2*) gene sensitized stably transfected cells to 5-fluorocytosine (5-FC) after 16 h of hypoxia (Dachs et al., 1997). *In vivo*, the stably transfected tumor xenografts showed enhanced *CD2* expression in cells known to be hypoxic. Another cell line stably transfected with a plasmid carrying the full-length murine *PGK*1 promoter gave a two- to three-fold induction of erythropoietin expression *in vitro* and in encapsulated mouse cell implants *in vivo* (Rinsch et al., 1997).

However, to fully exploit the HRE/HIF-1 promoter system for regulated gene therapy it will be important that it can be transferred between a variety of effective vector delivery platforms. Although significant advances have been made in plasmid/DNA-mediated gene therapy, the hypoxia-regulated promoters would have increased utility if they could be transferred between different virus vector platforms while maintaining functionality. Promoter elements have been known to respond differently when transferred to different vector platforms. For example, negative elements in viral vector genomes or promoter interference from the target cell genome could limit the use of certain regulated promoters (Shi et al., 1997). These considerations are particularly important because the primary way of investigating and developing new promoters is through the use of plasmid-based expression systems, which are later transferred to virus vector platforms.

The HRE promoter system has already successfully been transferred to both episomal and integrating viral vector platforms. Within an adenoviral vector, the enhancer present in the 3′ untranslated region (UTR) of the *EPO* gene boosted expression from a constitutive promoter (Setoguchi, Danel, and Crystal, 1994). In this case, the constitutive promoter gave rise to high-level expression in normoxic conditions. The synthetic 3×*PGK*1 HRE promoter was successfully incorporated into adenoviral, retroviral, and lentiviral vectors while maintaining the high level of induced expression seen within the plasmid setting (Fig. 23.4) (Boast et al., 1999; Binley et al., 1999). In a retroviral vector the 3×*PGK*1 HRE was incorporated directly into the 3′ long terminal repeat (LTR) in place of the natural U3 regulatory elements. Since retroviral transduction results in duplication of the 3′ LTR to the 5′ LTR, hypoxia-induced expression in the target cell benefited from both the 5′ and 3′ copies of the

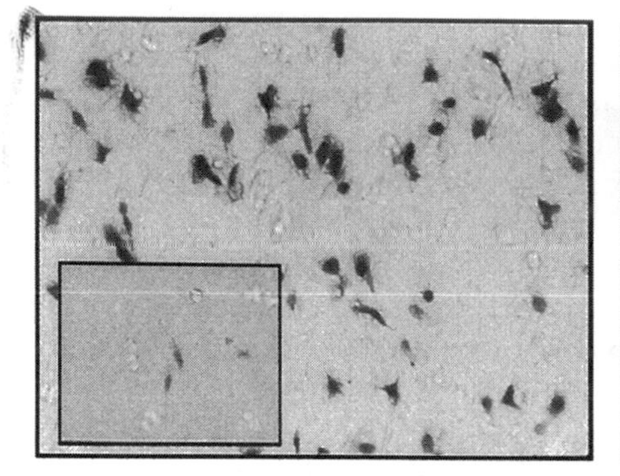

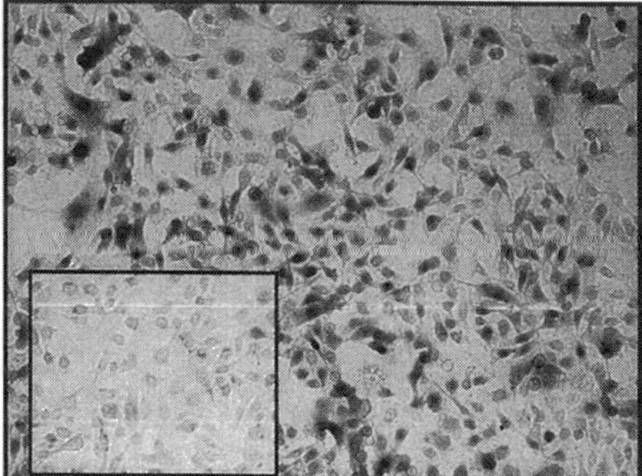

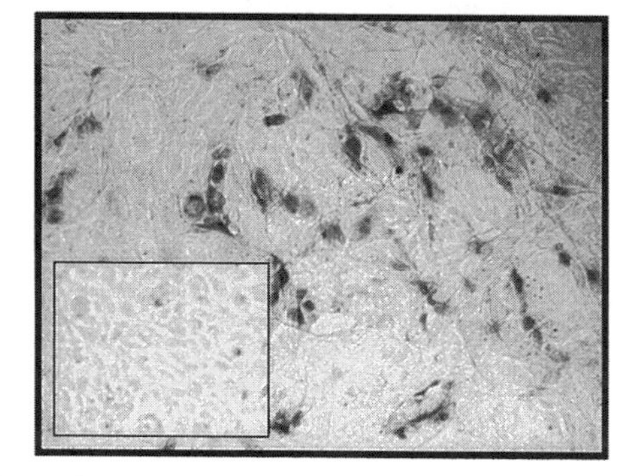

Figure 23.4 The hypoxia response element (HRE) promoter driving hypoxia stimulated β-galactosidase reporter gene expression in an Mo-MLV retrovirus vector, a first generation adenovirus vector, and an EIAV lentivirus in transduced HT1080 human fibrosarcoma cells. Transduced cells were incubated for 16 h in hypoxia (0.1% oxygen, main picture) or were left in normoxia (21% oxygen, inset picture) prior to X-gal staining.

hypoxia enhancer. More recently, an HRE promoter has been shown to function in an adeno-associated virus vector (Ruan et al., 2001).

Incorporation of the HRE promoter system into these viral vectors has extended the study of hypoxia signaling pathways to different primary cell types. Such cell types, including macrophages and skeletal muscle cells, are often refractory to plasmid transfection, so studying their utility for hypoxia-targeted gene delivery relies on virus-based systems (Griffiths et al., 2000; Binley et al., 1999). The ease with which the synthetic 3×PGK1 HRE promoter was transferred between the different vector platforms diversifies the utility of this promoter system for many disease targets where hypoxia is a major feature.

Modifying Hypoxic Gene Regulation

Posttranscriptional Control In addition to an increase in transcription, the expression of certain hypoxically regulated genes is regulated at the mRNA level. Posttranscriptional stability of mRNA plays an important role in increasing the expression of the angiogenic factor *VEGF* during hypoxia. The half-life of *VEGF* mRNA increases two- to eightfold in response to hypoxia due to the presence of an AU-rich element in the 3′ UTR of *VEGF* termed the *hypoxia stability region* (HSR) (Levy, Levy, and Goldberg, 1996; Claffey et al., 1998). The expression of chimeric reporters containing the *VEGF* HSR was significantly increased following hypoxia (Claffey et al., 1998). A minimal (145 bp) HSR improved the hypoxia-induced gene expression twofold with no apparent impact on the basal level of expression (Boast et al., 1999). The HSR may therefore prove to be a useful addition to hypoxia-regulated gene therapy products by further increasing the level of therapeutic gene expression.

Tissue Specificity Tissue-specific promoters often give relatively poor levels of expression, which may be increased by combining them with hypoxia regulatory elements. In this way, the hypoxia-driven gene expression could be restricted to a particular subset of cells using a tissue-specific promoter. To investigate this possibility, four copies of the *EPO* HRE were fused upstream of a minimal muscle specific promoter and hypoxic induction was obtained in skeletal and cardiac myocytes but not in HeLa cells *in vitro*. However, *in vivo* the chimeric promoter showed only modest induction in the postischemic rabbit heart (Prentice et al., 1997). In another example, two copies of the *PGK*1 HRE were combined with endothelial cell specific promoters, *KDR* and *E-selectin* (Modlich, Pugh, and Bicknell, 2000). The HRE conferred hypoxia inducibility to the *KDR* promoter with an increase in basal normoxic expression. However, the HRE overrode the tissue specificity of the *E-selectin* promoter.

It is clear that combining tissue-specific promoters with HREs while maintaining both hypoxia and tissue specificity is not a straightforward process. However, it is likely to be an area of future interest, because increasing tissue-specific expression in this way may be a useful development for targeted vectors.

GENERAL STRESS-MEDIATED GENE REGULATION

Heat shock or stress genes are present in every organism studied to date. Although some of these genes are expressed constitutively and appear to be involved in basic cellular processes such as protein synthesis and maturation, assembly of protein complexes, and intracellular trafficking, others are normally silent or expressed at low levels (Voellmy, 1994). Expression of the latter genes is enhanced when cells are subjected to stressful conditions that include elevated temperature, chronic hypoxia, low pH (acidosis), and glucose deprivation. Such conditions increase the number of malfolded or improperly processed proteins. In order to cope, cells have developed mechanisms to increase the production of certain molecular chaperones of the stress gene family. The chaperones promote folding and assembly of nascent polypeptides and facilitate repair or degradation of unfolded proteins. Two physiologically important chaperones are HSP70 and GRP78 (also known as BiP) (Feige and Polla, 1994; Little et al., 1994). HSP70 and GRP78 share 60% sequence identity at the amino acid level but reside in different subcellular locations (Munro and Pelham, 1986). GRP78 is present in the lumen of the endoplasmic reticulum and HSP70 is in the cytoplasm/nucleus. These proteins transiently bind to nascent polypeptides to facilitate proper folding and assembly and prevent formation of aggregates. The transcriptional control mechanisms that regulate the expression of the HSP70 and GRP78 genes have been explored for their potential utility in gene therapy applications and are discussed in the following sections.

Harnessing the *GRP78* Promoter for Gene Therapy

GRP78 is well characterized with regard to its up-regulation in response to glucose deprivation and endoplasmic reticulum (ER) stress (Lee, 1987). It is also elevated by other physiological conditions commonly found in the tumor microenvironment such as hypoxia, calcium depletion, and acid pH (Roll et al., 1991). *GRP78* expression is also elevated in virally, chemically, and radiation transformed cells (Patierno et al., 1987).

GRP78 performs its normal function as a molecular chaperone under moderate expression levels that increase 10–25-fold when the cell experiences stress conditions such as exist in poorly vascularized tumors (Lee, 1992). The GRP78 promoter is therefore a strong candidate for driving physiological regulation of gene expression in tumors. The role of GRP78 in other ischemic diseases is still unclear although several GRPs are elevated in kidney ischemia (Kuznetsov et al., 1996). To this end, much effort has been directed toward identifying the *cis*-and *trans*-components involved in the *GRP78* transcriptional control mechanism.

The rat *GRP78* promoter consists of *cis*-regulatory elements for both basal and inducible expression (Chang et al., 1987). Whereas most of the elements required for basal expression are located upstream, three stress-inducible elements, termed ESRE (endoplasmic reticulum stress response element), are clustered within a 200 bp region proximal to the TATA box. The ESRE gives rise to an evolutionarily conserved tripartite structure that interacts with a multiprotein stress inducible complex that includes ATF6, YY1, NF-Y, and TFII-I/ERSF (Table 23.1 and Fig 23.1).

The utility of a truncated rat *GRP78* promoter harboring ERSE elements was examined in a retrovirus vector. The promoter fragment gave rise to an eightfold elevation in neomycin expression in response to glucose starvation in B/C10ME mouse fibrosarcoma cells (Gazit et al., 1995). The regulation was preserved *in vivo* such that the tumors derived from stably transduced B/C10ME cells showed highest expression levels in the center near areas of necrosis.

The 5′ UTR of the *GRP78* gene contains a natural internal ribosome entry site (IRES) (Macejak and Sarnow, 1991). These structures, more commonly found in the 5′ UTR of picornaviral mRNAs, allow cap-independent protein translation and in some cases increase translation efficiency (Stein et al., 1998; Vivinus et al., 2001). A portion of the rat *GRP78* promoter (−520 to +175) including the natural IRES enhanced the sensitivity of tumor cells to the prodrug ganciclovir (GCV) due to twofold higher levels of HSV *TK* expression when compared to a constitutive retrovirus LTR promoter (Gazit et al., 1999; Chen et al., 2000). This sensitivity translated *in vivo* to complete tumor regression correlating with strong *TK* expression throughout the tumor tissue. In contrast, GCV treatment of tumors expressing HSV *TK* via the retrovirus LTR promoter did not reduce in size, indicating that the *GRP78* promoter gave higher expression levels in the tumor environment (Chen et al., 2000).

The construction of synthetic *GRP78*-based promoters may improve the expression profile in response to conditions of stress. The stress response of the native *GRP78* promoter requires at least two ERSE units (Roy and Lee, 1999). Although the duplicate copies of the ERSE confer full stress inducibility to an heterologous promoter *in vitro*, this promoter configuration has not yet been utilized in a gene therapy setting. Further multimerization of the ESRE elements could lead to the development of new synthetic stress responsive promoters with an advanced expression profile.

Harnessing the *HSP70* Promoter for Gene Therapy.

A number of physiological stresses that cause an accumulation of misfolded proteins also cause an increased production of HSP70. Most of these conditions are common to those that up regulate GRP78 but there is one important exception, heat shock. HSP70 is a member of a large family of heat shock proteins, each with its own characteristic size and location within the cell (Leppa and Sistonen, 1997). Many of the HSPs are essential for the day-to-day operation of all cells acting as housekeeping molecular chaperones and morticians for other proteins. However, *HSP70* expression is very low under normal physiological conditions but is strongly induced when the temperature in the cell increases by only a few degrees (Morimoto et al., 1997). HSP70 directly protects against ischemic damage. Overexpression of HSP70 has been shown to increase resistance to ischemic injury in a number of different cell types (Brar et al., 1999; Kubo et al., 2001; Lee et al., 2001).

The heat-shock transcription factor 1 (HSF1) is the primary mediator of the heat shock response (Wu, 1995). In unstressed conditions, HSF1 is sequestered as an inactive monomer within an hetero-oligomeric complex that includes HSP90 and HSP70 (Zou et al., 1998). Physiological stress disrupts this complex and promotes the formation of an HSF1 homotrimer which binds to triplet repeats of a nucleotide recognition motif called the *heat shock element* (HSE) present in target genes (Table 23.1 and Fig. 23.1). The induction of HSF1 and binding to its cognate DNA consensus occurs within 15 min of ischemic stress (Tacchini, Radice, and Bernelli-Zazzera, 1999).

The HSE can drive temperature-regulated expression of a heterologous gene. HSEs derived from rat, drosophila, and xenopus *HSP70* promoters all conferred heat inducibility on heterologous genes such that mRNA levels increased 15–20-fold following a 4 h 42°C heat shock (Pelham and Bienz 1982; Bienz and Pelham 1986; Wysocka and Krawczyk, 2000).

Because general stress factors influence the activity of the *HSP70* promoter, the response is limited to heat shock in many studies by excluding the major part of the upstream regulatory elements. Temperature dependent activity of a truncated human *HSP70B* promoter has been characterized using a variety of different reporter genes (Gerner et al., 2000; Borrelli et al., 2001; Huang et al., 2000). In each case, the magnitude of promoter activity increased proportionally with stronger heat shocks, with maximal induced expression similar to that driven by the strong CMV/IE promoter. A modest threefold induced expression was seen after 40°C heat shock with maximum induction occurring at 42°C. In addition, heat treatments at higher temperatures for shorter durations (48°C for 30 sec) induced expression equivalent to that of longer heat shocks at lower temperatures (43°C for 20 min) (Vekris et al., 2000).

The *HSP70B* promoter displays a gradient of reporter gene expression that decreases with increasing distance from the heat source (Vekris et al., 2000). Although there may be an inherent increase in the core temperature of a solid tumor, an exogenous heat source needs to be applied to give sufficient transgene expression. The heat sources used in the *in vivo* models include using needles traversing the tumors with hot water flowing through or simply immersing the tumor-bearing region in a hot water bath (Vekris et al., 2000, Huang et al., 2000).

The utility of the *HSP70B* promoter has been demonstrated by incorporating it into expression cassettes in plasmid-, retrovirus-, and adenovirus-based vector systems. Cells transduced with a retrovirus containing the *HSP70B* promoter driving expression of the HSV *TK* gene were 50,000 times more sensitive to GCV following a heat shock of 45°C for 30 min compared to cells not treated with heat shock (Braiden et al., 2000). This hyperthemia and suicide gene combination therapy gave significant tumor regression concurrent with an increased mouse survival rate compared to *HSP70B-TK*-treated mice with no hyperthermia treatment.

Heat-directed targeting of reporter genes such as GFP, encoding therapeutic protein genes such as interleukin2 (IL-2), IL-12, and tumor necrosis factor α (TNFα), as well as genes encoding proteins with good bystander effects such as TK and CD2, have shown efficacy in glioma and melanoma tumor models transduced with adenoviral vectors.

The sensitivity of prostate cells to 5-FC and GCV was dramatically increased by transducing them with an adenovirus where the *HSP70B* promoter regulated the expression of the *TK-CD* fusion gene. In this case, the decrease in survival rate was not only dependent on the severity of the heat shock but also on the multiplicity of infection (moi) (Blackburn et al., 1998). Higher levels of therapeutic gene expression were achieved as the moi increased. However, it is important that increasing the moi to give higher levels of therapeutic gene expression does not force expression in nontarget tissues. This is particularly important if the therapeutic protein is known to be toxic. Circulating concentrations of IL-12 exceeding a certain critical level are known to cause toxicity. Expression of IL-12 driven by the *HSP70B* promoter in tumors injected with recombinant adenovirus and exposed to hyperthermia (42.5°C for 30 min) were as high as the CMV/IE promoter. However, systemic IL-12 levels were significantly lower with the *HSP70B* rather than the CMV/IE promoter (Huang et al., 2000; Lohr et al., 2000).

The *HSP70B* promoter shows great promise for driving expression of tumor-specific gene therapy. However, like the hypoxia-regulated promoter, it may have utility in other diseases where general ischemic stress is a feature (Arai et al., 1999).

Improving Heat Directed Therapy Several studies indicate that the induced activity of the *HSP70B* promoter is similar to that of the CMV/IE promoter. Although expression from the CMV/IE promoter is generally considered to be high, expression may be low or switched off in certain tissues (Hersh and Stopeck, 1997). Driving expression of adequate amounts of therapeutic protein is important, so improving hyperthermia induced expression while maintaining minimal basal expression would be a useful addition to the *HSP70B* promoter system.

Addition of three extra HSE sequences upstream of the *HSP70B* promoter increased the responsiveness to heat treatment such that the threshold of activation reduced by 1–2°C (Brade et al., 2000). This is particularly important because it is known that the *HSP70B* promoter is less responsive in thermotolerant cells (cells preexposed to stress). Increasing the responsiveness of *HSP70B* in this way could therefore have great utility in cancer gene therapy applications where thermotolerance is likely to be a problem.

A different strategy to increase hyperthermia responsiveness is to drive therapeutic gene expression from a highly active promoter whose activity is dependent on a factor regulated by the *HSP70B* promoter. The basal expression of the therapeutic product in this system is reliant on the dependency of the highly active promoter on the factor regulated by the *HSP70B* promoter. For example, the activity of the HIV-1 LTR promoter was made temperature sensitive by expressing *tat* under the control of the *HSP70B* promoter (Gerner et al., 2000). Using this system, levels of IL-2 expression were significantly higher than the minimal *HSP70B* promoter alone in response to heat shock. However, basal expression at 37°C was high. It remains to be seen whether the strategy can be improved to give a high therapeutic index such that it could be useful for clinical applications.

The data discussed so far demonstrate the utility of the heat-shock promoter system for regulated cancer gene therapy applications. Human tissues can be heated relatively quickly to temperatures capable of activating the *HSP70B* promoter. In addition, heat rapidly diffuses from the tissue once the heat source is removed. Hyperthermia is already being investigated as an adjuvant modality for cancer radiation therapy and devices exist to produce local heating in tissues. Therefore, heat-activated expression of a therapeutic gene could be administered concomitantly with hyperthermia and ionizing radiation treatments already in use in some cancer therapies.

EVOLUTION OF PHYSIOLOGICAL GENE REGULATION

Exploiting Other Physiological Control Mechanisms

In the future, it may be possible to exploit other physiological conditions in addition to those discussed in this chapter for controlling therapeutic gene regulation. Gene expression profile analysis in diseased versus normal tissue could identify new promoters that may be useful in directing disease state-restricted gene expression. For example, the poor vascularization of arthritic joints and tumors increases anaerobic glycolysis and reduces efflux of acidic metabolites such that the microenvironment is acidic as well as hypoxic (Parak et al., 2000). Some genes show upregulation in expression at pH 6.3–6.7 that is reversible on returning to normal pH (Griffiths et al., 1997; Shi et al., 2000; Xu and Fidler, 2000). Identification of the promoter elements responsible for this change in expression could lead to the development of new physiologically responsive promoters. It may even be possible to combine hypoxia, heat, glucose, and acidic responsive elements in order to tailor tumor specific expression levels.

Improving Physiologically Regulated Promoters

Overall, protein synthesis is significantly inhibited in stressful conditions and complexes involved in cap dependent translation become limited. The 5′ UTR regions of the *VEGF* and *GRP78/BIP* genes include an IRES that ensures efficient cap-independent translation, thereby securing efficient production of the protein, even under unfavorable stress conditions. Therefore, expression of the therapeutic protein may be improved in adverse physiological conditions by incorporating an IRES upstream of the therapeutic transgene.

CONCLUSION

This chapter focused on harnessing the transcriptional response to a physiological condition to provide disease-related control of gene expression. In diseased tissue this expression can be spatially and temporally controlled by the presence of the abnormal condition. If this environment were important in the pathology of the disease state, gene expression would only be terminated once the disease is resolved.

Many disease states exhibit changes in a number of physiological parameters. However, the outcome of an environmental change depends on the tissue examined. Whereas ischemia may cause changes in normal tissues such as the myocardium and brain that may ultimately lead to cell death, such adverse environmental conditions are a prerequisite for, and have an inherent role in, the progression of many tumors. Deregulation in the response to hypoxia, heat shock, and glucose have been detected in tumor cell lines *in vivo* and *in vitro* such that HIF1, GRP78, and HSP70 are constitutively active (Zhong et al., 1999; Fuller et al., 1994; Fernandez et al., 2000; Gazit, Lu, and Lee, 1999). The overexpression of these proteins in tumor cells is not only due to the physiological nature of the tumor environment. Adaptation to adverse conditions appears to be an essential part of tumor progression. Tumor cells overexpressing these factors have a selective advantage and increased chance of survival.

Overexpression of these factors may be a cause of resistance to radiotherapy and chemotherapy in some tumors. For example, disrupting HIF-1 has been shown to reduce the development and progression of experimental tumors (Kung et al., 2000; Sun et al., 2001). Therefore, a novel treatment for malignant neoplasia could be an inhibitor of HIF-1, GRP78, or HSP70, the expression of which is tightly regulated via a physiologically responsive promoter.

ACKNOWLEDGMENTS

My thanks go to Leigh Martin and Stuart Naylor for help in preparing this chapter.

REFERENCES

Arai Y, Kubo T, Kobayashi K, Ikeda T, Takahashi K, Takigawa M, Imanishi J, Hirasawa Y (1999): Control of delivered gene expression in chondrocytes using heat shock protein 70B promoter. *J Rheumatol* 26:1769–1774.

Bienz M, Pelham HR (1986): Heat shock regulatory elements function as an inducible enhancer in the Xenopus hsp70 gene and when linked to a heterologous promoter. *Cell* 45:753–760.

Binley KM, Iqball S, Kingsman A, Kingsman S, Naylor S (1999): An adenoviral vector regulated by hypoxia for the treatment of ischaemic disease and cancer. *Gene Ther* 6:1721–1727.

Blackburn RV, Galoforo SS, Corry PM, Lee YJ (1998): Adenoviral-mediated transfer of a heat-inducible double suicide gene into prostate carcinoma cells. *Cancer Res* 58(7):1358–1362

Blanchard KL, Acquaviva AM, Galson DL, Bunn HF (1992): Hypoxic induction of the human erythropoietin gene: cooperation between the promoter and enhancer, each of which contains steroid receptor response elements. *Mol Cell Biol* 12:5373–5385.

Boast K, Binley K, Iqball S, Price T, Spearman H, Kingsman S, Kingsman A, Naylor S (1999): Characterisation of physiologically regulated vectors for the treatment of ischemic disease. *Hum Gene Ther* 10:2197–2208.

Borrelli MJ, Schoenherr DM, Wong A, Bernock LJ, Corry PM (2001): Heat-activated transgene expression from adenovirus vectors infected into human prostate cancer cells. *Cancer Res* 61(3):1113–1121.

Bowie A, O.Neill LA (2000): Oxidative stress and nuclear factor-kappaB activation: a reassessment of the evidence in the light of recent discoveries. *Biochem Pharm* 59:13–23.

Brade AM, Ngo D, Szmitko P, Li PX, Liu FF, Klamut HJ (2000): Heat-directed gene targeting of adenoviral vectors to tumor cells. *Cancer Gene Ther* 7:1566–1574.

Braiden V, Ohtsuru A, Kawashita Y, Miki F, Sawada T, Ito M, Cao Y, Kaneda Y, Koji T, Yamashita S (2000): Eradication of breast cancer xenografts by hyperthermic suicide gene therapy under the control of the heat shock protein promoter. *Hum Gene Ther* 11:2453–2463.

Brar BK, Stephanou A, Wagstaff MJ, Coffin RS, Marber MS, Engelmann G, Latchman DS (1999): Heat shock proteins delivered with a virus vector can protect cardiac cells against apoptosis as well as against thermal or hypoxic stress. *J Mol Cell Cardiol* 31:135–146.

Chang SC, Wooden SK, Nakaki T, Kim YK, Lin AY, Kung L, Attenello JW, Lee AS (1987): Rat gene encoding the 78-kDa glucose-regulated protein GRP78: its regulatory sequences and the effect of protein glycosylation on its expression. *Proc Natl Acad Sci USA* 84:680–684.

Chen X, Zhang D, Dennert G, Hung G, Lee AS (2000): Eradication of murine mammary adenocarcinoma through HSVtk expression directed by the glucose-starvation inducible grp78 promoter. *Breast Cancer Res Treat* 59:81–90.

Claffey KP, Shih SC, Mullen A, Dziennis S, Cusick JL, Abrams KR, Lee SW, Detmar M (1998). Identification of a human VPF/VEGF 3′ untranslated region mediating hypoxia-induced mRNA stability. *Mol Biol Cell* 9:469–481.

Cockman ME, Masson N, Mole DR, Jaakkola P, Chang GW, Clifford SC, Maher ER, Pugh CW, Ratcliffe PJ, Maxwell PH (2000): Hypoxia inducible factor-alpha binding and ubiquitylation by the von Hippel-Lindau tumor suppressor protein. *J Biol Chem* 18:25733–25741.

Dachs GU, Patterson AV, Firth JD, Ratcliffe PJ, Townsend KM, Stratford IJ, Harris AL (1997): Targeting gene expression to hypoxic tumor cells. *Nat Med* 3:515–520.

Feige U, Polla BS (1994): Heat shock proteins: the hsp70 family. *Experientia* 50:979–986.

Fernandez PM, Tabbara SO, Jacobs LK, Manning FC, Tsangaris, TN, Schwartz AM, Kennedy KA, Patierno SR (2000): Overexpression of the glucose-regulated stress gene GRP78 in malignant but not benign human breast lesions. *Breast Cancer Res Treat* 59:15–26.

Firth JD, Ebert BL, Pugh CW, Ratcliffe PJ (1994): Oxygen-regulated control elements in the phosphoglycerate kinase 1 and lactate dehydrogenase A genes: similarities with the erythropoietin 3′ enhancer. *Proc Natl Acad Sci USA* 91:6496–6500.

Fuller KJ, Issels RD, Slosman DO, Guillet JG, Soussi T, Polla BS (1994): Cancer and the heat shock response. *Eur J Cancer* 30(12):1884–1891.

Gazit G, Lu J, Lee AS (1999): De-regulation of GRP stress protein expression in human breast cancer cell lines. *Breast Cancer Res Treat* 54:135–146.

Gazit G, Kane SE, Nichols P, Lee AS (1995): Use of the stress-inducible grp78/BiP promoter in targeting high level gene expression in fibrosarcoma in vivo. *Cancer Res* 55:1660–1663.

Gazit G, Hung G, Chen X, Anderson WF, Lee AS (1999): Use of the glucose starvation-inducible glucose-regulated protein 78 promoter in suicide gene therapy of murine fibrosarcoma. *Cancer Res* 59:3100–3106.

Gerner EW, Hersh EM, Pennington M, Tsang TC, Harris D, Vasanwala F, Brailey J (2000): Heat-inducible vectors for use in gene therapy. *Int J Hyperthermia* 16:171–181.

Griffiths L, Dachs GU, Bicknell R, Harris AL, Stratford IJ (1997): The influence of oxygen tension and pH on the expression of platelet-derived endothelial cell growth factor/thymidine phosphorylase in human breast tumor cells grown in vitro and in vivo. *Cancer Res* 57(4):570–572.

Griffiths L, Binley K, Iqball S, Kan O, Maxwell P, Ratcliffe P, Lewis C, Harris A, Kingsman S, Naylor S (2000): The macrophage—a novel system to deliver gene therapy to pathological hypoxia. *Gene Ther* 7(3):255–262

Gu YZ, Moran SM, Hogenesch JB, Wartman L, Bradfield CA (1998): Molecular characterization and chromosomal localization of a third alpha-class hypoxia inducible factor subunit, HIF3alpha. *Gene Expr* 7:205–213.

Hersh EM, Stopeck AT (1997): Advances in the biological therapy and gene therapy of malignant disease. *Clin Cancer Res* 3:2623–2629.

Huang LE, Arany Z, Livingston DM, Bunn HF (1996): Activation of hypoxia-inducible transcription factor depends primarily upon redox-sensitive stabilization of its alpha subunit. *J Biol Chem* 271:32253–32259.

Huang Q, Hu JK, Lohr F, Zhang L, Braun R, Lanzen J, Little JB, Dewhirst MW, Li CY (2000): Heat-induced gene expression as a novel targeted cancer gene therapy strategy. *Cancer Res* 60(13):3435–3439.

Ivan M, Kondo K, Yang H, Kim W, Valiando J, Ohh M, Salic A, Asara JM, Lane WS, Kaelin Jr. WG (2001): HIFalpha targeted for VHL-mediated destruction by proline hydroxylation: implications for O2 sensing. *Science* 292(5516):464–468.

Jaakkola P, Mole DR, Tian YM, Wilson MI, Gielbert J, Gaskell SJ, Kriegsheim AV, Hebestreit HF, Mukherji M, Schofield CJ, Maxwell PH, Pugh CW, Ratcliffe PJ (2001): Targeting of HIF-alpha to the von Hippel-Lindau ubiquitylation complex by O2-regulated prolyl hydroxylation. *Science* 292(5516):468–472.

Jain S, Maltepe E, Lu MM, Simon C, Bradfield CA (1998): Expression of ARNT, ARNT2, HIF1alpha, HIF2alpha and Ah receptor mRNAs in the developing mouse. *Mech Dev* 73:117–123.

Jiang BH, Semenza GL, Bauer C, Marti HH (1996): Hypoxia-inducible factor 1 levels vary exponentially over a physiologically relevant range of O2 tension. *Am J Physiol* 271:C1172–80.

Kimura H, Weisz A, Ogura T, Hitomi Y, Kurashima Y, Hashimoto K, D'Acquisto F, Makuuchi M, Esumi H (2001): Identification of hypoxia-inducible factor 1 ancillary sequence and its function in vascular endothelial growth factor gene induction by hypoxia and nitric oxide. *J Biol Chem* 276(3):2292–2298.

Kubo T, Arai Y, Takahashi K, Ikeda T, Ohashi S, Kitajima I, Mazda O, Takigawa M, Imanishi J, Hirasawa Y (2001): Expression of transduced HSP70 gene protects chondrocytes from stress. *J Rheumatol* 28(2):330–335.

Kung AL, Wang S, Klco JM, Kaelin WG, Livingston DM (2000): Suppression of tumor growth through disruption of hypoxia-inducible transcription. *Nat Med* 6(12):1335–1340.

Kuznetsov G, Bush KT, Zhang PL, Nigam SK (1996): Perturbations in maturation of secretory proteins and their association with endoplasmic reticulum chaperones in a cell culture model for epithelial ischemia. *Proc Natl Acad Sci USA* 6:93.

Lee AS (1987): Coordinated regulation of a set of genes by glusose and calcium ionophores in mammalian cells. *Trends Biochem Sci* 12:20–23.

Lee AS (1992): Mammalian stress response: induction of the glucose regulated protein family. *Curr Opin Cell Biol* 4:267–273.

Lee JE, Yenari MA, Sun GH, Xu L, Emond MR, Cheng D, Steinberg GK, Giffard RG (2001): Differential neuroprotection from human heat shock protein 70 overexpression in in vitro and in vivo models of ischemia and ischemia-like conditions *Exp Neurol* 170(1):129–139.

Leppa S, Sistonen L (1997): Heat shock response—pathophysiological implications. *Trends in Molec Med* 29:73–78.

Levy, AP, Levy NS, Goldberg MA (1996). Hypoxia-inducible protein binding to vascular endothelial growth factor mRNA and its modulation by the von Hippel-Lindau protein. *J Biol Chem* 271:25492–25497.

Little E, Ramakrishnan M, Roy B, Gazit G, Lee AS (1994): The glucose-regulated proteins (GRP78 and GRP94): functions, gene regulation, and applications. *Crit Rev Eukaryoti Gene Expr* 4:1–18.

Lohr F, Hu K, Huang Q, Zhang L, Samulski TV, Dewhirst MW, Li CY. (2000): Enhancement of radiotherapy by hyperthermia-regulated gene therapy. *Int J Radiat Oncol Biol Phys* 48:1513–1518.

Macejak DG, Sarnow P (1991): Internal initiation of translation mediated by the 5′ leader of a cellular mRNA. *Nature* 353:90–94.

Maxwell PH, Wiesener MS, Chang GW, Clifford SC, Vaux EC, Cockman ME, Wykoff CC, Pugh CW, Maher ER, Ratcliffe PJ (1999): The tumor suppressor protein VHL targets hypoxia-inducible factors for oxygen-dependent proteolysis. *Nature* 399:271–275.

Modlich U, Pugh CW, Bicknell R (2000): Increasing endothelial cell specific expression by the use of heterologous hypoxic and cytokine-inducible enhancers. *Gene Ther* 7(10):896–902.

Morimoto RI, Kline MP, Bimston DN, Cotto JJ (1997): The heat-shock response: regulation and function of heat-shock proteins and molecular chaperones. *Essays in Biochem* 32:17–29.

Munro S, Pelham HRB (1986): An Hsp70-like protein in the ER: identity with the 78 kd glucose regulated protein and immunoglobulin heavy chain binding proteins. *Cell* 46:291–300.

Murphy BJ, Andrews GK, Bittel D, Discher DJ, McCue J, Green CJ, Yanovsky M,

Giaccia A, Sutherland RM, Laderoute KR, Webster KA (1999): Activation of metallothionein gene expression by hypoxia involves metal response elements and metal transcription factor-1. *Cancer Res* 59:1315–1322.

Parak WJ, Dannohl S, George M, Schuler MK, Schaumburger J, Gaub HE, Muller O, Aicher WK (2000): Metabolic activation stimulates acid production in synovial fibroblasts. *J Rheumatol* 27(10):2312–2322.

Parker R, Phan T, Baumeister P, Roy B, Cheriyath V, Roy AL, Lee AS (2001): Identification of tfii-i as the endoplasmic reticulum stress response element binding factor ersf: its autoregulation by stress and interaction with atf6. *Mol Cell Biol* 21:3220–3233.

Patierno SR, Tuscano JM, Kim KS, Landolph JR, Lee AS (1987): Increased expression of the glucose-regulated gene encoding the Mr 78, 000 glucose-regulated protein in chemically and radiation-transformed C3H 10T1/2 mouse embryo cells. *Cancer Res* 47:6220–6224.

Pelham HR, Bienz M (1982): A synthetic heat-shock promoter element confers heat-inducibility on the herpes simplex virus thymidine kinase gene. *Embo J* 1:1473–1477.

Prentice H, Bishopric NH, Hicks MN, Discher DJ, Wu X, Wylie AA, Webster KA (1997): Regulated expression of a foreign gene targeted to the ischaemic myocardium. *Cardiovasc Res* 35:567–574.

Rinsch C, Regulier E, Deglon N, Dalle B, Beuzard Y, Aebischer P (1997). A gene therapy approach to regulated delivery of erythopoietin as a function of oxygen tension. *Hum Gene Ther* 8:1881–1889.

Roll DE, Murphy BJ, Laderoute KR, Sutherland RM, Smith HC (1991): Oxygen regulated 80 kDa protein and glucose regulated 78kDa protein are identical. *Mol Cell Biochem* 103(2):141–148.

Roy B, Lee AS (1999): The mammalian endoplasmic reticulum stress response element consists of an evolutionarily conserved tripartite structure and interacts with a novel stress-inducible complex. *Nucleic Acids Res* 27(6):1437–1443.

Ruan H, Su H, Hu L, Lamborn KR, Kan YW, Deen DF (2001): A hypoxia-regulated adeno-associated virus vector for cancer-specific gene therapy. *Neoplasia* 3(3):255–263.

Setoguchi Y, Danel C, Crystal RG (1994). Stimulation of erythropoiesis by in vivo gene therapy: physiologic consequences of transfer of the human erythropoietin gene to experimental animals using an adenovirus vector. *Blood* 84:2946–2953.

Shi Q, Wang Y, Worton R (1997): Modulation of the specificity and activity of a cellular promoter in an adenoviral vector. *Hum Gene Ther* 8:803–815.

Shi Q, Le X, Wang B, Xiong Q, Abbruzzese JL, Xie K (2000): Regulation of interleukin-8 expression by cellular pH in human pancreatic adenocarcinoma cells. *J Interferon Cytokine Res* 20(11):1023–1028.

Shibata T, Giaccia AJ, Brown JM (2000): Development of a hypoxia-responsive vector for tumor-specific gene therapy. *Gene Ther* 7:493–498.

Shibata T, Akiyama N, Noda M, Sasai K, Hiraoka M (1998): Enhancement of gene expression under hypoxic conditions using fragments of the human vascular endothelial growth factor and the erythropoietin genes. *Int J Radiat Oncol Biol Phys* 42:913–916.

Stein I, Itin A, Einat P, Skaliter R, Grossman Z, Keshet E (1998): Translation of vascular endothelial growth factor mRNA by internal ribosome entry: implications for translation under hypoxia. *Mol Cell Biol* 18:3112–3119.

Sun X, Kanwar JR, Leung E, Lehnert K, Wang D, Krissansen GW (2001): Gene transfer of antisense hypoxia inducible factor-1 alpha enhances the therapeutic efficacy of cancer immunotherapy. *Gene Ther* 8(8):638–645.

Tacchini L, Radice L, Bernelli-Zazzera A (1999): Differential activation of some transcription factors during rat liver ischemia, reperfusion, and heat shock. *J Cell Physiol* 180(2):255–262.

Tian H, McKnight SL, Russell DW (1997): Endothelial PAS domain protein 1 (EPAS1), a transcription factor selectively expressed in endothelial cells. *Genes Dev* 11:72–82.

Vaupel PW (1993): Oxygenation of solid tumours. In *Drug Resistance in Oncology*, Teicher BA, ed. New York: Marcel Dekker, pp 53–85.

Vekris A, Maurange C, Moonen C, Mazurier F, De Verneuil H, Canioni P, Voisin P (2000): Control of transgene expression using local hyperthermia in combination with a heat-sensitive promoter. *J Gene Med* 2:89–96.

Vivinus S, Baulande S, Zanten M, Campbell F, Topley P, Ellis JH, Dessen P, Coste H (2001): An element within the 5′ untranslated region of human Hsp70 mRNA which acts as a general enhancer of mRNA translation. *Eur J Biochem* 268:1908–1917.

Voellmy R (1994). Transduction of the stress signal and mechanisms of transcriptional regulation of heat shock/stress protein expression in higher eukaryotes. *Crit Rev Euk Gene Exp* 4(4):357–401.

Waleh NS, Calaoagan J, Murphy BJ, Knapp AN, Sutherland RM, Laderoute KR (1998): The redox-sensitive human antioxidant responsive element induces gene expression under low oxygen conditions. *Carcinogenesis* 19:1333–1337.

Wang GL, Semenza GL (1993): Characterization of hypoxia-inducible factor 1 and regulation of DNA binding activity by hypoxia. *J Biol Chem* 268:21513–21518.

Wang GL, Jiang BH, Rue EA, Semenza GL (1995): Hypoxia-inducible factor 1 is a basic-helix-loop-helix-PAS heterodimer regulated by cellular O2 tension. *Proc Natl Acad Sci USA* 92:5510–5514.

Wenger RH, Gassmann M (1997): Oxygen(es) and the hypoxia-inducible factor-1. *Biol Chem* 378:609–616.

Williams GT, Morimoto RI (1990): Maximal stress-induced transcription from the human HSP70 promoter requires interactions with the basal promoter elements independent of rotational alignment. *Mol Cell Biol* 10(6):3125–3136.

Wu C (1995): Heat shock transcription factors: structure and regulation. *Annu Rev Cell Dev Biol* 11:441–469.

Wysocka A, Krawczyk Z (2000): Green fluorescent protein as a marker for monitoring activity of stress-inducible hsp70 rat gene promoter. *Mol Cell Biochem* 215:153–156.

Xu L, Fidler IJ (2000): Acidic pH-induced elevation in interleukin 8 expression by human ovarian carcinoma cells. *Cancer Res* 60(16):4610–4616.

Yan SF, Lu J, Zou YS, Won JS, Cohen DM, Buttrick PM, Cooper DR, Steinberg SF, Mackman N, Pinsky DJ, Stern DM (1999): Hypoxia-associated induction of early growth response-1 gene expression. *J Biol Chem* 274(21):15030-15040.

Zhong H, De Marzo AM, Laughner E, Lim M, Hilton DA, Zagzag D, Buechler P, Isaacs WB, Semenza GL, Simons JW (1999): Overexpression of hypoxia-inducible factor 1alpha in common human cancers and their metastases. *Cancer Res* 59:5830–5835.

Zou J, Guo Y, Guettouche T, Smith DF, Voellmy R (1998): Repression of heat shock transcription factor HSF1 activation by HSP90 (HSP90 complex) that forms a stress-sensitive complex with HSF1. *Cell* 94(4):471–480.

24

CLOSTRIDIUM-MEDIATED TRANSFER OF THERAPEUTIC PROTEINS TO SOLID TUMORS

PHILIPPE LAMBIN, M.D., PH.D., JAN THEYS, PH.D.,
SANDRA NUYTS, M.D., WILLY LANDUYT, PH.D.,
LIEVE VAN MELLAERT, PH.D., AND JOZEF ANNÉ, PH.D.

INTRODUCTION

During the past two decades, major advances in understanding the genetic basis of cancer development have created new opportunities for treatment. One of the most promising novel approaches is the possibility of using gene therapy to selectively target and destroy tumor cells. A variety of strategies and vector delivery systems have been developed in attempts to deliver high doses of therapeutic agents to the tumor or its microenvironment. To achieve selectivity, therapeutic genes must be introduced into cells in such a way as to ensure uptake and expression in as many cancer cells as possible, while limiting acquisition and expression by normal cells. Many different approaches have been conceived to increase the selectivity of vector systems, because strict targeting is necessary to ensure that only malignant cells will be killed. Nevertheless, the inability to mediate targeted and specific delivery of therapeutics to cancer cells, together with the poor distribution of the vector throughout

Vector Targeting for Therapeutic Gene Delivery, Edited by David T. Curiel and Joanne T. Douglas
ISBN 0-471-43479-5

the tumor mass, remain serious hurdles to the application of gene therapy in cancer treatment.

The transfer of a therapeutic gene product to the tumor by apathogenic clostridia has been proposed as an alternative vector system as it is highly tumor-specific and avoids many of the obstacles of other gene delivery systems. The anaerobe-mediated targeting strategy is based on several observations.

- An aberrant capillary network exhibiting many shunts and torsions characterizes solid tumors. As a consequence, a considerable part of the tumor becomes hypoxic. Severe hypoxia is not found in healthy, normal tissues.
- The lack of a balance between the process of angiogenesis and the tumor growth leads to nutrient deprivation, subsequently to cell death, and finally to the development of necrotic areas.
- Infiltration of anaerobic bacteria in human tumors has been observed in the clinic.

Because these hypoxic/necrotic regions are unique to solid tumors, and because proliferation of the obligate anaerobes can occur only in these hypoxic regions, there is potential to use anaerobes as a selective gene transfer system in anticancer treatment. Moreover, such a system will have several additional advantages. In contrast to systems that require the integration of the relevant gene into the chromosome of the host cell, the therapeutic gene will be expressed in *Clostridium* and its gene product will be secreted into the local tumor microenvironment. Hence, several limitations of other gene transfer systems are avoided, including transduction efficiency, the problem of the expression of a heterologous gene in the tumor cell, and the genetic instability of recombinant tumor cells (resulting in the eventual loss of the therapeutic gene). Although this approach may seem unusual at first glance, the attempt to treat cancer using bacterial agents dates from long ago. Recently however, this approach has attracted renewed interest because of the current availability of recombinant DNA technology for clostridia and the possibility to transform *Clostridium* and express heterologous genes in this host. Our group has been involved in the development of a tumor-specific clostridial delivery vehicle and in the assessment in vivo of this novel therapy for treating solid tumors. We have used nonpathogenic *Clostridium* species to deliver two types of therapeutic proteins specifically to the tumor: (1) the suicide gene cytosine deaminase (CD), which converts the nontoxic prodrug 5-fluorocytosine (5-FC) to the toxic drug 5-fluorouracil (5-FU) resulting in cell death, and (2) the cytokine tumor necrosis factor α (mTNFα). Both 5-FU and mTNFα are also well-known enhancers of the antineoplastic response following radiation treatment. A schematic overview of this concept is presented in Figure 24.1.

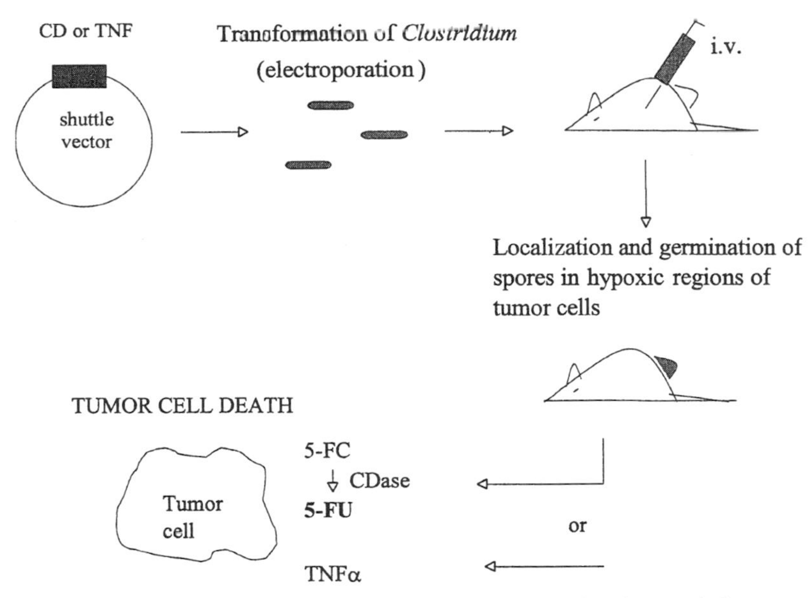

Figure 24.1 Schematic overview of the *Clostridium*-mediated gene delivery system for transfer of therapeutic proteins to hypoxic/necrotic regions of solid tumors.

CLOSTRIDIUM: INHERENT TUMOR SPECIFICITY DUE TO HYPOXIA

The genus *Clostridium* comprises a heterogeneous group of strictly anaerobic, Gram-positive endospore-forming bacteria, of which only 35 are considered to be pathogenic. They form spores located in either a central, subterminal, or terminal position. Most clostridial species are motile and are peritrichously flagellated. In the last decade, significant advances have been made in the development of molecular techniques to manipulate clostridia. Numerous clostridial genes have been cloned and characterized, procedures have been developed for the introduction of plasmid vectors by either protoplast transformation, conjugation, or electroporation, and an array of specialized cloning vectors have been constructed to introduce genes into various clostridial strains (Minton and Oultram, 1988; Young, Minton, and Staudenbauer, 1989).

Hypoxia as an Opportunity for Cancer Treatment

Normal tissues are characterized by a well-defined and structured vascular tree, which enables a spatial and temporal homeostatic flow and oxygen equilibrium. In contrast however, numerous investigations of experimental tumors have documented severe disturbances of the microcirculation, which occur already at an early stage of tumor growth (Horsman, 1993). This consequently leads to inefficient delivery of oxygen and other nutrients to many of the cells in tumors. Heterogeneities in the metabolic milieu have been demonstrated not only for experimental tumors but also for malignant disease in patients. Cells in this aberrant environment can remain viable and are often chemo- and radioresistant. It has been shown that hypoxia induced DNA overreplication and enhanced metastatic potential of murine tumor cells (Young, Marshall, and Hill, 1988). Moreover, several findings provide evidence that hypoxia, by selecting for mutant p53, might predispose tumors to a more malignant, proangiogenic phenotype (Giaccia, 1996; Graeber et al., 1996). Clinical data support this consideration (Brizel et al., 1996). Hence, hypoxia is considered a major hindrance to therapy.

However, the low oxygen levels in tumors can also be used as an opportunity for cancer treatment. The use of hypoxia-activated drugs, resulting in the tumor-specific killing of cells that are resistant to conventional therapy illustrates the principle of "complementary cytotoxicity" (Brown and Giaccia, 1998). In the same context, a strategy to exploit hypoxia in solid tumors would be to identify a promoter, highly responsive to the so-called hypoxia inducible factor (HIF-1) that would drive the expression of a therapeutic gene specifically in the tumor. Proof of this concept has been obtained using the CD enzyme (Dachs et al., 1997). Another approach involves the targeting of anaerobic bacteria such as *Clostridium* to solid tumors due to the occurrence of hypoxic/necrotic areas.

Clostridium as a Tumor-Specific Vector System: A Historical Review

Clostridial spore germination in necrotic tissue is most commonly associated with wound infections. However, as already mentioned, hypoxic/necrotic regions are also found in most solid tumors. The first data on the use of nonpathogenic clostridia for cancer treatment were reported in 1935 (Connell, 1935). Treatment with sterile filtrates of *C. histolyticum* resulted in clinical beneficial effects that were attributed to the production of proteolytic enzymes. Spore germination in tumors was first demonstrated in 1947 when it was shown that marked lysis of tumor tissue occurred following injection of *C. histolyticum* spores into mice sarcomas (Parker, 1947). The selectivity of *Clostridium* spores to germinate in hypoxic/necrotic regions was further clearly demonstrated through the injection of tumor-bearing mice with spore suspensions of *C. tetani* (Malmgren and Flanigan, 1955). The mice all died within 48 h, due to tetanus toxin produced by germinated spores in the tumor. A control group of tumor-free mice also injected with *C. tetani* spores remained healthy, clearly indicating that *Clostridium* was incapable of establishing itself in the animal in the absence of appropriate anaerobic conditions. Möse and Möse (Möse and Möse, 1964) reported that nonpathogenic clostridia (*C. butyricum* M55, later renamed *C. oncolyticum*) localized, germinated, and multiplied in solid and ascites Ehrlich tumors, causing oncolysis of the tumor.

Similar results were obtained with tumors of other rodents (Gericke and Engelbart, 1964; Thiele et al., 1964). Möse and Möse further demonstrated the nonpathogenicity of this procedure by injecting themselves with spores, and these studies were extended to human subjects with neoplastic disease (Carey et al., 1967; Heppner and Möse, 1978) particularly to patients with glioblastoma who received doses of 10^9 to 10^{10} spores injected either intravenously or intra-tumorally. Lysis was demonstrated only in the tumors, not in surrounding normal tissue and with the exception of a mild fever, patients suffered no ill effects. However, clinical trials were discontinued because no clinical benefit could be demonstrated. Despite the destruction of a large area of the tumor, an outer rim of viable tumor tissue remained from which regrowth occurred.

The fact that clostridial oncolysis was insufficient for complete tumor elimination led to the development of combined therapies in order to enhance the therapeutic effect. Even though coadministration of spores with several drugs (e.g., thioguanine, actinomycin D) in general increased lysis, it did not bring about complete tumor responses (Carey et al., 1967; Schlechte et al., 1982; Thiele et al., 1964). Increasing the degree of hypoxia has also been an option to improve therapeutic outcome. The application of microwaves, raising the temperature and the hypoxic fraction in the tumor, in combination with additional irradiation and spore administration led to a significant increase of survival rates in different mouse tumor models (Dietzel, 1978; Gericke, Dietzel, and Ruster, 1979). In another approach, the reduction of the oxygen concentration in respiratory air supplied to tumor-bearing rats resulted in improved antitumor effects (Möse, 1979). Based on these observations, clostridia have also been used in the field of cancer diagnosis. The detection in sera of antibodies directed against vegetative cells, is indicative for the presence of actively growing, vegetative clostridial cells, and thus for the existence of a tumor. This system has been extensively studied both in animals and humans with neoplastic disease (Fabricius et al., 1987; Wittmann et al., 1990).

The observations from animal and patient studies indicated that spore treatment was well tolerated but not sufficient for complete tumor regression. However, the development of recombinant DNA technology for clostridia and the possibility to transform *Clostridium* and to express heterologous genes in this host allows the construction of recombinant *Clostridium* with additional properties that may be potentially effective in tumor control. Schlechte and Elbe (1988) first tested this by introducing the gene encoding colicin E3, an *E. coli*-derived bacteriocin that was reported to have cancerostatic properties, into *C. butyricum* M55. The results were, however, contestable. The feasibility of generating recombinant strains that may be useful in antitumor therapy was later demonstrated by two separate research teams, through the construction of clostridial strains producing prodrug converting enzymes (Fox et al., 1996; Lemmon et al., 1997; Theys et al., 2001a) and cytokines (Theys et al., 1999).

EFFECTIVITY OF *CLOSTRIDIUM* FOR TUMOR-SPECIFIC COLONIZATION

The principal strain used in previous tumor studies was *C. oncolyticum* M55. However, no reliable transformation procedures are available for this strain, making the introduction of heterologous DNA impossible. In recent years, other strains for which recombinant DNA technology have been described have been examined for their ability to specifically colonize tumors. In one study, the ability of different saccharolytic *Clostridium* strains to colonize tumors was investigated and compared to the colonization properties of *C. oncolyticum* (Lambin et al., 1998). Using WAG/Rij rats with syngeneic rhabdomyosarcomas as a tumor model, the colonization efficiency of the various *Clostridium* strains was as follows: *C. oncolyticum* DSM754 ≈ *C. sporogenes* > *C. acetobutylicum* NI4082 ≈ *C. acetobutylicum* DSM792 >>> *C. beijerinkii* ATCC 17778 > *C. limosum* DSM1400. Quantitative analysis at different time intervals after systemic administration of spores at concentrations of at least 10^7 colony-forming units (cfu) ml^{-1} revealed that spores circulated throughout the whole body, but that only in tumors did *Clostridium* spores germinate to metabolically active, dividing cells. The number of cfu found in tumors was up to 10^9 per gram of tissue. The ratio of vegetative cells to spores in tumors differed at least by a factor of 100, indicating the presence of vegetative cells in the tumor. Colonization remained high during the entire follow-up period (Fig. 24.2). Vegetative cells could not be detected in normal tissues, nor was there any evidence of clostridia in the urine at either 4 or 8 days following administration of spores. This was not surprising, because healthy tissues are well

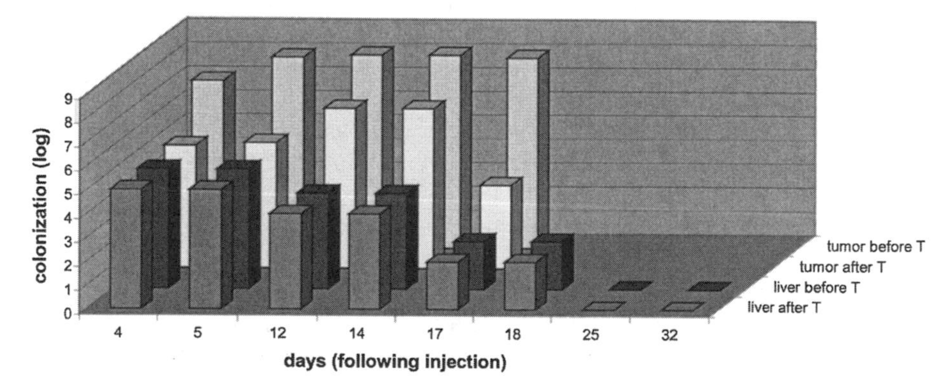

Figure 24.2 Colony-forming unit counts in tumors and liver as a function of time (up to 32 days) following systemic injection of *C. acetobutylicum*. Longer periods could not be investigated due to the fact that tumors grew to the maximum size permitted. Heat treatment at 73°C for 20 min (T) kills vegetative cells but not spores. Before T, the cfu represent the combined amount of spores and vegetative cells. After T, only spores are present.

oxygenated and clostridial spores do not germinate under these circumstances. The results were quantitatively confirmed: the number of cfu in samples of liver, spleen, kidney, brain, or eye was not changed after heating. Moreover, the number of spores present in normal tissues decreased as a function of time. After 32 days, spores could not be detected in any of the investigated normal tissues (Fig. 24.2). The presence of vegetative clostridial cells, specifically and exclusively in the hypoxic/necrotic regions of the tumors, was confirmed using histochemical staining techniques. Similarly, using EMT6 tumor-bearing mice, Lemmon et al. (1997) found that after administration of *C. beijerinckii* NCIMB8052 spores, high numbers of vegetative cells were present specifically in the hypoxic/necrotic regions of the tumor. No vegetative rods could be isolated from normal tissues.

Concentrated efforts in recent years have demonstrated the critical importance of angiogenesis in the development of tumors (Folkman, 1992). Angiogenesis is a complex process leading to the growth of new blood vessels. The interactions between the pro- and antiangiogenic molecules and the way they affect vascular structure are beginning to be elucidated. This knowledge has led to a number of new exciting strategies for cancer treatment. In the context of the *Clostridium*-mediated transfer system, vascular targeting of solid tumors is an attractive adjuvant treatment aimed at improving bacterial colonization by increasing the degree of hypoxia in the tumor. The tumor-specific antivascular activity of vascular targeting agents is based on the presence of morphologically and functionally abnormal blood vessels that are required for continuous tumor expansion (Denekamp, 1993). Combretastatin A-4 phosphate (CombreAp) is a vascular targeting agent interfering with tubulin polymerization. We showed that when CombreAp is given systemically at nontoxic doses (25 mg kg^{-1}) to WAG/Rij rats with rhabdomyosarcomas, rapid and severe vascular shutdown was induced within 3–6 h after injection, followed by necrosis within 1–3 days without obvious side effects (Landuyt et al., 2000). Very similar effects are documented in several rodent tumor models (Chaplin, Pettit, and Hill, 1999; Dark et al., 1997; Horsman et al., 1998). When tumors were divided into groups of predetermined volumes (< 1 cm^3, 1–3 cm^3, >3 cm^3) and *Clostridium* spores were administered systemically 4 h prior to CombreAp, the difference in colonization of small tumors (<3 cm^3) was striking (Theys et al., 2001b). Without CombreAp treatment, the amount of bacteria varied between nondetectable and 10^3 cfu per gram of tissue, while combined with CombreAp, more than 10^7–10^8 cfu per gram of tumor tissue were observed—an increase of at least a factor of 10^4 to 10^6 ($p < 0.001$ for tumors of 1–3 cm^3; $p < 0.0001$ for tumors <1 cm^3). Sham-treated tumors always revealed a solid well-vascularized tissue at transsection, with few scattered foci of necrosis. When treated with CombreAp, tumors were mostly soft on palpation, with an extensive core of necrosis and a viable-looking rim of tumor tissue surrounding the necrotic center. The result of the histopathological examinations and the gross appearance of the tumor at the time of transsection, led to the conclusion that a strong relationship exists between the necrosis induced by CombreAp and the increased presence of *Clostridium*. These results therefore support the combination of the vascular targeting approach and the proposed *Clostridium*-mediated delivery system.

To demonstrate that systemically applied clostridia (a class I hazardous agent according to the European list of infectious agents) could be removed from the tumor if desired, the effectiveness of antibiotic treatment with metronidazole (Flagyl®) was evaluated. This antibiotic is used routinely in the clinic and it appears to reach the poorly perfused hypoxic areas, making it ideal for treating anaerobic infections. Metronidazole was given twice a day starting at day 5 following spore administration. The number of cfu decreased as a function of time, and after 9 days of treatment no bacterial growth could be detected, while colonization in nontreated control animals remained high (Theys et al., 2001b). Another important factor to be considered is the eventual induction of an immune response after a single or repeated administration of *Clostridium* and its consequences on tumor colonization. We showed that a severe host immune response is not induced following repeated administration of clostridial spores. Neither fever nor any loss of body weight was observed when 10^8 to 10^{10} *Clostridium* spores were systemically administered to rhabdomyosarcoma-bearing WAG/Rij rats, regardless of the injected bacterial load. Also, repeated (3×) systemic injections of *Clostridium* spores did not induce measurable side effects. Experiments were also designed to investigate whether *Clostridium* could still colonize tumors following repeated (2×) administration of spores, after eradication of previously administered clostridia with antibiotics. Interestingly, regardless of the immune response status of the host, colonization efficiency of tumors was not affected following repeated spore administration, and always occurred to the same extent compared to animals treated only once. This could have important implications because it implies that long-term colonization of *Clostridium*, and hence long-term expression of the therapeutic genes introduced into *Clostridium* is possible.

GENETIC ENGINEERING OF *CLOSTRIDIUM* TO EXPRESS AND SECRETE THERAPEUTIC PROTEINS

To date, two types of therapeutic agents have been chosen for delivery to tumors: drugs that exert a direct cytotoxic effect, such as the cytokine mTNFα and proteins such as CD or nitroreductase, encoded by so-called suicide genes that convert a nontoxic prodrug into a toxic therapeutic drug. mTNFα, a trimeric protein consisting of 17 kDa monomers, was initially identified as a protein released by endotoxin-stimulated macrophages, although several other types of cells are able to synthesize small amounts (Seung et al., 1995; Weichselbaum et al., 1994). mTNFα has pleiotropic effects, including selective action on the neovasculature of tumors, stimulation of T cell-mediated immunity, and direct cytotoxicity to tumor cells. It can induce both necrotic and apoptotic forms of cell death. Preclinical in vitro studies in cell cultures and in vivo studies in animal models have demonstrated the antitumor capacities of mTNFα. *E. coli* CD converts nontoxic 5-FC to its toxic anabolite 5-FU, which is further metabolized and ultimately interferes with the synthesis of DNA and RNA. In the CD/5-FC system, the effectiveness of 5-FC in killing tumor cells transfected with the CD gene has been shown both in vitro and in vivo. Anti cancer effects of the CD/5-FC system have been observed for a wide variety of solid tumors, including colorectal (Nanni et al., 1998), gastric (Ge et al., 1997), hepatocellular (Ichikawa et al., 2000), breast cancers (Trinh et al., 1995), and glioma (Aghi, Hochberg, and Breakefield, 2000). The action of these suicide genes is associated with a so-called bystander effect: Eradication of tumor cells transduced with the suicide gene elicits a killing effect on the surrounding nontransduced tumor cells. In the CD/5-FC system, less than 4% of transfected cells were proven to be sufficient to achieve a 60% cure rate (Trinh et al., 1995).

Endogenous plasmids encoding selectable genetic markers have not been identified in the biotechnically important clostridia. However, suitable shuttle vectors are available that contain antibiotic resistance genes expressed in Gram-positive bacteria, as well as functional replicons that allow replication in both Gram-positive and Gram-negative hosts. Since we were interested specifically in investigating the effects of introducing mTNFα or CD into the local tumor microenvironment, a vector was constructed, which after introduction in *Clostridium* enabled the secretion of the therapeutic agent. Only a limited number of clostridial signal sequences are described in literature. Sequences from the endo-$\beta_{1,4}$-glucanase (*eglA*) gene of *C. acetobutylicum* P262 (Zappe et al., 1988) and from the clostripain gene (coding for a cysteine endopeptidase) of *C. histolyticum* (Dargatz et al., 1993), were used for expression and secretion of the therapeutic genes. To obtain in-frame fusions between the selected clostridial signal sequences and the coding sequences of the therapeutic genes, appropriate mutations were introduced at the 5′-end of the coding sequences of mTNFα and the *E. coli* CD gene (*codA*) and at the 3′-end of the signal sequences. The fusion constructs were subsequently cloned in appropriate shuttle vectors (pKNT19, pIMP1, pMTL500E). Following transformation of *C. acetobutylicum* using strain-specific electroporation protocols, mTNFα expression in *C. acetobutylicum* recombinant cultures was monitored by Western blot analysis. In lysates of the recombinant cultures, mTNFα was detected both as a preprotein (21 kDa) and as the mature form (17 kDa). Only the latter was present in the supernatant. mTNFα could not be detected in lysates or supernatants of the control *Clostridium* cultures. The functionality of the *eglA* and clostripain regulatory sequences preceding the mTNFα coding sequence was thereby proven, not only for the expression, but also for the extracellular secretion of this protein by both *C. acetobutylicum* strains. Biologically active mTNFα was quantified in lysates and supernatants of recombinant *C. acetobutylicum* cultures grown for different time periods, using a cytotoxicity test toward WEHI164 clone 13 cells. In this test, the amount of mTNFα was spectrophotometrically determined by measuring the in situ reduction of the yellow-colored MTT to a blue formazan by mitochondrial dehydrogenases of metabolically active cells, thus measuring essentially the percentage of nonviable cells. The high sensitivity of WEHI164 clone 13 cells for mTNFα makes it possible to detect very low mTNFα concentrations. The mTNFα concentration in lysates and supernatants of recombinant clostridia increased as the cells grew exponentially to an optical density at 600 nm (OD_{600}) of approximately 0.6 (midlog phase). In lysates and supernatants, a maximum of 10^3 to 10^5 U ml^{-1} was found, depending on the recombinant plasmid and strain that was used. The amount of mTNFα present in the supernatant decreased below the detection limit (3.1 U ml^{-1}) after 12 h, whereas in lysates biologically active mTNFα was still detectable during the 20 h follow-up period. mTNFα activity was not detected in supernatants or lysates of cultures that were untransformed or transformed with a control plasmid that does not contain the mTNFα gene. To investigate whether the observed decrease in mTNFα activity was due to increased acidity of the medium due to acidic fermentation products accumulating during the stationary growth phase, the culture medium was buffered with 50 mM MOPS (pH 7.2) in order to diminish acidification. As a result, elevated levels of biologically active mTNFα were obtained both in lysates and supernatants, suggesting that the decrease of pH negatively affected mTNFα stability (Fig. 24.3). This was in accordance with previous observations describing the stability of mTNFα within a pH region of 5.5 to 10 (Van Ostade, Tavernier, and Fiers, 1994).

CD expression was monitored in lysates and supernatants of early logarithmic growth phase cultures of recombinant *C. acetobutylicum*[pKNT19-closcodA]. In all samples, a protein of 52 kDa could be observed following sodium dodecylsulfate-polyacrylamide gel electrophoresis (SDS-PAGE) and Western blotting using the 16D8F2 monoclonal antibody for detection of the *E. coli* CD protein. No immunoreactive proteins were detected in lysates or supernatants of plasmid-free or *C. acetobutylicum* cultures carrying the control plasmid (Fig. 24.4). These results clearly show the functionality of both the clostripain promoter and signal sequence preceding the CD cDNA. Moreover, a considerable amount of heterologous protein was expressed and efficiently secreted by *Clostridium,* notwithstanding the large size of the *E. coli* CD protein. We developed a thin layer chromatography approach that made quantitative analysis of CD activity possible without the need for radioactivity. Recombinant bacteria containing the pKNT19closcodA construct were cultured and sampled at various stages of growth. CD activity measurements in lysates and supernatants showed an increase in enzyme activity in both lysates and supernatants until early stationary growth phase (OD_{600} ~1.2). The level of activity in lysates of stationary phase cultures remained high within the subsequent 20 h follow-up period (e.g., maximum enzyme activity = 1084.5 ± 189.5 pmol 5-FC converted to 5-FU min^{-1} ml^{-1} cell lysate for recombinant *C. acetobutylicum* cultures). In supernatants, CD activity decreased after this time point. Maximum levels obtained in supernatants were slightly lower as compared to CD enzymatic activity in lysates (e.g., maximum enzyme activity = 701.9 ± 104.3 pmol 5-FC converted to 5-FU min^{-1} ml^{-1} supernatant for

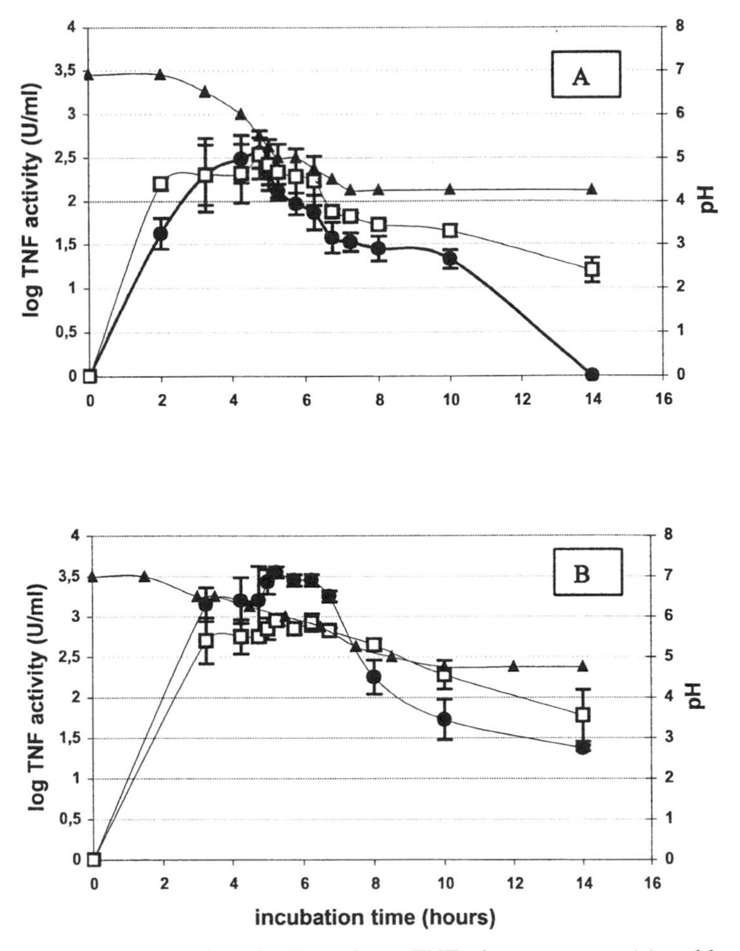

Figure 24.3 Amount of biologically active mTNFα in supernatant (•) and lysates (□) of *C. acetobutylicum* DSM792 transformed with pIMP1*eglA*mTNFα and evolution of pH (▲) in nonbuffered (A) and buffered (B) medium as a function of growth. Bars represent standard errors.

recombinant *C. acetobutylicum*). In other studies, recombinant plasmids capable of directing the expression of both the *cod*A and *nfn*B genes (encoding cytosine deaminase and nitroreductase, respectively) in *C. beijerinckii*, were generated (Fox et al., 1996; Lemmon et al., 1997). Using the pMTL500F and pMTL540FT expression vectors, the coding sequences were placed under regulatory control of the clostridial Ferredoxin (Fd) promoter. In all instances, lysates of cells carrying the recombinant plasmids encoding the heterologous genes were shown to contain the expected enzymatic activity and were able to convert the appropriate prodrug into the corresponding cytotoxic agent that is active against tumor cells.

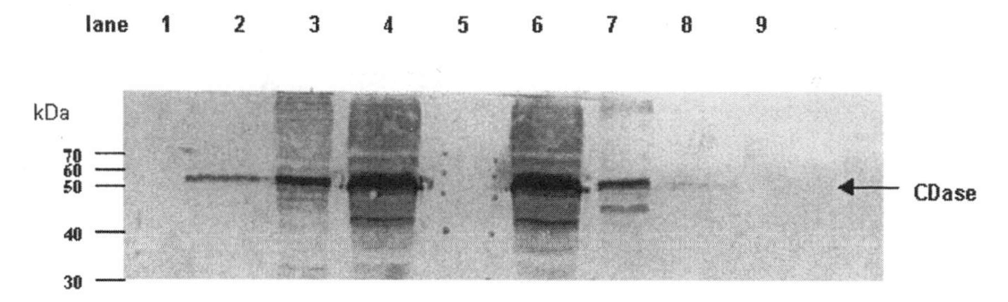

Figure 24.4 Western blot of supernatants and cell lysates of *C. acetobutylicum* strains expressing CD, after detection by mAb 16D8F2. Molecular weight standards are shown in kDa. Lanes: *1. C. acetobutylicum* NI4082 lysate (negative control); *2. C. acetobutylicum* NI4082 [pKNT19closcodA], supernatant; *3. C. acetobutylicum* N14082 [pKNT19closcodA], lysate; *4. E coli* PC0698 [pSD112] lysate (positive control); *5.* 10 kDa protein standard; *6. E. coli* PC0698 [pSD112] lysate (positive control); *7. C acetobutylicum* DSM792 [pKNT19closcodA], lysate; *8. C. acetobutylicum* DSM792 [pKNT19closcodA], supernatant; *9. C. acetobutylicum* DSM792, lysate (negative control).

CLOSTRIDIUM AS A SYSTEM TO TRANSFER THERAPEUTIC PROTEINS TO TUMORS

After validation that nonpathogenic *Clostridium* species could be genetically modified to express and secrete therapeutic proteins, the next step was to determine the ability of these recombinant strains to transfer the heterologous proteins specifically to the tumor site. *C. beijerinckii* engineered to produce nitroreductase was able to express this enzyme at the tumor site, as evidenced by Western blot analysis of tumor homogenates 4 days after recombinant spores were administered to EMT6 tumor-bearing mice (Lemmon et al., 1997). In our studies, spore suspensions prepared from recombinant *C. acetobutylicum* strains expressing cytosine deaminase were administered to rhabdomyosarcoma-bearing WAG/Rij rats. After 4 to 6 days, animals were sacrificed and tumors, as well as normal tissues (liver, spleen), were removed and minced. Liver and spleen were chosen as normal tissue controls because they contained the highest number of spores of the healthy tissues investigated (compare the foregoing). Gram staining revealed the presence of vegetative *Clostridium* cells in tumor specimens, but not in samples of liver or spleen. CD was never found in control tumors nor, more importantly, within the normal tissues investigated. Animals concomitantly treated with CombreAp showed higher incidence of CD-positive tumors (100% versus 58%). Moreover, the level of active CD in these tumor specimens was considerably higher (mean conversion efficiency of 5-FC to 5-FU ~11%) as compared to tumors not treated with the vascular targeting drug (mean conversion efficiency of 5-FC to 5-FU ~3%), as clearly evidenced by the intensity of the 5-FU spots observed in the analysis. These results illustrate that combining the administration of clostridia and CombreAp treatment increases the therapeutic dose intensity. Obviously, this can also be expected in combination with any other strategy that induces tumor necrosis, such as radiotherapy. In that context, 5-FU has been reported to be an effective radiosensitizer. Based on published data (Lambin *et al.*, 2000), it was calculated that a 1–3% conversion of 5-FC to 5-FU would be sufficient to achieve clinically significant radiosensitization. Based on the obtained in vitro and in vivo results, it is reasonable to believe that this is achievable with recombinant clostridia. *C. acetobutylicum* strains, recombinant for mTNFα, are currently being evaluated for the delivery of this therapeutic agent to the tumor. Preliminary results indicate the potential in vivo applicability of mTNFα-recombinant clostridia for the in situ production of mTNFα in tumor tissue.

TEMPORAL AND INCREASED SPATIAL CONTROL OF GENE EXPRESSION: USE OF RADIO-INDUCIBLE PROMOTERS IN *CLOSTRIDIUM*

More than 75% of all cancer patients are treated with radiotherapy during the progress of their disease. In spite of the enormous progress made over the last decennia, radiotherapy still encounters some limitations. One of the biggest problems is hypoxia, a known feature of solid tumors. Because oxygen fixates DNA damage after ionizing radiation, the absence of oxygen leads to radioresistance. Therefore, it seems promising to combine radiotherapy with an additional therapeutic modality, such as recombinant anaerobic *Clostridium*, which will specifically target these hypoxic cells.

To further increase the specificity of this tumor-directed delivery system, the gene of interest may be placed under the control of a radio-induced promoter. This will result in an activation of the promoter after irradiation of the tumor, leading to spatial and temporal control of gene expression (i.e., expression of the therapeutic genes will be limited to the irradiated tissues only). Radio-inducible promoters are being used in many viral vector systems to achieve spatial and temporal control of gene expression (e.g., Datta et al., 1992; Hallahan et al., 1995; Marples et al., 2000). We investigated if radiation-induced gene expression could also be attained in a bacterial vector system using anaerobic nonpathogenic clostridia.

We investigated whether radio-induced genes actually exist in *Clostridium* and at what dose of irradiation these genes are activated. Bacteria are known to be very radio-resistant, as a consequence of efficient DNA repair mechanisms. One of these mechanisms is the SOS repair system (Miller and Kokjohn, 1990), which consists of more than 20 genes that are all activated by the occurrence of single strand DNA breaks. We investigated if the central gene, *recA,* was activated by ionizing irradiation in *Clostridium.* Northern blot hybridizations using RNA extracted from irradiated and nonirradiated clostridia confirmed radio-induction of the *recA* gene, which was evident already at the clinically relevant dose of 2 Gy (Nuyts et al., 2001b, 2001c). When we quantified the degree of induction of the *recA* promoter using a reporter system, an overall 30% significant increase in activity was detected (Nuyts et al., 2001c).

Since this evidence supported the presence of a radio-induced gene in *Clostridium*, we were interested to see if we could increase the secretion yield of a therapeutic protein by *Clostridium* after irradiation, using this radio-inducible promoter. Therefore, a shuttle vector was constructed that contained the *recA* promoter upstream of the therapeutic cytokine mTNFα. To obtain secretion of mTNFα, the coding sequence was preceded by the signal sequence of the *eglA* gene. *Clostridium* was transformed with the construct via electroporation and subsequently irradiated with a dose of 2 Gy. At different time intervals, samples were taken to quantify the amount of secreted mTNFα using an enzyme-linked immunosorbent assay (ELISA). After a single dose of 2 Gray (Gy), we measured a significant 44% increase in mTNFα secretion, 3.5 h after irradiation compared to nonirradiated controls (Nuyts et al., 2001a) (Figure 24.5). However, in clinical practice, patients are not treated with single doses, but generally with daily small doses of 2 Gy. Therefore, we tested if we could reactivate the promoter activity by giving a second dose of irradiation. When we measured the induction of mTNFα, we found that indeed gene activation could be repeated and that the increase in mTNFα secretion was

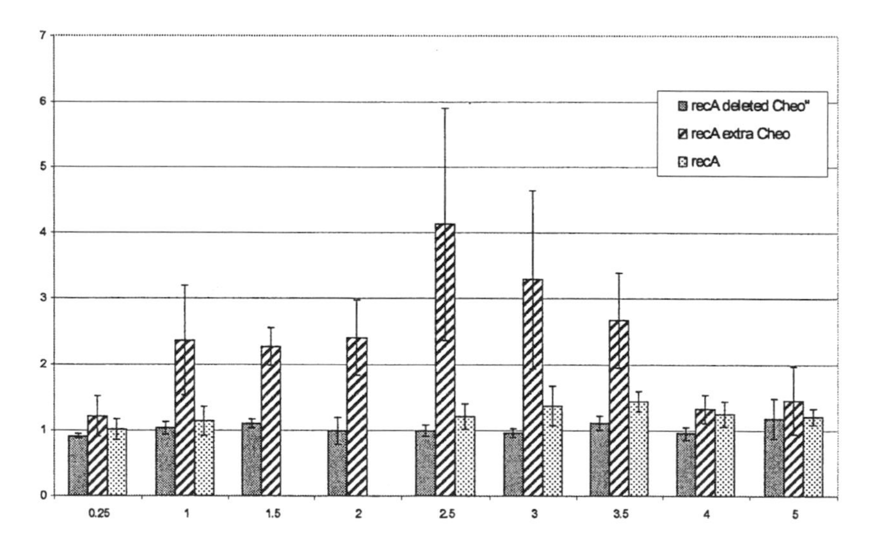

Figure 24.5 Fold increase of mTNFα secretion in *Clostridium acetobutylicum* DSM792 pIMP-*recA*-mTNFα (dotted bars), pIMP-*recA* deleted Cheo-mTNFα (grey bars) and pIMP-*recA* extraCheobox-mTNFα (hatched bars) 15 min, 1, 1.5, 2, 2.5, 3, 3.5, 4, and 5 h after a single dose of 2 Gy. The bars represent data from three independent experiments. Vertical bars represent standard deviations.

in the same range as after the first dose (Nuyts et al., 2001a). This proved that the radio-inducible promoter could be reactivated after each irradiation when using fractionated radiotherapy. Consequently, cell killing is increased by an exponent equal to the number of treatments.

However, when using the *recA* promoter we still have some expression of mTNFα under nonirradiated conditions, due to basal activity of the promoter. This is not optimal if temporal control is to be achieved. Therefore, we examined if we could suppress basal transcription by adding an extra repressor-binding site, or Cheo box, to the promoter region. Under basal conditions, the repressor DinR binds to this repressor-binding site, limiting transcription of the SOS genes. After activation by radiotherapy, both binding sites would become free and repression would be relieved. This would lead to an increase in transcription of the SOS genes, including *recA*, after radiotherapy. We first deleted the Cheo box sequence from the *recA* promoter to verify that this repressor-binding site was indeed responsible for the increase in promoter activity after radiotherapy. When bacteria containing this construct were irradiated, no increase in mTNFα- secretion was noticed (Fig. 24.5). However, addition of an extra Cheo box to the *recA* promoter resulted in a 412% increase of secreted mTNFα after irradiation, while only 44% yield increase was obtained using the wild-type promoter. As confirmed by reverse transcriptase-polymerase chain reaction (RT-PCR), the increase in secretion after irradiation was evidently the consequence of an increased promoter activity. Hence, the Cheo box sequence is the radio-responsive element and can be used to decrease basal transcription or to increase transcription upon induction (Nuyts et al., 2001d). The next step was to test if the Cheo box sequence could be used to bring a constitutive promoter under the control of irradiation. To this end, the Cheo box sequence was incorporated in the promoter region of the constitutive *eglA* promoter. This led to a 242% increase in mTNFα secretion after irradiation with 2 Gy. However, the wild-type promoter did not exhibit an increase in mTNFα secretion after irradiation. These data prove that the Cheo box is functional outside its natural sequence and can be used to bring other promoters under the control of ionizing irradiation (Nuyts et al., 2001d). Taken together, these data provide the proof of principle that radio-induced promoters can be used to control expression of therapeutic proteins by recombinant clostridia. In contrast to eukaryotic promoters tested so far, the *recA* promoter is activated at the clinically relevant dose of 2 Gy. This promoter gave rise to a significant yield increase of secreted mTNFα after irradiation. Moreover, we could enhance radio-responsiveness of the promoter by the insertion of an additional repressor-binding site to the promoter sequence. We also proved that strong, constitutive promoters can be brought under the control of ionizing irradiation if repressor-binding sites are incorporated into their promoter region.

The combination of radiotherapy, which preferentially kills well-oxygenated cells, with *Clostridium*-mediated protein delivery to target the hypoxic fraction, opens new possibilities for the future of cancer therapy.

CONCLUSIONS

Gene therapy was initially conceptualized as an approach for inherited genetic disease. Nevertheless, it is currently finding its widest use for treating neoplastic disorders. In this regard, more than 70% of patients treated to date in human clinical gene therapy protocols have been in the context of anticancer regimens. For successful gene vaccination and therapy of cancer, it is essential, however, to develop gene delivery vectors that can meet clinical and social requirements. The problem with current vector systems in most instances remains a lack of specificity and targeting therapeutic gene expression to tumor cells, representing a major challenge for cancer gene therapy. As a consequence, other strategies to improve specificity must be found. This can partially be overcome by the conditional expression of lethal genes in tumor cells and by the use of promoters that are preferentially active in cancer cells.

A completely different approach to specifically target tumors is the use of anaerobic bacteria as described in this overview. Spontaneous infection of tumors in cancer patients by anaerobic bacteria is an often-occurring problem. This phenomenon was noticed more than six decades ago and these observations, together with the effect it could cause on a tumor, constituted the onset of attempts to use anaerobic bacteria, and in particular *Clostridium*, to combat tumors in cancer patients. Ample experiments since then have shown that *Clostridium* and other anaerobic bacteria including *Bifidobacterium* (Yazawa et al., 2000) can specifically colonize tumors after intravenous injection, the specificity for tumors residing in their obligate requirement for anaerobic conditions. The efficiency of colonization could be further increased by the administration of antivascular targeting agents following which even small tumors could be colonized. As a consequence, these bacteria can be considered as tumor-specific vector systems to transport proteins of interest to tumors.

Whereas it was initially hoped that hydrolytic enzymes of the delivered strains would lead to tumor growth control, a major breakthrough came when it turned out to be possible to genetically modify *Clostridium* using recombinant DNA technology. As such *Clostridum* strains could be produced that express and secrete proteins of interest that either directly (e.g., by TNFα) or indirectly (e.g., by nitroreductase or CD) can lead to tumor cell death. Using this approach, clostridia have several advantages compared to classical gene therapy for cancer including high tumor-specificity, since in mammals hypoxia is a feature of solid tumors, not of normal tissues. The heterologous gene is not transduced into the genome of the tumor cell, because the anticancer gene will be expressed and secreted from the bacteria. Moreover, this approach can be considered as safe because targeted gene expression can be stopped at any time by administration of suitable antibiotics and no adverse effects have been observed (in experimental animals).

Selectivity and safety can be improved even further by using inducible promoters as described where radio-inducible promoters have been shown to be very interesting, not only because they are induced at a clinically relevant dose and they can be reactivated, but also the target proteins chosen have a synergistic or additive effect in combination with X rays, thus increasing the potentiality of the system.

In conclusion, the presented data show the proof of principle that *Clostridium* is an interesting host to specifically deliver therapeutic proteins to the tumor. If further improvement can be obtained with respect to the secretion of proteins to therapeutically relevant doses, an objective that certainly can be obtained when using stronger promoters and improved signal sequences, this system will without doubt be an invaluable alternative to the current classical gene therapy approaches tested in the clinic.

ACKNOWLEDGMENTS

We acknowledge the financial support from Het Fonds voor Wetenschappelijk Onderzoek-Vlaanderen, De Vlaamse Kankerliga, Verkennende Internationale Samenwerking, Het KULeuven Onderzoeksfonds, and Sportvereniging tegen Kanker of Belgium.

REFERENCES

Aghi M, Hochberg F, Breakefield XO (2000): Prodrug activation enzymes in cancer gene therapy. *J Gene Med* 2:148–164.

Brizel DM, Scully SP, Harrelson JM, Layfield LJ, Bean JM, Prosnitz LR, Dewhirst MW (1996): Tumor oxygenation predicts for the likelihood of distant metastases in human soft tissue sarcoma. *Cancer Res* 56:941–943.

Brown JM, Giaccia AJ (1998): The unique physiology of solid tumors: opportunities (and problems) for cancer therapy. *Cancer Res* 58:1408–1416.

Carey RW, Holland JF, Whang HY, Neter E, Bryant B (1967): Clostridial oncolysis in man. *Eur J Cancer* 3:37–46.

Chaplin DJ, Pettit GR, Hill SA (1999): Anti-vascular approaches to solid tumour therapy: evaluation of combretastatin A4 phosphate. *Anticancer Res* 19:189–195.

Connell H (1935): The study and treatment of cancer by proteolytic enzymes. A preliminary report. *Can Med Ass J* 33:364–370.

Dachs GU, Patterson AV, Firth JD, Ratclife PJ, Townsend KM, Stratford IJ, Harris AL (1997): Targeting gene expression to hypoxic tumor cells. *Nat. Med* 3:515–520.

Dargatz H, Diefenthal T, Witte V, Reipen G, von Wettstein D (1993): The heterodimeric protease clostripain from *Clostridium histolyticum* is encoded by a single gene. *Mol Gen Genet* 240:140–145.

Dark GG, Hill SA, Prise VE, Tozer GM, Pettit GR, Chaplin DJ (1997): Combretastatin A-4, an agent that displays potent and selective toxicity toward tumor vasculature. *Cancer Res* 57:1829–1834.

Datta R, Rubin E, Sukhatme V, Qureshi S, Hallahan D, Weichselbaum RR, Kufe DW (1992): Ionizing radiation activates transcription of the EGR1 gene via CArG elements. *Proc Natl Acad Sci USA* 89:10149–10153.

Denekamp J (1993): Review article: angiogenesis, neovascular proliferation and vascular pathophysiology as targets for cancer therapy. *Br J Radiol* 66:181–196.

Dietzel F (1978): Hyperthermia and radiotherapy. *Strahlentherapie [Sonderb]* 75:63–69.

Fabricius EM, Schmidt W, Schneeweiss U (1987): The serological *Clostridium* tumour test—application of discriminance analysis. *Zentralbl Bakteriol Mikrobiol Hyg [A]* 263:552–560.

Folkman J (1992): The role of angiogenesis in tumor growth. *Semin Cancer Biol* 3:65–71.

Fox ME, Lemmon MJ, Mauchline ML, Davis TO, Giaccia AJ, Minton NP, Brown JM (1996): Anaerobic bacteria as a delivery system for cancer gene therapy: in vitro activation of 5-fluorocytosine by genetically engineered clostridia *Gene Ther* 3:173–178.

Ge K, Xu L, Zheng Z, Xu D, Sun L, Liu X (1997): Transduction of cytosine deaminase gene makes rat glioma cells highly sensitive to 5-fluorocytosine. *Int J Cancer* 71:675–679.

Gericke D, Engelbart K (1964): Oncolysis by clostridia II. Experiments of a tumour spectrum with a variety of clostridia in combination with heavy metal. *Cancer Res* 24:217–221.

Gericke D, Dietzel F, Ruster I (1979): Further progress with oncolysis due to local high frequency hyperthermia, local x-irradiation and apathogenic clostridia. *J Microw Power* 14:163–166.

Giaccia AJ (1996): Hypoxic stress proteins: Survival of the fittest. *Semin Radiat Oncol* 6:46–58.

Graeber TG, Osmanian C, Jacks T, Housman DE, Koch CJ, Lowe SW, Giaccia AJ (1996): Hypoxia-mediated selection of cells with diminished apoptotic potential in solid tumours. *Nature* 379:88–91.

Hallahan DE, Mauceri HJ, Seung LP, Dunphy EJ, Wayne JD, Hanna NN, Toledano A, Hellman S, Kufe DW, Weichselbaum RR (1995): Spatial and temporal control of gene therapy using ionizing radiation. *Nat Med* 1:786–791.

Heppner F, Möse JR (1978): The liquefaction (oncolysis) of malignant gliomas by a non pathogenic *Clostridium. Acta Neurochir* 42:123–125.

Horsman MR, Ehrnrooth E, Ladekarl M, Overgaard J (1998): The effect of combretastatin A-4 disodium phosphate in a C3H mouse mammary carcinoma and a variety of murine spontaneous tumors. *Int J Radiat Oncol Biol Phys* 42:895–898.

Ichikawa T, Tamiya T, Adachi Y, Ono Y, Matsumoto K, Furuta T, Yoshida Y, Hamada H, Ohmoto T (2000): In vivo efficacy and toxicity of 5-fluorocytosine/cytosine deaminase gene therapy for malignant gliomas mediated by adenovirus. *Cancer Gene Ther* 7:74–82.

Lambin P, Theys J, Landuyt W, Rijken P, van der Kogel A, van der Schueren E, Hodgkiss R, Fowler J, Nuyts S, de Bruijn E, Van Mellaert L, Anné J (1998): Colonisation of *Clostridium* in the body is restricted to hypoxic and necrotic areas of tumours. *Anaerobe* 4:183–188.

Landuyt W, Verdoes O, Darius DO, Drijkoningen M, Nuyts S, Theys J, Stockx L, Wynendaele W, Fowler JF, Maleux G, Van den Bogaert W, Anné J, van Oosterom A, Lambin P (2000): Vascular targeting of solid tumours: a major 'inverse' volume-response relationship following combretastatin A-4 phosphate treatment of rat rhabdomyosarcomas. *Eur J Cancer* 36:1833–1843.

Lemmon MJ, van Zijl P, Fox ME, Mauchline ML, Giaccia AJ, Minton NP, Brown JM (1997): Anaerobic bacteria as a gene delivery system that is controlled by the tumor microenvironment. *Gene Ther* 4:791–797.

Malmgren RA, Flanigan CC (1955): Localisation of the vegetative forms of *Clostridium tetani* in mouse tumours following intravenous spore administration. *Cancer Res* 15:473–478.

Marples B, Scott SD, Hendry JH, Embleton MJ, Lashford LS, Margison GP (2000): Development of synthetic promoters for radiation-mediated gene therapy. *Gene Ther* 7:511–517.

Miller RV, Kokjohn TA (1990): General microbiology of *recA:* environmental and evolutionary significance. *Annu Rev Microbiol* 44:365–394.

Minton NP, Oultram JD (1988): Host: vector systems for gene cloning in *Clostridium. Microbiol Sci* 5:310–315.

Möse JR (1979): [Experiments to improve the oncolysis-effect of clostridial-strain M55 (author's transl)]. *Zentralbl Bakteriol [Orig A]* 244:541–545.

Möse JR, Möse G (1964): Oncolysis by clostridial activity of *Clostridium butyricum* (M-55) and other non-pathogenic clostridia against the Ehrlich carcinoma *Cancer Res* 24:212–216.

Nanni P, DeGiovanni C, Nicoletti G, Landuzzi L, Rossi I, Frabetti F, Giovarelli M, Forni G, Cavallo F, DiCarlo E, Musiani P, Lollini PL (1998): The Immune response elicited by mammary adenocarcinoma cells transduced with interferon-gamma and cytosine deaminase genes cures lung metastases by parental cells. *Hum Gene Ther* 9:217–224.

Nuyts S, Van Mellaert L, Theys J, Landuyt W, Bosmans E, Anné J, Lambin P, (2001a): Radio-responsive *recA* promoter significantly increases mTNFα production in recombinant clostridia after 2 Gy irradiation. *Gene Ther* 8:1197–1201.

Nuyts S, Theys J, Landuyt, van Mellaert L, Lambin P, Anné J (2001b): Increasing specificity of anti-tumor therapy: cytotoxic protein delivery by non-pathogenic clostridia under regulation of radio-induced promoters. *Anticancer Res* 21:857–861.

Nuyts S, Van Mellaert L, Theys J, Landuyt W, Lambin P, Anné J (2001c): The use of radiation-induced bacterial promoters in anaerobic conditions: a means to control gene expression in *clostridium*-mediated therapy for cancer. *Radiat Res* 155:716–723.

Nuyts S, Van Mellaert L, Barbé S, Lammertyn E, Theys J, Landuyt W, Anné J, Lambin P (2001d): Insertion or deletion of the Cheo box modifies radio-inducibility of *Clostridium* promoters. *Appl Environ Microbiol*, in press.

Schlechte H, Elbe B (1988): Recombinant plasmid DNA variation of *Clostridium oncolyticum*—model experiments of cancerostatic gene transfer. *Zentralbl Bakteriol Mikrobiol Hyg [A]* 268:347–356.

Schlechte H, Schwabe K, Mehnert WH, Schulze B, Brauniger H (1982): Chemotherapy for tumours using clostridial oncolysis, antibiotics and cyclophosphamide: model trial on the UVT 15264 tumour. *Arch Geschwulstforsch* 52:414–418.

Seung LP, Mauceri HJ, Beckett MA, Hallahan DE, Hellman S, Weichselbaum RR

(1995): Genetic radiotherapy overcomes tumor resistance to cytotoxic agents. *Cancer Res* 55:5561–5565.

Theys J, Nuyts S, Landuyt W, Van Mellaert L, Dillen C, Böhringer M, Dürre P, Lambin P, Anné J (1999): Stable *Escherichia coli-Clostridium acetobutylicum* shuttle vector for secretion of murine tumor necrosis factor alpha. *Appl Environ Microbiol* 65:4295–4300.

Theys J, Landuyt W, Nuyts S, Van Mellaert L, van Oosterom A, Lambin P, Anné J (2001a): Specific targeting of cytosine deaminase to solid tumors by engineered *Clostridium acetobutylicum. Cancer Gene Ther* 8:294–297.

Theys J. Landuyt W, Nuyts S, Van Mellaert L, Bosmans E, Rijnders A, Van Den Bogaert W, van Oosterom A, Anné J, Lambin P (2001b): Improvement of *Clostridium* tumour targeting vectors evaluated in rat rhabdomyosacromas. *FEMS Immunol Med Microbiol* 30:37–41.

Thiele EH, Arison RN, Boxer GE, (1964): Oncolysis by clostridia III. Effects of clostridia and chemotherapeutic agents on rodent tumours. *Cancer Res* 24:222–233.

Trinh QT, Austin EA, Murray DM, Knick VC, Huber BE (1995): Enzyme/prodrug gene therapy: comparison of cytosine deaminase/5-fluorocytosine versus thymidine kinase/ganciclovir enzyme/prodrug systems in a human colorectal carcinoma cell line. *Cancer Res* 55:4808–4812.

Van Ostade X, Tavernier J, Fiers W (1994): Structure-activity studies of human tumour necrosis factors. *Protein Eng* 7:5–22.

Weichselbaum RR, Hallahan DE, Beckett MA, Mauceri HJ, Lee H, Sukhatme VP, Kufe DW, (1994): Gene therapy targeted by radiation preferentially radiosensitizes tumor cells. *Cancer Res* 54:4266–4269.

Wittmann W, Fabricius EM, Schneeweiss U, Schaepe C, Benedix A, Weissbrich C, Schwanbeck U (1990): Application of microbiological cancer test to cattle infected with bovine leucosis virus. *Arch Exp Veterinarmed* 44:205–212.

Yazawa K, Fujimori M, Amano J, Kano Y, Taniguchi S (2000): *Bifiodbacterium longum* as a delivery system for cancer gene therapy: selective localization and growth in hypoxic tumors. *Cancer Gene Ther* 7:269–274.

Young SD, Marshall RS, Hill RP (1988): Hypoxia induces DNA overreplication and enhances metastatic potential of murine tumor cells. *Proc Natl Acad Sci USA* 85:9533–9537.

Young M, Minton NP, Staudenbauer WL (1989): Recent advances in the genetics of the clostridia. *FEMS Microbiol Rev* 5:301–325.

Zappe H, Jones WA, Jones DT, Woods DR (1988): Structure of an endo-beta-1,4-glucanase gene from *Clostridium acetobutylicum* P262 showing homology with endoglucanase genes from *Bacillus* spp. *Appl Environ Microbiol* 54:1289–1292.

PART IV

TARGET DEFINITION

25

SELECTION OF PEPTIDES ON PHAGE

MICHAEL A. BARRY, PH.D., SATOSHI TAKAHASHI, M.D., AND M. BRANDON PARROTT, B.S.

INTRODUCTION

Chapters 1–20 have described numerous examples of the ability to engineer viral and nonviral vectors to modify their cell specificity for transductional targeting. In most cases, this technology is based on the ability of investigators to introduce novel cell-targeting ligands into the vectors to target vector binding to new cell surface receptors. While this technology is now quite robust, its application has been fundamentally limited by the relative lack of useful cell-specific ligands to add to the vectors. This dearth of targeting ligands has historically been related to the fact that most ligands used for drug or gene therapy targeting have been identified coincidentally during basic research into the biology of various cells. For many cells of interest to gene therapy, there have been few cell-specific ligands known because they have not yet been fortuitously discovered. Therefore, cell-targeting gene therapy has not only been limited by engineering functional vectors, but has also been limited by the inability to find new ligands to target the vectors to the cells of interest.

Cell-Targeting Peptide Ligands for Gene Therapy

Given this lack of cell-targeting ligands, in 1992 we began work in Stephen Johnston's laboratory at University of Texas Southwestern Medical Center in

Vector Targeting for Therapeutic Gene Delivery, Edited by David T. Curiel and Joanne T. Douglas
ISBN 0-471-43479-5

Dallas to develop new technologies to supply these missing cell-binding ligands. Our goals in developing a ligand discovery system were:

- To identify ligands without any prior knowledge of the biology or receptors of the target tissue to speed the discovery process and make it generally useful for many cell targets.
- To develop a technology that would identify ligands that bind directly to the cells of interest for direct transduction.
- To identify cell-binding ligands that would be compatible with genetic engineering into viral gene delivery vectors.

Monoclonal antibodies were an obvious choice as efficient targeting ligands, but these are quite large when compared to viral proteins and we predicted these would likely be difficult to engineer into viruses. By contrast, peptide ligands have the attraction that they are relatively small for genetic engineering into viruses, and can be easily produced by chemical synthesis for targeting nonviral vectors. The possibility of identifying peptide ligands for vector targeting was suggested at the time by the early demonstrations using peptide-presenting phage libraries to select peptides against proteins in vitro (Cwirla et al., 1990; Devlin, Panganiban, and Devlin, 1990; Scott and Smith, 1990). These peptide-presenting phage libraries had been developed by engineering filamentous M13 and fd bacteriophage to display random peptide sequences by inserting semi-random DNA into their simple 11 protein 7 kilobase pair genomes. This peptide library technology was a uniquely powerful approach for the discovery of peptides that bind proteins, because the actual ligand is physically tethered to the DNA that encodes the random peptide ligand. This fact allowed one to simply sequence the DNA of a selected peptide-presenting phage to infer the protein-binding peptide sequence, rather than perform more laborious peptide sequencing as was required in synthetic peptide libraries. In addition, these libraries are produced in bacterial cells where libraries of 10^9 to 10^{10} members can be generated readily, allowing an exceptional repertoire of potential ligands to be screened by simple techniques.

At the time we began our work, peptide-presenting phage libraries had been used to select peptides against single proteins in vitro (Cwirla et al., 1990; Devlin, Panganiban, and Devlin, 1990; Scott and Smith, 1990) and it had not been applied to select peptides against the surfaces of mammalian cells. Given this, we developed the phage peptide library technology toward its specific application of finding targeting ligands for gene therapy vectors. The original proof of principle for this approach used phage libraries representing ~10^9 different 12 or 20 random amino acids on the amino-terminus of the phage pIII protein (Fig. 25.1) to select cell-binding and cell-entry peptides from random peptide (Barry, Dower, and Johnston, 1996; Barry and Johnston, 1995).

As we pursued this work, several articles were published describing selection of peptides against purified cell surface receptors (Gui et al., 1996;

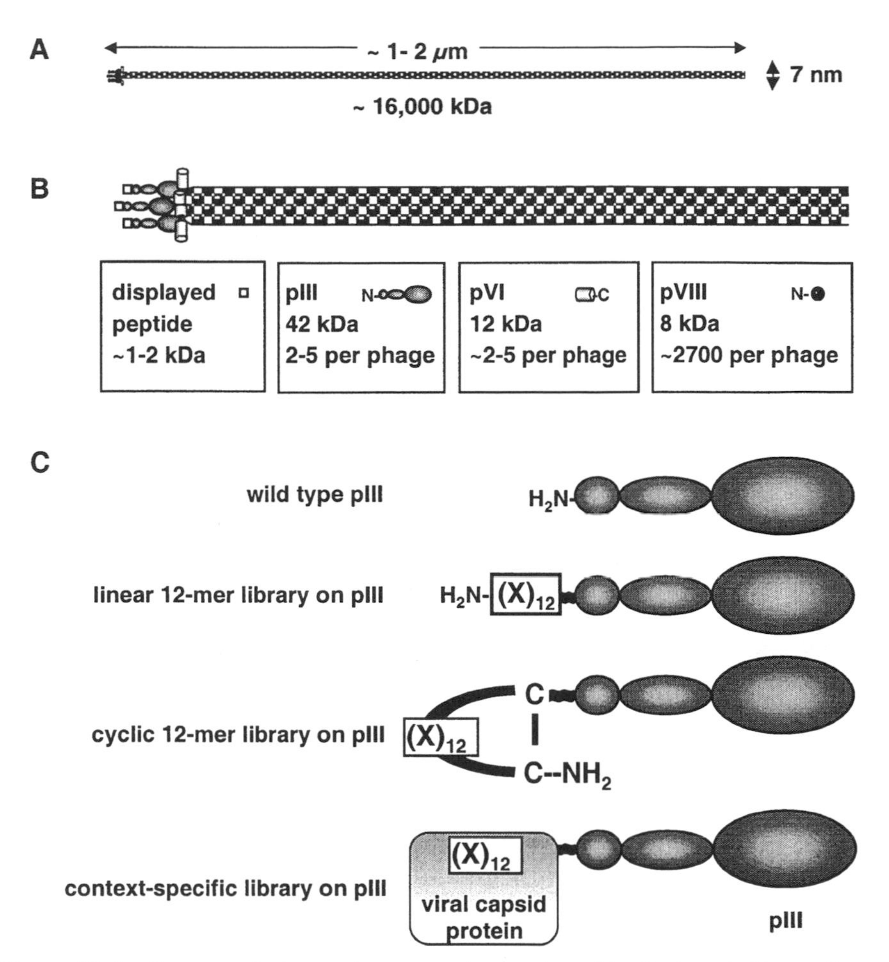

Figure 25.1 Cartoon of phage, phage proteins, and peptide display. *A*. Phage dimensions and mass. *B*. Phage proteins used for peptide display. *C*. Peptide display on pIII including linear, constrained, and context-specific libraries. Display on full-length pIII proteins is shown. Some phagemid display systems display peptides on a truncated pIII protein.

Koivunen, Gay, and Ruoslahti, 1993) and against intact platelets (Doorbar and Winter, 1994; Fong et al., 1994). Most notable of this work was the selection of the RGD peptide by Ruoslahti's group which has become the gold standard for vector targeting to date (Koivunen, Gay, and Ruoslahti, 1993). Since 1996, an increasing number of papers have been published describing selection of random peptide libraries in vivo and against mammalian cells in vitro. Examples of in vitro applications include selection against neutrophils (Mazzucchelli et al., 1999), airway epithelial cells (Romanczuk et al., 1999), vascular endothelial cells (Nicklin et al., 2000), prostate carcinoma (Romanov,

Durand, and Petrenko, 2001), and sperm (Eidne, Henery, and Aitken, 2000). In these cases, peptides were selected against the whole cells to identify any cell-binding ligands for those cells. More recent work has demonstrated the ability to select random peptides against specific receptors in their native format using cells engineered to express known receptors (White et al., 2001). Beyond these ligand discovery demonstrations, a few investigators have gone the full distance and translated selected peptides onto gene therapy vectors. These efforts include genetic translation of selected peptides into adenovirus and adeno-associated virus (AAV) (Dmitriev et al., 1998; Grifman et al., 2001), by chemical cross-linking to adenovirus (Romanczuk et al., 1999), and by incorporation of targeting peptides into vector-specific antibodies (Nicklin et al., 2000; Trepel et al., 2000).

The increasing number of papers showing the ability to select peptides against whole mammalian cells provides proof of principle for applying this technology to identify cell-targeting ligands for gene therapy applications for a number of disease systems. For a general review of phage technology, we refer the reader to excellent reviews (Burritt et al., 1996) and recent books (e.g., Barbas et al., 2001). In particular, we forward the reader to a recent chapter from the Affymax group that provides an excellent description of receptor ligand discovery with phage and other libraries as well as protocols for library construction (Cwirla et al., 2001). In the following section we describe our observations regarding the parameters of peptide library selection on whole mammalian cells in vitro. This work has primarily focused on the selection of long, linear peptides from fdTET filamentous phage libraries generated by Affymax against cells in vitro to identify peptides for translation into viral and nonviral gene delivery vectors.

CONSIDERATIONS IN CHOOSING A PEPTIDE LIBRARY

The Phage Protein Used for Peptide Display

The first random peptide-presenting phage libraries were constructed by insertion of random DNA codons in between the secretory leader and the N-terminus of the mature pIII phage protein (Cwirla et al., 1990; Devlin, Panganiban, and Devlin, 1990; Scott and Smith, 1990) and (Fig. 25.1)). Subsequent work demonstrated display of peptides on pVIII (Felici et al., 1991) and pVI of filamentous phage (Jespers et al., 1995). Two primary parameters determine which protein is used for display: (1) the valency or number of peptides per phage and (2) N-terminal or C-terminal display of peptides and proteins. The number of peptides displayed per phage is thought to influence whether peptides will be selected by affinity versus avidity and whether peptides that form multimers will be selected. pIII and pVI are present on each phage in approximately two to five copies. On wild-type phage pVIII is present in approximately 2700 copies per phage. However, this number will vary depending on the size of packaged DNA (i.e., phage versus phagemid), because pVIII coats the length of the filament of packaged DNA. Peptides displayed on both pIII and pVIII are fused to the N-termini of the mature proteins, whereas peptides displayed on pVI are displayed on its C-terminus. The display of peptides on N- or C-termini of phage proteins has implications for translation into viral capsid proteins. For example, it may be easier to translate a peptide isolated by C-terminal display on pVI onto the C-terminus of the adenovirus fiber protein than translating an N-terminal peptide identified on pIII. The issues of peptide selection, protein context, and vector context are discussed more fully in the Context section of this chapter below.

Phage Versus Phagemid Vectors

The first peptide-presenting phage libraries were constructed by insertion of random DNA into the pIII gene of an intact filamentous phage genome. From this, all copies of a pIII on the phage displayed a peptide. This display of two to five copies of peptide per phage allows protein binding by both affinity and avidity interactions. Because multivalent avidity interactions can theoretically prevent the selection of high affinity monovalent peptides (Dower, 1992), phage proteins were engineered into phagemid vectors with the goal of reducing the number of modified proteins to one per phage (Bass, Greene, and Wells, 1990). Whether monovalent display actually occurs is debatable and one is more likely to generate a population of phage without modified pIII, phage with one modified pIII, and phage with perhaps a few more modified pIII proteins. For pVIII modifications, display of peptides with five or more amino acids on all 2700 pVIII proteins from phage vectors largely interferes or ablates phage morphogenesis (Burritt et al., 1996). Therefore phagemid pVIII libraries are the vector of choice to reduce the number of modified pVIII proteins per phage to approximately 150 to 300.

On a more practical note, if you are performing library selections, phage are easier to use than phagemids, since one does not need to add helper phage at each round during selection to rescue the noninfectious phagemid clones. By contrast, if you are involved in constructing libraries, cloning into phagemid vectors is substantially easier than into phage vectors, because phagemids are manipulated as plasmids, whereas phage vectors are manipulated as the more cumbersome replicative form of DNA from the phage. By using phagemids, library sizes of 10^{11} clones or greater can be generated. The use of filamentous phage expressing antibiotic resistance markers is favored over wild-type phage libraries, because this allows phage clones to be collected as colonies rather than diffuse plaques.

Peptide Size

Libraries displaying peptides with lengths from 6 to 20 amino acids have been used for selection against mammalian cells (Barry, Dower, and Johnston, 1996;

Mazzucchelli et al., 1999; Pasqualini and Ruoslahti, 1996; Romanov, Durand, and Petrenko, 2001; Szecsi et al., 1999). In many, but not all, cases, library selection has generated peptides that bear consensus amino acids consisting of three to six amino acids in common. In most cases, buildup of this consensus is associated with the use of small-length peptide libraries encoding 6- or 7-mer peptides. By contrast, when using longer 12- or 20-mer peptide libraries, we rarely observe consensus between peptides. These selected long peptides are in most cases entirely divergent and presumably target different cellular receptors. The ability to generate peptides with consensus motifs is largely dependent on (1) the number of amino acids displayed by the library; (2) the library size; and (3) the number of amino acids actually involved in cell binding. For example, in considering a 6-mer peptide library, there are approximately 6×10^7 possible combinations of 6 amino acids that need to be represented. An average 6-mer library made up of 10^9 members should cover every combination of 6 amino acids approximately 10 times (assuming a perfect library—a large assumption). The same library will cover every combination of three amino acids approximately 100,000 times. Therefore, if the optimal binding motif is a 3-mer (e.g., RGD), it should be relatively easy to select it from a 6-mer library and 3-mer consensus motifs may be observed in other peptides. However, the other hand, if the optimal binding motif is a 10-mer, you are, of course, out of luck with this library.

Once the peptide length of a library reaches seven or eight amino acids, standard libraries of 10^9 to 10^{10} members cannot represent all possible combinations of amino acids effectively. For example, if one tests an average 10^9 member 20-mer library, this library would represent only one out of every 10^{17} of the 10^{26} possible combinations of 20 amino acid peptides. Similarly, a 10-mer library would represent approximately one in 10,000 possible 10-mers. From this, the likelihood of observing the "perfect" cell-binding 10-mer or even a 10 amino acid consensus is remote. By contrast, if the best motif were a 3-mer like RGD, the motif should be present in any short or long library more than 100,000 times and should be observed in a number of peptides. In fact, longer peptide libraries actually present more small peptides than an equivalent small peptide library. For example, each 20-mer peptide also contains eighteen 3-mer peptides within it. Thus, a 20 amino acid library should represent any given 3-mer approximately four times as frequently as a 6-mer library and in approximately 100,000 different sequence and structural contexts. Therefore, the use of libraries with longer peptides has several advantages, since these allow presentation of long cell-binding motifs while also presenting smaller motifs in more contexts.

In our original work, we used phage libraries presenting relatively large peptide sequences (12- and 20-mers), reasoning that these would not only display amino acids to bind receptors, but also might form structures that could be important for display and cell binding (Barry, Dower, and Johnston, 1996). To test the role of peptide size, we have performed several selections in which different libraries have been competed against each other. In all cases, the final peptides selected have always been from the longer peptide library, suggesting that the longer peptides are favored. By contrast, early work comparing peptide libraries for selection against antibodies in vitro (in parallel, not in competition) did not demonstrate any obvious advantage of larger peptides (Bonnycastle et al., 1996). This is not surprising, because it is estimated that the surface recognized by most antibodies can be represented by about six linear amino acids. Indeed, the recognition motifs of the peptides selected in (Bonnycastle et al., 1996) ranged from 3 to 10 amino acids with most being smaller than 6 amino acids. By contrast, most 20-mers we have selected appear to require at least nine to ten amino acids for cell binding (unpublished observations). Whether this peptide size phenomenon is specific to the selection of linear peptides on cells remains to be demonstrated.

Assuming the advantage of long peptide libraries on cells is not due to trivial effects (e.g., influence on phage biology versus cell selection), one rationale for this observation is that larger peptides out-select smaller peptides because they have more amino acids to generate higher affinity. Given that most proteins interact using many amino acid contacts, it is actually somewhat surprising that cell-binding peptides like the prototype RGD peptide can bind with high affinity using only 3 amino acid interactions. Since high affinity antibody interactions are thought to involve approximately 6 amino acids, it is likely that future cell-binding ligands from peptides will involve longer peptide motifs than 3 amino acids. An alternate explanation for the peptide length advantage is that the larger peptides can form secondary structures to stabilize interactions with receptors (see next section). A third explanation is that longer peptides internally present more small peptides than an equivalent small peptide library. A fourth explanation is that larger peptides, particularly with structure, are more stable and resistant to general degradation and cellular protease activity. This stability issue is significant, because most peptide libraries lose binding ability even after an overnight incubation at 4°C. When exposed to serum and cellular proteases, peptides with potentially useful targeting sequences may be rapidly degraded on cells or in vivo such that they are never selected during panning.

In summary, we believe there a number of beneficial features related to the use of longer rather than shorter peptide libraries including presentation of long motifs, better presentation of short motifs, and the potential to form structures. Arguments for the use of small libraries include higher likelihood of observing small amino acid consensus domains and a lower likelihood of disrupting phage protein function (particularly when displayed on pVIII).

Linear and Constrained Peptide Libraries

The earliest peptide libraries were constructed by insertion of linear peptides onto the mature N-termini of the pIII or pVIII proteins (Cwirla et al., 1990; Devlin, Panganiban, and Devlin, 1990; Scott and Smith, 1990; and Fig. 25.1). An alternate approach has been to generate constrained peptides in which the random peptide is flanked by two or more cysteine residues (McLafferty et al.,

1993). In this case, the intent is to form disulfide bonds between the flanking cysteine residues to form a semirigid loop to increase the stability of peptide interaction with target proteins. It has largely been assumed that constrained libraries will generate peptides with higher affinity than linear libraries due to free energy calculations of flexible peptides. This has been further supported by the observation that disulfide-constrained RGD peptides are preferentially selected even out of linear libraries and that these constrained peptides have 10-fold higher activity than their linear counterparts (Koivunen, Gay, and Ruoslahti, 1993).

The need for structure in the peptides is also suggested for some (but not all) peptides derived from linear peptide libraries. For example, one fibroblast-selected peptide from our original work was predicted to be an amphipathic α helix (Barry, Dower, and Johnston, 1996). Similarly, four promising cancer-targeting linear peptides we are developing are all predicted to form β-sheet-β-turn structures (unpublished observations). Therefore selection from linear peptide libraries may not only involve the selection of specific amino acids, but may also impose a second selection requirement to fold into a stable structure on which to display these amino acids.

If one were able to generate two libraries with the exact same repertoire of peptides in which one displayed linear peptides and the other displayed disulfide-constrained peptides, one might predict that each library would select a largely nonoverlapping set of peptides. This type of experiment has been approximated in work comparing constrained and linear peptides from eleven separate libraries for selection against proteins in vitro (Bonnycastle et al., 1996). In this case, no better peptide selection was observed from constrained or nonconstrained peptide libraries. Rather, antibodies and proteins with different specificities favored different types of peptide structures, either linear or constrained. Similarly, comparison of linear and constrained peptide libraries for selection on neutrophils demonstrated no great bias for disulfide-constrained peptides (Mazzucchelli et al., 1999). For at least one neutrophil-binding peptide consensus, linear peptides are favored, because this consensus was selected from both linear and disulfide libraries, but only observed from the constrained libraries when compensatory mutations ablated one of the cyclizing cysteines. Therefore, unlike RGD that is preferentially selected in constrained format (Koivunen, Gay, and Ruoslahti, 1993), this neutrophil peptide is preferentially selected in linear format (Mazzucchelli et al., 1999).

In practice, forcing one structural conformation on the peptides may either enable or disable ligand binding and selection of a particular peptide motif, depending on the specific peptide–receptor interaction. Therefore, as shown by Bonnycastle et al. (1996) and (Mazzucchelli et al. (1999), screening of several peptide libraries, from both constrained and linear libraries, may be the optimal protocol to ensure the identification of useful peptides. In our work, we have chosen to use primarily linear peptide libraries, because our goal was to engineer these into viruses like adenovirus or AAV. Since these viruses are assembled in the reducing environment of the nucleus, we want to avoid selecting peptides that were dependent on disulfide formation for activity. Regardless of whether a linear or disulfide constrained library is used, both suffer from the problem that peptides are selected in the context of a phage protein and are not selected in the context of the ultimate protein or vector that they will be used for in gene therapy targeting. Issues of context and peptide translation are discussed later in the section about context.

PEPTIDE SELECTION ON CELLS IN CULTURE

The basic peptide selection protocol we currently use is essentially the same as that described in (Barry, Dower, and Johnston, 1996) for selection of phage libraries with antibiotic markers against mammalian cells in culture.

Direct Selection on Target Cells

The basic protocol with specific comments consists of the following eight steps:

1. *Clear cell surface receptors of ligands.* This clearing is simply performed by incubation of target or competitor cells in serum-free media for 2 h prior to phage addition. Other investigators have selected peptides against serum proteins rather than against cells when selections have been performed without clearing. Removing all serum proteins from the cells is likely critical to avoid this problem.
2. *Add phage library to target cells.* Typically 10 to 100 library equivalents of phage are added in the first round of selection to a 60 mm dish of cells at approximately 80% confluence. Sick cells make for poor peptides from a selection, so ensure target cells are in optimal health on selection day. Incubation of the library is performed in serum-free Hanks Balanced Salt Solution to provide calcium and magnesium and avoid bicarbonate buffers. This solution may include protective agents such as chloroquine and protease inhibitors. In later rounds, one wants to reduce the amount of phage used at each round to prevent generating a "lawn" of phage clones on the plates. This is largely empirical, but typically at round 2, 50% of the round 1 phage are used, at round 3, 25% of round 2 phage are used, at round 4, 10% of round 3 phage are used, and so forth. This will vary from cell to cell and with different libraries. Binding is generally performed at 37°C. Binding at 4°C or at room temperature in our hands generally produces peptides with lower binding activity, particularly at physiologic temperatures.
3. *Wash nonspecific phage from the cells.* The step should be treated as a biochemical purification requiring that the background levels of phage in the final washes ultimately decline below the number of phage recovered

from the cells. Six washes for 5 min each is usually sufficient, but should be determined directly.

4. *Acid elute phage from cells by incubation at pH 2.2.* This is a typical recovery method for phage on proteins in vitro. It presumably works by protonating peptide side chains and receptors to reverse charge interactions to release the phage from their target surface. On cells, acid elution typically results in recovery of less than half of phage actually associated with the cells. In our hands, acid-eluted phage generally produce selected peptides with somewhat lower binding activity on the cells than cell-associated peptides.
5. *Recover cell-associated phage.* Phage are recovered by hypotonic lysis of the target cells followed by competitive binding and infection into F′-positive bacteria. More phage are generally recovered in this fraction than the acid fraction. Phage recovered in this fraction are either strongly attached to the cells by nonacid-sensitive interactions or by hydrophobic interactions. Alternately, these phage may include those that are internalized into cells. Recovery of phage from the mammalian cells rather than in an acid elution may also avoid selection of peptides against tissue culture plastic or other noncellular components in the culture. In many, but not all cases, the acid and cell-associated fractions select different peptides. In other cases, the same peptide will be selected from both fractions indicating that the two fractions are not fundamentally distinct, but likely provide a bias to the character of peptides. In our hands phage selected in this fraction tend to have higher binding activity as lead peptides on cells than acid-eluted phage.
6. *Amplify phage in bacteria.* Overnight growth on plates helps to reduce artifacts of selection due to more rapid growth of some phage versus others in liquid solution. Many phagemid libraries use liquid growth, because helper phage must be added.
7. *Purify phage population enriched for cell-binding peptides.*
8. *Recycle phage* for the next round of selection (go to step 1 if not clearing on nontarget cells) or add a cell clearing in all rounds after the initial round (see following instructions).

One round can be performed each day provided target cells are available and that phage from the previous round are purified (step 7) concurrently with receptor clearing (step 1) for the next round. We typically observe selection on mammalian cells within five rounds of panning as demonstrated by sequencing of individual phage clones as in Barry et al. (1996). Overall binding by the population can be monitored at each round as an alternate assessment. For libraries displaying short peptides (e.g., 6 amino acids), observation of consensus peptides between multiple phage clones may also indicate that effective cell-binding peptides have been selected. For libraries displaying longer peptides (e.g., 20-mers), observation of consensus is unusual. The most obvious end point is when the population collapses down to one or a few individual peptides. It is important to sequence bacterial clones produced at low multiplicities of infection, because bacteria can be superinfected with multiple phage clones to generate poor sequence data in the peptide-encoding DNA. If you observe good sequence in the constant regions of pIII or pVIII and the sequence signal falls in the variable region, it indicates that you are sequencing more than one phage clone.

Selection with Preclearing on Nontarget Cells

A number of cell-specific peptides have been selected by straight selection of peptide libraries on target cells (Stephen Johnston, personal commun.; Mazzucchelli et al., 1999). While this may be the case, simple selection on target cells also has generated a number of good cell-binding peptides that are quite promiscuous and bind many other cell types (Barry, Dower, and Johnston, 1996; Nicklin et al., 2000). These promiscuous peptides do not bind all cell types, but select sets (Barry, Dower, and Johnston, 1996 and unpublished observations). This, and their ability to bind cells in the presence of heparin or chondroitin or on cells treated with heparinase or chondroitinase (unpublished observations), suggests that these peptides were not selected against ubiquitous or common cellular targets. This is perhaps not surprising, because peptide selection theoretically selects out those peptides with highest affinity or avidity combined with those that target more abundant receptors. Therefore, simple selection on a target cell is perhaps more likely to select a non-cell-specific peptide that binds a common cellular protein than one that fortuitously binds a cell-specific receptor.

Selection of cell-specific peptides on cells also appears to be highly dependent on the target cells, perhaps related to viability issues or phenotypic variation. For example, when we have selected libraries against primary chronic lymphocytic leukemia (CLL) cells, we obtained six peptides from one patient that were highly specific for CLL cells and that do not bind normal B cells or other blood cells (Takahashi, Brenner, and Barry, 2001). By contrast, selection of the same library on a different CLL patient's cancer cells generated six peptides that bound not only CLL cells, but also normal B and T cells and macrophages. Similarly, we have performed direct parallel selections of the same library at the same time on two different human breast carcinoma cell lines at the same density under the same culture conditions (Barry et al., 2001). In this case, panning on one of the cell lines led to the selection of one breast carcinoma-specific peptide in five rounds that could bind both breast cell lines. By contrast, parallel selection on the second cell line failed to select a finite set of peptides in five or eight rounds of selection. Therefore, in both cases, selection with the same libraries against two externally "equivalent" cells generated very different repertoires of peptides. Since no cancer cell or cell line is truly the same, this variation in selection success on different target cells is likely due to differences in the repertoire of receptors displayed by each cell type. To combat this unpredictable variation, it is important to select libraries against several representative cell targets to obtain useful peptides. Entire selections should be performed against a single cell sample rather than different samples at each round, because variation of different patient samples at each round appears to focus selection against very common cellular receptors rather than against cell-specific receptors (unpublished observations).

Given that the peptides you select may or may not be cell-specific by direct selection on target cells, there are three approaches to try to obtain cell-specific peptides:

1. Select many peptides on target cells and screen for cell-specific peptides.
2. Do in vivo selection using the mass of the body to clear promiscuous peptides. This type of selection has an inherent bias toward isolating peptides that bind cells in direct contact with the injection site/tissue (e.g., endothelium by venous routes, epithelial cells by lung, etc.). This type of selection is also easier to do for animal cells than for human cells, because obtaining approval for direct selections in humans is difficult. If selecting against a particular cell type with the goal of identifying a peptide to mediate direct transduction of that cell, it is probably more straightforward to do in vitro selection.
3. Clear promiscuous peptides from the phage library by adsorbing these out on nontarget cells prior or during to selection on target cells.

Approaches 1 and 3 have been directly compared in Nicklin et al., (2000) for selection of peptides against endothelial cells in vitro. In this case, direct selection of peptides on human umbilical vein endothelial cells (HUVECs) generated 58 peptides. Selection on HUVECs after preclearing on smooth muscle cells, HepG2 hepatoma cells, and peripheral blood cells generated only three peptides: two that were also selected without clearing (each only 1 of 58 peptides) and a third that was not observed without clearing (Nicklin et al., 2000). When tested for cellbinding, two of the peptides selected without clearing actually appeared more specific than those identified by clearing. Much of the cross-reactivity was specific to binding to receptor-rich HepG2 cells, which suggests that the clearing process for these cells was less effective than for the other cell types. We similarly find hepatocyte cells have particularly high backgrounds for binding phage regardless of any peptides presented.

In our original work selecting against mouse fibroblasts, we attempted to clear the phage libraries in the first round of selection by first binding the library on mouse Hepa1-6 hepatoma cells and then taking the unbound phage and panning these on the target fibroblasts. When the selected peptides were ultimately tested, they bound the target fibroblasts no better than control phage, suggesting that we had lost most of our repertoire of fibroblast-binding peptides in the clearing step (unpublished observation). Since there are only 10 to 100 copies of a given peptide applied to the cells in the first round of selection, the danger of losing individuals to nonspecific adsorption may be high (e.g., phage with irrelevant peptides can bind some cells anywhere from 0.1% to 30% as well as phage with cell-binding peptides [Barry, Dower, and Johnston, 1996]).

Given this apparent loss of binding candidates, we have modified our clearing protocol to apply it in every round after one round of direct selection on target cells. The rationale for this is to use the first round of selection against the target cells to amplify true cell-binding peptides out of the background of irrelevant peptides. Since only approximately 1 in 100,000 phage bind in round one, this represents a substantial enrichment of cell-binding peptides. This clearing strategy appears to be generally useful and has generated human breast carcinoma-specific peptides by clearing on hepatoma cells and the reciprocal hepatoma-specific cells by clearing on breast, prostate, CLL, and muscle cells (Barry et al., 2001).

Phage Endocytosis and Protective Agents

We originally devised our selection strategy with the hope of biasing the selection for peptides that not only bind cells, but also target receptors that undergo endocytosis (Barry, Dower, and Johnston, 1996). We felt these types of ligands would be critical to facilitate vector entry into target cells. To bias selection towards this, we generally propagate only the cell-associated fraction and discard the acid-eluted fraction containing cell surface phage. To better protect any endocytosed phage, all selections were performed in the presence of 100 μM chloroquine. Since phage are relatively stable to low pH and proteases, it is uncertain if chloroquine is really needed for protection of the phage. In retrospect, given the size of phage relative to mammalian cells, it is likely that only a subset of the phage actually enter the cells. Indeed, when phage are immunolocalized on cells at high magnification, the phage appear clustered on the cell surface after 24 h, perhaps jammed in endocytic vesicles (Fig. 25.2). However, when peptide-presenting phage are applied for gene delivery, these appear to be assisted by the presence of chloroquine (Fig. 25.3; Barry and Johnston, 1995). Therefore, it is unclear if chloroquine is really required for selection. However, chloroquine does not appear to have negative effects, so we continue to include the reagent in the selections.

While phage are very stable, their peptides are less so. This observation is suggested by the fact that most phage libraries lose substantial binding activity even after incubation overnight at 4°C. To counter peptide instability, particularly for in vivo selections, we performed a number of selections in the presence of a cocktail of protease inhibitors (Complete^{TM} protease inhibitors, Roche). While one might not normally be interested in unstable peptides, these lead peptides may still be of use when optimized for affinity or when translated into an alternate context. Furthermore, we have observed selection of peptides, particularly in the lung that appear to have specific protease sites (e.g., for furin), suggesting that some peptides could be aberrantly selected for proteolysis (which may increase the ability of the phage to infect bacteria) rather than selected for cell binding. When we performed selections in the

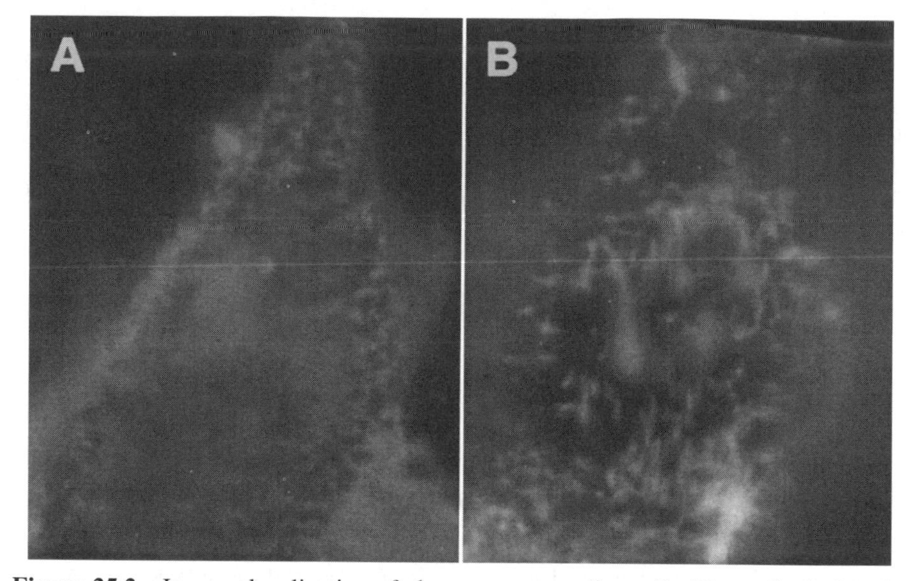

Figure 25.2 Immunolocalization of phage on mammalian cells. Phage displaying the 20.2 peptide from (Barry, Dower, and Johnston, 1996) were bound to fibroblasts for (*A*) 2 h or (*B*) 24 hours and were immunolocalized with anti-M13 antibody and fluorescent secondary antibody. Cells are observed at approximately 6000×. Note the filamentous or hairy appearance of the phage on cells.

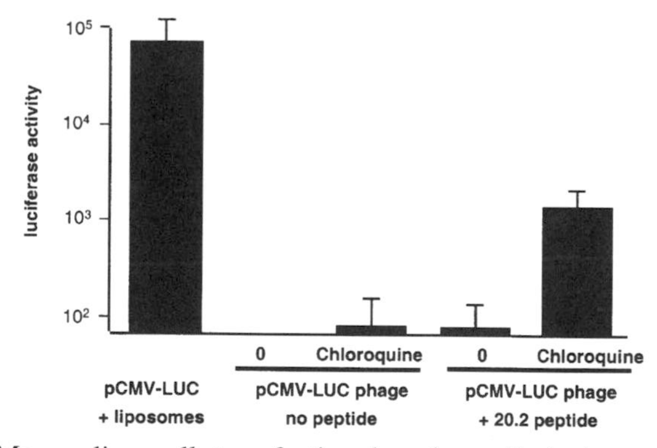

Figure 25.3 Mammalian cell transfection by phage-displaying selected peptides. Phagemid displaying the 20.2 peptide on pIII were constructed to also carry a CMV-luciferase expression cassette. These phage and controls without cell-binding peptides were packaged with helper phage and added to mouse fibroblasts for 24 h in the absence or presence of 100 μM chloroquine. Cells were assayed for luciferase activity and compared to transfection mediated by the same plasmid delivered with Lipofectamine. (Data presented at the 1995 Keystone Gene Therapy Meeting [Barry and Johnston, 1995] and described in (Barry, Dower, and Johnston, 1996).

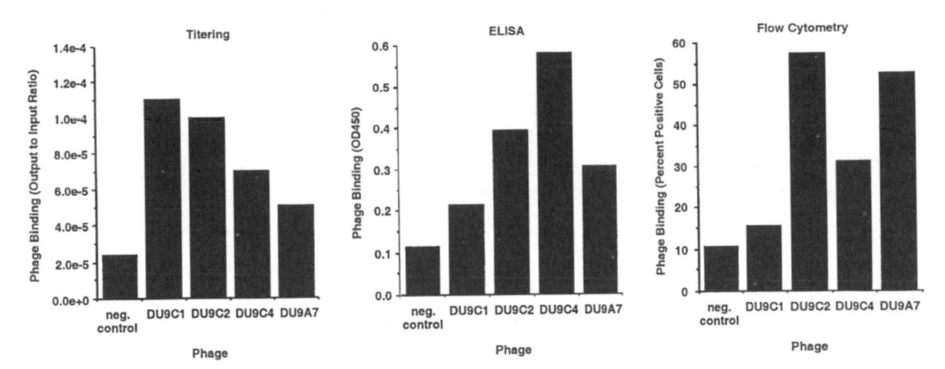

Figure 25.4 Peptide-presenting phage binding comparison by different assays. Phage DU9C1, DU9C2, DU9C4, and DU9A7 were selected on DU145 human prostate carcinoma cells. Negative control is phage selected against muscle. Phage were bound to target cells and assayed as indicated and described in the text.

presence of protease inhibitors, these selections produced legitimate cell-binding peptides (Fig. 25.4 for DU145 prostate carcinoma), but took substantially more rounds to complete (e.g., nine rounds). When the same selections were performed in the absence of protease inhibitors, selection occurred within four rounds and generated a completely different set of cell-binding peptides. In retrospect, the use of at least certain protease inhibitors is most likely a mistake, since some, such as PMSF, are alkylating agents that covalently modify specific amino acids. Therefore, in the presence of these alkylating agents, the amino acids of the peptides on the phage were also likely to be alkylated, modifying the binding activity of the phage. Other nonalkylating protease inhibitors may be of use to stabilize peptides, but could compete with phage for some cell surface proteases (e.g., aminopeptidase [Pasqualini et al., 2000]).

While some alkylating agents may be problematic for selection, some amino acid-specific alkylators may have utility for roughly assessing which residues are involved in peptide binding prior to large-scale site-directed mutagenesis or mutant library production. For example, the involvement of serines in peptide binding may be indicated by knockout of phage binding by pretreatment with PMSF. Or if disulfide formation is required, pretreatment with β-mercaptoethanol may ablate binding. Other amino acid-specific reagents exist (e.g., maleimides for cysteines, NHS succinamides for lysines and N-termini, iodoacetate for histidine and methionine, N-bromosuccinamide for tryptophan, etc.) but some alkylators will inactivate the phage as well (e.g., N-bromosuccinamide ablates phage propagation).

PHAGE-BINDING ASSAYS

Once peptides are selected, the next step is to screen these for their ability to bind their target cell and determine to what degree they bind other cell types. Given the expense of synthetic peptides, it is most economical to perform this preliminary screen of the selected peptides on phage for cell binding. The caveat with this is that some peptides may function well in the context of phage, but less so as synthetic peptides or in vectors after translation. While this may be a problem later in development, it is best to identify the promising cell-targeting candidates first on phage and then test later as free peptides or in vectors. In most cases, phage are selected on living cells at 37°C. Therefore, we attempt to perform all binding tests under similar conditions. Some protocols diverge quite a bit from the selection conditions (e.g., using fixed cells, using cells in suspension when selection was performed on monolayer cells, using substantially larger amounts of phage per cell). Beyond these differences at the front end of the assay, most assays differ in how the phage are detected and how sensitive the assay is. The following paragraphs describe and compare our experiences with a few typical phage binding assays.

Negative Controls

Normal phage are not good negative controls. Phage bearing any peptide tend to bind better than wild-type phage to cells. Phage presenting peptides selected against proteins or other cell types generally have higher background than wild-type phage or phage displaying random, unselected peptides.

Phage Input

Many phage assays use volumes of phage from an overnight culture for testing. The number of phage produced in such a culture is highly variable. To obtain accurate results, phage numbers need to be equalized at the input stage and verified in each binding assay.

Phage-Titering Assay

This assay is performed essentially the same as for selection except cell-associated phage are quantitated by infecting serial dilutions of phage into bacteria and counting the number of colonies as described by Barry, Dower, and Johnston, (1996). In this assay, phage inputs are first titered and an equal amount of each phage (10^7 to 10^9 phage) is added to each cell target and bound, washed, and recovered as for selection. Colonies from a plate with "countable" colonies from the serially-diluted bacterial plates are counted by hand or by an imager and the total number of phage bound to the cells is calculated from the titration (Fig. 25.4A). This method is the most laborious, but has the advantage of recapitulating the binding conditions used for selection. Under these conditions the multiplicity of infection (MOI) of phage on cells is relatively low, being 10 to 1000. This method is effective at comparing small differences in phage binding or weak binding peptides, particularly if phage binding is compared kinetically. Kinetic assays also correct for variations in the input phage between samples, since cell-binders increase with time, whereas background binding by phage remains relatively flat. While one might think weak-binding peptide may not be of interest, if this peptide is very cell-specific, it may be of considerable interest once its affinity is optimized.

Phage ELISA Assay

This assay is substantially easier to perform than titering, because one can use standard enzyme-linked immunosorbent assay (ELISA) techniques applied to whole cells. In this case, cells are plated in 96-well plates, a large amount of phage is added to each well, the wells are washed, and phage detected on a plate reader after binding of labeled phage-specific antibodies. One caveat to this approach is that one has to use large amounts of phage (e.g., 10^9 to 10^{11} phage), because the minimal level of detection by most phage antibodies in an ELISA is approximately 10^4 to 10^5 phage. Therefore, this type of assay usually runs at an MOI of 10,000 to 10,000,000 which is likely to be a less stringent test of peptide binding than titering assays. A second caveat is that if phage are internalized into the cells, binding may be underestimated (although this is not likely a large percent of the phage). A third caveat is that many cell types do not adhere to 96-well plates strongly enough to survive the multiple wash steps and the cells must be fixed onto the plates prior to phage binding. Therefore, the cells are no longer in the same format as used for selection and the conclusions are somewhat different for the same peptides (Fig. 25.4).

Phage Flow Cytometric Assay

This assay is substantially easier to perform than titering or ELISA, because the flow cytometer does the counting for you and quantitates at the single cell level. This assay also has the advantage that other cell-specific antibodies can be used in combination with phage antibodies to determine the cell-specificity of phage binding even in a complex mixture of cells such as peripheral blood. For this assay, a large amount of phage (10^{10} to 10^{11}) is added to each cell sample, and the cells are washed and stained with phage and cell-specific antibodies prior to flow cytometry. As with ELISA, large numbers of phage are used with MOIs of 100,000 to 10,000,000 driving avidity binding. A second caveat is that cells must be in suspension for assay on the flow cytometer. If phage were selected on monolayer or polarized cells and binding is performed on cells in suspension, the results may be somewhat misleading, because receptors may be internalized or relocalize when attached cells are disturbed. Therefore, it is best to perform binding and staining in monolayer and detach the cells for assay on the cytometer. A final issue is that the phage themselves are large enough to be detected on the flow cytometer and can bind antibodies nonspecifically resulting in false positive “cells.” Therefore, one needs to carefully select cells for analysis by scatter plot or other criteria.

PHAGE AS LIGAND DISPLAY SCAFFOLDS FOR CELL-BINDING ASSAYS AND GENE DELIVERY

Whereas phage are an excellent system to select random peptides on cells, they are less optimal in cell-binding assays when compared to other ligands like antibodies. When compared in vitro, control phage can have quite high backgrounds and when tested in vivo their large size makes them subject to pharmacologic filtration into tissues like the liver (Pasqualini and Ruoslahti, 1996). Typical binding backgrounds on cells by phage presenting irrelevant peptides can be as high as 10% (Fig. 25.4). Phage presenting any cell-selected peptide tend to have higher background binding to other cell types than wild-type phage. For the best peptides, only about 10% of input phage actually stay bound to the target (even for peptides selected against single proteins like antibodies). For many other legitimate cell-binding peptides, this output-to-input percentage may be as little as 0.1 to 0.01% (Fig. 25.4A and (Barry, Dower, and Johnston, 1996)).

This high background and poor performance stems in part from the fact that the actual cell-binding ligand represents less than 1/1,000 of the mass of the phage (i.e., a peptide is approximately 2 kDa, the phage displaying it is approximately 16,000 kDa) (Fig. 25.1A). The phage are also quite large relative to the cells they bind (i.e., phage is 1–2 μm long, an average fibroblast is 20 μm in diameter, an average lymphocyte is 8 μm in diameter). The phage are large enough that their filament structure can be directly observed at high magnification on mammalian cells giving the cells a hairy appearance (Fig. 25.2). This large size make them less optimal for binding assays than antibodies and is also likely to make their application as gene delivery vectors difficult (Fig. 25.3).

Phage proteins themselves also bear their own potential cell-binding ligands that may increase cell-binding background. For example, the pIII protein contains IDG, DGD, LDV-like, and NGR-like motifs, that could bind a number of integrin and nonintegrin receptors (Pasqualini et al., 2000; Yokosaki et al., 1998). It is an interesting possibility that weak binding by these cryptic motifs may actually enable low-affinity peptides to be selected by stabilizing phage interactions on cells. This might occur in a manner analogous to the multiple ligand interactions observed during lymphocyte "rolling" on endothelial layers. Indeed, given the necessity for multivalent interactions by lymphocytes, it may be that multivalent interaction of the presented peptides in combination with these cryptic motif may be fundamental to the success of in vivo selection.

Given the relative sizes of ligand, phage, and target cells, investigators need to recognize that phage are not antibodies and that they display less than optimal features when employed in ligand-binding assays. Given this, it is frequently useful to first identify lead peptides by phage binding assays and then to perform extensive ligand testing using synthetic peptides. Testing the ligands as synthetic peptides not only allows better ligand performance, but also importantly demonstrates the ability of the peptide to function out of the context of the phage. Peptides that do not function as synthetic peptides and that only function in the context of phage will be difficult to impossible to adapt for use on a heterologous scaffold in a nonviral or viral vector. This ability to bind cells out of context is therefore a critical demonstration needed before attempting to translate targeting peptides onto vectors. In addition, cell binding with synthetic peptides gives an estimate of ligand affinity to determine if this peptide will need to be optimized by mutant library approaches to increase the ligand's affinity to maximize its ability to target in vivo.

SYNTHETIC PEPTIDE ASSAYS

Synthetic peptides can be used to test phage selected ligands out of context in several different functional screens including: (1) phage inhibition assays; (2) cell binding by labeled ligands; and (3) for chemical conjugation to viral and nonviral vectors. One can synthesize unlabeled and labeled peptides for each application described in the following text. Because an average 20 amino acid peptide costs approximately $400 for 25–50 mg, producing peptides for a number of ligands can be cost prohibitive. A more frugal approach is to have one peptide synthesized representing the phage-encoded sequence with an additional C-terminal cysteine. This type of cysteine-labeled peptide can then be used for a variety of modifications using maleimide conjugates. Available modifications include biotin, fluorophores, or bifunctional cross-linking reagents. Cysteine-tagging the peptide is not feasible if the peptide contains cysteines elsewhere in its sequence (unless these are not involved in cell binding or cyclization). The ability to use cysteine tagging is one other advantage of avoiding disulfide-constrained peptides, because most selected peptides do not have critical cysteine residues. If cysteines are present in the peptide of interest, it is possible to label the peptide at lysine residues with NHS succinamide reagents (again provided there are not critical lysines in the peptide). Of course the other option is just to have the peptide synthesized with each modification as needed, which will generally add $200 to $300 to the cost of the peptide.

One caveat to the use of synthetic peptides is that some peptides are somewhat insoluble out of the context of the phage protein. This can be a problem for cell-binding assays, because these poorly soluble peptides may aggregate and can be taken up by nonspecific mechanisms in some cell types. In addition, aggregated peptides can actually be large enough to be observed as free particles by flow cytometry. These aggregates can also bind other antibodies nonspecifically to yield false double-positive cells. In many cases, filtration of a semisoluble peptide solution will remove most of the large aggregates, but one is still concerned by the presence of smaller aggregates during binding. Conjugation of the peptides to heterologous proteins such as albumin, to beads, or to gene delivery vectors may be an effective method to avoid solubility problems with synthetic peptides.

Synthetic Peptides to Block Phage Binding

One common test intended to demonstrate peptide-specificity is to use synthetic peptides to block binding of their cognate peptide-presenting phage. While this assay does work for a variety of phage (Pasqualini et al., 2000), there are also examples of effective cell-binding peptides that cannot be blocked by their synthetic peptide (Barry, Dower, and Johnston, 1996). The ability to block binding with peptides is likely related to the structure of the peptide, whether it acts as a monomer or multimer, and the biology of the receptor that is targeted. For example, a synthetic monomeric peptide may have difficulty in blocking a multivalent interaction of phage on a multimeric receptor. Because most lead peptides selected from phage libraries have low affinities in the micromolar range (Wrighton et al., 1996), one frequently needs to add high concentrations of peptide to observe competition. This test is simple to perform and is informative of specificity when it works. However, it does not work for every peptide. It is essential that competition of the specific peptide be compared to competition by another cell-binding peptide of the same size, because nonspecific effects can be observed. An ideal negative control peptide is one that is labeled identically and has the same pI as the test peptide.

Cell Binding by Labeled Synthetic Peptides

This assay is simple, particularly by flow cytometry. In this case, the peptide is either synthesized with a fluorophore or biotin, or it is synthesized with an additional cysteine and is conjugated by the investigator. Again, because many lead peptides isolated from libraries have lower affinity, these assays typically require the use of high concentrations of labeled peptide. As with blocking experiments, it is essential to use legitimate negative control cell-binding peptides for comparison to avoid artifactual results. A positive signal by this assay is very promising for the utility of the peptide, because this indicates the peptide can function out of context of the phage. Furthermore, direct binding by the labeled peptide suggests it can bind cells as a monomer or else can multimerize to facilitate cell binding.

Cell Binding by Synthetic Peptides on Scaffolds

Rather than directly fluorescently labeling the peptide, peptides can be conjugated to larger proteins such as albumin or fluorescent beads (e.g., Transfluor Beads, Molecular Probes). To use FluorBead or Transfluor beads, one can biotinylate the peptide of interest and use this to coat avidin-conjugated bead.

This approach may bias binding toward multivalent peptide interactions. This type of binding may mimic multivalent display on vectors and may be critical to the formation of clathrin-coated pits to enable vector entry.

Cell Binding by Synthetic Peptides Coupled to Gene Delivery Vectors

The assays previously are a few steps away from real modification of vector targeting. Recent work by a few groups has moved this screening process closer to real vector manipulation by either covalently or genetically coupling phage-selected peptides to vectors (Dmitriev et al., 1998; Grifman et al., 2001; Nicklin et al., 2000; Romanczuk et al., 1999; Trepel et al., 2000). As a robust peptide-screening method, the use of bifunctional polyethylene glycol (PEG) molecules to covalently cross-link peptides to vectors like adenovirus (Romanczuk et al., 1999) is a clever method to rapidly add ligands to vectors. Difficulties in this screening approach can be the inherent tropism of the vector itself and inactivation of the vector by covalent modification (unpublished observations). In our experience with adenovirus using bifunctional PEG or simple bifunctional cross-linkers, we find that many selected or positive control peptides (polylysine) will adsorb to the virus without cross-linking to the extent that the peptides cannot be removed even by extensive dialysis (unpublished observations). Therefore, peptides added to the virus modify transduction, but in a manner independent of a cross-linker. While there is a learning curve to modifying vectors covalently with synthetic peptides, once established, this will likely be a robust method to prescreen ligands prior to affinity optimization and prior to genetic translation of the ligand into viral vectors.

SELECTION OF PEPTIDE MONOMERS AND MULTIMERS

Although there are examples of selection of structured peptides, most selected peptides bear no obvious structural motifs. Some of these unstructured peptides may acquire structural stabilization by forming dimers or multimers during selection (Barry, Dower, and Johnston, 1996; Livnah et al., 1996). Because most cell surface receptors are themselves multimers, selection of multimeric peptides may reflect that the strongest binding by peptides may be mediated by those that can match receptor valency. Indeed, it has been speculated that, in general, peptide ligands will also have to multimerize to target these multimeric receptors effectively (Livnah et al., 1996). We would also predict that the ability to multimerize receptors by the ligand alone or by the ligand on the vector will likely be an acid test for vector targeting, given that many cell entry processes (e.g., the formation of clathrin-coated pits) require assembly of receptor complexes. If this proves true, we will likely observe ligands that can legitimately target vectors to cells, but that cannot enable vector entry. If this is the case, then there is value in identifying peptides that enable cell binding and those that enable cell binding and entry. These two classes of peptides could be applied in combination and would be analogous to the separate cell-binding and cell entry functions of adenovirus fiber and penton base proteins.

MOVING FROM PHAGE-SELECTED LEAD PEPTIDES TO HIGHER AFFINITY PEPTIDES

Due to library size and peptide length considerations, most peptide libraries do not represent all combinations of amino acids possible. Therefore many promising peptide candidates isolated from phage libraries are unlikely to be identified in their optimal sequence. The net effect of not being the perfect sequence is that the peptides isolated from a library will generally have affinities in the micromolar range which will be inadequate for in vivo targeting applications. Peptides isolated from phage libraries should therefore generally be considered lead compounds requiring further affinity optimization. An excellent more detailed chapter on lead ligand optimization can be found in Cwirla et al., 2001.

Mutant Library Selection

One effective approach to improve the activity of phage-selected peptides is to generate mutant peptide libraries based on of the lead peptide sequences (Wrighton et al., 1996). By applying these mutant libraries for selection, one should be able to optimize the affinity and specificity of candidate peptides by selecting mutant derivatives in which suboptimal amino acids in the original peptide are replaced by amino acids that can enhance the interaction of the peptide with its receptor. This process is exemplified by work to develop an erythropoietin peptide mimic where erythropoietin receptor-binding peptides were initially selected directly from phage libraries (Wrighton et al., 1996). Although, these peptides did bind the receptor, these lead ligands only had affinities in the micromolar range. To increase their affinity, a second peptide-presenting phage library was generated in which all peptides were based on the original candidate, but each peptide of the library was mutated at several residues in the peptide. By constructing this mutant library and selecting this against the receptor, secondary peptides were isolated that now had binding affinities in the nanomolar range.

When creating mutant peptide libraries, it is common practice to transfer the peptide from its original phage protein into an alternate phage protein to identify and remove context issues (Barry, Dower, and Johnston, 1996; Wrighton et al., 1996). For example, peptides selected in the context of the phage pIII protein are transferred to pVIII and peptides identified in pVIII are transferred to pIII to create the mutant library. In general, better affinity is thought to be achieved by reducing the number of peptides per phage, so many investigators favor selection on pVIII followed by later optimization in pIII phagemid vectors. Alternately, if context does not appear to be an issue, the peptide can be left in the same phage protein, but with the goal of displaying the peptide at low valency (i.e., 1–5 copies by phagemid vectors) to increase affinity binding and decrease binding by avidity.

Switching from pIII to pVIII or vice versa can be problematic if the peptides multimerize or bind better in one protein than the other. This is best demonstrated with the 20.2 peptide we originally selected against mouse fibroblasts (Barry, Dower, and Johnston, 1996). In this case, the 20.2 peptide originally selected on pIII failed to bind well when translated into the context of pVIII. In this case, peptide binding by the pVIII-displayed peptide could be rescued by adding additional exogenous synthetic 20.2 peptide, since the peptide appears to multimerize and this multimerization appears involved in effective cell binding (Barry, Dower, and Johnston, 1996). Hence, it is not always guaranteed that a peptide will work in any phage protein. Therefore, before building a peptide library in the other phage protein, it is prudent to first test the wild-type peptide for its binding ability in this new protein context prior to generating new peptide libraries. This also provides the lead peptide in the same vector backbone to directly compare binding of new peptides to the original.

Mutant libraries are constructed using an oligonucleotide encoding the original peptide sequence that is synthesized under conditions that give a 30% chance of placing a mutation in the first two bases and a G or T in the wobble position of each codon of the peptide within the oligonucleotide. This is typically done by synthesizing the peptide codons using "dirty" phosphoramidite base reagents that have been contaminated with 10% of each of the other three bases (i.e., for [Barry, Dower, and Johnston, 1996; Wrighton et al., 1996]). The wobble base of each codon is synthesized as a G or T to partially reduce favoring the display of amino acids represented by more codons. The net effect of 30% chance of mutation in the first two bases is to give an approximate 50% chance that any codon of the new peptide will be mutant with respect to the original selected peptide sequence. Once the oligonucleotide is synthesized with nonmutated flanking regions, it is cloned into a standard pIII or pVIII phagemid vector to display the peptide in low valency. The new constructs are then transformed into bacteria and these are infected with helper phage to create a secondary phage library in which the peptides are randomly derived from the first successful peptide (Cwirla et al., 2001). The resulting phage library is then selected against the target cells as in the first selection and peptide candidates are analyzed as described previously.

Screening Site-Directed Mutants

An alternate approach to random mutant libraries is to generate more rational mutants at each amino acid position of the peptide and screen for improved binding. This approach has the advantage of direct knowledge of individual mutation identities, but has the disadvantage of only making one mutation at a time. This approach can be used to first identify critical cell binding residues by observation of loss of cell-binding or to identify gain of function mutations. The labor associated with this approach is directly proportional to the size of the peptide ligand, because twice as many mutants will need to be constructed for a 12-mer as a 6-mer. While laborious, one is able to address every residue of the peptide, whereas a random peptide library will never represent every variant of large peptides. Generating these types of site-directed mutants has historically been complex, because one would need to generate one mutant at each codon for every different amino acid. This approach may now be feasible for addressing phage-displayed peptides with the development of long inverse PCR site-directed mutation methods and powerful bacterial systems to screen mutants (Kleina et al., 1990; Normanly et al., 1990; Rennell et al., 1991). In these mutant screening systems, a single (or multiple) codon is mutated to an amber (TAG) stop codon in the protein of interest. The resulting mutant is then tested for functionality in a panel of bacteria engineered to express different suppressor tRNAs that will insert a different amino acid at this stop codon. This system is available commercially (INTERCHANGE™ System, Promega) such that twelve different amino acids can be tested for each mutant construct. This type of system has been used to rapidly test more than 163 mutation sites resulting in more than 2000 amino acid substitutions in T4 lysozyme (Rennell et al., 1991). Similar approaches should be effective for rapid site-directed mutation of targeting peptides in bacteria constructed for phage infection. The practicality of this approach will largely be determined by how many amino acids are involved in binding and to what degree flanking interactions are needed.

Ligand Competition to Increase Peptide Affinity

If peptides are being selected against known receptors or if ligands are already in hand, one approach to drive high-affinity selection is to use these known ligands to compete with phage-displayed peptides during selection (Barrett et al., 1992; Wrighton et al., 1996; Yanofsky et al., 1996). By this competitive method, the peptides presented on the phage have to bind target receptors at similar or higher affinity than the known ligand. This approach may also be useful for peptides already selected on phage where these lead ligands can be used as synthetic peptides to compete with peptides derived from mutant libraries during affinity optimization. In either case, the approach is to begin competition after the first round of selection of the mutant library and to increase the concentration of the competitor ligand at each round thereafter to titrate up the stringency of peptide binding by the phage. By these protocols, affinities in the low nanomolar range have been achieved (Barrett et al., 1992; Wrighton et al., 1996; Yanofsky et al., 1996).

THE PROBLEM OF PROTEIN-LIGAND CONTEXT AND COMPATIBILITY IN TRANSLATING TARGETING LIGANDS INTO VECTORS

While peptide-presenting phage libraries have produced a wide variety of cell-binding ligands, these peptides bind cells in the structural context of a bacteriophage protein and may not bind in the same manner when translated into the different structural context of a viral protein. For example, a fibroblast-binding peptide we selected normally multimerizes and its 1000-fold improved binding to cells is ablated when it is displayed only as a monomer (Barry, Dower, and Johnston, 1996). Genetic translation of this amphipathic α helical fibroblast peptide into the HI loop of adenovirus fiber is unlikely to result in a tropism-modified virus, because the HI loop may not support an α helix structure and the peptide will not bind cells when displayed as a monomer in the spatially-isolated HI loops that project from the fiber protein. This ligand dimerization or multimerization requirement is not specific to this one peptide, but is the case for many ligands from nature or from peptide libraries.

Many ligands also form discrete structures for binding that may not be compatible with normal viral protein structures. For example, the fibroblast peptide is an amphipathic α helix (Barry, Dower, and Johnston, 1996), the erythropoietin peptide mimic selected from phage libraries contains a disulfide-constrained type I β turn (Livnah et al., 1996), and selected RGD peptides bind best when displayed as disulfide-constrained cyclic peptides (Pasqualini, Koivunen, and Ruoslahti, 1997). Given the structural constraints of peptide and protein ligands found in nature or from phage libraries, it is significant that these structures may not function correctly when translated into the heterologous structure of viral vector proteins or into nonviral vector components.

Although the context of vector structure may disrupt cell binding by targeting ligands, it is now well known that even short peptides can reciprocally disrupt viral protein function. For example, insertion of a variety of peptide or hormone ligands into retroviral envelopes disrupts the ability of the envelope to mediate cell fusion and cell entry (Fielding et al., 1998). Similarly, insertion of some peptides onto the C-terminus of the adenovirus fiber protein ablates fiber trimerization and vector production, even though other larger peptides are tolerated (Wickham et al., 1997). The context issues affecting the compatibility of ligands and vectors poses a significant problem for introducing any cell-targeting ligand into vectors, because an excellent targeting ligand may destroy vector function or the structure of the vector protein or component may destroy ligand function in an unpredictable fashion. Given these issues, investigators need to recognize that not every phage-selected peptide will function when translated into the heterologous context of their favorite gene therapy vector. This context issue is most problematic for viral vectors. Nonviral vectors will likely be compatible with most peptides that function off of the phage in synthetic form.

CIRCUMVENTING THE PROBLEM OF LIGAND CONTEXT

Given that the problems of ligand and vector context are only now being realized, methods to effectively circumvent this problem are in their infancy. To date, the approach has been simple screening where ligands are inserted into vector proteins or onto vectors and targeting functionality is determined empirically. This is largely a hit or miss process as best evidenced by work from Wickham's group to introduce peptides at the adenovirus fiber C-terminus (Wickham et al., 1997). In this case, the introduction of peptides of 17, 21, and 22 amino acids as well as two different 32 amino acid peptides were well tolerated and allowed viral encapsidation and vector production. By contrast, the introduction of a 26 and a 27 amino acid peptide into the same site ablated vector production. We have observed the same unpredictable size-independent effects of ligands on fiber trimerization where the addition of 18 and 83 amino acid peptides to the fiber C-terminus were tolerated and allowed trimerization, whereas insertion of 25 and 135 amino acid peptides into the same site ablated trimerization (unpublished observations). While these screening efforts will identify functional ligand vector combinations by hit or miss, more proactive approaches are currently under development by a number of laboratories.

Context-Specific Libraries

In this approach, the goal is to display random peptides already in the context that they will be applied in for gene therapy. The first step in developing this approach is to determine if the capsid proteins can be displayed on the phage. The second step will be to engineer random sequences into these capsids to generate context-specific peptide-presenting phage libraries. Toward this end, recent work has demonstrated the ability to display the adenovirus knob domain on pIII phagemid vectors (Pereboev, Pereboeva, and Curiel, 2001). We have similarly displayed the knob protein as well as AAV capsid proteins (Fig. 25.5). Other investigators have reported the display of retroviral envelopes on phage. In each case, display of capsids on phage can be difficult, because the size or structure of the capsid protein may be incompatible with secretion from the bacteria and may interfere with phage or bacterial function. For example, in our hands, the knob protein is exquisitely toxic to bacteria to the extent that only about 0.1% of phage displayed the capsid protein (unpublished observations). Other problems can occur particular to the function of the viral protein. For example, retroviral envelopes are difficult to display on phage in functional form, because these viral proteins require specific disulfide formation for cell binding. Getting the phage and bacteria to generate the same disulfides in envelope domains is difficult. Furthermore, viral cell-binding proteins are frequently trimeric, and it is difficult to display multiple proteins per phage. Although there are legitimate technical hurdles to be overcome, this approach holds promise by harnessing the power of phage while attempting to remove the problem of context change during ligand translation into vectors.

Using Phage as Gene Delivery Vectors

Phage displaying peptides or single chain antibodies that target endocytosing receptors also have the capacity to deliver genes into mammalian cells (Fig.

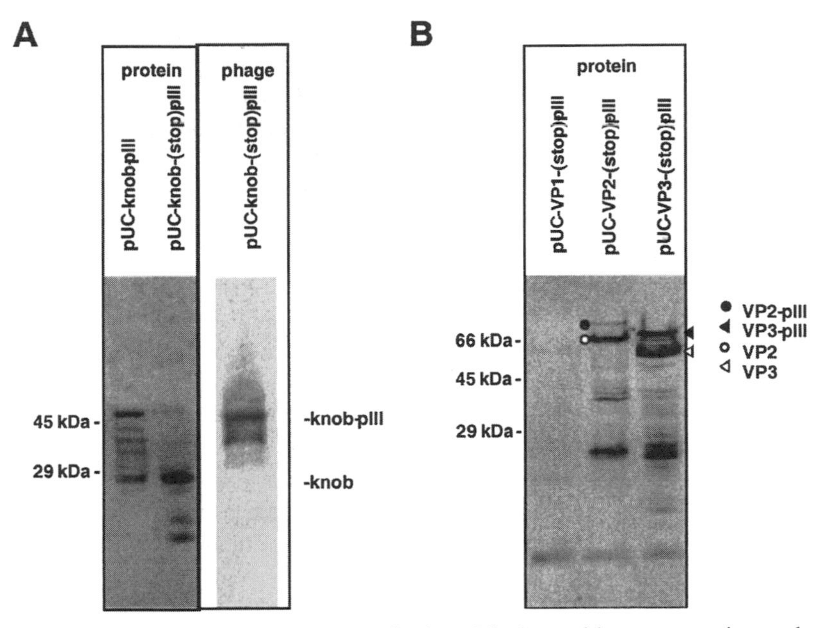

Figure 25.5 Context-specific phage display. Viral capsids were engineered onto phage pIII protein directly or by jun-fos dimerization domains (not shown). Bacteria or phage displaying the context-specific phage proteins were analyzed by Western blot with anti-pIII antibody. *A*. Display of adenovirus knob on pIII. *B*. Display of AAV VP capsid proteins on pIII.

25.3, (Kassner et al., 1999; Larocca et al., 2001; Larocca et al., 1999). With this approach, one can theoretically select peptides or antibodies from ligand libraries based on their ability to deliver a selectable plasmid into mammalian cells as described in Chapter 19. Recent developments have made the phage substantially better at gene delivery in vitro. Limitations in vivo may include phage size with respect to cells and cell entry and phage pharmacology as discussed earlier in this chapter.

Peptide-Presenting Vector Libraries

A highly pertinent question asked by incoming students is: "Why don't you just make peptide libraries in the vector?" The answer to this question has historically been simply that you cannot generate enough "clones" in vectors to make a library with a usefully large repertoire of different peptides. Phage libraries are powerful because they are built in bacteria, and *Escherichia coli* can be transformed to levels of 10^{10} transformants per microgram of DNA. By contrast, with older technology, one felt lucky to generate 10 to 1000 plaques of adenovirus with a microgram of DNA. Therefore, you might be able to "easily" generate an adenovirus library with 1000 clones. But, these 1000 library members could represent only a small number of peptide motifs. For example, 1000 members would represent about one in eight of the possible three amino acid combinations. So, phage have historically been used because of their large library sizes as well as the ease of growth in bacterial systems.

While there are definite benefits to phage libraries, the problem of context makes real vector libraries very attractive if the technology can be pushed to generate larger library numbers in mammalian systems. Recent advances toward this goal have been made in retroviral systems, where libraries of 10^6 have been generated that can cover all 4- to 5-mer combinations (Buchholz et al., 1998). While these peptide libraries were used to select protease sites, this work provides proof of principle for the application of retroviral libraries to select cell-targeting ligands. The ability to generate these moderate-sized libraries in retroviral vectors stems from the ability to easily transfect 293 cells with the retroviral vector backbone to produce virus. For other more complex systems like adenovirus or AAV, we can expect library sizes to be limited until the next breakthrough enables this technology. Since a number of investigators are at work toward this goal, we can expect significant advances in vector library development within the next five years.

CONCLUSION

Peptide-presenting phage libraries represent a potentially powerful tool for the selection of cell-binding and cell-targeting ligands for gene therapy vector targeting. Several issues are critical to the effective application of this technology for gene therapy vector targeting. The first issue is how best to select or screen to identify cell-specific peptides versus promiscuous cell-binding ligands. A second issue is the recognition that phage-selected peptides will generally be low-affinity lead ligands that will need affinity improvement prior to effective application in vivo. A third issue is the recognition of the potential for ligand-vector context problems where one cannot necessarily predict if a ligand will function. There are a number of biological hurdles to be faced, but new strategies and technologies hold great promise for circumventing these hurdles and advancing this already potent ligand technology for gene therapy vector targeting.

REFERENCES

Barbas CF, Burton DR, Scott JK, Silverman GJ (2001): *Phage Display : A Laboratory Manual*. Barbas CF, ed. Cold Spring Harbor, NY: Cold Spring Harbor Laboratory Press.

Barrett RW, Cwirla SE, Ackerman MS, Olson AM, Peters EA, Dower WJ (1992): Selective enrichment and characterization of high affinity ligands from collections of random peptides on filamentous phage. *Anal Biochem* 204:357–364.

Barry, MA, Johnston, SA (1995): Selection of cell-binding, endocytosing peptides using peptide-presenting phage. *J Cell Biochem Supplement 21A*, p. 389.

Barry MA, Dower WJ, Johnston SA (1996): Toward cell-targeting gene therapy vectors: selection of cell-binding peptides from random peptide-presenting phage libraries. *Nat Med* 2:299–305.

Barry ME, Menezes K, Takahashi S, Barry MA (2001): Selection of cancer-specific human tumor cell-targeting peptides from peptide-presenting phage libraries, in prep.

Bass S, Greene R, Wells JA (1990): Hormone phage: an enrichment method for variant proteins with altered binding properties. *Proteins* 8:309–314.

Bonnycastle LL, Mehroke JS, Rashed M, Gong X, Scott JK (1996): Probing the basis of antibody reactivity with a panel of constrained peptide libraries displayed by filamentous phage. *J Mol Biol* 258:747–762.

Buchholz CJ, Peng KW, Morling FJ, Zhang J, Cosset FL, Russell SJ (1998): In vivo selection of protease cleavage sites from retrovirus display libraries. *Nat Biotechnol* 16:951–954.

Burritt JB, Bond CW, Doss KW, Jesaitis AJ (1996): Filamentous phage display of oligopeptide libraries. *Anal Biochem* 238:1–13.

Cwirla SE, Peters EA, Barrett RW, Dower WJ (1990): Peptides on phage: A vast library of peptides for identifying ligands. *Proc Natl Acad Sci USA* 87:6378–6382.

Cwirla SE, Wagstrom CR, Gates CM, Dower WJ, Schatz PJ (2001): Identification of novel ligands for receptors using recombinant peptide. In *Phage Display: A Practical Approach*, Lowman HLCT, ed. London: Oxford University Press.

Devlin JJ, Panganiban LC, Devlin PE (1990): Random peptide libraries: a source of specific protein binding molecules. *Science* 249:404–406.

Dmitriev I, Krasnykh V, Miller CR, Wang M, Kashentseva E, Mikheeva G, Belousova N, Curiel DT (1998): An adenovirus vector with genetically modified fibers demonstrates expanded tropism via utilization of a coxsackievirus and adenovirus receptor-independent cell entry mechanism. *J Virol* 72:9706–9713.

Doorbar J, Winter G (1994): Isolation of a peptide antagonist to the thrombin receptor using phage display. *J Molec Biol* 244:361–369.

Dower W (1992): Phage power. *Curr Biol* 2:251–253.

Eidne KA, Henery CC, Aitken RJ (2000): Selection of peptides targeting the human sperm surface using random peptide phage display identify ligands homologous to ZP3. *Biol Reprod* 63:1396–1402.

Fielding AK, Maurice M, Morling FJ, Cosset FL, Russell SJ (1998): Inverse targeting of retroviral vectors: selective gene transfer in a mixed population of hematopoietic and nonhematopoietic cells. *Blood* 91:1802–1809.

Fong S, Doyle LV, Devlin JJ, Doyle MV (1994): Scanning whole cells with phage-display libraries: Identification of peptide ligands that modulate cell function. *Drug Development Res* 33:64–70.

Grifman M, Trepel M, Speece P, Gilbert LB, Arap W, Pasqualini R, Weitzman MD (2001): Incorporation of tumor-targeting peptides into recombinant adeno-associated virus capsids. *Mol Ther* 3:964–975.

Gui J, Moyana T, Malcolm B, Xiang J (1996): Identification of a decapeptide with the binding reactivity for tumor-associated TAG72 antigen from a phage displayed library. *Proteins* 24:352–358.

Jespers LS, Messens JH, De Keyser A, Eeckhout D, Van den Brande I, Gansemans YG, Lauwereys MJ, Vlasuk GP, Stanssens PE (1995): Surface expression and ligand-based selection of cDNAs fused to filamentous phage gene VI. *Biotechnology (NY)* 13:378–382.

Kassner PD, Burg MA, Baird A, Larocca D (1999): Genetic selection of phage engineered for receptor-mediated gene transfer to mammalian cells. *Biochem Biophys Res Commun* 264:921–928.

Kleina LG, Masson JM, Normanly J, Abelson J, Miller JH (1990): Construction of *Escherichia coli* amber suppressor tRNA genes. II. Synthesis of additional tRNA genes and improvement of suppressor efficiency. *J Mol Biol* 213:705–717.

Koivunen E, Gay DA, Ruoslahti E (1993): Selection of peptides binding to the alpha 5 beta 1 integrin from phage display library. *J Biol Chem* 268:20205–20210.

Larocca D, Kassner PD, Witte A, Ladner RC, Pierce GF, Baird A (1999): Gene transfer to mammalian cells using genetically targeted filamentous bacteriophage. *Faseb J* 13:727–734.

Larocca D, Jensen-Pergakes K, Burg MA, Baird A (2001): Receptor-targeted gene delivery using multivalent phagemid particles. *Mol Ther* 3:476–484.

Livnah O, Stura EA, Johnson DL, Middleton SA, Mulcahy LS, Wrighton NC, Dower WJ, Jolliffe LK, Wilson IA (1996): Functional mimicry of a protein hormone by a peptide agonist: The EPO receptor complex at 2.8Å. *Science* 273:464–471.

Mazzucchelli L, Burritt JB, Jesaitis AJ, Nusrat A, Liang TW, Gewirtz AT, Schnell FJ, Parkos CA (1999): Cell-specific peptide binding by human neutrophils. *Blood* 93:1738–1748.

McLafferty MA, Kent RB, Ladner RC, Markland W (1993): M13 bacteriophage displaying disulfide-constrained microproteins. *Gene* 128:29–36.

Nicklin SA, White SJ, Watkins SJ, Hawkins RE, Baker AH (2000): Selective targeting of gene transfer to vascular endothelial cells by use of peptides isolated by phage display. *Circulation* 102:231–237.

Normanly J, Kleina LG, Masson JM, Abelson J, Miller JH (1990): Construction of *Escherichia coli* amber suppressor tRNA genes. III. Determination of tRNA specificity. *J Mol Biol* 213:719–726.

Pasqualini R, Ruoslahti E (1996): Organ targeting in vivo using phage display peptide libraries. *Nature* 380:364–366.

Pasqualini R, Koivunen E, Ruoslahti E (1997): Alpha v integrins as receptors for tumor targeting by circulating ligands [see comments]. *Nat Biotechnol* 15:542–546.

Pasqualini R, Koivunen E, Kain R, Lahdenranta J, Sakamoto M, Stryhn A, Ashmun RA, Shapiro LH, Arap W, Ruoslahti E (2000): Aminopeptidase N is a receptor for tumor-homing peptides and a target for inhibiting angiogenesis. *Cancer Res* 60:722–727.

Pereboev A, Pereboeva L, Curiel DT (2001): Phage display of adenovirus type 5 fiber knob as a tool for specific ligand selection and validation. *J Virol* 75:7107–7113.

Rennell D, Bouvier SE, Hardy LW, Poteete AR (1991): Systematic mutation of bacteriophage T4 lysozyme. *J Mol Biol* 222:67–88.

Romanczuk H, Galer CE, Zabner J, Barsomian G, Wadsworth SC, O'Riordan CR (1999): Modification of an adenoviral vector with biologically selected peptides: a novel strategy for gene delivery to cells of choice [see comments]. *Hum Gene Ther* 10:2615–2626.

Romanov VI, Durand DB, Petrenko VA (2001): Phage display selection of peptides that affect prostate carcinoma cells attachment and invasion. *Prostate* 47:239–251.

Scott JK, Smith GP (1990): Searching for peptide ligands with an epitope library. *Science* 249:386–390.

Szecsi PB, Riise E, Roslund LB, Engberg J, Turesson I, Buhl L, Schafer-Nielsen C (1999): Identification of patient-specific peptides for detection of M-proteins and myeloma cells. *Br J Haematol* 107:357–364.

Takahashi S, Brenner MK, Barry MA (2001): Selection of chronic lymphocytic leukemia-targeting peptides from peptide-presenting phage libraries, in preparation.

Trepel M, Grifman M, Weitzman MD, Pasqualini R (2000): Molecular adaptors for vascular-targeted adenoviral gene delivery. *Hum Gene Ther* 11:1971–1981.

White SJ, Nicklin SA, Sawamura T, Baker AH (2001): Identification of peptides that target the endothelial cell-specific LOX-1 receptor. *Hypertension* 37:449–455.

Wickham TJ, Tzeng E, Shears LL, Roelvink PW, Li Y, Lee GM, Brough DE, Lizonova A, Kovesdi I (1997): Increased in vitro and in vivo gene transfer by adenovirus vectors containing chimeric fiber proteins. *J Virol* 71:8221–8229.

Wrighton NC, Farrell FX, Chang R, Kashyap AK, Barbone FP, Mulcahy LS, Johnson DL, Barrett RW, Jolliffe LK, Dower WJ (1996): Small peptides as potent mimetics of the protein hormone erythropoietin. *Science* 273:458–464.

Yanofsky SD, Baldwin DN, Butler JH, Holden FR, Jacobs JW, Balasubramanian P, Chinn JP, Cwirla SE, Peters-Bhatt E, Whitehorn EA, Tate EH, Akeson A, Bowlin TL, Dower WJ, Barrett RW (1996): High affinity type I interleukin 1 receptor antagonists discovered by screening recombinant peptide libraries. *Proc Natl Acad Sci USA* 93:7381–7386.

Yokosaki Y, Matsuura N, Higashiyama S, Murakami I, Obara M, Yamakido M, Shigeto N, Chen J, Sheppard D (1998): Identification of the ligand binding site for the integrin alpha9 beta1 in the third fibronectin type III repeat of tenascin-C. *J Biol Chem* 273:11423–11428.

26

ANTIBODY PHAGE DISPLAY LIBRARIES FOR USE IN THERAPEUTIC GENE TARGETING

P. ROHRBACH, PH.D. AND S. DÜBEL, PH.D.

INTRODUCTION

The increased interest in antibody-targeted therapeutics expresses a substantial change in the typical course of pharmaceutical development: one now chooses to use the body's own capabilities as a source for drug discovery rather than the usual chemist's reagent vessel.

Early in the 1970s, a method was developed to isolate monoclonal antibodies (Köhler and Milstein, 1975). Through hybridoma technology, one could take immunized mice and isolate antibody-producing B-lymphocytes. Fusing them with myeloma cell lines immortalized these cells. This method enabled the isolation of many murine monoclonal antibodies for a variety of different antigens. Unfortunately, the use of these antibodies became limited for therapeutics, as the human immune system was shown to initiate an immune response against the murine protein. The result of this so-called HAMA (human-anti-mouse antibody) response ranges from deactivation of the murine antibody to anaphylactic shock (Courtnay-Luck et al., 1986; Bach, Fracchia, and Chatenoud, 1993). Human antibodies for use in therapy were

Vector Targeting for Therapeutic Gene Delivery, Edited by David T. Curiel and Joanne T. Douglas
ISBN 0-471-43479-5

first isolated from the sera of immunized donors, whereas the monoclonal antibodies are produced using the hybridoma technique (Carson and Freimark, 1986). Another possibility of producing human monoclonal antibodies is the transformation of human B-lymphocytes with the Epstein-Barr virus (EBV) (Seigneurin et al., 1983). All of these methods are limited by various factors, including the instability of cell lines, low antibody production, limited access to compatible lymphocyte donors, and the difficulty of immunization in humans (Glassy and Dillman, 1988). As detailed information on both antibody structure and antibody generation in vivo became available, it was desired to transfer the basic principles of the humoral immune response to a bacterial system. The purpose of this undertaking was to generate a system able to overcome the limitations of murine antibodies generated by hybridoma technology.

It is not possible to produce functional full-length antibodies in bacteria, because glycosylation of the Fc region is fundamental for the activation of effector functions. However, various recombinant antigen-binding fragments of immunoglobulins have successfully been expressed in bacterial cells, including Fv-, single-chain Fv- (scFv), Fab-, and $F(ab)'_2$- fragments (Bird et al., 1988). The most common fragments used in research and therapy are the scFv and Fab fragments. In scFv fragments, the V_H and V_L domains are covalently linked with a short peptide linker, usually 15–20 amino acids long, which is introduced at the genetic level (Fig. 26.1).

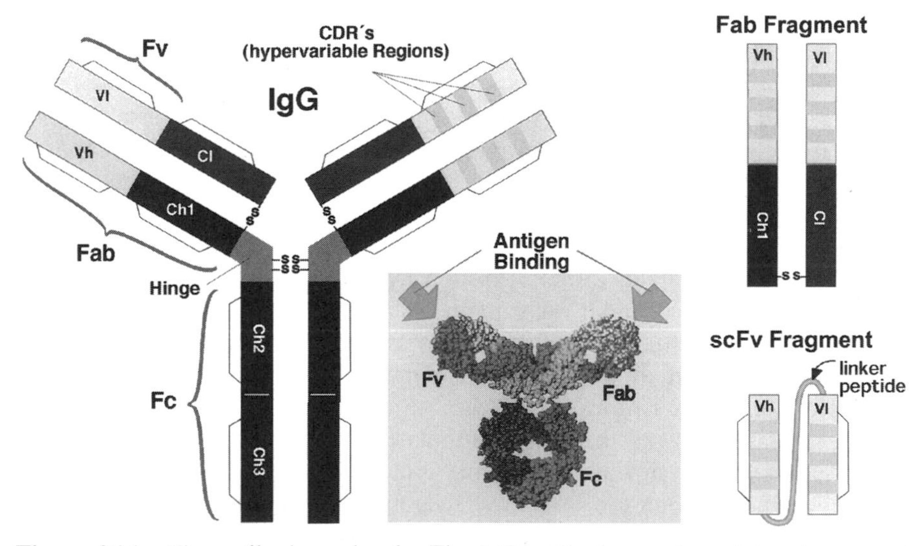

Figure 26.1 The antibody molecule. The IgG antibody consists of two heavy and two light chains. The variable domains, situated at the end of the IgG molecule, are responsible for antigen binding. The Fab fragment consists of both variable domains and the constant regions of the complete light chain and the first domain of the heavy chain. The scFv fragment contains only the variable domains, connected by a peptide linker.

The advantage of recombinant antibody fragments, especially in therapy, is their small size, facilitating tissue penetration, biodistribution, and blood clearance. In addition, the expression of recombinant antibody fragments in bacterial cells allows for the possibility of displaying the antibodies on the surface of filamentous phage. This phage display of antibodies is the most widespread method to date for in vitro selection of antibodies.

Phage display-derived antibodies are not yet listed among the Food and Drug Adminstration (FDA) approved products for the therapeutic targeted market. However, in contrast to hybridoma technology, which seems to have reached the endpoint of its technological development, research in antibody engineering is still ongoing. This can be seen from the dramatic increase of clinical studies employing antibodies for therapeutic targeting. From the vast variety of fusion proteins and targeting concepts employing recombinant antibodies that have emerged in the first decade, several have now reached clinical trials. These molecules are becoming by far the largest group of novel protein agents in the FDA approval pipeline.

Today, more than 200 laboratories worldwide use phage display as a means to isolate human antibodies. After the first decade of antibody engineering and phage display technology, the general methods required for harvesting nature's treasures had been established. The coming decade will most probably depend on specialized companies or central service groups in academic institutions for the targeted isolation of specific human antibodies. Efficient use of phage display and follow-up methods to generate human antibodies requires large libraries and further development in the area of robotics. One of the promising emerging technologies is ribosomal display, a method employing direct coupling of mRNA and the antibody fragment encoded on it (Hanes and Plückthun, 1997; Schaffitzel et al., 1999).

PHAGE DISPLAY OF ANTIBODIES

Today, the use of antibodies can be divided into two major areas. First, they can serve as molecular tools in research, where they are especially useful in the characterization and analysis of proteins and their functions. The increasing number of worldwide genome projects has lead to an ever-increasing demand for antibodies to newly identified proteins, a trend that will most definitely continue into the future. The conventional hybridoma technology, still being used to generate monoclonal antibodies, cannot meet these increased needs fast enough, as this method is not suited for high throughput. Second, antibodies are important for clinical use. Most diagnostic clinical testing used today is performed with antibody-based enzyme-linked immunosorbent assays (ELISAs). Furthermore, there is an increase in the numbers of antibodies being used in therapeutic trials. It is important to note that the vast majority of these antibodies are humanized to avoid the HAMA response in patients. The isolation of antibody candidates for therapeutic use is limited in hybridoma technology, because the generation of stable antibody-producing cell lines from humans is still a problem. Humanization of mouse antibodies is both a wearisome and time-consuming task. With the help of phage display, using recombinant antibody libraries, both central obstacles associated with hybridoma technology may be overcome. By using human lymphatic material for the generation of libraries, fully human antibodies can be isolated and are expected to perform well in therapeutic setting. Furthermore, the selection process applied in phage or ribosomal display mostly takes place in vitro and can therefore be automated. The various steps in which bacteria are used do not constitute a new bottleneck, as large amounts of these samples can be handled in parallel.

Filamentous Phage

Ever since the first demonstration of the display of foreign protein sequences on filamentous phage in 1985, this technology has been applied to the isolation of new ligands from phage-display libraries. Two advantages makes this technology so attractive: (1) Specific ligands can be selected from a vast number of different sequences by simple methods of affinity enrichment, and (2) the structural and functional information of the protein of interest displayed on the phage surface is physically linked to its genetic information contained within the phage genome. Further, rapid growth in the field of antibody engineering occurred after it was shown that functional antibody fragments could be secreted into the periplasmic space of *Escherichia coli* by fusing a bacterial signal peptide to the antibody's N-terminus (Better et al., 1988; Skerra and Plückthun, 1988).

The life cycle of filamentous phage (M13, f1, fd) is shown in Figure 26.2. These nonlytic bacteriophage have a single-stranded circular DNA genome coding for 10 different proteins involved in replication, morphogenesis, and formation of the phage coat. The phage infects F′ episome-bearing bacterial cells by binding to the tip of the F pilus. Once the pilus retracts, the phage genome penetrates the bacterial membrane. The bacterial cell replicates and translates the phage genome by producing all the proteins necessary for formation of new phage particles. pVIII is the major coat protein and is present in approximately 2700 copies. The proximal tip of the phage is formed by the minor coat proteins pIII and pVI present in 3 or 5 copies. The same number of the minor coat proteins pVII and pIX are found at the distal end. Functionally, the pIII can be divided into three domains. The tip of the N-terminal domain is responsible for penetration followed by a domain involved in recognition of the F pilus. The C-terminal domain is used for membrane anchorage and is hidden in the phage envelope. Assembly of the phage takes place in the cytoplasmic membrane, where the newly synthesized coat proteins are anchored. Approximately 150 to 300 phage may be produced during a bacterial life cycle.

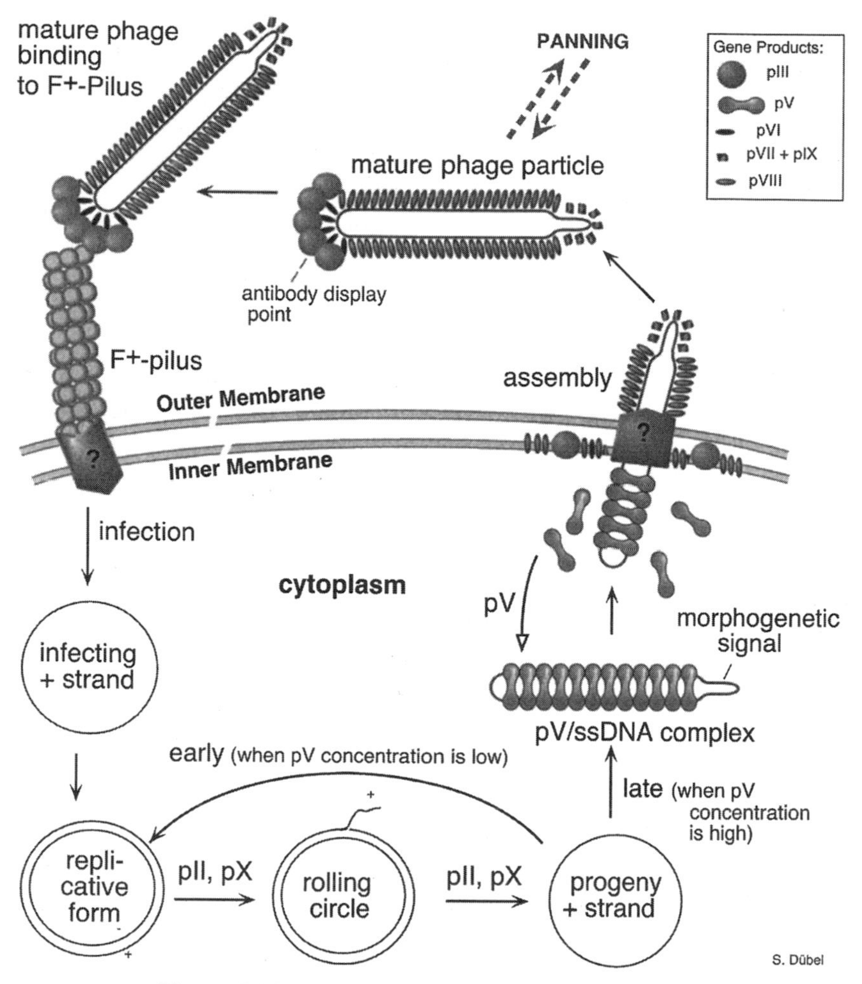

Figure 26.2 The life cycle of filamentous phages.

Phage Display

In the past, fusions to the N-terminus of the phage coat proteins pIII or pVIII have been used to display foreign sequences, showing no reduction in phage infectivity. In 1985, G. P. Smith first demonstrated that it is possible to display sequences encoding fragments of two different enzymes fused to pIII. This was followed by experiments showing that libraries displaying peptides of various lengths instead of gene fragments could be successfully applied for epitope mapping and identification of protein ligands. Finally, phages were used for the display of functional proteins such as antibody fragments (scFv, Fab′), hormones, enzymes, and enzyme inhibitors.

To display the antibody fragment on the phage surface, one needs to fuse it to one of the coat proteins. McCafferty et al. (1990) showed that the antibody gene can be directly inserted into the phage genome. Despite several reports describing the fusions of antibody fragments to the major coat protein pVIII, the vast majority of successful vectors are based on fusions to pIII.

Two different vector systems are used in phage display: the phage system and the phagemid system. In the phage system, the protein sequence of interest is cloned between the leader sequence and the coding region of pIII or pVIII gene of the phage DNA. This allows for expression of only pIII or pVIII fusion proteins, resulting in multivalent display. In the phagemid system, the vector contains only the genes for pIII or pVIII and bacterial and phage origins of replication (Fig. 26.3). In order to produce phage particles, cells containing the phagemid are rescued with helper phage, which provide all the remaining proteins required for the generation of phage particles. Because replication of the helper phage DNA is less efficient than that of the phagemid, the majority of the phagemid DNA is packaged into phage particles. Since the helper phage also provides wild-type pIII and pVIII, a mixture of wild-type and fusion proteins are displayed. These phagemid systems do not only improve infectivity and phage production but also result in mainly monovalent display, which is important for the isolation of high-affinity antibodies.

A recent breakthrough to enhance the performance of antibody phage display library was the development of hyperphage technology (Rondot et al., 2001). This approach uses a novel helper phage design to improve the presen-

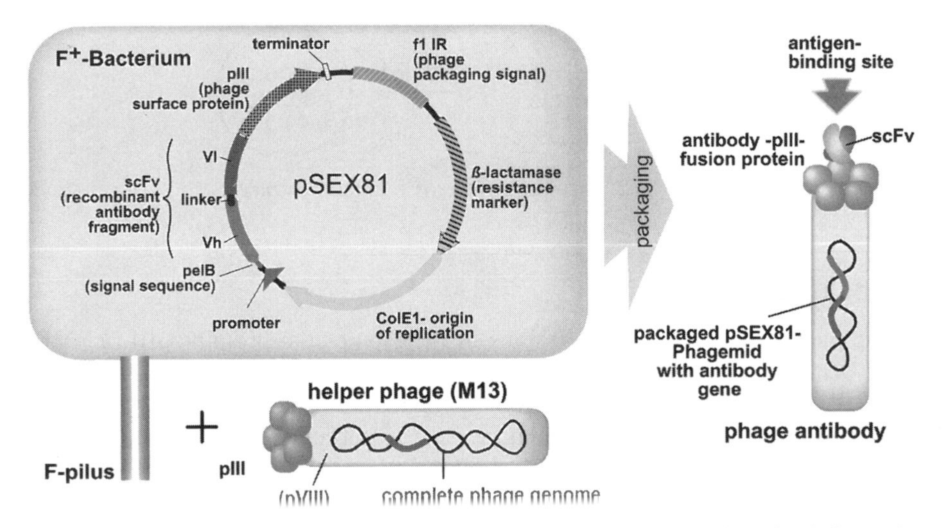

Figure 26.3 The phagemid plasmid contains a packaging signal derived from the filamentous phage. In the presence of helper phages, phagemid DNA is packaged into the phage particle. When the fusion protein scFv-pIII is induced, phage particles are produced carrying an antibody fragment at the tip.

tation of antibody fragments on the phage surface by more than two orders of magnitude. It was observed that in commonly used phagemid-based systems, the majority of the phage population carries one antibody fragment. This problem was overcome by avoiding the delivery of wild-type pIII during the phage antibody packaging. Consequently, a novel helper phage (hyperphage) was constructed having a wild-type pIII phenotype and capable of infecting F^+ *E. coli* cells with high efficiency. These phage, however, do not carry a functional pIII and render the phagemid-encoded antibody-pIII fusion as the only source of pIII in phage assembly (Fig. 26.4). By using this method, antigen-binding activity was increased approximately 400-fold by enforced oligovalent antibody display on every phage particle. The use of hyperphage for packaging a universal human scFv library improved the specific enrichment factor. After two rounds of panning, more than 50% of the isolated antibody clones bound to the antigen, compared to 3% when conventional M13KO7 helper phages were used. Thus, hyperphage are particularly useful in stoichiometrical situations, where the chances of a single phage locating the wanted antigen are known to be low. In particular, new tumor markers may be detected by allowing panning on cell surfaces with higher sensitivity. In the search for novel

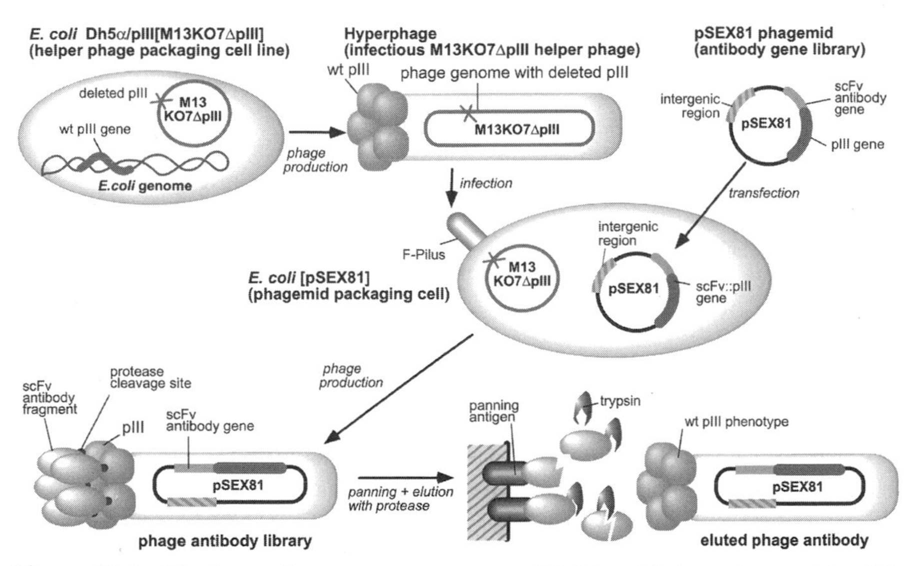

Figure 26.4 The hyperphage concept: a gene pIII-deleted helper phage with wild-type infection phenotype. Hyperphage are made by using an *E. coli* packaging cell line capable of producting pIII from a gene integrated into its genome. When hyperphage are used for pSEX81 phagemid packaging, phages carrying several copies of the scFv-antibody are produced. The trypsin cleavage site, linking the scFv-antibody fragment and the pIII, can be used for protease elution after panning. This allows for physiological buffer conditions that do not affect phage integrity and restores the wild-type pIII phenotype for reinfection.

targets for delivering genes or drugs tissue- or cell-specific, in vivo panning can be expected to benefit from hyperphage packaging.

ANTIBODY LIBRARIES

As mentioned, the generation of high-affinity monoclonal antibodies has traditionally involved the production of hybridomas from spleen cells of immunized animals. Now, the use of phage antibodies offers a new method for the generation of antibodies, including antibodies of human origin, which previously could not easily be isolated by the conventional hybridoma technology. Using phage display, antibodies can be made by bypassing the immune system and the immunization procedure and allowing in vitro modifications of the affinity and specificity of the antibody (Winter and Milstein, 1991). One advantage of phage antibody libraries is the generation of fully human monoclonal antibodies for medical and diagnostic use and, in the process, replacing hybridoma technology. With phage display, humanization strategies for antibodies become unnecessary.

The primary immune response consists of a large set of immunoglobulin M (IgM) antibodies that recognize a variety of antigens. These IgMs can be cloned as a naive repertoire of rearranged genes by harvesting the variable region (V) genes from the IgM mRNA of nonimmunized human donors, isolated from peripheral blood lymphocytes (PBLs), bone marrow, spleen, or tonsils. The V genes are amplified via polymerase chain reaction (PCR) from the cDNA using oligonucleotides binding to the framework 1 (FR1)- and constant heavy (CH1)- or constant light (CL)-encoding regions of the antibody genes. The amplified heavy and light chains are assembled randomly in a phagemid vector for expression and display (McCafferty et al., 1990; Marks et al., 1991b; Barbas et al., 1991, Dübel et al., 1992) (Fig. 26.5).

The main advantages of using large naive repertoires are:

- One library can be used to screen any given antigen.
- Human antibodies are isolated.
- Antibodies to self, nonimmunogenic, or toxic substances can be isolated, as the library is not preselected against self-binders.
- Antibody generation requires 3–5 rounds of screening (only approx. 3–4 weeks).

The disadvantages are the time and effort required to construct a reliable antibody library and the largely unknown contents of the naive libraries.

In a naive antibody library, the repertoire quality and content is influenced by unequal expression of the different V-gene families and the unknown history of the donor. Therefore an alternative method is sometimes used that contains the use of a semisynthetic or synthetic antibody library, where the

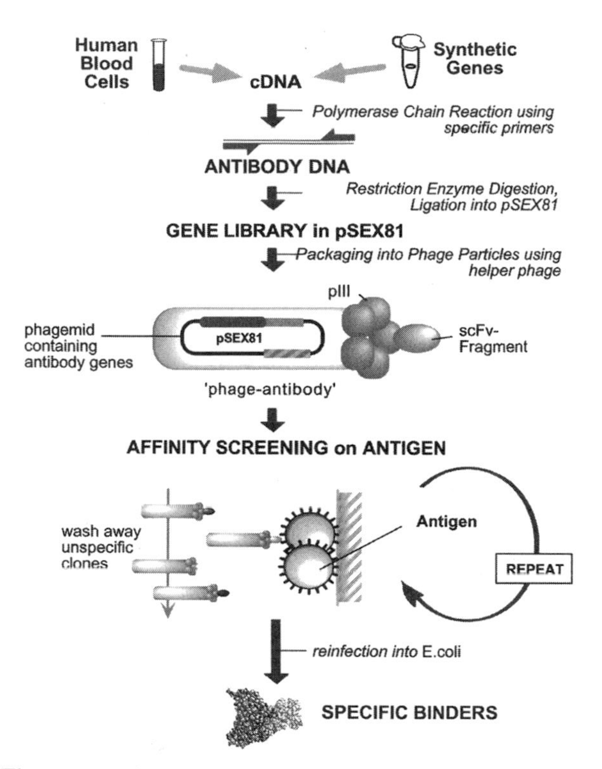

Figure 26.5 The contruction of antibody libraries and the selection procedure used for obtaining specific antibody binders.

complementarity-determining regions (CDRs) are randomly mixed (Hoogenboom and Winter, 1992; Barbas et al., 1992; Griffiths et al., 1994; Braunagel and Little, 1997). Furthermore, immunized antibody libraries may be used when a specific antibody is required and the donor has a titer for antibodies after viral infection, or for patients with autoimmune diseases or cancer. In this case, primers specifically amplifying IgG antibodies are used for the PCR (Rapoport, Portolano, and McLachlan, 1995, Welschof et al., 1995). This display technology and the construction of high-efficiency libraries of antibodies are constantly being refined.

Construction of Antibody Libraries

In order to generate scFv-antibody phage display libraries, the antigen-binding regions of the variable heavy and light chain genes are amplified from lymphatic tissues, resulting in large and complex scFv or Fab antibody repertoires. The creation of large diverse phage antibody libraries from human sources relies on primers that are able to amplify as many variable genes as possible. Many libraries have previously been constructed with primers designed at the amino acid and nucleotide level, using the sequences known at the time (Larrick et al., 1989a, 1989b; Marks et al., 1991a; Campbell et al., 1992; Tomlinson et al., 1992; Williams and Winter, 1993; Welschof et al., 1995; Sblattero and Bradbury, 1998). The number of variable antibody genes isolated and made available to the public in the last few years has dramatically increased. More and more antibodies are being characterized and the isolation of new variable genes is reaching an end point. In the Kabat database 4883 V_H and 1869 V_κ and 1738 $V\lambda$ sequences have been submitted and classified into the various gene families (Kabat et al., 1991). Furthermore, all known functional germline variable genes have been catalogued in the V-Base database, containing 638 different sequences (V-Base, MRC Cambridge, UK, CB2 2QH). Using this knowledge, the design of primer sets containing all relevant genes can be optimized. The quality of the primers increases the quality of the libraries produced today (Fig. 26.6).

The actual library is produced by cloning these antibody genes as fusion proteins with the minor coat protein, pIII, of bacteriophage (Marks et al., 1991b; Breitling et al., 1991; Hoogenboom et al., 1991; Nissim et al., 1994). By using this method, each phage particle expresses a functional recombinant antibody on its surface while holding the gene encoding the antibody within the phage genome. To isolate high-affinity binders from these libraries, both high complexity and quality of the library are important factors. In general, the affinity of the antibody isolated is proportional to the initial size of the library used for selection. Antibody libraries with complexities of over 10^9 individual clones have been constructed and successfully used for the selection of a number of specific antibodies (Sheets et al., 1998; Griffiths et al., 1993; Vaughan et al., 1996). Thus large antibody libraries have become very important sources for the isolation of high-affinity antibodies to virtually all types of antigens (Fig. 26.7).

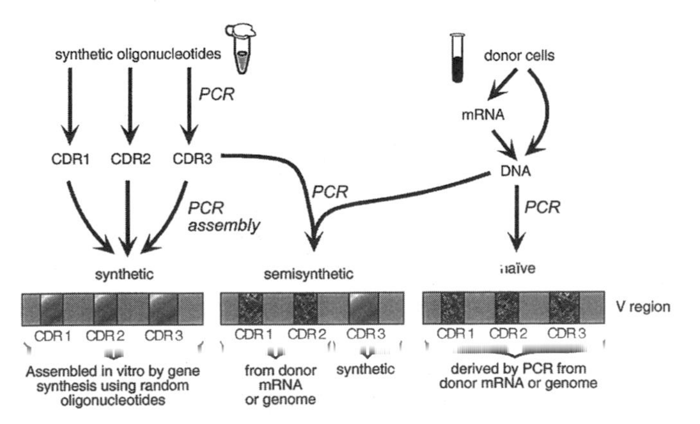

Figure 26.6 The various possibilities used to generate the variable region genes of antibodies using specific premers and PCR.

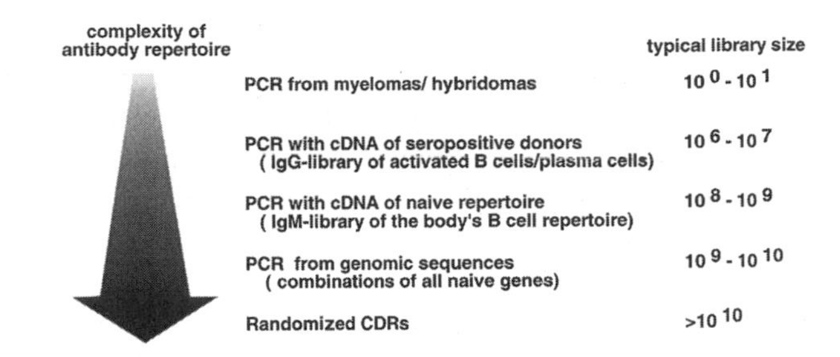

Figure 26.7 Recombinant antibody libraries of increasing complexity.

Another means by which high-affinity antibodies can be isolated is through the construction of specific Ig-class libraries. For example, with the help of specific IgE-class libraries, antibodies directed against a given allergen can be isolated with higher specificity and affinity and be directly used for therapy (Steinberger, Kraft, and Valenta, 1996).

Recombinant Technology

As mentioned, phage display was introduced as a means of making antibodies in vitro, and the quality of the antibodies isolated from the antibody libraries is seen to be proportional to the size of the library used for selection. The bottleneck of the antibody library complexity arises from the cloning steps required to produce scFv- or Fab-antibodies. Recombination has been proposed as an alternative to cloning for creation of large libraries. Lambda recombinase has been used to recombine Fabs, and Cre recombinase to recombine both Fabs and scFvs. For making large phage antibody libraries, however, there have only been two reports describing the use of Cre to recombine Fabs (Griffiths et al., 1994; Sheets et al., 1998). The system is difficult to use because deletion of antibody genes occurs at relatively high frequencies, as the library is formed in phage. In all recombination systems, the V_H genes are cloned in one vector and the V_L in another. Recombination is used to create a third vector capable of displaying functional antibodies. When the products of recombination are selected by the use of newly created antibiotic resistances, all plasmids with the appropriate resistance should be recombined correctly and render recombination irreversible. Other methods that have been employed include the ribosome display method (Schaffitzel et al., 1999) and selectively infective phage display (Krebber et al., 1997). With ribosome display, a method was developed in which whole functional proteins can be enriched in a cell-free system for their binding function, without the use of any cells, vectors, phages or transformation. This technology is based on in vitro translation, in which both the mRNA and the protein product do not leave the ribosome. This results in two fundamental advantages: first, the diversity of a protein library is no longer restricted by the transformation efficiency of the bacteria, and second, because of the large number of PCR cycles, errors can be introduced, and by the repeated selection for ligand binding, improved molecules are selected. Correctly folded proteins can be selected if the folding of the protein on the ribosome is secured. In selectively infective phage display, phage missing the N-terminal penetration domain of pIII are no longer infective. However, infectivity can be restored to a limited amount if this domain is reintroduced to the phage. An antigen–antibody interaction, for example, can be used to bring back some infectivity. The scFv fragment is displayed on phage by fusion to the truncated pIII on phage. Using repertoires of antibody fragments on phage in combination with a fusion protein of antigen with the N-terminal pIII domain, this method can be applied in situations where high-affinity interaction partners are to be selected from a limited number of clones.

Recently, several novel approaches to further improve library quality were developed. Sblattero and Bradbury (2000) have modified the Cre-lox system and have found a way to allow for multiple phage infection of a single bacterium. In this way, they have been able to increase the diversity of the library up to 10^{11} clones and select for antibodies with higher affinities.

Tomlinson has described another method (Recombinant Antibody Technologies Workshop in Heidelberg, Nov. 1999), in which protein A and protein L affinity columns are used to eliminate truncated and inproperly folded scFv gene fragments. The advantages of this system are that over 90% of the scFv phage antibodies obtained are full length and show correct folding. The disadvantages of this system may be the restrictions to a single human framework used for building the heavy and light chain libraries, VH3 (DP-47) and kappa (DPK-9). Thus, only the VH3 gene family is captured by protein A, and protein L only binds the DPK-9 kappa light chain. All other gene families are omitted.

MorphoSys has shown that fully synthetic human combinatorial antibody libraries (HuCAL) can successfully be constructed. By analyzing the human antibody repertoire, it was observed that seven heavy chain, four kappa-, and three lambda-light chain variable regions account for more than 95% of the human antibody diversity. The sequences of these variable genes were analyzed and a consensus sequence was derived for each family. Further diversity of the library was obtained by creating CDR3 library cassettes generated from mixed trinucleotides for both heavy and light chains. With this method, it was shown that the frequency of correct and potentially functional antibody sequences could be increased to approximately 60% (Knappik et al., 2000).

Another novel approach relies on molecules mimicking the antibody-binding pocket. An entire new class of targeting molecules functionally mimicking the antigen-binding site of antibodies are the so-called anticalins. These proteins were engineered by reshaping the four randomized loops of the core structure of lipocalins, producing a library of very stable small molecules being far less complex than antibodies. This library can be screened for binders in the usual manner using the phage display method (Beste et al., 1999). It was further possible to produce bispecific anticalins (duocalins). However, the high immunogenicity of anticalins could be a potential drawback for their use in targeting approaches. Various other approaches have recently emerged, mainly using Ig superfamily frameworks.

CONCLUSION AND FUTURE ASPECTS

Within the last one and a half decades, advances in gene technology have greatly facilitated the genetic identification, manipulation, production, and conjugation of recombinant antibody fragments. Both production and panning of recombinant antibody fragments from antibody libraries have evolved to become one of the major sources for novel antibodies. The key technological development for screening antibody libraries containing millions of different clones was the use of a selection system showing efficiencies comparable to our own immune system. By displaying antibodies on the surface of microorganisms containing the genetic information of the antibody's gene within its genome, an analogue was achieved for the expression of the IgM antigen receptor on the surface of inactivated B-lymphocytes. Further, the clones expressing antigen-specific binding are easily amplified. This phage display technology has proven to be essential for the isolation of various human antibodies against, for example, human target antigens, pathogenic infectious agents, or antigens where an immunized human donor is not available. Thus antibody libraries consisting of more than 10 billion different members allow for the isolation of human antibody fragments against almost every possible antigen, offering a magnificent source of high affinity human immunocompatible molecules for gene therapy.

The general agreement is that, while the making of antibodies is getting easier, there is still much to learn when considering the biological background of the targeting setups we place these antibodies in. It is still not clear what defined properties make the various antibodies good therapeutic agents. Many opinions are currently being evaluated, however concrete data is lacking. For example, an immune response against a therapeutic antibody may not necessarily influence its therapeutic effect (W. Wels and A. Magener, pers. commun.). Even the influence of affinity, which is currently considered to be quite important for the isolation of therapeutic antibodies, is not evaluated to a point allowing reliable predictions below a certain threshold. Affinity constants reaching picomolar levels have been achieved by using genetic engineering, however, it still remains unclear whether these ultrahigh affinities provide any advantage in the final biological setup when compared to the "conventional" nanomolar candidates provided by the native immune system. To solve these problems, two points must be addressed. First, one has to understand the influence of the antibodies' biochemical properties and ultrastructure details on pharmacology and activity in vivo. Second, one should develop methods where the selection of antibody or peptide libraries is comparable to the physiological situation faced by the assigned therapeutic protein. For example, a screening system that directly provides target cell survival if the antibody expressed by this cell binds to the right target (e.g., by blocking a surface receptor that induces apoptosis) could provide antibodies capable of fulfilling their dedicated task in a "real" physiological environment. Antibodies selected from libraries for high affinity are not necessarily suitable for the therapeutic environment.

Various selection strategies and their applications in therapeutic gene targeting are discussed in detail in Chapter 27. Here it is evident that there is still ample work to be done for further development of recombinant antibody technologies. Selection of new antibodies and design of fusion proteins should occur in close relation with the physiological setup, thus providing gene therapy targeting agents fulfilling all in vivo requirements. In vivo panning, in particular by using the enhanced antibody presentation with hyperphage technology, could be a crucial step in this direction.

REFERENCES

Bach JF, Fracchia GN, Chatenoud L (1993): Safety and efficacy of therapeutic monoclonal antibodies in clinical therapy. *Immunology Today* 14:421.

Barbas CF, Kang AK, Lerner RA, Benkovic SJ (1991): Assembly of combinatorial antibody libraries on phage surfaces: the gene III site. *Proc Natl Acad Sci USA* 88:7978.

Barbas CF, Bain JD, Hoekstra DM, Lerner, RA (1992): Semisynthetic combinatorial antibody libraries: A chemical solution to the diversity problem. *Proc Natl Acad Sci USA* 89:4457.

Beste G Schmidt FS, Stibora T, Skerra A (1999): Small antibody-like proteins with prescribed ligand specificities derived from the lipocalin fold. *Proc Natl Acad Sci USA* 96:1898.

Better M, Chang CP, Robinson RR, Horwitz AH (1988): *Escherichia coli* secretion of an active chimeric antibody fragment. *Science* 240(4855):1041–1043.

Bird RE, Hardman KD, Jacobson JW, Johnson S, Kaufman BM, Lee SM, Lee T, Pope SH, Riordan GS, Whitlow M (1988): Single-chain antigen-binding proteins. *Science* 242:423.

Braunagel M, Little M (1997): Construction of a semisynthetic antibody library using trinucleotide oligos. *Nucleic Acids Res* 25:4690.

Breitling F, Dübel S, Seehaus T, Klewinghaus I, Little M (1991): A surface expression vector for antibody screening. *Gene* 104:147.

Campbell MJ, Zelenetz AD, Levy S, Levy R (1992): Use of family specific leader region primers fro PCR amplification of the human heavy chain variable region gene repertoire. *Mol Immunol* 29:193.

Carson DA, Freimark BD (1986): Human lymphocyte hybridomas and monoclonal antibodies. *Adv Immunol* 38:275.

Courtenay-Luck NS, Epenetos AA, Moore R, Larche M, Pectasides D, Dhokia D, Ritter MA (1986): Development of primary and secondary immune responses to mouse monoclonal antibodies used in the diagnosis and therapy of malignant neoplasms. *Cancer Res* 46:6489.

Dübel S, Breitling F, Seehaus T, Little M (1992): Generation of a human IgM expression library in E. coli. *Meth Mol Cell Biol* 3:47–52.

Glassy MC, Dillman RO (1988): Molecular biotherapy with human monoclonal antibodies. *Mol Biother* 1:7.

Griffiths AD, Malmquist M, Marks JD, Bye JM, Embleton MJ, McCafferty J, Baier M, Hollinger KP, Gorick BD, Hughes-Jones NC, Hoogenboom HR, Winter G (1993): Human anti-self antibodies with high specificity from phage display libraries. *EMBO J* 12:725.

Griffiths AD, Williams SC, Hartley O, Tomlinson IA, Waterhouse P, Crosby WL, Kontermann RE, Jones PT, Low NM, Allison TJ, Prospero TD, Hoogenboom HR, Nissim A, Cox JPL, Harrison JL, Zaccolo M, Gherardi E, Winter G (1994): Isolation of high affinity human antibodies directly from large synthetic repertoires. *EMBO J* 13:3245.

Hanes J, Plückthun A (1997): In vitro selection and evolution of functional proteins by using ribosome display. *Proc Natl Acad Sci USA* 94:4937.

Hoogenboom HR, Griffiths AD, Johnson KS, Chiswell DJ, Hudson P, Winter G (1991): Multi-subunit proteins on the surface of filamentous phage: methodologies for displaying antibody (Fab) heavy and light chains. *Nucl Acids Res* 19:4133.

Hoogenboom HR, Winter G (1992): By-passing immunisation. Human antibodies from synthetic repertoires of germline V_H gene segments rearranged in vitro. *J Mol Biol* 227:381.

Kabat EA, Wu TT, Perry HM, Gottesman KS, Foeller C (1991): *Sequences of Proteins of Immunological Interest*, Fifth Ed. U.S. Department of Health and Human Services, Public Health Service, National Institutes of Health. NIH Publication No. 91-3242.

Knappik A Ge L, Honegger A Pack P, Fischer M Wellnhofer G, Hoess A, Wolle J, Plückthun A, Virnekas B (2000): Fully synthetic human combinatorial antibody libraries (HuCAL) based on modular consensus frameworks and CDRs randomised with trinucleotides. *J Mol Biol* 296:57–86.

Köhler G, Milstein C (1975). Continuous culture of fused cells secreting antibody of predefined specificity. *Nature* 256:495.

Krebber C, Spada S, Desplancq D, Krebber A, Ge L, Plückthun A (1997): Selectively-infective phage (SIP): a mechanistic dissection of a novel in vivo selection for protein-ligand interactions. *J Mol Biol* 268:607.

Larrick JW, Danielsson L, Brenner CA, Wallace EF, Abrahamson M, Fry KE, Borrebaeck CAK (1989a): Polymerase chain reaction using mixed primers: cloning of human monoclonal antibody variable region genes from single hybridoma cells. Bio/Technology 7,9:34.

Larrick JW, Danielsson L, Brenner CA, Abrahamson M, Fry KE, Borrebaeck CAK (1989b): Rapid cloning of rearranged immunoglobulin genes from human hybridoma cells using mixed primers and the polymerase chain reaction. *Biochem Biophys Res Commun* 160:1250.

Marks J (2000): Lecture at the 10th IBC meeting on antibody Engineering, San Diego, Dec. 2000.

Marks JD, Tristem M, Karpas A, Winter G(1991a): Oligonucleotide primers for polymerase chain reaction amplification of human immunoglobulin variable genes and design of family-specific oligonucleotide probes. *Eur J Immunol* 21:985.

Marks JD, Hoogenboom HR, Bonnert TP, McCafferty J, Griffiths AD, Winter, G

(1991b): By-passing immunization. Human antibodies from V-gene libraries displayed on phage. *J Mol Biol* 222:581.

McCafferty J, Griffiths AD, Winter G, Chiswell DJ (1990): Phage antibodies: filamentous phage displaying antibody variable domains. *Nature* 348:552.

Nissim A, Hoogenboom HR, Tomlinson IM, Flynn G, Midgley C, Lane D, Winter G (1994): Antibody fragments from a 'single pot' phage display library as immunochemical reagents. *EMBO J* 13:692.

Rapoport B, Portolano S, McLachlan SM (1995): Combinatorial libraries: new insights into human organ-specific autoantibodies. *Immunol Today* 16:43.

Rondot S, Koch J, Breitling F, Dübel S (2001): A helper phage to improve single-chain antibody presentation in phage display. *Nature Biotech* 19:75.

Sblattero D, Bradbury A (1998): A definitive set of oligonucleotide primers for amplifying human V regions. *Immunotechnology* 3:271.

Sblattero D, Bradbury A (2000): Exploiting recombination in single bacteria to make large phage antibody libraries. *Nat Biotechnol* 18:75.

Schaffitzel C, Hanes J, Jermutus L, Plückthun A (1999): Ribosome display: an in vitro method for selection and evolution of antibodies from libraries. *J Immunol Methods* 231:119.

Seigneurin JM, Desgranges C, Seigneurin D, Paire J, Renversez JC, Jacquemont B, Micouin C (1983): Herpes simplex virus glycoprotein D: Human monoclonal antibody produced by bone marrow cell line. *Science* 221:173.

Sheets MD, Amersdorfer P, Finnern R, Sargent P, Lindqvist E, Schier R, Hemingsen G, Wong C, Gerhart JC, Marks JD (1998): Efficient construction of a large nonimmune phage antibody library: The production of high-affinity human single-chain antibodies to protein antigens. *Proc Natl Acad Sci USA* 95:6157.

Skerra A, Plückthun A (1988): Assembly of a functional immunoglobulin Fv fragment in Escherichia coli. *Science* 240:1038.

Smith GP (1985): Filamentous fusion phage: novel expression vectors that display cloned antigens on the virion surface. *Science* 228:1315.

Steinberger P, Kraft D, Valenta R (1996): Construction of a combinatorial IgE library from an allergic patient. *J Biol Chem* 271:10967.

Tomlinson IM, Walter G, Marks JD, Llewelyn MB, Winter G (1992): The repertoire of human germline VH sequences reveals about fifty groups of VH segments with different hypervariable loops. *J Mol Biol* 227:776.

Vaughan TJ, Williams AJ, Pritchard K, Osburn JK, Pope AR, Earnshaw JC, McCafferty J, Hodits RA, Wilton J, Johnson K (1996): Human antibodies with subnanomolar affinities isolated from a large non-immunized phage display library. *Nature Biotec* 14:309.

Welschof M, Terness P, Kolbinger F, Zewe M, Dübel S, Dörsam H, Hain C, Finger M, Jung M, Moldenhauer G, Hayashi N, Little M, Opelz G (1995): Amino acid sequence based PCR primers for amplification of rearanged heavy and light chain immunoglobulin variable region genes. *J Immunol Methods* 179:203.

Williams SC, Winter G (1993): Cloning and sequencing of human immunoglobulin V lambda gene segments. *Eur J Immunol* 7:1456.

Winter G, Milstein C (1991): Man-made antibodies. *Nature* 349:293.

27

SINGLE-CHAIN Fv FRAGMENTS FROM PHAGE DISPLAY LIBRARIES

ROLAND E. KONTERMANN, PH.D.

INTRODUCTION

Single-chain Fv (scFv) molecules combine the coding sequences of the variable heavy and light chain domains of an antibody in a single-gene encoded format. The resulting polypeptides, with the variable light and heavy chain domains connected by a flexible peptide linker, assemble into functional antigen-binding sites. In contrast to dimeric antigen-binding molecules such as Fab fragments, the single-chain format simplifies fusion of additional polypeptides either to the N- or C-terminus. For phage display, scFvs are fused at their C-terminus to the gene III protein of filamentous bacteriophage (see Chapter 26). The single-chain scFv format not only facilitates fusion to other proteins but also improves the physicochemical stability of the recombinant antibody molecules (Glockshuber et al., 1990).

Various strategies have been developed over the past decade to generate scFv molecules for targeted delivery of gene therapeutic vectors (Pelegrin et al., 1998). In order to efficiently deliver a gene-therapeutic vector, scFvs have to bind to structures present on the surface of the target cells. The success of using scFvs for vector targeting depends mainly on the ability to isolate scFvs with the desired binding characteristics and to incorporate them into the surface

Vector Targeting for Therapeutic Gene Delivery, Edited by David T. Curiel and Joanne T. Douglas
ISBN 0-471-43479-5

of the gene-therapeutic vectors without interfering with target cell recognition. Furthermore, they have to induce internalization of the vector on binding to the target cell, either directly by processes mediated by the target molecule or indirectly by secondary interactions between the vector and the target cell.

The direct scFv-induced internalization requires binding to a cell surface receptor and the induction of receptor-mediated processes leading to internalization and endosomal release. Generally, this is caused either by receptor dimerization or by conformational changes of the receptor, both induced by binding of a ligand. In order to mediate internalization, scFv molecules have to mimic these processes. Consequently, not all scFv molecules which bind to a cell surface receptor will have the desired properties. Therefore, these requirements have to be considered in selecting a scFv. In contrast, if internalization is mediated by secondary interactions between vector and target cell, it is important that binding of the scFv to the target cell does not interfere with these processes.

SELECTION OF SINGLE-CHAIN Fv FRAGMENTS FROM PHAGE LIBRARIES

Selections with Purified Antigens

One of the most extensively applied methods for the isolation of specific scFvs is selection of scFv phage display libraries using purified antigens. By immobilization of the antigens on a polystyrene surface, e.g., using immunotubes or microtiter plates, bound phage can be enriched by a simple procedure involving removal of unbound phage and elution of bound phage (biopanning) (Kontermann, 2001). However, since binding of the antigen to the plastic surface can result in a partial or even complete denaturation of the protein, isolated scFvs often recognize only epitopes exposed by the denatured molecule. Nevertheless, numerous scFvs have been isolated by this approach, which are able to bind to cell surface displayed target antigens. This includes selection of anticarcinoembryonic antigen (CEA) scFvs from a large naïve library (Osbourn et al, 1996), antiepithelial glycoprotein 2 (EGP-2) scFvs from a semisynthetic scFv library (de Kruif et al., 1995), scFvs directed against high molecular weight-melanoma-associated antigen (HMW-MAA) isolated from a synthetic library (Desai et al., 1998), and antimesothelin scFvs selected from a murine immune library (Chowdhury et al., 1998).

In order to circumvent denaturation of the purified antigens on binding to the plastic surface, various indirect immobilization methods have been developed. These methods include capturing of the antigen with specific antibodies or using streptavidin and biotinylated antigen. The latter approach is also applied to immobilize antigens on paramagnetic beads (Chames, Hoogenboom, and Henderikx, 2001). The advantage of this method is that the antigen is kept in its native conformation and that much lower amounts of antigen are required. In addition, the interaction between phage-displayed antibodies and the antigen can also be performed in solution with subsequent capturing of the antibody–antigen complexes on streptavidin-coated beads. This allows for the selection under equilibrium conditions. By decreasing the concentrations of antigen during the rounds of selections, high-affinity scFvs are enriched (Hawkins, Russell, and Winter, 1992; Nielsen and Marks, 2001a). In conjunction with chain shuffling and random mutagenesis of the heavy and light chain complementarity-determining region 3 (CDR3) regions, Schier and co-workers (1996a, 1996b) were able to isolate scFv molecules against erbB2 with picomolar affinities. Interestingly, selections with erbB2 immobilized on polystyrene surfaces resulted in spontaneously dimerizing scFvs, which were selected due to their increased functional affinity and which had only a slightly improved affinity compared to the parental scFv. High-affinity scFvs were only selected in solution using limiting concentrations of biotinylated antigen and screening of soluble scFvs by measurement of k_{off} by surface plasmon resonance in a BIAcore (Schier et al., 1996a).

Selections employ either native proteins purified from target cells or recombinant proteins. Recombinant proteins can be either produced in prokaryotic or in eukaryotic systems. Because most target cells are human, it has to be considered that target molecules present on the cell surface are often glycoproteins consisting of one or more polypeptides, which quite often are stabilized by disulfide bonds. Such proteins or parts of them are normally difficult to express in a soluble and native conformation in prokaryotic systems, that is, bacterial cells, which lack glycosylation and disulfide-bond formation in the cytoplasm. Thus, selections with bacterially expressed proteins may result in antibodies not reacting with the native protein. Indeed, selections with the recombinant extracellular domain of human endoglin (CD105), a homodimeric disulfide-linked glycoprotein overexpressed by proliferating endothelial cells, resulted in scFvs recognizing endoglin on the surface of live cells, but which did not bind to the bacterially expressed protein (Nettelbeck et al., 2001).

Various studies have also used peptides derived from tumor-associated antigens (TAAs) to obtain scFvs which bind to tumor cells. These antigens are either coupled to a carrier protein, for example, bovine serum albumin (BSA), or are biotinylated and immobilized on streptavidin-coated magnetic beads. ScFvs recognizing a mutant epidermal growth factor receptor (EGFRvIII) found in glioblastoma and other carcinomas were selected from an immune library using a synthetic 17-mer biotinylated peptide (Lorimer et al., 1996). An immunotoxin produced from one of the selected scFvs fused to *Pseudomonas* exotoxin A was highly cytotoxic to cells expressing EGFRvIII. In a similar approach, scFvs directed against the tumor-associated antigen MUC-1 were selected from a naïve scFv library using a synthetic biotinylated 100-mer peptide containing five tandem repeats of the MUC-1 peptide core (Henderikx et al., 1998). In addition to peptides, other small antigens, such as tumor-associated carbohydrates, have been employed to obtain tumor-cell specific scFvs. In one such study, synthetic sialyl Lewis^X and Lewis^X coupled to BSA as carrier were used to isolate antibodies that bind to adenocarcinoma cells overexpressing these antigens (Mao et al., 1999).

ScFvs can not only be isolated against antigen but also using monoclonal antiidiotypic antibodies, which mimic cell surface receptors. The feasibility of this approach was demonstrated in a model system with an anti-CD30 scFv and antiidiotypic monoclonal antibodies that carry an internal image of an epitope of membrane bound CD30, an antigen overexpressed on Hodgkin's lymphoma cells (Hombach et al., 1998).

Selection with Proteins from Two-Dimensional Gel Electrophoresis

The previously described method of selection of scFv fragments using purified molecules is time consuming and elaborate in regard to obtaining sufficient amounts of antigen to perform selections. In addition, it would be advantageous to minimize the time between target identification and selection of a ligand. Two-dimensional gel electrophoresis is a powerful tool to identify molecules overexpressed by the target cells. The method can be applied to screen for differences in expression levels and patterns, for example, at different stages of the disease. However, research samples are often limited and the identified spots normally contain only minute amounts of material.

Pini and co-workers (1998) recently demonstrated that biotinylated proteins, which are eluted from a two-dimensional gel, can be used to select specific binders from a synthetic scFv library. As a model system they used a 91 amino acid residues long type III repeat of fibronectin (ED-B domain), a marker of angiogenesis. By using only 0.3 μg of antigen and three rounds of selection, they were able to identify specific scFvs with affinities in the 10–100 nM range. Of note, these antibodies recognized fibronectin in its native conformation. An affinity-matured version of one of these scFvs was recently applied for targeted delivery of tissue factor to the tumor vasculature overexpressing fibronectin. This scFv-tissue factor fusion protein mediated the infarction of three different kinds of solid tumours in mice (Nilsson et al., 2001). In a similar approach, antibodies were selected from a semisynthetic library against beta-actin eluted from a two-dimensional gel (Zhou et al., 1998). In contrast to the work of Pini et al. (1998), the protein was biotinylated after elution from the gel. Using four rounds of selection and only 50 ng biotinylated protein per round, they were able to isolate various independent clones reacting with beta-actin in an enzyme-linked immunosorbent assay (ELISA).

The method was further simplied by using protein spots blotted onto nitrocellulose membranes, thus omitting the biotinylation step (Ravn et al., 2000; Liu and Marks, 2000). However, when using nitrocellulose to immobilize the antigen one has to overcome the high background binding of phage caused by charge interactions between phage and the membrane (Liu and Marks, 2000). A comparative analysis of the optimal selection conditions revealed that the use of nitrocellulose membranes is superior to polyvinylidene difluoride (PVDF) membranes, and that blocking with 10% gelatin and the presence of 500 mM

NaCl in the binding buffer drastically reduced nonspecific binding, resulting in enrichment factors greater than 500-fold (Liu and Marks, 2000). In this study, a panel of antibodies was selected from a naïve scFv library against BSA and erbB2 with as little as 1–10 ng of protein. These antibodies worked in Western blotting and in ELISA. Interestingly, the minimal amount of antigen that had to be used to isolate antibodies depended on the kind of phage library applied. Multivalent display (see Chapter 26) yielded binders with as little as 1 ng antigen per round, compared to monovalent display, which required 10-fold more antigen.

Antibodies selected against proteins separated by two-dimensional gel electrophoresis are valuable tools for the analysis and identification of new targets in therapy. The major limitation for therapeutic gene delivery arises, however, from the possibility that antibodies selected against denatured proteins may not recognize the native antigen on the cell surface. Thus, additional screening or selection protocols have to be included in order to obtain antibodies with the desired properties.

Cell Selections

Ideally, one would like to select antibodies directly against the target cell or tissue. Thus, all antibodies selected are likely to bind to a component of the target cell membrane, a prerequisite for targeted gene delivery. This approach is especially applicable when purified antigen is not available or not stable in solution. As membrane proteins are embedded into the lipid bilayer, they are likely to interact with other membrane proteins. For example, target molecules are often part of larger complexes, for example, receptor complexes composed of various subunits. These complexes are easily disrupted by biochemical manipulation, which might also change the conformation of single subunits. However, the problems with cell selections arise from the fact that a large number of different antigens are present at different concentrations on a cell surface. Thus, an extensive analysis of the selected scFvs may be necessary and antibodies with broad cell type specificity may be isolated by this method.

Numerous cell selections were performed either on freshly isolated live cells (de Kruif et al., 1995; Siegel et al., 1997; Palmer, George, and Ritter, 1997; Lekkerkerker and Logtenberg, 1999; Mutuberria et al., 1999; Rossig, Nuchtern, and Brenner, 2000; Edwards et al., 2000), cell lines (Cai and Garen, 1995; Pereira et al., 1997; Noronha et al., 1998; Winthrop, DeNardo, and DeNardo, 1999; Ridgway et al., 1999; Kupsch et al., 1999; Tordsson et al., 2000; Shadidi and Sioud, 2001), stably transfected cell lines (Andersen et al., 1996; Hoogenboom et al., 1999; Marget et al., 2000; Peipp et al., 2001), fixed cells (van Ewijk et al., 1997; Schmidt et al., 1999; Roovers et al., 2001), or on cell extracts (Sanna et al., 1995; Sawyer, Embleton, and Dean, 1997).

Inital experiments used intact red blood cells to demonstrate the feasibility of the approach (Marks et al., 1993). ScFv fragments against various red blood cell group antigens were isolated from a naïve library, including antibodies against epitopes of the ABO and I blood group system (B and HI), of the Rh system (D and E), and the Kell system (Kp^b) present at a density of only 2000 to 5000 molecules per cell. Selections were performed by incubating cells in suspension with the phage and enrichment by pelleting and resuspending cells. This work also established preadsorption of the phage library with cells lacking the desired antigen in order to deplete for phage antibodies against common antigens.

In the case of tumor cells, preadsorption can be performed on normal cells of the same origin. By this approach, antimelanoma scFvs were isolated from a phage library generated from peripheral blood lymphocytes of two melanoma patients (Cai and Garen, 1995). In this work, selections were performed on attached monolayers of autologous melanoma cell lines and phage were preadsorbed against normal melanocytes. Isolated clones could be divided into three classes: (1) melanoma-specific scFvs, (2) tumor-specific scFvs also reacting with other tumor cell lines, and (3) lineage-specific scFvs reacting with melanocytes, melanoma, and glioma cell lines, but not with other cell lines. These experiments demonstrated that antibodies with different specificities can be selected from cells.

Further studies reported isolation of antibodies from the B-cell repertoire of patients immunized with autologous neuroblastoma (Rossig et al., 2000), anti-MUC-1 scFv fragments selected from a murine immune library against MCF-7 breast cancer cells (Winthrop, DeNardo, and DeNardo, 1999), anti-leukemic scFvs selected from a semisynthetic library on a premyelocytic leukemia cell line (HL60) (Shadidi and Sioud, 2001), and anti-CD55 scFvs selected by subtractive panning of a naïve library on a lung adenocarcinoma cell line (1264) (Ridgway et al., 1999). Thus, various kinds of antibody phage libraries have been successfully applied to isolate antitumor scFvs.

A simultaneous positive- and negative-selection strategy was employed to optimize the capture of antigen-specific phage and minimize the binding of irrelevant phage antibodies (Siegel et al., 1997). In this strategy, target cells were precoated with magnetic beads and were then diluted into an excess of antigen-negative cells, followed by incubation with the phage library. Target cell-specific antibodies were then isolated by magnetically activated cell sorting (MACS). Using this approach, anti-Rh(D) antibodies were selected with magnetically labelled Rh(D)-positive red blood cells and unlabelled Rh(D)-negative cells.

Selection on stably transfected cells enables preadsorption on untransfected cells. This has the advantage that all antibodies against common cell-surface antigens are efficiently eliminated and the selection of target antigen-specific antibodies is favored (Hoogenboom et al., 1999). By this approach, anti-CD13 scFvs were selected from a murine immune library on intact stably transfected NIH/3T3 cells expressing human CD13 (Peipp et al., 2001). Antibodies recognizing specific peptide/MHC complexes were isolated from a library of immunized mice selected on cells transfected with MHC class I molecule K^k, which were pulsed with a K^k-restricted influenza virus-derived peptide ($Ha_{255-262}$)

(Andersen et al., 1996). However, these approaches require cloning of the antigen of choice and the generation of stably transfected cell lines.

Most cell selections are performed on target cells in suspension or grown as adherent monolayers. It is also possible to select target cell-specific antibodies from a mixture of different cells including an isolation step during selection. In one such study, de Kruif and Logtenberg (1995) used selection by fluorescence-activated cell sorting (FACS) with target cell-specific fluorescently labeled antibodies to isolate from a semisynthetic scFv library antibodies specific for a subset of peripheral blood leukocytes, including $CD3^+$ T cells and $CD20^+$ B cells. The same approach was successfully applied to select for antibodies against subpopulations of blood dendritic cells (Lekkerkerker and Logtenberg, 1999).

The success of cell selections not only depends on the availability of suitable target cells and cells for negative selection by preadsorption, but is also influenced by the antigen itself. Hoogenboom and co-workers (1999) found that cell selections with CHO cells stably transfected with a receptor for somatostatin resulted in a large panel of pan-specific antibodies, but none of these was receptor-specific. In contrast, selections with CD36-expressing CHO cells resulted in a high frequency of antigen-specific antibodies. Such an immunodominance in vitro was also observed in selections of a synthetic scFv library on human melanoma cell lines, with all isolated antibodies reacting with HMW-MAA (Noronha et al., 1998). These findings indicate that the nature of the antigen can have a profound effect on the outcome of the selection.

In a recent study, Huie and co-workers (2001) efficiently selected a panel of specific antibodies to fetal nucleated red blood cells (NRBCs) from a naïve multivalent display scFv library. It was concluded that multivalent display overcomes the limitations of cell selections caused by a low concentration of a given binding phage in a naïve library.

ProxiMol and Step-Back Selections

As just discussed, selections with whole cells often led to a large panel of different antibodies with only a fraction of the antibodies, if any, possessing the desired specificity. In order to enrich for antibodies that bind in close proximity to an antigen of choice, one can apply the ProxiMol selection method (Osbourn et al., 1998a; Osbourn, 2001). ProxiMol was designed to increase the probability of finding the desired antibody. The method is based on catalyzed reporter enzyme deposition (CARD), originally developed as a method of signal amplification in immunoassays (Bobrow et al., 1989). The success of the method relies on the availability of a ligand, such as an antibody, specifically reacting with the target cell. The ligand, which has to be conjugated to horseradish peroxidase (HRP), is used as a guide molecule to target the antigen. Target cells are incubated with the ligand and a phage library and subsequently biotin tyramine and hydrogen peroxide are added, which results in enzyme-catalyzed formation of biotin tyramine free radicals. These radicals react with proteins in close vicinity of the enzyme, including phage particles, thus leading to biotinylation of the phage bound nearby. The biotinylated phage are then enriched by the use of streptavidin-coated magnetic beads. Biotinylation is restricted to an area of an approximate radius of 25 nm from the ligand-binding site. Thus, the selected antibodies bind close to the site of ligand binding, but do not overlap with it. Using this method, scFv molecules directed against CEA, E-selectin, P-selectin, and transforming growth factor $\beta1$ (TGF-$\beta1$) were selected from a naive library (Osbourn et al., 1998a).

In order to isolate antibodies which bind to the epitope recognized by the ligand, one can perform an additional round using phage from the first round to guide selection. In this step-back selection method (Osbourn et al., 1998b; Osbourn, 2001), cells are incubated with the biotinylated phage from the first round and subsequently with an unselected phage library. Streptavidin-HRP and biotin tyramine/hydrogen peroxide are then added, which results in biotinylation of phage bound near the guiding phage. This method was, for example, successfully applied to isolate anti-CCR-5 receptor scFv molecules from a naïve phage library using MIP-1α as a guide molecule and $CD4^+$ cells for selection (Osbourn et al., 1998b). Of note, none of the antibodies selected by conventional cell selection bound CCR-5 whereas approximately 50% of the $CD4^+$ clones isolated by the inital ProxiMol selection were CCR-5 positive. The step-back selection resulted then in scFvs that block the binding of MIP-1α to $CD4^+$ cells. Instead of defined ligands, it is also possible to use HRP-conjugated polyclonal sera as a guide in ProxiMol selections, as shown for the isolation of adipocyte-specific scFvs (Edwards et al., 2000). Thus, if target cell specific antibodies or natural ligands, such as growth factors or chemokines, are available, these methods are powerful tools to drive the selection to a specific antigen.

Selection for Cell Binding and Internalization

Internalization of the gene therapeutic vehicles after binding to the target cell is a prerequisite for gene delivery. After binding to the target cell, viral delivery systems, such as adenoviruses, can be internalized and directed to the endocytotic pathway through interaction of the virus envelope with natural receptors on the cell surface. In contrast, nonviral systems, such as liposomes or DNA complexes, are normally not internalized per se. However, internalization can be accomplished by the use of antibodies that access receptor-mediated uptake mechanisms of the target cell. As discussed before, this receptor-mediated internalization requires homo- or heterodimerization through a bivalent ligand or conformational changes of the receptor caused by monovalent ligands (Ullrich and Schlessinger, 1990). These processes can be mimicked by antibodies and antibodies with these properties can be identified by screening available receptor-specific antibodies.

As has been shown recently, it is also possible to directly select antibodies, which are internalized, from phage libraries by recovering phage particles from within the cell (Becerril, Poul, and Marks, 1999; reviewed in Nielsen and Marks, 2000). This method has already been applied to isolate novel internalizing scFvs directed against erbB2, transferrin receptor, and EGFR (Poul et al., 2000; Heitner et al., 2001). These selections were carried out either using tumor cells overexpressing these receptors or cells transfected with the receptor gene. Binding of the phage is done at 4°C on adherent cells in the presence of non-target cells (e.g., fibroblasts or untransfected cells) in suspension to deplete the library of phage that bind to common antigens. Internalization is then induced by shifting the temperature to 37°C.

This incubation step is restricted to 15–30 min in order to allow sufficient endocytosis of cell-bound phage but to avoid degradation of phage within the cell. Recovery of internalized phage involves stripping of cells with a low pH buffer to remove phage bound to the cell surface, incubation with trypsin to remove phage bound to the extracellular matrix and cell culture dish, and finally lysis of cells with a high pH buffer (Nielsen and Marks, 2001b). In comparison to selection of cell surface-bound phage, recovery of internalized phage antibodies increases the enrichment of specific antibodies more than 10- to 30-fold (Becerril, Poul, and Marks, 1999). Selection of a naïve scFv library on the breast tumor cell line SKBR3 for internalized antibodies resulted in 40% clones that bound to tumor cells but not to normal cells (Poul et al., 2000). Of the selected clones, two were identified to react with erbB2 and one with the transferrin receptor. These antibodies were efficiently endocytosed into SKBR3 in the form of monomeric scFvs. However, they also induced downstream signaling through the erbB2 receptor or growth inhibition through the transferrin receptor, which might limit the application of the approach. One of the scFvs directed against erbB2 was used to generate doxorubicin-containing immunoliposomes and was shown to be superior in delivering the drug to erbB2-expressing cells compared to unconjugated liposomes (Heitner et al., 2001).

Monovalent display selects for scFvs that induce conformational changes of the receptor. However, most antibodies will require di- or multivalent binding to induce receptor cross-linking and endocytosis. Thus, libraries displaying antigen-binding sites multivalently (see Chapter 26) or in the form of bivalent molecules, for example, as diabodies (Becerril, Poul and Marks, 1999), should extend the probability of finding internalizing antibodies.

Tissue Selections

Cell selections do not require purified antigen but depend on the availability of cells or cell lines derived from the target tissue. However, in some cases such cells might be difficult to isolate and some cells might also change their phenotype in culture. In addition, isolated cells do not reflect the exact situation in vivo, that is, some target antigens may be expressed in a site-specific manner. This is especially true for invasive and metastatic tumor cell populations, which are the main targets in cancer gene therapy, and which are in cell culture primarily selected for proliferation. Thus, it would be advantageous to perform selections in situ or directly in vivo, which takes into account developmental, organ-specific, and stage-specific molecular differences. Selection of an scFv phage library generated from melanoma-immunized nonhuman primates on metastatic melanoma tissue sections resulted in scFv fragments reacting with different components of melanoma and normal tissues (Tordsson et al., 1997), including antigens expressed on the cell surface, the nucleus, or the extracellular stroma. The tissue selections were performed on small acetone-fixed cryosections (approx. 5 mm in diameter) of resected tissue samples (Tordsson, Brodin, and Karlstöm, 2001).

In Vivo Selections

The selection of phage in vivo has been successfully performed with peptide libraries, but difficulties in translating these protocols to the selection of antibody fragments have been reported. Nevertheless, recent work by Johns, George, and Ritter (2000) describes the isolation of organ-specific scFv fragments selected in vivo from a semisynthetic phage library. The analysis of unselected ^{125}I-labeled phage after intravenous injection into CBA mice showed that many phage become trapped nonspecifically in the lungs, but also in the liver and spleen. These phage are cleared from the organs within 24 h. In contrast, very few phage are trapped within the majority of the organs at any time. As the localization of phage to the liver was also reported for phage displaying peptides, this is likely to be a property of the whole phage rather than the displayed polypeptide (Pasqualini, Koivunen, and Ruoslahti, 1997). Interestingly, nonviable phage were mainly localized in the liver whereas phage recovered from other organs were predominantly viable. The recovery of viable phage reached its maximum after 1 h. However, the authors chose 2 h as the point to select phage in order to reduce the probability to enrich for phage still nonspecifically trapped in the tissue. Applying these conditions, scFvs which bind to the endothelium in a tissue-specific manner were isolated from the thymus (Johns, George, and Ritter, 2000).

PREPARATION OF SINGLE-CHAIN Fv FRAGMENTS

Various systems have been developed for the expression and purification of single-chain Fv fragments. The most frequently used method is the expression of scFv fragments in bacterial cells. In order to obtain properly folded scFv molecules containing a disulfide bond in both the V_H and V_L domains, an oxidizing environment is necessary. In bacteria this is achieved by secretion of the polypeptide into the periplasmic space adding a signal peptide to the N-terminus of the scFv molecule (Skerra and Plückthun, 1988). In addition, for purification short sequences, such as a FLAG tag or a hexahistidyl stretch, are added to the N- or C-terminus of the scFv fragment (Fiedler and Skerra, 2001). By these methods it is possible to obtain milligram amounts of pure scFv molecules from shake flask cultures. Similarly, scFv fusion proteins, for example bifunctional or bispecific molecules applied for vector targeting (see next section), can be expressed in these systems. However, the fusion of additional sequences to an scFv molecule often leads to insoluble products in prokaryotic expression systems. In such cases, either refolding strategies (Kipriyanov et al., 1995) or eukaryotic expression systems, that is, *Pichia pastoris*, insect cells, plant cells, or mammalian cells, have to be applied in order to obtain soluble material (Gram and Ridder, 2001; Liang & Dübel, 2001; Conrad and Fiedler, 2001; Bradbury, 2001).

VECTOR TARGETING USING SINGLE-CHAIN Fv FRAGMENTS

As described in previous chapters, various nonviral and viral systems have been developed for gene delivery. In principle, there are two ways to use scFvs for targeting. They can be either directly incorporated into the gene-therapeutic vehicle or they can be used as part of a molecular adaptor, for example in the form of bispecific or bifunctional molecules.

Targeting of Nonviral Vectors

Nonviral vectors can be divided into two major types: molecular conjugates (polyplexes) and liposomes (lipoplexes) (Felgner et al., 1997). Polyplexes are composed of DNA and DNA-condensing polycations that assemble into stable complexes. For targeted DNA delivery, ligands, such as antibodies, have been chemically coupled to polycations. This includes coupling to polylysine (Gupta et al., 2001), polyethylenimine (PEI) (O'Neill et al., 2001), or cationic amphiphiles, such as cholesterol spermine (Mohr et al., 1999). In case of short cationic peptides, these sequences can be directly incorporated into recombinant antibodies by means of genetic engineering. In addition, fusion proteins consisting of an scFv fragment, a DNA-binding domain, and a translocation domain have been employed for targeted delivery of DNA (Fominaya and Wels, 1996; Uherek and Wels, 2001). The scFv fusion protein is indirectly complexed with the DNA by the DNA-binding domain, which binds to a recognition site present in the DNA. Whereas the scFv fragment is responsible for targeted delivery, the translocation domain is required for efficient release from the endosome after internalization. These acid-activatable translocation domains are derived, for example, from *Pseudomonas* exotoxin A or diphtheria toxin, which retain their endosomolytic activity even in an unrelated context. A bacterially expressed fusion protein consisting of an anti-erbB2 scFv, the yeast GAL4-DNA-binding domain and the translocation domain of diphtheria toxin was employed for selective gene delivery to erbB2-expressing carcinoma cells (Uherek, Fominaya, and Wels, 1998).

ScFvs can also be incorporated into liposomes resulting in immunoliposomes. This can be achieved by chemical coupling to components of the liposome surface. In order to facilitate and direct coupling, Heitner and co workers (2001) added a cysteine residue to the C-terminus of an scFv directed against erbB2, which was selected for internalization. The scFv was coupled to maleimide-terminated polyethyleneglycol phosphatylethanolamine (M-PEG-PE) already incorporated into the liposome surface (PEG-MAb linkage) or was subsequently inserted into the liposome surface by micellar incorporation. These immunoliposomes efficiently delivered genes or drugs to erbB2-expressing tumor cells in vitro and in vivo (Park et al., 2001).

Another way to incorporate scFvs into liposomes is by the use of biosynthetically lipid-tagged molecules (Laukkanen, Teeri, and Keinänen, 1993; Keinänen and Laukkanen, 1994). These molecules are generated by genetically fusing the signal peptide and the nine N-terminal amino acid residues of the LPP lipoprotein of *Escherichia coli* to the N-terminus of an scFv. The lipid-tagged scFv is directed to the outer membrane from where it is purified after solublization with detergent. The purified proteins are then incorporated into liposomes by detergent dilution (Laukkanen, Alfthan, and Keinänen, 1994). Immunoliposomes generated from in vivo lipid-tagged anti-CD22 scFv were shown to bind to $CD22^+$ cell lines and B lymphocytes and to accumulate in intracellular compartments (de Kruif et al., 1996).

Targeting of Viral Vectors Displaying Single-Chain Fv Fragments

Several strategies have been described to use scFvs to alter the tropism of viruses in gene therapy (Russell and Cosset, 1999). One way is to incorporate a scFv-encoding gene into the viral genome, which becomes thus an integral part of the virus. In order to display the scFv on the virus surface, it has to be fused to a component of the viral envelope. Importantly, this should not abolish production of functional virus particles and should result in binding of the vector to the target cell and the internalization upon binding.

Extensive studies have been performed with retroviruses (see Chapters 13 and 14). Initial experiments with ecotropic Moloney murine leukemia virus (MLV) and an scFv directed against a hapten demonstrated that scFvs can be incorporated into the viral genome and displayed in active form by inserting the scFv between residues 6 and 7 of the viral SU envelope protein (Russell, Hawkins, and Winter, 1993). Subsequently, using scFvs specific for the low density lipoprotein receptor (Somia, Zoppe, and Verma, 1995) or human major histocompatibility complex class I molecules (Marin et al., 1996), gene transfer to cells expressing these receptors was reported. Of note, the coincorporation of unmodified envelope proteins was found to be indispensible for high-titer virus infection (Somia, Zoppe, and Verma, 1995). In further studies, specific gene transfer was demonstrated for MLVs codisplaying anti-CEA scFv-envelope chimeric proteins in conjunction with unmodified envelope proteins (Konishi et al., 1998; Kuroki et al., 2000; Khare et al., 2001).

Other studies showed that the type of receptor used for targeting also plays a critical role. For example, studies with chimeric ecotropic MLV envelope proteins containing an anti-CD34 scFv showed that binding to $CD34^+$ cells did not catalyze virus–cell fusion (Benedict et al., 1999). Similar results were obtained with an anti-EGFRvIII-specific scFv (Lorimer and Lavictoire, 2000). Additional experiments indicated that the insertion site and the length of the peptide spacer between the scFv fragment and the envelope protein can play a critical role for target cell binding and in enhancement of infectivity (Ager et al., 1996; Kayman et al., 1999).

Further experiments have used scFvs to extend the tropism of amphotropic MLV. In one such experiment, the tropism was extended to melanoma cells by generating chimeric envelopes through fusion of an HMW-MAA-specific scFv to the N-terminus of the envelope protein (Martin et al., 1998). The transduction efficiency of these viruses could be increased by incorporating a protease cleavage site between the scFv and the envelope protein. Thus, after binding to the target cell mediated by the scFv, tumor cell-specific proteinases, such as matrix metalloproteinases, cleave the scFv from the virus envelope allowing infection of the cells (Martin et al., 1999). Other studies have successfully employed spleen necrosis virus (SNV), an avian reticuloendotheliosis virus, for targeting by fusing scFv fragments to the transmembrane (TM) subunit of the viral envelope protein (Env) (Chu et al., 1994; Chu and Dornburg, 1995, 1997; Jiang et al., 1998; Engelstädter et al., 2000). Depending on the scFv applied, wild-type envelope proteins had to be present to infect human cells with significant efficiencies (Jiang et al., 1998). Interestingly, display of an anti-Her2neu scFv led to high levels of infectivity even in the absence of wild-type envelope proteins, but coexpression of wild-type Env further increased the efficiency of infection. Using this virus, selective gene transfer to human cells expressing this receptor was also observed in vivo (Jiang and Dornburg, 1999). Recently, specific targeting has also been reported for other viruses displaying scFv fragments on the viral envelope, including adeno-associated virus (AAV) (Yang et al., 1998), a subgroup A avian leukosis virus (Snitkovsky et al., 2000), and measles virus (Hammond et al., 2001).

Targeting of Viral Vectors with Bifunctional and Bispecific Adaptor Molecules

Bifunctional and bispecific molecules combine a virus-specific binding moiety with a target cell-specific ligand. They are employed as molecular adaptors, which bind to the gene-therapeutic vector and direct it selectively to a target cell. Thus, they form noncovalently conjugated complexes. ScFvs can be used as target cell-specific ligands, but can also be applied for vector binding. Watkins and co-workers (1997) isolated scFvs directed against the adenovirus serotpye 5 knob domain from a murine immune library. One of the scFvs (S11), which was able to neutralize adenovirus infection, was fused to EGF. This fusion protein ablated the native adenovirus tropism and redirected infection to EGF receptor-expressing tumor cells. In a similar approach, the same antiadenovirus scFv was fused to a peptide selected against human umbilical vein endothelial cells (HUVEC) (Nicklin et al., 2000). The fusion protein mediated specific retargeting of adenoviral gene delivery to endothelial cells in vitro with enhanced efficiency compared to untargeted adenoviruses. Two further studies have described the use of scFv S11 for the construction of recombinant bispecific antibodies. Haisma and co-workers (2000) generated a tandem scFv molecule fusing scFv S11 to an anti-EGF receptor scFv. This molecule markedly enhanced the infection of EGF receptor-expressing tumor cell lines. To overcome the problem of heterogeneous expression of the EGF receptor, this bispecific molecule was combined with a genetically modified αv-integrin-specific adenovirus vector containing an RGD-4C peptide in the fiber knob domain. This combination further enhanced gene transfer into glioma cells in an additive manner (Grill et al., 2001). In another study, scFv S11 was combined with an antiendoglin scFv in a bispecific single-chain diabody (Nettelbeck et al., 2001). In contrast to tandem scFvs, which are often difficult to express in prokaryotic systems, single-chain diabodies (scDbs) can be produced in bacterial cells (Brüsselbach et al., 1999). The single-chain diabody mediated enhanced transduction of endoglin-expressing proliferating endothelial cells in vitro.

Bacteriophage Displaying ScFvs As Gene-Therapeutic Vectors

As described previously, bacteriophage displaying ligands can be selected for cell-surface receptor-mediated internalization into mammalian cells. Of note, internalization can lead to the expression of a transgene. Thus, it has been proposed to directly use bacteriophage as gene-therapeutic vehicle (see chapters 19 and 20). Initial experiments described the successful gene delivery with noncovalently attached fibroblast growth factor 2 (FGF2) (Larocca et al., 1998). Subsequently, phage displaying ligands by direct fusion to gene III protein were applied. In addition to various natural ligands, such as FGF2 and EGF, a recent study reported the use of phage displaying a scFv fragment directed against erbB2, which were selected for internalization (Poul and Marks et al., 1999). The current limitation of this system is the low transduction efficiency, with only a few percent of the cells transduced in vitro. Improvements of the system include the use of multivalent phagemid systems, which combine increased internalization and transduction efficiency with the advantage of larger insert size and vector stability (Larocca et al., 2001).

TARGETING MOIETIES DERIVED FROM IMMUNOGLOBULIN ANTIGEN-BINDING SITES

Various approaches have been described to reduce the size of the immunoglobulin-derived antigen-binding sites, as smaller molecules should be more stable and easier to produce and to incorporate into gene therapeutic vectors. One such approach represents the generation of single-domain antibodies, utilizing either V_H or V_L domains (Nuttall, Irving, and Hudson, 2000). The problem to be overcome is the reduced solubility and nonspecific interaction of these single domains exposing hydrophobic residues of the normally occupied V_H/V_L interface. Solubility can be increased by replacing these residues by hydrophilic residues. An interesting alternative was derived from the discovery of natural antibodies in camelids consisting only of heavy chains (Muyldermans, Cambillau, and Wyns, 2001). Camelizing human V_H domains by substituting human interface residues with those used in camel V_H domains resulted in soluble fragments. Camelized V_H domains were used as scaffolds to generate antibody libraries, for example, containing randomized CDR loops, and to select specific binders (Tanha et al., 2001). Single-domain antibody libraries were also generated from human V_L domains. Van den Beucken and co-workers (2001) generated a V_L library from human B cells, which was spiked with the CDR3-like loop of CTLA-4 and further diversified by DNA shuffling. From this library, novel B7.1- and B7.2-specific V_L domains were isolated which were expressed as high soluble monomers in bacteria. A further reduction in size is achieved by using only parts of the antigen-binding site. This site is formed by the six CDRs of the V_H and V_L domains (see also Chapter 26). Various studies have shown that only a limited number of CDR residues interact with an epitope. By identifying these residues and keeping them in a context that conserves their orientation it should be possible to obtain small antigen-binding peptides. Compared to antigen-binding proteins such as scFvs or single-domain antibodies, large amounts of these peptides can be produced synthetically. In addition, they can be easily fused or coupled to gene therapeutic vectors and should have the advantage of being less immunogenic. Several studies have shown that peptides derived from CDR sequences indeed possess antigen-binding activity, although with much lower affinities compared to the parental antibodies (Dougall, Peterson, and Greene, 1994). The affinities of such peptides can be increased by dimerization or by constraining the confirmation through the introduction of cysteine bridges (Williams et al., 1991). Furthermore, conformationally restricted cyclic organic peptides containing the relevant contact residues can serve as mimetics of antibody CDRs (Saragovi et al., 1991). The feasibility of CDR-derived peptides for vector targeting was demonstrated for a peptide derived from the CDR-2-like region of CD4 (Slepushkin et al, 1996). Covalent coupling of these peptides to liposomes resulted in specific binding to human immunodeficiency virus (HIV)-infected cells by binding to the HIV envelope protein gp120.

CONCLUSION

The combination of single-chain Fv fragments as target cell-binding molecules with the phage display technology has led to powerful strategies to obtain scFv molecules which can be applied for vector targeting in gene therapy. Most of these strategies involve selection of scFvs against the target antigen or cell and subsequent screening for those scFvs appropriate for vector targeting. Several newly developed methods also allow for the direct selection of scFv molecules with the desired targeting properties. The incorporation of these scFvs into gene therapeutic vectors results in targeted delivery systems. In addition, single-chain Fv fragments as antigen-binding entities can easily be modified and used in a modular way to combine them with additional moieties, either by genetic engineering or by chemical coupling to generate targeting molecules. Thus, single-chain Fv fragments offer numerous possibilities for application in vector targeting for both viral and nonviral systems.

REFERENCES

Ager S, Nilson BH, Morling FJ, Peng KW, Cosset FL, Russell SJ (1996): Retroviral display of antibody fragments: Interdomain spacing strongly influences vector infectivity. *Hum Gene Ther* 7:2157–2164.

Andersen PS, Stryhn A, Hansen BE, Fugger L, Engberg J, Buus S (1996): A recombinant antibody with the antigen-specific, major histocompatibility complex-restricted specificity of T cells. *Proc Natl Acad Sci USA* 93:1820–1824.

Becerril B, Poul M-A, Marks JD (1999): Toward selection of internalizing antibodies from phage libraries. *Biochem Biophys Res Comm* 255:386–393.

Benedict CA, Tun RY, Rubinstein DB, Guillaume T, Cannon PM, Anderson WF (1999): Targeting retroviral vectors to CD34-expressing cells: binding to CD34 does not catalyze virus-cell fusion. *Hum Gene Ther* 10:545–557.

Bobrow MN, Harris TD, Shaughnessy KJ, Litt GJ (1989): Catalyzed reported deposition, a novel method of signal amplification. *J Immunol Meth* 125:279–285.

Bradbury A (2001): Expression of antibodies in mammalian cells. In *Antibody Engineering*, Kontermann, RE, Dübel S, eds. Heidelberg: Springer-Verlag, pp. 357–366.

Brüsselbach S, Korn T, Völkel T, Müller R, Kontermann RE (1999): Enzyme recruitment and tumor cell killing in vitro by a secreted bispecific single-chain diabody. *Tumor Target* 4:115–123.

Cai X, Garen A (1995:) Anti-melanoma antibodies from melanoma patients immunized with genetically modified autologous tumor cells: Selection of specific antibodies from single-chain Fv fusion phage libraries. *Proc Natl Acad Sci USA* 92:6537–6541.

Chames P, Hoogenboom HR, Henderikx P (2001): Selections on biotinylated antigens. In *Antibody Engineering*, Kontermann RE, Dübel S, eds. Heidelberg: Springer-Verlag, pp. 149–166.

Chowdhury PS, Viner JL, Beers R, Pastan I (1998): Isolation of a high-affinity stable single-chain Fv specific for mesothelin from DNA-immunized mice by phage display and construction of a recombinant immunotoxin with anti-tumor activity. *Proc Natl Acad Sci USA* 95:669–674.

Chu T-HT, Martinez I, Sheay WC, Dornburg R (1994): Cell targeting with retroviral vector particles containing antibody-envelope fusion proteins. *Gene Ther* 1:292–299.

Chu T-HT, Dornburg R (1995): Retroviral vector particles displaying the antigen-binding site of an antibody enable cell-type-specific gene transfer. *J Virol* 69:2659–2663.

Chu T-HT, Dornburg R (1997): Toward highly efficient cell-type-specific gene transfer with retroviral vectors displaying single-chain antibodies. *J Virol* 71:720–725.

Conrad U, Fiedler U (2001): Expression of antibody fragments in plant cells. In *Antibody Engineering*, Kontermann RE, Dübel S, eds. Heidelberg: Springer-Verlag, pp. 367–382.

de Kruif JD, Logtenberg T (1995): Selection and application of human single-chain Fv antibody fragments from a semi-synthetic phage antibody library with designed CDR3 regions. *J Mol Biol* 248:97–105.

de Kruif J, Storm G, van Bloois L, Logtenberg T (1996): Biosynthetically lipid-modified human scFv fragments from phage display libraries as targeting molecules for immunoliposomes. *FEBS Lett* 399:232–236.

de Kruif JD, Terstappen L, Boel D, Logtenberg T (1995): Rapid selection of cell subpopulation-specific human monoclonal antibodies from a semisynthetic phage antibody library. *Proc Natl Acad Sci USA* 92:3938–3942.

Desai SA, Wang X, Noronha EJ, Kageshita T, Ferrone S (1998): Characterization of human anti-high molecular weight-melanoma-associated antigen single-chain Fv fragments isolated from a phage display antibody library. *Cancer Res* 58:2417–2425.

Dougall WC, Peterson NC, Greene MI (1994): Antibody-structure-based design of pharmacological agents. *Trends Biotechnol* 12:372–379.

Edwards BM, Main SH, Cantone KL, Smith SD, Warford A, Vaughan TJ (2000): Isolation and tissue profiles of a large panel of phage antibodies binding to the human adipocyte cell surface. *J Immunol Meth* 245:67–78.

Engelstädter M, Bobkova M, Baier M, Stitz J, Holtkamp N, Chu T-HT, Kurth R, Dornburg R, Buchholz CJ, Cichutek K (2000): Targeting human T cells by retroviral vectors displaying antibody domains selected from a phage display library. *Hum Gene Ther* 11:293–303.

Felgner PL, Barenholz Y, Behr JP, Cheng SH, Cullis P, Huang L, Jessee JA, Seymour L, Szoka F, Thierry AR, Wagner E, Wu G (1997): Nomenclature for synthetic gene delivery systems. *Hum Gene Ther* 8:511–512.

Fiedler M, Skerra A (2001): Purification and characterisation of His-tagged antibody fragments. In *Antibody Engineering*, Kontermann RE, Dübel S, eds. Heidelberg: Springer-Verlag, pp. 243–256.

Fominaya J, Wels W (1996): Target cell-specific DNA transfer mediated by a chimeric multidomain protein. *J Biol Chem* 271:10560–10568.

Glockshuber R, Malia M, Pfitzinger I, Plückthun A (1990): A comparison of strategies to stablize immunoglobulin Fv-fragments. *Biochemistry* 29:1362–1367.

Gram H, Ridder R (2001): Expression of scFv antibody fragments in the yeast *Pichia pastoris*. In *Antibody Engineering*, Kontermann RE, Dübel S, eds. Heidelberg: Springer-Verlag, pp. 321–333.

Grill J, van Beusechem VW, van der Valk P, Dirven CM, Leonhart A, Pherai DS, Haisma HJ, Pinedo HM, Curiel DT, Gerritsen WR (2001): Combined targeting of adenoviruses to integrins and epidermal growth factor receptors increases gene transfer into primary glioma cells and spheroids. *Clin Cancer Res* 7:641–650.

Gupta S, Eastman J, Silski C, Ferkol T, Davis PB (2001). Single chain Fv: a ligand in receptor-mediated gene delivery. *Gene Ther* 8:586–592.

Haisma HJ, Grill J, Curiel DT, Hoogeland S, van Beusechem VW, Pinedo HM, Gerritsen WR (2000): Targeting of adenoviral vectors through a bispecific single-chain antibody. *Cancer Gene Ther* 7:901–904.

Hammond AL, Plemper RK, Zhang J, Schneider U, Russell SJ, Cattaneo R (2001): Single-chain antibody displayed on a recombinant measles virus confers entry through the tumor-associated carcinoembryonic antigen. *J Virol* 75:2087–2096.

Hawkins RE, Russell SJ, Winter G (1992): Selection of phage antibodies by binding affinity. Mimicking affinity maturation. *J Mol Biol* 226:889–896.

Heitner T, Moor A, Garrison JL, Marks C, Hasan T, Marks JD (2001): Selection of cell binding and internalizing epidermal growth factor receptor antibodies from a phage display library. *J Immunol Meth* 248:17–30.

Henderikx P, Kandilogiannaki M, Petrarca C, von Mensdorff-Pouilly S, Hilgers JK, Krambovitis E, Arends JW, Hoogenboom HR (1998): Human single-chain Fv antibodies to MUC1 core peptide selected from phage display libraries recognize unique epitopes and predominantly bind adenocarcinoma. *Cancer Res* 58:4324–4332.

Hombach A, Pohl C, Heuser C, Sircar R, Diehl V, Abken H (1998): Isolation of single chain antibody fragments with specificity for cell surface antigens by phage display utilizing internal image anti-idiotypic antibodies. *J Immunol Meth* 218:53–61.

Hoogenboom HR, Lutgerink JT, Pelsers MM, Rousch MJ, Coote J, van Neer N, de Bruine A, van Niewenhoven FA, Glatz JF, Arends JW (1999): Selection-dominant and nonaccessible epitopes on cell-surface receptors revealed by cell-panning with a large phage antibody library. *Eur J Biochem* 260:774–784.

Huie MA, Cheung M-C, Muench MO, Becerril B, Kan YW, Marks JD (2001): Antibodies to human fetal erythroid cells from a nonimmune phage antibody library. *Proc Natl Acad Sci USA* 98:2682–2687.

Jiang A, Chu T-HT, Nocken F, Cichutek K, Dornburg R (1998): Cell-type-specific gene transfer into human cells with retroviral vectors that display single-chain antibodies. *J Virol* 72:10148–10156.

Jiang A, Dornburg R (1999): In vivo cell type-specific gene delivery with retroviral vectors that display single chain antibodies. *Gene Ther* 6:1982–1987.

Johns M, George AJT, Ritter MA (2000): In vivo selection of sFv from phage display libraries. *J Immunol Meth* 239:137–151.

Kahre PD, Shao-Xi L, Kuroki M, Hirose Y, Arakawa F, Nakamura K, Tomita Y, Kuroki M (2001): Specifically targeted killing of carcinoembryonic antigen (CEA)-expressing cells by a retroviral vector displaying single-chain variable fragmented antibody to CEA and carrying the gene for inducible nitric oxide synthase. *Cancer Res* 61:370–375.

Kayman SC, Park H, Saxon M, Pinter A (1999): The hypervariable domain of the murine leukemia virus surface protein tolerates large insertions and deletions, enabling development of a retroviral particle display system. *J Virol* 73:1802–1808.

Keinänen K, Laukkanen M-L (1994): Biosynthetic lipid-tagging of antibodies. *FEBS Lett* 346:123–126.

Kipriyanov SM, Dübel S, Breitling F, Kontermann RE, Little M (1994): Recombinant single-chain Fv fragments carrying C-terminal cysteine residues: production of bivalent and biotinylated miniantibodies. *Mol Immunol* 31:1047–1058.

Kipriyanov SM, Dübel S, Breitling F, Kontermann RE, Heynann S, Little M (1995): Bacterial expression and refolding of single-chain Fv fragments with C-terminal cysteines. *Cell Biophys* 26:187–204.

Konishi H, Ochiya T, Chester KA, Begent RH, Muto T, Sugimura T, Terada M (1998): Targeting strategy for gene delivery to carcinoembryonic antigen-producing cancer cells by retrovirus displaying a single-chain variable fragment antibody. *Hum Gene Ther* 9:235–248.

Kontermann RE (2001): Immunotube selections. In *Antibody Engineering*, Kontermann RE, Dübel S, eds. Heidelberg: Springer-Verlag, pp. 137–148.

Kupsch JM, Tidman NH, Kang NV, Truman H, Hamilton S, Patel N, Newton Bishop JA, Leigh IM, Crowe JS (1999): Isolation of human tumor-specific antibodies by selection of an antibody phage library on melanoma cells. *Clin Cancer Res* 5:925–931.

Kuroki M, Arakawa F, Khare PD, Kuroki M, Liao S, Matsumoto H, Abe H, Imakire T (2000): Specific targeting strategies of cancer gene therapy using a single-chain variable fragment (scFv) with a high affinity for CEA. *Anticancer Res* 20:4067–4071.

Larocca D, Jensen-Pergakes K, Burg MA, Baird A (2001): Receptor-targeted gene delivery using multivalent phagemid particles. *Mol Ther* 3:476–484.

Larocca D, Witte A, Johnson W, Pierce GF, Baird A (1998): Targeting bacteriophage to mammalian cell surface receptors for gene delivery. *Hum Gene Ther* 9:2393–2399.

Laukkanen M-L, Alfthan K, Keinänen K (1994): Functional immunoliposomes harboring a biosynthetically lipid-tagged single-chain antibody. *Biochemistry* 33:11664–11670.

Laukkanen M-L, Teeri TT, Keinänen K (1993): Lipid-tagged antibodies: bacterial expression and characterization of a lipoprotein-single-chain antibody fusion protein. *Protein Eng* 6:449–454.

Lekkerkerker A, Logtenberg T (1999): Phage antibodies against human dendritic cell subpopulations obtained by flow cytometry-based selection on fresh isolated cells. *J Immunol Meth* 231:53–63.

Liang M, Dübel S (2001): Production of recombinant human IgG antibodies in the baculovirus expression system. In *Antibody Engineering*, Kontermann RE, Dübel S, eds. Heidelberg: Springer-Verlag, pp. 334–356.

Liu B, Marks JD (2000): Applying phage antibodies to proteomics: selecting single chain Fv antibodies to antigens blotted on nitrocellulose. *Anal Biochem* 286:119–128.

Lorimer IAJ, Keppler-Hafkemeyer A, Beers RA, Pegram CN, Bigner DD, Pastan I (1996): Recombinant immunotoxins specific for a mutant epidermal growth factor receptor: targeting with a single-chain antibody variable domain isolated by phage display. *Proc Natl Acad Sci USA* 93:14815–14820.

Lorimer IAJ, Lavictoire SJ (2000): Targeting retrovirus to cancer cells expressing a mutant EGF receptor by insertion of a single chain antibody variable domain in the envelope glycoprotein receptor binding lobe. *J Immunol Meth* 237:147–157.

Marget M, Sharma BB, Tesar M, Kretzschmar T, Jenisch S, Westphal E, Davarnia, P, Weiss E, Ulbrecht M, Kabelitz D, Krönke M (2000): Bypassing hybridoma technology: HLA-C reactive human single-chain antibody fragments (scFv) derived from a synthetic phage display library (HuCAL) and their potential to discriminate HLA class I specificities. *Tissue Antigens* 56:1–9.

Marin M, Noël D, Valsesia-Wittman S, Brockly F, Etienne-Julan M, Russell S, Cosset FL, Piechaczyk M. (1996): Targeted infection of human cells via major histocompatibility complex class I molecules by Moloney murine leukemia virus-derived viruses displaying single-chain antibody fragment-envelope fusion proteins. *J Virol* 70:2957–2962.

Marks JD, Ouwehand WH, Bye JM, Finnern R, Gorick BD, Voak D, Thorpe SJ, Hughes-Jones NC, Winter G (1993): Human antibody fragments specific for human blood group antigens from a phage display library. *Bio/Technol* 11:1145–1149.

Mao S, Gao C, Lo C-HL, Wirsching P, Wong CH, Janda KD (1999): Phage-display library selection of high-affinity human single-chain antibodies to tumor-associated carbohydrate antigens sialyl Lewis^x and Lewis^x. *Proc Natl Acad Sci USA* 96:6953–6958.

Martin F, Kupsch J, Takeuchi Y, Russell S, Cosset FL, Collins M (1998): Retroviral vector targeting to melanoma cells by single-chain antibody incorporation in envelopes. *Hum Gene Ther* 9:737–746.

Martin F, Neil S, Kupsch J, Maurice M, Cosset F, Collins M (1999): Retrovirus targeting by tropism restriction to melanoma cells. *J Virol* 73:6923–6929.

Mohr L, Schauer JI, Boutin RH, Moradpour D, Wands JR (1999): Targeted gene transfer to hepatocellular carcinoma cells in vitro using a novel monoclonal antibody-based gene delivery system. *Hepatology* 29:82–89.

Muyldermans S, Cambillau C, Wyns L (2001): Recognition of antigens by single-domain antibody fragments: the superfluous luxury of paired domains. *Trends Biochem Sci* 26:230–235.

Mutuberria R, Hoogenboom HR, van der Linden E, de Bruine AP, Roovers RC (1999): Model systems to study the parameters determining the success of phage antibody selections on complex antigens. *J Immunol Meth* 231:65–81.

Nettelbeck DM, Miller DW, Jérôme V, Zuzarte M, Watkins SJ, Hawkins RE, Müller R, Kontermann RE (2001): Targeting of adenovirus to endothelial cells by a bispecific single-chain diabody directed against the adenovirus fiber knob domain and human endoglin (CD105). *Mol Ther* 3:882–891.

Nicklin SA, White SJ, Watkins SJ, Hawkins RE, Baker AH (2000): Selective targeting of gene transfer to vascular endothelial cells by use of peptides isolated by phage display. *Circulation* 102:231–237.

Nielsen UB, Marks JD (2000): Internalizing antibodies and targeted cancer therapy: direct selection from phage display libraries. *Pharm Sci Technol Today* 3:282–291.

Nielsen UB, Marks JD (2001a): Affinity maturation by chain shuffling and site directed mutagenesis. In *Antibody Engineering*, Kontermann RE, Dübel S, eds. Heidelberg: Springer-Verlag, pp. 515–539.

Nielsen UB, Marks JD (2001b): Selection of phage antibody libraries for binding and internalization into mammalian cells. In *Antibody Engineering*, Kontermann RE, Dübel S, eds. Heidelberg: Springer-Verlag, pp. 234–240.

Nilsson F, Kosmehl H, Zardi L, Neri D (2001): Targeted delivery of tissue factor to the ED-B domain of fibronectin, a marker of angiogenesis, mediates the infarction of solid tumors in mice. *Cancer Res* 61:711–716.

Noronha EJ, Wang X, Desai SA, Kageshita T, Ferrone S (1998): Limited diversity of human scFv fragments isolated by panning a synthetic phage-display scFv library with cultured human melanoma cells. *J Immunol* 161:2968–2976.

Nuttall SD, Irving RA, Hudson PJ (2000): Immunoglobulin VH domains and beyond: design and selection of single-domain binding and targeting reagents. *Curr Pharm Biotechnol* 1:253–263.

O'Neill MM, Kennedy CA, Barton RW, Tatake RJ (2001): Receptor-mediated gene delivery to human peripheral blood mononuclear cells using anti-CD3 antibody coupled to polyethylenimine. *Gene Ther* 8:362–368.

Osbourn JK (2001): Proximity (ProxiMol) and step-back selections. In *Antibody Engineering*, Kontermann RE, Dübel S, eds. Heidelberg: Springer-Verlag, pp. 184–192.

Osbourn JK, Derbyshire EJ, Vaughan TJ, Field AW, Johnson KS (1998a): Pathfinder selection; in situ isolation of novel antibodies. *Immunotechnology* 3:293–302.

Osbourn JK, Earnshaw JC, Johnson KS, Parmentier M, Timmermans V, McCafferty J (1998b): Directed selection of MIP-1α neutralizing CCR5 antibodies from a phage display human antibody library. *Nat Biotechnol* 16:778–781.

Osbourn JK, Field A, Wilton J, Derbyshire E, Earnshaw JC, Jones PT, Allen D, McCafferty J (1996): Generation of a panel of related human scFv antibodies with high affinity for human CEA. *Immunotechnology* 2:181–196.

Palmer DB, George AJ, Ritter MA (1997): Selection of antibodies to cell surface determinants on mouse thymic epithelial cells using a phage display library. *Immunology* 91:473–478.

Park JW, Kirpotin DB, Hong K, Shalaby R, Shao Y, Nielsen UB, Marks JD, Papahadjopoulos D, Benz CC (2001): Tumor targeting using anti-her2 immunoliposomes. *J Control Rel* 74:95–113.

Pasqualini R, Koivunen E, Ruoslahti E (1997): Alpha$_V$-integrins as receptors for tumor targeting by circulating ligands. *Nat Biotechnol* 15:542–546.

Peipp M, Simon N, Loichinger A, Baum W, Mahr K, Zunino SJ, Fey GH (2001): An improved procedure for the generation of recombinant single-chain Fv antibody fragments reacting with human CD13 on intact cells. *J Immunol Meth* 251:161–176.

Pelegrin M, Marin M, Noël D, Piechaczyk M (1998): Genetically engineered antibodies in gene transfer and gene therapy. *Hum Gene Ther* 9:2165–2175.

Pereira S, Maruyama H, Siegel D, van Belle P, Elder D, Curtis P, Herlyn D (1997): A model system for detection and isolation of a tumor cell surface antigen using antibody phage display. *J Immunol Meth* 203:11–24.

Pini A, Viti F, Santucci A, Carnemolla B, Zardi L, Neri P, Neri D (1998): Design and use of a phage display library: Human antibodies with subnanomolar affinity against a marker of angiogenesis eluted from a two-dimensional gel. *J Biol Chem* 273:21769–21776.

Poul M-A, Becerril B, Nielsen UB, Morisson P, Marks JD (2000): Selection of tumor-specific internalizing human antibodies from phage libraries. *J Mol Biol* 301:1149–1161.

Poul MA, Marks JD (1999): Targeted gene delivery to mammalian cells by filamentous bacteriophage. *J Mol Biol* 288:203–211.

Ravn P, Kjaer S, Jensen KH, Wind T, Jensen KB, Kristensen P, Brosh RM, Orren DK, Bohr VA, Clark BF (2000): Identification of phage antibodies toward the Werner protein by selection on Western blots. *Electrophoresis* 21:509–516.

Ridgway JBB, Ng E, Kern JA, Lee J, Brush J, Goddard A, Carter P (1999): Identification of a human anti-CD55 single-chain Fv by subtractive panning of a phage library using tumor and nontumor cell lines. *Cancer Res* 59:2718–2723.

Roovers RC, van der Linden E, de Bruine AP, Arends J-W, Hoogenboom HR (2001): Identification of colon tumour-associated antigens by phage antibody selections on primary colorectal carcinoma. *Eur J Cancer* 37:542–549.

Rossig C, Nuchtern JG, Brenner MK (2000): Selection of human antitumor single-chain Fv antibodies from the B-cell repertoire of patients immunized against autologous neuroblastoma. *Med Pediatr* 35:692–695.

Russell SJ, Cosset F-L (1999): Modifying the host range properties of retroviral vectors. *J Gene Med* 1:300–311.

Russell SJ, Hawkins RE, Winter G (1993): Retroviral vectors displaying functional antibody fragments. *Nucleic Acids Res* 21:1081–1085.

Sanna PP, Williamson RA, de Logu A, Bloom FE, Burton DR (1995): Directed selection of recombinant human monoclonal antibodies to herpes simplex virus glycoproteins from phage display libraries. *Proc Natl Acad Sci USA* 92:6439–6443.

Saragovi HU, Fitzpatrick D, Raktabutr A, Nakanishi H, Kahn M, Greene MI (1991): Design and synthesis of a mimetic from an antibody complementarity-determining region. *Science* 253:792–795.

Sawyer C, Embleton J, Dean C (1997): Methodology for selection of human antibodies to membrane proteins from a phage-display library. *J Immunol Meth* 204:193–203.

Schier R, Bye J, Apell G, McCall A, Adams GP, Malmqvist M, Weiner LM, Marks JD (1996a): Isolation of high-affinity monomeric human anti-c-erbB-2 single chain Fv using affinity-driven selection. *J Mol Biol* 255:28–43.

Schier R, McCall A, Adams GP, Marshall KW, Merritt H, Yim M, Crawford RS, Weiner LM, Marks C, Marks JD (1996b): Isolation of picomolar affinity anti-c-erbB-2 single-chain Fv by molecular evolution of the complementarity determining regions in the center of the antibody binding site. *J Mol Biol* 263:551–567.

Schmidt S, Braunagel M, Kürschner T, Little M (1999): Selection of an anti-CD20 single-chain antibody by phage ELISA on fixed cells. *BioTechniques* 26:697–702.

Shadidi M, Sioud M (2001): An anti-leukemic single chain Fv antibody selected from a synthetic human phage antibody library. *Biochem Biophys Res Commun* 280:548–552.

Siegel DL, Chang TY, Russell SL, Bunya VY (1997): Isolation of cell surface-specific human monoclonal antibodies using phage display and magnetically activated cell sorting: applications in immunohemtalogy. *J Immunol Meth* 296:73–85.

Skerra A, Plückthun A (1988): Assembly of functional immunoglobulin Fv fragments in *Escherichia coli. Science* 249:1038–1040.

Slepushkin VA, Salem II, Andreev SM, Dazin P, Düzgünes, N (1996): Targeting of liposomes to HIV-1-infected cells by peptides derived frm the CD4 receptor. *Biochem Biophys Res Comm* 227:827–833.

Snitkovsky S, Niederman TM, Carter BS, Mulligan RC, Young JA (2000): A TVA-single-chain antibody fusion protein mediates specific targeting of a subgroup A avian leukosis virus vector to cells expressing a tumor-specific form of epidermal growth factor receptor. *J Virol* 74:9540–9545.

Somia NV, Zoppe M, Verma IM (1995): Generation of targeted retroviral vectors by using single-chain variable fragment: An approach to in vivo gene delivery. *Proc Natl Acad Sci USA* 92:7570–7574.

Tanha J, Xu P, Chen Z, Ni F, Kaplan H, Narang SA, MacKenzie CR (2001): Optimal design features of camelized human single-domain antibody libraries. *J Biol Chem* 276:24774–24780.

Tordsson J, Abrahmsén L, Kalland T, Ljung C, Ingvar C, Brodin T (1997): Efficient selection of scFv antibody phage by adsorption to in situ expressed antigens in tissue sections. *J Immunol Meth* 210:11–23.

Tordsson J, Brodin T, Karlstöm PJ (2001): Selections on tissue sections. In *Antibody Engineering*, Kontermann RE, Dübel S, eds. Heidelberg: Springer-Verlag, pp. 193–205.

Tordsson J, Lavasani S, Ohlsson L, Karlström P, Svedberg H, Abrahmsen L, Brodin T (2000): A3—a novel colon and pancreatic cancer reactive antibody from a primate phage library selected using intact tumour cells. *Int J Cancer* 87:559–568.

Uherek C, Fominaya J, Wels W (1998): A modular DNA carrier protein based on the structure of diphtheria toxin mediates target cell-sepcific gene delivery. *J Biol Chem* 273:8835–8841.

Uherek C, Wels W (2001): Antibody fusion proteins for targeted gene delivery. In *Antibody Engineering*, Kontermann RE, Dübel S, eds. Heidelberg: Springer-Verlag, pp. 710–723.

Ullrich A, Schlessinger J (1990): Signal transduction by receptors with tyrosine kinase activity. *Cell* 61:203–212.

van den Beucken T, van Neer N, Sablon E, Desmet J, Celis L, Hoogenboom HR, Hufton SE (2001): Building novel binding ligands to B7.1 and B7.2 based on human antibody single variable light chain domains. *J Mol Biol* 310:591–601.

van Ewijk W, de Kruif J, Germeraad WTV, Berendes P, Röpke C, Platenburg PP, Logtenberg T (1997): Subtractive isolation of phage-displayed single-chain antibodies to thymic stromal cells by using intact thymic fragments. *Proc Natl Acad Sci USA* 94:3903–3908.

Watkins SJ, Mesyanzhinov VV, Kurochkina LP, Hawkins RE (1997): The 'adenobody' approach to viral targeting: specific and enhanced adenoviral gene delivery. *Gene Ther* 4:1004–1012.

Williams WV, Kieber-Emmons T, VonFeldt J, Greene MI, Weiner DB (1991): Design of bioactive peptides based on antibody hypervariable region structures. *J Biol Chem* 266:5182–5190.

Winthrop MD, DeNardo SJ, DeNardo GL (1999): Development of a hyperimmune anti-MUC-1 single chain antibody fragments phage display library for targeting breast cancer. *Clin Cancer Res* 5:3088s–3094s.

Yang Q, Mamounas M, Yu G, Kennedy S, Leaker B, Merson J, Wong-Staal F, Yu M, Barber JR (1998): Development of novel cell surface CD34-targeted recombinant adenoassociated virus vectors for gene therapy. *Hum Gene Ther* 9:1929–1937.

Zhou JN, Linder S, Franzen B, Auer G, Hochstrasser DF, Persson MA (1998): Rapid isolation of phage displayed antibodies to beta-actin eluted from two-dimensional electrophoresis gel. *Electrophoresis* 19:1808–1810.

28

RETROVIRAL PARTICLE DISPLAY FOR COMPLEX GLYCOSYLATED AND DISULFIDE-BONDED PROTEIN DOMAINS

SAMUEL C. KAYMAN, PH.D.

INTRODUCTION

A major limitation of current efforts to apply gene therapeutic approaches to widespread clinical practice is the difficulty in directing gene delivery to specific target cells in vivo (Curiel, 1999). One of the levels at which such targeting might be accomplished is via a specific binding interaction between the gene delivery vehicle and the appropriate cell type. Such an approach requires methods for identifying moieties having the desired binding characteristics. In recent years, a variety of systems have been developed for displaying libraries of peptide sequences such that those with specific binding properties can be isolated and identified by enrichment procedures. Systems using bacteriophage, and bacterial and yeast cell surfaces are reviewed elsewhere in this volume. This chapter describes efforts to use a retrovirus, murine leukemia virus (MLV), for particle display (Kayman et al., 1999).

This retroviral system utilizes a hypervariable, proline-rich region (PRR) of

Vector Targeting for Therapeutic Gene Delivery, Edited by David T. Curiel and Joanne T. Douglas
ISBN 0-471-43479-5

the MLV surface protein (SU) that tolerates large insertions, including domains containing multiple glycosylation signals and complex disulfide-bonded structures. When the recombinant retroviral genome is expressed in mammalian cells, heterologous protein domains inserted into the PRR of SU are glycosylated and fold efficiently into native structures. Even structures containing multiple disulfide bonds form correctly. Use of fusions to the N-terminus of MLV SU for enrichments based on retroviral particle display has also been described (Buchholz et al., 1998). The extensive processing of N-linked glycans that occurs in mammalian cells should allow the MLV retroviral particle display system to explore a larger range of structures than is available with yeast or baculovirus particle display systems (Grabherr et al., 2001). Thus, the retroviral system should be particularly well suited for identifying variants of protein domains that cannot be expressed in simpler systems or that need to be fine-tuned by taking advantage of the size and charge properties of complex N-linked glycans. The PRR of the MLV SU carries O-linked glycans (Linder et al., 1992), suggesting that protein domains that are normally O-glycosylated and perhaps those that carry other posttranslational modifications may also be accurately presented. The retroviral display system could also be adapted to present ligands for receptors in which lectinlike domains provide specificity based on the presence of particular carbohydrate structures that depend on the array of glycosyl transferases expressed by particular cell types (Vestweber and Blanks, 1999). Thus, the retroviral particle display system based on insertions into the PRR of SU provides a promising approach for the exploration of complex protein structures as ligands for specific binding interactions.

STRUCTURE AND FUNCTION OF MLV SU

MLV envelope proteins consist of a trimer of heterodimers of SU and a transmembrane protein (TM) (Hunter and Swanstrom, 1990; Kamps, Lin, and Wong, 1991). SU is a peripheral membrane protein that associates with the virion via interactions with the integral membrane protein TM; these interactions include a metastable disulfide bond between the subunits (Opstelten, Wallin, and Garoff, 1998; Pinter and Fleissner, 1977; Pinter et al., 1997). The fusion activity of MLV Env is believed to reside in TM, which contains a hydrophobic N-terminal putative fusion domain and heptad repeats thought to form coiled coil structures, both of which are characteristic of viral fusion proteins (Chambers, Pringle, and Easton, 1990). The cell surface receptors for a number of classes of MLV and other C-type mammalian retroviruses have been identified. All of these receptors span the membrane multiple times and have transporter functions (Weiss and Tailor, 1995), suggesting that properties of this class of membrane protein may be important in the fusion process (Valsesia-Wittmann et al., 1997).

The initial envelope expression product is a single precursor protein carrying high mannose N-linked glycans in the endoplasmic reticulum; during transport through the Golgi apparatus to the cell surface, N-linked glycans are processed to complex forms, O-linked glycans are added, and the precursor is cleaved into SU and TM subunits by a cellular protease (Pinter, 1989). Removal of a small peptide from the C-terminus of the cytoplasmic domain of TM, by an additional proteolytic cleavage that occurs during particle maturation, greatly enhances the fusogenic activity of the Env complex (Ragheb and Anderson, 1994; Rein et al., 1994).

The secondary structure of SU of ecotropic Friend MLV, which is used in the display system, consists of two internally disulfide-bonded globular domains connected by a cysteine-free, central domain that includes a PRR (Fig. 28.1).

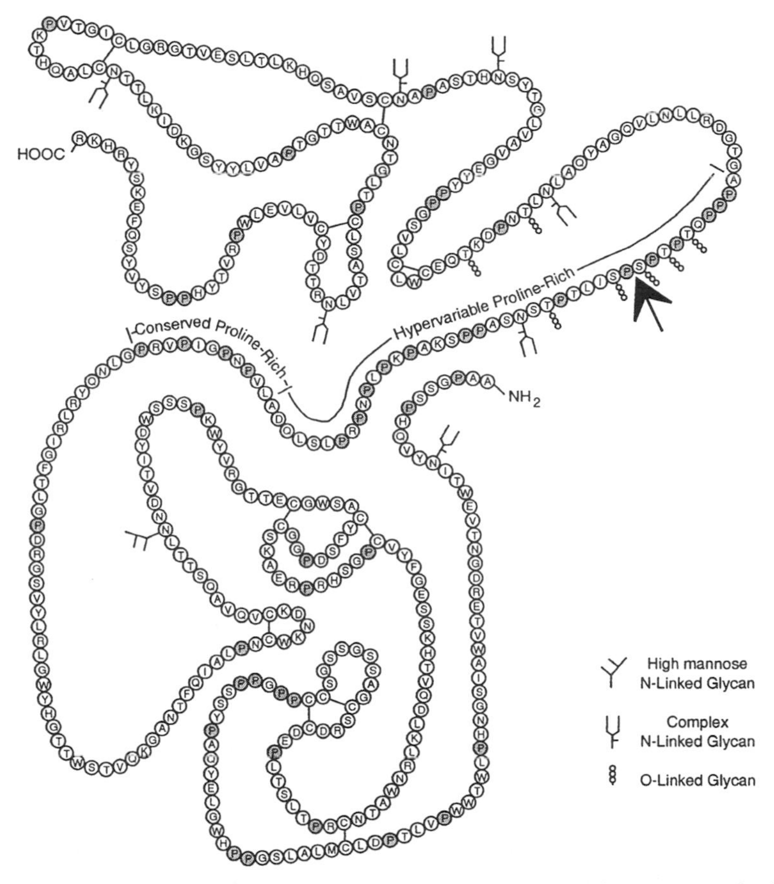

Figure 28.1 Secondary structure of Friend ecotropic SU. The sequence is that of clone 57 (Koch, Hunsmann, and Friedrich, 1983), used in the particle display vector; the disulfide bonding and O-linked glycosylation is according to Linder et al. (1992); assignment of the high mannose glycan is according to Kayman et al. (1991). The conserved and hypervariable regions of the PRR are labeled, and prolines are shaded; the 273/274 insertion site is indicated by an arrow.

The globular domains correspond to the products observed following limited proteolysis, during which cleavage occurs primarily in the central, cysteine-free region of SU (Pinter and Honnen, 1983; Pinter et al., 1982). The N-terminal domain of SU contains the receptor binding activity, the C-terminal domain is responsible for binding to TM, and the majority of the PRR appears to be a passive linker between the globular domains. The domain structure of SU and the high variability of the C-terminal portion of the PRR suggested that insertions into this region of SU might result in the display of heterologous sequences on the surface of infectious MLV particles.

The earliest indication that receptor binding activity resides in the N-terminal domain of SU came from sequence comparisons among MLV Envs (Battini, Heard, and Danos, 1992; Stoye and Coffin, 1987). MLV isolates fall into several different classes based on the species whose cells they infect, which in turn reflects the use of different cell surface molecules as receptors for attachment to cells. The C-terminal domain of SU and all of TM are highly conserved among all MLVs, while the C-terminal region of the PRR was highly variable even within receptor classes. However, the N-terminal domain of SU has three regions that are highly divergent between receptor classes but highly conserved within receptor classes, suggesting that these regions of the N-terminal domain determine receptor specificity. The primary role of the N-terminal domain in controlling receptor specificity was confirmed by construction of chimeras between Envs of different receptor classes. Receptor specificity correlates with the source of the N-terminal domain (Battini, Heard, and Danos, 1992; Morgan et al., 1993), although in some cases effects from the PRR are also seen (Battini, Danos, and Heard, 1995; Ott and A.Rein, 1992). Direct demonstration that receptor-binding activity resides in the N-terminal domain came from studies with N-terminal truncation fragments of ecotropic (Heard and Danos, 1991) and amphotropic (Battini et al., 1996) Envs. These N-terminal domain fragments bind to cells in a receptor-specific fashion, and they establish superinfection exclusion when expressed in target cells. Mutational analysis (Davey, Zuo, and Cunningham, 1999; MacKrell et al., 1996) and the crystal structure of an N-terminal truncation fragment of Friend ecotropic SU allowed identification of a putative receptor-binding pocket (Fass et al., 1997). Additional support for this receptor pocket comes from a neutralizing monoclonal antibody that blocks receptor binding and reacts with a conformational epitope expressed by N-terminal domain truncation fragments and involves Ser84 of the putative binding pocket (Burkhart, Kayman and Pinter, in prep.).

The central PRR can be divided into an N-terminal portion that is conserved among MLV Envs and a hypervariable C-terminal portion (Fig 28.1). Env function is sensitive to modification in the conserved region of the proline-rich domain (Wu et al., 1998), possibly due to disruption of a pattern of β-turns (Lavillette et al., 1998). In contrast, the hypervariable regions of the Friend ecotropic (Kayman et al., 1999) and the 4077A amphotropic (Wu et al., 1998) MLVs were shown to tolerate large insertions and deletions with only minor effects on Env function. Thus, the hypervariable portion of the PRR appears to be under little or no structural or functional constraint.

The C-terminal domain of SU plays a major role in the binding of SU to TM. MLV SU and TM are joined by a disulfide bond that is labile upon detergent lysis of virions (Pinter et al., 1997; Pinter, Lieman-Hurwitz, and Fleissner, 1978). This linkage resides in the C-terminal domain, because a C-terminal fragment of SU coprecipitates with TM under nonreducing conditions following limited proteolytic digestion of virions (Pinter et al., 1982). The cysteine(s) of SU involved in this disulfide have been mapped to one or both of the cysteines in the CWLC motif found near the beginning of the C-terminal domain (Pinter et al., 1997). This motif is conserved across a very broad group of retroviral Env proteins (Kayman et al., 1991; Sitbon et al., 1991). Mutation of either of these cysteines prevents folding and processing of the Env precursor (unpublished results), while mutations between the cysteines attenuate the pathogenicity of Friend ecotropic MLV (Sitbon et al., 1991). This motif is similar to the active site motif in disulfide isomerase enzymes, suggesting that MLV and related Envs may possess a disulfide exchange activity that is responsible for the lability of the intersubunit disulfide linkage and that rearrangement of the disulfide linkage between SU and TM may be involved in triggering fusion (Pinter et al., 1997).

COMPLEX INSERTS IN THE HYPERVARIABLE PRR OF MLV SU

A variety of complex domains from heterologous proteins have been inserted into the MLV SU at several sites within the hypervariable PRR. The inserted domains fold into native conformation and are displayed on the surface of virus particles. For some of the insertion sites tested, the chimeric Envs are processed efficiently and function normally during virus infection (Kayman et al., 1999). Among the domains successfully expressed in this way are single-chain antigen-binding domains (scFvs), the V1/V2 domain of HIV-1 gp120, and the V3 domain of HIV-1 gp120.

ScFvs contain approximately 250 amino acids that encompass the Fv domains of IgG light and heavy chains joined by a $(Gly_4Ser)_3$ linker. Because they fold efficiently in many contexts and expression systems, they provide a low-stringency test for folding of inserts into the PRR while demonstrating the large size of tolerated insertions. Particle-associated chimeric MLV SU carrying an scFv insert directed against a CD4 binding site epitope of human immunodeficiency virus type 1 (HIV-1) gp120 binds gp120 (Kayman et al., 1999), demonstrating that the scFv folds properly and is accessible on the surface of virions.

The V1/V2 domain of HIV-1 gp120 is a complex structure that involves 2 disulfide bonds and, depending on the viral isolate, contains 80 to 105 amino acids and four to eight signals for N-linked glycosylation (Kuiken et al., 1999). Efficient formation of the correct disulfide bonds requires addition of particular glycans (Wu et al., 1995). Thus, the V1/V2 domain provides a stringent test for the ability of insertions into the PRR to undergo proper glycosylation and correct folding. This domain folds into its native conformation and is exposed on the surface of virions when inserted into the PRR (Kayman et al., 1999), demonstrated by reactivity with a monoclonal antibody directed against a conformational epitope expressed only by V1/V2 domains possessing the correct pattern of disulfide bonds (Wu et al., 1995).

The V3 domain of HIV-1 gp120 is also expressed in native form when inserted into the hypervariable PRR of MLV SU. Immune reagents specific for the native conformation of this domain are not available, so biochemical criteria were used to characterize V3 domain inserts (Fig. 28.2). When a 46-amino acid V3 loop sequence containing the two cysteines that form the loop and 3 signals for N-linked glycosylation was taken from a laboratory-adapted X4-tropic isolate (lanes B), recovery of SU in virions was poor compared to the parental MLV Env (lanes A), a chimera carrying the same V3 domain with one of the cysteines mutated to serine (lanes C), or a chimeric Env carrying the equivalent sequence from an R5-tropic primary isolate HIV-1 (lanes D). The electrophoretic mobility of wild-type X4 V3 chimeric SU in nonreducing sodium dodecylsulfate polyacrylamide gel electrophoresis (SDS-PAGE) is similar to that of the other chimeric SUs, but this protein dissociates into two proteolytic fragments in the presence of DTT. Thus, the poor recovery of this chimera appears to be due to high sensitivity to an endogenous protease that cleaves it into disulfide-linked fragments.

This cleavage is reminiscent of in vivo V3 loop cleavage by a cellular protease that has been reported for viral gp120 and that can be modeled in vitro by

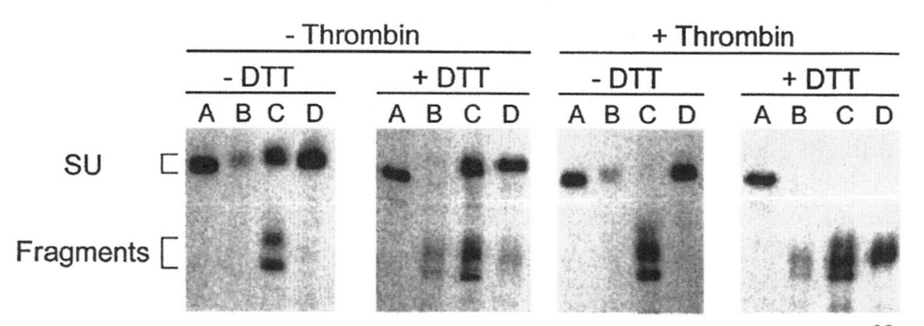

Figure 28.2 Properties of V3 domain insertions. Virus particles labeled with [^{35}S]-cysteine were concentrated by centrifugation and separated from soluble proteins by Sepharose CL4B column chromatography. Samples were immunoprecipitated with polyclonal antiserum against MLV SU, with and without pretreatment with thrombin, analyzed by SDS-PAGE in the presence and absence of reducing agent, and visualized by autoradiography. *A:* wild-type Env; *B:* chimeric Env carrying wild-type V3 domain from HIV-1 HXB2, a TCLA isolate; *C:* chimeric Env carrying Cys → Ser mutant V3 domain from HIV-1 HXB2; *D:* chimeric Env carrying wild-type V3 domain from HIV-1 JR-CSF, an R5 primary isolate.

cleavage with thrombin (Clements et al., 1991). Treatment of virions bearing each of the V3 chimeras with thrombin in vitro resulted in complete cleavage of the chimeric SU into fragments similar to those seen for the X4 V3 chimera (+Thrombin, + DTT lanes in Fig. 28.2), consistent with the cleavage of the V3 chimeras being analogous to that reported for HIV-1 viral gp120. The sizes of the proteolytic fragments are also consistent with cleavage occurring within the V3 inserts, between the N- and C-terminal domains of SU. There are no disulfide bonds linking the N- and C-terminal domains of MLV SU. Therefore, if these cleavages occur within the V3 loop inserted between the globular domains of SU, the V3 loop disulfide bond would provide the only linkage between the fragments. The cleavage products of the mutant X4 V3 chimera in which the loop was linearized by mutation of one of the cysteines migrates as the two fragments even without reduction by DTT. Thus, the cleavage site is within the V3 loop. The formation of the V3 disulfide bond was highly efficient as indicated by the observation that no detectable fragments were released from the wild-type X4 chimera, or from the R5 V3 chimera following thrombin cleavage, without reduction by DTT.

Although essentially all of the wild-type X4 V3 chimera is cleaved by the endogenous cellular protease, only a small fraction of the Cys → Ser X4 V3 chimera is cleaved. The decreased proteolysis of the mutant chimera indicates that the disulfide bond at the base of the V3 loop constrains this domain into a conformation that is more efficiently cleaved. This is consistent with modeling that indicates that some of the possible conformations of the V3 loop are more compatible than others with binding at the active site of thrombin (Johnson et al., 1994). The R5 V3 chimera is the most resistant to cleavage, both by the endogenous protease in vivo (Fig. 28.2) and by thrombin in vitro (data not shown). This pattern parallels the relative sensitivity of viral gp120s from laboratory-adapted X4 isolates compared to that of R5 gp120s (Gu, Westervelt, and Ratner, 1993; Werner and Levy, 1993). Taken together, these data demonstrate that the V3 loop disulfide forms with high efficiency in the MLV SU chimeras and suggest that subtle aspects of V3 domain conformation are faithfully reproduced.

PRR INSERTION CHIMERAS AS A RETROVIRAL PARTICLE DISPLAY SYSTEM

The data summarized previously show that complex protein domains bearing N-linked glycans and multiple disulfide bonds are accurately expressed and presented on the surface of virus particles when inserted into the PRR of MLV SU. We are therefore exploring the utilitity of such chimeras for enrichments based on binding properties of the inserts, according to the scheme diagrammed in Fig. 28.3. Insertions following residue 273 in the PRR of Friend MLV SU (see Fig. 28.1) are used because this position is the most tolerant of large insertions (Kayman et al., 1999). The plasmid vector pDisplay273 encodes a colin-

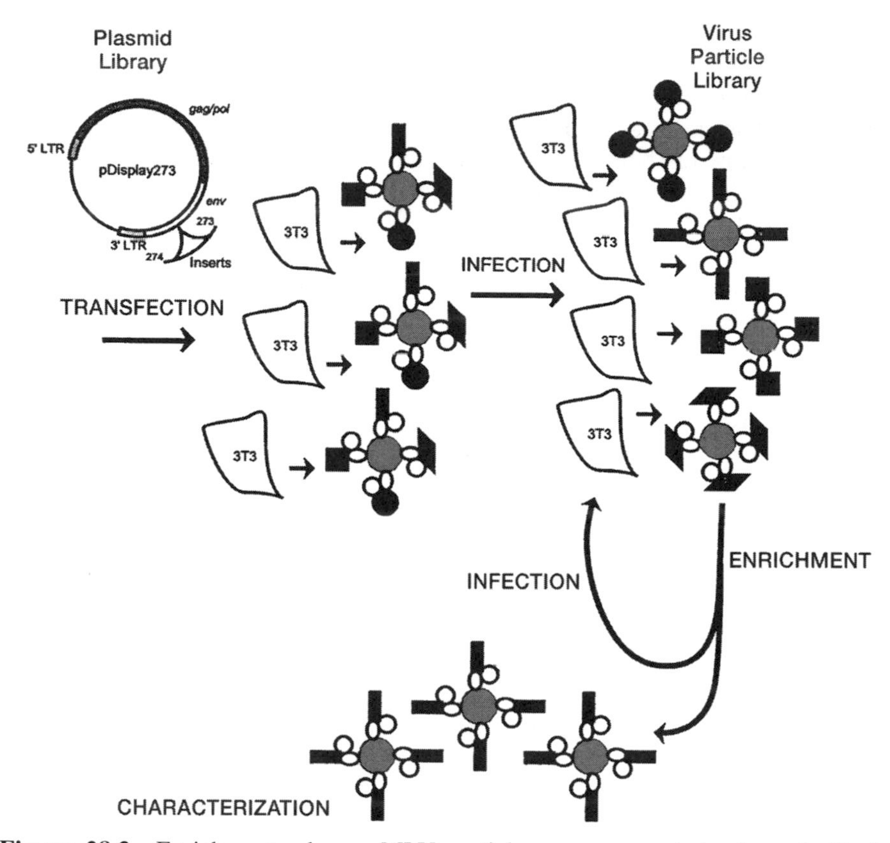

Figure 28.3 Enrichment scheme. MLV particles are represented schematically by gray circles; MLV Env complex by open ovals; inserted protein domains by filled shapes, with different shapes indicating different inserts.

ear 2 LTR Friend ecotropic MLV genome containing unique *Nhe*I and *Ngo*MI restriction sites for insertion of foreign sequences between residues 273 and 274 of SU, which result in AlaSer and GlyAla dipeptides at the N- and C-terminal junctions, respectively.

To use this system to screen structures, a library of variant sequences would be prepared in pDisplay273 by insertion between the *Nhe*I and *Ngo*MI sites. A plasmid stock of this library would then be transfected into 3T3 cells under conditions that yield a large number of transfectants, each of which will have taken up and expressed multiple copies of the viral plasmid. The virus produced by the transfected cells would then be amplified by being allowed to spread through the cell culture by infection, which would also introduce additional variation into the library due to the high error rate of reverse transcriptase. Because expression from multiple viral genomes is required to establish effective superinfection interference (Odawara et al., 1998), most cells in an outgrown culture would express many different chimeric Envs. Thus, the resulting stock of the viral display library would be unsuitable for use in enrichments because any given virus particle would display multiple different chimeric Envs. In order to obtain virions with only a single insert sequence displayed per particle and strong linkage between the insert displayed and the insert encoded by the viral genome, the stock of particles used for enrichment must derive from a controlled, single round of infection initiated with the outgrown library stock at a multiplicity of infection less than one.

The library of retroviral particles expressing single inserts would be subjected to positive and/or negative enrichment procedures based on the binding properties of the inserted domains. The recovered virus would be expanded by outgrowth in 3T3 cells, and a fresh particle library would be prepared by single round infection for each additional cycle of enrichment. For characterization, the enriched insert sequences would be recovered directly from enriched virions by RT-PCR or from the DNA of infected cells by PCR.

TESTING THE DISPLAY SYSTEM WITH MODEL ENRICHMENTS

A model system was established to test this scheme for a retroviral particle display system. A chimera bearing a V1/V2 domain insert was used with a monoclonal antibody reactive with a conformational epitope expressed by this insert (Kayman et al., 1999). Mixtures were prepared containing particles bearing the chimeric Env and particles bearing wild-type Env, with the species to be enriched present as a minority of the particles. The mixtures were then incubated with the monoclonal antibody. Positive enrichment, that is, the recovery of virions that express the epitope, was performed by adding Protein A carrying a six histidine tag prebound to nickel-nitrilotriacetic acid (Ni-NTA) agarose, followed by low-speed centrifugation to sediment virions carrying bound antibody. Bound virions were recovered by suspension of the pelleted resin in 10 mM EDTA to elute the His6-tagged Protein A from the Ni-NTA resin, followed by a second low speed centrifugation and immediate neutralization of the EDTA by addition of balancing magnesium chloride to the supernatant fluid. Negative enrichment, that is, the removal of virions that express the epitope, was performed by adding Pansorbin followed by low-speed centrifugation; the supernatant fluid contained the desired virions that had not bound antibody. Recovered virus were amplified by infection of 3T3 cells, and the ratio of chimeric Env to wild-type Env, based on their distinct molecular masses, was determined by radioimmunoprecipitation and SDS-PAGE. A single round of these enrichment procedures achieved over 30-fold positive enrichment and 400-fold negative enrichment. When more extreme ratios of input virus were used, this method enriched for nonreactive virions present at 10^{-5} within two rounds of enrichment (Fig. 28.4). However, enrichment for reactive virions present at low initial levels was not achieved by multiple rounds of enrichment (data not shown). It appears that transient treatment with EDTA results

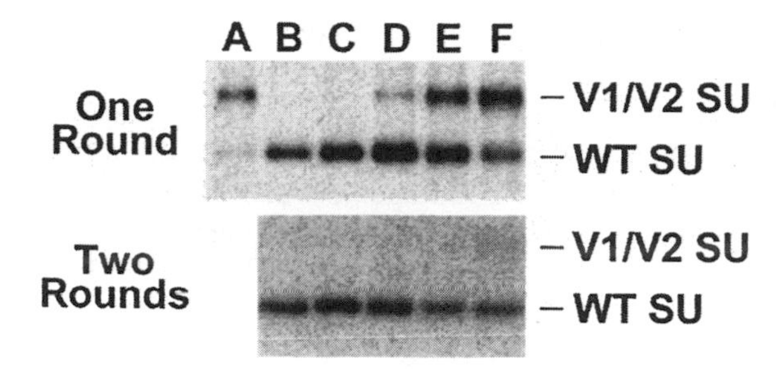

Figure 28.4 Multiple rounds of negative enrichment. V1/V2 insert virus were mixed with wild-type virus at various ratios and subjected to one or two rounds of negative enrichment with monoclonal antibody SC258. Cells infected with the viral mixture before (lane A) or after (lanes B–F) enrichment were labeled with [^{35}S]-cysteine, virus in culture supernatants were precipitated with polyclonal anti-SU serum and analyzed by SDS-PAGE and autoradiography. Lane A: $10^1:1$ mixture, no enrichment; Lane B: $10^1:1$ mixture; Lane C: $10^2:1$ mixture; Lane D: $10^3:1$ mixture; Lane E: $10^4:1$ mixture; Lane F: $10^5:1$ mixture.

in sufficient toxicity to create a population bottleneck that eliminates the reactive virions that are still rare after the first round of enrichment.

FUTURE DEVELOPMENT

The studies just described provide an encouraging indication of the ability to use MLV virions as a particle display system. However, enrichment methodologies suitable to retroviral particles and specific targets require further development. For molecularly defined targets, methods for recovery of bound virions that are less toxic to MLV need to be developed. Alternative elution conditions for His6-tagged material from Ni-NTA agarose and/or methods for use of alternative affinity tags need to be developed. Fusions to maltose binding protein (di Guan et al., 1988) would be one promising alternative. Although labor intensive, the need to elute infectious virions bound to the target could be bypassed by using reverse transcriptase-polymerase chain reaction (RT-PCR) to recover the sequences of the inserts carried by the bound virions for preparation of fresh plasmid libraries for additional cycles of enrichment. Even if RT-PCR were used for recovery of enriched sequences, the infectivity of the chimeric constructs would remain important for establishing clonality and linkage between the genome and chimeric Env on the surface of the particles used for enrichment.

For purposes of defining targeting moieties for gene therapy, it would be desirable to use whole cells as the basis for enrichments. This should be achievable since the ecotropic Env used in this retroviral particle display system does not bind to human cell types, and assays of MLV particle binding to cells show receptor specificity (Battini et al., 1996; Kadan et al., 1992). If the need is to enhance the binding specificity of a ligand that shows a preference for the desired target cell, identifying variants that bind more poorly to nontarget cells might be sufficient. This might be accomplished using only negative enrichment, and thus it would be straightforward. If positive enrichments are required, either as the primary enrichment from a library or to identify variants that retain binding activity following a primary negative enrichment, RT-PCR should allow recovery of the enriched variants.

This retroviral particle display system might prove particularly useful for developing ligands for cell surface receptors for which specific complex carbohydrate structures contribute to binding specificity. For example, both the presence of specific carbohydrate moieties and their arrangement on the glycoprotein ligand appear to play roles in the binding specificity of C-type lectin receptors such as the selectins (Vestweber and Blanks, 1999). A particle display system based on a mammalian virus is ideally suited to production of the display library in cells from the appropriate tissue in order to obtain the proper glycan modifications. If cell lines or primary cultures of the proper cell type were available, a retroviral display library could be passaged through a mouse cell line expressing the amphotropic MLV Env or the vesicular stomatitis virus G protein to produce a library stock capable of a single round of infection into essentially any desired human cell line or dividing culture. This final passage would generate a library of virions carrying the cell-type specific modifications suitable for enrichment. Modifying the viral display vector to use either an LTR with a broader pattern of cell type expression than that of the Friend MLV or a particular MLV LTR that is highly active in the appropriate cell type might also be required to achieve sufficient viral replication.

ACKNOWLEDGMENTS

I thank Abraham Pinter for helpful discussions during the course of this work, and Karl Drlica for helpful comments on this manuscript. This work was supported by the National Institutes of Health (AI44410).

REFERENCES

Battini JL, Danos O, Heard JM (1995): Receptor-binding domain of murine leukemia virus envelope glycoproteins. *J Virol* 69:713–719.

Battini JL, Heard JM, Danos O (1992): Receptor choice determinants in the envelope glycoproteins of amphotropic, xenotropic, and polytropic murine leukemia viruses. *J Virol* 66:1468–1475.

Battini JL, Rodrigues P, Muller R, Danos O, Heard JM (1996). Receptor-binding properties of a purified fragment of the 4070A amphotropic murine leukemia virus envelope glycoprotein. *J Virol* 70:4387–4393.

Buchholz CJ, Peng KW, Morling FJ, Zhang J, Cosset F, Russell SJ (1998): In vivo selection of protease cleavage sites from retrovirus display libraries. *Nat Biotechnol* 16:951–954.

Chambers P, Pringle CR, Easton AJ (1990). Heptad repeat sequences are located adjacent to hydrophobic regions in several types of virus fusion glycoproteins. *J Gen Virol* 71:3075–3080.

Clements GJ, Price-Jones MJ, Stephens PE, Sutton C, Schulz TF, Clapham PR, McKeating JA, McClure MO, Thomson S, Marsh M, et al. (1991): The V3 loops of the HIV-1 and HIV-2 surface glycoproteins contain proteolytic cleavage sites: a possible function in viral fusion? *AIDS Res Hum Retroviruses* 7:3–16.

Curiel DT (1999): Considerations and challenges for the achievement of targeted gene delivery. *Gene Ther* 6:1497–1498.

Davey R, Zuo Y, Cunningham JM (1999). Identification of a receptor-binding pocket on the envelope protein of friend murine leukemia virus. *J Virol* 73:3758–3763.

di Guan C, Li P, Riggs PD, Inouye H (1988): Vectors that facilitate the expression and purification of foreign peptides in *Escherichia coli* by fusion to maltose-binding protein. *Gene* 67:21–30.

Fass D, Davey RA, Hamson CA, Kim PS, Cunningham JM, Berger JM (1997): Structure of a murine leukemia virus receptor-binding glycoprotein at 2.0 Angstrom resolution. *Science* 277:1662–1666.

Grabherr R, Ernst W, Oker-Blom C, Jones I (2001): Developments in the use of baculoviruses for the surface display of complex eukaryotic proteins. *Trends Biotechnol* 19:231–236.

Gu R, Westervelt P, Ratner L (1993): Role of HIV-1 envelope V3 lop cleavage in cell tropism. *AIDS Res Hum Retroviruses* 9:1007–1015.

Heard JM, Danos O (1991): An amino-terminal fragment of the Friend murine leukemia virus envelope glycoprotein binds the ecotropic receptor. *J Virol* 65:4026–4032.

Hunter E, Swanstrom R (1990): Retrovirus envelope glycoproteins. *Curr Top Microbiol Immunol* 157:187–253.

Johnson ME, Lin Z, Padmanabhan K, Tulinsky A, Kahn M (1994): Conformational rearrangements required of the V3 loop of HIV-1 gp120 for proteolytic cleavage and infection. *FEBS Lett* 337:4–8.

Kadan MJ, Sturm S, Anderson WF, Eglitis MA (1992): Detection of receptor-specific murine leukemia virus binding to cells by immunofluorescence analysis. *J Virol* 66:2281–2287.

Kamps CA, Lin YC, Wong PK (1991): Oligomerzation and transport of the envelope protein of Moloney murine leukemia virus-TB and of ts1, a neurovirulent temperature-sensitive mutant of MoMuLV-TB. *Virology* 184:687–694.

Kayman S, Kopelman R, Kinney D, Projan S, Pinter A (1991): Mutational analysis of N-linked glycosylation sites of the Friend murine leukemia virus envelope proteins. *J Virol* 65:3323–3332.

Kayman SC, Park H, Saxon M, Pinter A (1999). The hypervariable domain of the murine leukemia virus surface protein tolerates large insertions and deletions, enabling development of a retroviral particle display system. *J Virol* 73:1802–1808.

Koch W, Hunsmann G, Friedrich R (1983). Nucleotide sequence of the envelope gene of Friend murine leukemia virus. *J Virol* 45:1–9.

Kuiken CL, Foley B, Hahn B, Korber B, McCutchan F, Marx PA, Mellors JW, Mullins JL, Sodroski J, Wolinsky S, eds (1999): *Human Retroviruses and AIDS 1999: A compilation and analysis of nucleic acid and amino acid sequences* (Los Alamos, NM, Theoretical Biology an Biophysics Group, Los Alamos National Laboratory).

Lavillette D, Maurice M, Roche C, Russell SJ, Sitbon M, Cosset FL (1998): A proline-rich motif downstream of the receptor binding domain modulates conformation and fusogenicity of murine retroviral envelopes. *J Virol* 72:9955–9965.

Linder M, Linder D, Hahnen J, Schott H-H, Stirm S (1992): Localization of the intra-chain disulfide bonds of the envelope glycoprotein 71 from Friend murine leukemia virus. *Eur J Biochem* 203:65–73.

MacKrell AJ, Soong NW, Curtis CM, Anderson WF (1996): Identification of a subdomain in the Moloney murine leukemia virus envelope protein involved in receptor binding. *J Virol* 70:1768–1774.

Morgan RA, Nussbaum O, Muenchau DD, Shu L, Couture L, Anderson WF (1993): Analysis of the functional and host range-determining regions of the murine ecotropic and amphotropic retrovirus envelope proteins. *J Virol* 67:4712–4721.

Odawara T, Oshima M, Doi K, Iwamoto A, Yoshikura H (1998): Threshold number of provirus copies required per cell for efficient virus production and interference in moloney murine leukemia virus-infected NIH 3T3 cells. *J Virol* 72:5414–5424.

Opstelten D-JE, Wallin M, Garoff H (1998): Moloney murine leukemia virus envelope protein subunits, gp70 and Pr15E, form a stable disulfide-linked complex. *J Virol* 72:6537–6545.

Ott D, Rein A (1992): Basis for receptor specificity of nonecotropic leukemia virus surface glycoprotein gp70(Su). *J Virol* 66:4632–4638.

Pinter A (1989): Functions of murine leukemia virus envelope products in leukemogenesis. In *Retroviruses and Disease*, Hanafusa H, Pinter A, Pullman M, eds. San Diego, CA: Academic Press, pp. 20–39.

Pinter A, Fleissner E (1977): The presence of disulfide-linked gp70-p15(E) complexes in AKR MuLV. *Virology* 83:417–422.

Pinter A, Honnen WJ (1983). Comparison of structural domains of gp70s of ecotropic Akv and its dualtropic recombinant MCF-247. *Virology* 129:40–50.

Pinter A, Honnen WJ, Tung J-S, O'Donnell PV, Hammerling U (1982): Structural domains of endogenous murine leukemia virus gp70s containing specific antigenic determinants defined by monoclonal antibodies *Virology* 116:499–516.

Pinter A, Kopelman R, Li Z, Kayman SC, Sanders DA (1997): Localization of the labile disulfide bond between SU and TM of the murine leukemia virus envelope protein complex to a highly conserved CWLC motif in SU the resembles the active site sequence of thiol-disulfide exchange enzymes. *J Virol* 71:8073–8077.

Pinter A, Lieman-Hurwitz, Fleissner E (1978): The nature of the association between the murine leukemia virus envelope proteins *Virology* 91:345–351.

Ragheb JA, Anderson WF (1994): pH-independent murine leukemia virus ecotropic envelope-mediated cell fusion: implications for the role of the R peptide and p12E TM in viral entry. *J Virol* 68:3220–3231.

Rein A, Mirro J, Haynes JG, Ernst SM, Nagashima K (1994): Function of the cytoplasmic domain of a retroviral transmembrane protein: p15E-p2E cleavage activates the membrane fusion capability of the murine leukemia virus Env protein. *J Virol* 68:1773–1781.

Sitbon M, D'Auriol L, Ellerbrok H, Andre C, Nishio J, Perryman S, Pozo F, Hayes SF, Wehrly K, Tambourin P, et al. (1991): Substitution of leucine for isoleucine in a sequence highly conserved among retroviral envelope surface glycoproteins attenuates the lytic effect of the Friend murine leukemia virus. *Proc Natl Acad Sci USA* 88:5932–5936.

Stoye JP, Coffin JM (1987): The four classes of endogenous murine leukemia virus: structural relationships and potential for recombination. *J. Virol* 61:2659–2669.

Valsesia-Wittmann S, Morling FJ, Hatziioannou T, Russell SJ, Cosset FL (1997): Receptor co-operation in retrovirus entry: recruitment of an auxiliary entry mechanism after retargeted binding. *EMBO J* 16:1214–1223.

Vestweber D, Blanks JE (1999): Mechanisms that regulate the function of the selectins and their ligands. *Physiol Rev* 79:181–213.

Weiss RA, Tailor CS (1995): Retrovirus receptors. *Cell* 82:531–533.

Werner A, Levy JA (1993): Human immunodeficiency virus type 1 envelope gp120 is cleaved after incubation with recombinant soluble CD4, *J Virol* 67:2566–2574.

Wu BW, Cannon PM, Gordon EM, Hall FL, Anderson WF (1998): Characterzation of the proline-rich region of murine leukemia virus envelope protein. *J Virol* 72:5383–5391.

Wu Z, Kayman SC, Revesz K, Chen HC, Warrier S, Tilley SA, McKeating J, Shotton C, Pinter A (1995): Characterization of neutralization epitopes in the V2 region of HIV-1 gp120: role of conserved glycosylation sites in the correct folding of the V1/V2 domain. *J. Virol* 69:2271–2278.

29

CELL SURFACE DISPLAY AND CYTOMETRIC SCREENING FOR PROTEIN LIGAND ISOLATION AND ENGINEERING

PATRICK S. DAUGHERTY, PH.D.

INTRODUCTION

Since the development of phage display technology, an expanded set of display technologies has been developed that improve on existing methods and enable entirely new applications in genetics, clone isolation, and protein engineering. The term *display* is used to indicate the use of a physical link between a protein and encoding nucleic acid, which enables isolation of genes of interest from a library. Display technologies for protein engineering have been reviewed previously (Shusta, VanAntwerp, and Wittrup, 1999). See Chapters 25–28 in this volume for discussions of phage and retroviral display. Cell surface display is unique among display technologies because quantitative flow cytometry (FCM) and cell sorting instrumentation can be used for high-throughput library screening (Boder and Wittrup, 1997; Georgiou et al., 1997; Daugherty et al., 1998). Potential applications of cell surface display include identification and isolation of target receptors, isolation of receptor-binding ligands, affinity and specificity improvement of known ligands, and improvement of phage display selection efficiency.

Vector Targeting for Therapeutic Gene Delivery, Edited by David T. Curiel and Joanne T. Douglas
ISBN 0-471-43479-5

Overview of Methodology

The display of proteins on the surface of microorganisms for the purpose of protein engineering can be accomplished using a wide variety of cell surface display vectors. Typically, an expression plasmid is designed to produce a chimeric protein composed of the protein of interest, fused in frame to a surface-exposed or anchored membrane protein. A wide variety of surface display vectors are available for a range of hosts (Georgiou et al., 1997). Typical methods for bacterial, yeast, and mammalian display are illustrated in Fig. 29.1.

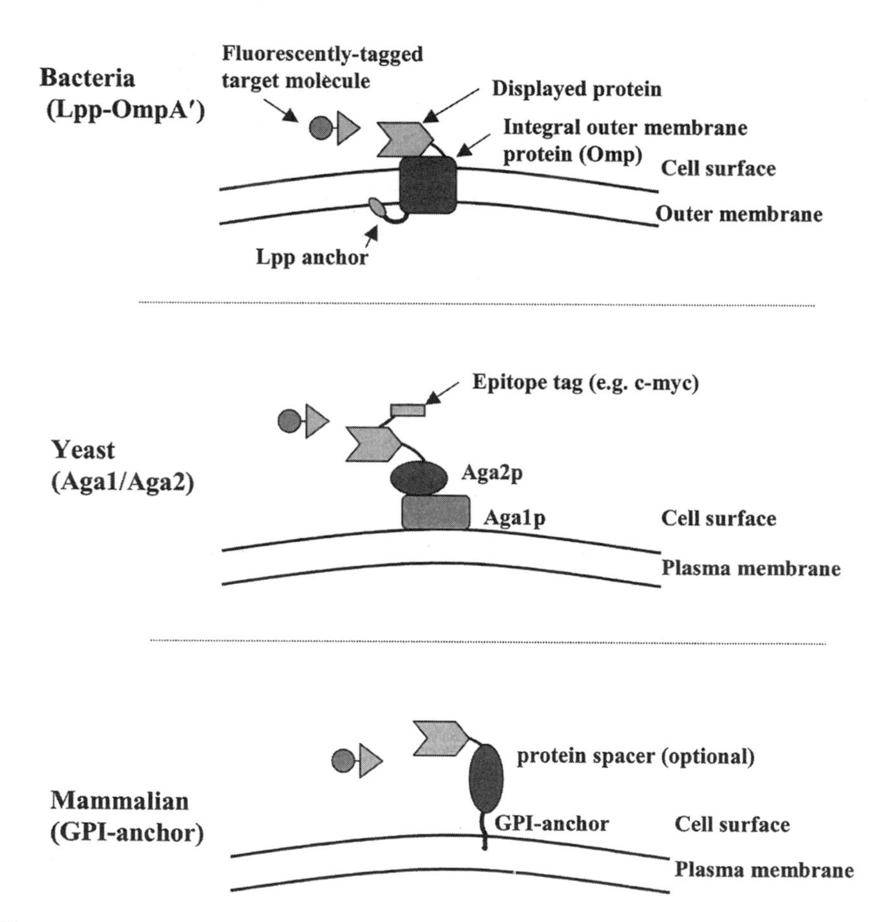

Figure 29.1 Cell surface display methods. In bacteria, display has been accomplished using fusions and insertions into integral outer membrane proteins, as well as via the proteins of pili and flagella structures (Georgiou et al., 1997). Yeast display typically relies on a fusion to the yeast mating protein Aga2p which is disulfide bonded to the membrane anchored Aga1p protein. In mammalian cells, the C-terminal signal sequence from glycosyl-phosphatidylinositol-glycan (GPI) linked proteins has also been used to display proteins at the cell surface.

The general strategy for screening cell surface-displayed libraries is illustrated in Fig. 29.2. A gene library is created using recombination and mutagenesis methods or from genomic or cDNA. The library is then introduced into the host cells for expression (bacteria, yeast, or mammalian cells) using an appropriate transformation method. Each cell in the population will display a unique protein on its outer surface, thereby allowing extracellular access to targets including fluorescently tagged proteins, viruses, or whole cells. The labeled target ligand is allowed to equilibrate with an aliquot of the library for a sufficient period of time, and the cells are washed and sorted, or sorted directly to isolate cells displaying binding proteins. The isolated cell population can then be amplified by growth and sorted again (if needed) to further enrich clones having desired properties. When the sorted population becomes highly

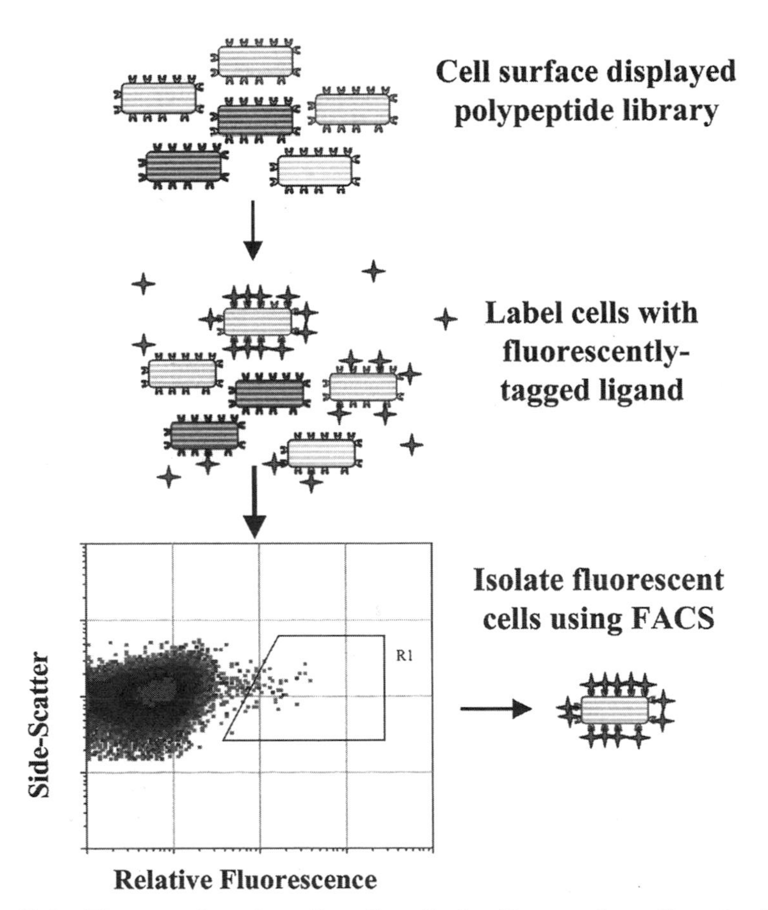

Figure 29.2 The screening of a cell surface display library using cell sorting instrumentation (FCM). The library is incubated with a fluorescently tagged target ligand. Data shown is actual data from an scFv library screening experiment, using BOPDIPY-conjugated digoxin probe as described previously (Daugherty et al., 1998).

enriched for target cells, individual clones can be isolated and characterized directly using FCM analysis and DNA sequencing.

The use of flow cytometry offers some potentially important advantages for isolating targeting ligands from protein libraries and subsequently improving their function by directed evolution. Probably the most important feature is the ability to visualize and optimize the screening process. For example, it is possible to directly measure the efficiency of display for the protein of interest as well as the affinity of the interaction with the target. The system screening parameters such as ligand concentration, length of the dissociation phase, the fraction of the population to be collected, and number of rounds needed can all be calculated using a relatively simple mathematical analysis (Boder and Wittrup, 1998). Also, libraries can be analyzed directly for the presence of target clones by FCM analysis of a suitable number (10^5–10^7) of clones. Thus, if multiple libraries were available, each could be quickly analyzed for the presence or absence of candidate target proteins prior to investing effort in library screening.

General Advantages of Cell Surface Display

Cell surface display can provide distinct advantages for particular protein engineering problems. However, the development of targeted vectors for gene therapy is certainly a demanding application, irrespective of the display technology employed. Thus, it is worth considering whether cell surface display may provide advantages as compared to alternative display technologies. To begin with, cell surface display (particularly with bacteria) is simple, fast, and nonlabor intensive, relative to typical phage methods. A cycle of display and selection, using *Escherichia coli* as a host, requires about 24 h, with only about 4 h of preparation and sorting time per cycle. Upon optimizing the cell sorting, fewer rounds of screening are required for the direct isolation of binding ligands. Cell surface display formats also offer an expanded range of expression capabilities because yeast and mammalian cells can be used as hosts. As a result, it is possible to engineer proteins that require posttranslational modifications including glycosylation. Given the important role of envelope glycosylation for retroviral immune evasion, eukaryotic machinery is likely to be important for many targeting applications. Finally, cell surface display systems have enabled the engineering of receptor–ligand interactions from nanomolar to subpicomolar affinity (Boder, Midelfort, and Wittrup, 2000). This is truly impressive when compared to noteworthy examples of affinity maturation using phage display, yielding antibodies in the 10 pM range (Yang et al., 1995; Schier et al., 1996). Such powerful affinity maturation may be important for certain applications such as selective targeting of tumor cells.

Cell surface display further differs from common phage methods by being stochastic. In other words, the analysis of a single cell expressing tens of thousands of copies of a protein of interest represents an average measurement for all molecular binding events on the cell surface. This contrasts with monovalent phage display protocols, which require expression and functional measurement for a perhaps a single copy per phage. Thus, averages can only be obtained by processing thousands of copies of a given phage clone in a single selection. In addition, the quantitative screening capability afforded by FCM allows fine discrimination of ligand specificity and affinity (Daugherty et al., 1998; Kieke et al., 1999; VanAntwerp and Wittrup, 2000). In certain cases, the enrichment of clones with less than 10% differences in the dissociation rate constant has been demonstrated (Daugherty et al., 1998), as measured by flow cytometry and surface plasmon resonance. Such fine affinity discrimination provides improved resolution during screening, a feature beneficial for affinity maturation studies where improved clones may differ only subtly from the wild-type. Lastly, cell surface display has been shown to coselect for high expression levels, along with improved affinity (Shusta et al., 1999a). Expression mutants are often an unexpected consequence of the inability to uncouple expression and affinity during selection, but can be valuable when an application requires eventual protein expression and purification.

APPLICATIONS

Improving the Affinity of Ligands

Cell surface display has been used most extensively for improving ligand binding affinity, because cell surface display coupled with flow cytometry provides a very convenient selection for improved affinity. The affinity maturation of peptides, including a cystine knot protein (Christmann et al., 1999), single-chain Fvs (scFvs) (Boder and Wittrup, 1997; Daugherty et al., 1998), and single-chain T-cell receptors (scTCRs) (Holler et al., 2000) has been demonstrated. In one spectacular case, an scFv specific to fluorescein was affinity improved by 10,000-fold (Boder, Midelfort, and Wittrup et al., 2000). Fewer studies have demonstrated improvement of antibodies specific toward protein antigens, perhaps due in part to the relative difficulty of preparing fluorescent protein conjugates and the inaccessibility of some surface displayed proteins to large proteins. In one recent study, the affinity of a T-cell receptor was improved toward a peptide-MHC complex (Holler et al., 2000). Ligand affinity has most often been improved by polymerase chain reaction (PCR)-based random mutagenesis of the entire gene (Boder, Midelfort, and Wittrup, 2000; Daugherty et al., 2000), or targeted mutagenesis within the antibody hypervariable domains (Daugherty et al., 1998). Several cycles of mutagenesis and selection can be performed, allowing potentially large improvements in affinity.

Isolation Receptor Binding Proteins

Although binding is not generally sufficient for gene delivery, phage-based selections for peptides enabling entry has shown that entry, probably via endocytosis, can be achieved from random peptide libraries (Larocca et al., 2001). Cell surface display can also be used to isolate proteins that bind to cell surface receptors. At least three different approaches could be used to isolate and engineer receptor ligands. First, if the receptor contains a soluble cell surface domain, such as that from CD4, a fluorescently tagged version of this fragment could be prepared and used for screening. Alternatively, entire mammalian cells could be used as targets in a manner analogous to panning phage on cells (de Kruif et al., 1995). Intracellular expression of suitable fluorescent proteins within the display host can be used to fluorescently label entire cells, providing a simple and cost-effective signal amplification scheme. A third approach, reverse display, could also be envisioned, wherein viral particles pseudotyped with a target receptor are used to infect cells displaying the corresponding viral envelope. Such mammalian display of a viral envelope may become a useful tool for enhancing specificity of viral entry when targeting of one particular receptor is desired.

In addition to being useful for isolating and improving ligands, the screening of cDNA expression libraries holds promise for the identification of target receptors for gene delivery. Already, retroviral vector-based cDNA libraries have been used to clone several retroviral receptors (Battini and Miller, 1999; Rasko et al., 1999). The increasing availability of cDNA, peptide, and scFv antibody libraries should promote further application of this powerful method.

Improvement of Expression Level

Cell surface display methods can also be used to increase expression and display level, as well as thermal stability of proteins (Shusta et al., 2000). Improved scFv expression has been found to occur much more frequently than improved affinity in mutant libraries (Daugherty et al., 1998; Kieke et al., 1999). Consequently, mutants with improved expression and display levels can be isolated from relatively small random mutant libraries (Fig. 29.3). The protein of interest should probably be displayed at some detectable level in the chosen host, before attempting to increase the expression level by mutation and selection. Libraries are created by random PCR mutagenesis and screened for highly fluorescent cells. Several rounds of mutagenesis have been used to create scTCRs that can be displayed at high levels on the surface of yeast, allowing further engineering (Kieke et al., 1999; Shusta et al., 1999a).

Improving the Quality of Phage Selections

Phage display remains a highly useful tool for isolating and engineering ligands. Given the labor-intensive nature of the screening process, it is of interest to improve the efficiency when possible. A recent report has demonstrated that display of a target antigen on the surface of a cell, followed by phage panning can provide highly efficient selections (Benhar et al., 2000). The authors

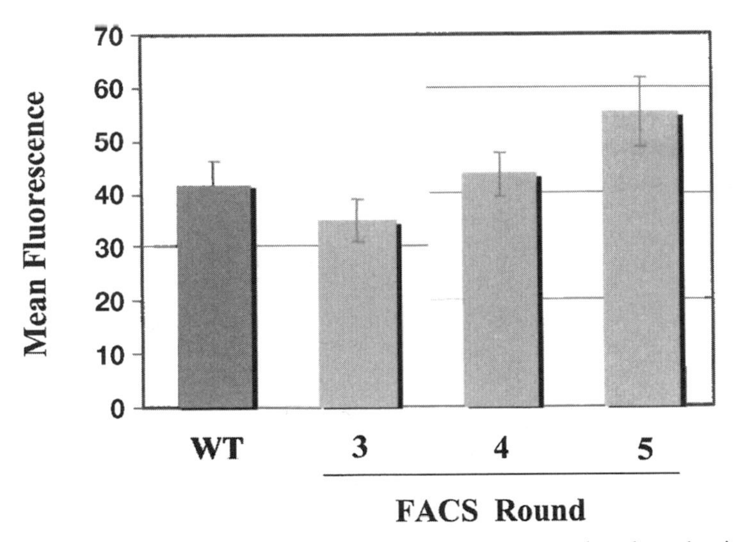

Figure 29.3 Improvement in expression level for mutant pools selected using FCM during rounds 3–5. Error bars are calculated from the standard deviation of triplicate measurements.

displayed various proteins on the surface of *Escherichia coli* using the Lpp-OmpA' system (Francisco, Earhart, and Georgiou, 1992) and panned a phage scFv library on these cells. Using this method the authors were able to select antibodies that bound with high affinity to several different antigens.

SELECTION OF AN APPROPRIATE HOST CELL AND DISPLAY VECTOR

Appropriate Protein Expression Capabilities

Clearly, one must consider the proteins to be displayed in relation to the expression capabilities of the host. Bacteria, while being relatively simple to manipulate in the laboratory, have relatively poor abilities to form disulfide bonds and are unable to perform glycosylation, with special exceptions (Schaffer, Graninger, and Messner, 2001). Thus, if glycosylation is thought to be required for function, one should use a yeast or mammalian host. However, if the protein of interest is small, has only a few disulfide bonds, and does not require glycosylation, the use of a bacterial host could provide advantages as discussed in this section. Yeast systems are advantageous for gycosylation and are somewhat easier to work with than mammalian cells. However, yeast strains have been shown to hypergylcosylate some proteins including scFvs, resulting in a modification of their binding properties (Haidaris et al., 1998). Increasing evidence suggests that glycosylation is seldom a strict requirement for function, while hyperglyosylation may interfere with function in unpredictable ways (Wester et al., 2000).

Regulated Protein Expression

The regulation of protein expression from surface display vectors is often overlooked despite being an important consideration for success. Problems with unregulated protein expression occur when potentially desired mutants are outcompeted by cells producing less toxic variants or less total protein during expansion in culture. The result is a decrease in the representation of potentially desirable clones within the library. For example, the frequency of a clone initially present at $1:10^6$ may fall to $1:10^9$ after many generations in culture, potentially preventing successful isolation. For this reason, it is desirable to regulate protein expression using a tightly controlled promoter. Under tightly repressed conditions, individual clones will grow at nearly the same rate. Thus, the *tet* promoter in mammalian cells (Baron and Bujard, 2000) or the *araBAD* promoter in bacteria (Guzman et al., 1995) are better choices than constitutive expression from common promoters including those from *lac*, *tac*, and *trc* or CMV and SV40. In an *E. coli* surface display system, it has been shown that library diversity is well conserved using the *araBAD* promoter on a low-copy (p15A) plasmid (Daugherty et al., 1999). In contrast, the use of a leaky uninduced *lac* promoter resulted in loss of desired active clones after about 50 generations of growth. In addition, unregulated expression can lead to clone depletion as a result of cell death (Christmann et al. 1999). Display in yeast is generally regulated by the *GAL1,10* promoter (Boder and Wittrup, 2000), which has proven effective, although it is not clear whether the use of a more tightly regulated promoter might yield improved results.

General Host and Display System Considerations

Several other factors should be considered when choosing a display format for a particular problem. First the cell should display a sufficient number of copies of the protein of interest at the surface to provide high signal-to-noise ratios of target to nontarget cells. For example, in bacteria the expression of about 3×10^4 copies of the protein of interest has resulted in a 10-fold signal-to-noise ratio between fully stained positive cells and nonbinding cells, when using a BODIPY-FL probe from Molecular Probes (Eugene, Oregon). Second, the display host should possess a sufficiently high transformation efficiency to allow the construction of the libraries of the needed size. Third, the host cells should remain viable after the rigors of displaying the target protein, the labeling and dissociation period, and the process of being sorted. Of course, viability depends on various factors, including the toxicity of the protein being displayed, total expression level, and the mode of display. Finally, the ideal display system should be relatively simple to manipulate in the laboratory and should permit the shortest screening time possible (i.e., 24 h cycles of sorting and regrowth for a bacterial host). Having said this, none of the current systems meet all of these criteria. Hence, the investigator should consider all available options with each of these criteria in mind.

Choice of an Appropriate Host Cell

Bacterial Display: Features Bacterial display has proven effective for the display of a wide range of proteins and peptides (Georgiou, et al., 1997). Despite the absence of glycosylation capabilities and a limited protein-folding machinery, bacterial display libraries offer the following advantages. First, the rapid growth rate of *E. coli* (about 30 min/doubling) enables short, 24 h cycles of sorting and regrowth. Second, the transformation of bacteria by electroporation is both simple and efficient, allowing roughly 10^{10} transformants per microgram of plasmid DNA. The high transformation efficiency of *E. coli* makes this host suitable for the construction of very large libraries containing more than 10^9 members. Third, *E. coli* remains a widely used host for large-scale protein production. Thus, proteins expressed and displayed in *E. coli* can generally be produced at higher levels, with lower costs relative to yeast or mammalian systems. Finally, the well understood genetics and reduced complexity of *E. coli* make this host suitable for improvement by genetic engineering. If necessary, a wide array of available strains, modified for improved expression of recombinant proteins (Meerman and Georgiou, 1994), can be tested with relatively little effort.

Bacterial Display Vectors Most early display vectors for a bacterial host typically allowed peptide insertions into outer membrane proteins such as LamB (Lang 2000). Other peptide display systems were subsequently developed to allow better access to larger proteins or peptide presentation in a defined manner. One such system, commercially available from Invitrogen (Carlsbad, California), utilizes thioredoxin as a scaffold to constrain displayed peptides, which itself is inserted into the major flagellin protein FliC (Lu et al., 1995). While proven effective as a means to map antibody epitopes, this system is not well suited to FCM screening because flagella are sensitive to shear forces in the flow stream (unpublished observations). Recently, the bacterial fimbrial protein FimH has been used as a display vehicle, again for peptide insertions (Klemm and Schembri, 2000). Fimbria have the advantage of being located at a distance from the cell surface and thus may provide improved binding to larger targets, including whole cells. In one case, a fimbrial peptide library was screened for binding to metal oxide surfaces (Kjaergaard et al., 2000). Unfortunately, fimbrial display of proteins larger than about 60 amino acids has not yet been demonstrated. Another interesting system utilizing the ice-nucleation protein has been shown capable of displaying larger proteins (Shimazu, Mulchandani, and Chen, 2001), though has not yet been applied for library screening.

The Lpp-OmpA' system (Francisco et al., 1992) was the first cell surface display system demonstrated to be effective for the screening of protein libraries, including scFv antibodies (Georgiou et al., 1997; Daugherty et al., 1998). This system has since been used for screening cystein knot protein libraries (Christmann et al., 1999) and for displaying antigen targets for hybrid cell surface/phage selections (Benhar et al., 2000). This vector is composed of a fusion of a nine-amino acid signal sequence from the *E. coli* major lipoprotein (Lpp), to amino acids 46–159 of the abundant outer membrane protein A (OmpA). This segment of OmpA was subsequently verified to constitute five transmembrane strands in the native β-barrel structure (Pautsch and Schulz, 2000). The display of a wide range of soluble proteins has been demonstrated with this system (Georgiou et al., 1997). Some proteins displayed by this system were not fully accessible to large proteins (Francisco et al., 1993a), whereas others have allowed adhesion to microstructures (Francisco et al., 1993b; Christmann et al., 1999).

Yeast Display: Features Yeast display offers an expanded range of protein processing capabilities, including improved disulfide bond formation and glycosylation. Thus, yeast display may be advantageous for displaying relatively complex proteins. For example, the display and evolution of functional single-chain T-cell receptors has been demonstrated in *Saccharomyces cerevisia* (Kieke et al., 1999; Holler et al., 2000). However, improved expression capabilities come at the expense of smaller libraries and slower growth rates and screening cycles. Yeast display libraries typically contain around 10^5 unique mutants, although this could be improved by 10-fold after scaling up. The size constraint stems from the relatively low efficiency of lithium chloride transformation of yeast (Griffin et al., 2001).

Yeast Display Vectors The most widely used yeast display vector employs fusions to the agglutinin mating protein Aga2p, which is covalently linked to the surface-expressed Aga1p through disulfide bridges. Wittrup and co-workers have constructed such display plasmids (pCT302) (Boder and Wittrup, 2000), under the control of the *GAL1,10* promoter, inducible by growth in the presence of galactose. This vector is used in combination with a yeast strain (EBY100) engineered to express inducible Aga1p from the genome. The library is grown in glucose for repression, and then switched to galactose during exponential growth. The expression level of the fusion protein in different cells can be normalized by labeling cells with anti-epitope (i.e., cMyc or haemagglutin [HA]-directed antibodies (Holler et al., 2000).

Mammalian Display: Features The primary potential advantage of using a mammalian cell surface display format is improved processing of complex proteins. In particular, the production of a mammalian-derived protein in a native-like environment should likely allow the display of a wider range of proteins. Posttranslational modifications such as proteolytic cleavage or glycosylation may be required for certain aspects of protein function. For example, if one wished to display the human immunodeficiency virus (HIV) envelope, which has more than thirty N-linked glycosylation sites, the absence of glycosylation could dramatically alter this protein's immunogenicity or result in undesired amino acid changes in regions of the protein surface ordinarily masked by glycosylation.

Despite these potential advantages, mammalian cell surface display has been used only in a few special cases (Rode et al., 1996; Kitamura, 1998), likely owing to the complexity and labor-intensive nature of such a system. Because mammalian cells typically have doubling times of eight hours or longer, screening cycles are on the order of weeks. Thus, a single experiment designed to select the best mutant from a pool of 10^6 might require a month from start to finish. An affinity maturation project could be particularly labor intensive if several mutagenesis and selection cycles are required. In addition, it been difficult to construct larger libraries, since methods generally require transfection, or preferably retroviral infection, to avoid multiple distinct proteins in a single cell. Also, the recovery of integrated sequences from isolated clones is a time-intensive process, particularly when compared to a plasmid mini-prep from a bacterial culture. For mammalian cells, genomic DNA is prepared from individual clones or pools and nested PCR is performed to recover sequences.

Mammalian Cell Surface Display Vectors In mammalian cells, the C-terminal signal sequence from glycosyl-phosphatidylinositol glycan (GPI)-linked proteins can be used to display a variety of proteins at the cell surface (Scallon et al., 1992). In particular, the signal sequence from human placental alkaline phosphatase and a low affinity IgG receptor have been used to target heterologous proteins to the cell surface (Whitehorn et al., 1995). Protein display can also be accomplished by producing the protein of interest in soluble form as a fusion to a GPI anchor. An aliquot can then be added to the target cells *in-vitro*, whereupon fusion proteins can incorporate into the membrane and retain function (Medof, Nagarajan, and Tykocinski, 1996).

Retroviral expression vectors are being used increasingly for cloning surface antigens from cDNA libraries (Kitamura, 1998; Wong et al., 1994). The MoMLV-derived pMX retroviral vector possesses a full-length packaging signal including a portion of the *gag* gene followed by a multiple cloning site. The vector is subcloned into a high-copy number, pUC-based plasmid for efficient library construction (Kitamura 1998). In principle, this technique could be used to engineer surface antigens in mammalian cells through mutagenesis and FCM selection.

LIBRARY SCREENING USING FLOW CYTOMETRY

The use of flow cytometry and cell sorting to screen cell surface-displayed libraries provides a powerful approach for mutant analysis and isolation. Technically, cell sorting is a screen rather than a selection, because during library processing, the accept and reject criteria are evaluated in series rather than in parallel. Given that the maximum throughput rate is upward of 100,000 cells/s, roughly 5×10^8 cells can be processed per hour. Less expensive bench-top sorting instruments (Partec PAS-III, BD FACSCalibur, etc.) are also capable of processing at least 5×10 cells per h. Since most typical libraries are composed of about 10^6–10^9 members, including phage libraries, this throughput rate does not impose a bottleneck for most applications.

Flow Cytometry Instrumentation

There are a number of factors to take into consideration when evaluating a flow cytometer for library screening applications. Since cytometers were originally used in the greatest numbers by immunologists, many current instruments do not simultaneously provide all features desirable for library applications. First, the screening of large libraries can be limited by the maximum throughput speed that allows high-purity sorting. Bench-top instruments such as the FACSCalibur are stated to allow a maximum analysis of 10,000 s^{-1}. In practice, this instrument has allowed processing rates of 4,000 s^{-1}, when the desired target cells are relatively rare. While sort speeds could be potentially increased, the sorter design makes this less practical. Specifically, the maximum capture rate for sorted cells is specified at 300 s^{-1} (Chapman, 2000). On the other end of the throughput scale are the stream-in-air instruments including Cytomation's MoFlo (Fort Collins, Colorado), which reportedly allows high-purity sorting at input speeds greater than 30,000 s^{-1} and capture rates equally high (Chapman, 2000).

The instrument sensitivity to both fluorescence and light scatter should also be considered. As will be discussed, the resolution one obtains from a mixed population of target and nontarget cells can make a large difference in the sorting efficiency. Generally speaking, target cells should have a 10-fold greater signal than nontarget cells on the instrument used for sorting. Stream-in-air instruments are at a disadvantage when it comes to sensitivity. Sensitivity largely depends on the light-gathering capacity of the collection lens (Shapiro, 1995). Light-gathering capacity depends on the maximum numerical aperture (N.A. 1.0 for stream-in-air instruments), determining the range of angles over which light is collected. Some bench-top instruments including the Calibur and PAS-III (Chapman, 2000), utilize excitation within a quartz flow cell, which can be optically coupled to detection optics using an oil-immersion collection lens, resulting in a high numerical aperture and excellent sensitivity. One can calculate that it is possible to achieve a two- to threefold improved sensitivity for a system with an oil-immersion objective (N.A. = 1.25) relative to a typical air objective (N.A. ≈ 0.85) for a given laser configuration. In addition to improved fluorescence sensitivity, the improved forward angle and 90-degree light scatter signals obtained with immersion coupled systems can be important for bacterial applications requiring population resolution based on light scatter (unpublished observations).

The light source may also be an important issue to consider (Shapiro, 2001).

Most cytometers are equipped with an excitation source for fluorescein/GFP excitation and, increasingly, a red-diode laser emitting around 630 nm, although occasionally, the experiment could benefit from the use of non-standard wavelengths including ultraviolet (UV), violet, green, orange, or orange-red. For instance, cellular autofluorescence resulting from flavins in mammalian cells can limit the resolution between target and nontarget cells (Shapiro, 1995). Thus, excitation of suitable green- or yellow-absorbing probes (phycoerythrin [PE], red fluorescent protein, propidium iodide) with a green laser such as the solid-state Nd:YAG emitting at 532 nm (Shapiro, 1995), could potentially provide enhanced resolution of target cells and thus more efficient library screening.

Another issue to consider is the volume of fluid collected during the sorting process. Stream-in-air cytometers collect very small volumes of fluid along with sorted cells, however, certain fluidic sorters such as the BD Calibur and Partec PAS-III may collect a few hundred cells in a volume of 30 ml. If one desires to plate the cells on solid media, the sample must be centrifuged or filtered onto membranes and transferred to agar plates. If the population is to be amplified by growth, large volumes require more doublings and thus slightly longer growth periods. Aside from longer growth periods, the growth of heterogeneous pools of cells over many generations can result in undesirable Darwinian selection of accidentally cocaptured nontarget cells (Daugherty et al., 1999). Nevertheless, the collection of large fluid volumes has not prevented successful application of these instruments in library screening (Daugherty et al., 1998).

Fluorescent Probe Selection

The fluorescent–ligand conjugate used for screening a cell surface display library is an important factor for creating a system amenable to efficient sorting. A large number of suitable probes are commercially available or can be synthesized readily (Haugland 1996). Some desired properties include optimal excitation by the cytometer light source, strong fluorescence emission compatible with the cytometer optics, tight emission spectrum, neutral charge, and low levels of nonspecific binding. The level of purity is not particularly important unless the impurities give rise to nonspecific staining or compete with the primary probe for specific binding. The probe should be relatively free of diffusion barriers imposed by the host cell surface environment. Preferably, the probe should bind monovalently to the desired target ligand. The use of a monovalent ligand avoids the selection of clones having only increased apparent affinity due to an avidity effect (Boder and Wittrup, 1997). Collectively, the characteristics of the probe can determine the success or failure of the system.

Fluorescent conjugates of proteins can be used in some cases to achieve very strong fluorescent signals either directly or through signal amplification strategies. One relatively simple method is the use of a fluorescently labeled antibody specific to the target antigen. If commercial antibodies are not available, polyclonal sera can be labeled with small fluorochromes such as fluoroscein isothiocyanate (FITC), Texas Red, or the BODIPY and Alexa probes from Molecular Probes (Eugene, Oregon). Alternatively, PE-labeled streptavidin has been used to label biotinylated target antigens, providing improved signals (Christmann et al., 1999; Wentzel et al., 1999). The commercial availability of genetically encoded fluorescent reporters including blue, cyan, green, yellow, and red fluorescent proteins, should allow the preparation of fluorescent protein fusions to target antigens by expression and purification from a suitable host. Viability probes can also be useful for improving the efficiency and purity of selections by reducing the capture of viable nontarget cells along with nonviable target cells during sorting.

Experimental Setup and Target Cell Definition

Resolving Target from Nontarget Cells In a well-optimized system, it should be possible to enrich cells occurring at frequencies less than one in ten million, allowing the thorough screening of libraries composed of more than 10^7 clones. The use of newer instrumentation for high-speed cell sorting may improve this number by 10-fold, for some applications. In general, target cells should possess a 10-fold greater fluorescence than nontarget cells to provide enrichment ratios as high as 10^5 in a single round of sorting. However, for affinity maturation studies, the separation of mutants with only twofold improved affinity may be desired. Here, the mean fluorescence intensity of wild-type and improved mutants may differ by less than threefold, resulting in overlapping populations. Thus, it is best to consider methods to improve the resolution of target cells from nontarget cells before initiating such studies.

Several different strategies can be used to enhance target cell resolution: (1) the total number of displayed receptors can be increased via changes in the expression system; (2) brighter fluorescent probes can be used (e.g., BODIPY-FL instead of fluorescein); (3) one can attempt to reduce the autofluorescence of nontarget cells by changing the excitation light sources and collection optics; (4) a fluorescent signal amplification strategy can be used (e.g., streptavidin-PE).

Eliminating False Positives Since the frequency of false positive target events dictates the maximum possible enrichment when sorting, false positives should be minimized (Leary, 1994; Gross et al., 1995; Daugherty, Iverson, and Georgiou, 2000). Experience suggests that the primary sources of false positives in cell surface library screening experiments arise from nonspecifically stained cells, experimental carryover, and impure sheath or sample buffers. Particle-free buffer should be prepared by filtration with a filter pore size of <0.2 μm, because 0.2 μm particles can be detected by most cytometers. Experimental carryover is usually best minimized by washing the instrument with a 10% solution of bleach that both sterilizes the instrument and inactivates any fluorochromes in the flow stream. The instrument should then be rinsed with sterile, filtered water for sufficient time until the frequency of events occurring within the target cell sort gate is low (<1 event/minute). Nonspecifically stained cells frequently are nonviable or posseses irregular light scatter signals (Leary 1994). Thus, a gate can be set in a forward versus 90 degree light scatter plot to include only cells with average light scatter (i.e., the top 90% of a peak, as identified in a density or three-dimensional plot). Another method to reduce nonspecifically stained cells is to use one or more additional nonspecific exclusion probes (Gross et al., 1995; Leary, 1994), which like the primary probe, are likely to bind to such irregular cells. Viability indicators may also be useful for this purpose (Daugherty, Iverson, and Georgiou, 2000). Care should be taken when using forward-scatter (FSC) or side-scatter (SSC) as a threshold parameter when sorting for purity since, if a poorly illuminated or small cell is undetected, it will not be considered in coincident event detection/rejection.

Screening Based on Equilibrium Affinity

Target cells displaying ligands of a given affinity can be isolated quantitatively using equilibrium-binding conditions. In this case, cells are incubated with a ligand concentration that maximizes the expected resolution between target and nontarget cells. For affinity maturation, the optimum ligand concentration $[L]_{\text{opt}}$ can be calculated using the expression (Boder and Wittrup 1998),

$$\frac{[L]_{\text{opt}}}{K_{D,\text{wt}}} = \frac{1}{\sqrt{S_r \cdot \dfrac{K_{D,\text{wt}}}{D_{D,\text{mut}}}}} \tag{29.1}$$

The parameter S_r is the ratio of maximum fluorescence to background autofluorescence, and the dissociation constant of the target mutant $K_{D,\text{mut}}$ is estimated. This method is effective when screening for ligands with low micromolar or better affinity, since high concentrations of the fluorescent probe in the sample stream results in substantial background fluorescence. Using this approach, cells are labeled and sorted directly without a wash step. The usefulness of Eq. 29.1 has been tested experimentally. Remarkably, the use of $0.2 \cdot K_D$ resulted in the isolation of scFv clones with three-fold improved affinity in a single sorting step (Daugherty et al., 1998).

Screening Based on Ligand Dissociation Kinetics

The screen can also be performed using the ligand dissociation rate as the selection parameter. Using a simple mathematical analysis, the optimal duration of the dissociation period can be calculated (Fig. 29.4). Since association rate constants typically do not change substantially (Schier et al., 1996; Daugherty et al., 1998), the dissociation rate constant (k_{off}) usually correlates directly with affinity. For kinetic selection, cells are labeled with saturating concentrations of the target ligand, washed, and resuspended in buffer containing a nonfluorescent ligand as a competitor. After an optimal period of time t_{opt} predicted

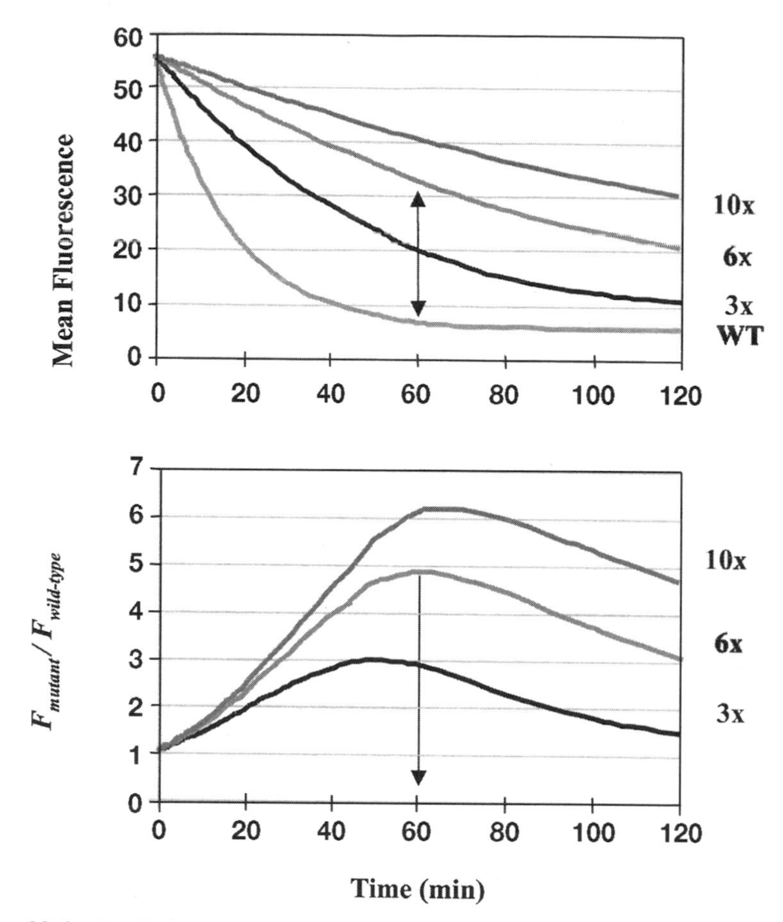

Figure 29.4 Prediction of optimal dissociation time before maximal fluorescence resolution of mutants improved by a factor of 3-, 6-, or 10-fold relative to the wildtype (WT), with appropriate parameters for a bacterial display system (Daugherty et al., 1999). *Top*. Mean fluorescence verses time for cells displaying wild-type ($K_D = 1$ nM) or 3-, 6-, or 10-fold improved mutants. Arrow is drawn at time where flourescence ratio of 6-fold improved mutant to background is maximized. *Bottom*. Ratio of fluorescence signal of improved mutants to that of the wild-type (arrow at ratio peak).

by Eq. 29.2 (Boder and Wittrup 1998),

$$k_{\text{of wt}} t_{\text{opt}} = 0.293 + 2.05 \log k_r + \left(2.30 - 0.759 \cdot \frac{1}{k_r} \right) \log S_r \qquad (29.2)$$

sorting is started. The ratio of the dissociation rate constants between wild-type and mutant proteins k_r can be estimated as two to three for affinity maturation. For sorting, the dissociation reaction can be slowed by placing the sample on ice. If the dissociation reaction is not fully stopped, the target cells can be collected during a particular time window. This approach thus provides for convenient fractionation of a library based on affinity during sorting.

Sorting

As discussed previously, the use of multiple logical gates has provided good results in practice (Daugherty et al., 2000; Leary, 1994). Gated parameters typically include forward and 90 degree light scatter, fluorescence due to the labeled ligand (FL1), and fluorescence due to nonspecific stain controls and nonviability indicators (FL2). Thus, one gate is set in the FSC-SSC window. A second gate is set in a 90 degree or FSC versus FL1 (Fig. 29.5), and a third gate is set in a FSC versus FL2 plot to include cells that are not FL2-positive. The sort gate is then defined as a logical combination of gates (Gate 1 AND Gate 2 AND Gate 3). The gates drawn in FSC-FL plots may often be diagonal because both autofluorescence and specific fluorescence correlate with cell volume, and hence with roughly with FSC. After acquiring more than 10^5 events, the shape of the nontarget population can be visualized in a two parameter histogram and the shape of the target cell population can be inferred.

The probability of recovering a given target cell during sorting is dependent on several factors including the frequency of the target cell within the population, the probability of detecting that cell, the probability that it falls into all of the gates, and whether that cell is viable on returning to culture. Thus, each of these factors should be considered independently (Daugherty, Iverson, and Georgiou, 2000).

Typically, two to four rounds of sorting may be required depending on the experimental goals. The fraction of gated cells depends on multiple factors including the type of library being screened, maximum resolution between target and nontarget cells, and various other parameters. In general, the fraction of fluorescent cells gated is determined by balancing the desired enrichment factor per round of sorting and the desired probability that all target clones will be collected. For the special case of affinity maturation of an existing protein, roughly the most fluorescent 1% of the cell population is recovered in the first round (Boder and Wittrup, 1998). In subsequent rounds, increasingly smaller numbers of cells are collected because their representation will have been increased in the initial round. Typical enrichment results from a kinetic-based screen are shown in Figure 29.6.

Magnetic Cell Sorting for Preenrichement Magnetically activated cell sorting (MACS) offers potential benefits for screening very large cell surface libraries. In particular, MACS allows parallel processing of more than 10^9–10^{10} or more cells in only minutes. Cells are incubated with magnetically tagged reagents, washed, and applied to the magnetic column. Cells simply pass through the magnetic matrix material by gravity flow. Thus, this method provides a means to screen larger libraries, using MACS as a preenrichment

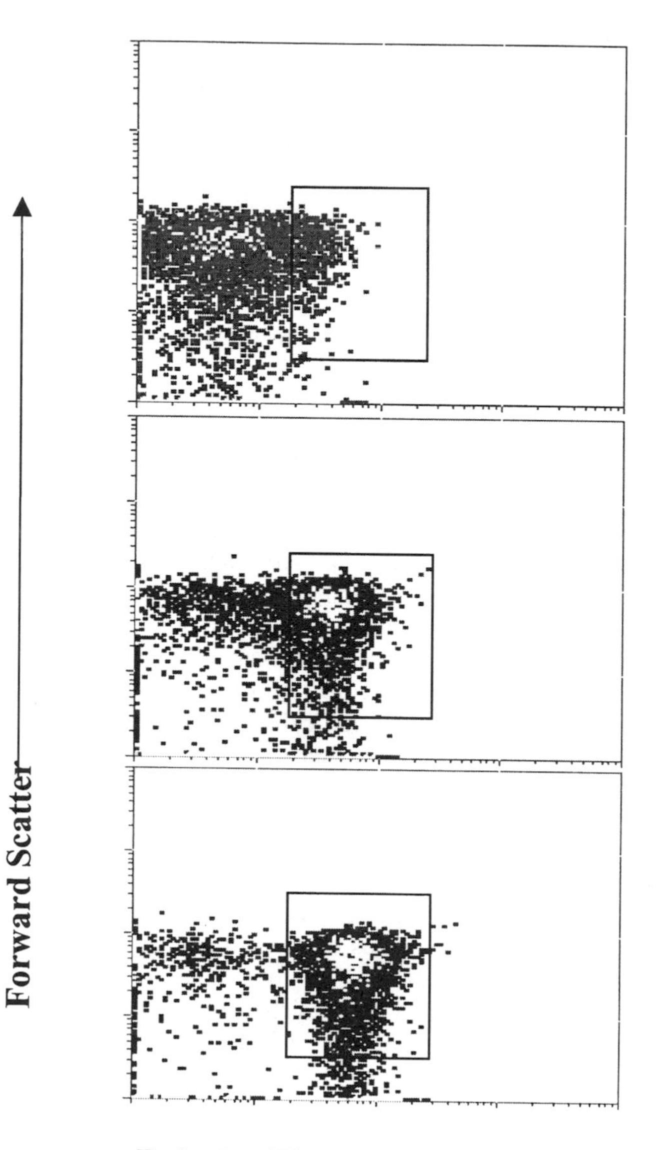

Figure 29.5 Enrichment of target clones from a library of randomly mutated scFv, for the purpose of affinity maturation. Forward scatter verses fluorescence of heterogeneous populations from presort (*top*), postsort two (*middle*), and postsort four (*bottom*).

before using FCM. Kolmar and co-workers used a MACS-based protocol for the preenrichment of a large library of cystein knot proteins displayed on bacteria (Christmann et al., 1999). Unfortunately, despite the low initial cost of magnetic separation columns, reagents can be quite expensive.

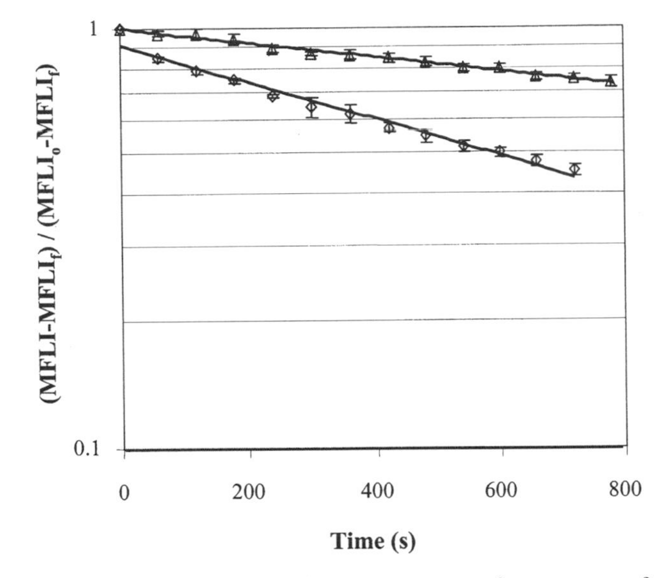

Figure 29.6 Improvement of dissociation rate constant of a mutant, surface displayed scFv: Decay in mean fluorescence as BODIPY-digoxin dissociates from whole cells displaying wild-type scFv (dig) (◇, botttom), and the mutant EP1-1(△, top) isolated after 5 rounds of FACS. Standard deviations were determined from quadruplicate measurements. Experimental details are exactly as described (Daugherty et al., 2000).

Analysis of Isolated Clones

Another benefit of surface display coupled with FCM is the ability to directly measure relative binding parameters of isolated library members without further experimental manipulation, including protein expression and purification. Enriched cell populations can be plated onto solid growth media to isolate individual clones for further analysis. Clones are then grown in liquid medium, labeled, and analyzed using FCM. Two different methods can be used to determine binding parameters with good precision. First, individual clones can be separately labeled with a series of probe concentrations, allowing estimation of the equilibrium binding affinity. Second, the relative dissociation rate constant can be determined by recording the decay in fluorescence of a population as a function of time (Fig. 29.6). In practice, these rate constants correlate strongly with those measured using surface plasmon resonance (Daugherty et al., 1998).

CONCLUSIONS

The use of flow cytometry to screen cell surface-displayed protein libraries is a relatively new yet powerful technology with significant potential for assisting the production of targeted vectors for gene therapy applications. Whereas bacterial systems may allow efficient selection and engineering of ligands, yeast systems offer an expanded range of processing for mammalian proteins requiring posttranslational modifications. It also appears likely that mammalian surface display and retroviral expression libraries will become increasingly used for target identification and ligand engineering. But perhaps the most important benefit of displaying proteins on cell surfaces is the ability to visualize, quantify, and optimize the screening process using FCM. The fine affinity discrimination capability of FCM should allow a more precise engineering of existing ligands and improved selection of new and useful receptors and ligands.

REFERENCES

Baron U, Bujard H (2000): Tet repressor-based system for regulated gene expression in eukaryotic cells: principles and advances. *Methods Enzymol* 327:401–421.

Battini JL, Rasko JE, Miller AD (1999): A human cell-surface receptor for xenotropic and polytropic murine leukemia viruses: possible role in G protein-coupled signal transduction. *Proc Natl Acad Sci USA* 96(4):1385–1390.

Benhar I, Azriel R, Nahary L, Shaky S, Berdichevsky Y, Tamarkin A, Wels W (2000): Highly efficient selection of phage antibodies mediated by display of antigen as Lpp-OmpA' fusions on live bacteria. *J Mol Biol* 301:893–904.

Boder ET, Midelfort KS, Wittrup KD (2000): Directed evolution of antibody fragments with monovalent femtomolar antigen-binding affinity. *Proc Natl Acad Sci USA* 97:10701–10705.

Boder ET, Wittrup KD (1997): Yeast surface display for screening combinatorial polypeptide libraries. *Nat Biotechnol* 15:553–557.

Boder ET, Wittrup KD (1998): Optimal screening of surface-displayed polypeptide libraries. *Biotechnol Prog* 14:55–62.

Boder ET, Wittrup KD (2000): Yeast surface display for directed evolution of protein expression, affinity, and stability. *Methods Enzymol* 328:430–444.

Chapman GV (2000): Instrumentation for flow cytometry. *J Immunol Meth* 243:3–12.

Christmann A, Walter K, Wentzel A, Kratzner R, Kolmar H (1999): The cystine knot of a squash-type protease inhibitor as a structural scaffold for *Escherichia coli* cell surface display of conformationally constrained peptides. *Protein Eng* 12:797–806.

Daugherty PS, Chen G, Iverson BL, Georgiou G (2000): Quantitative analysis of the effect of the mutation frequency on the affinity maturation of single chain Fv antibodies. *Proc Natl Acad Sci USA* 97:2029–2034.

Daugherty PS, Chen G, Olsen MJ, Iverson BL, Georgiou G (1998): Antibody affinity maturation using bacterial surface display. *Protein Eng* 11:825–832.

Daugherty PS, Iverson BL, Georgiou G (2000): Flow cytometric screening of cell-based libraries. *J Immunol Meth* 243:211–227.

Daugherty PS, Olsen MJ, Iverson BL, Georgiou G (1999): Development of an optimized expression system for the screening of antibody libraries displayed on the *Escherichia coli* surface. *Protein Eng* 12:613–621.

de Kruif J, Terstappen L, Boel E, Logtenberg T (1995): Rapid selection of cell subpopulation-specific human monoclonal antibodies from a synthetic phage antibody library. *Proc Natl Acad Sci USA* 92:3938–3942.

Francisco JA, Earhart CF, Georgiou G (1992): Transport and anchoring of beta-lactamase to the external surface of *Escherichia coli. Proc Natl Acad Sci USA* 89:2713–2717.

Francisco JA, Campbell R, Iverson BL, Georgiou G (1993): Production and fluorescence-activated cell sorting of *Escherichia coli* expressing a functional antibody fragment on the external surface. *Proc Natl Acad Sci USA* 90:10444–10448.

Francisco JA, Stathopoulos C, Warren RA, Kilburn DG, Georgiou G (1993b): Specific adhesion and hydrolysis of cellulose by intact *Escherichia coli* expressing surface anchored cellulase or cellulose binding domains. *Bio/Technology* 11:491–495.

Georgiou G, Stathopoulos C, Daugherty PS, Nayak AR, Iverson BL, Curtiss R 3rd (1997): Display of heterologous proteins on the surface of microorganisms: from the screening of combinatorial libraries to live recombinant vaccines. *Nat Biotechnol* 15:29–34.

Griffin MD, Holman PO, Tang Q, Ashourian N, Korthauer U, Kranz DM, Bluestone JA (2001): Development and applications of surface-linked single chain antibodies against T-cell antigens. *J Immunol Methods* 248:77–90.

Gross HJ, Verwer B, Houck D, Hoffman RA, Recktenwald D (1995): Model study detecting breast cancer cells in peripheral blood mononuclear cells at frequencies as low as 10(-7). *Proc. Natl Acad Sci USA* 92:537–541.

Guzman LM, Belin D, Carson MJ, Beckwith J (1995): Tight regulation, modulation, and high-level expression by vectors containing the arabinose PBAD promoter. *J Bacteriol.* 177:4121–4130.

Haidaris CG, Fisher DJ, Gigliotti F, Simpson-Haidaris PJ (1998): Antigenic properties of recombinant glycosylated and nonglycosylated Pneumocystis carinii glycoprotein A polypeptides expressed in baculovirus-infected insect cells. *Mol Biotechnol* 9:91–97.

Haugland RP (1996): *Handbook of Fluorescent Probes and Research Chemicals.* Eugene, OR: Molecular Probes, Inc.

Holler PD, Holman PO, Shusta EV, O'Herrin S, Wittrup KD, Kranz DM (2000): In vitro evolution of a T cell receptor with high affinity for peptide/MHC. *Proc Natl Acad Sci USA* 97:5387–5392.

Kieke MC, Shusta EV, Boder ET, Teyton L, Wittrup KD, Kranz DM (1999): Selection of functional T cell receptor mutants from a yeast surface-display library." *Proc Natl Acad Sci USA* 96:5651–5656.

Kitamura T. (1998): New experimental approaches in retrovirus-mediated expression screening. *Int J Hematol* 67:351–359.

Kjaergaard K, Sorensen JK, Schembri MA, Klemm P (2000): Sequestration of zinc oxide by fimbrial designer chelators. *Appl Environ Microbiol* 66:10–14.

Klemm P, Schembri MA (2000): Fimbrial surface display systems in bacteria: from vaccines to random libraries. *Microbiology* 146:3025–3032.

Lang H (2000): Outer membrane proteins as surface display systems. *Int J Med Microbiol* 290:579–585.

Larocca D, Jensen-Pergakes K, Burg MA, Baird A (2001): Receptor-targeted gene delivery using multivalent phagemid particles. *Mol Ther* 3:476–484.

Leary JF (1994): Strategies for rare cell detection and isolation. *Meth Cell Biology* 42:331–358.

Lu Z, Murray KS, Van Cleave V, LaVallie ER, Stahl ML, McCoy JM (1995): Expression of thioredoxin random peptide libraries on the *Escherichia coli* cell surface as functional fusions to flagellin: a system designed for exploring protein-protein interactions. *Biotechnology* 13:366–372.

Medof ME, Nagarajan S, Tykocinski ML (1996): Cell-surface engineering with GPI-anchored proteins. *FASEB J* 10:574–586.

Meerman HJ, Georgiou G (1994): Construction and characterization of a set of *E. coli* strains deficient in all known loci affecting the proteolytic stability of secreted recombinant proteins. *Biotechnology* 12:1107–1110.

Pautsch A, Schulz GE (2000): High-resolution structure of the OmpA membrane domain. *J Mol Biol* 298:273–282.

Rasko JE, Battini JL, Gottschalk RJ, Mazo I, Miller AD (1999): The RD114/simian type D retrovirus receptor is a neutral amino acid transporter." *Proc Natl Acad Sci USA* 96:2129–2134.

Rode HJ, Little M, Fuchs P, Dorsam H, Schooltink H, de Ines C, Dübel S, Breitling F (1996): Cell surface display of a single-chain antibody for attaching polypeptides. *Biotechniques* 21:650, 652–653, 655–656, 658.

Scallon BJ, Kado-Fong H, Nettleton MY, Kochan JP (1992): A novel strategy for secreting proteins: use of phosphatidylinositol-glycan-specific phospholipase D to release chimeric phosphatidylinositol-glycan anchored proteins. *Biotechnology* 10:550–556.

Schaffer C, Graninger M, Messner P (2001): Prokaryotic glycosylation. *Proteomics* 1:248–261.

Schier R, Bye J, Apell G, McCall A, Adams GP, Malmquist M, Weiner LM, Marks JD (1996): Isolation of high-affinity monomeric human anti-c-erbB-2 single chain Fv using affinity-driven selection. *J Mol Biol* 255:28–43.

Shapiro H. (1995): *Practical Flow Cytometry*. New York: Wiley-Liss.

Shapiro HM (2001): Optical measurements in cytometry: light scattering, extinction, absorption, and fluorescence. *Meth Cell Biol* 63:107–129.

Shimazu M, Mulchandani A, Chen W (2001): Cell surface display of organophosphorus hydrolase using ice nucleation protein. *Biotechnol Prog* 17:76–80.

Shusta EV, Holler PD, Kieke MC, Kranz DM, Wittrup KD (2000): Directed evolution of a stable scaffold for T-cell receptor engineering. *Nat Biotechnol* 18:754–759.

Shusta EV, Kieke MC, Parke E, Kranz DM, Wittrup KD (1999): Yeast polypeptide fusion surface display levels predict thermal stability and soluble secretion efficiency. *J Mol Biol* 292:949–956.

Shusta EV, VanAntwerp J, Wittrup KD (1999): Biosynthetic polypeptide libraries. *Curr Opin Biotechnol* 10:117–122.

VanAntwerp JJ, Wittrup KD (2000): Fine affinity discrimination by yeast surface display and flow cytometry. *Biotechnol Prog* 16:31–37.

Wentzel A, Christmann A, Kratzner R, Kolmar H (1999): Sequence requirements of the GPNG beta-turn of the Ecballium elaterium trypsin inhibitor II explored by combinatorial library screening. *J Biol Chem* 274:21037–21043.

Wester L, Fast J, Labuda T, Cedervall T, Wingardh K, Olofsson T, Akerstrom B (2000): Carbohydrate groups of alpha1-microglobulin are important for secretion and tissue localization but not for immunological properties. *Glycobiology* 10:891–900.

Whitchorn EA, Tate E, Yanofsky SD, Kochersperger L, Davis P, Mortensen RB, Yonkovich S, Bell K, Dower WJ, Barrett RW (1995): A generic method for expression and use of "tagged" soluble versions of cell surface receptors. *Biotechnology* 13:1215–1219.

Wong BY, Chen H, Chung SW, Wong PM (1994): High-efficiency identification of genes by functional analysis from a retroviral cDNA expression library. *J Virol* 68:5523–5531.

Yang WP, Green K, Pinz-Sweeney S, Briones AT, Burton DR, Barbas CF 3rd (1995): CDR walking mutagenesis for the affinity maturation of a potent human anti-HIV-1 antibody into the picomolar range. *J Mol Biol* 254:392–403.

PART V

MONITORING OF TARGETING

30

MONITORING GENE THERAPY BY POSITRON EMISSION TOMOGRAPHY

HARVEY R. HERSCHMAN, PH.D., JORGE R. BARRIO, PH.D.
NAGICHETTIAR SATYAMURTHY, PH.D., QIANWA LIANG, PH.D.,
DUNCAN C. MACLAREN, PH.D., SHARIAR YAGHOUBI, B.S.,
TATSUSHI TOYOKUNI, PH.D., SIMON R. CHERRY, PH.D.
MICHAEL E. PHELPS, PH.D., AND SANJIV S. GAMBHIR, M.D., PH.D.

INTRODUCTION

The fundamental goals of gene therapy are easy to state and easy to grasp: to express, from somatically transferred genes, gene products that produce a therapeutic benefit to the recipient. Gene therapists might replace a damaged gene with a functional gene, introduce a gene whose product will lead to the death of an unwanted cell population, or introduce a new function into a target cell population.

The devil, of course, is in the details; a wide variety of pitfalls to successful therapeutic gene administration immediately come to mind. How can therapeutic genes be targeted to the cells/tissues/organs that require therapeutic modification? How can one make certain that only the appropriate target cells express the therapeutic gene? How can the time, in biological terms, of therapeutic gene expression be regulated? How can the magnitude of therapeutic gene expression be regulated? How can the expression of the therapeutic gene be altered as the recipient responds to the therapy?

Vector Targeting for Therapeutic Gene Delivery, Edited by David T. Curiel and Joanne T. Douglas
ISBN 0-471-43479-5

Animal model systems have been used to address these important and difficult practical questions. Because of their short breeding time and well-characterized genetics, mice have been the primary species used for these basic mechanistic studies. To investigate changes in vector design, routes of administration, timing of administration, and the many other variables that might play a role in successful somatic gene transfer and expression of the desired gene product, researchers need an assay(s) to evaluate levels of ectopic gene expression following somatic transfer. To develop technological improvements in somatic gene delivery, "reporter genes" that can rapidly, easily and effectively be monitored have been used.

Reporter Gene Analysis

The basic concept of reporter gene analysis is predicated on the idea that an inexpensively, rapidly, and sensitively assayed reporter gene protein product can be used to monitor the efficacy of gene delivery vehicles by incorporating the reporter gene coding region into the therapeutic gene delivery system. Two common reporter genes that were used during the early development of gene therapy were chloramphenicol acetyl transferase (Westphal et al., 1985; Leite, Niel, and D'Halluin, 1986) and β-galactosidase (Forss-Petter et al., 1990; Naciff et al., 1999). The disadvantage of chloramphenicol acetyl transferase and β-galactosidase as reporter genes is that each assay requires sacrifice of the animal; consequently one gets only a snapshot—an assay at a single point in time—of their expression from somatically transferred DNA. Consequently, to do time-dependent analyses of gene expression, groups of animals for each time point must be sacrificed and assayed and statistical variations among groups of animals at each time point must be analyzed. Clearly, such reporter genes would not be useful for monitoring somatic gene delivery in patients. Reporter genes whose products are secreted (e.g., alkaline phosphatase) permit the investigator serially to monitor recipients of somatically transferred DNA by repeated blood sampling and analyses, but sacrifice the benefit of determining the site of reporter gene expression.

In Vivo Monitoring of Reporter Gene Expression

To both advance research in somatic gene transfer and reduce to clinical practice this new form of therapy, it would be of great advantage to develop techniques to noninvasively, quantitatively, and repetitively monitor the location, magnitude, and duration of therapeutic gene expression. Recent developments in molecular and medical imaging modalities in the past several years have now made it possible to achieve these goals. While it is currently not practical to develop *in vivo* imaging probes for each therapeutic gene under investigation, reporter gene technologies have, since 1996, been developed that permit noninvasive, repetitive, and—in some cases—quantitative imaging. A variety of radionuclide, magnetic resonance imaging, and optical detection systems are now under development to monitor the expression of reporter genes following somatic gene transfer.

Genes whose protein products can be assayed by optically based techniques, such as green fluorescent protein (Chalfie et al., 1994; Naylor, 1999; Yang et al., 2000), β-lactamase (Zlokarnik et al., 1998), and firefly luciferase (Contag et al., 2000) have been introduced as reporter genes to monitor somatic gene transfer. Expression of these reporter genes, like chloramphenicol acetyl transferase and β-galactosidase, can be analyzed in tissue extracts following sacrifice or biopsy. As a result of sensitive new technologies, the expression of these optically assayed reporter genes can be monitored in vivo in small animals, using video cameras, digital cameras, and sensitive charged coupled device cameras. The convenience and relative inexpensiveness of noninvasive optical imaging provides advantages for some applications in small animals. However, optical techniques are often limited by fluorophore quenching effects, depth-dependent photon attenuation, scatter, and resolution issues. Moreover, without endoscopic analysis, optical imaging of reporter gene expression is likely to be quite limited in patient applications.

We have recently reviewed a variety of methods for in vivo monitoring of reporter gene expression (Ray et al., 2001). Our laboratories have focused primarily on reporter gene technologies that utilize positron emission tomography (PET) to detect reporter gene expression; we review PET procedures for imaging reporter gene expression here.

PRINCIPLES OF POSITRON EMISSION TOMOGRAPHY

Although PET has made enormous contributions to clinical oncology, neurology, cardiology, and other medical specialties (Phelps, 2000), it is still not a household word or a technology familiar to most cell biologists, virologists, and biochemists. PET utilizes compounds labeled with positron-emitting radioisotopes to measure biological processes such as enzyme reactions and ligand binding to receptors in living individuals. Because tracer amounts of the radiolabeled probes are used, PET can measure biological processes without producing mass disturbances of biochemical processes. When positrons are emitted from nuclei of proton-rich radioisotopes (e.g., ^{18}F, $t\frac{1}{2}$ = ~110 min; ^{11}C, $t\frac{1}{2}$ = ~20 min), the positrons interact with electrons in the surroundings. Annihilation results, and the mass of the positron and electron is converted into two 511 keV gamma rays that travel outward from the annihilation site at ~180° from one another. The positron emission tomograph, or PET scanner, is a circular array of scintillation detectors coupled to photomultiplier tubes. The scanner records coincident gamma ray emissions from the chemical probes that contain positron-emitting isotopes.

Molecules labeled with positron-emitting radionuclides, injected intravenously in tracer quantities, are retained in tissues as a result of either (1) the labeled ligand binding to receptor or (2) conversion of the labeled enzyme substrate to a trapped product. The kinetics of probe entry into tissue, clearing from tissue, and receptor- or enzyme-dependent retention of the positron-emitting probe in tissue are detected with the tomograph. It should be emphasized that PET uses relatively small mass amounts of high specific activity probes. Concentrations of PET imaging probes (generally in the range of nano- to picomole amounts) are typically orders of magnitude lower than the concentrations of the corresponding endogenous or pharmacologic agents required to elicit a biological response.

An Example: PET Analysis of Enzyme Reactions

To measure enzyme reactions with PET, a positron-emitting substrate that can be delivered to tissue via the bloodstream and can move into and out of cells is selected. The substrate is converted by the target enzyme to a product that is retained in cells. For example, the positron-emitting glucose analogue 2-deoxy-2-[^{18}F]fluoro-D-deoxyglucose (FDG), following intravenous injection, is transported across the vascular endothelium and then is transported into and out of cells. The enzyme hexokinase converts FDG to FDG-6-phosphate, which is trapped inside cells. Thus, FDG is transported into, phosphorylated, and retained in tissues in proportion to the rate of hexokinase-mediated phosphorylation of glucose. When the unphosphorylated FDG has been eliminated from the tissues and circulation, the accumulation of phosphorylated FDG can be imaged and quantitatively measured by the PET scanner (Phelps, 2000). The principle of PET assay of glucose metabolism with FDG is illustrated in Figure 30.1.

An Example: PET Measurement of Receptors

A positron-emitting dopamine D2 receptor (D2R) antagonist, 3-(2-[^{18}F]fluoroethyl)spiperone (FESP), binds to the D2R with a very slow off rate (~10 h). Following intravenous FESP injection, tissues expressing D2R (e.g., the striatum) bind and retain the positron-emitting labeled ligand. When images are collected in the tomograph, D2R-rich tissues can be identified and D2R levels can be quantitated, at ng/mg tissue levels, in living subjects. PET can detect 2–20 nM D2R concentrations in the striatum of living individuals (Phelps, 1991). Using PET, properties such as receptor density and metabolic rate can be repeatedly and noninvasively quantitated in living subjects.

PRINCIPLES OF PET REPORTER GENE/PET REPORTER PROBE ANALYSIS OF REPORTER GENE EXPRESSION

Analysis of PET reporter genes employs the same principles used to analyze enzyme reactions or receptor–ligand interactions by conventional PET analysis. PET reporter genes encode proteins that are either (1) receptors that can

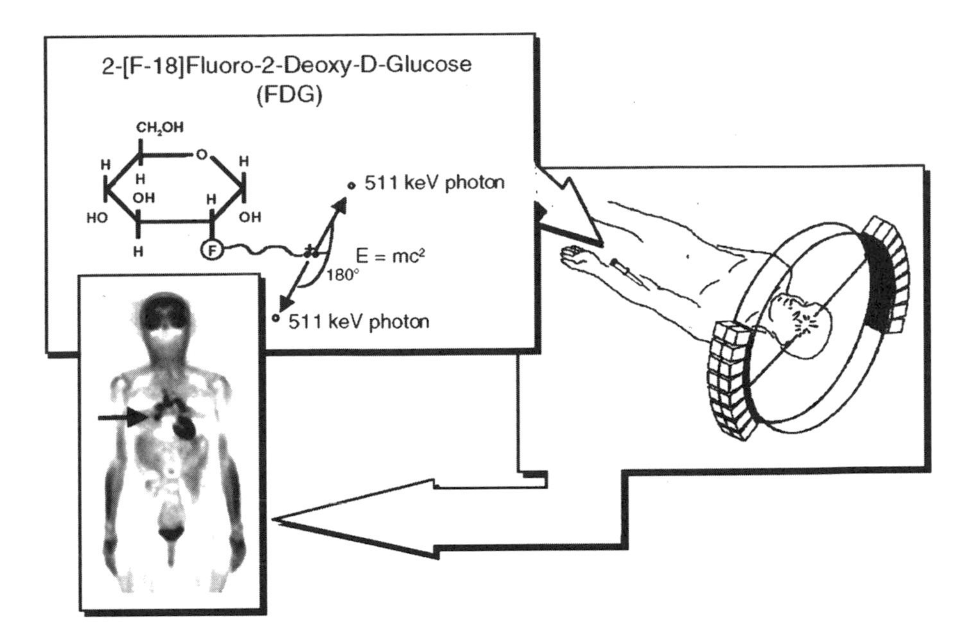

Figure 30.1 Principles of PET imaging. A biologically active molecule such as FDG, labeled with a positron emitting radioisotope, is injected intravenously. The tracer distributes throughout the vasculature, where it can move into and out of tissues. In cells/tissues with elevated glucose transport and/or hexokinase II activity, the FDG is phosphorylated and cannot escape. Radiolabeled product accumulates in cells/tissues with active hexokinase, while the bolus of unphosphorylated FDG clears from the body. Positrons emitted from the trapped FDG combine with electrons, where annihilation occurs. The masses of the two particles are converted to two 511-keV photons 180° apart, according to the relationship $E = mc^2$. The two emitted photons that (nearly) simultaneously strike opposing detectors are recorded as a coincidence event. Although the figure shows a single coincidence event, there are many detector pair combinations recording events. After correction for attenuation, tomographic images of probe concentrations are reconstructed. This figure shows a section (6 mm thick) from a woman with previously diagnosed ovarian cancer. The arrow points to bilateral lung metastases. The brain and heart, which have high metabolic rates and consume large amounts of glucose, can also be easily seen in the illustration. (From Phelps 2000, by permission.)

bind, and therefore sequester, positron-emitting PET reporter ligand probes or (2) enzymes that convert nonsequestered, positron-emitting PET reporter substrate probes to sequestered products (Fig. 30.2). When individuals expressing a PET reporter gene that encodes a receptor are injected with the corresponding PET reporter ligand probe, cells expressing the reporter receptor will sequester the reporter ligand probe as a ligand-receptor complex detectable in the PET scanner. When animal or human subjects expressing a PET reporter gene that encodes an enzyme are injected with the corresponding PET reporter substrate probe, cells expressing the reporter enzyme will sequester the reporter sub-

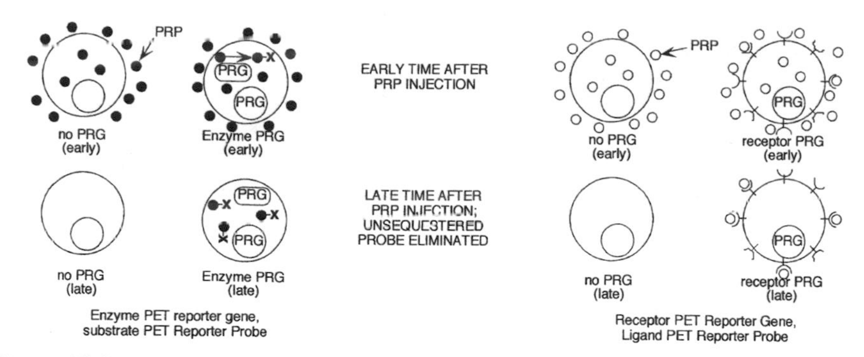

Figure 30.2 Principles of two PET reporter gene/PET reporter probe systems. On the left is illustrated a PRG/PRP system that utilizes an enzyme as a PET reporter gene. A positron-emitting substrate PRP is converted by the PRG to a sequestered product that is subsequently detected by tomography. On the right is illustrated a PRG/PRP system that utilizes a receptor as a PET reporter gene. A positron-emitting ligand PRP is trapped by the PRG. The trapped ligand is subsequently detected by tomography.

strate probe as a positron-emitting product of the enzymatic reaction catalyzed by the PET reporter enzyme. Cells that do not express PET reporter genes will not retain PET reporter probes. Tomographic imaging then demonstrates PET reporter gene-dependent sequestration of the positron-emitting reporter probe. For enzyme-based PET reporter gene systems, the reporter probe must be transported into the target cell.

MICROPET, A SMALL ANIMAL PET SCANNER THAT PROVIDES ACCESS TO MICE

Current clinical PET scanners have a resolution of $(5–8)^3$ mm^3, a substantial portion of the volume of a mouse. Our UCLA colleagues developed a PET scanner, microPET, dedicated to small animal imaging. MicroPET can image rodents at a resolution of 1.8^3 mm^3 (Cherry et al., 1997; Chatziioannou et al., 1999); a one-to-two order of magnitude improvement in volumetric resolution over existing clinical PET scanners. MicroPET II, currently in development, will provide resolution of $\sim 1^3$ mm^3, that is, volumetric pixels of 1 μl will be distinguishable with this instrument. Other small animal PET scanners are also being developed (Correia et al; 1999; Weber et al., 1999; Jeavons, Chandler, and Dettmar, 1999; Chatziioannou, 2001).

THE DOPAMINE D2 RECEPTOR AS A PET REPORTER GENE

The D2R is expressed primarily in the striatum, in the central nervous system. When bound by agonist ligands, the D2R activates a G-protein coupled signal transduction system and elicits a reduction in cyclic adenosine monophosphate (cyclic AMP) accumulation in cells. Several positron-emitting labeled probes derived from neuroleptics and related structures have been developed to image and measure the D2 receptor levels in living individuals. One of these probes, FESP, has been used to monitor the D2 receptor levels in vivo in rodents, nonhuman primates, and humans (reviewed in Phelps, 1991). For one of our PET reporter gene/PET reporter probe systems, we simply used the D2R gene as a PET reporter gene. Ectopic expression of the D2R reporter gene is imaged by using FESP as the PET reporter probe.

We used a replication-defective adenovirus delivery system to demonstrate the noninvasive, repetitive, and quantitative ability of the D2R/FESP PET reporter gene/PET reporter probe combination as a noninvasive in vivo imaging system. As discussed extensively in other chapters in this volume, when adenovirus is injected intravenously into mice, infection and expression of both viral genes and reporter genes is overwhelmingly restricted to the liver. The dominant hepatic expression of adenovirus-directed genes is a consequence both of hepatic filtration properties and the extensive hepatic expression in mice, relative to other tissues, of the coxsackie and adenovirus receptor.

To create a replication-deficient adenovirus expressing the D2R PET reporter gene, we first placed the rat D2R coding sequence behind the cytomegalovirus (CMV) early promoter. The CMV-D2R cassette was then inserted into the E1 region of a replication-deficient type 5 adenovirus to create ad.D2R (MacLaren et al., 1999). To determine whether ad.D2R could effectively deliver the D2R as an imagable PET reporter gene, mice were injected by tail vein with ad.D2R. Ad-βGal, a replication-defective adenovirus expressing the β-galactosidase gene as a reporter, was injected into control mice. Two days after injection of the adenoviral vectors, the mice were injected by tail vein with FESP. After three hours, the D2R-dependent sequestration of FESP was measured by microPET imaging. Whole body coronal PET images of living mice injected with ad.βGal and ad.D2R, after systemic FESP administration, are shown in the left panel of Figure 30.3. Substantial retention of fluorine-18 activity can be observed in the liver of the mouse injected with ad.D2R. In contrast, relatively little accumulation of fluorine-18 activity is present in the liver of the mouse injected with ad.βGal.

MicroPET analysis can quantitatively monitor hepatic D2R expression in living ad.D2R infected mice. Mice were first injected with varying amounts of ad.D2R virus, expressing D2R from the CMV promoter. The mice were subsequently injected intravenously with FESP, then imaged by microPET. FESP concentrations in the liver were determined by region of interest (ROI) analysis of the microPET scans (MacLaren et. al, 1999). The mice were then killed and liver samples were analyzed for fluorine-18 retention in a well counter. Additional liver samples were assayed for D2R activity by [^{3}H]spiperone binding, using a conventional in vitro D2 receptor binding assay (Bunzow et al., 1988). The relationship between the concentration of fluorine-18 retained in liver, determined by ROI measurements of PET images, was then plotted against

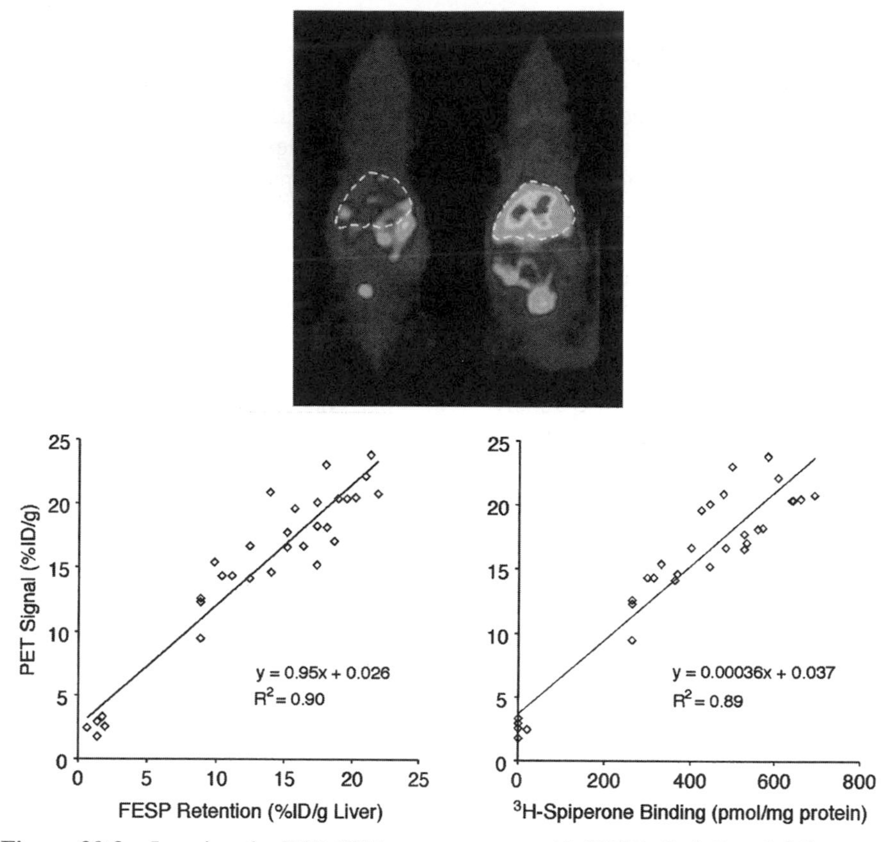

Figure 30.3 Imaging the D2R PET reporter gene with FESP. (*Left Panel*) Mice were injected intravenously with ad.βGal or ad.D2R (2×10^9 pfu). Two days later the mice were injected with FESP then imaged by microPET. Whole body coronal projections are shown, with the livers outlined in white. Red indicates the greatest intensity; purple the least. (*Right Panel*). Mice were injected with varying titers of ad.D2R ranging from 5×10^6 to 9×10^9 pfu, subjected to microPET analysis with FESP and then sacrificed. Liver samples were counted by well counting to determine the level of retained [^{18}F]-FESP and labeled metabolites. Additional liver samples were assayed for [^{3}H] spiperone binding. The graph on the left plots in vivo ^{18}F retention determined by region of interest analysis from the PET scan data versus the in vivo hepatic ^{18}F retention determined by well counting. The right graph shows in vivo ^{18}F retention determined by region of interest analysis from the PET scan data plotted versus the level of hepatic D2R receptor activity determined by [^{3}H] spiperone binding. (From Herschman et al. 2000, by permission.) Figure also appears in Color Figure section.

either the well counter data or the D2R-dependent [^{3}H]spiperone binding (i.e., versus hepatic D2R levels). The PET ROI values for FESP retention are proportional to the D2R levels, determined by receptor binding assays (Fig. 30.3). MicroPET analysis of the levels of hepatic D2R reporter gene expression in living animals accurately reflect in vitro measurements of hepatic D2R expression levels.

We have also analyzed the specificity of the D2R PET reporter gene/FESP PET reporter probe system with pharmacologic tools. Two mice were injected with ad.D2R. The expression of hepatic D2R was then assayed with FESP, either in the presence (mouse B, Fig. 30.4) or absence (mouse A, Fig. 30.4) of excess unlabelled (+)-butaclamol (a D2R antagonist). The excess butaclamol effectively blocks specific binding of FESP to the D2R and, therefore, retention of FESP in the liver. Because the half-life of ^{18}F is only 110 minutes, a second microPET analysis can be performed relatively quickly. After the initial FESP activity had decayed and the butaclamol had cleared from the tissue, the mouse that initially received (+)-butaclamol and FESP (mouse B) was reinjected with FESP a second time and reimaged by microPET. Hepatic FESP retention, that is, D2R PET reporter gene expression, can now be observed. These data demonstrate (1) the pharmacologic specificity of the FESP probe for D2R reporter gene imaging and (2) the noninvasive and repetitive nature of PET reporter gene/PET reporter probe analyses.

The D2R/FESP reporter gene/reporter probe system has the advantage that FESP can be routinely produced by well-characterized procedures (Barrio et al., 1989; Satyamurthy et al., 1990). In addition, other positron-emitting ligands such as [^{11}C]raclopride (Ehrin et al., 1985; Hall et al., 1988; Hume et al., 1992) and N-[^{11}C]methylspiperone (Wagner et al., 1983) may also be useful as probes

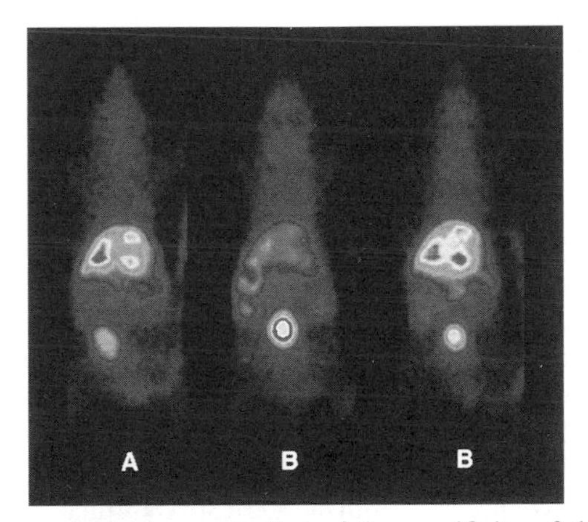

Figure 30.4 Pharmacologic demonstration of the specificity of the D2R/FESP PET reporter gene/PET reporter probe imaging system. Two mice were injected intravenously with ad.D2R. After 2 days, mouse B was injected intraperitoneally with (+)-butaclamol (2 mg/kg). After an additional hour, both mice were injected with FESP and subsequently imaged with microPET. Mouse B was injected a second time with FESP after another 3 days (5 days after virus injection) and again imaged with microPET. Summed whole body coronal images are shown. (From Herschman et al. 2000, by permission.) Figure also appears in Color Figure section.

to detect the D2R reporter gene by PET. Finally, [^{123}I]iodobenzamide derivatives (Kessler et al., 1991) should be useful for analyzing the D2R reporter gene by SPECT. PET offers the advantage of higher sensitivity (>10-fold) and better spatial resolution as compared to SPECT. The higher sensitivity is particularly important in the reporter gene approaches when low levels of reporter gene expression are encountered.

A SECOND GENERATION D2R PET REPORTER GENE

One potential problem with D2R as a PET reporter gene is the possibility that ectopic D2R expression might change the physiology of target cells. Agonist binding to the D2R elicits activation of a G-protein complex leading to inactivation of adenylyl cyclase. As a result, cellular cyclic AMP levels decrease. FESP administration is not likely to elicit a biological effect, because of the low mass levels used for D2R reporter gene detection. However, circulating *endogenous* ligands for ectopically expressed D2R might provide a chronic stimulus, inducing undesired biological responses.

Two mutations in the D2R are able to uncouple ligand binding from activation of the G-protein complex signaling pathway leading to suppressed cyclic AMP synthesis (Neve et al., 1991; Woodward et al., 1996; Cox et al., 1992). We have investigated the utility of these two mutated D2R genes as PET reporter genes (Liang et al., 2001). We created ad.D2R80A and ad.D2R194A by standard site-directed mutagenesis and adenovirus construction protocols. In ad.D2R80A the aspartic acid at residue 80 of the D2R has been replaced by alanine; in ad.D2R194A the serine at residue 194 has been replaced with alanine. C6 glioma cells infected with ad.D2R, ad.D2R80A, and ad.D2R194A all bind equivalent amounts of [^{3}H]spiperone; mutating these residues has no effect on spiperone binding.

Stimulation of cells with forskolin activates cyclic AMP synthesis. When cells expressing the D2 receptor are exposed to dopamine, the forskolin stimulated decrease in cyclic AMP levels can be blocked by the dopamine/D2R activation of the G-protein system. Thus, when C6 glioma cells are infected with ad.D2R and treated with forskolin, the cyclic AMP accumulation induced by forskolin is attenuated by dopamine (Fig. 30.5). In contrast, cyclic AMP accumulation in cells infected with ad.D2R194A is only slightly reduced following dopamine treatment, and the level of forskolin-stimulated cyclic AMP accumulation in cells infected with ad.D2R80A is unaffected by dopamine treatment.

Because of its essentially complete uncoupling of ligand binding and modulation of cyclic AMP levels in cultured cells, we characterized the effectiveness of the D2R80A mutant D2R as a PET reporter gene. When mice are injected with adenovirus expressing either the wild-type D2R gene or the mutant D2R80A gene, for which ligand binding and ligand activation of the G-protein system are uncoupled, no difference in hepatic FESP accumula-

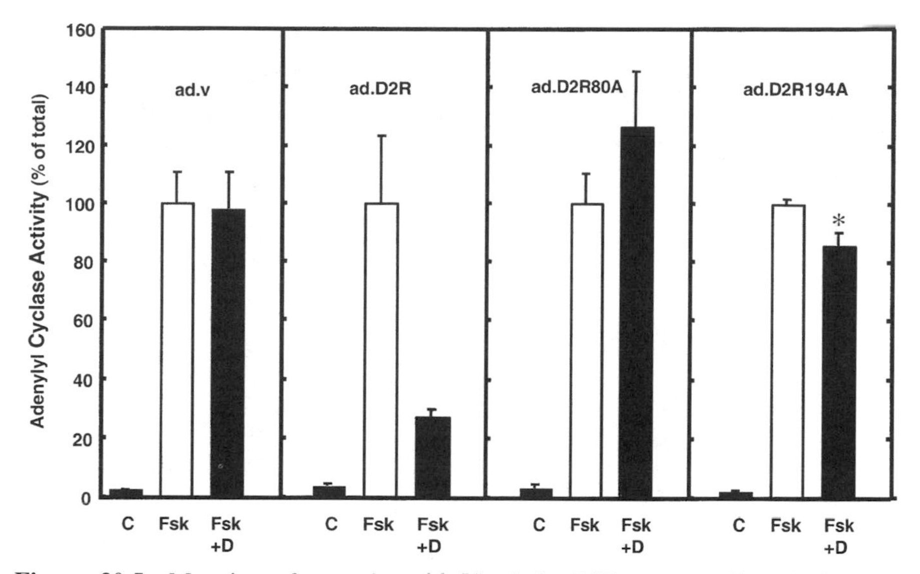

Figure 30.5 Mutation of aspartic acid 80 of the D2R prevents ligand-stimulated cyclic AMP attenuation. C6 cells grown in 12 well plates (3×10^5 cells per 3.8 cm^2 well) were infected with 1×10^6 pfu of ad.D2R, ad.D2R80A, or ad.D2R194A. After 24 h, the cells were treated with 10 μM forskolin, either in the presence or absence of dopamine (10 μM) according to the protocol described by Liang et al. (2001). After 15 min the cells were lysed and cyclic AMP levels were determined. Results of four independent experiments are expressed as a percentage of total cyclic AMP accumulation induced by 10 μM forskolin in the absence of dopamine. The * indicates that for ad.D2R194A-infected cells, there is a significant reduction ($p < 0.05$) of cAMP levels in Fsk+D treated cells versus cells treated with Fsk alone. (From Liang et al. 2001, by permission.)

tion could be detected (Fig. 30.6). The D2R80A protein is as effective a PET reporter as the wild-type D2R protein and does not activate signal transduction in response to ligand binding.

THE HERPES SIMPLEX VIRUS 1 THYMIDINE KINASE (HSV1-TK) GENE AS A PET REPORTER GENE

Like murine and human thymidine kinase (TK), HSV1-TK phosphorylates thymidine. However, acycloguanosines (e.g., acyclovir, ACV; ganciclovir, GCV; penciclovir, PCV) are much more effectively phosphorylated by HSV1-TK than by mammalian TKs. Following phosphorylation of the acycloguanosines by HSV1-TK, the acycloguanosine monophosphates are converted by cellular kinases to di- and triphosphates. If phosphorylated acycloguanosines are present in high enough concentrations they can kill cells, either by chain termination or by inhibition of DNA polymerase. The ability

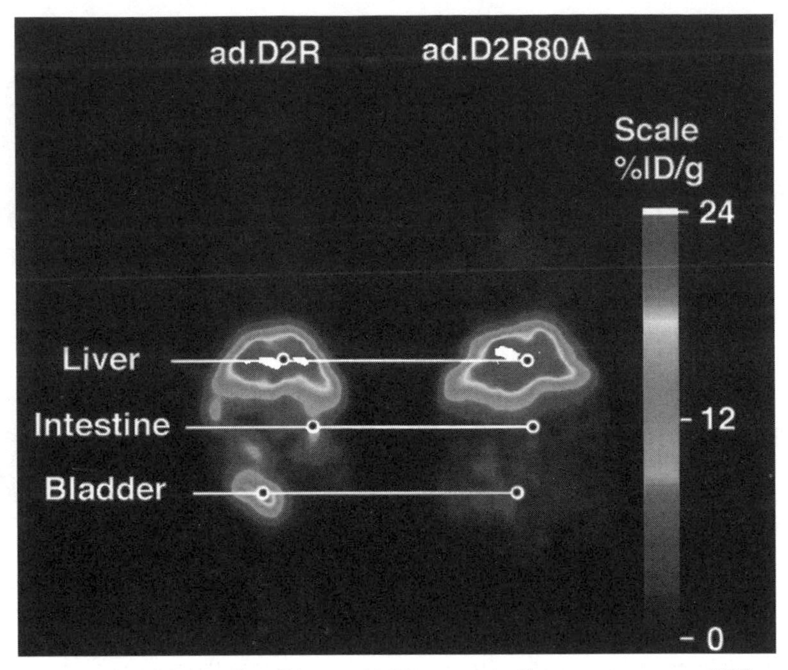

Figure 30.6 D2R80A and wild-type D2R are equally competent as PET reporter genes. Mice were injected intravenously with 2×10^9 pfu of ad.D2R or ad.D2R80A. Two days later, the mice were injected with FESP and imaged with microPET. Figure also appears in Color Figure section.

of HSV1-TK to convert acycloguanosine prodrugs to toxic compounds is used in cancer gene therapy protocols to target tumor cells for destruction; HSV1-*tk* is utilized as a suicide gene. If HSV1-*tk* expression can be restricted to tumor cells, then the acycloguanosine prodrugs can be used to kill the tumor cells expressing HSV1-TK. HSV1-*tk* is also used in therapeutic somatic gene transfer protocols as a safety device. Should cells expressing HSV1-*tk*, along with a potential therapeutic gene product, elicit an adverse effect, the cells can be killed by systemic administration of pharmacologic levels of an acycloguanosine (Reviewed in Herschman et al., 2000; Gambhir et al., 2000b).

We prepared a replication-deficient adenovirus, ad.HSV1-tk, similar to ad.D2R and ad.D2R80A. The wild-type HSV1-*tk* gene is expressed from the CMV promoter in ad.HSV1-tk (Gambhir et al., 1998). To investigate the ability of HSV1-tk to act as a PET reporter gene, we synthesized the positron-emitter labeled HSV1-tk substrate 8-[^{18}F]fluoro-9-[(1,3-dihydroxy-2-propoxy)methyl]guanine (fluoroganciclovir, or FGCV) (Namavari et al., 2000).

MicroPET imaging of hepatic FGCV retention following ad.HSV1-tk administration was used to examine the ability of HSV1-*tk* to serve as a PET reporter gene (Gambhir et al., 1999). Mice were injected via tail vein with ad.βgal or ad.HSV1-tk. Two days after viral administration the mice were

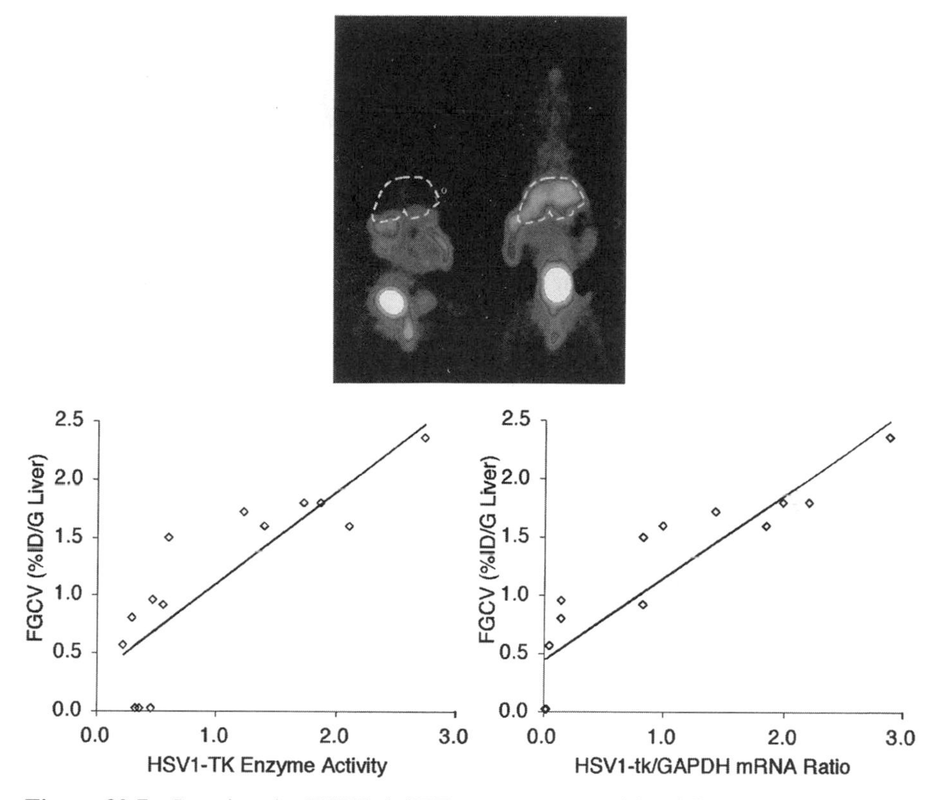

Figure 30.7 Imaging the HSV1-tk PET reporter gene with FGCV. (*Left Panel*) Mice were injected intravenously with 1×10^9 pfu of ad.βgal or ad.HSV1-tk. Two days later mice were injected with FGCV and imaged with microPET after 1 h. (*Right Panel*). Mice were injected intravenously with varying numbers of ad.HSV1-tk. After 2 days, the mice were injected with FGCV and sacrificed after 1 h. Liver samples were analyzed for FGCV retention (by well counting), HSV1-tk mRNA (by Northern blot), and HSV1-TK enzyme (by measuring conversion of [^{3}H]GCV to [^{3}H]GCV-P). (From Herschman et al., 2000, by permission.) Figure also appears in Color Figure section.

injected via tail vein with FGCV, then subjected to microPET analysis (Fig. 30.7). Mice injected with ad.HSV1-tk had substantial fluorine-18 accumulation in their livers. In contrast, mice receiving ad.βgal had no detectable fluorine-18 retention. It should be emphasized that the FGCV levels used for microPET imaging are well below the pharmacologic levels of GCV required for HSV1-TK dependent toxicity.

We analyzed the quantitative relationship between HSV1-*tk*-dependent FGCV retention and HSV1-*tk* PET reporter gene expression in ad.HSV1-tk infected mice, much as we did in our analysis of the D2R/FESP PET reporter gene/PET reporter probe in vivo imaging system. Mice were injected intravenously with varying numbers of ad.HSV1-tk. Two days later the mice were injected intravenously with FGCV. After 1 h, the mice were sacrificed, and hepatic fluorine-18 tissue concentrations were analyzed using a well counter. Liver samples were also analyzed both for HSV1-*tk* mRNA levels and HSV1-TK enzyme levels (Fig. 30.7). The hepatic retention of the FGCV PET reporter probe is proportional to expression of the HSV1-*tk* PET reporter gene. An excellent correlation between the FGCV %ID/gm liver and both HSV1-*tk* mRNA ($r^2 = 0.81$) and HSV1-TK enzyme activity ($r^2 = 0.71$) is observed (Fig. 30.7).

Penciclovir is a more effective HSV1-TK substrate than ganciclovir. We thought that a fluorinated form of penciclovir might, therefore, be a better substrate than FGCV for imaging the HSV1-*tk* PET reporter gene. We synthesized 8-[^{18}F]fluoro-9-[4-hydroxy-3-(hydroxymethyl)but-1-yl]guanine (fluoropenciclovir, or FPCV) and investigated the utility of FPCV for imaging HSV1-tk expression. FPCV is two- to threefold more effective than is FGCV in detecting hepatic HSV1-TK expression in mice injected with ad.HSV1-tk (Iyer et al., 2001).

Several laboratories have described the synthesis of acycloguanosines labeled with fluorine-18 in the side chain, rather than in the C8 position. Alauddin et al. (1996) first suggested that labeling of the side chain of ganciclovir, rather than the ring, might provide a more effective substrate for HSV1-TK. They described the synthesis of 9-[3-[^{18}F]-fluoro-1-hydroxy-2-propoxy)methyl]guanine (FHPG) (Alauddin et al., 1996), and tested this probe in vivo as an imaging probe for HSV1-*tk* (Alauddin et al., 1999). Both Hospers et al. (2000) and Hustinx et al. (2001) demonstrated substantial uptake of FHPG in HSV1-*tk* expressing tumors following intravenous injection of the tracer.

Alauddin and Conti (1998) synthesized 9-[(4-[^{18}F]-fluoro-3-hydroxymethyl-butyl)guanine (FHBG), the side-chain fluorinated analogue of penciclovir. Although they did not describe PET imaging studies with FHBG, comparative cell culture studies with FHPG (Alauddin et al., 1996) and FHBG (Alauddin and Conti, 1998) suggested that FHBG would be a better imaging probe for HSV1-*tk*.

Positron-emitter labeled thymidine analogues have also been developed to image the HSV1-*tk* PET reporter gene in living animals. Radiodine labeled 5-iodo-2′-fluoro-2′deoxy-1-β-D-arabinofuranosyluracil (FIAU) and (E)-5-(2-iodovinyl)-2′-fluoro-2′-deoxyuridine have been used to image both stable HSV1-*tk* gene expression in xenografted tumors and HSV1-*tk* genes delivered in viral vectors to xenografted tumors (reviewed in Gambhir et. al., 2000b). In particular, [^{124}I]FIAU has been developed as an imaging substrate for HSV1-TK and utilized as a probe to image HSV1-*tk* expression in vivo by PET analysis. Tjuvajev et al. (1998), using [^{124}I] FIAU, imaged HSV1-*tk* expression in tumors by PET. PET imaging of HSV1-TK reporter gene expression with [^{124}I]FIAU should be a useful clinical tool for monitoring somatic gene transfer and has been validated in a number of experimental models (reviewed in Gambhir et al., 2000b; Sadelain and Blasberg, 1999). Because of the longer half-life (4.2 days) of the positron emitter iodine-124, [^{124}I]FIAU may be a preferred alternative to image HSV1-TK in applications where systemic clearance of substrate limits imaging resolution or sensitivity. A recent report compared [^{124}I]FIAU and FHPG as substrates for the wild-type HSV1-TK enzyme, using xenografted tumors stably expressing HSV1-*tk*, and concluded that [^{124}I]FIAU may be the preferred substrate (Brust et al., 2001) for imaging the HSV1-TK reporter.

A SECOND GENERATION HSV1-TK PET REPORTER GENE

The HSV1-tk gene is under intense investigation as a therapeutic gene for cancer treatment. HSV1-tk can convert pharmacologic levels of acycloguanosine prodrugs such as acyclovir, ganciclovir, and penciclovir to toxic forms. Thus, if ectopically overexpressed in target cells, HSV1-TK can kill these cells, following delivery of pharmacologic levels of these compounds. To improve the therapeutic efficacy of HSV1-TK, Black et al. (1996) created mutant HSV1-TK enzymes and screened for HSV1-TK mutants that are more effective than wild-type HSV1-TK at phosphorylating ganciclovir and less effective than wild-type TK at utilizing endogenous thymidine. We reasoned that mutant HSV1-TK enzymes that more effectively utilize acycloguanosines as substrates might also be more effective than wild-type TK in phosphorylating positron emitter-labeled acycloguanosines, making them more effective PET reporter genes. Using both cell culture models and stably transformed tumor cell lines carrying the transfected HSV1-*sr39tk* mutant gene, we demonstrated that FGCV, FPCV, FHPG, and FHBG are more effectively phosphorylated by HSV1-sr39TK than by wild-type TK. As expected, FHBG is the most effective substrate for HSV1-sr39TK.

We constructed a replication-defective adenovirus, ad.TKm, in which the HSV1-sr39*tk* PET reporter gene is expressed from the CMV promoter. When equivalent titers of ad.TKm (expressing HSV1-sr39TK) and ad.TK (expressing HSV1-TK) are injected intravenously into mice, the mutant enzyme is substantially more effective than the wild-type enzyme, utilizing FPCV as substrate, as a reporter gene (Gambhir et al., 2000a). Modifying both the PET reporter gene and the PET reporter probe for the HSV1-TK imaging system has resulted in a substantial improvement in imaging capability; the sensitivity of the ad.TKm/FHBG system is about an order of magnitude greater than the sensitivity of the ad.TK/FGCV system described in our original studies. Our data suggest that the HSV1-sr39tk/FHBG imaging system should provide an excellent PET reporter gene/PET reporter combination with which to noninvasively monitor somatic gene transfer in living individuals. FHBG distribution in human volunteers is consistent with its potential as a PET reporter probe to image HSV1-*tk* based PET reporter genes (Yaghoubi et al., 2001a). It will be of great interest to compare the sensitivities of the HSV1-TK/[^{124}I]FIAU and HSV1-sr39TK/FHBG imaging systems.

The HSV1 and HSV2 thymidine kinase genes have been recombined in vitro, using DNA family shuffling, and chimeras with an enhanced ability to phosphorylate AZT (3′-azido 3′deoxythymidine) have been isolated (Christians et al., 1999). Similar procedures could be used to develop HSV-*tk* reporter genes that will more effectively use either FIAU or FHBG, further improving the imaging capability of the HSV1-*tk* PET reporter gene system.

ADVANTAGES AND DISADVANTAGES OF THE D2R AND HSV1-TK PET REPORTER GENE SYSTEMS

Our current HSV1-sr39tk/FHBG and D2R80A/FESP PET reporter gene/PET reporter probe imaging systems now have essentially equivalent sensitivities. Each reporter gene system, however, has distinct advantages and disadvantages. The positron emitter-labeled probes for HSV1-TK do not significantly penetrate the intact blood–brain barrier. Consequently, no matter what the level of reporter gene expression, *in vivo* imaging is not possible within the brain, if the blood–brain barrier is not breached (e.g., in the case of some brain tumors). In contrast, FESP—the probe used for imaging D2R—has access to all tissues.

The HSV1-sr39*tk* gene is a foreign gene. HSV1-sr39TK expression following somatic gene transfer may lead to an immune response. Because the D2R gene is an endogenous gene, present in the mammalian genome, ectopic expression should not cause an immune response. However, endogenous D2R expression in the striatum and, to a lesser extent, in other tissues increases the background signal for FESP retention. In some contexts, endogenous localized D2R expression could be used to clinical advantage; striatal FESP sequestration could potentially be used for internal normalization when measuring ectopic D2R reporter gene expression in other tissues.

HSV1-TK expression has little or no effect on cells and tissues, as long as acycloguanosine derivatives are not present. In contrast, occupancy by endogenous agonists of the wild-type D2R receptor ectopically expressed from gene therapy delivery vehicles might have physiological consequences, since ligand-activated D2R regulates intracellular cyclicAMP levels (Strange, 1990). The development of the D2R80A mutant, in which ligand binding and activation of the G-protein mediated modulation of cyclic AMP levels are uncoupled (Liang et al., 2001), should alleviate this concern. However, it will be necessary to examine other signal transduction pathways normally stimulated by activation of the D2R in cells expressing D2R80A.

MONITORING OF COORDINATELY EXPRESSED THERAPEUTIC GENES

Ideally, it would be of great utility if the expression of each therapeutic gene could be analyzed directly by an *in vivo* imaging technique, using a gene-specific imaging probe sequestered either by the protein product or the mRNA product of the therapeutic gene. Because this goal has not yet been obtained, imaging technology has turned to indirect or inferential measurements of therapeutic gene expression. The basic concept is to coexpress a therapeutic gene product and a reporter gene product in a coordinately regulated fashion. By noninvasively measuring changes in expression of the reporter gene product, one can infer correlative changes in the expression of the therapeutic gene product. Our laboratories have investigated several alternative methods to administer coregulated therapeutic and imaging genes.

Bicistronic Vectors

The most common way in which to coordinately express two genes in a common cell is to utilize bicistronic vectors. Many DNA viruses express polycistronic messages, from which several proteins are translated (Jang et al., 1989; Schmid and Wimmer, 1994). The translation of internally initiated proteins, in a cap-independent fashion, is facilitated by internal ribosomal entry sites (IRES) in the viral messages. In a bicistronic reporter gene expression system, two coding regions are translated from a common message. One coding sequence (e.g., encoding a therapeutic protein) is placed in a cap-dependent position proximal to an IRES. A second coding sequence (e.g., encoding a reporter protein) placed distal to the IRES can be translated by the cap-independent mechanism. Since the two gene products are translated from a common message, expression of the two gene products should be proportional to one another even if the expression level of the bicistronic message varies substantially over time following DNA transfer. Thus, by measuring changes in the level of one of the protein products of the bicistronic message, investigators can infer similar changes in the relative level of the second protein product.

To determine whether coordinate expression from a bicistronic message can be observed with PET reporter genes, we took advantage of our ability to monitor two different PET reporter systems. Our initial studies used xenografted tumors to demonstrate proof of principle. We constructed a plasmid containing the CMV promoter driving a message in which the D2R PET reporter gene is placed proximal to the encephalomyocarditis virus IRES and the HSV1-sr39*tk* PET reporter gene is placed distal to the IRES (Fig. 30.8). C6 glioma cells were transfected with this construct and clones stably expressing differing levels of the D2R-IRES-HSV1-sr39*tk* message were grown as xenograft tumors on nude mice. The levels of D2R and HSV1-sr39*tk* reporter gene expression were measured by microPET, using FESP and FHBG respectively as imaging probes (Fig. 30.9). The level of D2R expression (measured as FESP retention)

Figure 30.8 The dual PET reporter gene bicistronic plasmid expressing D2R and HSV1-sr39TK proteins from a common message.

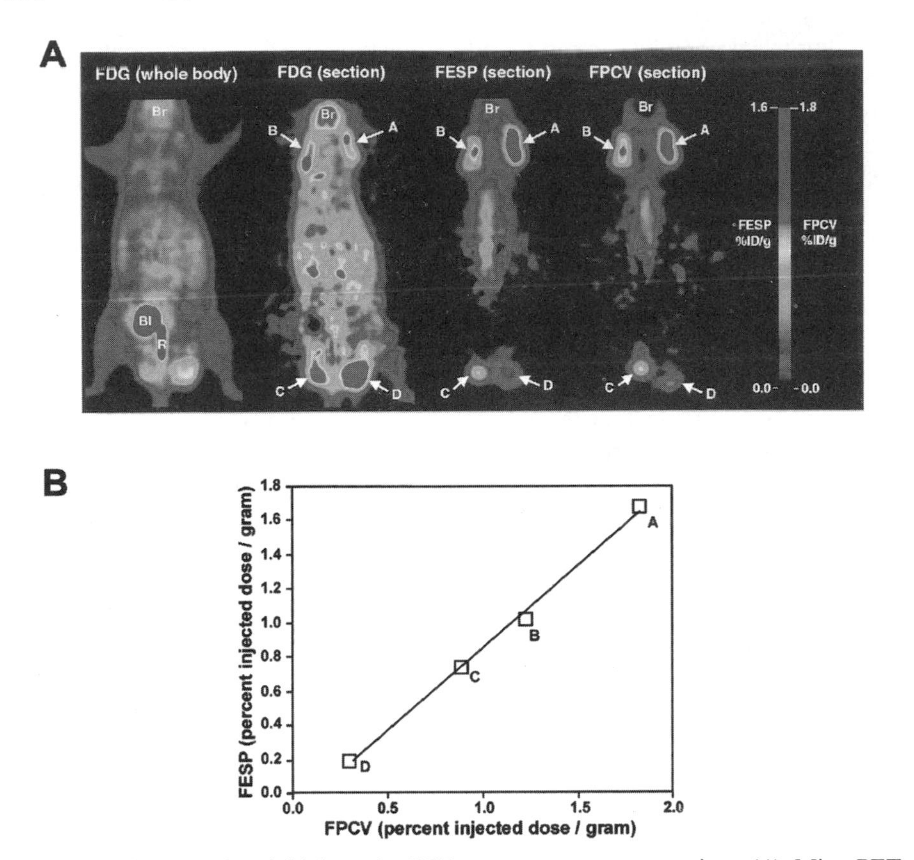

Figure 30.9 Correlated bicistronic PET reporter gene expression. (*A*) MicroPET imaging of xenograft tumors expressing differing levels of the pCMV-D2R-IRES-HSV1-sr39tk bicistronic plasmid. Three stably transfected C6 cell lines (A, B, and C) and C6 cells (D) were injected into four distinct sites on a nude mouse. Ten days later, the mouse was injected with FDG and analyzed by microPET. A microPET analysis was performed with FESP 24 h later. The following day a third microPET analysis was performed with FPCV. The FDG whole body image on the left averages all coronal (horizontal) planes. As a result, the tumors are not well visualized. The second FDG image, a set of coronal images that passes through all the tumors, shows FDG accumulation in each tumor. Br: brain; Bl: Bladder; R. Rectum. (*B*) Plot of FESP %ID/gm tissue versus FPCV %ID/gm of tissue. The data are obtained from regions of interest drawn on the images in panel A. The error bars represent the standard deviations for three regions of interest placed on the tumor images. The correlation between FESP and FPCV microPET signals is $r^2 = 0.99$. (From Yu et al., 2000, by permission.) Figure also appears in Color Figure section.

and the level of HSV1-sr39*tk* expression (measured as FHBG retention) are proportional (Yu et al., 2000). Tjuvajev et al. (1999), using SPECT imaging for in vivo analysis of HSV1-TK coupled with sacrifice and analysis of β-galactosidase, have also demonstrated that noninvasive imaging of HSV1-TK can reflect the location and magnitude of expression for a *cis*-linked second coding region.

We have recently created the virus ad.DTm, D2R in which the D2R PET reporter gene is placed proximal to the EMCV IRES and the HSV1-sr39tk PET reporter gene is placed distal to the IRES. Expression of the bicistronic message in ad.DTm is driven by the CMV promoter. When ad.DTm is injected intravenously into mice, both the D2R and HSV1-sr39tk PET reporter genes can be noninvasively and quantitatively monitored in living mice following somatic gene transfer (Fig. 30.10).

The use of bicistronic vectors is not without problems. Attenuation of expression for the distal gene is quite common, potentially reducing the sensitivity of a reporter gene placed in this position. Moreover, attenuation of distal gene expression can vary for the same bicistronic message expressed in different tissues, providing potentially misleading results for relative levels of therapeutic gene expression. Experiments with "super IRES" sequences (Chappell, Edelman, and Mauro, 2000) are underway in several laboratories, in attempts to ameliorate these problems.

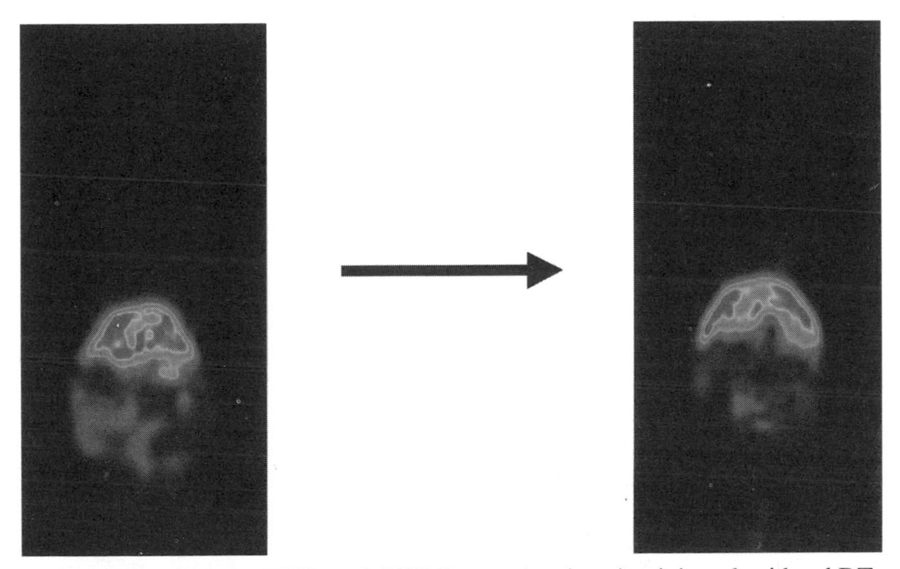

Figure 30.10 Hepatic FESP and FHBG retention in mice injected with ad.DTm, expressing the D2R and HSV1-sr39tk PET reporters from a bicistronic message. Ad.DTm (2×10^9 pfu), in which the D2R-IRES-HSV1-sr39tk message is expressed from the CMV promoter, was injected via the tail vein into a Swiss-Webster mouse. Two days later, the mouse was injected with FHBG and scanned by microPET to image HSV1-sr39tk reporter gene expression. The following day, the mouse was injected with FESP and scanned by microPET to image expression of the D2R reporter gene. Data are shown as % ID/gm of FESP and FHBG, determined from region of interest analysis.

Covector Administration

A relatively simple approach to analyzing the location and expression of a therapeutic gene is to administer, at the same time, a second vector identical in all respects, with the single exception that the therapeutic gene is replaced by the imaging gene. Clearly, a large number of assumptions are made in such an approach. Although on an individual cell basis substantial differences in infection, selection, and so forth might occur, at the macrosopic/organ level such individual cell variation may not play a substantial role. To test this approach to imaging, we coadminstered ad.D2R and ad.HSV1-sr39tk, distinct adenoviruses expressing D2R and HSV1-sr39tk, to nude mice and compared FESP and FHBG retention. We examined D2R and TK expression by microPET following tail vein injection, intramuscular injection, and intratumoral injection of the two viruses. The expression of the D2R and HSV1-sr39*tk* genes, as measured by FESP and FHBG retention, is proportional both with respect to viral dose and with respect to time after injection of the two viruses (Yaghoubi et al., 2001b). Although this is, in principle, a very simple approach to indirect imaging of ectopic gene expression, coadministration is an alternative that might be useful, certainly in experimental situations.

Bidirectional Transcription Vectors

There exist bidirectional expression vectors in which two messages can be expressed from a common promoter (Krestel et al., 2001). These vectors might be thought of as expressing two genes independently from the same delivery system. We have used this type of expression cassette in xenografts of stably transfected tumor cells to express our two PET reporter genes from a tetracycline-inducible, bidirectional promoter (Sun et al., 2001). Prior to doxycycline administration, both HSV1-sr39tk expression (measured by microPET analysis of FHBG accumulation) and D2R expression (measured by microPET analysis of FESP accumulation) were minimal. In response to doxycycline administration in the drinking water, to activate the tet-responsive bidirectional promoter element, HSV1-sr39TK and D2R reporter gene expression were coordinately induced, as measured by microPET analyses of these same tumor-bearing mice. Following doxycycline withdrawl, microPET scanning for a third time demonstrated the coordinate reduction of both D2R and HSV1-sr39*tk* gene expression. This dual gene expression cassette experiment demonstrates both (1) another mechanism by which correlated expression of therapeutic and reporter genes can potentially be expressed and (2) the ability of PET reporter gene analyses to monitor pharmacologically regulated alterations in gene expression in living individuals.

CONCLUSION

The imaging technologies discussed in this chapter are generalizable, both to all methods of somatic gene transfer and gene targeting and to cell trafficking and cell therapies. The laboratories that have developed the in vivo imaging technologies discussed in this chapter are not, for the most part, laboratories that previously worked in the area of gene therapy research. Instead, the development of in vivo reporter gene imaging procedures has come primarily from laboratories whose major interest has been either in the area of regulation of gene expression or in the area of biomedical imaging technology.

We are often asked whether we plan to build a specific type of vector—a lentivirus, retrovirus, plasmid DNA or other vector-incorporating in vivo reporter genes such as D2R or HSV1-*tk*. We are also often asked whether we plan to apply the new molecular imaging technologies to specific problems in oncology, cardiology, neurology, infectious diseases, or to any problem in which somatic gene transfer is under investigation as a potential therapeutic modality. We do, of course, have our own areas of biological interest in which we are using noninvasive reporter gene imaging as a tool. However, the goal of our molecular imaging group at UCLA is to develop a new set of tools that can be integrated by other researchers into their own scientific objectives. We anticipate that noninvasive in vivo imaging will be added into vector design both by laboratories committed to the development of specific types of vectors and by laboratories committed to the somatic DNA therapy approach to specific diseases.

The ability to noninvasively and quantitatively monitor targeted gene expression following somatic DNA exposure should provide an enormous benefit to the goals of vector targeting studies. In addition, by stably transfecting with PET reporter genes cells used for therapeutic purposes, investigator/physicians can monitor both initial trafficking of cell transplants as well as the subsequent viability, proliferation, and location of transplanted cells. Now that PET scanners are becoming more widely distributed and radiopharmaceutical companies are making positron-labeled probes more widely available, PET approaches to monitoring reporter genes in humans will become much more accessible.

REFERENCES

Alauddin MM, Conti PS (1998): Synthesis and preliminary evaluation of 9-(4-[^{18}F]-fluoro-3-hydroxymethylbutyl)guanine ([^{18}F]FHBG): a new potential imaging agent for viral infection and gene therapy using PET. *Nucl Med Biol* 25:175–180.

Alauddin MM, Conti PS, Mazza SM, Hamzeh FM, Lever JR (1996): Synthesis of 9-[(3-[^{18}F]fluoro-1-hydroxy-2-propoxy)methyl]guanine ([^{18}F]FHPG): A potential imaging agent of viral infection and gene therapy using PET. *Nucl Med Biol* 23:787–792.

Alauddin MM, Shahinian A, Kundu RK, Gordon EM, Conti PS (1999): Evaluation of 9-[(3-^{18}F-fluoro-1-hydroxy-2-propoxy)methyl] guanine ([^{18}F]FHPG) in vitro and in vivo as a probe for PET imaging of gene incorporation and expression in tumors. *Nucl Med Biol* 26:371–376.

Barrio JR, Satyamurthy N, Huang SC, Keen RE, Nissenson CH, Hoffman

JM, Ackermann RF, Bahn MM, Mazziotta JC, Phelps ME (1989): 3-(2′-[^{18}F]fluoroethyl)spiperone: in vivo biochemical and kinetic characterization in rodents, nonhuman primates, and humans. *J Cereb Blood Flow Metab* 9:830–839.

Black ME, Newcomb TG, Wilson H-MP, Loeb LA (1996): Creation of drug-specific Herpes Simplex Virus type 1 thymidine kinase mutant for gene therapy. *Proc Natl Acad Sci USA* 93:3525–3529.

Brust P, Haubner R, Friedrich A, Scheunemann M, Anton M, Koufaki ON, Hauses M, Noll S, Noll B, Haberkorn U, Schackert G, Schackert HK, Avril N, Johannsen B (2001): Comparison of [^{18}F]FHPG and [$^{124/125}$I]FIAU for imaging herpes simplex virus type 1 thymidine kinase gene expression. *Eur J Nucl Med* 28:721–729.

Bunzow JR, Van Tol HH, Grandy DK, Albert P, Salon J, Christie M, Machida CA, Neve KA, Civelli O (1988): Cloning and expression of a rat D2 dopamine receptor cDNA. *Nature* 336:783–787. Comment in *Nature* (1989): 342:865.

Chalfie M, Tu Y, Euskirchen G, Ward WW, Prasher DC (1994): Green fluorescent protein as a marker for gene expression. *Science* 263:802–805.

Chappell SA, Edelman GM, Mauro VP (2000): A 9-nt segment of a cellular mRNA can function as an internal ribosome entry site (IRES) and when present in linked multiple copies greatly enhances IRES activity. *Proc Natl Acad Sci USA* 97:1536–1541.

Chatziioannou A (2001): PET scanners dedicated to molecular imaging of small animal models. *Mol Imaging Biol* 4:47–63.

Chatziioannou AF, Cherry SR, Shao Y, Silverman RW, Meadors K, Farquhar TH, Pedarsani M, Phelps, ME (1999): Performance evaluation of microPET: A high resolution Leutetium Oxyorthosilicate PET scanner for animal imaging. *J Nuc Med* 40:1164–1175.

Cherry SR, Shao Y, Silverman RW, Meadors K, Siegel S, Chatziioannou A, Young JW, Jones W, Moyers JC, Newport D, Boutefnouchet A, Farquhar TH, Andreaco M, Paulus MJ, Binkley DM, Nutt R, Phelps ME (1997): MicroPET: a high resolution PET scanner for imaging small animals. *IEEE Trans Nucl Sci* 44:1161–1166.

Christians FC, Scapozza L, Crameri A, Folkers G, Stemmer WP (1999): Directed evolution of thymidine kinase for AZT phosphorylation using DNA family shuffling. *Nat Biotechnol* 17:259–264.

Contag CH, Jenkins D, Contag PR, Negrin RS (2000): Use of reporter genes for optical measurements of neoplastic disease in vivo. *Neoplasia* 2:41–52.

Correia JA, Burnham CA, Kaufman D, Fischman AJ (1999): Development of a small animal PET imagnng device with resolution approaching 1 mm. *IEEE Trans Nucl Sci* 46:631–635.

Cox BA, Henningsen RA, Spanoyannis A, Neve RL, Neve KA (1992): Contributions of conserved serine residues to the interactions of ligands with dopamine D_2 receptors. *J Neurochem* 59:627–635.

Ehrin E, Farde L, de Paulis T, Eriksson L, Greitz T, Johnstrom P, Litton JE, Nilsson JL, Sedvall G, Stone-Elander S, et al. (1985): Preparation of ^{11}C-labelled Raclopride, a new potent dopamine receptor antagonist: preliminary PET studies of cerebral dopamine receptors in the monkey. *Int J Appl Radiat Isot* 36:269–273.

Forss-Petter S, Danielson PE, Catsicas S, Battenberg E, Price J, Nerenberg M, Sutcliffe JG (1990): Transgenic mice expressing beta-galactosidase in mature neurons under neuron-specific enolase promoter control. *Neuron* 5:187–200.

Gambhir SS, Barrio JR, Wu L, Iyer M, Namavari M, Satyamurthy N, Parrish C, MacLaren DC, Borghei AR, Bauer E, Green LA, Sharfstein S, Berk AJ, Cherry SR, Phelps ME, Herschman, HR (1998): Imaging of adenoviral directed herpes simplex virus Type 1 thymidine kinase reporter gene expression in mice with ganciclovir. *J Nucl Med* 11:2003–2011.

Gambhir SS, Barrio JR, Phelps ME, Iyer M, Namavari M, Satyamurthy N, Wu L, Green LA, Bauer E, MacLaren DC, Nguyen K, Berk AJ, Cherry SR, Herschman HR (1999): Imaging adenoviral-directed reporter gene expression in living animals with positron emission tomography. *Proc Nat Acad Sci USA* 96:2333–2338.

Gambhir SS, Bauer E, Black ME, Liang Q, Kokoris MS, Barrio JR, Iyer M, Namavari M, Satyamurthy N, Green LA, Nguyen K, Cherry SR, Phelps ME, Herschman HR (2000a): A mutant herpes simplex virus Type 1 thymidine kinase reporter gene shows improved sensitivity for imaging reporter gene expression with positron emission tomography. *Proc Nat Acad Sci* 97:2785–2790.

Gambhir SS, Herschman HR, Cherry SR, Barrio JR, Satyamurthy S, Toyokuni T, Phelps ME, Balatoni J, Finn R, Tjuvajev J, Blasberg R (2000b): Imaging transgene expression with radionuclide imaging technologies. *Neoplasia* 2:118–138.

Hall H, Kohler C, Gawell L, Farde L, Sedvall G (1988): Raclopride, a new selective ligand for the dopamine-D2 receptors. *Prog Neuropsychopharmacol Biol Psychiatry* 12:559–568.

Herschman HR, MacLaren DC, Iyer M, Namavari M, Bobinski K, Green LA, Wu L, Berk AJ, Toyokuni T, Barrio JR, Cherry SR, Phelps ME, Sandgren EP, Gambhir SS (2000): Seeing is believing: Non-invasive, quantitative and repetitive imaging of reporter gene expression in living animals, using positron emission tomography. *J Neurosci Res* 59:699–705.

Hospers GA, Calogero A, van Waarde A, Doze P, Vaalburg W, Mulder NH, de Vries EF (2000): Monitoring of herpes simplex virus thymidine kinase enzyme activity using positron emission tomography. *Cancer Res* 60:1488–1491.

Hume SP, Myers R, Bloomfield PM, Opacka-Juffry J, Cremer JE, Ahier RG, Luthra SK, Brooks DJ, Lammertsma AA (1992): Quantitation of carbon-11 labeled raclopride in rat striatum using positron emission tomography. *Synapse* 12:47–54.

Hustinx R, Shiue CY, Alavi A, McDonald D, Shiue GG, Zhuang H, Lanuti M, Lambright E, Karp JS, Eck SL (2001): Imaging in vivo herpes simplex virus thymidine kinase gene transfer to tumour-bearing rodents using positron emission tomography and. *Eur J Nucl Med* 28:5–12.

Iyer M, Barrio JR, Namavari M, Bauer E, Satyamurthy N, Green LA, Nguyen K, Cherry SR, Toyokuni T, Phelps ME, Herschman HR, Gambhir SS (2001): 8-[^{18}F]-Fluoropenciclovir: an improved reporter probe for imaging HSV1-tk reporter gene expression in vivo using positron emission tomography. *J Nuc Med* 42:96–105.

Jang SK, Davies MV, Kaufman RJ, Wimmer E (1989): Initiation of protein synthesis by internal entry of ribosomes into the 5′ nontranslated region of encephalomyocarditis virus RNA in vivo. *J Virol* 63:1651–1660.

Jeavons AP, Chandler RA, Dettmar CAR (1999): A 3D HIDAC-PET camera with sub-millimetre resolution for imaging small animals. *IEEE Trans Nucl Sci* 46:468–473.

Kessler RM, Ansari MS, de Paulis T, Schmidt DE, Clanton JA, Smith HE, Manning RG, Gillespie D, Ebert MH (1991): High affinity dopamine D2 receptor radioligands. 1. Regional rat brain distribution of iodinated benzamides. *J Nucl Med* 32:1593–1600.

Krestel HE, Mayford M, Seeburg PH, Sprengel R (2001): A GFP-equipped bidirectional expression module well suited for monitoring tetracycline-regulated gene expression in mouse. *Nucleic Acids Res* 29:E39.

Leite JP, Niel C, D'Halluin JC (1986): Expression of the chloramphenicol acetyl transferase gene in human cells under the control of early adenovirus subgroup C promoters: effect of E1A gene products from other subgroups on gene expression. *Gene* 41:207–215.

Liang Q, Satyamurthy N, Barrio JR, Toyokuni T, Phelps MP, Gambhir SS, Herschman HR (2001): Noninvasive, quantitative imaging, in living animals, of a mutant dopamine D2 receptor reporter gene in which ligand binding is uncoupled from signal transduction. *Gene Ther*, 8:1490–1498.

MacLaren DC, Gambhir SS, Satyamurthy N, Barrio JR, Sharfstein S, Toyokuni T, Wu L, Berk AJ, Cherry SR, Phelps ME, Herschman HR (1999): Repetitive, non-invasive imaging of the dopamine D2 receptor as a reporter gene in living animals. *Gene Ther* 6:785–791.

Naciff JM, Behbehani MM, Misawa H, Dedman JR (1999): Identification and transgenic analysis of a murine promoter that targets cholinergic neuron expression. *J Neurochem* 72:17–28.

Namavari M, Barrio JR, Toyokuni T, Gambhir SS, Cherry SR, Herschman HR, Phelps ME, Satyamurthy N (2000): Synthesis of 8-[^{18}F]fluoroguanine derivatives: in vivo probes for imaging gene expression with positron emission tomography. *Nucl Med Biol* 27:157–162.

Naylor LH (1999): Reporter gene technology: the future looks bright. *Biochem Pharmacol* 58:749–757.

Neve KA, Cox BA, Henningsen RA, Spanoyannis A, Neve RL (1991): Pivotal role for aspartate-80 in the regulation of dopamine D2 receptor affinity for drugs and inhibition of adenylyl cyclase. *Mol Pharmacol* 39:733–739.

Phelps ME (1991): PET: A biological imaging technique. *Neurochem Res* 16:929–940.

Phelps ME (2000): Positron emission tomography provides molecular imaging of biological processes. *Proc Nat Acad Sci USA* 97:9226–9223.

Ray P, Bauer E, Iyer M, Barrio JR, Satyamurthy N, Phelps ME, Herschman HR, Gambhir SS (2001): Monitoring gene therapy with reporter gene imaging. *Sem. Nuc Med*, 31:312–320.

Sadelain M, Blasberg RG (1999): Imaging transgene expression for gene therapy. *J Clin Pharmacol Suppl*, 39:34S–39S.

Satyamurthy N, Barrio JR, Bida GT, Huang SC, Mazziotta JC, Phelps ME (1990): 3-(2'-[^{18}F]fluoroethyl)spiperone, a potent dopamine antagonist: synthesis, structural analysis and in-vivo utilization in humans. *Int J Rad Appl Instrum* 41:113–129.

Schmid M, Wimmer E (1994): IRES-controlled protein synthesis and genome replication of poliovirus. *Arch Virol Suppl* 9:279–289.

Strange PG (1990): Aspects of the structure of the D2 dopamine receptor. *TINS* 13:373–378.

Sun X, Annala AJ, Yaghoubi S, Barrio JR, Nguyen K, Toyokuni T, Satyamurthy N,

Namavari M, Phelps ME, Herschman HR, Gambhir SS (2001): Quantitative imaging of gene induction in living animals. *Gene Ther* 8:1572–1579.

Tjuvajev JG, Avril N, Oku T, Sasajima T, Miyagawa T, Joshi R, Safer M, Beattie B, DiResta G, Daghighian F, Augensen F, Koutcher J, Zweit J, Humm J, Larson SM, Finn R, Blasberg R (1998): Imaging herpes virus thymidine kinase gene transfer and expression by positron emission tomography. *Cancer Res* 58:4333–4341.

Tjuvajev JG, Joshi A, Callegari J, Lindsley L, Joshi R, Balatoni J, Finn R, Larson SM, Sadelain M, Blasberg RG (1999): A general approach to the non-invasive imaging of transgenes using *cis*-linked herpes simplex virus thymidine kinase. *Neoplasia* 1:315–320.

Wagner HN Jr, Burns HD, Dannals RF, Wong DF, Langstrom B, Duelfer T, Frost JJ, Ravert HT, Links JM, Rosenbloom SB, Lukas SE, Kramer AV, Kuhar MJ (1983): Imaging dopamine receptors in the human brain by positron tomography. *Science* 221:1264–1266.

Weber S, Herzog H, Cremer M, Engels R, Hamacher K, Kehren F, Muehlensiepen H, Ploux L, Reinartz R, Reinhart P, Rongen F, Sonnenberg F, Coenen HH, Halling, H. (1999): Evaluation of the TierPET system. *IEEE Trans Nucl Sci* 46:1177–1183.

Westphal H, Overbeek PA, Khillan JS, Chepelinsky AB, Schmidt A, Mahon KA, Bernstein KE, Piatigorsky J, de Crombrugghe B (1985): Promoter sequences of murine alpha A crystallin, murine alpha 2(I) collagen or of avian sarcoma virus genes linked to the bacterial chloramphenicol acetyl transferase gene direct tissue-specific patterns of chloramphenicol acetyl transferase expression in transgenic mice. *Cold Spring Harb Symp Quant Biol* 50:411–416.

Woodward R, Coley C, Daniell S, Naylor LH, Strange PG (1996): Investigation of the role of conserved serine residues in the long form of the rat D2 dopamine receptor using site-directed mutagenesis. *J Neurochem* 66:394–402.

Yaghoubi SS, Barrio J, Dahlborn M, Iyer M, Namavari M, Satyamurthy N, Goldman R, Herschman HR, Phelps ME, Gambhir SS (2001a): Human pharmacokinetic and dosimetry studies of [^{18}F]-FHBG, a reporter probe for imaging Herpes Simplex Virus Type 1 thymidine kinase (HSV1-tk) reporter gene expression. *J Nucl Med* 42:1225–1234

Yaghoubi SS, Wu L, Liang Q, Toyokuni T, Barrio JR, Namavari M, Satyamurthy N, Phelps ME, Herschman HR, Gambhir SS (2001b): Direct correlation between positron emission tomographic images of two reporter genes delivered by two distinct adenoviral vectors. *Gene Ther* 8:1072–1080.

Yang M, Baranov E, Jiang P, Sun FX, Li XM, Li L, Hasegawa S, Bouvet M, Al-Tuwaijri M, Chishima T, Shimada H, Moossa AR, Penman S, Hoffman RM (2000): Whole-body optical imaging of green fluorescent protein-expressing tumors and metastases. *Proc Natl Acad Sci USA* 97:1206–1211.

Yu Y, Annala AJ, Barrio JR, Toyokuni T, Satyamurthy N, Namavari M, Cherry SR, Phelps ME, Herschman HR, Gambhir SS (2000): Quantification of target gene expression by imaging reporter gene expression in living animals. *Nature Med* 6:933–937.

Zlokarnik G, Negulescu PA, Knapp TE, Mere L, Burres N, Feng L, Whitney M, Roemer K, Tsien RY (1998): Quantitation of transcription and clonal selection of single living cells with beta-lactamase as reporter. *Science* 279:84–88.

INDEX